www.wadsworth.com

wadsworth.com is the World Wide Web site for Wadsworth Publishing Company and is your direct source to dozens of online resources.

At *wadsworth.com* you can find out about supplements, demonstration software, and student resources. You can also send e-mail to many of our authors and preview new publications and exciting new technologies.

wadsworth.com
Changing the way the world learns®

SOCIOLOGY

*The United States
in a Global Community*

FOURTH EDITION

SOCIOLOGY

*The United States
in a Global Community*

JOAN Ferrante

NORTHERN KENTUCKY UNIVERSITY

Australia • Canada • Denmark • Japan • Mexico • New Zealand • Philippines • Puerto Rico
Singapore • South Africa • Spain • United Kingdom • United States

Publisher: Eve Howard
Assistant Editor: Ari Levenfeld
Editorial Assistant: Bridget Schulte
Marketing Assistant: Kelli Goslin
Project Editor: Jerilyn Emori
Print Buyer: Karen Hunt
Permissions Editor: Susan Walters
Production: Andrea Cava
Interior and Cover Designer: Paul Uhl/Design Associates
Photo Researcher: Sue McDermott

Copyeditor: Jill Hobbs
Illustrations: Seventeenth Street Studios
Cover Images: © Corbis/Walter Hodges; © Jeff Noble,
Noble Stock Inc.; © Jack McConnell; © Michael
Moschogianis/ImageQuest; Digital Stock; © Dana
Schuerholz Impact Visuals/PNI; © Phillip Jarrell;
© David Wagenaar; © Corbis/Ron Watts; © Corbis/
Mark Stephenson; Digital Vision
Compositor: New England Typographic Service
Printer: World Color Book Services/Versailles

For permission to use material from this
text, contact us by
web: www.thomsonrights.com
fax: 1-800-730-2215
phone: 1-800-730-2214

**Library of Congress
Cataloging-in-Publication Data**
Ferrante-Wallace, Joan
    Sociology: the United States in a global
community/Joan Ferrante.—4th ed.
        p.   cm.
    Includes bibliographical references and index.
    ISBN 0-534-57060-7 (alk. paper : paperback)
    1. Sociology.   2. Social history Cross-
cultural studies.
    I. Title.
HM585.F47   1999
301—dc21   99-28522

Wadsworth/Thomson Learning
10 Davis Drive
Belmont, CA 94002-3098
USA
www.wadsworth.com

International Headquarters
Thomson Learning
290 Harbor Drive, 2nd Floor
Stamford, CT 06902-7477
USA

UK/Europe/Middle East
Thomson Learning
Berkshire House
168-173 High Holborn
London WC1V 7AA
United Kingdom

Asia
Thomson Learning
60 Albert Street #15-01
Albert Complex
Singapore 189969

Canada
Nelson/Thomson Learning
1120 Birchmount Road
Scarborough, Ontario M1K 5G4
Canada

To my mother,

  Annalee Taylor Ferrante

and in memory of my father,

  Phillip S. Ferrante
  (March 1, 1926 – July 8, 1984)

Brief Contents

Contents

Economics and Politics 360
With Emphasis on the United States

Population and Family Life 390
With Emphasis on Brazil

Preface

As a professor who teaches undergraduates every semester, I am constantly reminded of our shared mission—to find the best way to introduce students to the discipline of sociology and to help them see that sociological concepts and theories are powerful tools for thinking about any issue. My research and writing are guided by the assumption that the introduction to sociology course must be an eye-opening experience that challenges readers to see their world in a new light. I will try to explain how this textbook offers such an experience.

Key Features of This Textbook

First, students will learn from this textbook that sociology offers a coherent perspective for analyzing any social event or issue. To illustrate this I use the concepts and theories covered in each chapter to address a central question or theme. For example, in Chapter 7, "Social Organizations," I present the same concepts and theories covered in the mainstream introduction to sociology textbooks but instead of using an encyclopedia-like overview, I weave concepts and theories together to analyze the well-known organization McDonald's, a U.S.-based multinational corporation with operations in 111 countries. I use the concepts and theories to specifically show how this organization coordinates the activities of over 1 million employees in corporate offices and franchises across the globe to serve an estimated 38 million customers per day or 14 billion customers a year in 111 countries. McDonald's achieves this by not only coordinating its employees' actions but also the activities of millions of people employed by the meat, potato, produce, bread, and condiment suppliers.

Second, students are exposed to *meaningful* examples to illustrate concepts and theories. This is the only way to grab the students' interest and convince them that sociology is relevant to their lives. Just like any mainstream textbook, this textbook covers the ideas of people considered important to the discipline's development, such as Marx, Durkheim, Weber, DuBois, Martineau, and others. But it also is obligatory to show why ideas that are decades to centuries old have survived the test of time and are still relevant today. One such idea was voiced by W. E. B. DuBois in 1903 when he wrote about the "strange meaning of being 'black' in America at the dawning of the 20th century." This phrase is brought to life when readers learn that DuBois' interest in this concept was surely affected by his being both American-born and of Dutch, French, and African ancestry. Yet DuBois has always been labeled a "black" sociologist, and no one thinks to question why he is not labeled as a Dutch, French, or an American one. Student voices are used throughout the text to clarify ideas even further. For example, one student writes about "the strange meaning of being black": "I can't be anything but what my skin color tells people I am. I am 'black in America' because I look 'black.' It does not matter that my family has a complicated biological heritage and that I am mixture of Cherokee, French, and African descent."

Third, like most mainstream textbooks, this textbook addresses issues central to American society. Not ignored, however, is the fact that the United States is part of a global community. I cannot pretend that people in the United States interact only with those inside its borders. I cannot act as if problems such as unemployment, inequality, disease, and illegal drug use can be understood and solved by reference only to domestic factors. I cannot forget that the United States has the world's most powerful and diverse economy; that it has a military presence in at least 140 countries; or that it especially known around the world for its entertainment-related exports (films, television programs, videos, radio programming, and recorded music). Finally, I cannot pretend that the individual biographies of those who have come to live in the United States are unaffected by events and processes in foreign locations. Simply consider the number of people in the United States who are descendants of immigrants who crossed paths with native-born people or other immigrants to produce offspring. The offsprings' existences depended on events in one or more countries "pushing" their parents together.

Because the United States is connected to and affected by its place in a global community, sociological concepts and theories are used to explore its global place. Specifically discussed are the transfer of labor-intensive manufacturing operations out of the United States to countries such as Mexico (Chapter 2); the meaning of and explanations for the U.S. trade deficit with Japan (Chapter 3); the influence of culture, especially as it relates to those influenced by more than one culture such as Korean students studying in the United States and U.S. servicemen and servicewomen stationed in South Korea (Chapter 4); the socialization mechanisms by which long-standing ethnic conflicts are passed from one generation to the next (Chapter 5); the connection between the HIV/AIDS problem and a complex set of intercontinental, international, and intrasocietal interactions (Chapter 6); changing conceptions of what constitutes conformity and deviance in two societies (Chapter 8); stratification systems and the massive effort it takes to dismantle inequality (Chapter 9); the U.S. system of racial classification as it compares with Germany's system (Chapter 10); the distinction between sex and gender and the role gender plays in affecting life chances in the United States and in American Samoa (Chapter 11); the political and economic systems of the largest and most diverse economy in the world (Chapter 12); the demographic events that shape the structure of family life in the United States and Brazil (Chapter 13); the U.S. system of public education and its standing in the international community (Chapter 14); the ways religion is and has been used to justify the best and worst behaviors (Chapter 15); and the Internet as a tool for connecting people on a global scale (Chapter 16).

You may wonder why the global emphasis is listed third especially when many consider it the key feature distinguishing my textbook from others. I do this purposely because I consider the other two features—(1) the coordinated and integrated presentation of sociological concepts and theories and (2) meaning and memorable examples—as more critical. At first glance the global nature of this book may seem intimidating to students, but I think sociology professors underestimate their familiarity with global issues and their knowledge of the countries covered in this text. Consider the emphasis on the Democratic Republic of the Congo (formerly Zaire) in Chapter 6 (Social Interac-

tion and the Social Construction of Reality). The chapter is really not so much about the Congo but about larger social issues with which any sociologist is already familiar. I simply use the Congo as a vehicle for showing how sociological concepts and theories related to interaction and reality construction help us answer an important question: In light of the fact that the earliest sample of HIV-infected blood is an unidentified sample taken in 1959 and stored in a Congo blood bank, how is it that AIDS has grown from a few cases to a monumental health problem in the United States and around the globe? It is obvious that the transmission of HIV is connected to a complex set of intimate interactions (intimate enough to allow for the exchange of bodily fluids) between people from different nations. All sociologists know that the forces behind global-scale interactions are related to changes in the division of labor, colonization, industrialization, and to specific inventions such as the jet engine and advances in preserving and storing blood products. Similarly, all sociologists are familiar with the principles of reality construction and would not be surprised to learn that the unidentified blood sample is not necessarily that of a "black" African. The simple fact that the Congo was a Belgian colony opens up the possibility that the blood sample belonged to a "white" European.

An organizing principle is present in all chapters. If instructors or readers feel, however, that they need more information about the country emphasized, they can refer to the study guide, the instructor's manual, or to the Wadsworth Sociology homepage: http://sociology.wadsworth.com.

Several additional features help achieve the goals outlined above. Each chapter opens with a "Why Focus on" discussion as a way of showing how the chapter emphasis offers a good vehicle for integrating the concepts and theories covered in that chapter. For example, Chapter 2 (Theoretical Perspectives) opens with "Why Focus on U.S. Manufacturing Operations in Mexico?" In Chapter 2 special attention is given to Mexico because, although the United States shares a 1,952-mile border with Mexico (which millions cross each year to shop and visit vacation spots), most Americans know little about what binds the two countries together. In truth, few people know that 2,800 U.S.-headquartered corporations have manufacturing operations on the Mexican side of the border or understand their economic significance. The three major sociological perspectives offer us quite different ways to assess the relationship between Mexico and the United States and to put what we hear, read, or experience about Mexico into a broader context. In addition, the three perspectives offer some constructive ways to think about not just this relationship but also a larger global trend: the transfer of labor-intensive manufacturing or assembly operations out of the United States to countries with lower wage rates.

Each chapter also contains "U.S. in Perspective" boxes that further illustrate the place of the United States in the global community. In Chapter 1, for example, a box titled "Top 20 Global Grossers of '98" ranks U.S. films in terms of gross worldwide incomes. Students are reminded that the United States dominates the world's film industry. Furthermore, U.S. films offer a vehicle by which people in other countries gain impressions of what life is like in the United States. Students are asked to study the list and determine what impressions of the United States someone might acquire from watching these films.

In each chapter, key concepts are boxed and highlighted on the pages where they are first introduced. This reinforces my goal of showcasing sociology as a discipline offering a powerful vocabulary for thinking about virtually any event or situation. I have also included Internet exercises at the end of each chapter. URL addresses were included in the Third Edition but most of them changed even as the book went to press. In this edition, I chose instead to give enough information so that students can search for the relevant Web sites.

New to This Edition

A number of *major* changes in the Fourth Edition strengthen the way the discipline of sociology is introduced and strengthen coverage of the United States as part of a global community.

- Chapter 1 includes new sections "Why Study Sociology?" and "The Importance of a Global Perspective." The "why-study-sociology?" change is important because students, especially students thinking of majoring in sociology, are often asked to explain why sociology is a useful area of study. Understanding the meaning of a global perspective also is important because although *global* is a word students hear daily (e.g., global economy, global airlines, the Internet as a global connector, global problems, the U.S. as a global leader, global competition), it nevertheless is a vague term in need of elaboration.

- In Chapter 7, "Social Organizations," the emphasis has changed from "The Multinational Corporation in India" to "McDonald's, a U.S.-Based Multinational Corporation with Operations in 111 Countries." In previous editions I gave emphasis to the U.S.-headquartered multinational corporation Union Carbide. I had chosen Union Carbide as a way of exploring the organizational failures behind the largest industrial accident in world history. That accident occurred in 1984 in Bhopal, India, and although the case against Union Carbide has yet to be resolved, few students remember this event. The focus on McDonald's (its organizational successes and failures) highlights an organization with which almost every student is familiar.

- The emphasis of Chapter 11, "Gender," has changed from "the former Yugoslavia" to "American Samoa." It made sense to drop coverage of the former Yugoslavia because it is now four distinct countries. Using American Samoa as a replacement gives coverage to one of the many "geographic spaces" designated as a U.S. territory. American Samoa is a particularly good choice because, as we will learn, it has been used as a point of reference and comparison for thinking about gender in the United States since the 1920s.

- Chapter 12, "Economics and Politics: With Emphasis on the United States," is new to the Fourth Edition. Although economic and political issues are covered in other chapters, this chapter offers a coherent overview of these key institutions and gives special attention to the United States, the country with the world's largest and most diverse economy and the country considered to be the world's only "superpower."

- One of the new features to this edition for which I am most proud is the inclusion of student voices throughout. In the sociology classes that I teach I ask students to respond in writing to the concepts and examples I present. Many times their responses open my eyes to applications I would never have considered on my own. As an example, consider one student's association upon learning that social facts are ideas, feelings, and ways of behaving that possess the remarkable property of existing outside the individual; that is, social facts do not originate with the people experiencing them. "After hearing this information about social facts, I remembered my experiences while traveling in Africa. In the United States it is typically unacceptable for two men to hold hands; immediately the men would be labeled homosexual. I spent some time in Ghana, Africa, several years ago and one of the first cultural differences I noticed was that men, including the men I was with, hold hands. This cultural difference definitely hit home when one day one of the men I was with took my hand as we walked. In order not to offend him, I followed through with this until an appropriate opportunity allowed me to disengage our hands. Even though I was in a country where this was perfectly acceptable, I still felt extremely uneasy with this tradition."

There are too many less dramatic changes to list here, of course. But I can say that I revised each chapter with an eye toward improving its clarity, updating information, and replacing outdated examples. As a case in point, since the publication of the Third Edition, the South African Truth and Reconciliation Commission has completed the bulk of its investigations into the human rights' violations that occurred under apartheid. The South African government, under the leadership of Nelson Mandela, established the Truth and Reconciliation Commission as a vehicle for dealing with the country's past. The postapartheid government had taken the position that dealing with the past is uncomfortable and painful but it is essential for building a new future and that all South Africans, especially those who benefited, must confront the reality of apartheid, express remorse, and commit to a new multiracial system. Thus in Chapter 9 we consider the lessons the United States might learn from South Africa's decision to confront its past head on. As with this chapter, every other chapter contains an important change of this nature.

Ancillary Materials

The *Study Guide,* which I also wrote, contains study questions, concept applications, applied research questions, Internet sites, InfoTrac College Edition sources, and movie recommendations. There is also a Web site that offers an interactive learning tool, including on-line quizzes and chapter-by-chapter links to related Internet sites. Visitors to the site are encouraged to participate in ongoing discussions on current hot topics in sociology and to take advantage of the additional sociology and Internet-related resources in Virtual Society, the Wadsworth Sociology Resource Center (http://sociology.wadsworth.com).

Acknowledgments

The Fourth Edition builds on the efforts of those who helped me with the first three editions. Five people stand out as particularly influential: Sheryl Fullerton (the editor who signed this book), Serina Beauparlant (the editor who saw the first and second editions through to completion), Maggie Murray (the developmental editor for the first edition), John Bergez (the developmental editor for the second edition), and Alan Venable (the developmental editor for the third edition).

I give special thanks to my editor, Eve Howard, who gave me fresh insights about how to revise the Fourth Edition and suggested that I change the subtitle to highlight my longstanding emphasis on the United States and its role in the global community. Of course, a revision plan depends on thoughtful, constructive, and thorough reviewer critiques. In this regard I wish to extend my deepest appreciation to those who have reviewed this edition:

Walter Dean, Baltimore City Community College
Kevin Early, Oakland University
Mark G. Eckel, McHenry County College
Jim Fraser, University of Tennessee at Chattanooga
Lee Hamilton, New Mexico State University
James D. Jones, Mississippi State University
Jerry Karst, Aims Community College
Shirlee Owens, Northeast Louisiana State University
Robert L. Perry, Johnson County Community College
Rob Reynolds, Weber State University
Stacy G. Ruth, Jones Junior College
Jacqueline Scherer, Oakland University
Jenny Stuber, Normandale Community College
Sarah Whitmer Foster, Florida Agricultural and Mechanical University

I am also grateful to the reviewers of past editions:

Deborah A. Abowitz, Bucknell University
John Brenner, Southwest Virginia Community College
R. Carlos Cavazos, Texas State Technical College
Paul Ciccantell, Kansas State University
Mary A. Cook, Vincennes University
Peter W. Cookson, Jr., Adelphi University
Carol Copp, California State University, Fullerton
Norman A. Dolch, Louisiana State University, Shreveport
Joseph S. Drew, University of the District of Columbia
Kevin Early, Oakland University
John Ehle, Jr., Northern Virginia Community College

J. Lynn England, Brigham Young University
Juanita Firestone, University of Texas
Kimberly Folse, Southwest State University
Pat Gadban, Monterey Peninsula College
T. Neal Garland, University of Akron
P. Kreutzer Garman, DeKalb College, North
Thomas Gold, University of California, Berkeley
Franklin Goza, Bowling Green State University
Jack Harkins, College of DuPage
Janet Hilowitz, University of Massachusetts, Dartmouth
Jen Hlavacek, University of Colorado
Gary Hodge, Collin County Community College
Moon H. Jo, Lycoming College
Jane I. Johnson, Southwest Texas State University
Nancy Kleniewski, State University of New York–Geneso
Gary R. Lemons, Scottsdale Community College
Robert Liebman, Portland State University
Clinton M. Lipsey, Tennessee State University
Patricia McNamara, University of New Mexico
Martin Marger, Michigan State University
M. Cathy Maze, Franklin University
Michael V. Miller, University of Texas, San Antonio
Bronislaw Misztal, Indiana University at Fort Wayne
Fenno Ogutu, Diablo Valley College
Kristin Park, Emporia State University
Robert Perry, Johnson County Community College
Terry C. Rodenberg, Central Missouri State University
James Carroll Simms, Virginia State University
Bhavani Sitaraman, The University of Alabama in Huntsville
William L. Smith-Hinds, Lock Haven University
Richard Sweeney, Modesto Junior College
Susan Tiano, University of New Mexico
Walter C. Veit, Burlington County College
Suzan Waller, University of Central Oklahoma
Leslie Wang, University of Toledo
Richard Wood, De Anza College

For the third edition, I have had the privilege of working with Leigh Cherni (NKU, Class of 2000) and Michael Vaughn (NKU, Class of 2001), my research assistants on this book. They worked behind the scenes helping me gather hundreds of articles, Internet documents, and statistics required to revise and update this book. I formulate the questions to guide the research; they search for the answers. Sometimes the searching is like trying to find a needle in a haystack. The working relationship I have with Leigh and Michael is one of the most satisfying, rewarding, and enjoyable aspects of my life as a professor (and I might add, there are many other rewarding aspects to my profession).

I am grateful to Kim Bo-Kyung and Kevin Kirby for their thoughtful reading and critique of Chapter 4, "Culture: With emphasis on South Korea." I thank Ari Levenfeld for consulting with me about America Samoa and for the work he has done to keep the book on schedule. I also appreciate the help of librarians at the University of Cincinnati, the Public Library of Cincinnati, and, especially, Northern Kentucky University. One of the most underrecognized roles in producing a textbook is that of production editor. I was fortunate to work with Andrea Cava. No one can imagine the coordination and overwhelming amount of detailed work that is required to get a book ready to go to press.

As always, I wish to express my appreciation to my best friend and mother, Annalee Taylor Ferrante, who not only files, clips newspaper articles, and writes for permissions but also cooks meals for my husband and me several times a week. The care with which she prepares food and the exquisite results are the subject of movies such as *Big Night* and *Like Water for Chocolate*. Thanks also go to my father-in-law, Walter D. Wallace, who sent me newspaper clippings several times a week, many of which gave me ideas for examples. He died on December 27, 1998. He took the time to clip articles and mail them to me even in the last days of his life when he was very ill.

In closing, I acknowledge, as I have done in all editions of this and other books, the tremendous influence of Horatio C. Wood IV, M.D., on my philosophy of education. As time passes my feelings of warmth and gratitude toward him only deepen. Finally, I express my love for the most important person behind this book, my husband, colleague, and greatest supporter Robert K. Wallace. I dedicate this book to the memory of my father, Phillip S. Ferrante, and to my mother, Annalee Taylor Ferrante.

SOCIOLOGY

*The United States
in a Global Community*

The Sociological Imagination

[*]Harriet Martineau and W. E. B. DuBois are covered in this section.

A plasma center in the United States. (Courtesy of ABRA, The International Authority of the Source Plasma Collection Industry.)

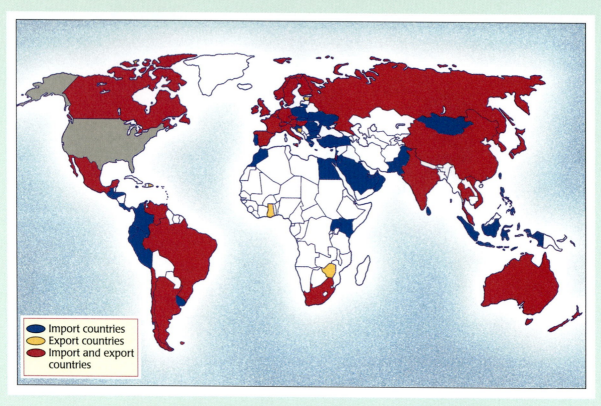

Source: International Trade Administration (1997a, 1997b)

The map shows the countries *to which* the United States exported blood and blood products in 1997 and the countries *from which* the United States imported blood and blood products. The United States is the world's largest exporter of blood, shipping these products to 44 countries. Millions of people from different places in the world interact intimately with one another through the exchange of blood without ever meeting. As one indicator of the amount of blood exchanged, consider that the United States supplies 60 percent of the world's demand for plasma (Krever Commission Report 1997).

- On May 12, 1998, Dr. Noel Buskard, the president of the Canadian Society for Transfusion Medicine, announced that Canada "had only one and a half days supply of blood for transfusions, and some vital blood products are already out of stock. . . . We face the very real prospect that Canada could run out of blood. We are currently importing more than 50 percent of our blood and blood products, mainly from the U.S. But shortages in the U.S. mean we can't increase imports to make up for a drop off in local donors. If anything, our U.S. supply will be reduced" (Canada News Wire 1998).

- On February 27, 1998, the U.K. government authorized its National Health Service to import blood plasma as a way of protecting the public against the theoretical risk of contracting the human equivalent of mad cow disease. Health Secretary Frank Dobson said the move came after three recalls of blood products in November because donors contributing to British-made plasma developed Creutzfeldt-Jakob disease [CJD] (Reuters News Service 1998).

- At the end of October 1997, 1,644 people in Japan had AIDS or had died from AIDS and 4,935 were HIV-infected (not AIDS). Thirty percent of HIV-infection cases and 40 percent of AIDS cases have been traced to contaminated blood products imported from the United States in the 1980s (Japan Surveillance Report 1998).

- Three plasma centers in El Paso, Texas, placed an ad in the Spanish-language newspaper, *El Diario* (1998), to recruit donors from Juárez, Mexico. The ad indicated that new donors could earn $40 for two plasma "donations," and that it was possible to earn as much as $120 per month donating plasma.

Notice that these four accounts do not focus on specific individuals; the people affected remain nameless and faceless. Yet, for whatever reason, not enough people living in Canada, Japan, and the United Kingdom donate enough blood to sustain the needs of their respective countries. The United States is somehow able to mobilize enough people to give blood (including those who cross the border into the United States to give blood) in order to supply 60 percent of the world's demand for plasma. Can we assume that people living in the United States and in Mexican border towns are more generous or altruistic than people living in Canada, Japan, and the United Kingdom? Or are other factors responsible for this phenomenon? These are examples of the kinds of questions sociologists ask. In this chapter we introduce you to the discipline of sociology. In particular, we focus on the sociological imagination, an approach to analyzing any social issue. We also explore the historical origins of this approach and some of the key vocabulary words sociologists draw upon to help them think about a whole range of social issues.

Troubles, Issues, and Opportunities: The Sociological Imagination

The discipline of **sociology** offers us an approach to analyzing social issues known as the sociological imagination. In his now-classic book *The Sociological Imagination*, sociologist C. Wright Mills (1959) explains that the key element of the sociological imagination is an ability to make a distinction between troubles and issues. Mills defines **troubles** as personal problems and difficulties that can be explained in terms of individual characteristics such as motivation level, mood, personality, or ability. The resolution of a trouble, if it can be resolved, lies in changing an individual's character or immediate relationships. Mills states that when only one man or woman is unemployed in a city of 100,000, that is his or her personal trouble, and for its relief we properly look to that person's character, skills, and immediate opportunities (that is, "She is lazy," "He has a bad attitude," "She has always been a moody person," "He didn't try very hard in school," "She had the opportunity but didn't take it").

An **issue**, on the other hand, is a matter that can be explained only by factors outside an individual's control and immediate environment. When 15 million men and women are unemployed in a nation of 50 million employees, that is an issue, and we may not hope to solve this unemployment problem by focusing on the character flaws of 15 million unemployed individuals. Likewise, we cannot explain Canada's and Japan's reliance on the United States for blood products by pointing to character flaws in Japanese and Canadian people that make them unlikely to donate blood. Rather, the **structure of opportunities**—the chances available in a society to achieve a valued goal such as employment or to give blood—is flawed or has collapsed. In the case of large-scale unemployment, not enough opportunities are available to accommodate

everyone who needs a job. Likewise, no opportunities to "sell" blood exist in Canada and Japan; donors are unpaid. If we compare the percentages of Japan's and Canada's blood needs met by unpaid volunteers with the percentage of U.S. blood needs met by volunteers, we would find similar donation rates in the three countries. The United States manages to meet both its own blood needs and 60 percent of the world's demand for blood because its plasma centers pay "donors."

Mills argues that people often feel "trapped" by historical forces that affect them because, even though they respond to their personal plight, it is as if they remain spectators feeling little control over the larger forces shaping their lives. When a war breaks out, for example, an ROTC student becomes a rocket launch technician, a teenager buys and wears a sweatshirt that displays his or her support (or lack of support) for the war, an automobile owner thinks about the role of oil in the war while pumping gas, and a child whose parents are in the military has to live without them for a while—or forever. The personal experience of war "may be how to survive it or how to die in it with honor; how to make money out of it; how to climb into the higher safety of the military apparatus; or how to contribute to the war's termination" (Mills 1959:9). Whatever the experience, the origins of their personal problems lies outside the individuals' control. To understand these origins, we can examine the issue of war that may be caused by a threat to the resources (such as oil) on which a country depends (see Table 1.1), a threat to a country's major trading partners, or a threat to a country's credibility or image in the world if it does not act. Such threats are significant because they affect the structure of opportunities. In the case of a threat to oil, the opportunity to drive a vehicle and to drive it at a certain price per gallon is threatened.

Likewise, when the CEOs of corporations based in the United States decide that if they are to compete in a global economy, they must downsize their operation or move some of their daily operations to a foreign country or to a lower-wage area of the United States, those affected react in many ways. The laid-off employees may file for unemployment, return to college, open a business, take on third-shift work, choose early retirement, or move to another city to find work. The personal response connected with this kind of economic restructuring may include finding a way to live on a reduced budget, fighting drowsiness while driving home from third-shift work, trying to stay cheerful while hunting for a job, or finding ways to earn "extra" money (selling blood) while on unemployment. Such

Sociology The systematic study of social interaction.

Troubles Personal problems and difficulties that can be explained in terms of individual characteristics such as motivation level, mood, personality, or ability.

Issue A matter that can be explained only by factors outside an individual's control and immediate environment.

Structure of opportunities The chances available in a society to achieve a valued goal.

Table 1.1   1997 U.S. Reliance on Foreign Supplies of Critical Nonfuel Mineral Materials

U.S. dependence on oil makes the headlines, but the United States also relies on other nonfuel mineral materials. This chart shows the extent to which the United States relied on foreign countries for critical materials and the major uses for which those materials are employed.

Commodity	Percentage	Major Sources (1993–1996)	Major Uses
Arsenic	100	China, Japan, Hong Kong, Mexico	Wood preservatives, agricultural chemicals, nonferrous alloys
Bauxite and alumina	100	Australia, Guinea, Jamaica, Brazil	Aluminum production, abrasives, chemicals, refractories
Columbium (niobium)	100	Brazil, Canada, Germany	Steelmaking, superalloys
Graphite (natural)	100	Mexico, Canada, China, Madagascar, Brazil	Refractories, brake linings, packings, lubricants, foundry dressings, molds
Thorium	100	France	Ceramics, alloys, welding electrodes
Fluorspar	100	China, South Africa, Mexico	Hydrofluoric acid production, aluminum fluoride, steelmaking
Manganese	100	South Africa, Gabon, Australia, France	Steelmaking, batteries, agricultural chemicals
Mica, sheet (natural)	100	India, Brazil, Belgium, China	Electronic and electrical equipment
Strontium (celestite)	100	Mexico	Television picture tubes, ferrite magnets, pyrotechnics
Thallium	100	Belgium, Canada, Mexico	Superconductor materials, electronics, alloys, glass
Yttrium	100	China, Japan	TV manufacturing, fluorescent lights, temperature sensors
Gemstones	99	Israel, India, Belgium	Jewelry, carvings, gem and mineral collections
Tin	85	Brazil, Bolivia, Indonesia, China	Cans and containers, electrical, transportation, construction
Tungsten	85	China, Russia, Germany, Bolivia	Cemented carbide parts, electronic components, steel
Platinum	84	South Africa, Russia, UK, Germany	Catalysts, jewelry
Tantalum	80	Australia, Germany, Thailand, Brazil	Electronic components

Source: U.S. Department of the Interior, Bureau of Mines (1998).

responses do not change the issues associated with global competition, which is driven largely by an economic system that measures success in terms of profit.[1]

Mills asks, "Is it any wonder that ordinary people feel they cannot cope with the larger worlds with which they are so suddenly confronted?" (pp. 4–5). He believes that to gain some sense of control over their lives, people need "a quality of mind that will help them to use information" in a way that they can think about "what is going on in the world and of what may be happening within themselves" (p. 5). Mills equates this quality of mind with the sociological imagination. Those who possess the sociological imagination can view their inner life and human career in terms of opportunity structures and the larger historical forces that affect those structures. The opportunity structure may be related to almost any valued goal—the opportunity to attend a high school prom, the opportunity to become an effective reader, the opportunity to work, the opportunity to drive, the opportunity to live a long life, or the opportunity to receive money for blood donated. The sociological imagination empowers us to

[1]Profit is achieved by lowering production costs, hiring employees who will work for lower wages, introducing labor-saving technologies (computerization), and moving production out of high-wage zones inside or outside the country. Such strategies affect employment opportunities and leave workers vulnerable to unemployment.

see how opportunities, whatever they may be, are shaped by the way in which behavior is patterned or organized in society (that is, structured). In the United States, the opportunity to attend a high school prom, for example, is constrained for many students by the fact that they (or school officials) define a date as a social occasion planned in advance between two people of the opposite sex. Ideally one partner (the male) is taller than the other (the female). Some school officials and students have recognized how this definition limits students' opportunities to participate and have taken steps to widen the opportunity structure by encouraging and accepting other dating arrangements, such as group dates whose members may involve any combination of males and females.

Mills maintains that the payoff for those who possess this quality of mind is that they will understand the forces shaping their lives and thus be able to respond in ways that benefit their lives as well as those of others. In summary, sociology offers a perspective that allows people to know when personal problems are troubles or are connected to issues. The same perspective also discourages us from thinking that we can solve large-scale societal issues by changing individual characteristics or traits. It does not promise that people who possess the sociological imagination will solve their personal problems easily. The sociological imagination, however, does help people to see how opportunity structures shape their lives. The ability to see the structure of opportunities becomes most evident to people when they can grasp two interrelated concepts: social relativity and the transformative powers of history.

Social Relativity as a Key to Analyzing Opportunity Structures

Social relativity is the view that ideas, beliefs, and behavior vary according to time and place. People who have lived in more than one culture can teach us much about social relativity. Consider the following excerpts from diaries kept by college students in my classes who were asked to interview foreign students about their adjustment to life in the United States:

- I asked François (from France) what was the biggest adjustment he had to make coming to America. He answered in this way. When he came to America he took a job. The first day on the job

Social relativity The view that ideas, beliefs, and behavior vary according to time and place.

he expected to take a lunch break around 12 o'clock. What seemed like several hours passed before his boss told him to take a half-hour lunch break. Lunch is not taken so lightly in France. François was accustomed to a larger meal and to taking as long as two hours to eat.

- I asked Irma if she had any problems readjusting to her native culture when she visited Indonesia. She said that she did have problems. Irma mentioned that in Indonesia everyone she knows drops in on a relative or friend without advance notice. In the United States, on the other hand, she had to learn to call people first. When she was home over Christmas break, she began dialing the phone number of a relative to let the person know that she and her mother were coming over. Her mother asked her why she was behaving so strangely. Irma replied that she wanted to check to see if this was a convenient time. Her mother thought Irma was going crazy. In Indonesia calling before a visit is equivalent to asking someone if they will go out of their way to make special preparations. Indonesians view such a request as rude.

- I asked Hannah what were some of the hardest things for her to get used to since coming to the United States. She explained that in Jordan a girl would never introduce a boy as a boyfriend. If that were to happen, the girl would be disowned not only by the family, but by the whole society. In general, the nature of family ties is very different in the two countries. In Jordan people do not always do what they would like. They have a deep respect for family. This respect is so strong that family comes first and individual achievement second. Hannah described how marriage partners are chosen. The first and most important consideration is not the marriage partner per se, but the person's family. If someone was from a bad family, you would not marry the person for two reasons. First, the prevailing assumption is that if the family is bad, so is an individual raised in that family. Second, you would not disgrace the family by marrying a person of questionable character.

These accounts show that ideas about eating, visiting relatives, and dating vary across regions and cultures and that these ideas affect opportunities—in these cases, structured opportunities to relax in the middle of the day, to visit friends and relatives, and to select potential mates. Furthermore, they help us recognize and understand that many of our ideas and behaviors do not originate from any one individual but are products of the environment into which we were born.[2] The concept of social relativity helps us to see that things "'could be otherwise.' That is, social life—the way school, work, families, daily experience are organized—may feel permanent and given, but the arrangements are socially constructed, have changed over time [and place], and can be changed" (Thorne 1993:xxxi).

Transformative Powers of History and Opportunity Structures

According to the concept of the **transformative powers of history,** the most significant historical events have dramatic consequences on people's lives and on the structure of opportunities. To understand the transformative power of an event, a person must have some idea of the way people thought or behaved before and after that event.

Consider the tremendous transformative power of the Industrial Revolution of the mid- to late nineteenth century (1850–1903). Many would call this time the "greatest period of technological change in terms of things that affected huge numbers of people's lives in basic ways" (Post 1997). The new inventions included a revolution in steel making, the light bulb, the phonograph, the telephone, radio, automobiles, subways, elevated trains, diesel fuel, refrigeration, and the airplane (Lohr 1997). All of these inventions had a dramatic effect on the structure of opportunities, including the opportunity to travel long distances, to communicate with those in distant locations, and to hear the "news" as it happens. Before the Industrial Revolution, most people's lives varied little from one generation to the next. As a result, parents could assume that their children's opportunities would be much like their own and that their children would face the same challenges and problems that they faced. Consequently, societies embarked on long-term projects that extended to subsequent generations. For example, in the Middle Ages people built cathedrals that took several lifetimes to complete, "presumably never doubting that such edifices would be used and appreciated by their great-grandchildren when construction was complete" (Ornstein and Ehrlich 1989:55).[3]

During the Industrial Revolution, however, the pace of change increased to such a degree that parents could no longer assume that their children's lives and opportunities would be like their own. "Imagine the reactions of an American today if asked to contribute to a building that would take 150 years to finish: 'We don't want to tie up our capital in something with no return for one hundred and fifty years!' 'Won't a new design and construction process make this one obsolete long before it's finished?'" (Ornstein and Ehrlich 1989:55).

The sociological imagination permits those who adopt it to see how biography and history intersect—to understand that even some of the most personal experiences are shaped by time and place and by the transformative powers of history. This type of imagination is characteristic of all the great sociologists, including three of the most influential: Emile Durkheim, Max Weber, and Karl Marx. All three had a tremendous impact on the discipline of sociology; all three spent much of their professional careers attempting to understand the nature and consequences of one event: the Industrial Revolution and its effect on the structure of opportunities.

The Industrial Revolution and the Emergence of Sociology

Between 1492 and 1800, an interdependent world began to emerge. Europeans learned of, conquered, or colonized much of North America, South America, Asia, and Africa and set the tone of international relations for centuries to come. During this time, colonists forced local populations to cultivate and harvest crops and to extract minerals and other raw materials for export to the colonists' home countries. When the Europeans' labor needs could not be met by indigenous populations, they imported slaves from Africa or indentured workers from Asia. In fact, an estimated 11.7 million

[2]Two other examples from student journals follow:

Alex, from Brazil, observed that the Americans he has encountered seem more individualistic than Brazilians. He thinks this comes from how people are raised. For example, when U.S. parents buy a toy, it is usually for a particular child—that child does not have to share. In Brazil, when parents buy a toy, it belongs to all their children to share.

Monika, a Swedish-born student, told me that in her country, love is a very strong word. Swedes use the word more selectively than do Americans. People in the United States might say, "I love the outdoors." Based on her experiences, if a Swedish person said this, people would literally think something was wrong with the person.

[3]This is a "romantic" example of the transformative powers of the Industrial Revolution appropriate for this point in the chapter. As the material that follows will show, the Industrial Revolution was a disruptive, even violent, event that affected people all over the world.

Transformative powers of history The concept that most significant historical events have dramatic consequences on people's opportunities and that events should be viewed in the context of time and place.

Africans survived their journey to the "New World"[4] between the mid-fifteenth century and 1870 (Chaliand and Rageau 1995; Holloway 1996; Conrad 1996).

In light of these events, one can argue that the world has been interdependent for at least 500 years. The scale of social and economic interdependence changed dramatically, however, with the Industrial Revolution, which gained momentum in England in about 1850 and soon spread to other European countries and the United States. The Industrial Revolution drew people from even the most remote parts of the world into a process that produced unprecedented quantities of material goods, primarily for the benefit of the colonizing countries.

Between 1880 and 1914, European annexation and colonization expanded to meet Europe's growing demand for raw materials and labor. This period in history is known as the Age of Imperialism. The Age of Imperialism represents the most rapid colonial expansion in history, in which rival European powers (for example, the United Kingdom, France, Germany, Belgium, Portugal, the Netherlands, Italy) competed to secure colonies and spheres of influence in Asia, the Pacific, and especially Africa. By 1914, all of Africa had been divided into European colonies. In that year, 84 percent of the world's land area had been affected by colonization, and an estimated 500 million people lived as members of European colonies (*Random House Encyclopedia* 1990).[5]

One fundamental feature of the Industrial Revolution was **mechanization**, the addition of external sources of power such as oil or steam to hand tools and modes of transportation. This innovation turned the spinning wheel into a spinning machine, the hand loom into a power loom, and the blacksmith's hammer into a power hammer. It replaced wind-powered sailboats with steamships and horse-drawn carriages with trains. On a social level, it changed the nature of work and the ways in which people interacted with one another.

The Nature of Work

Before the mechanization brought on by the Industrial Revolution, goods were produced and distributed at a human pace, as illustrated by the effort required to bake bread or to make glass:

- Bakeries produced bread almost entirely by manual labor, the hardest operation being that of

preparing the dough, "usually carried on in one dark corner of a cellar, by a man, stripped naked down to the waist, and painfully engaged in extricating his fingers from a gluey mass into which he furiously plunges alternately his clenched fists."

- In glassmaking, everything was done by hand, "the gatherers taking the metal from the furnace at the end of an iron rod, the blower shaping the body of the bottle with his breath, while the maker who finished the bottle off . . . tooled the neck with a light spring-handled pair of tongs. Each bottle was individually made no matter what household, shop, or tavern it was destined for." (Zuboff 1988:37–38)

Workers paid a price for the reduction of the physical requirements necessary to produce goods, because skills and knowledge were intertwined with the physical effort.

> Before mechanization, knowledge was inscribed in the laboring body—in hands, fingertips, wrists, feet, nose, eyes, ears, skin, muscles, shoulders, arms, and legs—as surely as it was inscribed in the brain. It was knowledge filled with intimate detail of materials and ambiance—the color and consistency of metal as it was thrust into a blazing fire, the smooth finish of the clay as it gave up its moisture, the supple feel of the leather as it was beaten and stretched, the strength and delicacy of glass as it was filled with human breath. (Zuboff 1988:40)

Industrialization transformed individual workshops into factories, craftspeople into machine operators, and hand production into machine production. Products previously designed as unique entities and assembled by a few people were now standardized and assembled by many workers, each performing only one function in the overall production process. This divi-

---

[4]The European name for the Americas.

[5]These figures do not include those areas of the world imperialist powers divided according to "spheres of influence" such as China. Compared with the European countries, the United States and Japan were latecomers to colonialism and thus were labeled as "minor" powers. During the Age of Imperialism, the U.S. government declared Oklahoma and other territories that were home to Native Americans open to white settlement, entered its last major military conflict with Native Americans resisting resettlement, annexed the independent republic of Hawaii, purchased the Virgin Islands from Denmark, and established military governments in the Dominican Republic and Haiti. In addition, the United States fought and won the Spanish-American War and acquired Puerto Rico, Guam, and the Philippines from Spain. The United States also gained temporary control over Cuba, turning it into a protectorate. Japan annexed Korea, Taiwan, Karafuto (the southern half of the former Soviet island Sakhalin), and the southern half of the Pacific Islands.

Mechanization The addition of external sources of power, such as oil or steam, to hand tools and modes of transportation.

Lewis Hine's remarkable photograph "The Steamfitter" exemplifies many people's concern about the molding of human beings to machines brought about by the Industrial Revolution.

International Museum of Photography at George Eastman House

sion and standardization of labor meant that no one person could say, "I made this; this is a unique product of my labor." The artisans yielded power over the production process to the factory owner because their skills were rendered obsolete by the machines. Now people with little or no skill could do the artisan's work, and at a faster pace.

The Nature of Interaction

Between 1820 and 1860, a series of developments—the railroad, the steamship, gas lighting, running water, central heating, stoves, iceboxes, the telegraph, and mass-circulation newspapers—transformed the ways in which people lived their daily lives. Although all of these developments were important, the railroad and the steamship were perhaps the most crucial. These new modes of travel connected people to one another in reliable, efficient, less time-consuming ways. These inventions caused people to believe they had "annihilated" time and space (Gordon 1989). They permitted

people to travel day and night; in rain, snow, or sleet; across smooth and rough terrain.

Before the steam engine train was introduced, a trip by coach between Nashville, Tennessee, and Washington, D.C., took one month. Before the advent of the steamboat, a 150-mile trip on a canal took 32 hours (Gordon 1989). An overseas trip from Savannah, Georgia, to Liverpool, England, took months. Each year following the discovery of steam power, transportation became faster and more reliable. In 1819, the year of the first ocean crossing by a steam-driven vessel, the trip took one month. By 1881, the time needed to make that trip had been reduced to seven days (*New Columbia Encyclopedia* 1975).

Railroads and steamships increased opportunities for personal mobility as well as the freight traffic between previously remote areas. These modes of transportation facilitated an unprecedented degree of economic interdependence, competition, and upheaval. Now people in one area could be priced out of a livelihood if people in another area could provide goods and materials at a lower price (Gordon 1989).

Industrialization did more than change the nature of work and interactions; it affected virtually every aspect of daily life. "Within a few decades a social order [that] had existed for centuries vanished, and a new one, familiar in its outline to us in the late twentieth century, appeared" (Lengermann 1974:28). The changes triggered by the Industrial Revolution, and especially by mechanization, are incalculable; they are still taking place.

Sociological Perspectives on Industrialization

Sociology emerged as an effort to understand the dramatic and almost incalculable effects of the Industrial Revolution on human life across the globe. The perspective sociology offers for understanding human affairs can be traced to those who first tried to make sense of this dramatic event: Karl Marx, Emile Durkheim,[6] and Max Weber.

Karl Marx (1818–1883)

Karl Marx was born in Germany but spent much of his professional life in London, working and writing in

---

[6]In 1888, Durkheim began an opening lecture for a course in sociology with the following words: "Charged with the task of teaching a science born only yesterday, one [that] can as yet claim but a small number of principles to be definitely established, it would be rash on my part not to be awed by the difficulties of my task" (p. 43).

Karl Marx
German Information Center

collaboration with Friedrich Engels. Two of Marx and Engels's most influential treatises are *The Communist Manifesto* and *Das Kapital*. *The Communist Manifesto* is a 23-page pamphlet that was issued in 1848 and has since been translated into more than 30 languages (Marcus 1998). Upon reading it today, more than 150 years later, one is "struck by the eerie way in which its 1848 description of capitalism resembles the restless, anxious and competitive world of today's global economy" (Lewis 1998:A17). The *Manifesto* begins, "A specter is haunting Europe—the specter of communism," and includes these famous lines: "The workers have nothing to lose but their chains; they have a whole world to gain. Workers of all countries unite."

Das Kapital, a massive multivolume work published in 1867, 1885, and 1894, is critical of the capitalist system and predicts its defeat by a more humane and more cooperative economic system—socialism. Marx's cri-

tique of capitalism as outlined in these treatises has profoundly influenced economic, social, and political life around the world, although Marx would not necessarily have approved of its effects. In an essay marking the 150-year anniversary of *The Communist Manifesto*, John Cassidy (1997) wrote that "in many ways, Marx's legacy has been obscured by the failure of Communism, which wasn't his primary interest. In fact, he had little to say about how a socialist society should operate, and what he did write, about the withering away of the state and so on, wasn't very helpful . . . Marx was a student of capitalism, and that is how he should be judged" (p. 248).

At his funeral, Engels ([1883] 1993) spoke of Marx as the best-hated and most-calumniated[7] man of his time.

> Governments, both absolutist and republican, deported him from their territories. Bourgeois, whether conservative or ultra-democratic, vied with one another in heaping slanders upon him. All this he brushed aside as though it were a cobweb, ignoring it, answering only when extreme necessity compelled him. And he died beloved, revered and mourned by millions of revolutionary fellow workers—from the mines of Siberia to California, in all parts of Europe and America—and I make bold to say that, though he may have had many opponents, he had hardly one personal enemy. His name will endure through the ages, and so also will his work.

According to Marx, the sociologist's task is to analyze and explain **conflict**, the major force that drives social change. The character of conflict is shaped directly and profoundly by the **means of production**, the resources (land, tools, equipment, factories, transportation, and labor) essential to the production and distribution of goods and services. Marx viewed every historical period as characterized by a system of production that gave rise to specific types of confrontation between an exploiting class and an exploited class. For Marx, class conflict was the vehicle that propelled people from one historical epoch to another. Over time, free people and slaves, nobles and commoners, barons and serfs, guildmasters and trade workers have confronted one another.

The Industrial Revolution was accompanied by the rise of two distinct classes or a fundamental divide:

Conflict According to Karl Marx, the major force that drives social change.

Means of production The resources (land, tools, equipment, factories, transportation, and labor) essential to the production and distribution of goods and services.

---

[7]*Calumniated* means to be a victim of unjust and false accusations intended to damage one's reputation. In an interview with the *Chicago Tribune* in 1879, Marx said if he took the time to deny untrue and unjust statements made and written about him that he "would require a score of secretaries."

the **bourgeoisie**, the owners of the means of production, and the **proletariat**, those individuals who must sell their labor to the bourgeoisie. Marx expressed profound moral outrage over the plight of the proletariat, who at the time of his writings were unable to afford the products of their labor and suffered from deplorable living conditions. Marx devoted his life to documenting and understanding the causes and consequences of this inequality, which he connected to a fatal flaw in the organization of production (Lengermann 1974).

Karl Marx believed that an economic system —capitalism—ultimately caused the explosion of technological innovation and the enormous, unprecedented increase in the amount of goods and services produced during the Industrial Revolution. In a capitalist system, profit is the most important measure of success. To maximize profit, the successful entrepreneur reinvests profits to expand consumer markets and obtain technologies that allow products and services to be produced in the most cost-effective way.

The capitalist system is a vehicle of change in that it requires the instruments of production to be revolutionized constantly. Marx believed that capitalism was the first economic system capable of maximizing the immense productive potential of human labor and ingenuity. He also felt, however, that capitalism ignored too many human needs and that too many people could not afford to buy the products of their labor. Marx stated that capitalism already had unleashed "wonders far surpassing Egyptian pyramids, Roman aqueducts, and Gothic cathedrals—[and] expeditions that put in the shade all former Exoduses of nations and crusades" ([1881] 1965:531). He believed that if this economic system were in the right hands—those of socially conscious people motivated not by a desire for profit or self-interest, but by an interest in the greatest benefit to society—public wealth would be more than abundant and would be distributed according to need.

Instead, according to Marx, capitalism survived and flourished by sucking the blood of living labor. The drive for profit (which Marx maintained is derived from the labor of those directly involved in the production process) is a "boundless thirst—[a] werewolflike hunger—[that] takes no account of the health and the length of life of the worker unless society forces it to do so" (Marx 1987:142). The thirst for profit "chases the bourgeoisie over the whole surface of the globe" ([1881] 1965:531).

Emile Durkheim (1858–1918)

The Frenchman Emile Durkheim believed that the sociologist's task is to study social facts. **Social facts** are

Emile Durkheim
Corbis-Bettmann

ideas, feelings, and ways of behaving "that possess the remarkable property of existing outside the consciousness of the individual" (Durkheim 1982:51). That is, social facts do not originate with the people experiencing them. From the time we are born, the people around us seek to impose upon us ways of thinking, feeling, and acting that we had no hand in creating. For example, the words and gestures people use to express thoughts, the monetary and credit system they use to pay debts, the rules governing games such as baseball, tennis, soccer, and basketball, the beliefs and rituals of the religions we follow—all were created before we came on the scene. Thus social facts have a life that extends beyond individuals.

Not only do social facts exist outside individuals, but they also have compelling and coercive powers. When people freely conform to social facts, this power "is not felt or felt hardly at all, since it is unnec-

Bourgeoisie The owners of the means of production.

Proletariat Those individuals who must sell their labor to the bourgeoisie.

Social facts Ideas, feelings, and ways of behaving "that possess the remarkable property of existing outside the consciousness of the individual" (Durkheim 1982:51).

essary" (Durkheim 1982:55). Only when we resist do we come to know and experience the power of social facts. As Durkheim writes, "I am not forced to speak French with my compatriots, nor to use the legal currency, but it is impossible for me to do otherwise. If I tried to escape the necessity, my attempt would fail miserably. . . . Even when in fact I can struggle free from these rules and successfully break them, it is never without being forced to fight against them" (Durkheim 1982:51). That is, even when people manage to rebel and go their own ways, social facts make their power known in the outside resistance they experience. Durkheim argues that "There is no innovator, even a fortunate one, whose ventures do not encounter opposition of this kind" (Durkheim 1982:51). As one example, see "The Power of Social Facts: Facial and Body Hair on Women."

A second situation where people experience the coercive power of social facts is when they are introduced to social facts that conflict with those they know and have come to take for granted. One student in my introduction to sociology class offered this example:

> In the United States it is typically unacceptable for two men to hold hands; immediately the men would be labeled homosexual. I spent some time in Ghana, Africa, several years ago and one of the first cultural differences I noticed was that men, including the men I was with, hold hands. This cultural difference definitely hit home when one day one of the men I was with took my hand as we walked. In order not to offend him, I followed through with this until an appropriate opportunity allowed me to disengage our hands. Even though I was in a country where this was perfectly acceptable, I still felt extremely uneasy with this tradition.

Durkheim's interest in social facts can be traced to the changes he witnessed in connection with the Industrial Revolution. In particular, the social facts governing people's interactions with others began to give way to new ones. Durkheim observed that, as society became industrialized, the nature of the ties binding people together changed in profound ways. Before the Industrial Revolution, most people earned their livelihoods in the same way as their ancestors, and most "spent their whole lives within a few miles of

Max Weber
AKG Photo/London

where they had been born . . . [and] nearly all lived much as their parents and grandparents had lived before them" (Gordon 1989:106).

Durkheim maintained that new kinds of ties based on differences and interdependence among people emerged in conjunction with the Industrial Revolution. Industrialization led to a more complex division of labor, with people earning their livelihoods in a variety of ways. People became united because they could not survive apart from one another; as a result people had to cooperate with each other, even though they might feel no emotional ties with most of the people they encountered in the course of a day.

Max Weber (1864–1920)

The German-born scholar Max Weber has had a monumental influence not only on sociology, but also on political science, history, philosophy, economics, and anthropology. Like Marx and Durkheim, Weber was preoccupied with the Industrial Revolution. In particular, he focused on how this event changed thought and action and how it brought about a process he called rationalization.

Weber defined **rationalization** as a process whereby thought and action motivated by emotion,

Rationalization A process whereby thought and action motivated by emotion, superstition, respect for mysterious forces, and tradition are replaced by thought and action grounded in the logical assessment of the most efficient ways to achieve a valued goal or end (known as value-rational action).

The Power of Social Facts: Facial and Body Hair on Women

Like many females I've talked to about this, I couldn't wait to become old enough to tweeze my eyebrows and shave my legs and underarms. Whenever I reached the age when my mom would let me do these things, I thought, I'd be a woman, no longer a little kid. I had seen my mom do these things, and to me they were exciting rituals into womanhood.

Finally, around the age of 13 or so, my mom gave me permission to start removing offensive hairs. She showed me how to numb the eyebrows by applying an ice cube to each eyebrow. Supposedly, this trick would lessen the pain of ripping out hairs. However, even with the ice cubes, I still found this plucking to be extremely painful. My eyes watered fiercely from the pain. It felt like some form of torture, not the exciting adult ritual I'd anticipated when I was younger.

Likewise, I discovered that shaving was a painful chore. No matter how careful I tried to be with the razor, I always seemed to end up with nicks and cuts. So often the water in the tub would turn red from the blood flowing out of one of my newest cuts. And it seemed like I would never be able to achieve the smooth legs our society wants females to have.

Shaving under my arms was equally problematic. It seemed like shaving my underarms was yet another exercise in frustration, futility, and pain. As with my legs, I could never achieve the smoothness shown in magazines. I would always have some stubble, even if I shaved daily. When I was finished shaving, my underarms would itch. Putting deodorant on top of freshly shaved underarms burned.

Despite the physical pain of shaving and tweezing, I continued to do it because I accepted it as a part of life. I didn't really know why I did these things; it was just what women did, which I picked up on in childhood. The images of women I had seen in the media showed women with smooth, completely hairless legs and underarms and eyebrows that were plucked and shaped. Plus, the women I knew—the women in my family—shaved and tweezed, too.

One day, when I was 19, I decided to abandon shaving my underarms. I was tired of the pain and tired of spending so much time trying to capture the smooth underarm look that seemed impossible anyway. Around the same time, I quit tweezing my eyebrows.

The first time my mom caught sight of my hairy underarms marked the beginning of a tumultuous period in our relationship. We were at our lake cabin. It was July and we were all swimming, my family and our neighbors at the lake. Mom was sitting in a lounge chair when I got out of the water and sat down next to her. Even though I tried to hide my underarms, she saw them. "Rachel," she hissed, "go shave your pits! That's disgusting!" I told her it wasn't disgusting and that I wasn't going to shave them. She has always been very concerned with how others perceive us, not wanting her children to embarrass her with dirty nails or wrinkled clothes. Seeing her only daughter having hair under her arms was mortifying. She was afraid the neighbors would see.

From that day forward, every time I called her she would nag me about my underarms. I would get aggravated with her, telling her over and over again that I

was not going to shave them. She would get angry with me, and I with her. She just couldn't understand why I wouldn't want to shave my underarms. I couldn't understand why it was such a big deal to her. So for a long time, our relationship was tense. I was excluded from summertime gatherings unless I promised not to wear sleeveless shirts. I felt hurt that she could be so embarrassed.

My mother was also bothered when I quit tweezing my eyebrows, but not mortified as she was with my underarm hair. She'd just tell me that if I would tweeze them, it would really open up my eyes and make me look better. I told her it hurt too badly to tweeze them and that I was confident in the way I was.

As we've discussed in class, there are penalties for deviating from social facts. Because I refuse to shave, I have to pay a price. On a personal level, this price was my mom's hostility. It sounds ridiculous but it is true that our relationship suffered because I quit shaving. On a public level, the price is dealing with the stares of strangers when they catch sight of the hair under my arms. (Once, a child, about six, saw my hair and pointed at me and yelled, "Look! THAT LADY HAS HAIR UNDER HER ARMS!!" We were in a crowded restaurant and this was embarrassing.) It is definitely true that when you practice certain behaviors that go against social facts, you have to pay the price. You have to decide if it is worth it or not. For me, freeing myself from shaving is worth occasional public embarrassment, and was even worth a rocky relationship with my mom for a while.

Source: Rachel Shelton, student, Northern Kentucky University (1998).

superstition, respect for mysterious forces, and tradition become replaced by thought and action grounded in the logical assessment of the most efficient ways to achieve a valued goal or end (known as **value-rational action**). That valued goal may be anything—earning a profit, achieving a state of health, earning a college degree, or moving from one location to another. To understand how thought and action guided by tradition, superstition, or emotion differ from value-rational

action, consider the traditional Hmong culture and its "clear and time-honored" views on epilepsy and its treatment.

> **Value-rational action** Thought and action grounded in the logical assessment of the most efficient ways to achieve a valued goal or end.

A person who has seizures is considered to be both burdened and blessed by the condition known as *qaug dab peg* (pronounced "kow da pay")—which roughly translated means "the spirit catches you and you fall down." The epileptic suffers but is also spiritually gifted. According to Hmong culture, epileptics should be protected from injury during seizures and should receive attention from a *txiv neeb* (pronounced "tsi neng")—that is, a Hmong healer who uses "chants, amulets, and chicken and pig sacrifices" (Konner 1997:28).

By contrast, the value-rational thought that guides "modern medicine" views epilepsy not as a sign of spiritual giftedness, but as a brain condition that causes people to lose consciousness for a short period of time and move in violent and uncontrolled ways. The most efficient way to treat this condition is via medications such as phenobarbital. Even though many people may still suffer seizures and side effects while on medication, this treatment option is viewed as a better and safer (more efficient) alternative than finding people to protect the epileptic during his or her seizures.

Weber used the term "rationalization" to refer to the way in which daily life is organized to accommodate large numbers of people. For example, large industries make products such as cars, pesticides, clothes, prescriptions, fast foods, and televisions that enable millions of people to use them. Companies advertise such products as efficient ways of moving from one place to another; killing bugs in the house, in the garden, and on pets; covering the body; achieving a state of health; going from a state of hunger to being satisfied; or being entertained and informed. Yet most people who buy and use these products have no idea how they work, where they come from, or what consequences they bring—they simply know that the products help us to satisfy some goal in a quick and cost-efficient manner. For example, each day millions of people eat chicken sandwiches or hamburgers prepared by fast-food restaurants. Yet few people have ever killed a chicken or cow or know how to prepare the meat once the animal is killed.

Rationalization, therefore, does not assume a better understanding or greater knowledge. People who live in rational environments typically know little about their surroundings (nature, technology, the economy). "The consumer buys any number of products in the grocery without knowing how they are made or of what substances they are made. By contrast, 'primitive' man in the bush knows infinitely more about the conditions under which he lives, the tools he uses and the food he consumes" (Freund 1968:20). Most people are not troubled by such ignorance, but

are content to let specialists or experts understand how things work and make the necessary corrections when something goes wrong. That is, most people assume they do not have to know how things work.

Weber was most concerned about the adverse consequences of such rationalization. He believed that when people in rational environments determine a goal, such as making a profit or achieving a state of health, and then determine the means needed to achieve that goal, they rarely consider other ways of achieving that goal beyond using the most efficient means. In addition, they rarely consider competing goals that might be sacrificed or neglected to achieve that valued goal.

How the Discipline of Sociology Evolves

Even today, the ideas of Marx, Durkheim, and Weber continue to influence the discipline of sociology because they have survived the test of time. Over many decades, people from a variety of backgrounds have found the ideas of these three scholars useful for thinking about a wide array of situations. (Ideas lose their usefulness if they cannot explain the situations and events that people consider important.)

Yet, despite the importance of Marx, Durkheim, and Weber, we must keep in mind that some voices and experiences have been left out of the record of sociology's history; still others were left out initially but were "discovered" later. Harriet Martineau (1802–1836) is one example. This Englishwoman began writing in 1825,[8] but only in recent years has her name been included in some sociology textbooks. From a sociological viewpoint, Martineau's most important book is *Society in America*.[9] Sociologists can learn much from the way she conducted her research on the United States. Martineau made it a point to see the country in all its diversity, and she believed it was important to hear "the casual conversation of all kinds of people" ([1837] 1968:54):

---

[8]In her day, Harriet Martineau was a popular author. One measure of her popularity is that her first book sold more copies in one month than John Stuart Mill's *Principles of Economics* sold in four years. Martineau's success with the general public can be attributed in part to her style: She developed a plot, a setting, and a cast of characters and used them to illustrate economic principles at work in a community (Fletcher 1974).

[9]Martineau receives the most credit for introducing sociology to England; she translated Auguste Comte's six-volume work *Positive Philosophy* into English. Many people consider Comte to be the father of sociology (Terry 1983; Webb 1960).

Harriet Martineau's Society in America *was an important contribution to sociology, but only recently have her name and ideas been included in sociology textbooks.*

Archive Photos

I visited almost every kind of institution. The prisons of Auburn, Philadelphia, and Nashville: the insane and other hospitals of almost every considerable place: the literary and scientific institutions, the factories of the north, the plantations of the south, the farms of the west. I lived in houses that might be called palaces, in log-houses, and in a farm house. I traveled much in wagons, as well as stages; also on horseback, and in some of the best and worst of steamboats. I saw weddings, and christenings; the gatherings of the richer at watering places, and of the humbler at country festivals. I was present at orations, at land sales, and in the slave market. I was in frequent attendance on the Supreme Court and the Senate; and witnessed some of the proceedings of state legislatures.

[10]DuBois began "The After-Thought" to *The Souls of Black Folk* ([1903] 1996) with these words: "Hear my cry, O God the Reader; vouchsafe that this my book fall not still-born into the world-wilderness. Let there spring, Gentle One, from out of its leaves vigor of thought and thoughtful deed to reap the harvest wonderful." These words indicate that DuBois was not confident that his ideas would be heard or taken seriously.

I traveled among several tribes of Indians; and spent months in the southern States, with Negroes. (pp. 52–53)

Perhaps most useful are the methods that Martineau chose to make sense of all this information. First, she wanted to communicate her observations without expressing her judgments of the United States. Second, she gave a focus to her observations by asking the reader to compare the actual workings of the society with the principles on which the country was founded, thus testing the state of affairs against an ideal standard. Third, with this focus in mind, Martineau asked her reader "to judge for themselves, better than I can for them . . . how far the people of the United States lived up to, or fell below, their own theory" (pp. 48, 50).

Another voice that initially was ignored and later was "discovered" as important to sociology is W. E. B. DuBois (1868–1963).[10] Two sentences in the "Forethought" to *The Souls of Black Folk* ([1903] 1996) summarize DuBois' preoccupation and conceptual contribution to sociology: "Herein lie buried many things which if read with patience may show the strange meaning of being black here in the dawning of the Twentieth Century. The meaning is not without interest to you, Gentle Reader; for the problem of the Twentieth Century is the problem of the color-line." The strange meaning of being black in America includes a **double consciousness** that DuBois defined as "this sense of always looking at one's self through the eyes of others, of measuring one's soul by the tape of a world that looks on in amused contempt and pity." The double consciousness includes a sense of twoness: "an American, a Negro; two souls, two thoughts, two unreconciled strivings; two warring ideals in one dark body, whose dogged strength alone keeps it from being torn asunder."

DuBois' preoccupation with the "strange meaning of being black" was no doubt affected by the fact that his father was a Haitian of French and African descent and his mother was an American of Dutch and African descent (Lewis 1993). Historically in the United States, a "black" person is a black even when his or her parents

Double consciousness According to DuBois, "this sense of always looking at one's self through the eyes of others, of measuring one's soul by the tape of a world that looks on in amused contempt and pity." The double consciousness includes a sense of two-ness: "an American, a Negro; two souls, two thoughts, two unreconciled strivings; two warring ideals in one dark body, whose dogged strength alone keeps it from being torn asunder."

W. E. B. DuBois
Corbis-Bettmann

through this world but once. Few tragedies can be more extensive than the stunting of life, few injustices deeper than the denial of an opportunity to strive or even to hope, by a limit imposed from without, but falsely identified as lying within" (1981:28–29).

Gould's statements can be illustrated by considering the case of Josh Gibson, an African American baseball player whose talents can be equated to those of Babe Ruth. Legend has it that Gibson hit a thousand home runs while in the Negro Leagues—leagues for players of color who were not permitted to play in "organized baseball" of the time. (Baseball was integrated with the arrival of Jackie Robinson in 1947.) Spectators said that Gibson's home runs were "quick smashing blows that flew off the bat and rushed out of the stadium" (Charyn 1978:41). In 1943, Gibson suffered a series of nervous breakdowns and was institutionalized at St. Elizabeth's Hospital in Washington, D.C., leaving only to play baseball on the weekends. He began to hear voices; his teammates found him "sitting alone engaged in a conversation with [Yankee star] Joe DiMaggio," a man he was never permitted to play with or against: "'C'mon, Joe, talk to me, why don't you talk to me? . . . Hey, Joe DiMaggio, it's me, you know me. Why don't you answer me? Huh, Joe? Huh? Why not?'" (p. 41). The case of Josh Gibson reminds us that people's physical characteristics (race, gender, age) cannot be the criteria for judging the worth of their ideas and contributions.

Today the discipline of sociology owes much to Marx, Durkheim, Weber, Martineau, and DuBois. Although they wrote in the nineteenth and early twentieth centuries, their observations remain relevant today. In fact, many insights into the character of contemporary society can be gained by reading their writings, because those who witness and adjust to a significant event are intensely familiar with its consequences in daily life. Because most of us living today know only an industrialized life, we lack the insights

are of different "races."[11] To accept this idea, we must act as if whites and blacks do not marry or produce offspring, and as if one parent—the "black" one—contributes a disproportionate amount of genetic material, so large that it negates the genetic contribution of the other parent. In *The Philadelphia Negro: A Social Study*, DuBois ([1899] 1996) wrote about popular ideas of race and reality. Ironically, almost 100 years after this book was published, the flaws of the U.S. system of racial classification, which rests on the assumption that people can be classified into clear-cut racial categories, are just now becoming a standard topic in many sociology texts. Yet DuBois documented that blacks and whites married and paired off, despite laws prohibiting marriage, and that they did have children. He reminded us that race amalgamation took place "largely under the institution of slavery and for the most part, though not wholly, outside the bonds of legal marriage" ([1899] 1996:359). This kind of data, which DuBois painstakingly collected, is used today to discredit the idea of race as a valid way of categorizing humanity and to remind us that "multiracial" people are not a recent phenomenon.

Ignoring some people's ideas not only weakens the discipline of sociology but also does a disservice to those who have been ignored. In *The Mismeasure of Man*, Stephen Jay Gould writes about the agony of being denied the opportunity to participate: "We pass

[11]As one of my students wrote, "I can't be anything but what my skin color tells people I am. I am black because I look black. It does not matter that my family has a complicated biological heritage. One of my great-great-grandmothers looked white but she was of French and African American descent. Another great-great-grandmother looked Indian but she was three-fourths Cherokee and one-quarter black. My great-grandfather looked white but his sister was so black she looked purple. My coloring is a middle shade of brown, but I have picked up a lot of red tones in my hair from my Indian heritage. My family is a good example of how classifying people according to skin color is ridiculous." In the United States, Tiger Woods is a "black" golfer and Colin Powell is a "black" general. We ignore the fact that Woods is a mixture of Chinese, Native American, Thai, white, and African. We ignore that Powell's ancestry is African, English, Irish, Scottish, Jewish, and Arawak Indian (Gates 1995; Page 1996).

In spite of his great athletic abilities, Josh Gibson ("the black Babe Ruth") could play professional baseball only in the Negro Leagues. Gibson was profoundly troubled by the fact that he never got to play with or against a player like Joe DiMaggio.

UPI/Corbis-Bettmann

that come from living in transitional times. Consequently, the words of witnesses can be revealing, as the following anecdote suggests.

Recently a scientist was interviewed on the radio. He maintained that scientists were close to understanding the mechanisms that govern the aging process and that people might soon be able to live to be 150 years old. If the aging mechanisms in fact are controlled, the first people to witness the change will have to make the greatest adjustment. In contrast, people born after this discovery will know only a life in which they can expect to live 150 years. If these postdiscovery humans are curious, they may wish to understand how living to age 150 shapes their lives. To learn about this subject, they will have to look to those who recorded life before the change and who described their adjustments to the so-called advancement.

So it is with industrialization: To understand how it shapes human life, we can look to some of the early sociologists, who laid the foundations of the discipline of sociology. Because the Industrial Revolution (and all that it encompassed) was so important in the professional and personal lives of the early sociologists, the sociological perspective may be regarded as a "necessary tool for analyzing social life and the recurring issues that confront human beings caught up in the intensity and uncertainty of the modern era" (Boden,

Giddens, and Molotch 1990:B1). The insights of early sociologists teach us that "The fascination of sociology lies in the fact that its perspective makes us see in a new light the very world in which we have lived all our lives. . . . It can be said that the first wisdom of sociology is this—things are not what they seem" (Berger 1963:21, 23).

What Is Sociology?

In the classic book *Invitation to Sociology*, Peter L. Berger (1963) presents sociology as a form of consciousness or as a perspective that gives us the theories and concepts to look below the surface of popular meanings and interpretations. Sociologists assume that all human activity has several levels of meaning, some of which remain hidden from view. Sociologists are most interested in those hidden levels. Berger equates the sociologist with a curious observer walking the neighborhood streets of a large city, fascinated with what he or she cannot see taking place behind the building walls. The facades of the buildings offer few clues beyond the architectural tastes of the people who build the structures and who may no longer inhabit them. According to Berger, the wish to penetrate the facade is analogous to the sociological perspective.

Berger offers the following example to illustrate his point. In Western countries, and especially in the

Lennon and McCartney sang, "All you need is love," but when people do attach themselves to someone else, many considerations are quietly at work in their minds. Usually, we have checked out numerous social indicators about the other person before we "fall" in love.

© Anne Dowie

United States, we assume that people marry because they are in love. Popular belief states that love is a violent, irresistible emotion that strikes at random. Upon investigating which people actually marry, we find that the so-called lightning shaft of Cupid seems to be guided by considerations of age, height, income, education, race, and so on. Thus it is not so much the emotion of love by itself that causes us to marry; rather, when certain conditions are met (for example, a person is the right height in relation to another—usually the male is taller), we allow ourselves to "fall in love."

"The sociologist investigating our patterns of 'courtship' and marriage soon discovers a complex web of motives related in many ways to the entire institutional structure within which an individual lives his life—class, career, economic ambition, aspirations of power and prestige" (Berger 1963:35–36). For example, people meet potential spouses in the circles within which they move (work, school, church, neighborhoods, and so on). Considerations of class and education are obviously present if someone meets their spouse at a college campus where the tuition is $1,600 per year versus $20,000. "The miracle of love now begins to look somewhat synthetic" (p. 36). This does not mean that sociologists dismiss the emotion of love as irrelevant. Rather, they look beyond the popular meanings and publicly approved interpretations. In summary, the sociological viewpoint requires those who use it to go beyond official, popular, and widely accepted interpretations and look for other levels of reality.

What exactly do sociologists study? Berger argues that the distinctiveness of the sociological perspective lies with its focus on society. Sociologists use the term **society** to mean a system of **social interaction**, or everyday events in which at least two people communicate[12] and respond to affect one another's behavior and thinking. In the process, the parties involved define, interpret, and attach meaning to the encounter. Sociologists may study a *society* that includes millions, or even billions, of people interacting to various degrees, or they may study a society of people interacting on a much smaller scale, such as a basketball team, a sociology class, or two people meeting for the first time. From a sociological point of view, three people studying together for an exam as well as an exchange student meeting his host family in an airport terminal for the first time qualify as societies. A society is not decided on numerical grounds. A society is a system of interaction analyzed in terms of its social characteristics.

Sociologists use the term **social**[13] to refer to a quality of interaction that involves ways of seeing, thinking, acting, and responding to others that are shaped by two factors: (1) forces outside the individuals or (2) the presence of other people who notice what is going on. Thus two strangers passing on a street corner who remain unaware of one another's presence do not constitute a *social* relationship. If they acknowledge one another, then what transpires between them is social, if only in the sense that the two communicate with gestures they did not invent or speak to one another in a language they did not create.

Why Study Sociology?

A major or minor in sociology, or even a series of sociology courses, can lead to many careers. The American Sociological Association (1995) names the following employment sectors as areas with job opportunities for graduates with a B.A. or B.S. in sociology: social services, community work, corrections, business, health services, publishing/journalism/public relations, government, and teaching. Because sociology is not connected with a specific job track (unlike accounting, management, psychology, or public relations), sociology students must be able to explain to employers that they have acquired, through their study of sociology (and other college courses), a basic set of analytical skills that can be applied to many of the tasks associated with the desired position. When employers, parents, friends, and other outsiders to the discipline of sociology ask, "Why did you major in sociology?" or "Why take sociology classes?," the reply must be convincing. Responses such as "I like people"

Society A large complex of human relationships; a system of interaction.

Social interactions Everyday events in which at least two people communicate and respond to affect one another's behavior and thinking. In the process, the parties involved define, interpret, and attach meaning to the encounter.

Social A quality of interaction that involves the ways of seeing, thinking, acting, and responding to others that are shaped by two factors: (1) forces outside the individuals or (2) the presence of other people who notice what is going on.

[12]The content of that communication may include messages of conflict, hostility, sexual attraction, friendship, loyalty, or the intent to sell or purchase goods or services.

[13]In popular speech, *social* can refer to the informal quality of a certain gathering ("This is a *social* event, not a business meeting"), an altruistic attitude ("He had a strong sense of *social* responsibility"), or something resulting from contact with people ("a *social* disease").

or "Sociology is about people and I want to work with people" are too vague and will lead to puzzled looks and responses such as "So what can you do with that kind of degree?"

Replies must point to the distinctive aspects of the sociological perspective. Sociology is a discipline that looks beyond the individual. "The [basic] insight of sociology is this: human behavior is largely shaped by the groups to which people belong and by the social interaction that takes place within these groups" (Texas A&M 1998). A pattern to interaction exists, and sociologists seek to determine how that pattern emerged, can be maintained, or can be changed. Thus sociologists believe that "problematic" behavior can be modified by changing the pattern or structure of social interactions.

As one example of how the sociological perspective informs decision making, imagine you work for a company that employs thousands of workers. Because of the company's size, employees know only those people working in their unit. Company executives are concerned that the units know little about the company's overall operations. A company picnic is arranged to help employees meet each other. At this event, however, everyone talks only with those they know. How might you change this pattern of interaction? A person informed by the sociological perspective would think of ways to "make" people break out of their limited social circles—for example, separating the crowd into 12 groups according to birthday month and asking each group to arrange themselves in a line according to birth date. This exercise may sound like a corny solution, but it forces people to talk to one another and to solve problems. While still in their birthday groups, employees could be asked to move five steps away from their group, stop, and introduce themselves to the closest person by telling them something memorable ("I love to eat green peppers," "I played tennis in college," "I was at the ballpark when Mark McGwire broke the homerun record").

As a second example of how the sociological perspective informs decision making, imagine a situation where company executives are seeking to promote physical fitness among employees who work at their desks in front of computers all day. One strategy might involve asking workers to take the stairs instead of the elevator. A few people may respond to such a request. A better strategy might be to give workers who take the stairs one Friday off each month. More workers will follow through on such a request. Executives informed by the sociological perspective will also recognize when policies favor one group of employees over others. Not all employees are capable of taking

the stairs (those in wheelchairs, with heart trouble, and so on). Nevertheless, such employees can do other things to maintain their physical fitness. Consequently, executives must devise a fitness plan that includes everyone.

These examples show how the sociological perspective, with its focus on interaction patterns, informs decision making. This approach is valuable because most work-related tasks and issues revolve around coordinating interactions among employees, clients, consumers, suppliers, and other interested parties. In particular, a degree, minor, concentration, or strong interest in sociology will give you the following distinct analytical skills:

- To anticipate the intended and unintended consequences that policies, practices, and technologies might have on interaction patterns and structure of opportunities (see Chapters 2, 9, 10, and 11).

- To see the connections between individual behavior and the structure of social relationships and opportunity structures (all chapters).

- To identify and project population trends—births, deaths, migration, marriage, divorce, family size (see Chaper 12).

- To appreciate and consider viewpoints other than your own (see Chapter 4).

- To recognize and create quality information (see Chapter 3).

- To collect information via interviews, questionnaires, participant observation, case studies, secondary data analysis, and content analysis and to analyze the results (see Chapter 3).

- To avoid making decisions and recommendations that affect others when they are based on superficial knowledge and/or derived from biased perspective about those being affected.[14]

The Importance of a Global Perspective

The classic sociologists—Marx, Durkheim, Weber, Martineau, and DuBois—were wide-ranging and

---

[14]As an Iranian education reformer wrote, "unless we have seen the school's environment and surrounding community, unless we have lived among the people, unless we have been friends with the people, we have not heard their voices and have not known their desires, it is not even proper to have sympathy for the environment, or impose policies, or even write stories or textbooks for them" (Behrangi 1994).

comparative in their outlook. They did not limit their observations to a single academic discipline, a single period in history, or a single society. They were particularly interested in the transformative powers of history, and they located the issues they studied according to time and place. All five lived at a time when Europe was colonizing much of Asia and Africa; when Europeans were migrating to the United States, Canada, South Africa, Australia, New Zealand, and South America; and when slaves and indentured servants were moving to areas to fill demands for cheap labor. Sociologist Patricia M. Lengermann believes that European expansion and movement of peoples had significant consequences for the discipline of sociology. She writes:

> Explorers, traders, missionaries, administrators, and anthropologists recorded and reported more or less accurately the details of life in the multitudes of new social groupings which they encountered. Westerners were deluged with the flood of ethnographic information. Never had man more evidence of the variety of answers which his species could produce in response to the problems of living. This knowledge was built into the foundations of sociology—indeed, one impulse behind the emergence of the field must surely have been Western man's need to interpret this evidence of cultural variation. (1974:37)

This textbook continues this tradition by incorporating a global perspective throughout. A global perspective is guided by the following assumptions:

- Globalization is not new, although the scale of global interdependence changed dramatically with the Industrial Revolution, which gained momentum in England about 1850 and soon spread to other European countries and the United States. The Industrial Revolution contributed to global interdependence because it drew people from even the most remote corners of the world into a process that produced unprecedented quantities of material goods, primarily for the benefit of the colonizing powers.

- The lives of people around the world are intertwined; social relationships do not stop at national borders (see Figure 1.1).

- One country's problems—unemployment, drug abuse, pollution, AIDS, inequality—are part of a larger global situation and cannot be explained by reference only to domestic conditions (Albrow 1990).

- Seemingly local events are shaped by events taking place in foreign countries (Giddens 1990).

The "now-undeniable global extensions of social life" force people to engage with others outside the boundaries of their country (Martin and Beittel 1998).

- The individual biography is affected by events and processes in foreign locations (Moore 1966). Simply consider the number of people in the United States who are descendants of immigrants who crossed paths with native-born people or other immigrants to produce offspring. The offspring's existence depended on events in one or more countries "pushing" their parents together.

- Established social arrangements that we "never see" among people around the globe nevertheless deliver to us products and services we use on a daily basis (Lemert 1995). For example, most of us are not familiar with social arrangements that deliver blood and blood products on a global scale.

- The exchange of goods, services, and influences on a global scale is uneven with some countries—most notably, the United States is generally the more dominant trading partner (Donnelly 1996). U.S. dominance can be observed in the country's military presence in 140 countries and every region of the world (U.S. Department of Defense 1996).[15] In addition, the United States has the most powerful and diverse economy in the world. Approximately 2,000 of the top 8,000 corporations in the world are headquartered in the United States (Center for International Financial Analysis and Research 1992).[16] The United States is also the largest exporter of goods and services, and it is especially known for its entertainment-related exports: films, television programs, videos, radio programming, and recorded music (see "Top 20 Global Grossers of '98").[17, 18]

- The movement of goods, services, and influences is not a one-way process, because people "have some freedom to interpret and reinterpret those products in any way they choose" (Donnelly 1996:243).

---

[15]Other evidence can be found in the amount of money it spends on defense. The $284.4 billion that the United States spends each year represents 37.9 percent of the approximately $750 billion that all governments spend on defense (U.S. Central Intelligence Agency 1995).

[16]The significance of this fact is evident when we consider the evolving global economy is "spearheaded by a few hundred corporate giants," many of which have revenues larger than most countries in the world (Barnet and Cavanagh 1994:14).

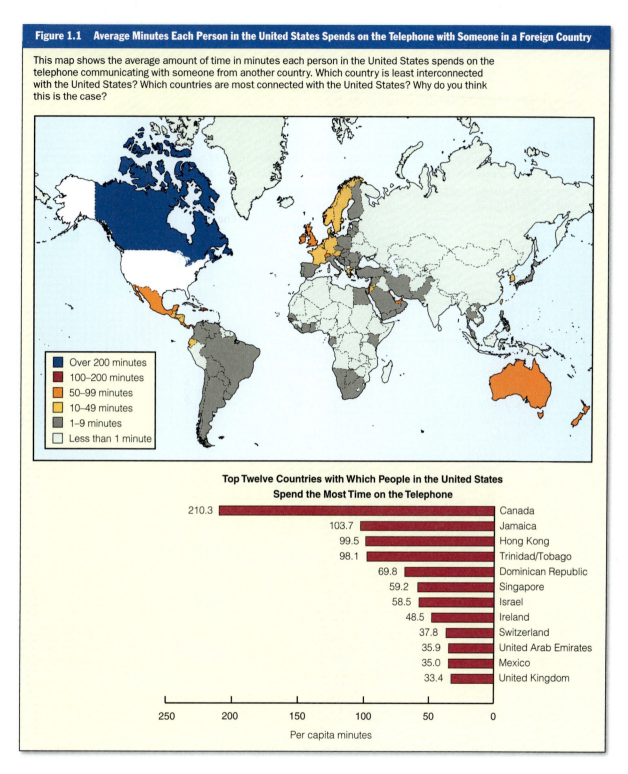

Figure 1.1   Average Minutes Each Person in the United States Spends on the Telephone with Someone in a Foreign Country

This map shows the average amount of time in minutes each person in the United States spends on the telephone communicating with someone from another country. Which country is least interconnected with the United States? Which countries are most connected with the United States? Why do you think this is the case?

Legend:
- Over 200 minutes
- 100–200 minutes
- 50–99 minutes
- 10–49 minutes
- 1–9 minutes
- Less than 1 minute

Top Twelve Countries with Which People in the United States Spend the Most Time on the Telephone

Per capita minutes	Country
210.3	Canada
103.7	Jamaica
99.5	Hong Kong
98.1	Trinidad/Tobago
69.8	Dominican Republic
59.2	Singapore
58.5	Israel
48.5	Ireland
37.8	Switzerland
35.9	United Arab Emirates
35.0	Mexico
33.4	United Kingdom

[17]In this regard, Australian writer Peter Carey (1995) maintains:

Americans have no understanding of the power that their culture has on everybody else; so even when there are people who are your political enemies, they are people who have taken on and internalized your popular music, your sitcoms, and there's a part in their heart which loves this country and its culture. I don't know whether it's Mary Tyler Moore or Bruce Springsteen or what it is, but these parts of this country are out there in the most unexpected places. (pp. 124–125)

[18]The U.S. Department of Commerce (1993) estimates that films produced by United States–based studios are shown in more than 100 countries and that U.S. television programs have audiences in more than 90 countries. As the executive summary of a Department of Commerce (1993) special report, *Globalization of the Mass Media*, states, "the products of the U.S. electronic mass media industry are vehicles through which ideas, images, and information are dispersed across the United States and throughout the world. As such these mass media can be a powerful agent for political and social change."

U.S. in Perspective

Top 20 Global Grossers of '98

These are the 20 top U.S. films of 1998 ranked in terms of gross worldwide incomes. U.S. films are an important factor in how people in other countries view life in the United States. Based on this list, what impressions do you think they get of us? What impressions might be accurate? Which might be just a little off the mark?

Table 1.2 The 20 Top-Grossing U.S. Films (in millions of dollars)			
Title/Distributor	**Domestic**	**Foreign**	**World**
1. *Titanic* (Par/Fox)	$488.2	$1,209.0	$1,697.2
2. *Armageddon* (BV)	201.5	262.8	464.3
3. *Saving Private Ryan* (Dreamworks/Par)	190.8	232.0	422.8
4. *Godzilla* (Sony)	136.2	239.7	375.9
5. *Deep Impact* (Par)	140.5	208.3	348.8
6. *There's Something About Mary* (Fox)	174.1	153.6	327.7
7. *As Good As It Gets* (Sony)	123.3	165.6	288.9
8. *Dr. Dolittle* (Fox)	144.1	132.8	276.9
9. *Lethal Weapon 4* (WB)	129.8	141.7	271.5
10. *Mulan* (BV)	120.6	149.5	270.1
11. *The Truman Show* (Par)	125.6	118.0	243.6
12. *Good Will Hunting* (Mrmx)	134.1	87.0	221.1
13. *The Mask of Zorro* (Sony)	93.6	112.0	205.6
14. *Tomorrow Never Dies* (MGM)	52.0	133.9	185.9
15. *The Horse Whisperer* (BV)	75.4	109.3	184.7
16. *The X-Files* (Fox)	83.9	100.3	184.2
17. *City of Angels* (WB)	78.7	105.1	183.8
18. *The Man in the Iron Mask* (MGM)	57.0	123.1	180.1
19. *Rush Hour* (New Line)	134.8	25.1	159.9
20. *Six Days, Seven Nights* (BV)	74.3	84.2	158.5

Source: *Reprinted with permission of VARIETY Magazine, Reed Elsevier Inc. January 4–10, 1999.* © *by Cahners Publishing.*

- Multinational corporations, international organizations, and special-interest groups are significant forces in structuring social relationships that transcend national boundaries (Harvey, Rail, and Thibault 1996). Consider that the blood industry, which is dominated by a handful of multinationals, facilitates "interaction" between millions of people each year who never meet. As a second example of interactional organizations structuring relationships that transcend national boundaries, consider the World Federation of Hemophilia, which joins together 88 national hemophilia associations.

- Efforts to open and erase national boundaries are accompanied by simultaneous efforts to protect and enforce boundaries. As an example, airports around the world are seeking to process passengers and cargo from around the world more rapidly. At the same time, airport officials are developing systems to preselect passengers considered to be high-risk threats (such as drug traffickers, terrorists, and illegal aliens) for special screening (White House Press Release 1995).

- The production process increasingly draws labor and raw materials from around the world. We can no longer say with certainty that a product is

Globalization from above Connects people from around the world with educational, economic, and political advantages, excluding or pushing to the sidelines those who are not so advantaged.

Globalization from below Involves interdependence at the grassroots level that aims to protect, restore, and nurture the environment and to enhance ordinary people's access to the basic resources they need to live a dignified existence.

"made in the U.S.A." or produced by a U.S. company (Schmitt 1997). "One of the ironies of our age is that an American who buys a Ford automobile or an RCA television is likely to purchase less American workmanship than if he bought a Honda or Matsushita TV" (Reich 1988:78).

- Globalization occurs on at least two levels. **Globalization from above** connects people from around the world with educational, economic, and political advantages, excluding or pushing to the sidelines those who are not so advantaged. **Globalization from below** involves interdependence at the grassroots level that aims to protect, restore, and nurture the environment and to enhance ordinary people's access to the basic resources they need to live a dignified existence (Brecher, Childs, and Cutler 1993).

In this book, we use a global perspective and its assumptions to introduce sociological concepts and theories. We apply those concepts and theories to a wide range of critical issues and events affecting the United States that cannot be separated from a larger global context.

The growth of international airports and the relatively inexpensive cost of rapid international travel are but two signs of how globally oriented our lives and society are becoming. Consider also the many ways we may now communicate with or affect the lives of people halfway around the globe while never leaving home.

© Kevin R. Morris/Tony Stone Images

Key Concepts

Use this outline to organize your review of key chapter ideas.

Sociological imagination
    Troubles
    Issues
    Structure of opportunities
    Social relativity
    Transformative powers of history
Means of production
    Bourgeoisie
    Proletariat

Rationalization
Social facts
Sociology
    Society
    Social
    Social interaction
Globalization from above
Globalization from below

internet assignment

The following is a list of international, national, and regional sociological associations with Web sites.

- International Sociological Association
- American Sociological Association
- Mid-South Sociological Association
- North Central Sociological Association
- Pacific Sociology Association
- Society for Applied Sociology

- Society for the Study of Symbolic Interaction (SSSI): Papers of Interest
- Southern Sociological Association

Use a search engine to find several of the sites listed here, following any links that look interesting. Do any of these sites provide information or links that make sociology appear attractive as a field of study? Explain.

2

Theoretical Perspectives

With Emphasis on U.S. Manufacturing Operations in Mexico

United States–Mexico border crossing at Tijuana. (Lee Celano/Gamma Liaison.)

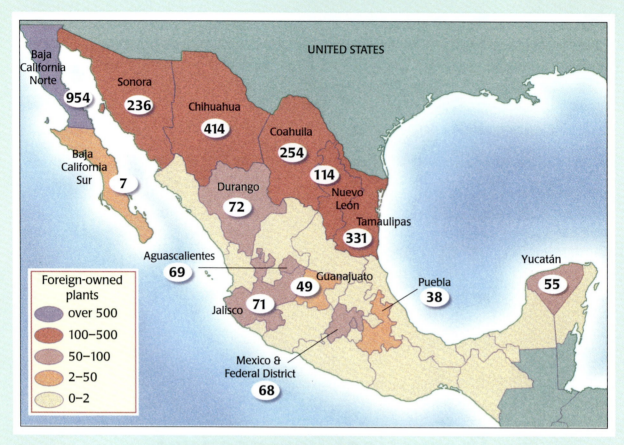

Source: Secretariat of Commerce and Industrial Development, Mexico; *Twin Plant News* (1998d).

The Number of Foreign-Owned Assembly Plants in Mexico

Approximately 2,900 foreign-owned manufacturing plants operate in Mexico. Sixty-four percent of *maquilas* are located along the Mexican side of the U.S.–Mexico border; the rest are located in the interior states of Mexico (Vargas 1998).

Is the presence of so many U.S.-owned factories in Mexico just below the U.S. border good or bad for the people of Mexico? Is it good or bad for the people of the United States? Three basic sociological theories—functional, conflict, and interactionist—provide three ways to examine the issues.

Consider the following facts:

- Approximately 2,900 foreign-owned manufacturing plants exist in Mexico, with an estimated 938,438 employees. Thirty-eight percent are U.S.-owned and another 13 percent have joint U.S.-Mexican ownership (Brayshaw 1998, Twin Plant News 1998d).
- General Motors (GM) is the largest private employer in Mexico, with approximately 65,000 Mexican workers. Delphi Automotive Services, a subsidiary of General Motors, employs 12,000 Mexican workers (Dillon 1998).
- In 1997, 1.2 million loaded trucks and 14.3 million cars and buses traveled into Mexico through the port of Laredo in Texas (Laredo 1997).

Why Focus on Mexico?

In this chapter, we pay special attention to Mexico because, although the United States shares a 1,952-mile border with Mexico (which millions cross each year to shop and visit vacation spots), most Americans[1] know little about what binds the two countries together. In truth, few people know about the large number of U.S. corporations with manufacturing operations on the Mexican side of the

U.S.–Mexico border or understand their economic significance. In addition, most of us are unaware that Mexico is the United States' second largest trading partner (after Canada) or that the United States is Mexico's largest trading partner. In 1997, Mexico purchased 77 percent of its imports from the United States (U.S. Department of State 1998).

Depending on the health of the U.S. economy, many Americans have interpreted the presence of U.S. corporations in Mexico as an economic advantage or a threat to the United States. When unemployment is high, many argue that Mexico, a major source of cheap labor, lures companies out of the United States, takes away jobs, and drives down U.S. workers' wages. Based on this interpretation, the natural impulse is to revoke trade agreements such as the North American Free Trade Agreement (see "What Is NAFTA?"), to pass laws to stop the flow of Mexicans into the United States, to prevent U.S. companies from investing in Mexico, and to limit the flow of Mexican-assembled products into the United States. In contrast, when unemployment is low, concern over illegal immigrations and U.S. corporate presence in Mexico is low.

In light of these narrow, yet very common, interpretations of events, the three major sociological perspectives offer us quite different ways to

assess the relationship between Mexico and the United States and to put what we hear, read, or experience about Mexico into a broader context. In addition, the three perspectives offer us some constructive ways to think about not just this relationship, but also a larger global trend: the transfer of labor-intensive manufacturing or assembly operations out of Western European countries, the United States, and Japan into low-wage, labor-abundant countries such as Mexico, China, Malaysia, Singapore, the Philippines, Taiwan, South Korea, Haiti, Brazil, the Dominican Republic, Bangladesh, Honduras, and Colombia. In this chapter, we use U.S. manufacturing operations in Mexico as a vehicle for thinking about this global trend.

[1]In the broadest sense, the word *Americans* applies to people living in North, South, and Central America, and all of the islands in the surrounding waters. Yet *American* is almost always used in reference to the people, culture, government, history, economy, or other activities of the United States of America. Some critics argue that using *Americans* as synonymous with residents of the United States is an example of "American" arrogance. However, there is no succinct word such as Mexican, Canadian, Brazilian, Central American, South American, or North American that can be applied easily (*United Statesian* is not an accepted word and it is awkward). For this reason I use the word *American* as synonymous with "pertaining to the United States."

In the most general sense, a **theory** is a framework that can be used to comprehend and explain events. In every science, theories serve to organize and explain events going on around us. A **sociological theory** is a set of principles and definitions that tell how societies operate and how people in them relate to one another and respond to the environment.

Three theories dominate the discipline of sociology: the functionalist, the conflict, and the symbolic interactionist perspectives. This chapter outlines the basic assumptions and definitions of each of these theories, showing how each one offers a distinct framework or angle that can be used to interpret any event. Each perspective offers central questions to help guide thinking and provides a vocabulary for answering those questions. We turn first to an overview of the functionalist perspective. The central questions that functionalists ask are, "Why does a particular arrangement exist?" and "What are the consequences of this arrangement for society?" Following this overview, we use the functionalist perspective to explain why U.S.-owned manufacturing plants exist in Mexico and to determine the consequences of this arrangement for both countries.

The Functionalist Perspective

Functionalists focus on questions related to order and stability in society. They define *society* as a system of interrelated, interdependent parts. To illustrate this vision, early functionalists used the human body as an analogy for society. The human body is composed of parts that include bones, cartilage, ligaments, muscles, a brain, a spinal cord, nerves, hormones, blood, blood vessels, a heart, a spleen, kidneys, lungs, and chemicals, all of which work together in impressive harmony. Each body part functions in a unique way to maintain the entire body, but it cannot be separated from other body parts that it affects and that in turn help it function.[2]

Society, like the human body, is made up of parts, such as schools, automobiles and other modes of trans-

portation, sports, medicine, bodily adornments such as tattoos, funeral rites, ways of greeting people, religious rituals, laws, language, household appliances, and tools. Like the various body parts, each of the society's parts *functions* to maintain a larger system. Functionalists consider a **function** to be the contribution of a part to order and stability within the larger system. Consider sports teams—whether they be Little League, grade school, high school, college, city, Olympic, or professional teams. Sports teams function to draw people together who are often extremely different from one another economically, culturally, linguistically, politically, religiously, and in other ways. Loyalty to a sports team transcends individual differences and fosters a sense of belonging to the school, a company, a city, or a country associated with it.

In the most controversial form of this perspective, functionalists argue that all parts of society—even those that seem not to serve a purpose, such as poverty, crime, illegal immigration, and drug addiction—contribute in some way to the larger system's stability. Functionalists maintain that a part would cease to exist if it did not serve some function. Thus they strive to identify how any parts—even seemingly problematic ones—contribute to the stability of the larger society.

Herbert Gans (1972) argues this point in his classic analysis of the functions of poverty. In this analysis Gans asked, "Why does poverty exist?" He answered that poverty performs at least 15 functions, several of which are described below:

- The poor have no choice but to take on the unskilled, dangerous, temporary, dead-end, undignified, menial work of society at low pay. Hospitals, hotels, restaurants, factories, and farms draw their employees from a large pool of workers who are forced to work at minimum or below-minimum wages. This hiring policy keeps the costs of their services reasonable and increases the employer's profits.

- Affluent persons contract out and pay low wages for many time-consuming activities, such as housecleaning, yard work, and child care. This practice gives them time for other, more "important" activities.

Theory A framework that can be used to comprehend and explain events.

Sociological theory A set of principles and definitions that tell how societies operate and how people relate to one another and respond to the environment.

Function The contribution of a part to the order and stability within the larger system.

---

[2]Consider eyelids, for example. When they blink, they work in conjunction with tear fluid, tear ducts, the nasal cavity, and the brain to keep the corneas (the transparent coat over the eyes) from drying out and clouding over. Furthermore, some scientists speculate that blinking functions in some way to activate the brain, which controls the body and performs thought processes (Rose 1988).

What Is NAFTA?

Many Americans have fiercely opposed NAFTA because they believe that companies will take advantage of the agreement to export U.S. jobs to Mexico.

J. Patrick Forden/Sygma

The North American Free Trade Agreement (NAFTA) is a 2,000-page document written with the purpose of eventually eliminating all trade barriers between the United States, Canada, and Mexico. The details of the agreement have been worked out and approved by the U.S. Congress. The agreement went into effect on January 1, 1994. NAFTA sets guidelines for the gradual elimination of tariffs and duties on goods that are traded between the three countries. In its original form, it called for the immediate elimination of many tariffs and the subsequent reduction of other tariffs over a period of up to 15 years. For example, half of U.S. farm goods exported to Mexico immediately became duty-free. Throughout North America, all other tariffs on agricultural goods will be phased out over the next 15 years. Within five years of implementation of the agreement, 65 percent of U.S. goods gained duty-free status (White and Maier 1992).

In general, the goals of the agreement are to eliminate or decrease trade barriers so as to increase trade between the three countries. Designers of the pact believe that this increase will force the three economies to specialize in whatever they do best and, therefore, become more efficient. The increased efficiency, they theorize, will lead to lower consumer prices, the creation of new jobs, and an increased standard of living for the people of all three countries.

One should not conclude that NAFTA will bring about changes without causing problems to some segments of the three countries' economies. After all, it has created the largest market in the world: 396 million consumers with a total GDP of $6 trillion. It will unite three countries with very diverse cultures and levels of economic development. The United States has a population of 267.9 million and a per capita GDP of $28,600. Mexico, on the other hand, has a population of 97.6 million and a per capita GDP of $8,100. Canada's population is only about one-third of Mexico's, at 30.3 million, with a per capita GDP of $25,000 (U.S. Central Intelligence Agency 1998). One thing is clear: "the overall benefits and losses are impossible to predict because there is no historic evidence of what happens when two countries with such different living standards and political systems attempt to integrate" (Fallows 1993). Because the Mexican economy is about 5 percent of the size of the U.S. economy, we can expect that NAFTA will have a greater impact on the Mexican economy.

NAFTA has the potential to "leave virtually no aspect of the economy untouched" (Risen 1992:A1). The significance and strength of the agreement's effects will vary depending on where the individual, corporation, or agency is positioned within the economy of its particular country. The potential effects of the agreement on jobs, wages, and the environment seem to have drawn the most attention. The U.S. government has instituted a retraining program for American workers who lose their jobs because of NAFTA. As of 1998, 150,000 workers were eligible for this program (Smith 1998).

Although some economic analysts predicted that the *maquila* industry would disappear under NAFTA (Lee and Kraul 1993; Essential Organization 1996), NAFTA has had a positive economic impact on most *maquila* operations because "export quotas on originating textile goods and wearing apparel have been eliminated; U.S. Custom duties on originating goods have been decreased or eliminated altogether; and *maquiladoras* are now allowed to sell more goods into the Mexican national market" (Pina 1996:39).

Source: Kevin Steuart, Northern Kentucky University (August 1996); updated July 1998.

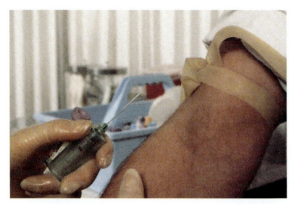

Low-income people often volunteer for over-the-counter and prescription drug tests. This volunteer is testing an AIDS vaccine.

Custom Medical Stock Photo

- The poor often volunteer for over-the-counter and prescription drug tests. Most new drugs, ranging from AIDS vaccines to cat allergy medicine, must eventually be tried on healthy subjects to determine their potential side effects (for example, rashes, headaches, vomiting, constipation, drowsiness) and appropriate dosages. Money motivates subjects to volunteer. Because payment is relatively low, however, the tests attract a disproportionate share of low-income, unemployed, or underemployed people as subjects (Morrow 1996).

- The occupations of some middle-class workers—police officers, psychologists, social workers, border patrol guards, and so on—exist to serve the needs or to monitor the behavior of poor people. For example, approximately 7,300 agents patrol the U.S.–Mexican border (Puente 1997). Similarly, physicians and grocery store owners do not give groceries and medical care away for free to poor people. The poor use food stamps and medical cards to pay providers.

- Poor people purchase goods and services that otherwise would go unused. Day-old bread, used cars, and secondhand clothes are purchased by or donated to the poor. In the realm of services, the labor of many less competent professionals (teachers, doctors, lawyers), who would not be hired in more affluent areas, is absorbed by low-income communities.

Gans outlines the functions of poverty to show how a part of the society that everyone agrees is problematic and should be eliminated remains intact: It contributes to the stability of the overall system. Based on this reasoning, the economic system would be strained seriously if we completely eliminated poverty; industries, consumers, and occupational groups that benefit from poverty would be forced to adjust.

Critique of Functionalism

As you may have realized by now, the functionalist perspective has a number of shortcomings. First, critics argue that the functionalist perspective is by nature conservative in that it defends existing arrangements (Merton 1967). In other words, when functionalists identify how a problematic part of society such as poverty contributes to the system's stability, by definition they are justifying its existence and legitimating the status quo. Functionalists reject this criticism, claiming that they are not justifying poverty's existence, but rather simply illustrating why such controversial practices or "parts" continue to exist despite efforts to change or eliminate them.

Second, critics take issue with the functionalist claim that "parts" exist because they serve a function. Critics argue that a part may not serve any function when it is first introduced. Often people have to work to make parts useful. When the automobile was invented, for example, no paved roads existed, so the automobile's use remained limited. Only later did it realize its transportation function:

> The availability of cheap energy and an inexpensive mass [-produced] car soon transformed the American landscape. Suddenly there were roads everywhere, paid for, naturally enough, by a gas tax. Towns that had been too small for the railroads were reached now by roads, and farmers could get to once-unattainable markets. Country stores that sat on old rural crossroads and sold every conceivable kind of merchandise were soon replaced by specialized stores, for people could drive off and shop where they wanted to. Families that had necessarily been close and inwardly focused in part because there was nowhere else to go at night became somewhat weaker as family members got in their cars and took off to do whatever they wanted to do. The car stimulated the expansiveness of the American psyche and the rootlessness of the American people; a generation of Americans felt freer than ever to forsake the region and the habits of their parents and strike out on their own. (Halberstam 1986:78–79)

The example of the automobile complicates the functionalist argument that a part *exists* because it contributes to order and stability in society. First, the case of the automobile suggests that "necessity is not the mother of invention," but rather that "invention is the mother of necessity." The automobile, for example, did not become a necessity until sometime after it was invented. Second, while the automobile functioned to connect markets, people, and towns, it also weakened ties to family and the community.

Third, although the automobile connects people with one another, it is not the most environmentally sound means of doing so. In fact, the widespread use of and reliance on the automobile hinder the development of more environmentally efficient modes of transportation, such as electric cars, buses, and trains.

Because the functionalist perspective assumes that every part functions in some way to support the smooth operation of society, the theory has difficulty accounting for the origins of social instability. This assumption also leads functionalists to overlook the fact that stability and order are frequently achieved at a cost to some segment of the society, such as poor and powerless individuals.

To address some of this criticism, sociologist Robert K. Merton (1967) introduced a few concepts to the functionalist perspective that help us think about a part's overall effect on the social system and not just its contribution to stability. These concepts are manifest, latent, and dysfunction.

Merton's Concepts

Merton distinguished between two types of functions that contribute to the smooth operation of society: manifest functions and latent functions. **Manifest functions** are a part's intended, recognized, expected, or predictable effects on order and stability in the larger society. **Latent functions** are the unintended, unrecognized, unanticipated, or unpredicted effects on order and stability.

To illustrate this distinction, consider the manifest and latent functions associated with annual community-wide celebrations such as fireworks displays on the Fourth of July or during concerts in the park. Often corporate sponsors join with the city to mount such events. Three expected, or manifest, functions associated with such community celebrations readily come to mind: (1) the celebration functions as a marketing and public relations event for the city and for corporate sponsors, (2) the event provides an occasion to plan activities with family and friends, and (3) the celebration unifies the community through a shared experience.

At the same time, however, several unexpected or latent functions are associated with community celebrations. First, such celebrations often give a visible role to public transportation systems as people take buses or ride trains to avoid traffic jams. Second, such

events break down barriers across neighborhoods. People who do drive may find that they must park some distance from the event, often in neighborhoods that they would not otherwise visit. Consequently, after they park, people have the opportunity to walk through such neighborhoods and observe life up close instead of at a distance.

Merton also points out that parts of a social system can have **dysfunctions;** that is, they can sometimes have disruptive consequences to society or to some segment of society. Like functions, dysfunctions can be manifest or latent. **Manifest dysfunctions** are a part's expected or anticipated disruptions to order and stability. Expected disruptions that seem to go hand in hand with community-wide celebrations include traffic jams, closed streets, piles of garbage, and a shortage of clean public toilets.

In contrast, **latent dysfunctions** are unintended, unanticipated disruptions to order and stability. For instance, community-wide celebrations often have some unanticipated negative consequences. Sometimes police departments and other city workers choose to negotiate contracts with the host city just before the celebration, thereby using the event as a bargaining tool to secure a good contract. (Actually, one might argue that this development is a latent function for the police and a latent dysfunction for the city.) In addition, many people celebrate so vigorously that they miss class or work the next day to recover.

You can see from just this brief analysis of community-wide celebrations that the concepts of manifest and latent functions and dysfunctions provide a more balanced perspective than does the concept of function alone. The addition of these concepts eliminates many of the criticisms leveled at the functionalist perspective. This broader functionalist approach also introduces a new problem: It gives us "no techniques . . . for adding up the pluses and the minuses and coming out with

---

Manifest functions A part's intended, recognized, expected, or predictable effects on order and stability.

Latent functions The unintended, unrecognized, unanticipated, or unpredicted effects on order and stability.

Dysfunctions Disruptive consequences to society or to some segment in society.

Manifest dysfunctions A part's expected or anticipated disruptions to order and stability.

Latent dysfunctions Unintended, unanticipated disruptions to order and stability.

some meaningful overall calculation . . . of the net effect" (Tumin 1964:385). With regard to the community-wide celebration, its positive effects on society do seem to outweigh the negative ones. A part's overall effect is not always so easily determined, however, as we will see when we examine the case of U.S. manufacturing operations in Mexico.

The Functionalist Perspective on U.S. Manufacturing Operations in Mexico

To see how the functionalist perspective can be applied to a specific event, let us consider how functionalists would view U.S. manufacturing operations in Mexico, known as *maquiladoras*[3] (pronounced mah-kee-la-doras). In Mexico, approximately 2,900 *maquiladoras* or foreign-owned manufacturing plants operate. Although there are Asian-, European-, and Canadian-owned *maquiladoras*, about one-third are U.S.-owned and another 13 percent are joint U.S.–Mexican-owned. Well-known companies with *maquila* operations include Xerox, Johnson & Johnson, GTE, Texas Instruments, Black and Decker, Uniroyal, Singer, Clark Equipment, Ford, General Motors, Chrysler, Honeywell, Kellogg, Foster Grant, and Westinghouse.

Maquiladoras work in this way: Foreign companies ship tools, machinery, raw materials, and components duty-free into Mexico. Workers use the tools and machinery to process and assemble the raw materials and components into finished or semifinished products, which are then exported back to the country of the contracting company. No tariff is charged on the finished product; only the cost of labor to assemble the product is subject to tax (Dowla 1997). Mexico, because of its proximity to the United States and because of its abundant, cheap labor force, is the United States' most important partner with regard to assembly operations. Mexican workers perform a variety of labor-intensive tasks, ranging from assembling TV receivers to sorting U.S. retail store coupons. Assembly operations are concentrated, however, in the electronics, automotive, and textile sectors.[4]

In assessing *maquiladoras*, functionalists ask two questions: Why do *maquiladoras* exist, and what consequences do they have for the United States and Mexico? We can apply Merton's concepts to analyze the overall

consequences of the *maquiladora* program on the United States and Mexico. The concepts of manifest and latent functions and dysfunctions remind us that in assessing *maquiladoras*, we must look for intended and unintended consequences that lead to both order and disorder.

MANIFEST FUNCTIONS   To identify the intended, planned functions of the *maquiladoras*, we need to understand why they were created in the first place. The *Maquiladora* Program, also known as the Border Industrialization Program (BIP),[5] was launched in 1965 after the *Bracero* Program was terminated. The *Bracero* Program, which began in 1942, allowed Mexicans to work legally in the United States to relieve labor shortages in rural areas and to bolster the American work force during World War II. The *braceros* replaced American workers who were needed at that time to work in defense plants or to serve in the armed forces. After World War II, the *Bracero* Program was extended several times to supplement the American work force during the Korean War and to provide laborers for agricultural and other low-wage work. With the termination of the *Bracero* Program, the Mexican government established the Border Industrialization Program.

One purpose of the BIP program was to encourage foreign investors to locate plants along the border so that employment opportunities could be created for returning *braceros*. A second purpose was to give U.S. corporations access to a low-wage labor pool that would cut their production costs and allow them to compete in the increasingly global economy. A third purpose was to increase the employment opportunities in Mexican border cities, which were experiencing a high level of population growth because of a large influx of migrants from the interior of Mexico. In addition, the program's creators envisioned that border industrialization would increase economic ties between the border region and the rest of Mexico, provide foreign exchange, and stimulate technology transfer from the United States to Mexico.

Some advocates of the *Maquiladora* Program assess its contribution to stability by estimating the effect on Mexico and the United States if it were eliminated. They project that under such a scenario, Mexico's gross

---

[3]*Maquila* is derivative of the verb *maquilar*, "to do work for another." Originally, *maquila* signified the toll that a farmer paid to a miller for processing grain (Magaziner and Patinkin 1989:319).

[4]This economic arrangement also goes by the name *export processing zones*, of which there are at least 200 in the world (Amirahmadi and Wu 1995).

[5]A foreign company can take part in the BIP in several ways: It can (1) establish its own plant, (2) subcontract with an existing firm in Mexico, or (3) participate in a shelter plan. A shelter plan is an arrangement with a liaison company that provides facilities (buildings and warehouses), contracts with a work force, and handles paperwork and red tape. The contracting company provides raw materials and components, production equipment, and on-site management.

The physical border separating Tijuana, Mexico, from the United States can be marked with signs. The social and economic "border" cannot be marked so easily. A person walking along the beaches of the Pacific Ocean, the Gulf of Mexico, or along most of the border itself would be unable to tell where Mexico begins and the United States ends. Although the Rio Grande might be considered a natural boundary, it marks only about two-thirds of the 1,952-mile border (Verhovek 1997). The map shows that the man-made border between the two countries has changed three times since 1845.

© Peter Johansky/FPG International

national product (GNP) would drop by $3.1 billion because *maquilas* are directly responsible for an estimated 1 million jobs and because two additional "support" jobs are created for every one *maquila* job (*Twin Plant News* 1996a). The U.S. economy would be affected similarly. In addition, if the *Maquiladora* Program disappeared, prices for "American-made" goods would increase by an estimated 36 percent. The higher prices would mean a reduced demand for U.S. products and a corresponding loss of manufacturing, management, marketing, and retail jobs (Cornejo 1988). This projected job loss would occur because *maquilas* function in two important ways to save and create U.S. jobs.

First, U.S.-based companies offer services[6] and supply raw materials and components for assembly by Mexican *maquila* workers. In 1994, for example, the *maquiladoras* in Baja California, Mexico, imported $5.4 billion in goods and services from the United States (California's share alone was $810 million; Kraul 1996). Second, the low-cost, labor-intensive *maquila* output is sent to sister plants in the United States for final assembly. In this way, the Mexican contribution helps to create other manufacturing jobs and permits U.S. workers to receive higher wages for their work.[7]

LATENT FUNCTIONS   On a latent level, the growth of *maquila* plants along the border has the unintended consequence of intensifying not just economic ties, but also political and social ties between the United States and Mexico.[8]

This integration is symbolized by the bridges and border crossings that connect border cities with their respective sister cities across the Rio Grande. These bridges increase not only the flow of goods, but also the exchange of persons, information, and services between the United States and Mexico. For example, in 1997 an estimated 40,000 people from both sides of the border crossed the San Diego–Tijuana line every day legally to work, shop, eat, or go to school, making it one of the two busiest land-border crossings in the

[6]Examples of companies located on the U.S. side of the border that supply services include Border Restaurant Supply (designs, furnishes, and installs cafeterias and dining rooms for *maquilas*), Language Plus (language services to the *maquila* industry), and Latin Moves (manages machinery moves between *maquila* plants).

[7]This economic arrangement is typified by the *maquila* plant Delredo, a division of Delco, one of the largest producers of magnets. Delredo is located in Nuevo Laredo, Mexico. Iron components needed to make magnets are purchased from companies in Oklahoma. The magnets are produced at Delredo and then shipped to a sister plant in Rochester, New York. There the magnets are placed in electric motors that operate power windows and windshield wipers (Jacobson 1988).

[8]Although the *maquila* industry *clearly* has increased this social and economic integration, especially along the border separating the two countries, the United States and Mexico have always been interconnected. This fact becomes obvious when we simply consider that in 1848 the U.S. government annexed (in the name of manifest destiny) Mexican territory, along with many of its inhabitants (who lived in what is now California, Nevada, Texas, Utah, and parts of Arizona, Colorado, New Mexico, and Wyoming). Just because a line was drawn to separate the two countries politically does not mean that the social and economic ties between people on either side ceased to exist.

In a world seemingly committed to freer trade among countries, borders become more and more paradoxical. Even as the U.S. Immigration and Naturalization Service puts up more new security fences to block illegal human crossings from Mexico, governments on both sides want to increase the flow of goods, information, and money.

© Jeffry D. Scott/Impact Visuals

world (Krause 1996). As noted earlier in this chapter, the port of Laredo (in Texas) handled 1,227,000 loaded trucks and 14,300,000 cars and buses in that same year (Laredo 1997). This integration between the two countries is also reflected in recent changes to Mexico's citizenship laws, which now permit anyone born in Mexico or born to Mexican nationals who acquired citizenship elsewhere to officially claim dual nationality. These changes, which resulted from pressures on the Mexican government from Mexican-born U.S. citizens, now allow those with dual-nationality status to work and invest in Mexico and to own property along the border and coast as well as in the interior of Mexico. An estimated 3 million Americans are expected to claim Mexican nationality (Verhovek 1998b).

As a final indicator of Mexican–U.S. interdependence, consider that a significant number of border residents understand both worlds and act to mediate, facilitate, and smooth exchanges between the two countries. Bilingual consultants who are comfortable with both cultures help U.S. businesses establish assembly operations in Mexico. Academic centers that facilitate the blending of cultures include the American

Center for Mexican Studies at the University of California in Los Angeles, the Center for Frontier Studies at the University of Texas in El Paso, and Mexico's Colegio de la Frontera Norte in Tijuana (Sanders 1987). On its Web page, the city of McAllen, Texas, advertises itself as a city "Bridging the Americas for Trade and Manufacturing" (McAllen Economic Development Corporation 1995–1997).

Although the latent functions described here are positive, it does not mean that no cultural frictions exist at the border. In *Days of Obligation: An Argument with My Mexican Father*, Richard Rodriguez (1992) reminds us that on the U.S. side of the border "are petitions to declare English the official language of the United States [and] the Ku Klux Klan nativists posing as environmentalists, blaming illegal immigration for freeway congestion" (p. 84). Rodriguez's observations suggest that this interdependence between the two countries brings some dysfunctional consequences.

MANIFEST DYSFUNCTIONS   Although the *Maquiladora* Program contributes to stable relations between Mexico and the United States, it has several expected (manifest) dysfunctions that are associated with disorder and instability. Job displacement is one obvious manifest dysfunction: Whenever an event alters the way in which significant numbers of people earn their livelihoods, we can anticipate that some adjustments will have to be made. In the United States, for example, an unknown number of workers have been laid off or fired because their jobs were transferred to Mexico. Since 1992, GM's U.S. work force has declined yet during that time the company has built 15 new component plants along the border or in Mexico (Meredith 1997).

The disappearance of entry-level and other manufacturing jobs has left many unskilled workers without jobs and lacking the qualifications to fill newly created jobs. As one example, at its El Paso, Texas, plant Johnson & Johnson laid off 100 workers who cut fabric for surgical gowns, drapes, and other materials. These jobs were transferred to the company's Juarez, Mexico, plant, which employed 3,000 workers (Myerson 1997a). On a larger scale, consider that the last television factory in the United States closed down in 1995, when Zenith laid off 430 workers. Zenith is one of the largest employers of *maquila* workers (Bacon 1995). Workers in rural U.S. counties are particularly vulnerable to this form of job displacement, because manufacturing jobs in these areas tend to be concentrated in routine manufacturing industries (such as food, textiles, furniture production, and apparel). These industries require less-skilled labor (such as garment inspection, repetitive assembly, and simple machine operation) and can be shifted to foreign-based plants (O'Hare 1988).[9]

A second "expected" dysfunction relates to employment in Mexico. As you might expect, the U.S. decision to terminate the *Bracero* Program compounded unemployment problems across the border, which the *Maquiladora* Program was expected to alleviate. Although we have little information about the fate of the *braceros*, we know that the BIP alone could not possibly function to absorb all of them into the Mexican economy. More than 4 million permits to work in the United States were issued to *braceros* between 1951 and 1964 (Garcia 1980), whereas the *maquila* industry has generated only about 1 million jobs over a 30-year period.

LATENT DYSFUNCTIONS Several unexpected or latent dysfunctions are also associated with the growth of *maquilas* along the border. First, because the BIP has increased interdependence between the United States and Mexico, problems in one country affect the other. For example, the 1974–1975 and 1981–1982 recessions in the United States contributed to *maquila* plant layoffs and closings. When a $100 billion debt crisis hit Mexico in the early 1980s and again in 1995, and when Mexico's stock market crashed in 1987, the peso subsequently was devalued. Each time the peso was devalued, the president of Mexico announced that he had to make the people poorer, at least for a time (Golden 1995). These economic crises devastated retail businesses on the U.S. side of the border dependent on cross-border shoppers (Myerson 1995). (Under normal economic conditions, an estimated 40 to 70 percent of the spending by border-area Mexicans takes place on the U.S. side.) Although *maquila* owners benefit from a weak Mexican economy and a weak peso, the adjustments that the Mexican government imposes on its people—forced austerity, currency devaluation, curtailment of social programs, decline of real wages, privatization of government-owned enterprises—make

life very difficult[10] for the masses, increasing the flow of illegal immigrants to the United States and sowing the seeds for political upheaval.

Another latent dysfunction involves the transfer of significant numbers of white-collar jobs from the United States into Mexico. When the *maquila* industry started in 1965, most of the white-collar work that was necessary for these operations was performed by employees in the United States. Since then, however, many Mexican citizens have participated in high school equivalency programs offered by *maquilas* or graduated from one of Mexico's technical schools, whose missions are to prepare young people for skilled work (Uchitelle 1993). File clerks, receptionists, stenographer typists, tool and die makers, and offset press operators can now be drawn from the Mexican population. This alternative labor force is attractive for *maquila* owners because salaries for Mexican workers are much less than those of their American counterparts (*Twin Plant News* 1998a; U.S. Bureau of Labor Statistics 1998; see Table 2.1). Although this situation is in some ways functional for Mexico, it represents a latent dysfunction for the white-collar segment of the labor force in the United States.

The population growth in cities on both sides of the border has been rapid and unregulated, resulting in another latent dysfunction: the establishment of large human settlements characterized by substandard or nonexistent housing, without proper sanitation, transportation, water, electricity, or social services. More than 200,000 people in Texas and New Mexico live in such settlements, known as *colonias* (NAFTA 1993; see "Frequently Asked Questions About *Colonias* in the United States"). In addition, unemployment along the border is high, reaching 28 percent in Starr County, Texas, which has a per capita income of $7,233. While great economic activity clearly occurs along the border, much of it takes the form of warehouses, distribution centers, and truck depots (Verhovek 1998c). The factory jobs that were once located on the U.S. side of the border have moved to the Mexican side. Holders of the factory jobs that remain on the U.S. side receive $7.71 per hour, while their counterparts on the Mexican side earn $1.36.

Problems such as severe and even health-threatening pollution, traffic congestion, flooding, and disease

[9]It would be very misleading, however, if we were to attribute job displacement and loss in the United States to just the *Maquiladora* Program. Mexico is one of many locations in the world to which U.S.-based corporations have moved assembly and other operations. Another often-overlooked reason for job displacement and loss is the failure of U.S. corporations to invest in the technologies that would allow them to compete in a global economy.

[10]During the 1995 economic crisis in Mexico, the value of the peso was devalued 55 percent against the dollar. In personal terms, that meant that a worker earning 50 pesos per day before the devaluation earned the equivalent of $15.40. After the devaluation, that same worker earned 70 pesos per day but earned only the equivalent of $10 (Levi 1995).

Table 2.1   Average Hourly Wages for Selected Occupations in Reynosa, Mexico, Versus U.S. Counterpart

Rates are subject to daily Mexican peso exchange rates and changes in Mexican minimum wage (8.953 pesos = $1.00).

	Pesos	Dollars	U.S. Counterpart ($ U.S.)
General worker	3.50	0.39	10.29
File clerk	4.89	0.55	8.22
Seamstress (factory)	4.53	0.51	7.25
Truck driver (freight)	5.24	0.59	14.07
Warehouse clerk	4.62	0.52	10.73
Nurse (licensed)	5.79	0.65	17.43
Tool and die maker	4.62	0.52	18.00
Offset press operator	5.12	0.57	12.10
Receptionist (bilingual)	4.57	0.51	8.33
Stenographer typist (Spanish)	4.79	0.54	10.25
Watchman	4.53	0.51	8.32
Bartender	4.64	0.52	7.75
Cashier	4.55	0.51	6.18
Chef	5.19	0.58	7.00
Baker	5.12	0.57	6.50
Bulldozer operator	5.38	0.60	12.25
Plumber	4.90	0.55	14.78
Automotive painter	4.93	0.55	10.50
Jeweler	4.76	0.53	12.63
Newspaper reporter	10.52	1.17	11.20
Carpenter	4.76	0.53	11.90

Sources: U.S. Bureau of Labor Statistics (1998); *Twin Plant News* (1998a); Xenon Laboratories (1998).

affect community life on both sides of the border (Herzog 1985). Hence, these problems cannot be solved by one side alone. Problems do not stop at the border, as illustrated by the automobile and truck exhaust generated by traffic slowdowns and delays on the many bridges and other border crossings between Mexico and the United States. It is estimated that vehicles waiting to cross the Cordova Bridge (Bridge of the Americas), which joins El Paso, Texas, and Ciudad Juarez, Mexico, emit 1,280 tons of carbon monoxide, hydrocarbons, and nitrogen oxides annually (Roderick and Villalobos 1992). Obviously, emissions from commercial vehicles affect both sides of the border.

Table 2.2 summarizes the manifest and latent functions and dysfunctions associated with the *maquiladora* industry. These concepts help us answer these questions: Why do *maquiladoras* exist, and what consequences do *maquilas* have on the United States and Mexico? The strength of the functionalist perspective is that it gives us a balanced overview of a part's

contribution—negative, positive, intended, and unintended. Its weakness is that it leaves us wondering whether the overall impact of *maquilas* contributes to stability or instability in the United States and Mexico. Nevertheless, the functionalist perspective does help us see that some people and industries in each country benefit more than others from this labor transfer arrangement. At the same time, it leads us to believe that the negative consequences are simply the costs of overall order and stability. As we will learn in the next section, such a conclusion would receive no support from a conflict theorist.

The Conflict Perspective

In contrast to functionalists, who emphasize order and stability, conflict theorists focus on conflict as an inevitable fact of social life and as the most important agent for social change. Conflict can take many forms, including physical confrontation, manipulation,

U.S. in Perspective

Frequently Asked Questions About Colonias in the United States

WHAT ARE *COLONIAS*?

Colonias are housing clusters with substandard infrastructures occupied predominantly by Hispanic American citizens. They are notable for the absence of one or more of the following: paved streets, numbered street addresses, sidewalks, storm drainage, sewers, electricity, potable water, or telephone services. The quality of housing varies from brick ranch-style homes to packing sheds.

WHERE ARE THEY LOCATED?

Colonias are found in Texas, New Mexico, Arizona, and California, along the border with Mexico.

WHO LIVES IN *COLONIAS*?

Almost all of the residents of *colonias* are of Hispanic origin. There are currently approximately 1 to 1.5 million residents. Eighty-five percent are U.S. citizens. Some are fifth-generation American citizens.

WHAT IS THE FAMILY SIZE?

Colonias average 3.6 persons per household, compared with 2.7 persons in urban areas and 2.6 persons in rural areas. The percentage of households headed by single females is lower in the *colonias* (7.5 percent) than in rural

areas overall (21.4 percent), urban areas (22.1 percent), and the state of Texas (21.1 percent). Also, 36.6 percent of the population are age 17 or younger. In urban areas, 27.8 percent are in that age group.

WHAT IS THE AVERAGE ANNUAL INCOME?

According to the 1990 census, the median income for residents of *colonias* was $16,608. In comparison, the median income for residents of the state of Texas was $26,000, and the average for residents in rural areas was $21,000. Approximately 43 percent of all *colonia* residents live in poverty, whereas only 18 percent of all residents of Texas are below the poverty level.

HOW DID *COLONIAS* BECOME ESTABLISHED?

The lots in the *colonias* are typically sold to low-income migrant farm workers. Thousands of low-income border families are not able to find affordable rental units or obtain home financing from traditional sources; thus they must rely on financing provided by *colonia* developers. Generally, the property does not transfer until the final payment is made.

The lots are small—many only 60 by 100 feet—and sell for $3,000 to $12,000. Lots in *colonia* areas within the corporate limits of border cities cost more. Lots in El Paso and Laredo can cost $10,000 to $20,000.

WHAT ARE THE LIVING CONDITIONS IN THE *COLONIAS*, AND ARE THERE HEALTH RISKS?

In Texas, 26 percent of *colonia* residents live in crowded conditions. Overall, 8 percent of the state's residents live in crowded conditions. "Crowded" is defined as more than one person per room. In *colonias*, 8.9 percent of residents live in housing units with seven or more persons. In urban areas, 1.7 percent live in smaller situations; in rural areas, 1.9 percent live in units with seven or more people.

Only 35 percent of the homes in *colonias* have public sewage disposal systems (81 percent of the homes in Texas have public sewage). *Colonias* also have a high percentage of homes with septic tanks, cesspools, and no sewage removal system.

Source: U.S. Department of Housing and Urban Development (1996).

---

disagreement, dominance, tension, hostility, and direct competition. Conflict theorists emphasize the role of competition in producing conflict. In any society, dominant and subordinate groups compete for scarce and valued resources (access to material wealth, education, health care, well-paying jobs, and so on). Those who gain control of these resources strive to protect their own interests against the resistance of others.

Conflict theorists ask this basic question: Who benefits from a particular pattern or social arrangement and at whose expense? In answering this question, they strive to identify (1) dominant and subordinate groups and (2) practices that the dominant groups have established, consciously or unconsciously, to promote and

protect their interests. Exposing these practices helps explain why access to valued and scarce resources remains unequal.

Conflict theorists draw their inspirations from Karl Marx, who focused on class conflict. Marx maintained that two major social classes exist and that class membership is determined by relationship to the means of production. **Means of production** refers to the land,

Means of production The land, machinery, buildings, tools, and other technologies needed to produce and distribute goods and services.

Table 2.2	**A Functional Analysis of the *Maquiladora* Program**

Manifest Functions

- Creates employment opportunities for returning *braceros* and border residents
- Creates a low-wage labor pool from which U.S. corporations can draw
- Increases economic ties between border region and the rest of Mexico
- Provides a major source of foreign exchange for Mexico
- Saves and creates jobs in the United States
- Allows U.S. corporations to compete in a global economy
- Stimulates technology transfer from the United States to Mexico

Latent Functions

- Integrates Mexico and the United States socially and politically
- Increases exchange of persons, information, and services between the United States and Mexico
- Transfers some white-collar jobs from the United States into Mexico (a plus for Mexico)

Manifest Dysfunctions

- A mismatch between skills needed for jobs lost and jobs created in the United States
- Many workers lost jobs in the United States
- Returning *braceros* were not hired

Latent Dysfunctions

- Environmental and economic problems on one side of the border affect the other
- Transfer of some white-collar jobs from the United States into Mexico (a minus for the United States)
- Rapid growth of *maquila* cities contributes to problems on both sides of the border
- Economic adjustments imposed on Mexican people make their lives very difficult
- Increased numbers of illegal immigrants to the United States
- Establishment of *colonias*

Bourgeoisie The owners of the means of production (land, machinery, buildings, tools) who purchase labor.

Proletariat A less powerful class composed of workers who own nothing of the production process and who sell their labor to the bourgeoisie.

Facade of legitimacy An explanation that members in dominant groups give to justify their actions.

machinery, buildings, tools, and other technologies needed to produce and distribute goods and services. The more powerful class is the **bourgeoisie,** or the owners of the means of production and the purchasers of labor. The bourgeoisie, motivated by a desire for profit, need constantly to expand markets for their products. In an effort to increase profits, they search for ways to make the production process more efficient and less dependent on human labor (using machines, robots, and automation, for example), and they strive to find the cheapest labor and raw materials (see Figure 2.1 for a comparison of labor costs in the apparel industry). This need for profit spread "the bourgeoisie over the whole surface of the globe. It must nestle everywhere, settle everywhere, establish connections everywhere" (Marx [1888] 1961:531). According to Marx, in less than 100 years of existence, the bourgeoisie "has created more massive and more colossal productive forces than have all preceding generations together" (p. 531).

The less powerful class, the **proletariat,** consists of the workers who own nothing of the production process except their labor. The bourgeoisie view the proletariat's labor no differently than they see machines or raw materials. Mechanization combined with the specialization of labor leaves the worker with no skills, according to Marx; the worker is an "appendage of the machine, and it is only the most simple, most monotonous, and most easily acquired knack that is required of him" (Marx [1888] 1961:532). As a result, workers produce goods that have no individual character and no sentimental value to either the worker or the consumer.

Conflict exists between the bourgeoisie and the proletariat, because those who own the means of production exploit workers by "stealing" the value of their labor. They do so by paying workers only a fraction of the profits they make from the workers' labor and by pushing workers to increase output. Increased output without a commensurate pay raise shrinks wages to an even smaller fraction of the profit.

The capitalists' exploitation of the proletariat is disguised by a **facade of legitimacy**—an explanation that members in dominant groups give to justify their actions. On close analysis, however, these explanations are based on "misleading arguments, incomplete analyses, unsupported assertions, and implausible premises" that ultimately support the interests of the dominant group (Carver 1987:89–90).

To illustrate, consider that the bourgeoisie's exploitation of the proletariat is justified by the explanation that members of the proletariat are free to take

Figure 2.1   Hourly Wage Estimates for Apparel Workers in the United States and Selected Countries

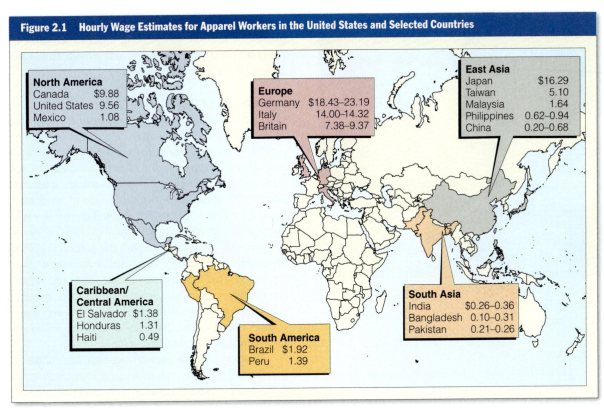

North America
Canada	$9.88
United States	9.56
Mexico	1.08

Europe
Germany	$18.43–23.19
Italy	14.00–14.32
Britain	7.38–9.37

East Asia
Japan	$16.29
Taiwan	5.10
Malaysia	1.64
Philippines	0.62–0.94
China	0.20–0.68

Caribbean/Central America
El Salvador	$1.38
Honduras	1.31
Haiti	0.49

South America
Brazil	$1.92
Peru	1.39

South Asia
India	$0.26–0.36
Bangladesh	0.10–0.31
Pakistan	0.21–0.26

Sources: Werner International as reported by Greenhouse (1997) and Sweatshop Watch (1997).

their labor elsewhere if they are not satisfied with their working conditions, salary, or benefits. On closer analysis, we see that this is not the case. "The Capitalist, if he cannot agree with the Labourer, can afford to wait, and live upon his capital. The workman cannot. He has but wages to live upon, and must therefore take work when, where, and at what terms he can get it. The workman has no fair start. He is fearfully handicapped by hunger" (Engels [1881] 1996).

On the most basic level, employers have considerably more leverage over workers than workers enjoy over their employers: If workers make too many demands or do not produce—or if business is slow or in need of restructuring—employers can fire or lay off their workers. Workers have no comparable leverage against unreliable and overdemanding employers. As a case in point, consider the United Automobile Workers strike against GM in the summer of 1998. During the strike GM threatened to close some of its factories and eliminate some car models. GM had $13.6 billion in cash reserves to fight the strike. Auto workers, on the other hand, received $150 per week in strike benefits (Bradsher 1998a, 1998b).

The most common methods of justifying exploitative practices are (1) blaming the victims by proposing that character flaws impede their chances of success and (2) emphasizing that the less successful benefit from the system established by the powerful. Consider the argument a Denver woman gave *MacNeil/Lehrer Newshour* correspondent Tom Bearden (1993) for hiring an illegal immigrant to care for her children. In particular, pay attention to how the woman justifies the economic relationship with the woman she employs:

MR. BEARDEN: Does the employer of the undocumented worker have too much power over that person? There are some that believe that people who hire undocumented aliens gain an unfair power over them, it gives them influence over them because they're, in a sense, collaborating in something that's against the law. Do you agree with that, or have any thoughts about that?

DENVER WOMAN: I guess I would disagree with that. The one thing that you get in undocumented child care or the biggest thing that you probably get, my woman from Mexico was available to me

This maquila *worker is paid about $5 per day to sew pants for export to the United States.*
© Louis DeMatteis

twenty-four hours a day. I mean, her cost of living in Mexico and quality of life in Mexico compared to what she got in my household were two extremes. When we hired her, she said, "I'll be available all hours of the day, I'll clean the house, I'll cook." They do everything. And if you hire someone from here in the States, all they're going to do is take care of your children. So not only do you have a differentiation in price, you have a differentiation in services in your household. I have to admit that was, at that point, with a newborn infant, wonderful to have someone who was so available. . . .

MR. BEARDEN: And it's not like indentured servitude?

DENVER WOMAN: That crossed my mind, and after she had been here for six months or so, we went to a schedule where she finished at 6 or 7 o'clock at night. And I don't think I ever really took advantage of her. Once a week I'd have her get up with the baby, so I didn't. . . . [S]he was available to me, but I don't feel like I really took advantage of her, other than the fact that I paid her less and she was certainly more available. But she got paid more here than she would have gotten paid if she'd stayed where she was. (p. 8)

Conflict theorists would take issue with the logic that this Denver woman uses to justify hiring an undocumented worker at a low salary. When it comes right down to it, the Denver woman is protecting and promoting her interests (having someone available all hours of the day to cook, clean, and provide child care) at the expense of the illegal worker.

Critique of Conflict Theory

Like the functionalist perspective, conflict theory has its shortcomings. A major criticism is that it overemphasizes the tensions and divisions between dominant and subordinate groups and underemphasizes the stability and order that do exist within societies. It tends to assume that those who own the means of production are all-powerful and impose their will on workers who have nothing to offer except their labor. The theory also assumes that the owners exploit the natural resources and cheap labor of poor countries at will and without resistance. This idea is a somewhat simplistic view of the employer–employee relationship and of relationships between corporations and host countries. It also ignores the real contributions of industrialization in improving people's standard of living.

For example, the president of New Age Intimates, a company that has manufactured brassieres in the Dominican Republic, Jamaica, Haiti, and Colombia for 25 years, points out that although his company pays sewers between $0.75 and $1.25 (a competitive wage, comparable to other sewing factories in the area), it also provides medical care and breakfast. Given that the real unemployment rate in these countries is 50 percent, these jobs are desperately needed and eagerly sought after (Goldberg 1995).

Moreover, some owners of production, such as the CEO of Sequins International in Queens, New York, believe that educated, healthy workers who are not limited to mindless repetitive work tasks will work more effectively, exhibit higher commitment, and produce higher-quality products. The employer can therefore both charge more and cut costs and compete against "500,000 women and children with needle and thread sequins" in China, India, and in other low-wage regions of the world (Sexton 1995). Likewise, many companies such as Allstate Insurance, DuPont, Ford, Procter and Gamble, and Texas Instruments deserve attention for their favorable employee policies, as they were named by *Working Mother* magazine (1997) as being among the "100 Best Companies for Working Mothers."

Finally, conflict theorists tend to neglect situations in which consumers, citizen groups, or workers use economic and other incentives to modify or control the way capitalists pursue profit. For example, they give little attention to cross-border organizing aimed at improving working conditions, establishing unions, and monitoring the environment (Merideth and Brown 1995; Bacon 1997). One such organization is the Coalition for Justice in the *Maquiladoras* (based in San Antonio), "a tri-national coalition of religious, environmental, labor, Latino and women's organizations that seek to pressure U.S. transnational corporations to adopt socially responsible practices within the *maquiladora* industry, to ensure a safe environment along the U.S./Mexican border, safe work conditions inside the *maquila* plants, and a fair standard of living for the industries workers" (Coalition for Justice in the *Maquiladoras*, home page).

The Conflict Perspective on U.S. Manufacturing Operations in Mexico

From a conflict perspective, *maquilas* represent the pursuit of profit. The means of production (machinery, raw materials, tools, and other components) are owned by capitalists in the United States and other foreign countries. The Mexican workers own only their labor and must sell it to the owners of production at a low price. They cannot demand higher wages and better working conditions because the employers can easily replace them from a large pool of Mexican workers. If labor problems emerge, foreign companies can move their operations to the interior of Mexico (or to another country) where the labor is less expensive and they face less competition with other employers for workers (Patten 1996).

As for the facade of legitimacy, some maintain that the *Maquiladoras* Program benefits both the United States and Mexico. Mexico gains because its growing work force gets jobs, and the United States gains because its industries become competitive in the world market. From a conflict point of view, however, the facade of legitimacy masks the real purpose of the *maquila* industry, which is to increase profits by exploiting a vulnerable and low-cost work force. Businesspeople in the United States describe this labor arrangement as being of mutual benefit, but on closer analysis their economic success derives from the fragility of the Mexican economy: the weaker the peso, the lower the wages that company owners have to pay workers (Parra, Fernandez, and Osmond 1996). In fact, the editors of *Twin Plant News*, the leading magazine covering the *maquila* industry, argue that "the *maquila* industry got its start in the mid-1960s, but it wasn't until the peso devaluation of 1988 that it exploded into the giant industry it is today" (*Twin Plant News* 1996a:37). It is no accident that after a second major peso devaluation in 1994, which caused real

wages to drop by 40 percent, Mexico attracted "a record amount of new investment in the *maquila* program as foreign assembly operations benefited from lower labor costs and the improved price competitiveness of Mexican imports to world markets" (Mata 1998).

Conflict theorists argue that, compared with the employer, the Mexican worker gains very little (see Table 2.3). *Maquila* jobs are characterized by insecurity, lack of advancement, and exceedingly low wages. Most of the work is mind-numbing and outrageously repetitive. In electronics plants, for example, workers peer "all day through a microscope, bonding hair-thin gold wires to a silicon chip destined to end up inside a pocket calculator" (Ehrenreich and Fuentes 1985:373). In the *maquilas* where manufacturers' coupons are sorted, workers classify as many as 1,300 coupons each hour. Tens of thousands of times per day, workers drag coupons across an electronic bar code scanner, which flashes a numeric code that identifies the slot where the worker is to file the coupon (Glionna 1992).

Clearly, conflict theorists focus on exploitive conditions, of which plenty prevail. However, not all assembly plants organize the production process and work environment in the manner described here. In *Exports and Local Development: Mexico's New Maquiladoras*, Patricia Wilson (1992) describes three different environments:

> In a hot, stifling, poorly lighted Quonset hut on the Mexican border where shoes are assembled for well-known U.S. department stores, the noxious fumes overwhelmed me as I watched the women and men work. Using sewing machines, thread, and leather from the United States, rows of women stitched tops to soles. Some men attached heels to shoes with hot glue in rapid succession, while others deftly removed excess rubber from the heel by turning the shoe around the sharp blade of a trimmer with their hands, rapidly, mechanically, under the gaze of the Virgin's image pasted on the machine to remind them that they are human.
>
> In another Mexican city I visited a well-lighted, air-conditioned electronics factory with row after row of women inserting scores of tiny colored pieces into circuit boards in just the right order. At break time they doffed the colored robes that indicate their seniority to reveal attractive dresses and high heels. "We come to get out of the house, to socialize, to meet men, not just to support our families," one young worker told me. On the other side of the factory floor was a new automated high-density double-sided insertion machine, attended by one man.
>
> In an auto parts assembly plant I saw young men working in teams aided by machines to stamp, bend,

Table 2.3   The Costs of Everyday Items for Mexican Versus U.S. Workers

According to the conflict theorist Karl Marx, workers exchange their labor for a wage that goes to purchase what they need to survive. A comparison of the labor time Mexican and American workers pay for
a number of everyday items is shown here. Prices of various items in Mexico and metropolitan Los
Angeles are presented below. The time worked to earn these goods is based on a 40-hour work week
and 1998 U.S. Labor Department statistics for the average wage of manufacturing workers in the United
States and Mexico ($13.08 and $0.83 per hour, respectively).

Item	Mexican Price ($ U.S.)	Time to Earn (minutes)	U.S. Price ($ U.S.)	Time to Earn (minutes)
NOKIA cell phone				
Model 636	93.53	6,756.0	300	1,374
Model 2160	83.32	6,018.0	300	1,374
Buffet—Chinese (All-You-Can-Eat)	5.38	390.0	3.99	18.3
Mobile HD40 (5 liters w/bonus funnel)	8.44	606.0	6.45	29.6
Movie theater tickets				
Regular adult prices	3.01	216.0	6.75	30.9
Discounted price	1.94	138.0	4.00	18.3
Ketchup—Hunt's 14 oz	0.63	45.6	1.15	5.3
Canned golden sweet corn	0.43	31.2	0.59	2.7
Pancake mix	0.61	43.8	2.49	11.4
Clorox bleach (1 gallon)	1.82	126.0	1.49	6.8
Pine Sol (28 oz)	0.89	66.0	1.79	8.2
Bottled water (1.5 liters)	0.29	20.4	0.69	3.2
Presidente brandy (750 ml bottle)	4.25	307.2	14.49	66.0

Sources: Albertson's (1998), Amex Liquor Store (1998), Bassett Cinema 6 (1998), Cellular One (1998),
Chinese Gourmet Express (1998), K-Mart (1998).

weld, and paint materials. Each team was responsible
for an entire subassembly. Not far away in a clearing
on the shop floor a soundproof room housed a blackboard where line workers are encouraged to meet
with their team managers to discuss production
problems and suggest alternative solutions. A hand-
written diagnosis was still on the board from the last
occupants. (pp. 1–2)

Conflict theorists would argue that U.S. and other
foreign employers have the upper hand, not only
because Mexico needs a seemingly endless number of
jobs, but also because the country faces intense pressures to generate foreign currency to pay back an estimated $15.3 billion in loans (Preston 1998).[11] Lending
institutions in countries like the United States make

profits from the high interest rates that they charge.
Maquila owners benefit from the conditions of the
loans, which stipulate that Mexico must promote
export manufacturing as a means of generating foreign
currency for loan repayments.

In the meantime, many U.S. companies gain
another measure of control over their domestic workers
and communities. Many local, state, and federal governments give land, special tax breaks, and wage conces-
sions to keep operations in their region or to entice
companies to locate new operations in their areas
(Lekachman 1985). Moreover, companies can threaten
relocation to Mexico or other off-shore locations if workers vote to unionize, complain about working conditions, or ask for pay raises. When Kate Bronfenbrenner
(1997), an Industrial Relations professor, surveyed a
random sample of lead union negotiators involved in
union organizing efforts in the United States between
1993 and 1995, she found that 62 percent of such campaigns were met with plant-closing threats, a significant

[11]Foreign companies cannot pay workers in dollars. They must
exchange dollars for pesos to ensure that the dollars go to the government to pay off debts rather than to the workers. In 1995 Mexico's
debt was $128 billion (U.S. Central Intelligence Agency 1995).

number of which were carried through in manufacturing, transportation, and warehouse/distribution industries. Two students in my 1998 Introduction to Sociology class offered personal examples:

- I can picture my brother-in-law holding a sign that reads "Don't Send My Job to Mexico." Just recently his job was eliminated because the company at which he worked for 30 years moved to Mexico. Because my brother-in-law only did one job for all of those years, his skills were not very marketable. His salary had topped out and for him to work somewhere else he would take a large cut in pay. He is very bitter about the move and has had some depression.

- My boyfriend once worked for a company that would threaten to move the operations to Mexico whenever contract negotiations came around (this company already had a plant located in Mexico). The company and the workers seemed to always come to an agreement right before the strike was set to begin. My boyfriend no longer works there, but we have heard that an upcoming contract negotiation with employees has led to another round of relocation threats.

In addition to focusing on those who exploit Mexican labor, conflict theorists focus on the *maquilas'* effect on the surrounding environment. The Mexican equivalent of the U.S. Environmental Protection Agency found that approximately 1,000 U.S.-owned *maquila* plants generate hazardous waste. Only about one-third comply with Mexican laws requiring them to file reports on the handling of that waste. Only one in five could document that it disposes of hazardous wastes properly[12] (Suro 1991).

Conflict theorists therefore ask, Who benefits from the transfer of labor-intensive manufacturing or assembly operations from the United States into low-wage, labor-abundant countries such as Mexico? Unlike the functionalist perspective, which does not reach a conclusion about the overall effect that an event or arrangement has on society, the conflict perspective zeros in on its exploitive consequences. For conflict theorists, the answer to their question is clear: the owners of produc-

tion, or the capitalists. "But if you stretch it a little, we're all a part of it—anyone who has anything to do with Ford, GM, GE, or if you own anything electronic that's not Asian. Anyone in Wisconsin or Michigan working for these firms or using their products does so at the expense of Latin America" (Weisman 1986:133). Figure 2.2 invites you to make your own comparison between the functionalist and conflict perspectives.

We turn now to the third theoretical perspective: symbolic interaction. It is distinct from the functionalist and conflict perspectives, which focus on the way social systems are organized. Instead, symbolic interactionists focus on how people experience and understand their environment.

The Symbolic Interactionist Perspective

In contrast to functionalists, who ask how parts contribute to order and stability, and to conflict theorists, who ask who benefits from a particular social arrangement, symbolic interactionists ask, How do people define reality? In particular, they focus on how people make sense of the world, on how they experience and define what they and others are doing, and on how they influence and are influenced by others. These theorists argue that something very important is overlooked if an analysis does not consider these issues.

Symbolic interactionists have drawn much of their inspiration from American sociologists George Herbert Mead, Charles Horton Cooley, and Hebert Blumer (who coined the term "symbolic interactionism"). Symbolic interactionists are concerned with how the self develops, how people attach meanings to their own and other people's actions, how people learn these meanings, and how meanings evolve. Consequently, they focus on people and their relationships with one another. Symbolic interactionists maintain that we learn meanings from others, that we organize our lives around those meanings, and that meanings are subject to change (Mead 1934).

According to symbolic interactionists, symbols play a central role in **social interaction,** situations in which two people communicate, interpret, and respond to each other's words and actions. A **symbol** is any kind

---

[12]An agreement between Mexico and the United States, known as Annex III, specifies that hazardous waste and materials (HWM) produced at *maquila* plants must be returned to the country of origin. An application must be filed 45 working days before every HWM shipment. The application specifies, among other things, a detailed description of the HWM, the route and final destination, and emergency measures to be taken in case of accidental spill (Partida and Ochoa 1990).

Social interaction Everyday events in which two people communicate, interpret, and respond to each other's words and actions.

Symbol Any kind of physical phenomenon to which people assign a name, meaning, or value.

Figure 2.2   Border Activity

"From the perspective of the border, borderlines are not lines of sharp demarcation, but broad scenes of intense interactions in which people from both sides work out everyday accommodations based on face-to-face relationships. Each crosser seeks something that exists on the other side of the border. Each needs to make herself or himself understood on the other side of the border, to get food or gas or a job" (Thelen 1992:437). This map shows the tremendous two-way traffic across the border as well as attempts on the part of the United States to control that traffic. How would a functionalist interpret that activity? What would a conflict theorist make of this activity?

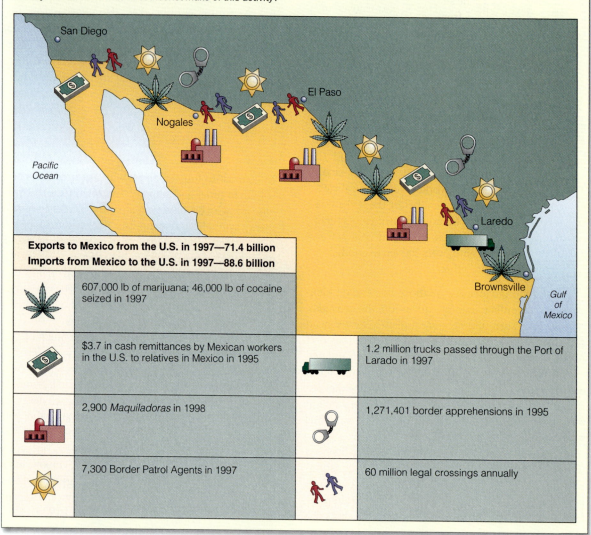

	Exports to Mexico from the U.S. in 1997—71.4 billion Imports from Mexico to the U.S. in 1997—88.6 billion		
	607,000 lb of marijuana; 46,000 lb of cocaine seized in 1997		
	$3.7 in cash remittances by Mexican workers in the U.S. to relatives in Mexico in 1995		1.2 million trucks passed through the Port of Larado in 1997
	2,900 *Maquiladoras* in 1998		1,271,401 border apprehensions in 1995
	7,300 Border Patrol Agents in 1997		60 million legal crossings annually

Sources: Case 1996; Hamashige 1995; Myerson 1997b; Puente 1997; *Twin Plant News* 1998b, 1998c; U.S. Department of Justice 1997; University of California, Davis Migration Dialogue 1998; Verhovek 1998a.

of physical phenomenon—a word, object, color, sound, feeling, odor, movement, taste—to which people assign a name, meaning, or value (White 1949). The meaning or value, however, is not evident from the physical phenomenon alone. This deceptively simple idea suggests

that people construct meaning—that is, they decide what something means.

As an example, consider the various meanings that people have assigned to a suntan. In the United States, a tan has at various times represented quite different ideas about social class, youthfulness, and health. The shift in meanings supports the symbolic interactionists' premise that people assign meanings. Around the turn of the century, wealthy persons purposely avoided tanning to distinguish themselves from members of the working class (farmers and laborers). Pale complexions showed that they did not have to make their living outdoors, laboring under the sun. Then, as the basis of the U.S. economy changed from agriculture to manufacturing, a large portion of the population moved indoors to work. The meaning attached to a pale complexion changed accordingly to represent unrelieved indoor labor; in turn, a tan came to symbolize abundant leisure time (Tuleja 1987).

The presence or absence of a suntan also has reflected ideas about health and youthfulness. Many Americans describe a tan as something that makes them look good and feel better. This meaning will likely change as the result of increasing reports about the connection between exposure to the sun and premature aging and skin cancer (Blakeslee 1994; *New York Times* 1996). In the face of such evidence, a tanned complexion symbolizes the skin's desperate attempt to protect the body from ultraviolet radiation. These changes in the meaning of a suntan underscore (1) the fact that people assign meanings to physical forms and (2) demonstrate that meanings are subject to change. "Because individuals are able to think, they can always put the meanings of objects together in new ways and design new lines of action" (Lal 1995:422).

Symbolic interactionists maintain that people must share a symbol system if they are to communicate with one another. Without some degree of mutual understanding, encounters with others would be ambiguous and confusing. How people come to share and attach meanings is of extreme interest to symbolic interactionists. Consequently, they are interested in how the media, schools, churches, and popular culture convey images and information that influence the process by which people come to learn and share meanings (Lal 1995). The importance of shared symbols frequently is overlooked unless a misunderstanding occurs. Several TV situation comedies—including *Third Rock from the Sun, Mork and Mindy, Perfect Strangers,* and *Beverly Hillbillies*—have emphasized such misunderstandings by featuring characters who do not share the same symbol system as do other characters. These comedies show us that communication problems arise when involved parties place different interpretations on the same event. In addition, they show that when people interact, the parties involved interpret others' actions, words, and gestures first and then respond on the basis of those interpretations (Blumer 1962). We tend to take the interpretation–response process for granted. Usually we are not conscious of how the meanings that we assign to objects, people, and settings shape our reactions. To make us aware of this fact, something must happen that challenges our interpretations.

On a more serious note, consider how physical and social scientists employ the principles of symbolic interaction as they work to design a "keep out sign" to warn people 10,000 years into the future about nuclear waste repository sites.

Critique of Symbolic Interaction

Symbolic interactionists inquire into factors that influence how we assign meanings to what people say and do. They are especially interested in how significant numbers of people come to agree on the meaning of something. Related topics of interest include origins of symbolic meaning, the way in which meanings persist, and the circumstances under which people question, challenge, criticize, or reconstruct meanings. Symbolic interactionists, however, have not established any systematic frameworks for predicting what symbolic meanings will be generated, for determining how meanings persist, or for understanding how meanings change.

Because of these shortcomings, the symbolic interactionist perspective does not provide precise guidelines about where to focus one's attention. For example, on whose interpretation should we focus when we analyze an event? The cast of characters involved in any situation is virtually endless. Even if we were able to consider every possible interpretation of an event, we may still be left with two questions: What really happened? and Whose interpretation best captures the reality of the situation?

The Symbolic Interactionist Perspective on U.S. Manufacturing Operations in Mexico

As we have seen, symbolic interactionists explore the meanings that people assign to words, objects, actions, and human characteristics. Such a broad focus allows considerable flexibility when it comes to applying the symbolic interactionist perspective to foreign manufacturing operations in Mexico. We can examine the different meanings assigned to *maquila* operations by people living on opposite sides of the U.S.–Mexico border, policy makers in the United States and Mexico,

The tourism industry often promotes images of a country and its people that do not correspond with reality or that focus on a small slice of reality. Such images foster and maintain stereotypes, even in the minds of those who never visit the country.

© Rick Strange/The Picture Cube

American and Mexican businesspeople, union leaders in both countries, or displaced workers. We can also analyze why U.S. corporations initially moved their assembly operations to Mexico only slowly. When the value-added system (in which only the labor costs associated with assembling the product are subject to tariff) first became available in the mid-1960s, Hong Kong—not Mexico—emerged as the most important foreign assembly partner of the United States. Longstanding images associated with Mexican workers and Asian workers may help explain why this happened.

Americans tend to stereotype Asians as hardworking, intelligent, and obedient (Yim 1989). In sharp contrast, Mexicans are often stereotyped as unambitious and lazy and as people who value leisure more than work (Noll 1992). The unflattering image of Mexicans that many Americans hold is that of the

"lazy peon asleep in the sun, his sombrero tipped forward over his eyes" (Dodge 1988:48). Many uninformed Americans also tend to associate Mexico with the afternoon siesta and believe that the country closes shop for a couple of hours after a big noontime meal. Given these contrasting conceptions, it is understandable that a label bearing "Assembled in Mexico" symbolizes a much different meaning for consumers and businesspeople than a label bearing "Assembled in Hong Kong."

These images of the siesta and the sombrero, although prevalent, are at odds with the fact that the Mexican people, through their powerful work ethic and sheer sacrifice, have pulled out of at least three economic crises over the past 12 years. Further, many who equate the afternoon siesta with lack of ambition and unwillingness to work probably do not know that for many Mexicans the workday begins at 5:30 A.M. and ends at 7:00 or 8:00 P.M. When we consider that Mexico is now the United States' most important foreign assembly partner and that other foreign corporations also capitalize on the strengths of the Mexican labor force, we might expect such negative images to fade. Yet, the results of a survey of 2,800 Japanese youths by the Mexican government suggest that these conceptions are still strong and are shared by people other than Americans. When Japanese youths were asked, "What comes to mind when you hear the word *Mexico*?" some 30 percent answered "sombrero," 20 percent said "dirt," 10 percent said "crime," and 30 percent said "desert" (Pearce 1987).

Although the origins of these images of lazy Mexicans are not clear, they are rooted in part in American views toward poverty and its causes and are perpetuated by the mass media. Much poverty exists among the Mexican masses, and Americans typically define the poor as a drain on society. They tend to attribute poverty to inferior traits of the people themselves—laziness, lack of discipline, lack of skill, and so on. For some reason, many Americans find it difficult to accept that people can work hard and yet remain poor.

In spite of the negative images of Mexico, this country has emerged over the past decade as a world leader in international assembly operations. To see Mexico as a viable production site, owners of American and other foreign companies had to be convinced that popular conceptions of Mexicans were inaccurate. Working together, Mexican communities trying to attract American investors, corporations offering shelter plans, and U.S. border communities that might benefit from *maquila* operations advertised a new image of Mexico. What attributes did they emphasize to offset the negative stereotypes of Mexicans? Let's examine some of the strategies they used (and continue to use) to market (or define) Mexico.

First, advertisers list the names of U.S. companies with established plants in Mexico to assure potential clients that reputable companies are succeeding there. Second, they emphasize the proximity of *maquila* cities to the United States to remind potential clients that the border is "close to home" and that transportation and communication costs are lower than in Asia. Third, they describe the size and cost of the Mexican labor force relative to that of other countries. Finally, they show pictures of happy, middle-class-type workers to counter images of exploited foreign workers and to

suggest that the workers' standard of living benefits from *maquila* employment. Here are some examples of advertisements from *Twin Plant News* (1990a, 1990b, 1996b) that illustrate these strategies:

- "What Industrial City Is 3 Times the Size of Dallas and 45 Times Closer to You Than Taiwan?" (The advertisement promotes Monterrey, Nuevo Leon.)

- "Shopping the Interior [of Mexico]? Picture This . . . ABUNDANT LABOR . . . Worker Housing for 180,000 is in close proximity to the park."

- "Yucatan is only 90 minutes from the USA. . . . Yucatan has an abundant and efficient labor pool, easy to train and with very low turnover."

Summary and Implications

We began this chapter with some facts that describe ways in which Mexico and the United States are interconnected. Taken alone, the facts tell us little about the relationship between the two countries. At first glance, many Americans see these facts as proof that Mexico threatens their economic well-being. The three theoretical perspectives give us a strategy for thinking about these facts and tempering hasty and oversimplistic reactions. The strategy is reflected in the questions and vocabulary of each perspective (see Table 2.4).

Now that we have seen how each perspective guides analysis through its central question and key terms, it becomes apparent that no single sociological

perspective can give us a complete picture of any situation. Each of the three basic perspectives offers only one way of looking at something. Because no single perspective can capture all aspects of a situation, we can know more of a given situation if we apply more viewpoints. All three perspectives are useful, however, in that each makes a distinct contribution to our understanding.

With regard to *maquilas*, functionalists emphasize how the arrangements between the United States and Mexico create a pattern that contributes to the *overall* economic stability and the well-being of both countries. Although this overall stability comes at a price,

Table 2.4   Overview of the Three Theoretical Perspectives

	Functionalist Perspective	Conflict Perspective	Symbolic Interactionist Perspective
Focus	order and stability	conflict over scarce and valued resources	shared meaning
Vision of Society	system of interrelated parts	dominant and subordinate groups in conflict over scarce and valued resources	interaction is dependent on shared symbols
Key Terms	function, dysfunction, manifest, and latent	means of production, facade of legitimacy	symbols
Central Question	How does a part contribute to overall stability of a society?	Who benefits from a particular pattern or social arrangement, and at whose expense?	How are symbolic meanings generated?
Major Criticisms	defends existing social arrangements; offers no technique to establish a part's "net effect"	exaggerates tension and divisions in society	no systematic framework for predicting which symbolic meanings will be generated or for how meanings persist or change

Different theoretical perspectives highlight different observations on social life. The sign above this top seamstress at the Nova/Link maquiladora *in Matamoros, Mexico, means that she meets or exceeds 100 percent of her production goal. How would each theoretical perspective tend to evaluate that detail?*

© Joel Sartore/National Geographic Society

it must outweigh the disruptive consequences or *maquilas* would cease to exist. Conflict theorists examine how these arrangements benefit the owners of production and exploit workers in the United States, but especially Mexico. Symbolic interactionists consider the meanings assigned by various groups to *maquila* industries and explore how these meanings affect relationships between Mexicans and Americans. Significantly, despite the differences among these theories, none supports the claim that Mexico is a drain on the U.S. economy. In fact, all three support the notion that, for better or worse, the two countries are dependent on each other.

Very few sociologists adhere to only one perspective and maintain that it should be adopted, to the neglect of the other two perspectives. Ideally, the three perspectives should not be viewed as clashing or incompatible. In fact, they overlap considerably. The conflict perspective's focus on exploitation overlaps the functionalist perspective's focus on manifest and latent dysfunctions. Both perspectives recognize the loss of U.S. jobs, the low wages paid to Mexican workers, and the vulnerability of communities that rely on assembly jobs. But their emphasis differs as well. Conflict theorists concentrate on exploitation; functionalists view the negative consequences as the price some segments

in each society pay to support economic prosperity in both countries.

Symbolic interactionists can benefit from the insights of both functionalists and conflict theorists as they attempt to understand the various meanings that *maquilas* symbolize for different segments of the world population. The functionalist and conflict perspectives can be used to explain the origins of the various symbolic meanings assigned to *maquilas*. Among those segments of society that benefit or profit, *maquilas* are likely to evoke positive images. In contrast, *maquilas* are more likely to evoke negative images among exploited segments.

With regard to U.S. assembly operations in Mexico, the larger lesson of this chapter is that we must avoid hasty and oversimplistic reactions to the headline news, sound bites, and so-called facts we hear about issues such as illegal immigration and NAFTA. Consider these recent headlines:

> "U.S. Strengthens Patrols Along Mexican Border"

> "A Losing Battle Against Illegal Immigration"

> "After Peso Collapse, an Unemployed Population Looks North"

> "'Free Trade' Proves Costly to U.S. Jobs: A Somber Look at NAFTA"

Headlines such as these should raise questions rather than solidify opinions. For example, with regard to illegal immigration, three sets of questions associated with theoretical perspectives should immediately come to mind:

1. How does illegal immigration contribute to order and stability in Mexico and the United States? What are the manifest and latent functions and dysfunctions? (functionalist)

2. Who benefits from the existence of illegal immigrants, and at whose expense? (conflict)

3. Does everyone in the United States and Mexico see illegal immigrants in the same way? (symbolic interaction)

Knowing the answers to these questions will not necessarily make decisions about how to respond easier, but the answers will at least offer more information on which to base a constructive response. Although it may be depressing to learn that the answers to these questions may lead to agonizing choices about how to respond, keep in mind that the choices *should* be agonizing whenever other people's lives are affected by your response.

Key Concepts

Use this outline to organize your review of the key chapter ideas.

Sociological theory
    Functionalist perspective
        Function
            Latent function
            Manifest function
Dysfunction
    Latent dysfunction
    Manifest dysfunction

Conflict perspective
    Means of production
        Facade of legitimacy
    Bourgeoisie
    Proletariat
Symbolic interactionist
    Social interaction
        Symbols

internet assignment

Below is a list of twin cities on the U.S.–Mexican border. Twin cities are two geographically adjacent cities—one on the Mexican side of the border, the other on the U.S. side—that are interconnected economically, socially, and politically. Use a search engine to find the home page for at least two cities on each side of the border. What information on the home pages would a conflict theorist, functionalist, and symbolic interactionist find useful for understanding the border?

- San Diego, CA/Tijuana, BCN
- Tecate, CA/Tecate, BCN
- Calexico, CA/Mexicali, BCN
- Yuma (San Luis), AZ/San Luis Rio Colorado, SON
- Nogales, AZ/Nogales, SON
- Douglas, AZ/Agua Prieta, SON
- Columbus, NM/Palomas, CHIH
- El Paso, TX/Ciudad Juarez, CHIH
- Presido, TX/Ojinaga, CHIH
- Del Rio, TX/Ciudad Acuna, COAH
- Eagle Pass, TX/Piedras Negras, COAH
- Laredo, TX/Nuevo Laredo, TAMPS
- Laredo, TX/Columbia, NL
- McAllen (Hidalgo), TX/Reynosa, TAMPS
- Brownsville, TX/Matamoros, TAMPS

Research Methods in the Context of the Information Explosion

With Emphasis on Japan

Japanese cargo ship unloading imported cars at U.S. dock.
(Sylvain Coffie/Tony Stone Images.)

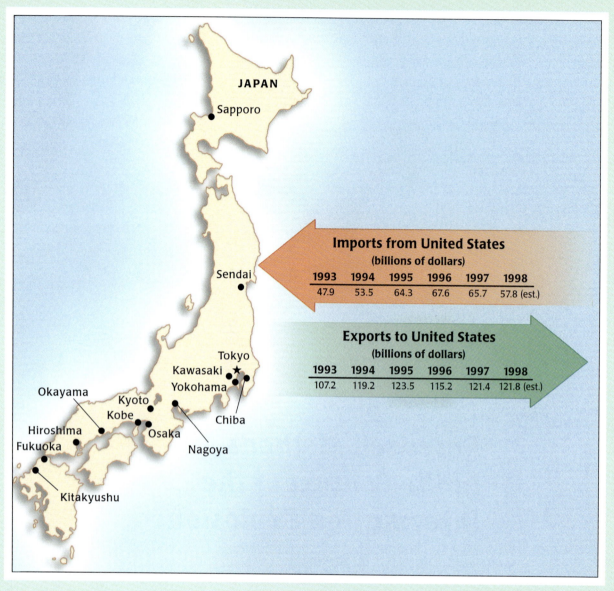

Imports from United States
(billions of dollars)

1993	1994	1995	1996	1997	1998
47.9	53.5	64.3	67.6	65.7	57.8 (est.)

Exports to United States
(billions of dollars)

1993	1994	1995	1996	1997	1998
107.2	119.2	123.5	115.2	121.4	121.8 (est.)

Source: International Trade Association (1998).

The U.S. Trade Deficit with Japan
The United States posts a trade deficit with almost every country in the world, but its largest deficit is with Japan. Many informed Americans worry about this deficit; others maintain that it is no cause for concern. What factors might explain the deficit? What does the trade deficit say about the United States? What does it say about Japan?

How can we begin to study and make comparisons between two societies? This chapter explores how sociologists study social phenomena scientifically. The goal is to gather reliable and valid data and to analyze those data according to the rules of the scientific method.

Consider that in the United States, as of March 1999, the amount of printed material about Japan included:

- 13,076 books in print
- 30,744 magazine articles (since 1960)
- 873,000 holdings in the United States Library of Congress
- 12,339 articles listed in *Social Science Research Index*
- 36,396 articles listed in *Periodical Abstracts*
- 12,329,303 Web sites (Alta Vista search engine)

A significant number of these articles and books present a holistic image of Japanese society, portraying it as a group-oriented society that places great emphasis on consensus and social harmony. Often this image is juxtaposed against an equally holistic image of the United States portrayed as an individual-oriented society that places high value on competition, personal freedom, and independence. These characteristics of Japanese and U.S. society have been used to explain economic growth *and* economic stagnation or recessions in both countries. Certainly each society possesses distinct qualities, but is it accurate to present the two as polar opposites? A knowledge of research methods can help us evaluate this practice.

Why Focus on Japan?

Most Americans know that of the 179 countries with which the United States trades, the largest trade deficit is with Japan and that U.S. economic policy toward Japan is aimed at opening the Japanese market to U.S. goods and services (see Figure 3.1). In fact, the trade imbalance with Japan is the single largest contributor to the record trade deficits that the United States incurred throughout the 1980s and 1990s. Although considerable debate prevails over what this large deficit means, it has increased public awareness about the interdependence between the United States and Japan.

Politicians, the media, and many social critics have used the trade deficit as a measure of the overall health of the U.S. economy, Japanese aggressiveness, the patriotism of U.S. consumers, the competitiveness of U.S. corporations, and the overall quality of U.S. workmanship. The attention given to the trade deficit has caused Americans at almost every level to question why they are not selling as many "U.S.-made" products abroad as they are purchasing from foreign countries (particularly Japan). In the 1980s and early 1990s, the large U.S. trade deficit with Japan caused social critics to question what quality in the Japanese made them so "successful" and what shortcomings in Americans contributed to their "failure." In the mid- to late 1990s, U.S. social critics pointed to the trade deficit as having "nothing to do with how productive" American workers are or with how competitive American businesses are (Miller 1998). Rather, the U.S. trade deficit with Japan reflected fundamental weaknesses in the Japanese economy.

No matter which interpretation of the trade deficit dominated, U.S. politicians, journalists, and social critics tried to explain it by pointing to such things as the special bond between Japanese mothers and their children, the Japanese work ethic (said to be so extreme for men that Japanese children grow up "fatherless"), the value that the Japanese place on the group, unfair trading practices, and the general unwillingness of Japanese consumers to purchase U.S.-manufactured products. The stated or unstated implication was that the opposite qualities exist in the United States. A basic understanding of research methods enables us not only to evaluate the accuracy of the things we read, hear, and view about Japan, but also to evaluate comparisons made between the United States and Japan.

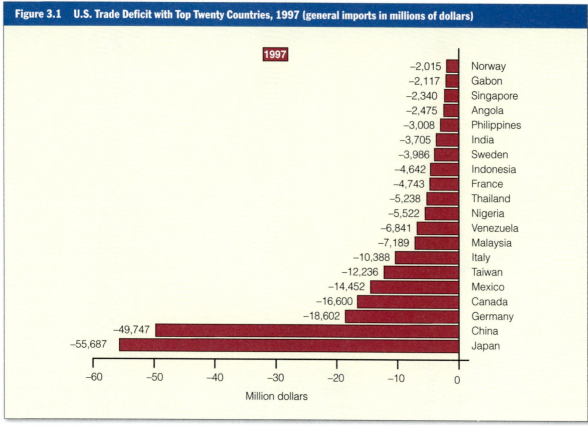

Figure 3.1   U.S. Trade Deficit with Top Twenty Countries, 1997 (general imports in millions of dollars)

1997

−2,015	Norway
−2,117	Gabon
−2,340	Singapore
−2,475	Angola
−3,008	Philippines
−3,705	India
−3,986	Sweden
−4,642	Indonesia
−4,743	France
−5,238	Thailand
−5,522	Nigeria
−6,841	Venezuela
−7,189	Malaysia
−10,388	Italy
−12,236	Taiwan
−14,452	Mexico
−16,600	Canada
−18,602	Germany
−49,747	China
−55,687	Japan

Million dollars

Source: U.S. Department of Commerce (1998), Table 13.

Research is a fact-gathering and fact-explaining enterprise governed by strict rules (Hagan 1989). **Research methods** are the various techniques that sociologists and other investigators use to formulate meaningful research questions and to collect, analyze, and interpret facts in ways that allow other researchers to check their results. We need to possess a working knowledge of research methods even if we do not plan to become sociologists or to do research of our own. One important reason is connected with a relatively new global phenomenon—the **informa-tion explosion.** This dramatic term describes the unprecedented increase in the volume of information made possible by the development of the computer and telecommunications.

In addition to coping with large quantities of information, we have to consider its quality as well. Most of what we hear, read, and see has been created by others. Consequently, we can never be sure that information is accurate. In his article "Too Much of a Good Thing? Dilemmas of an Information Society," Donald Michael (1984) argues that we cannot assume that more information will lessen uncertainty and increase feelings of control and security. In fact, the opposite may be true: More information can overwhelm us to the point that we conclude we cannot believe anything we hear, read, or see. We need not accept Michael's gloomy assessment, however, if we possess a working knowledge of social research methods. Such knowledge gives us the skills to identify and create high-quality information. In addition, this working knowledge enables us to evaluate the accuracy of the things we read, hear, and view and to cope with the consequences of the information explosion.

Research Fact gathering and explaining enterprise governed by strict rules.

Research methods The various techniques used to formulate meaningful research questions and to collect, analyze, and interpret facts in ways that allow other researchers to check the results.

Information explosion A term describing the unprecedented increase in the volume of data made possible by the development of the computer and telecommunications.

The Information Explosion

At least two technological innovations are responsible for the information explosion: computers and telecommunications. Both technologies help people create, store, retrieve, and distribute large quantities of printed, visual, and spoken materials at mind-boggling speeds. Comparing the size and capabilities of the first computers with those of the present suggests why the volume of information has increased so rapidly over the past 50 years or so. The computers of the 1940s weighed five tons, stood 8 feet tall and 51 feet long, and contained 17,468 vacuum tubes and 5,000 miles of wiring (Joseph 1982). Although they performed simple calculations in a few seconds, they tended to overheat and break down. Early computers also used so much power that the lights in nearby towns often failed when the machines were turned on. And, because of their size and cost, these machines were used only by the U.S. Defense Department and the U.S. Census Bureau. Today, in contrast, a single silicon chip only a quarter-inch thick can process millions of bits of information in a second. The chip has reduced computer size and cost and has made possible the widespread use of the personal computer.

Similarly, telecommunications have increased our ability to send information quickly across space. Although the telephone, radio, and television have existed in some form for as long as 100 years, methods of rapidly transmitting clear signals have changed considerably over that time. Fiber-optic cables and satellites have replaced wire cables as the means of transmitting images, voices, and data. In 1923, the cable connecting Great Britain and the United States contained 80,000 miles of iron and steel wire (enough to circle the Earth three times) and 4 million pounds of copper. It could transmit the equivalent of 1,200 letters of the alphabet per minute across the ocean. In contrast, when the capabilities of fiber optics are exploited fully, a single fiber the diameter of a human hair can carry the entire telephone voice traffic of the United States and transmit the contents of the Library of Congress anywhere in the world in a few seconds (Lucky 1985). This latter example is no small feat, considering that the Library of Congress houses 100 million items on 532 miles of shelves (Thomas 1992).

One software tool that has helped to increase the amount of data available at our fingertips is **hypertext.** Anyone who has used the World Wide Web as a research tool has encountered hypertext. Hypertext allows readers to pick and choose among highlighted keywords (that is, important ideas) and follow links to related documents that are stored in computers around the world. As readers select keywords and move from one linked document to another, they may wander off along tangential links, some of which may not be even remotely related to the topic at hand.

Likewise, such innovations as the personal computer and fiber optics have increased the ease with which people can publish, produce, gather, and distribute information and put information creation technologies into the hands of the general public. Writing about the information explosion in *Overload and Boredom: Essays on the Quality of Life in the Information Society,* sociologist Orrin Klapp uses a vivid metaphor to describe the dilemma of sorting through and keeping up with the massive amounts of information generated. Klapp envisions a person "seated at a table fitting [together] pieces of a gigantic jigsaw puzzle. From a funnel overhead, pieces are pouring onto the table faster than one can fit them. Most of these pieces do not match up. Indeed, they do not all belong to the same puzzle" (Klapp 1986:110). The pieces falling from overhead represent research accumulating at a pace that interferes with people's ability to organize it in meaningful ways. Klapp uses this analogy to show that the amount of and speed by which information is produced and distributed overwhelms the brain's capacity to organize and evaluate it.

Consequently, when we want information on any subject, we must sift through and select from a large quantity of data. To complicate matters further, the information is often distorted or exaggerated. Klapp explains why: New technologies permit the existence of large numbers of magazines, newspapers, journals, radio stations, and television channels. Message senders must therefore compete for our attention. Reporters, producers, and others in the media often devise ways to entice us to read and listen to their message. Common strategies include eye-catching headlines, misleading titles, and shocking stories. Some titillating headlines related to Japan, for example, include "What's the Weakest Link in the World Economy? Japan," "Collision Course: Confrontation Looms as Japan Persists with Trade Barriers," "In Japan, Even Toddlers Feel the Pressure to Excel," "Still in Diapers but Cramming for Those Competitive Exams," and "U.S. Trade Deficit Hits Decade-High Level" (Dornbusch 1998). Some examples of boldface excerpts boxed off and highlighted within newspaper articles include the following:

Hypertext Software that allows readers to pick and choose among highlighted keywords (that is, important ideas) and follow links to related documents.

- "When Japanese get fed up, they don't go out and kill their neighbors, as in America. Instead they kill themselves" (Kristoff 1997a:E3).

- "Health care in Japan is equalitarian, meaning you might get well without being treated well" (Kristoff 1997b:7).

- "Think of the global economy as a game of musical chairs, and cautious and reticent, the Japanese have been left without a seat" (Kristoff 1997a:7).

Too often such exaggerated headlines lure people into reading and listening to material that turns out to be trivial, repetitive, contradictory, or ultimately uninformative in any broader sense. The titles and headlines might catch our attention, but they usually mask rather than reveal a more complex reality.

Klapp also cites a **dearth of feedback** as a second factor in creating poor-quality information. What does he mean by "dearth of feedback"? Much of the televised and published information is not subjected to honest, constructive feedback, because there are too many messages and not enough critical readers and listeners to evaluate it before the material is released or picked up by the popular media. Without feedback, the creators cannot correct their mistakes; thus they produce information of diminished quality. Klapp believes that information overload, coupled with distortion, exaggeration, and triviality, is as problematic as a lack of information.

Although it is easy to become pessimistic about people's ability to organize, evaluate, comprehend, and trust the growing quantity of data, the situation has another, brighter side. For one thing, the information explosion also increases the chances that good and useful ideas that might have been overlooked or rejected by information gatekeepers will receive some exposure. For another, the variety translates into something for everyone. And, even though the data are not well organized, researchers still can draw from and organize this information in new and unexpected ways.

Dearth of feedback A situation in which not enough critical readers and listeners evaluate material before it is used by the popular media.

Research-methods literate Knowing how to collect data worth entering into a computer and then how to interpret the resulting data.

Scientific method An approach to data collection in which knowledge is gained through observation and its truth confirmed through verification.

The information explosion does not negate the need to be informed; rather, it simply increases the need to create, identify, and synthesize information. Decisions still must be made, actions still must be taken, and policies still must be formed. No constructive decision, action, or policy can be based on haphazard, misleading, or inadequate data. That situation would be equivalent to a physician's decision to perform heart surgery based only on the intuition that such action will solve the patient's health problems, without ordering medical tests, reviewing the patient's history, or interviewing the patient beforehand.

The larger point is that we live in a society in which people need to be more than computer literate (that is, able to operate a computer and use it to input, access, and print out data). They also need to be **research-methods literate**; that is, they must know how to create information that is worth putting into the computer and how to interpret what comes out of it. Unless computer-literate people also have research-methods literacy, they merely possess the skills to access and navigate, not evaluate (Kinnaman 1994). We turn now to basic techniques and strategies that sociologists (and all other researchers) use to evaluate and gather reliable data. We begin with the guiding principle: the scientific method.

The Scientific Method

Sociologists are guided by the scientific method when they investigate human behavior; in this sense they are scientists. The **scientific method** is an approach to data collection that relies on two assumptions: (1) knowledge about the world is acquired through observation, and (2) the truth of the knowledge is confirmed by verification—that is, by others making the same observations. Researchers collect data that they and others can see, hear, taste, touch, and smell (that is, observe through the senses). They must report the process by which they make their observations and present conclusions so that interested parties can duplicate or critique that process. If observations cannot be duplicated, or if repeating the study yields results that differ substantially from those of the original study, we consider the study to be suspect. Findings endure as long as they can withstand continued reexamination and duplication by the scientific community. "Duplication is the heart of good research" (Dye 1995:D5). No finding can be taken seriously unless other researchers can repeat the process and obtain the same results. When researchers know that others are critiquing and checking their work, it works to reinforce careful, thoughtful, honest, and conscien-

Japanese children visit the Hiroshima Peace Memorial Museum. Might a study of the clothing worn by Japanese children or other groups in Japan lead to insights about changes in Japan's relations with the United States and the world as part of its rise to economic power? What questions might such a study ask?

©Jodi Cobb/National Geographic Society

tious behavior. Moreover, this "checking" encourages researchers to maintain **objectivity**—that is, not to let personal and subjective views about the topic influence their observations or the outcome of the research.

Because of continued reexamination and revision, research is both a process and a dialogue. It is a process because findings and conclusions are never considered final. It is a dialogue because a critical conversation between researchers and readers leads to more questions and additional research. This description of the scientific method is an ideal one, because it outlines how researchers and reviewers *should* behave. "The research enterprise is based on core values—honesty, skepticism, fairness, collegiality, and openness . . . So long as they remain strong, science—and the society it serves—will prosper" (National Academy of Sciences 1995). In practice, though, questionable acts occur sometimes. Some reviewers may dismiss another's research as unimportant (often even before reading the results) and as unworthy of examination simply because the topic is controversial or departs from mainstream thinking, or because the results are reported by someone from a supposedly "inferior" group. Moreover, the ideal scientific method assumes that researchers are honest—that they do not make up data (fabrication) or change data to support personal, economic, and political agendas. The extent to which researchers actually are honest remains unknown.

In one survey sponsored by the American Association for the Advancement of Science (AAAS), 25 percent of the 1,500 professional researchers surveyed reported that in the past 10 years they had witnessed some faking, falsifying, or plagiarizing of data (Marsa 1992). In another survey of 2,000 doctoral candidates and 2,000 faculty[1] from 99 of the largest chemistry, civil engineering, microbiology, and sociology graduate schools in the United States, 22 percent of faculty reported instances where colleagues overlooked sloppy use of data. Another 15 percent knew of cases where researchers failed to report data that would have contradicted their previous research (Swazey, Anderson, and Lewis 1993).

Ideally, research is a carefully planned, multistep, fact-gathering, and fact-explaining enterprise (Rossi 1988) that involves a number of interdependent steps:

1. Choosing the topic for investigation/deciding on the research question
2. Reviewing the literature
3. Identifying core concepts
4. Choosing a research design, forming hypotheses, and collecting data
5. Analyzing the data
6. Drawing conclusions

Researchers do not always follow these six steps in sequence, however. Sometimes they do not define the topic (step 1) until they have familiarized themselves with the literature (step 2). Sometimes an opportunity arises to gather information (step 4), and a project is defined to fit that opportunity (step 1). Although the six steps need not be followed in sequence, all must be completed at some point to ensure the quality of the project.

In the sections that follow, we will examine each stage individually, making reference to a variety of research projects comparing people living in the United States with people living in Japan on some attribute, documenting some supposedly unique qualities about Japanese society, and investigating the U.S.–Japan trade deficit (see Figure 3.1).

Step 1: Defining the Topic for Investigation

The first step of a research project involves choosing a topic or deciding on a research question. It would be impossible to compile a comprehensive list of the topics that sociologists study, because almost any subject involving humans represents a potential target for

---

[1]Two thousand six hundred people returned the survey, for a response rate of 65 percent.

Objectivity A state in which personal, subjective views do not influence one's opinions or behavior.

The preschool obentōs *that inspired Anne Allison's study are a good example of how sociologists' curiosity may be aroused by almost any fact of social life.*

©Ulrike Welsch

investigation. Sociology is distinguished from other disciplines not by the topics it investigates, but by the perspectives it uses to study topics (see Chapter 2).

Good researchers explain to their readers *why* their topic or research question has significance. This explanation is vital because it clarifies the purpose and significance of the project, as well as the researcher's motivation for doing the work. If a researcher cannot say why he or she is conducting a study, the project will lack focus or purpose.

Researchers choose their topics for a number of reasons. Personal interest is a common and often understated motive. It is perhaps the most significant reason that someone picks a specific topic to study, especially if we consider how a researcher eventually chooses one topic from a virtually infinite set of possibilities. Consider the reasons why researcher Anne Allison (1991) studied *obentōs*, the lunch boxes that Japanese mothers make up for their children when they are in nursery school. An *obentō* is a "small box packaged with a five or six course miniaturized meal whose pieces and parts are artistically arranged, perfectly cut, and neatly arranged" (Allison 1991:196). Allison became interested in the *obentōs* while she was living in Japan and taking her child to a Japanese nursery school. Her son's nursery school teacher talked with Allison "daily about the progress he was making finishing his *obentōs*" (p. 200). As Allison tells it:

The intensity of these talks struck me at the time as curious. We had just settled in Japan and David, a highly verbal child, was attending a foreign school in a foreign language he had not yet mastered; he was the only non-Japanese child in the school. Many of his behaviors during this time were disruptive: for example, he went up and down the line of children during morning exercises hitting each child on the head. Hamada-sensei [the teacher], however, chose to discuss the *obentōs*. I thought surely David's survival in and adjustment to this environment depended much more on other factors, such as learning Japanese. Yet it was the *obentō* that was discussed with such recall of detail ("David ate all his peas today, but not a single carrot until I asked him to do so three times") and seriousness that I assumed her attention was being misplaced. (pp. 200–201)

On a personal level, Allison undertook this study to learn more about her son's new environment so she could help him adapt to it. Realizing that the *obentō* was somehow significant to his success in a Japanese preschool, Allison picked it as a research topic. The choice of a research topic, however, usually has some further significance that goes beyond personal interest. Allison's research on *obentōs*, for example, examines why Japanese mothers are expected to make such elaborate lunches for their children and why Japanese teachers pay so much attention to how well children eat lunch. In addition, her research offers readers some insights into the Japanese system of preschool education and allows us to see the educational significance of eating lunch—a seemingly routine activity. According to Allison's research, the elaborately prepared lunch indicates a Japanese woman's commitment as a mother, which inspires her child to be similarly committed as a student.

Step 2: Reviewing the Literature

All good researchers take existing research into account. They read what knowledgeable authorities have written on the chosen topic, if only to avoid repeating earlier work. More importantly, reading the relevant literature can generate insights that the researcher may not have considered. Even if a researcher believes that he or she has revolutionary ideas, the researcher must consider the works of past thinkers and show how the new research verifies, advances, and corrects what has been done in the past.

One example of research that "corrected" existing literature is Roger Goodman's (1993) work on *kikokushijo*—Japanese children who return to Japan after living overseas with their parents who work for Japanese corporations operating in foreign countries.

Goodman found that literature on this subject indicated that *kikokushijo*, especially those who have returned to Japan after staying in the United States or Europe, suffer from nonadaptation disease—"all kinds of mental and physical problems." They suffer these problems because they have difficulty readjusting to "the very nature of Japanese society," which is portrayed as "homogeneous, exclusivist, conformist, and harmonious" after living in Europe and the United States, which are portrayed as individualist, heterogeneous, independent, and argumentative (p. 3). According to the literature, the reentry problems are so great that special schools known as *ukeireko* have been established to offer "relief education" and to "rejapanize" returnees or "peel off their foreignness." Upon reviewing the literature, however, Goodman found many descriptions of these special schools but discovered that no one had investigated the process by which these special schools accomplish these goals.

Goodman's review of the literature shaped his research in that he decided to "see for himself" how these special schools address the physical and mental problems associated with reentry into Japanese society. After spending one year observing Japanese junior high schools, he spent a second year observing a *ukeireko*. Goodman found that this special school differed greatly from that described in the literature.

> The *ukeireko* proved to be more interested in the education of future leaders of society than in "decontaminating" *kikokushijo* of their overseas experience. The school modelled itself more on a British public school than on a refugee camp. The teachers sought to instill a combination of "Japaneseness" and "internationalness" in their education programmes that seemed to undermine the widely held image of Japan's innate exclusivity, and the problems that such exclusivity caused for *kikokushijo*. (p. 4)

Step 3: Identifying and Defining Core Concepts

After deciding on a topic and reading the relevant literature (not necessarily in that order), researchers typically state their core concepts. **Concepts** are powerful thinking and communication tools that enable researchers to give and receive complex information efficiently. The mention of a concept triggers in the minds of those who are familiar with its meaning a definition and a range of important associations that help to frame and give focus to observations. To illustrate, consider how the three following sociological concepts focus researchers' attention and observations:

- **Global interdependence:** a state in which the lives of people in any one country are intertwined closely with the lives of people outside their border.

- **Hidden curriculum:** all the "other things" students learn along with the formal subject matter.

- **Feeling rules:** general rules for behavior that specify appropriate ways to express emotions toward another person.

Notice how the meaning of the concept "global interdependence" focuses attention on interconnections among people residing in different countries. The meaning reminds the interested researcher that in order to "observe" global interdependence he or she must find evidence of ties between people in different countries. With this focus in mind, the researcher may decide to examine telephone conversations that cross national boundaries (see "Telephone Conversations: Evidence of Global Interdependence").

The concept "hidden curriculum" invites researchers to look beyond the official subject matter (math, English, reading) and focus on other "hidden" or less "planned" lessons. Anne Allison's research on *obentōs* is an example of such a study. She learned that the *obentō* is more than "lunch." Its real purpose is to "teach" children to do their best in school.

As a final example, the general concept "feeling rules" guided social psychologist Tomoko Hamada (1996) in her study of Euro-American and Japanese male managers who worked together at a Japanese multinational corporation located in the United States. She focused on the two groups of managers and the norms each held with regard to touching each other's bodies in a workplace setting.

All three examples show how researchers use sociological concepts or language of their discipline to give focus to their observations.

Step 4: Choosing a Research Design and Data-Gathering Strategies

Once researchers have clarified core concepts, they decide on a **research design**, a plan for gathering data on the topic they have chosen. A research design

Concepts Thinking and communication tools that are used to give and receive complex information efficiently and to frame and explain observations.

Research design A plan for gathering data that specifies the population and method of data collection.

Telephone Conversations: Evidence of Global Interdependence

The map and accompanying charts show (1) the number of phone calls made from the United States to people in 11 different countries and (2) the number of phone calls made from those 11 countries to people in the United States. The total phone calls are the sum of phone calls made to the United States and from the United States for each of these 11 countries. The chart also shows the total number of minutes that people talked. We cannot look at the total phone calls and assume that U.S. interdependence is highest with Israel and lowest with Afghanistan. Because the total number of calls will be influenced greatly by the population size of each country, we might want to use a per capita measure instead, such as total number of calls divided by total population of the country. The table shows the resulting per capita numbers. Do these results make sense to you, based on your knowledge about the countries and their relationship with the United States?

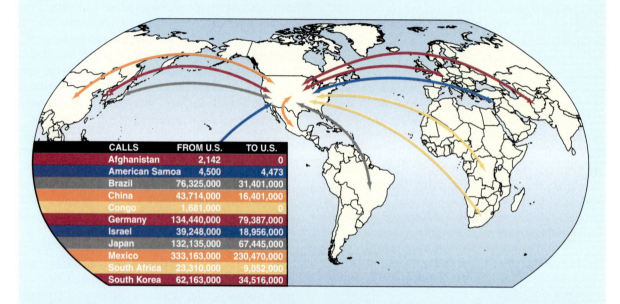

CALLS	FROM U.S.	TO U.S.
Afghanistan	2,142	0
American Samoa	4,500	4,473
Brazil	76,325,000	31,401,000
China	43,714,000	16,401,000
Congo	1,681,000	0
Germany	134,440,000	79,387,000
Israel	39,248,000	18,956,000
Japan	132,135,000	67,445,000
Mexico	333,163,000	230,470,000
South Africa	23,310,000	9,052,000
South Korea	62,163,000	34,516,000

Average Annual Telephone Messages and Population (by Country)

Country	Total Phone Calls (thousands)	Number of Minutes (thousands)	Population Size (thousands)	Per Capita Minutes	Per Capita Messages
Afghanistan	2	23	24,792	0.00	0.00
American Samoa	10	58	62,000	0.95	0.161
Brazil	107,726	496,269	161,087	3.08	0.67
China	60,007	356,210	1,232,083	0.29	0.0049
Germany	213,827	1,094,542	81,922	13.36	2.61
Israel	58,204	317,435	5,664	56.04	10.28
Japan	199,580	1,040,603	125,351	8.30	1.59
South Korea	96,679	537,046	45,314	11.85	2.13
Mexico	563,633	3,328,907	92,718	35.90	6.08
South Africa	32,362	126,716	42,393	2.99	0.76
Zaire	1,681	8,099	46,812	0.17	0.036

Sources: U.S. Federal Communications Commission (1997); United Nations (1998).

specifies the population to be studied and the **method of data collection,** or the procedures used to gather relevant data. One research design is not inherently better than another. Instead, researchers choose the design that best enables them to address their research question (Smith 1991).

THE POPULATION TO BE STUDIED   Researchers must decide who or what they are going to study. The most common "thing" social scientists study is individuals, but they may also decide to study traces, documents, territories, households, small groups, or individuals (Rossi 1988).

Traces   **Traces** are materials or other evidence that yield information about human activity, such as the items that people throw away, the number of lights on in a house,[2] or changes in water pressure.[3] One example of research that examines traces is the Garbage Project, a University of Arizona program directed by William L. Rathje (Rathje and Murphy 1992). Since the 1970s, Rathje and his team of researchers have been collecting garbage from landfills and selected neighborhood garbage cans. Among other things, Rathje has found that the hazardous waste generated by households across the United States equals the hazardous waste generated by commercial establishments.

Documents   **Documents** are written or printed materials, such as magazines, books, calendars, graffiti, birth certificates, and traffic tickets. For her research on *obentōs*, Anne Allison (1991) examined a variety of documents. She did a content analysis of *obentō* magazines and cookbooks and *obentō* guidelines that are issued biweekly by Japanese nursery schools. From examining the documents, Allison realized the cultural importance of the *obentō*. She found that the magazines, cookbooks, and guidelines were filled with recipes, hints, pictures, and ideas about how to prepare food so that children will eat it. One common suggestion was that the mother should carefully design the presentation of food items and the bag in which the

food is carried. In other words, the mother should not let someone else prepare the *obentō*.

Territories   **Territories** are settings that have borders or that are set aside for particular activities. Examples include countries, states, counties, cities, streets, neighborhoods, classrooms, and buildings. William H. Whyte's *City: Rediscovering the Center* (1988) is an example of research that examines a territory. For more than 16 years, Whyte visited and observed New York City and other cities across the United States and around the world. He was interested in identifying the conditions under which people use urban spaces. Whyte found that some city spaces simply are not designed to be used: "Steps too steep, doors too tough to open, ledges you cannot sit on because they are too high or too low, or have spikes on them" (p. 1). Often city streets are simply not very interesting places. Among other things, he found Tokyo's streets to be consistently more interesting than most city streets in the United States. According to Whyte, the Japanese "do not use zoning to enforce a rigid separation of uses. They encourage a mixture, not only side by side, but upwards. In the buildings you will see showrooms, shops, pachinko parlors, offices, all mixed together and with glass-walled restaurants rising one on top of the other—three, four, and five stores up" (p. 89).[4]

Households   **Households** include all related and unrelated persons who share the same dwelling. Economist Raymond A. Jussaume, Jr., and sociologist Yoshiharu Yamada surveyed households in Seattle, Washington, and Kobe, Japan, and found some important differences between the two. The very composition of Kobe and Seattle households is different: 3 percent of Kobe households have no children or elderly residents, in contrast to 15.7 percent of Seattle households. Also, although virtually every Kobe and Seattle household

---

[2]Sociologists Robert and Helen Lynd ([1929] 1956) observed the times at which lights were turned on during winter mornings to determine whether people from working-class households started their days earlier than middle-class people.

[3]Researchers have studied reductions in municipal water pressure to obtain estimates of the number of people watching commercials shown at breaks in prime-time programs. The assumption is that fewer people go to the bathroom when commercials are interesting than when they are uninteresting (Rossi 1988).

[4]Pachinko is a type of pinball game. Pachinko parlors are typically located near train stations and commercial districts (Kōji 1983).

Method of data collection The procedures used to gather relevant data.

Traces Materials or other evidence that yield information about human activity.

Documents Written or printed materials used in research.

Territories Settings that have borders or that are set aside for particular activities.

Households All related and unrelated persons who share the same dwelling.

A systematic look at specific social units such as the family reveals differences between the United States and Japan. For example, households without children or elderly residents are much more common in Seattle than in Kobe.

©Dave Bartruff/Artistry International

has at least one telephone, only 2.1 percent of Kobe households have unlisted telephone numbers, compared with approximately 20 percent of Seattle households (Jussaume and Yamada 1990).

Small Groups **Small groups** are defined as two to about twenty people who interact with one another in meaningful ways (Shotola 1992). Examples include father-child pairs, doctor-patient pairs, families, sports teams, circles of friends, and committees. An example of research conducted at this level of analysis is the

Small groups Two to about twenty people who interact with one another in meaningful ways.

Populations The total number of individuals, traces, documents, territories, households, or groups that could be studied.

Sample A portion of the cases from a larger population.

Random sample A sample in which every case in the population has an equal chance of being selected.

Representative sample A research sample with the same distribution of characteristics as the population from which it is selected.

Sampling frame A complete list of every case in the population.

work of sociologist Masako Ishii-Kuntz (1992). She studied father-child relationships in three countries (Japan, Germany, and the United States) to evaluate the popular belief that Japanese fathers are largely absent from the home and uninvolved in their children's lives. The results of Ishii-Kuntz's study are described later in this chapter.

Because of time constraints alone, researchers cannot study entire **populations**—the total number of individuals, traces, documents, territories, households, or groups that could be studied. Instead, they study a **sample**, or a portion of the cases from a larger population.

SAMPLES Ideally, a sample should be a **random sample**, with every case in the population having an equal chance of being selected. The classic, if inefficient, way of selecting a random sample is to assign every case a number, place the cards or slips of paper on which the numbers are written into a container, thoroughly mix the cards and pull out one card at a time until the desired sample size is achieved. Rather than follow this tedious system, most of today's researchers use computer programs to generate their samples. If every case has an equal chance of becoming part of the sample, then theoretically the sample should be a **representative sample**—that is, one with the same distribution of characteristics (such as age, gender, and ethnic composition) as the population from which it is selected. For example, if 56.4 percent of the population from which a sample is drawn is at least 30 years old, then approximately 56.4 percent of a representative sample should be that age. In theory, if the sample is representative, then whatever is true for the sample should also hold true for the larger population.

Obtaining a random sample is not as easy as it might appear. For one thing, researchers must begin with a **sampling frame**—a complete list of every case in the population—and each member of the population must have an equal chance of being selected. Securing such a complete list can be difficult. Campus and city telephone directories are easy to acquire, but lists of, for example, U.S. citizens, adopted children in the United States, U.S.-owned companies in Japan, or Japanese-owned companies in the United States are more difficult to obtain. Almost all lists omit some people (such as persons with unlisted numbers, members too new to be listed, or between-semester transfer students) and include some people who no longer belong (such as individuals who have moved, died, or dropped out). What is important is that the researcher consider the extent to which the list is incomplete and update it before drawing a sample. Even if the list is complete, the researcher also must think of the cost and time required to take random samples and consider the problems of inducing all sampled persons to participate.

Researchers sometimes select nonrandom samples to study people who they know are not representative of the larger population but who are easily accessible. For example, they often use high school and college students as a sample because they are a captive audience. The sample in Stella Ting-Toomey's (1991) study of intimacy in three cultures consisted of 781 volunteers from various college classes in Japanese, U.S., and French universities. The U.S. sample of 256, for example, came from students enrolled in a mid-sized East Coast university. The Japanese sample of 279 came from a university in the southwest region of Japan. Other researchers may choose nonrepresentative samples for other important reasons: (1) little is known about them, (2) they have special characteristics, or (3) their experiences clarify important social issues. (See "The Innovative Research of Erving Goffman.")

METHODS OF DATA COLLECTION   In addition to identifying who or what is to be studied, the design must include a plan for collecting information. Researchers can choose from a variety of data-gathering methods, including self-administered questionnaires, interviews, observations, and secondary sources.

---

[5]Achievers are those students who scored in the expected range for someone in that grade.

[6]Two students in my Introduction to Sociology class described how they responded to a questionnaire. Their accounts show how the quality of the response affects the quality of the results.

- Several months ago, I received a questionnaire in the mail to fill out. At the time, I was extremely busy so I put the questionnaire aside. Several weeks later, I received another questionnaire along with a letter chastising me for not returning the first one. I was very angry about this letter so I answered the questionnaire and returned it. I filled it out so quickly that I didn't give much thought to the questions. I wonder if everyone who received this letter answered the questions in the same way I had. If they did, then I don't believe the questionnaire was very helpful to this company.
- I remember being interviewed for a research project my freshman year in college. During my lunch break someone asked me if I wanted to earn three dollars for about five minutes of my time; I said "yes." The survey I filled out contained questions asking how I felt about controversial topics. I remember simply answering the questions, not as I thought, but how I thought the interviewer would want the questions to be answered. Looking back I realized I was trying to please the researcher. However, I think that many respondents answer questions as I did. My experience represents one of the obstacles researchers must overcome if they are to obtain unbiased and accurate data.

Self-Administered Questionnaire   A **self-administered questionnaire** is a set of questions given (or mailed) to respondents, who read the instructions and fill in the answers themselves. The questions may require respondents to write out answers (open-ended) or to select from a list of responses the one that best reflects their answer (forced choice). This method of data collection is probably most common. The questionnaires found in magazines or books, displayed on tables or racks in service-oriented establishments (hospitals, garages, restaurants, groceries, physicians' offices), and mailed to households are all self-administered questionnaires. This method of data collection has a number of advantages. No interviewers are needed to ask respondents questions; the questionnaires can be given to large numbers of people at one time; and an interviewer's facial expressions or body language cannot influence respondents, so they feel more free to give unpopular or controversial responses.

Researchers Paul Tuss, Jules Zimmer, and Hsiu-Zu Ho (1995) used a self-administered questionnaire with open-ended questions to study the beliefs that fourth-graders labeled as achievers[5] and underachievers in the United States, Japan, and China held about the reasons for successful and unsuccessful mathematics performance. Two of the questions were as follows:

1. People use different reasons to explain why they have done things well or poorly. Think of the last test you did poorly on. Why do you think you did so poorly? Write the reasons on the lines below.

2. Now think of the last math test you did well on. Why do you think you did so well? Write the reasons on the lines below. (p. 414)

Self-administered questionnaires pose some problems, however. Respondents can misunderstand or skip over some questions. Often questionnaires are mailed, set out on a table or counter, or published in a magazine or newspaper. Researchers must then wonder whether the people who volunteer or choose to fill out a questionnaire have different opinions than those who ignore the survey. The results of a questionnaire depend not only on respondents' decisions to fill it out, answer questions conscientiously and honestly, and return it, but also on the quality of the survey questions asked and a host of other considerations.[6]

Self-administered questionnaire   A set of questions given to respondents who read the instructions and fill in the answers themselves.

The Innovative Research of Erving Goffman

Erving Goffman made the study of talk, conversation, and interaction his life's work. There is no doubt that Goffman's social "methods for accomplishing this were, to say the least, singular" (Drew and Wootton 1988:6). His writings span 30 years, beginning in 1953 with the completion of his Ph.D. dissertation (*Communication Conduct in an Island Community*) and ending in 1983 with the publication of his presidential address to the American Sociological Association ("The Interaction Order"). Some sociologists believe that Goffman was "the greatest sociologist of the latter half of the twentieth century" (Collins 1988:41). Although other sociologists take issue with such a claim, "no one would question the claim that Erving Goffman was one of the leading sociological writers of the post–[World War II] period" (Giddens 1988:250). Few sociologists writing in the past five decades have made the impression that Goffman has on professionals and academics outside the discipline of sociology (Giddens 1988), including those in anthropology, linguistics, folklore, communication, political science, psychiatry, ethnology (Winkin 1989).

One of the many qualities that makes Goffman unique among sociologists is the sources from which he drew to form, support, and illustrate his ideas and insights about the largely taken-for-granted world of everyday interaction. These sources included newspaper clippings, comic strips, scenes from popular films, cartoons, personal observations and anecdotes, "Dear Abby" columns, etiquette manuals such as those by Emily Post, popular fiction, serious novels, autobiographies, the writings of "respectable" researchers, "informal memoirs written by colorful people" (Goffman 1959:xi), and contemporary

Erving Goffman
Courtesy American Sociological Association

and ancient theatrical writings. In the essay "Radio Talk: A Study of the Ways of Our Errors," Goffman (1981) tells readers that he draws on the following sources: "eight of the LP records and three of the books produced by Kermit Schafer from his recordings (Jubilee Records) of radio bleepers . . . ; twenty hours of taped programs from two local stations in Philadelphia and one in the San Francisco Bay area; a brief period of observation and interviewing of a classical DJ at work; and informal note-taking from broadcasts over a three-year period" (p. 197).

Goffman believes that any group of persons—subsistence farmers, prisoners, pilots, patients, disk jockeys, people with disabilities, the socially rejected—"develop a life of their own that becomes meaningful, reasonable, and normal once you get close to it, and that a good way to learn about any of these worlds is to submit oneself in the company of the members to the daily round of petty contingencies to which they are subject" (Goffman

1961:x). This idea was not idle talk on the part of Goffman. For his doctoral dissertation, Goffman spent one year studying a Shetland Island community of 300 people with a subsistence farming economy. For his book *Asylums: Essays on the Social Situation of Mental Patients and Other Inmates*, Goffman (1961) assumed the role of an assistant to the athletic director in a 7,000-bed federal mental hospital for one year to learn about "the social world of the hospital inmate, as this world is subjectively experienced by him" (p. ix).

Goffman avoided one of the most common methods of research—gathering and generating statistical evidence. He believed that creating statistics interfered with his goals of capturing and conveying to his readers the "tissue and fabric" of the interaction order. Goffman acknowledged that, in addition to the absence of statistical data in his writings, his approach to studying social encounters is biased in the direction of one category of participants over other participants. Goffman described this bias most openly in the preface to *Asylums*:

> The world view of a group functions to sustain its members and expectedly provides them with a self-justifying definition of their own situation and a prejudiced view of nonmembers, in this case, doctors, nurses, attendants, and relatives. To describe the patient's situation faithfully is necessarily to present a partisan view. (For this last bias I partly excuse myself by arguing that the imbalance is at least on the right side of the scale, since almost all professional literature on mental patients is written from the point of view of the psychiatrist, and he, socially speaking, is on the other side.) (Goffman 1961:x)

Interviews In comparison with self-administered questionnaires, **interviews** are more personal. In these face-to-face sessions or telephone conversations between an interviewer and a respondent, the interviewer asks questions and records the respondent's answers. As respondents give answers, interviewers must avoid pauses, expressions of surprise, or body language that reflects value judgments. Refraining from such conduct helps respondents feel comfortable and encourages them to give honest answers.

Interviews can be structured or unstructured, or some combination of the two. In a **structured interview**, the wording and sequence of questions are set in advance and cannot be altered during the course of the interview. In one kind of structured interview, respondents choose answers from a response list that the interviewer reads to them. In another kind of structured interview, respondents are free to answer the questions as they see fit, although the interviewer may ask them to clarify or explain answers in more detail.

In contrast, an **unstructured interview** is flexible and open-ended. The question-answer sequence is spontaneous and resembles a conversation in that the questions are not worded in advance and are not asked in a set order. The interviewer allows respondents to take the conversation in directions they define as crucial. The interviewer's role is to give focus to the interview, ask for further explanation or clarification, and probe and follow up interesting ideas expressed by respondents. The interviewer appraises the meaning of answers to questions and uses what was learned to ask follow-up questions. Talk show hosts, for instance, often use an unstructured format to interview their guests. Sociologists, however, have much different goals than talk show hosts do. For one thing, sociologists do not formulate questions with the goal of entertaining an audience. In addition, sociologists strive to ask questions in a neutral way, and no audience reaction is possible to influence how respondents answer the questions.

Observation As the term implies, **observation** involves watching, listening to, and recording behavior and conversations as they happen. This research technique may sound easy, but it involves more than seeing and listening. The challenge of observation lies in knowing what to look for while remaining open to other considerations; success results from identifying what is worth observing. "It is a crucial choice, often determining the success or failure of months of work, often differentiating the brilliant observer from the . . .

plodder" (Gregg 1989:53). Good observation techniques must be developed through practice to learn to recognize what is worth observing, be alert to unusual features, take detailed notes, and make associations between observed behaviors.

If observers come from a culture different from the one under study, they must be careful not to misinterpret or misrepresent what is happening. Imagine for a moment how an uninformed, naive observer might describe a sumo wrestling match: "One big, fat guy tries to ground another big, fat guy or force him out of the ring in a match that can last as little as three seconds" (Schonberg 1981:B9). Actually, for those who understand it, sumo wrestling is "a sport rich with tradition, pageantry, and elegance and filled with action, excitement, and heroes dedicated to an almost impossible standard of excellence down to the last detail" (Thayer 1983:271).

Observational techniques are especially useful for (1) studying behavior as it occurs, (2) learning things that cannot be surveyed easily, and (3) acquiring the viewpoint of the persons under observation. Observation can take two forms: participant and nonparticipant. **Nonparticipant observation** consists of detached watching and listening: the researcher only observes and does not interact or become involved in the daily life of the study subjects. A good example of nonparticipant observation is Catherine C. Lewis's (1988) research on Japanese first-grade classrooms. Lewis observed 15 Japanese classrooms each for a full day, "from morning greetings through the children's departure from school" (p. 161). She observed:

Interviews Face-to-face or telephone conversations between an interviewer and a respondent in which the interviewer asks questions and records the respondent's answers.

Structured interview An interview in which the wording and sequence of questions are set in advance and cannot be changed during the interview.

Unstructured interview An interview in which the question-answer sequence is spontaneous, open-ended, and flexible.

Observation A research technique involving watching, listening to, and recording behavior and conversations as they happen.

Nonparticipant observation A research technique involving detached watching and listening in which the researcher does not interact with the study participants.

[I]t was common for two [student] monitors to stand at the front of the class, ask the class to be quiet, announce what subject was about to be studied, and ask the children to rise and greet the teacher. In some classrooms, the monitors picked which group showed the best order, cautioned groups to improve their posture, granted check marks to groups that were quiet and ready to begin the class period, or chose which groups would receive lunch first. Often, the monitors assembled and quieted the class when the teacher was not yet present. Typically, monitors rotated each day, and all children became monitors in turn. (pp. 162–163)

In contrast, researchers engage in **participant observation** when they join a group and assume the role of a group member, interact directly with those whom they are studying, assume a position critical to the outcome of the study, or live in a community under study. Anne Allison's research on *obentōs* described earlier in this chapter provides an excellent example of participant observation. Her research revolved around conversations she had with other mothers, daily conversations with the teacher about her son's eating habits, Mothers' Association Meetings, and other school-related events.

In both participant and nonparticipant observation, researchers must decide whether to hide or to announce their identity and purpose. One major reason for choosing concealment is to avoid the Hawthorne effect,[7] a phenomenon whereby research subjects alter their behavior when they learn they are being observed.[8] If researchers announce their identity and purpose, they must give participants time to adjust to their presence. Usually, if researchers are present for a long enough time, the subjects eventually will display natural, uninhibited behaviors.

Secondary Sources or Archival Data Another data-gathering strategy relies on **secondary sources** or **archival data**—that is, data that have been collected by other researchers for some other purpose. Government researchers, for example, collect and publish

data on many areas of life, including births, deaths, marriages, divorces, crime, education, travel, and trade. "Every researcher who uses an existing data set or who does a literature review in which published research findings are taken out of their original context and applied to a different issue or question is involved in 'archival research'" (Horan 1995:423). For instance, for their research on changing norms and expectations related to the care of the elderly in Japan, Naohiro Ogawa and Robert D. Retherford (1993) used data from the National Survey on Family Planning, a biannual random survey conducted by the Population Problems Research Council of the Mainchi Newspapers since 1950. These surveys included a question asking women, "Are you planning to depend on your children in your old age (including adopted children, if any)?"[9]

Another kind of secondary data source consists of materials that people have written, recorded, or created for reasons other than research (Singleton, Straits, and Straits 1993). Examples include television commercials and other advertisements, letters, diaries, home videos, poems, photographs, artwork, graffiti, movies, and song lyrics.

IDENTIFYING VARIABLES AND SPECIFYING HYPOTHESES As researchers acquire a conceptual focus, identify a population, and determine a method of data collection, they also identify the variables they want to study. A **variable** is any trait or characteristic that

---

[7]A student in my Introduction to Sociology class wrote about the Hawthorne effect in a response paper:

"When I read about the Hawthorne effect, I immediately thought back to high school. During my senior year a group of extremely important officials from the state of Ohio gave notice that they were going to visit our school for a week. The janitors washed *and* waxed the floors (even in the locker rooms); students were sent outside during study hall to pull weeds and clean up litter; suddenly student work was displayed in the hallways. The week the officials were at the school, male teachers who generally wore casual shirts and khakis wore ties and our principal who always sat in his office was walking the halls talking to us and shaking our hands. This personal experience allows me to perfectly understand the Hawthorne effect."

[8]The term "Hawthorne effect" originates from a series of studies of workers' productivity conducted in the 1920s and 1930s, which involved female employees of the Hawthorne, Illinois, plant of Western Electric. Researchers found that no matter how they varied working conditions—bright versus dim lighting, long versus short breaks, frequent versus no breaks, piece rate pay versus fixed salary—workers' productivity increased. One explanation for these findings is that workers were responding positively to the fact that they had been singled out for study (Roethlisberger and Dickson 1939).

[9]The researchers found that in 1950, 65 percent of women expected to depend on their children in old age, and that percentage declined to 18 percent in 1990.

Participant observation A research technique in which researchers interact directly with study participants.

Secondary sources or **archival data** Data that have been collected by other researchers for some other purpose.

Variable Any trait or characteristic that can change under different conditions or that consists of more than one category.

consists of more than one category. The variable "sex," for example, is generally divided into two categories: male and female. The variable "marital status" is often separated into six categories: single, living together, married, separated, divorced, and widowed. The variable "comfort level making conversation with strangers" could be divided into four categories: very comfortable, comfortable, uncomfortable, and very uncomfortable.

Researchers strive to find associations between variables to explain and/or predict behavior. The behavior to be explained or predicted is the **dependent variable.** The variable that explains or predicts the dependent variable is the **independent variable.** Thus a change in the independent variable brings about a change in the dependent variable. A **hypothesis,** or trial explanation put forward as the focus of research, predicts the relationship between independent and dependent variables. It specifies which outcomes are expected to occur as the independent variable varies.

As one example, consider the hypotheses that researcher Stella Ting-Toomey (1991) advanced in her study of intimacy expressions in Japan and the United States as they relate to close, opposite-sex relationships. She hypothesized that the independent variable "respondent's culture"—Japanese or American—would affect four dependent variables: (1) love commitment (the amount of perceived interdependence in a relationship); (2) disclosure (the extent to which partners communicate in a relationship); (3) ambivalence (the extent of uncertainties about the relationship); and (4) conflict (the extent of overt arguments and disagreements). Four of Ting-Toomey's hypotheses follow:

1. Members of the U.S. culture would express a significantly higher degree of *love commitment* than members of the Japanese culture.

2. Members of the U.S. culture would express a significantly higher degree of *disclosure* than members of the Japanese culture.

3. Members of the U.S. culture would express a significantly higher degree of *ambivalence* than members of the Japanese culture.

4. Members of the U.S. culture would express a significantly higher degree of *conflict* than members of the Japanese culture.

In arriving at these hypotheses, Ting-Toomey reasoned that because Japanese society is less individualistic than American society, and because people in Japan have stronger connections to family and group than people in the United States, the answers to intimacy questions would reflect these cultural differences. Ting-Toomey argued that romantic love holds a central place in an individualist society such as the United States because a strong attachment to one individual "serves as the active solution to weak ties." On the other hand, romantic love "also induces a higher level of uncertainty and anxiety" because if a person loses the "love of their life," they are "all alone." Conversely, in Japan strong ties with family members "diffuse the intensity level of love commitment and uncertainty level" (pp. 32–33).

One major reason researchers collect data is to test hypotheses. If their findings are to matter, other researchers must be able to replicate the study. For this reason, researchers need to give clear and precise definitions and instructions about how to observe (and/or measure) the variables being studied.

OPERATIONAL DEFINITIONS   In the language of research, such definitions and accompanying instructions are called **operational definitions.** An analogy can be drawn between an operational definition and a recipe. Just as anyone with basic cooking skills can follow a recipe to achieve a desired end, anyone with basic research skills should be able to replicate a researcher's observations if he or she knows the operational definitions (Katzer, Cook, and Crouch 1991).

Operational definitions include precise explanations of how researchers decide that a behavior of interest has occurred. Suppose a researcher is interested in the question of who washes their hands after using public restrooms. An operational definition of handwashing must include an account of what must take place for a researcher to count someone as a handwasher. If people simply run water over their fingertips or rinse their hands quickly without using soap, should it count as handwashing? What if someone uses soap but only washes his or her fingertips? Should handwashing be counted as such only if it satisfies the

Dependent variable The behavior to be explained or predicted.

Independent variable The variable that explains or predicts the dependent variable.

Hypothesis A trial explanation put forward as the focus of research that predicts how independent and dependent variables are related and what outcome will occur.

Operational definitions Clear, precise definitions and instructions about how to observe and measure variables.

U.S. in Perspective

Is the Trade Balance a Valid Measure of the Economic Relationship Between Japan and the United States?

Consider the trade balance between two countries. It is calculated by adding up the dollar value of the goods and services that each country exports to the other and then subtracting the smaller amount from the larger. On the basis of this formula, the United States has sustained a trade deficit with Japan for approximately a decade and a half.

Some critics argue that this formula is not an accurate or valid measure of the amount of trade between the United States and Japan. It does not consider the dollar value of goods and services produced by American and Japanese companies located in the other's country. A more valid measure would be to add up the goods and services that each country exports to the other, plus what each country's firms produce and sell within the other country (Robinson 1985).

In 1985, Richard D. Robinson, a professor of international management, figured the U.S. trade deficit with Japan on the basis of two formulas. According to the formula that counts annual American exports to Japan ($25.6

How does this worker in a Honda factory in Marysville, Ohio, illustrate the difficulty of calculating the trade balance between two countries in a global economy?
©Andy Snow/Saba

billion) and Japanese exports to the United States ($56.8 billion), the trade deficit was $31.2 billion for the United States. According to the second formula, which includes sales by Japanese-owned businesses in the United States ($12.6 billion) and sales by American-owned businesses in

Japan ($43.9 billion), the trade surplus was $100 million for the United States, significantly different from the deficit figure of $31.2 billion.[1]

According to the second formula, then, the United States had a slight trade surplus with Japan. Even this formula, however, is not free of

---

guidelines issued by the American Society of Microbiology? Those guidelines specify using warm or hot running water and soap while washing for 10 to 15 seconds "all surfaces thoroughly, including wrists, palms, back of hands, fingers and under fingernails" (American Society of Microbiology Web site 1996). These are the kinds of questions researchers must address in creating operational definitions.

In her study of intimacy expression, Stella Ting-Toomey (1991) operationalized intimacy expression in this way: she asked participants to "think of a relationship you have known with an opposite-sex close friend." She then asked subjects to consider their communication relationship with this person, and circle the number on a scale of 1–5 (with "5" being "to a very great extent" and "1" being "to a very little extent") that best described the participant's response to each question. Ting-Toomey asked 25 questions to operational-

ize the four dimensions of intimacy expression. Examples of questions follow:

- Love commitment: "To what extent did you feel that your relationship was special compared to others?"

- Disclosure: "To what extent did you reveal or disclose very intimate things about yourself or personal feelings to the person?"

- Ambivalence: "To what extent did you feel 'trapped' or pressured to continue this relationship?"

- Conflict: "How often did you and the person argue with each other?"

Operational definitions do not have to be questions. Sometimes operational definitions are precise accounts or descriptions of what a researcher observed and a

problems. For one thing, it does not take into account that countries export goods and services *indirectly* to one another. For example, 70 percent of Japanese males use Schick razors. Schick, with headquarters in Connecticut, exports razors to Japan via Hong Kong (Totten 1990). Similarly, Japan exports cars to the United States from Canada, Mexico, and various countries in Asia.

On the basis of this information, we can argue that the best operational definition is a third formula: for each country, add up the dollar value of goods and services it exports to the other country, *plus* what that country's firms produce and sell within the other country, *plus* what that country exports indirectly to the other country. Then figure the difference between the two totals. Although in theory this third formula provides a more valid measure of trade, it is virtually impossible to find complete and accurate figures on indirect exports. In addition, anyone who tries to calculate indirect exports faces the challenge of determining the national identity or origin of the exported products. For example, should the approximately 164,000 vehicles

produced in Japan and exported to the United States as Chryslers, Dodges, and Chevrolets be considered Japanese exports (Sanger 1992)? Similarly, should the 50,000 Nissan Quest minivans produced at Ford's Avon, Ohio, production plant be considered Japanese exports? As you can see, the issue of trade balance is not simple or clear-cut.

Similarly, when we use the trade deficit with Japan as an operational definition of the openness of the Japanese market, Japanese aggressiveness, or the patriotism of American consumers, we must question whether it provides a valid measure of such behaviors. For example, the trade deficit as currently calculated cannot be used as a measure of the openness of Japanese markets to U.S. goods because it does not consider (1) the dollar value of goods and services produced by American and Japanese companies located in the other's country or (2) the dollar value of the indirect exports that each country sends to the other. Some examples of more valid measures of market openness are the number of trade barriers each country puts up against the other, the number of sug-

gestions each country offers the other for opening its markets, and per capita purchases of imports (see Figure 3.2).

To further complicate the meaning of a trade deficit, currency devaluation (without any change in the volume of imports or exports) can affect the size of the deficit. In addition, a currency devaluation may cause people in the country where it occurred to stop buying products from other countries. Conversely, people whose currency is strong may buy more goods from the country with the weakened currency. Finally, we must question the significance of a trade deficit when we learn that 40 percent of U.S. imports "represent intracompany transactions by American multinational corporations" (Tyson 1997:A17).

[1]To evaluate this difference ($31.2 billion versus $100 million), it is important to understand the difference between a million and a billion. "For example, knowing that it takes only about eleven and a half days for a million seconds to tick away, whereas almost thirty-two years are required for a billion seconds to pass, gives one a better grasp of the relative magnitude of these two common numbers" (Paulos 1988:10).

description of the context in which the observations were made. In Tomoko Hamada's (1996) study of feeling rules governing bodily contact at work, she observed "touching" and reactions to it. As you read her account, keep in mind that Hamada is offering her operational definitions of "touching" and "response to touching."

> Some Japanese men seemed to enjoy holding another man or [being] touched or held by another man. Japanese engineers slapped each other's back as a friendly or congratulatory gesture, and they sometimes leaned casually against each other. A Japanese engineer named Toru Ueno, who was a martial arts champion, actually liked to touch Japanese and Euro-American men to show his friendliness. Toru stated that his fraternal love toward fellow males had been nurtured through his sports activities. He said that touching was a positive thing that would intensify the feeling of connectedness among coworkers and help

> convey positive feelings, particularly when there was a linguistic barrier. Toru's English was not fluent. When a job was well done by his mates, he tapped their shoulders, touched their backs, and sometimes tried to put his arm around their shoulders.

> Euro-American male engineers, on the other hand, became very uptight about such bodily contact in business settings. Whenever Toru touched one of them, the Euro-American's whole body stiffened, his shoulders tightened, and an expression of acute embarrassment or anger surged on the Euro-American's face. Many jokes and changes of topics suddenly became necessary to ease their tension. A rumor began concerning Toru's "homosexual" tendency among Euro-American managers. (p. 165)

If the operational definitions are not clear or do not indicate accurately the behaviors they were designed to represent, they have questionable value. Good

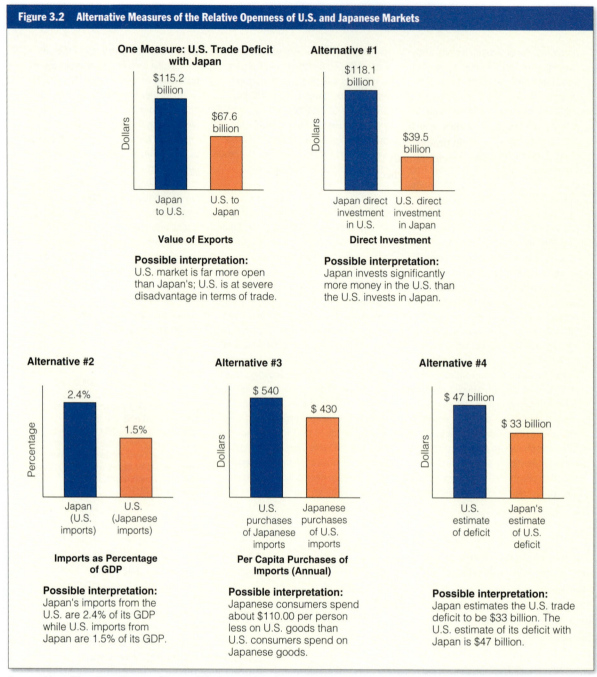

Figure 3.2   Alternative Measures of the Relative Openness of U.S. and Japanese Markets

One Measure: U.S. Trade Deficit with Japan

$115.2 billion — Japan to U.S.
$67.6 billion — U.S. to Japan

Dollars

Value of Exports

Possible interpretation:
U.S. market is far more open than Japan's; U.S. is at severe disadvantage in terms of trade.

Alternative #1

$118.1 billion — Japan direct investment in U.S.
$39.5 billion — U.S. direct investment in Japan

Dollars

Direct Investment

Possible interpretation:
Japan invests significantly more money in the U.S. than the U.S. invests in Japan.

Alternative #2

2.4% — Japan (U.S. imports)
1.5% — U.S. (Japanese imports)

Percentage

Imports as Percentage of GDP

Possible interpretation:
Japan's imports from the U.S. are 2.4% of its GDP while U.S. imports from Japan are 1.5% of its GDP.

Alternative #3

$540 — U.S. purchases of Japanese imports
$430 — Japanese purchases of U.S. imports

Dollars

Per Capita Purchases of Imports (Annual)

Possible interpretation:
Japanese consumers spend about $110.00 per person less on U.S. goods than U.S. consumers spend on Japanese goods.

Alternative #4

$47 billion — U.S. estimate of deficit
$33 billion — Japan's estimate of U.S. deficit

Dollars

Possible interpretation:
Japan estimates the U.S. trade deficit to be $33 billion. The U.S. estimate of its deficit with Japan is $47 billion.

Sources: Adapted from Japan Ministry of Finance (1998); Sanger (1992); Totten (1990); U.S. Central Intelligence Agency (1997); *The World Almanac and Book of Facts 1998* (1997).

operational definitions are reliable and valid. **Reliability** is the extent to which the operational definition gives consistent results. For example, the question "How

Reliability   The extent to which the operational definition gives consistent results.

many magazines do you read each month?" may not yield reliable answers because respondents may forget some magazines. Thus, if you asked the question at two different times, the respondent likely would give two different answers. One way to increase the reliability of this question is to ask respondents to list the magazines that they have read in the past week. The act of listing forces respondents to think harder about the question, and shortening the amount of time to the past week makes it easier to remember. Under what conditions might the question "How did you usually get to work last week?" yield unreliable answers? Suppose some respondent works at home 20 hours per week and at a corporate office to which he or she drives during the other 20 hours. Such respondents may answer "work at home" when queried at one point in time or "car" when asked at another time. The researcher can remedy this reliability problem by allowing respondents to check more than one response to the question.

Validity is the degree to which an operational definition measures what it claims to measure. Professors, for example, give tests to measure students' knowledge of a particular subject as covered in class lectures, discussions, reading assignments, and other projects. Students may question the validity of this measure if the questions on a test reflect only the material covered in lectures. In such instances, students may argue that the test does not measure what it claims to measure. As a second example, many critics claim that the trade balance as currently figured is not a valid measure of the economic relationship between countries (see "Is the Trade Balance a Valid Measure of the Economic Relationship Between Japan and the United States?" and Figure 3.2). Remember, when assessing validity, always ask: Is the operational definition really measuring what it claims to measure?

Steps 5 and 6: Analyzing the Data and Drawing Conclusions

When researchers reach the stage of analyzing collected data, they search for common themes, meaningful patterns, and/or links. The researcher must "pick and choose among the available numbers [and observations] and then fashion a format" (Hacker 1997:478). In presenting their findings, researchers may use graphs, frequency tables, photos, statistical data, and so on. The choice of presentation depends on what results are significant and how they might be best shown.

Researchers William T. Bailey and Wade C. Mackey (1989), for example, used percentages to summarize the data that they collected in their 1989 nonparticipant observation study of 18,272 child-adult interactions in public settings in Japan and the United States. They constructed tables that show the percentages of child-adult interactions in Japan and the United States involving women only, men only, and men and women. As the tables show, contrary to the popular image of Japanese men as people who work long hours and spend little or no time with their children, Japanese men are observed

Contrary to a common stereotype, the results of Bailey and Mackey's comparative study of parent-child interactions indicated that Japanese fathers spend about as much time with their children as U.S. fathers do.

©Sonia Katchian/Photo Shuttle: Japan

as frequently as American men interacting with their children in public settings (see Figure 3.3).

Of course, Bailey and Mackey's presentation represents only one example of the many ways in which researchers can present their findings. Table 3.1, for example, shows the ways in which Ting-Toomey chose to present her findings related to intimacy expressions in close, opposite-sex relationships. Table 3.1 also considers other statistical information that helps readers to evaluate and conveys findings.

In the final stage of the research process, sociologists comment on the **generalizability** of findings, the extent to which the findings can be applied to the larger population from which the sample is drawn. The sample used and the response rate are both important factors in determining generalizability, as is clear from the following example.

Validity The degree to which an operational definition measures what it claims to measure.

Generalizability The extent to which findings can be applied to the larger population from which the sample is drawn.

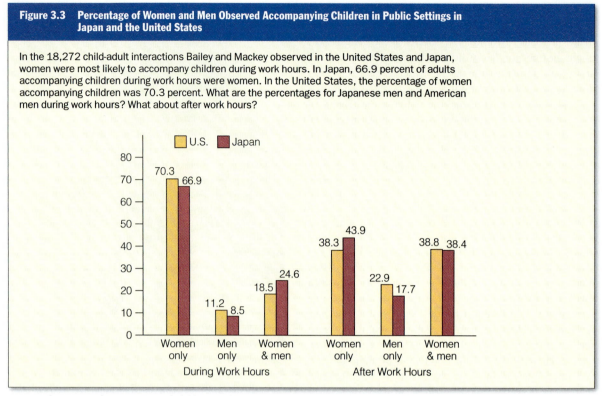

Figure 3.3   Percentage of Women and Men Observed Accompanying Children in Public Settings in Japan and the United States

In the 18,272 child-adult interactions Bailey and Mackey observed in the United States and Japan, women were most likely to accompany children during work hours. In Japan, 66.9 percent of adults accompanying children during work hours were women. In the United States, the percentage of women accompanying children was 70.3 percent. What are the percentages for Japanese men and American men during work hours? What about after work hours?

Source: "Observations of Japanese Men and Children in Public Places: A Comparative Study," by W. T. Bailey and W. C. Mackey. Page 733 in *Psychological Reports*, 65. Copyright © 1989 by *Psychological Reports*. Reproduced by permission of the authors and publisher.

In 1991, the U.S. Centers for Disease Control (CDC) abandoned its plans to conduct a nationwide survey of households to determine the prevalence of HIV infection after a pilot study involving 1,724 households in Pittsburgh and Allegheny County, Pennsylvania, showed that the researchers could not secure a high enough response rate to be confident in their estimate. Although 85 percent of the pilot study households agreed to take a blood test, CDC researchers wondered whether the 15 percent who refused to participate might have a higher risk of HIV infection than those who agreed. Researchers contacted those 15 percent a second time and convinced approximately half to participate. These "reluctant participants" were found to be twice as likely as the original participants to have used intravenous drugs and to have reported male-to-male sex. Was this 7.5 percent representative of the other 7.5 percent who did not participate at all? The fact that the CDC researchers knew nothing about the latter group prompted them to abandon the study rather than risk making inaccurate generalizations about rates of HIV infection among the general population (Hilts 1991).

If a sample is randomly selected, if almost all subjects agree to participate, and if the response rate for every question is high, we can say that the sample is representative of the population and that the findings theoretically are generalizable to that population. If a sample is chosen for some other reason—perhaps because it is especially accessible or interesting—then the findings cannot be generalized to the larger population. Keep in mind that even though one goal of drawing conclusions is to make generalizations about the larger population, generalizations are not statements of certainty that apply to everyone. Consider the comparative research on father-child interaction in three cultures conducted by sociologist Masako Ishii-Kuntz (1992).

In this case, knowing a respondent's culture did not mean that the researchers could predict the frequency with which fathers went home for dinner in the evenings to eat with their children. The study data show that Japanese fathers are significantly less likely than U.S. and German fathers to eat dinner with their children every day. Approximately 50 percent of Japanese fathers ate dinner every day with their children

Table 3.1   Basic Statistics Researchers Use to Convey Findings

In her study of intimacy expressions in the United States and Japan described on page 70, Stella Ting-Toomey summarized her research findings as follows:

	United States		Japan	
	Mean	Standard Deviation	Mean	Standard Deviation
Love commitment	4.11*	0.72	3.43	0.79
Disclosure	3.40*	0.81	2.86	0.82
Ambivalence	2.36*	0.95	2.11	0.78
Conflict	2.44	0.79	2.50	0.78

Mean—the sum of all individual scores divided by the total number of respondents. Recall there were 256 U.S. respondents and 279 Japanese respondents. In the case of love commitment, the average for U.S. respondents was 4.11; for Japanese respondents it was 3.43.

Standard Deviation (s.d.)—the standard deviation measures how far the data are spread from the mean. Most of the data (at least 75%) will fall within two standard deviations of the mean, and almost all of the data (90%) will fall within three standard deviations of the mean. To calculate the spread of data around the mean, multiply s.d. by two and add and subtract that number to and from the mean.

Minimum—lowest value that an individual respondent can score. We know that Ting-Toomey used a five-point scale. Therefore, the lowest possible score is 1.

Maximum—highest value that an individual respondent can score. The highest possible score is a 5.

Range—represents the numbers between which respondents can score. In this study the numbers are between 1 and 5.

Mode—the value that occurs most often. Ting-Toomey did not include this information in her research article. We can use a hypothetical example to show how the mode is determined. The table below shows the number of U.S. and Japanese respondents who scored "1" through "5" on love commitment. As you can see, 20 Japanese scored 1, while 16 U.S. respondents scored 1. The score that occurs most often is 3 for Japanese respondents and 4 for U.S. respondents.

Love Commitment

Score	Japan	United States
1	20	16
2	60	20
3	120	40
4	60	120
5	19	60
	$n = 279$	$n = 256$

n—a symbol that stands for number of respondents. There were 279 Japanese respondents and 256 American respondents.

Median—the number below which 50 percent of the cases fall above and 50 percent fall below. The median is not a very useful statistic when researchers are working with five-point scales. The median is useful when researchers are examining issues such as income, test scores on a scale of 1–100, and age.

Statistical significance—statistical significance is too complex a topic to discuss in an introduction to sociology textbook. However, note that Ting-Toomey places an "*" after three of the four sets of numbers to indicate that the differences in mean scores are statistically significant. It means that there is a significant statistical difference between Japanese and U.S. respondents with regard to love commitment, disclosure, and ambivalence.

compared with 70 percent of U.S. fathers and 60 percent of German fathers. Clearly, not all *individual* Japanese fathers behave in this way, but *as a group* they eat dinner with their children less frequently than U.S. and German fathers do.

Because the generalization does not apply to 50 percent of Japanese fathers, it is virtually impossible to claim that one independent variable (in this case, nationality) causes a dependent variable (in this case, the frequency that fathers are home for dinner). Conse-quently, instead of claiming cause, researchers search for independent variables that make significant contributions toward explaining the dependent variable.

At least three conditions must be met before a researcher can claim that an independent variable contributes significantly toward explaining a dependent variable. First, the independent variable must precede the dependent variable in time. Time sequence can be established easily when the independent variable is a predetermined factor, such as sex or birth date. These kinds of factors are fixed before a person becomes capable of any kind of behavior. Usually, however, time order cannot be established so easily.

Second, the two variables must be correlated. The strength of this contribution is often represented by a **correlation coefficient,** a mathematical representation of the extent to which a change in one variable is associated with a change in another (Cameron 1963). Correlation coefficients are numbers that fall between 0.00 and 1.00 and that are preceded by either a negative sign or a positive sign (for example, -0.23, $+0.54$). Researchers use these numbers to indicate two things about the relationship between two variables: the relationship's strength and its direction. The closer the correlation coefficient is to 1.0, the stronger the relationship or association between variables. A value of 1.0 represents a perfect association, meaning that if researchers know the value of the independent variable they can predict the value of the dependent variable with 100 percent certainty. A value of 0.00 indicates that no relationship exists between variables. The sign in front of the number shows the direction of the relationship. A positive sign means that as one variable increases or decreases, the second variable follows suit. A negative sign in front of the number means that as one variable increases or decreases, the other variable moves in the opposite direction. A correlation of $+0.90$ between grade-point average (GPA) and income, for example, means that the higher the GPA, the higher the income. Conversely, a correlation of -0.90 between grade-point average and income means that the higher the GPA, the lower the income.

Third, establishing a correlation is a necessary step but is not in itself sufficient to prove causation. A correlation shows only that the variables are related; it does not mean that one variable causes the other. For one thing, a correlation can be spurious. A **spurious correlation** is one that is coincidental or accidental; in reality, some third variable is related to both the independent and dependent variables and makes it seem that those two variables are related. For a researcher to claim that an independent variable helps explain a dependent variable, then, no evidence must indicate that another variable is responsible for a spurious correlation between the independent and the dependent variables. To check this possibility, sociologists identify **control variables,** variables suspected of causing spurious correlations.

A good example of a significant relationship between two variables that disappears when we control for a third variable involves the strong correlation between the number of fire trucks at a fire scene and the amount of damage done in dollars (the more fire trucks, the greater the dollar amount of fire damage). Common sense, however, tells us that this relationship is a spurious correlation. The number of fire trucks on the scene could not possibly be responsible for the dollar amount of fire damage. A third variable—the size of the fire—accounts for both the number of fire trucks sent and the amount of damage produced. Although the number of fire trucks called to the scene does help us predict the amount of damage, it is not the variable of cause.

A second example of a possible spurious correlation comes from the *Dictionary of Statistics and Methodology* by Paul W. Vogt (1996).

> If the students in a psychology class who had long hair got [significantly] higher scores on the midterm than those who had short hair, there would be a correlation between hair length and test scores. Not many people, however, would believe that there was a causal link and that, for example, students who wished to improve their grades should let their hair grow. The real cause might be gender: that is, women (who usually have longer hair) did better on the test. Or that might be a spurious relationship, too. The real cause might be class rank: Seniors did better on the test than sophomores and juniors, and, in this class, the women (who also had longer hair) were mostly seniors, whereas the men (with shorter hair) were mostly sophomores and juniors. (p. 217)

Thinking about the possibility of a spurious relationship is especially important when the independent variable is an **ascribed characteristic**—any physical trait that is biological in origin and/or cannot be changed but to which people assign overwhelming significance, such as hair texture and color, eye shape, and skin color. Other examples include age and country of birth. Such findings announcing an ascribed characteristic as a

---

Correlation coefficient A mathematical representation of the extent to which a change in one variable is associated with a change in another.

Spurious correlation A correlation that is coincidental or accidental because some third variable is related to both the independent and dependent variables.

Control variables Variables suspected of causing a spurious correlation.

Ascribed characteristic Any physical trait that is biological in origin and/or cannot be changed but to which people assign overwhelming significance.

"cause" can be used to stereotype and stigmatize some groups of people. For example, many Americans believe that Japanese workers are more loyal than Americans, that Japanese students are more studious, that Japanese criminals are more contrite, that Japanese people are less litigious than Americans, and that Japanese people in general are more group-oriented. Actually, however, such observed differences are beyond the control of individual Japanese and can be explained as due to some other factors:

> The "loyalty" of white-collar workers to their company, in contrast to the constant movement of employees in other countries, is one clear example. [In Japan] the major corporations tacitly agree never to hire someone who has left another firm. Japanese children are studious in large part because admission to the University of Tokyo, which is based on examination scores, is essentially their only hope for having an influential place in society.
>
> Japanese are "nonlitigious," not just because of their alleged love of consensus but also because of the acute shortage of lawyers. The Ministry of Justice controls the Legal Training and Research Institute, where future lawyers and judges must train, and it admits only 2 percent of those who apply. (Of the 23,855 who took the entrance examination in 1985, 486 were admitted.)
>
> Most criminals arrested by the police confess partly out of a sense of remorse but also because they know what a trial would mean: in 99 percent of criminal trials, the verdict is guilty. (Fallows 1989:28)

Summary and Implications

In this chapter, we have identified the technological forces (computer, telecommunications, the Internet, hypertext) that allow people to produce large quantities of information, and we have considered the quality-control problems—dearth of feedback and exaggerated, distorted headlines—that are associated with the information explosion. These issues require us to think critically about the research we encounter.

Despite such problems, a basic knowledge of research methods helps us sort through information in a way that allows us to distinguish media hype from balanced research. To illustrate the power of research methods to help us evaluate "facts," we focused on Japan; specifically, we examined the trade deficit between Japan and the United States, the meanings people attach to that deficit, and generalizations made about Japanese society (and, by implication, U.S. society) to explain the deficit. We also reviewed research that investigated some of the traits that many people assume are somehow connected with Japan's economic success or failure. Consider some of these highlights from research studies:

- In Japan, an elaborately prepared lunch is a sign of a Japanese woman's commitment as a mother, which inspires her child to be similarly committed as a student. This observation helps us understand the Japanese teacher's attention to eating habits. (By implication, U.S. mothers are more concerned with children's social adjustment than with their eating habits.)

- *Kikokushijo* are not "refugee camps" for returning Japanese students who have become individualistic, independent, and argumentative while abroad and must relearn the value of harmony, conformity, and group-mindedness.

- Only 2.1 percent of Kobe households have unlisted numbers compared with 20 percent of Seattle households.

- In Japan, student monitors play an important role in maintaining classroom order and discipline. (By implication, in the United States, teachers bear almost total responsibility for maintaining order and discipline.)

- The percentage of women who expect to depend on their children in old age has declined dramatically.

- The trade deficit is not a valid measure of economic exchange between the United States and Japan.

- American fourth graders tend to blame lack of ability and difficult tests as reasons for poor performance on math tests, whereas their Japanese and Chinese counterparts tend to blame lack of effort.

- Japanese fathers are less likely than their U.S. and German counterparts to eat dinner every day with their children.

- Japanese men are observed as frequently as American men interacting with their children in public settings.

Taken together, these findings suggest some interesting differences between Japan and the United States with regard to certain features: *obentōs*, *kikokushijo*, unlisted phone numbers, attribution for lack of success on math tests, methods of keeping classroom order. For the most part, however, these differences are not absolute. For example, people living in the United States and Japan share some qualities, such as the amount of time fathers appear and interact with their children in public. In addition, when we find significant differences between the people in two countries with regard to some attributes, "exceptions to the rule" almost always exist. In other words, although most residents of Japan do not have unlisted numbers, a small percentage do. Likewise, although a significant percentage of U.S. residents have unlisted phone numbers, most choose to make their phone numbers public. Likewise, some fourth graders in the United States do blame lack of effort for poor scores on math tests just as some fourth graders in Japan and China do blame difficult tests and lack of ability for poor performances.

Although some of the findings cited here suggest that some characteristics, such as the *kikokushijo* or *obentōs*, appear to be uniquely Japanese, we must be careful about making such claims. The world has thousands of cultures, and anyone who claims that some quality is unique to one place implies that they are an expert on all the world's culture. Today, the Internet or World Wide Web (WWW) can help in identifying unique characteristics. For example, Darryl Macer (1995), founder of the Eubios Ethics Institute, a Japan-based organization with the goal of promoting an integrated and cross-cultural approach to bioethics and building an international network of interested parties, used the WWW for this purpose when he asked readers of the on-line *Eubios Journal of Asian and International Bioethics* for comments on the following observation:

> On 27 October the University Animal Research Center held its annual memorial service for the research animals. In the past year 13,000 animals were used, and about 110 persons came to the shrine for experimental animals outside the animal research center. It was my first time to participate in the 20-minute service, during which time the people place a white chrysanthemum on the shrine while saying a short prayer. It is among trees, with birds flying overhead—overall a very interesting experience. It may serve for the relief of guilt of using animals, and recognizing their contribution to research. Earlier in the month the annual memorial service for the families who gave dead relatives' bodies for research, education, and autopsies was held. In the past year 214 bodies were given; included were about a dozen or so fetuses. This is a part of Japan that appears to be unique; I wonder if anyone can tell us of parallels. (Macer 1995)

The fact that everyone in Japan who participated in the research studies covered in this chapter did not respond in the same way suggests that we must reassess "the notion of Japan as a cohesive and tightly knit society based on a single value system" (Mouer and Sugimoto 1990:108). Likewise, we cannot assume that the United States is simply a society of individuals who value only self-expression and personal independence. The larger implication is that we cannot explain away differences between the two countries by labeling one society as group-oriented and the other as individualistic.

When researchers report cross-national, comparative findings, readers (and even the researchers) tend to treat the people in the country or society as one unit when, in actuality, the people within each society possess a wide range of characteristics. For example, in his research on the *kikokushijo*, Goodman (1993) noted that the experiences of Japanese children living abroad varies depending on

> where the children went, how long they were overseas, what type of education they received there, and how much contact they had with the local communities. There are also, of course, differences depending on their age, their gender, and their own individual personalities. Yet all such children, when they return to Japan, are classified as *kikokushijo*. They all tend to have the same qualities ascribed to them as a result of their overseas experience. (p. 2)

Goodman maintains that most researchers who study *kikokushijo* do not question this category as a valid way of classifying all returnees. Instead, they focus on and document the problems returnees supposedly have while overlooking evidence that suggests healthy outcomes. In a similar vein, many researchers study the trade deficit without considering whether it represents a valid measure of an economic relationship between two countries. The information in this chapter suggests that if we plan to conduct research aimed at understanding the economic relationship between the United States and Japan, then we cannot use the trade deficit as an operational definition of that relationship.

The fact that quality and conscientious research yields complex findings may leave some people feeling frustrated about how to approach problems believed to be related to the trade deficit or to cultural differences. On the other hand, the complex (rather than absolute) findings suggest that problems between groups are not insurmountable because differences are not as great as previously assumed.

Key Concepts

Use this outline to organize your review of the key chapter ideas.

Information explosion
- **Information**
- **Data**
- **Hypertext**
- **Dearth of feedback**

Research

Scientific method
- **Objectivity**

Concepts

Research design
- **Population**
  - **Traces**
  - **Documents**
- **Households**
- **Small groups**

Sample
- **Random sample**
- **Representative sample**
- **Sampling frame**

Method of data collection
- **Self-administered questionnaire**

Interviews
- **Structured interviews**
- **Unstructured interviews**

Observation
- **Participant observation**
- **Nonparticipant observation**
- **Hawthorne effect**

Secondary sources or archival data

Variables
- **Dependent variables**
- **Independent variables**
- **Control variables**

Operational definitions
- **Reliability**
- **Validity**

Generalizability

Correlations
- **Positive correlation**
- **Negative correlation**
- **Spurious correlation**

internet assignment

The Internet is a great tool for helping researchers to gather data from secondary sources. Search for the cremation rate for each of the 50 states. Find two variables such as "median income" or "percent of population that has graduated from college" that might help explain the cremation rate. Write two hypotheses specifying how each variable is related to the cremation rate.

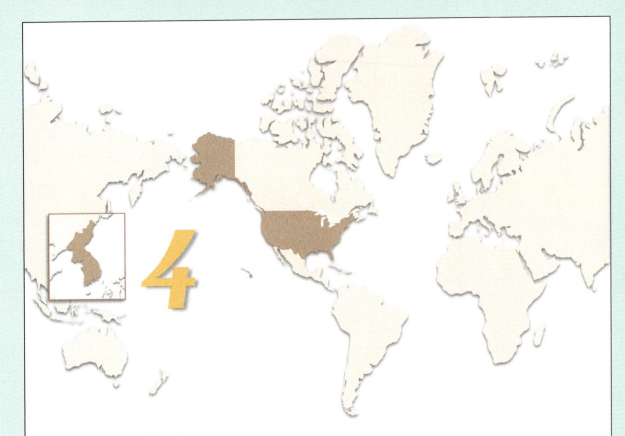

Culture

With Emphasis on South Korea

Demilitarized zone between North and South Korea. (© Nathan Benn/Woodfin Camp & Associates.)

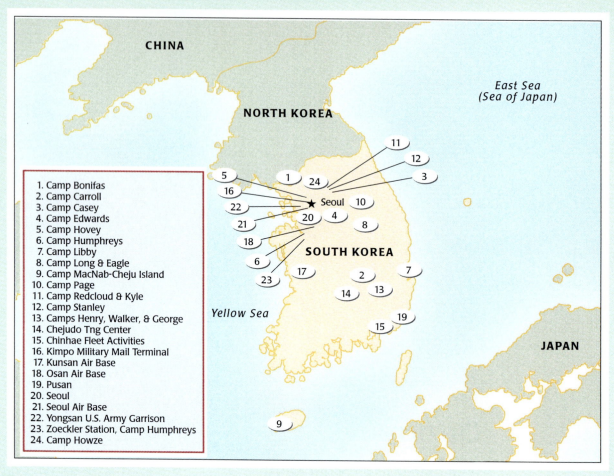

CHINA

NORTH KOREA

East Sea
(Sea of Japan)

1. Camp Bonifas
2. Camp Carroll
3. Camp Casey
4. Camp Edwards
5. Camp Hovey
6. Camp Humphreys
7. Camp Libby
8. Camp Long & Eagle
9. Camp MacNab-Cheju Island
10. Camp Page
11. Camp Redcloud & Kyle
12. Camp Stanley
13. Camps Henry, Walker, & George
14. Chejudo Tng Center
15. Chinhae Fleet Activities
16. Kimpo Military Mail Terminal
17. Kunsan Air Base
18. Osan Air Base
19. Pusan
20. Seoul
21. Seoul Air Base
22. Yongsan U.S. Army Garrison
23. Zoeckler Station, Camp Humphreys
24. Camp Howze

Seoul

SOUTH KOREA

Yellow Sea

JAPAN

Sources: Evinger (1995); U.S. Army (1998).

This map shows the locations of U.S. military bases in South Korea. Approximately 37,000 service women and men are stationed at these bases. The U.S. military presence in South Korea goes back more than 50 years when the peninsula was first divided into North and South Korea. In this chapter we ask how this division has affected Korean culture and what it means to be Korean.

Who Is Korean?

- Wayne Berry was born in Korea but was adopted and raised by a U.S. couple. [He] "thought the chances of finding his birth parents were just about zero. The adopted son of Minnesota dairy farmers, Berry, 26, was looking for two people among 45 million in South Korea. He didn't know their names and he didn't even speak Korean.

  After sending out 100 letters, [Berry] employed a translator, made a two-minute video and sent it with his baby picture to three TV stations in Seoul. As luck had it, his aunt was watching the one station that aired the tape—and she recognized him.

  Last year, [Berry] flew to Korea and met his parents, even though he learned he had been conceived during an illicit affair and they were now married to other people. Again, luck was on his side and they were overjoyed, not ashamed, to see him. 'I think they wanted to make up for lost time,' he said.

  What's more, Berry said he received 'the red carpet treatment' from about 30 other relatives he never knew he had. He is planning to stay in touch with his parents. And now, he said, his aunt is seeking a son she gave up for adoption in the United States" (Smith 1996:A1).

- [In 1965], at the age of 22, the future seemed bleak for Sang Hyun Kim, a poor farm boy from South Korea, so he packed his bags and headed for Argentina, in search of the great American dream. Today, it seems that Mr. Kim, who once worked on an Argentine peanut farm, has found that dream. Now 52 and an insurance entrepreneur, he is one of 35,000 South Korean immigrants who make up a prosperous and closely knit population here that is often the envy of other Argentines.

  Since arriving in 1965, Mr. Kim, who speaks Spanish like a native, has brought nine of his siblings from Korea to Argentina, and all of them now own thriving businesses (Sims 1995:A4).

- In 1951 at the age of nine, K. Connie Kang fled from her home in North Korea, and the Korean War, to live in Japan. Later she attended college in the United States and eventually became a *Los Angeles Times* reporter. While growing up in Japan, Kang recalls that she "covered the wall of [her] bedroom with pictures of James Dean, Elvis Presley, Sandra Dee and Natalie Wood and sang along with [her] record player to 'Love Me Tender,' 'Jailhouse Rock' and 'Hound Dog.' [She] was so wrapped up with [her] life in Tokyo that [she] did not think about Korea much" (Stephens 1995:2). Kang describes how she "became Korean" in college:

  "After spending a big chunk of my childhood and all of my adolescence in Japan, meeting Korean students with strong Korean identities gave me the oddest feeling. On one hand, it provided me with a connection to my ancestral land that had been missing during those formative years. But on the other hand, I knew how different I was from the other Korean students. My mannerisms were a mixture of the influences from my Japanese and American schooling. Despite our differences, though, they treated me like a younger sister. They taught me popular Korean songs, jokes, and idioms then in vogue. So, in Missouri, of all places, my Koreanization began. I felt like a little girl relearning Korea from the real teachers. As much as my parents had tried, I knew they had been out of touch from being away for so long" (Kang 1995:167).

- U.S. servicewoman Andrea "Simone" Bowers was stationed in South Korea for 13 months. During that time she became best friends with a Korean civilian who worked on her base. He introduced Simone to his country. Simone climbed mountains every weekend with him. On these weekends she met the "real" Korean people—not those who lived in and around the tourist towns next to the military bases catering to servicemen and women. She "sampled their cooking, coffee, and slept at their houses on the weekend." In the meantime she picked up enough of the Korean language to communicate basic information. Simone visited mountain villages so remote that the people did not have access to television and had never seen dark-skinned people of African ancestry. She recalls that some Koreans were so fascinated with her skin color that they rubbed her skin just to feel it. She still keeps in touch with her Korean friend and sometimes finds herself longing for Korean food and coffee (Cherni 1998).

Why Focus on South Korea?

The Korean peninsula, divided into North and South Korea, is located in a region of the world known as the Pacific Rim.[1] The peninsula is surrounded by three powerful neighbors—China, Russia, and Japan. Americans should know more about this area of the world if only because 37,213 U.S. military personnel are stationed in South Korea today (U.S. Department of Defense 1998). U.S. military involvement in Korea dates back to the end of World War II (1945), when Premier Joseph

Stalin of the Soviet Union, Prime Minister Winston Churchill of Great Britain, and President Franklin Roosevelt of the United States met and "without consulting even one Korean" agreed to chop the country in half (Kang 1995:75).[2] During this 50-year-plus involvement, an estimated 7.4 million U.S. servicemen and servicewomen have fought, died, and otherwise served in Korea. Yet most Americans have little idea how U.S. military interest in Korea developed or why it continues today. News about North and South Korea seems to flit "across the world's newspaper headlines and television screens, . . . only to disappear from view when the immediate dangers seemed to pass" (Oberdorfer 1997:xiii). These headlines may involve terrorism (such as the 1987 bombing by North Korean agents of a South Korean jetliner, killing 135 passengers), showdowns with North Korea over nuclear weapons, the sudden deaths of North or South Korean leaders, economic crises, famines in North Korea, revelations that the Japanese military forced 100,000 young Koreans to be "comfort women" to its troops, or South Korea's capture of North Korean spy submarines off its shores (Savada and Shaw 1990).

Although these news events are all connected in some way to the division of Korea into north and south regions, they are presented to viewers without historical context. Undoubtedly, this division affected the lives of the four people highlighted at the start of this chapter. In the case of K. Connie Kang, the Korean War was the reason she lived in Japan during her youth. Korean-born Wayne Berry, adopted by U.S. parents, was also affected by the Korean War, even though he was born 17 years after it ended. Harry and Bertha Holt, the founders of Holt International Children's Services, pioneered intercountry adop-

tion in 1955 upon learning about the large number of Korean children orphaned by that conflict (Holt International Children Services 1997). Sang Hyun Kim immigrated to Argentina from South Korea in 1965, 12 years after the Korean War ended. His decision to seek employment opportunities elsewhere was directly connected to the fact that the country was recovering from a "bitter and costly civil war, which . . . bled the whole peninsula" (Halberstam 1986:700). U.S. servicewoman Andrea "Simone" Bowers found herself stationed in South Korea because U.S. and Soviet leaders decided as long ago as 1945 that Korea was important to each country's national interests. As you can see from the examples, the Korean War and the subsequent division of the peninsula into north and south has had a profound effect on Korean culture and the meaning of being Korean.

A few words of caution are in order before delving into this subject. Although this chapter focuses on South Korea[3] and the United States, we can apply the concepts discussed here to understand *any* culture and frame other cross-cultural comparisons. As you read about culture, remember that South Korea is broadly referred to as a country possessing an Eastern (or Asian) culture and the United States is regarded as a country possessing a Western culture. Therefore, many of the patterns described here are not necessarily unique to Korea or the United States, but rather are shared with other Eastern or Western societies. At the same time, you should not overestimate the similarities among countries that share a broad cultural tradition. Do not assume, for example, that South Korea (or any other Pacific Rim country or city-state) is just like Japan. As we will see, much of Korean identity is intricately linked with the

idea of being "not Japanese" (Fallows 1988). To assume that South Korea is like Japan is equivalent to assuming that the United States is just like a Western European country, such as Germany or the United Kingdom. As we know, however, the United States is a country that celebrates its independence from European influence.

[1]The countries of the Pacific Rim are Japan, South Korea, China, Taiwan, Hong Kong, the Philippines, Vietnam, Thailand, Malaysia, Singapore, and Australia.

[2]The Korean people, who had been governed by the Japanese since 1910, could not have established a strong, stable government if left on their own. The United States and Russia gave themselves shared trusteeship of Korea. Japan surrendered to the Russians north of the 38th parallel and surrendered to the Americans south of the 38th parallel. In 1947, the United Nations (UN) called for elections and the establishment of a united and independent Korea. Russia and the United States supported candidates sympathetic to their respective governments. The Soviet Union refused to allow the UN to supervise elections in the north, so elections were held only in the south. The south elected Kim Il-Sung as president and established Pyongyang as its capital. As a consequence, two leaders and two governments emerged—a Soviet-dominated communist government in the north and an American-dominated democratic government in the south. The Korean War began when the North Korean government attempted to take control of South Korea in 1950. The civilian populations endured tremendous suffering as the north and the south pushed each other back and forth across the 38th parallel. During the fighting, family members became separated, children were orphaned, and entire cities, industrial plants, roads, bridges, and homes were reduced to rubble. An estimated 5 million Koreans were separated from their families when the 38th parallel was reestablished. To this day, separated families have a tremendous desire to reunite, although they have been allowed no contact for almost 50 years.

[3]Republic of Korea is the official name for South Korea. From this point on, we will use the terms *South Korea* and *Korea* interchangeably. Any reference to North Korea specifically is noted as such.

The Challenge of Defining Culture

Consider the entry for *culture* in the *Cambridge International Dictionary of English* (1995), which presents core definitions of words and the most common usage of the word among English speakers:

> **culture** WAY OF LIFE *n* the way of life, esp. general customs and beliefs of a particular group of people at a particular time • *youth/working-class/Russian/Roman/mass culture* • *She's studying modern Japanese language and culture.* • *The cultures of Britain and Nigeria are very different.* • *Thatcher's enterprise culture* (= way of thinking and behaving) *of the 1980s brought many changes.* • *There's a* **culture gap** (= difference in ways of thinking and behaving) *between many teenagers and their parents.* • *It was a real* **culture shock** *to find herself in London after living on a small island* (= She felt alone and was confused by the completely different way of life there).

This entry shows that we use the word *culture* in conjunction with specific places (Russia, Rome, Nigeria, Britain, small islands) and categories of people (the masses, teenagers, parents, Russians, Japanese). We also use the word in ways that emphasize differences ("The cultures of X and Y are very different"; "There is a culture gap between X and Y"; "It is culture shock to come from X and live in Y"). Our use of the word suggests that we think of culture as having clear boundaries, as explaining behavior, and as providing a blueprint for living that people follow in mechanical ways. In addition, it suggests that we think of interaction between people of different cultures as problematic.

In light of the seemingly clear way in which we use the word, we may be surprised to learn that the real challenges and the sources of endless debate among people who study culture include the following:

- *Describing a culture*. That is, is it possible to find words to define something so vast as the way of life of a people?

- *Determining who belongs to a group designated as a culture*. Does a person who "looks Korean" and who has lived in the United States most of his or her life belong to Korean or American culture? Can the four people described in the chapter introduction be classified as Korean?

- *Identifying the distinguishing characteristics that set one culture apart from others*. For example, is eating rice for breakfast a behavior that makes someone Korean? Is an ability to speak Korean a behavior that makes someone Korean? Are ethnic Koreans

Koreans have been part of the United States in small numbers for about 150 years, even though their presence has received little attention until recently. This photo of a Korean farmer and his son was taken around 1920.

Courtesy USC Korean Heritage Library

who speak English or Spanish not Korean? Can we make a case that Andrea "Simone" Bowers is more Korean than Wayne Berry?

Cynthia K. Mahmood and Sharon Armstrong (1992) faced these nagging questions when they traveled to Eastermar, a village in the Netherlands province of Friesland, to study the Frisian people's reactions to a book published about their culture. They found that the Frisian people could not agree on a single "truth" about them as described in the book. At the same time, the villagers could not come up with a list of features that would apply to all Frisians and that would distinguish them from other people living in Eastermar. Yet the Frisians remained "convinced of their singularity," and the villagers reacted emotionally to the suggestion that they might not constitute a culture.

The Frisian situation captures the conceptual challenges associated with the idea of culture: the paradox of recognizing a culture but being unable to define its **boundaries,** the characteristics determining where a culture begins and ends or the qualities marking some people off from others as a unified and distinctive group. This chapter offers a framework for thinking about culture that considers both its elusive nature and its importance in shaping human life. This framework includes nine essential principles defining the nature of culture.

Material and Nonmaterial Components

Principle 1: Culture consists of material and nonmaterial components. **Material culture** consists of all physical objects that people in one society have invented themselves or borrowed from members of other societies and to which they have attached meaning. Material culture includes natural resources such as plants, trees, and minerals or ores, as well as items that people have converted from natural resources into other forms for a purpose. Examples of the latter include cars and trucks to transport people, animals, and goods; microwave ovens to cook and heat food; computers to make calculations; video cameras equipped with devices that selectively soften facial features to make people look younger than they really are; radios to entertain, inform, and provide background sound; and so on.

Boundaries The qualities marking some people off from others as a unified and distinctive group.

Material culture All physical objects that people have borrowed, discovered, or invented and to which they have attached meaning.

Nonmaterial culture Intangible creations or things that we cannot identify directly through the senses.

Beliefs Conceptions that people accept as true, concerning how the world operates and where the individual fits in relationship to others.

In thinking about material culture, it is important to learn not only the most obvious and practical uses for which an object is designed, but also the meanings assigned to that object by the people who use it (Rohner 1984). As an example, consider the radio, a device for receiving and broadcasting sound messages. For many people, the radio takes on meaning beyond these obvious purposes. People use it to fill the void that can accompany boring tasks, daily routines, or loneliness; to sustain or create a mood (for example, upbeat, romantic, relaxed); and to provide a social lubricant in that people can talk with one another about what they have heard on the air. The importance of the radio in some people's lives becomes evident in statements like these: "To me, when the radio is off, the house is empty"; "I listen to the radio from the time I get up until I go to bed"; "Radio puts me in a better mood"; and "It makes driving easier" (Mendelsohn 1964:242–243). Learning the meaning that people assign to objects in the material culture helps sociologists grasp the significance of those objects in people's lives.

Nonmaterial culture consists of nonphysical creations. Thus a person cannot hold or see nonmaterial culture. Three of the most important of these creations are beliefs, values, and norms.

Beliefs

One type of nonmaterial culture is **beliefs,** conceptions that people accept as true, concerning how the world operates and where the individual fits in relationship to others. Beliefs can be rooted in blind faith, experience, tradition, or the scientific method. Whatever their accuracy or origins, they can exert powerful influences on actions as they are used to justify behavior, ranging from the most generous to the most violent. For example, some people share the following beliefs:

- Interruptions or imbalances in the flow of *qi* (pronounced chee)—the vital energy that flows through the body—cause illness.

- Very small organisms called germs cause disease.

- Continuous conversation, rather than silence, validates a relationship.

- Athletic talent is inherited.

- Athletic talent is essentially a product of hard work, practice, and persistence.

- After death the human spirit returns to earth in a different form.

Values

A second component of nonmaterial culture is **values,** general, shared conceptions of what is good, right, appropriate, worthwhile, and important with regard to conduct, appearance, and states of being. Perhaps the most significant study on values was made by social psychologist Milton Rokeach (1973). Rokeach identified 36 values that people everywhere share to differing degrees, including the values of freedom, happiness, true friendship, broadmindedness, cleanliness, obedience, and national security. He suggests that societies are distinguished from one another not on the basis of which values are present in one society and absent in another, but rather according to which values are the most cherished and dominant. Americans, for example, place considerable value on the individual as an individual; they stress personal achievement and unique style (free choice). In contrast, Koreans value

Dodgers pitcher Chan Ho Park bows to the umpire before throwing the first pitch. What values and beliefs might inspire this behavior?
© Ronald Modra/Sports Illustrated

the individual in relationship to the group (particularly the family); they stress self-discipline and respect toward older people.

In sports, for example, the value placed by Americans on the individual is demonstrated by the fact that they single out the most valuable player of a game, a season, a league, or a tournament. Furthermore, when most Americans view an outstanding athletic feat, they tend to give more credit to the individual's innate talent or desire to win than to his or her disciplined practice habits. In addition, American athletes work to find the style that is "right" for them and are willing to change this style if it does not bring success. Koreans, on the other hand, spotlight individual achievement but also place considerable value on discipline, particularly on form (that is, adhering to time-tested and efficient methods of accomplishing goals). From the Korean point of view, athletic achievement does not occur simply because a person wants to excel or because he or she possesses raw talent. Athletic competence develops over time, after the individual masters and appreciates the steps that combine to produce the intended result. Compared with the American system, the Korean system minimizes individual achievement because the achiever owes success to the mastery of technique.

Values, we should note, transcend any particular situation. For example, the American emphasis on individual achievement and unique style—and the Korean emphasis on the group, form, and discipline—are not confined to one single area of life, such as sports. As we will see later in this chapter, these cultural values permeate many areas of life, including studying (see "Group Study, Cheating, and the Korean Foreign Student Experience").

Norms

A third component of nonmaterial culture is **norms,** written and unwritten rules that specify behaviors appropriate and inappropriate to a particular social situation. Examples of written norms are rules that appear in college student handbooks, on signs in

Values General, shared conceptions of what is good, right, appropriate, worthwhile, and important with regard to conduct, appearance, and states of being.

Norms Written and unwritten rules that specify behaviors appropriate and inappropriate to a particular social situation.

U.S. in Perspective

Group Study, Cheating, and the Korean Foreign Student Experience

Even though it is very interesting to live in a foreign country, you cannot help but be frustrated almost every day. You always feel that you are missing something in the culture—whatever you do and wherever you go. At worst, you may even hurt other people's feelings unintentionally. Of course, yours can get hurt, too. To avoid those kinds of occasions as a foreigner living in America, you try to open your eyes and ears as wide as possible to collect any information about the culture, hoping you can use it sometime in the future. When the information seems to be hard to collect, you become even more alert. And even though you have accurate information, there are always cases when you use it at the wrong time and at the wrong place. There are even some cross-cultural nightmares that would never have taken place if anybody had taken the trouble to pass on to you only one sentence beforehand. In some of these nightmares, you may have to live with a mark of shame on your back for what you have done even though you did not intend to do it. The worst part of it is that you may not know that the mark is on your back. On a nice evening after dinner, long after I had graduated from an American university, I learned that the mark many unwitting Korean students were living with was "Master of Cheating."

When I was a graduate student in the late 1980s, with few exceptions the Korean students would associate almost exclusively with Korean students. Students who were married gathered with other families who lived in the same apartment complexes. Those who were not married shared apartments. Because most of us went to the engineering school and majored in a small number of different subjects, many of us took the same classes at the same time, or a little earlier, or a little later. Therefore, we were in a good position to help each other with our studies. We could naturally pool all the old homework assignments and exams. Of course, everyone took good care of the Old Exam Folder, as it was the most important reference. This folder was off-limits to other classmates who were not Koreans. However, any intent to keep it a secret was not mentioned outright by any of us. There were many foreign students at our school, and all of them worked desperately to get good grades. Seeing them studying with their own people in their own languages, we who had to compete with them just assumed that they were taking care of themselves just as we were taking care of ourselves. For Koreans, "racial identity" is a familiar concept, so we did not give this business a second thought. Anyhow, that kind of group study helped me at several points during my student years.

Some years later I married an American who worked for a while as a teaching assistant at the same school. At one point, he and I happened to be sipping a cup of coffee after dinner talking about our graduate student years. We started talking about my old days as a foreign student and Korean students in general. Then, suddenly, I could not believe what I was hearing from him. He said that all his American friends who were teaching assistants had a saying: "Beware of the Korean network; the Korean students are masters of cheating." This was very strange to hear. We were such a close-knit community (as we say in Korean, we knew the count of each other's silverware), so we naturally would have heard if some of us were known to be cheaters! What could this mean?

The answer was very short: our homework solutions, computer programming assignments, and written reports were very much the same, and we prepared for our tests with copies of old exams. I said with a shaking voice that students in Korea often prepare for tests with copies of old exams. But that is not cheating, is it? Well, maybe not precisely. But to American students it is not considered very upright, ethically. Let me try to explain.

American universities do not give a fixed course sequence to students in a given major. The university just gives lists of what courses the students need to take to graduate. Accordingly, every

restaurants ("No Smoking Section"), and on garage doors of automobile repair centers ("Honk Horn to Open"). Unwritten norms exist for virtually every kind of situation: wash your hands before preparing food; do not hold hands with a friend of the same sex in public; leave at least a 15 percent tip for waiters and waitresses; remove your shoes before entering the house. One unwritten rule followed by a majority of women in the United States is to shave or otherwise remove facial and body hair.

Some norms are considered more important than others—and so the penalties for their violation are more severe. Depending on the importance of a norm, punishment can range from a frown to death. In this regard we can distinguish between folkways and mores.

student plans her or his own classes independently.

Very often you find a very mixed class: sophomores sitting next to juniors and seniors, or seniors sitting next to master's students. There is not much class-year unity to be seen. Another difference from Korea is that at many U.S. schools students often have their own part-time jobs. At my school, many American students showed up on campus right before their classes began, then immediately after their classes ended they rushed out to home or work. Because of this, too, they did not have any strong bond with people who majored in the same subjects and did not know (and did not care) which classmate would graduate in what year. There was no notion at all of *sunpae* or *hupae*.[1] It was very natural that they could not think of getting precious old exams from people who took the class in earlier years. Sometimes they organized a study group on their own or at the recommendation of an instructor, but most of the time Americans did homework all alone and prepared tests without the benefit of any old exams.

To most Americans, studying for long hours alone at their desks is part of the college experience. Occasionally turning in a few incomplete or incorrect homework problems in a physics class, say, is no great shame. Instructors often give students some credit for their partial solutions, and indeed, they may respect the student's efforts even if it did not result in perfection. The American graduate assistants I knew in computer science, for example, considered difficult 1,000-line programming assignments part of a rite of passage for majors. When they saw students in their own classes taking the "easy route" by consulting the work of others, they were filled with disdain. Whereas we Koreans saw our goals as merely to learn the subject matter well and achieve a good grade, for Americans, individual solitary effort for its own sake seemed part of the formula.

Of course, all this applies to Asian foreign students in general, who make up large parts of science and engineering graduate programs in the United States. There was even a group of German students who studied together closely at my school. But as I now understand it, the problem was not that we studied together. The problem was that we made one shared project out of succeeding in school, literally sharing old exams and homework, and so the individual effort, not to mention the individual creativity, was simply not there for each person. The harsh reaction of the teaching assistants I have mentioned is not universal here, of course. However, it would be a good idea for students who are planning to study here to ask around about their university's customs or policies in this matter.

It is distressing to think about how many habits we bring from Korea that are judged as failings here. What is more troubling is that these judgments are often hidden from us and that we lead our personal and professional lives in this new country completely un-aware. Of course, there is a lesson from the reciprocal of this scenario also. The *American* habits that trouble us deeply should always be placed in context. How much will I really understand deeply while I live here? Maybe that is why I try to be kind to Americans. On occasions when they make me upset, I try to calm myself down first and try to tell them how I view the situation. This kind of direct talk is usually acceptable. In many cases, I end up learning to interpret the occasion in a way I never imagined before. In fact, sometimes I talk and talk about my culture to American listeners until they get sick and tired of hearing me, because I hope that through all my talk they find the false negatives about Korea fading away. I also hope that through my listening I will come to know better the real America, as the America in my mind is certainly misrepresented. However, the greatest hope I have is that when Americans remember my "bad behavior" at some occasions long past, they will be able to judge me anew, more clearly and fairly, after all these years.

[1]Translator's note: In Korea, a *sunpae* is a student at your school who began his or her studies one or more years before you did. A *hupae* is a student at your school who began after you. There is a strict respect hierarchy here. And even decades after graduation, recognizing that a stranger is in fact a *sunpae* or *hupae* establishes an instant bond.

Source: Kim Bo-Kyung, Northern Kentucky University (1997). This essay is adapted from *"Hankuk Yuhaksaeng: 'Khunning Tosa'ranun Numyengi Woeyn Malinka?"* to appear in a collection of essays published in Korea by Hanul Press, 1997.

Folkways are norms that apply to the mundane aspects or details of daily life: when and what to eat, how to greet someone, how long the workday should be, how many times each day caregivers should change babies' diapers. As sociologist William Graham Sumner (1907) noted, "Folkways give us discipline and support of routine and habit"; if we were forced constantly to make decisions about these details, "the burden would be unbearable" (p. 92). Generally, we go about everyday life without asking "Why?" until something reminds us or forces us to see that other ways are possible.

Folkways Norms that apply to the mundane aspects or details of daily life.

The Korean table has no clear place settings, a practice that reflects values about the relationships of the individual to the group. In addition, Korean diners use the same utensils for serving and eating.

© David Bartruff/Artistry International

Consider the folkways that govern how a meal typically is eaten at Korean and American dinner tables. In Korea, diners do not pass items to one another, except to small children. Instead, they reach and stretch across one another and use their chopsticks to lift small portions from serving bowls to individual rice bowls or directly to their mouths. The Korean norms of table etiquette—reaching across instead of passing, having no clear place settings, and using the same utensils to eat and to serve oneself food from platters and bowls—deemphasize the individual and reinforce the greater importance of the group.

Americans follow different dining folkways. They have individual place settings, marked clearly by place mats or blocked off by eating utensils. It is considered impolite to reach across another person's space and to use personal utensils to take food from the serving bowls. Diners pass items around the table and use special serving utensils. The fact that Americans have clearly marked eating spaces, do not trespass into other diners' spaces, and use separate utensils to take food reinforces values about the importance of the individual.

Koreans and Americans even have different folkways about how they should use resources such as notebook paper and the electricity needed to keep refrigerators cold. Koreans open the refrigerator door only as wide as necessary to remove an item, blocking the opening to minimize the amount of cold air that escapes. Americans open the refrigerator door wide and often leave it open while they decide what they want or until they move the desired item to a stove or countertop.

As for notebook paper, until about 15 years ago Koreans filled every possible space on a sheet of paper before throwing it away. Sometimes they used a ruler to draw extra lines between those already on the paper to double the writing space. In contrast, Americans will often throw away a sheet of paper with only one line of writing because they do not like what they wrote or their penmanship.

Another more visible folkway related to conservation in Korea can be observed when cars stop at red traffic lights in the city; drivers turn off their headlights. When the light turns green, they turn them on again (Kim and Kirby 1996).

Often travel guides and international business guides list folkways that foreign travelers should follow when visiting a particular country. The *Los Angeles Times* (1995:D4), for example, published tips for business executives traveling to South Korea advising them to observe the following folkways:

- Be punctual. Many Koreans are more sensitive than other Asians when it comes to deadlines.

- Don't talk excessively during a meal. Koreans may interpret that as a lack of appreciation for either the food or the service.

- Try to get a Korean colleague, the higher-ranking the better, to introduce you. Connections and referrals can greatly facilitate your entry into the South Korean business scene.[4]

Mores are norms that people define as essential to the well-being of a group. People who violate mores are usually punished severely—they are ostracized, institutionalized in prisons or mental hospitals, sentenced to physical punishment, condemned to die. In contrast to folkways, people consider mores to be unchangeable, regarding them as "the only way" and "the truth."

Mores Norms that people define as essential to the well-being of a group.

[4]The interesting point about these suggestions is that many apply to behavior and/or practices that are common in the United States, but are presented here as if they represent real differences between the two cultures.

In a 1994 case that received international attention, Singapore officials found Michael Faye, an 18-year-old American, guilty of violating their official mores, which place social order and citizens' general well-being ahead of individual rights. Faye was found guilty of vandalism because, over a 10-day period, he had damaged cars with eggs and spray paint. Singapore officials defined this behavior as "a calculated course of criminal conduct" and sentenced him to caning and 60 days in jail. President Bill Clinton appealed the sentence, and a Clinton administration official called the penalty excessive "for a youthful, nonviolent offender who pleaded guilty to reparable crimes against private property" (Wallace 1994:A5).

One way to explain differences in behavior is to point to differences in Korean and American values and norms. For example, one could argue that Koreans value conservation and Americans value consumption and that the two groups have devised standards of appropriate behavior that reflect these values. Yet to say that these differences are caused by values and norms offers few insights. We must instead investigate the geographic and historical circumstances that gave rise to specific norms and values.

The Role of Geographic and Historical Forces

Principle 2: Geographic and historical forces shape the character of culture. Sociologists operate under the assumption that culture is "a buffer between [people] and [their] habitat" (Herskovits 1948:630). That is, material and nonmaterial aspects of culture represent the solutions that people of a society have worked out over time to meet their distinctive historical and geographic challenges and circumstances.[5]

Part of the reason that Koreans and Americans use refrigerators and notebook paper differently has to do with the amount of natural resources available in each country. Korea has no oil, only moderate supplies of coal, and depleted forests. Relative to Korea, the United States possesses abundant supplies of oil, wood, and coal. Although Koreans can import these resources, they face pressures unknown to most peo-

ple in the United States, even as Americans come to realize that their resources are dwindling. Because Koreans depend on other nations for most resources, they are vulnerable to any world event that might disrupt the flow of resources into their country. This vulnerability reinforces the need to use resources sparingly and not to take them for granted. The relative lack of natural resources has affected the energy costs in Korea and may ultimately explain conservation-oriented behavior.[6]

In sum, conservation- and consumption-oriented values are rooted in circumstances of shortage and abundance. To understand this connection, recall a time when your electricity or water was turned off. Think about the inconvenience you experienced after a few minutes and how it increased after a few hours. The idea that one must conserve available resources takes root. People take care to minimize the number of times they open the refrigerator door.

Imagine how a permanent resource shortage or the dependence on other countries for resources can affect people's lives. Consider how a long-term resource shortage affected Californians when their state experienced a six-year drought (1987–1992). In some water districts, Californians cut their use of water by 25 to 48 percent. In Contra Costa County in the San Francisco Bay area, for example, customers cut water consumption below the 280 gallons of water per household per day recommended under the voluntary rationing plan to an average of 165 gallons (Ingram 1992).

In contrast, you can imagine how a greater abundance of resources breaks down conservation-oriented behaviors. After spending some time in the United States, many Koreans stop double-lining their paper because they have few incentives to do so. Supplies of paper are abundant, and no one else is conserving this way. Likewise, as Korea has improved its economic status, the frugal use of notebook paper seems to have disappeared (Kim and Kirby 1996).

For the most part, people do not question the origin of the values they follow and the norms to which they conform, "any more than a baby analyzes the atmosphere before it begins to breathe it" (Sumner 1907:76). Nor are they aware of alternatives, because many values and norms that people believe in and adhere to were established before they were born. Thus people behave as they do simply because they know no other way. And, because these behaviors seem so

---

[5]"All mankind shares a unique ability to adapt to circumstances and resolve the problems of survival. It was this talent which carried successive generations of people into the many niches of environmental opportunity that the world has to offer—from forest, to grassland, desert, seashore, and icecap. And in each case, people developed ways of life appropriate to the particular habitats and circumstances they encountered." (Reader 1988:7)

[6]For example, even wealthy Koreans rarely air-condition their homes but use ordinary room and window fans (Kim and Kirby 1996).

natural, we lose sight of the fact that culture (in this case, conservation and excessive consumption) is learned.

The Transmission of Culture

Principle 3: Culture is learned. Parents transmit to their offspring via their genes a biological heritage common to all humans but uniquely individual. The genetic heritage that we share with all humans gives us a capacity for language development, an upright stance, four movable fingers and an opposable thumb on each hand, and other characteristics. If these traits seem overly obvious, consider that they allow humans to speak innumerable languages, perform countless movements, and devise and use many inventions and objects. In fact, "most people are shaped to the form of their culture, because of the enormous malleability of their genetic endowment" (Benedict 1976:14).

Regardless of their physical traits (for example, eye shape and color, hair texture and color, skin color), babies are destined to learn the ways of the culture into which they are born and raised. That is, our genes endow us with our human and physical characteristics, but not our cultural characteristics. We cannot assume that someone comes from a particular culture simply because he or she looks like a person whom we expect to come from that culture. This fact becomes obvious to Korean American youth who participate in cultural immersion programs that involve study in Korea. "Many say they have never felt so American as when they are slurping noodles in Korea. Even their slurps have an American accent" (Kristof 1995:47).

An excerpt from a letter written by a first-generation Taiwanese-American mother to her daughter shows that even parents have a hard time accepting this idea when their children are raised in a country and culture different from their own:

> To you, Taiwan is just a fun but humid place that you visited one summer, and your grandparents are just fuzzy voices over a telephone. To me, there is a lifetime that sits thousands of miles away, tucked inside the navy blue suit you see on me now. And to me, Taiwan is still ohm. HOME. Yes, our white

stucco house with the orange door is home, too, but part of my blood still flows toward Taiwan. Can you understand that? Maybe that's why sometimes I expect you to understand Chinese culture without having experienced any of it firsthand. I think that you have the same blood, and that it pulses to the same beat. You ARE Chinese still, and I know that you have some interest and even some pride in it, but there's so much you don't know. It is your right to know. It is your right to know your family's experiences, even if you don't care about them. Maybe someday, you will care. (Yeh 1991:2)

The Role of Language

The development of language illustrates the relationship between genetic and cultural heritages. Human genetic endowment gives us a brain that is flexible enough to allow us to learn the language(s) that we hear spoken by the people around us. As children learn words and the meanings of words, they learn about their culture. They also acquire a tool that enables them to think about the world—to interpret their experiences, establish and maintain relationships, and convey information. Anyone who speaks only one language might not realize this property of language until he or she learns another language. To become fluent in another language is not merely a matter of reading and conversing in that language but of actually being able to think in that language.[7] Similarly, when young children learn the language of their culture, they acquire a thinking tool.

The following characteristics of language show the relationship between learning the meaning of words and learning the ways of a culture:

- *Language conveys important messages above and beyond the actual meaning of words.* Words have two levels of meaning—denotative and connotative. **Denotation** is literal definition; **connotation** is the set of associations that a word evokes. Idioms help us see the distinction between denotation and connotation. An

---

Denotation A literal definition.

Connotation The set of associations that a word evokes.

[7]In the 1920s, linguist Edward Sapir wrote a highly influential book, *Language: An Introduction to the Study of Speech.* Sapir (1949), who alerted us to the social issues and structure of language, believed, "No two languages are ever sufficiently similar to be considered as representing the same social reality. The worlds in which different societies live are distinct worlds, not merely the same world with different labels attached" (p. 162). These assumptions underlie the linguistic relativity hypothesis advanced by Sapir: languages are so different that it is nearly impossible to make translations in which words produce approximately the same effects in the speaker of language X as in the speaker of language Y. Today, most social scientists reject this position and argue that, although considerable work may be required, it is possible for people to discover one.

An American cultural norm is the idea that when a child is adopted, it typically leaves a biological family to join an unrelated family. Is there anything particularly "natural" about this? Adoption practices are quite different in Korean culture.

© Myrleen Ferguson/PhotoEdit

idiom is a group of words that when taken together have a meaning different from the internal meaning of each word understood on its own. Examples of idioms include "to bump into," "on the edge of my seat," "to be eating out of someone's hand," and "to be in hot water" (Comenius Group 1996). In a literal sense, "to bump into" means "to collide with or hit something" but when used in the sentence "Guess who I bumped into the other day?" the phrase connotes "to meet unexpectedly" (Comenius Group 1996). These examples show that the connotation of a word is as important—and sometimes even more important—in understanding meaning as is the literal definition.

• *Words refer to more than things; they also describe relationships.* The word *adoption,* for example, as in the adoption of a child, refers to more than the child or the taking in and raising of that child. It also implies the presence of biological and adoptive parents and evokes assumptions about the relationship among the child, the biological parents, and the adoptive parents. The norms that guide these relationships are important features of the word. In Korea, for example, most people think of adoption as "a system whereby a sonless couple may receive a son from one of the husband's brothers or male cousins" (Peterson 1977:28). An adopted son cannot come from the wife's family or the husband's sister. In contrast, in the United States adoption usually involves a situation in which a child's biological parents release him or her to responsible adults who agree to raise the child as their own.

• *Words mirror cultural values.* Language embodies values considered important to the culture. For instance, in Korean society age is an exceedingly important measure of status: the older a person is, the more status, or recognition, he or she has in the society. Korean language acknowledges the importance of age by its use of special age-based hierarchical titles for everyone. In fact, it is nearly impossible to carry on a conversation, even among siblings, without taking age into consideration. Every word referring to one's brother or sister acknowledges his or her age in relation to the speaker. Even twins are not equal, because one twin was born first. Furthermore, norms that guide Korean forms of address do not allow the speaker to refer to elder brothers or sisters by their first names. A boy addresses his elder brother as *hyung* and his elder sister as *muna;* a girl addresses her elder brother as *oppa* and her elder sister as *unni.* Regardless of gender, however, people always address their younger siblings by their first names (Kim and Kirby 1996).

Among Koreans, the importance of the group over the individual is reflected in rules governing the writing and speaking of one's name. That is, Koreans tend to identify themselves by stating the family name first and then the given name. In effect, the family is deemed more important than the individual. Likewise, a letter is addressed to the country, the province, the city, the street, the house number, and finally the recipient.

• *Common expressions embody the preoccupations of the culture.* Frequently used phrases and words serve as indicators of cultural preoccupations—stresses, strains, and values. For instance, Americans use the word *my* to express "ownership" of persons or things over which they do not have exclusive rights: my mother, my school, my country. The use of *my* reflects the American preoccupation with the needs of the individual over those of the group. In contrast, Koreans express possession as shared: our mother, our school, our country. The use of the plural possessive reflects the Korean preoccupation with the group's needs over the individual's interests.

Another example of cultural preoccupations involves the Korean response to a full moon—"Isn't

Idiom A group of words that when taken together have a meaning different from the internal meaning of each word understood on its own.

that sad?"—or to singing birds—"They are weeping!" These comments reflect patterns of response rooted in centuries of invasions and wars. In part because of its geographic location, Korea has experienced a continuous history of invasions—by the Japanese, the Chinese, and the Russians—that caused severe hardship, substantial loss of life, and widespread devastation. Generations of warfare and occupation by foreigners have created in the Korean people a sadness that is reflected in their responses to many natural phenomena.

These four characteristics of language demonstrate that children do more than learn words and meanings. They also acquire a perspective that reflects what is important to the culture. Nevertheless, learning about and acquiring a cultural perspective do not transform people into cultural replicas of one another.

The Importance of Individual Experiences

Principle 4: People are products of cultural experiences but are not cultural replicas of one another. The information presented thus far may suggest that culture is simply a blueprint that guides, and even limits, thought and behavior. If true, then everyone would be cultural replicas of one another. Why are people not cultural replicas? Although culture is a blueprint of sorts, it also functions as a "toolkit" that allows people to select from a menu of cultural options (Schudson 1989:155). How does this selection process work? A baby enters the world and, by exposure to an already-established set of human relationships, is introduced to many versions of the culture. Virtually every event the child experiences—being born, nursing, being cleaned, being talked to, weaning, toilet training, talking, playing, and so on—involves people. The people present in the child's life at any one time include various combinations of father, mother, grandparents, brothers, sisters, playmates, other adult relatives, neighbors, babysitters, and others (Wallace 1952). All of these people expose the child to their own "versions" of culture, which they have acquired in the same way and which they pass on to the child in modified terms. The following excerpt from the essay "Faculty Brat: A Memoir," by Emily Fox Gordon (1995), captures this selective and interpretive dynamic:

> My father's background was Jewish, my mother's Presbyterian. Both of them were agnostic rationalists, and I grew up hearing almost nothing of belief or doctrine. My mother preserved the aesthetic parts of her Christian heritage. We spent two weeks before Christmas, my mother, sister, brother and I at the kitchen table mixing food coloring into vanilla icing in small glass dishes—pale green, pink, a shade I called chocolate blue. We used toothpicks to paint striped frosting trousers on the rudimentary legs of gingerbread men, buttoned up their blurred pastel waistcoats with silvery sugar balls. We also collected pine cones and sprayed them, over newspaper, with silver and gold (the wonderful toxic reek of those cans, which were also preternaturally cold to the touch!); we saved the tops and bottoms of tin cans and used metal shears to cut them into stars and spirals for the Christmas tree. We made Santas, gluing triangles of cotton on the chins of walnuts and red felt hats on their foreheads. . . .
>
> We children learned nothing of Judaism, except a vague understanding that the pickles and corned beef sandwiches my father loved, and the demonstrative relatives from New York and Philadelphia we occasionally visited, were things from the Jewish side of the universe. . . .
>
> I am the only one of my siblings to marry a Jew, and from her birth my daughter has always been Jewish to me. My brother and sister consider themselves and their children to be unaffiliated, but they celebrate Christmas and Easter. I'm not sure what I call myself, but now, having a child, I find I cannot celebrate the Christian holidays, even though the memory of some carols—"It Came upon a Midnight Clear," and "Lo, How a Rose Ere Blooming"—brings tears to my eyes when I find myself humming them in December. I know more about Judaism now, and I have a great abstract respect for it, but my mother's holiday Christianity, its sweetness, the memories of food and music and the surfaces of familiar things embellished and glittering, is like a beloved country from which I have exiled myself. (pp. 6–8)

This excerpt shows that individuals are products and carriers of cultural experiences and that they pass on those experiences selectively with varying degrees of clarity and confusion. The people to whom they transmit these experiences then repeat the process. Because individuals can reject, manipulate, revive, and create culture, they cannot be viewed as passive agents who absorb one dominant version of culture. As a case in point, consider that Christmas and other "religious" holidays are celebrated in the United States as if everyone participates in the festivities. Businesses close on Christmas Eve and Day; public schools give children the week of Christmas off; stores, houses, and streets are decorated; and television commercials and shows run Christmas themes for a month or more. Yet many people in the United States reject this cultural "option." As one of my students noted in a response paper:

I grew up in a very legalistic religion. We did not celebrate Christmas or Easter (even though we called ourselves Christians), we did not go to doctors or take medicines like aspirin, we did not do any work or activities on Saturday, and we observed unusual, little-known holy days mentioned in the Old Testament. While growing up I never challenged these practices. I just did what I was told.

Yet this same student as an adult decided to celebrate Christmas and other holidays that her family rejected in her youth:

> Now that I am older and have a child, I have turned away from my parents' religion. . . . This year was the first year I put up a Christmas tree. I didn't know how to decorate it and I didn't know any Christmas carols. I decided that I don't want my children growing up the way I did.

Culture as a Tool for the Problems of Living

Principle 5: Culture is the tool that enables the individual to adjust to the problems of living. Although our biological heritage is flexible, it presents all of us with a number of challenges. As noted previously, all people everywhere are dependent on others for a relatively long time. Everyone feels emotions and experiences hunger, thirst, and sexual desire. All humans age and eventually die. All cultures have developed "formulas" to help their members respond to these biological inevitabilities. Formulas exist for caring for children; satisfying the need for food, drink, and sex; channeling and displaying emotions; segmenting the stages and activities of the life cycle; and eventually departing this world. In this section, we will focus on the differing cultural formulas for dealing with two biological events—hunger and social emotions.

Cultural Formulas for Hunger

All people become hungry, but the factors that stimulate and satisfy appetite vary considerably across cultures. One indicator of a culture's influence is that people define only a portion of the potential food available to them as edible. Culture determines not only what is defined as edible but also who prepares the food, how the food is served and eaten, what relationship exists among those eating together, how many meals are consumed in a day, and when during the day meals are taken. For example, dogs and snakes are among the foods defined by many Korean and other Asian peoples as edible, but most Ameri-

Americans may wonder how some Koreans can eat dogs. Koreans and others may wonder how an American can stand to let a dog lick her face. Much of the world would be dismayed at the resources that Americans "waste" on pet dogs and cats. Why feed dogs when people starve?

© Ann Cecil/Photo 20-20

cans would find it appalling that someone would eat dog meat. That reaction should not surprise us when we consider that more than one-third of U.S. households include at least one dog and that owners often treat their dogs as family members. The results of an American Animal Hospital Association survey showed that almost 80 percent of respondents said that they give their dogs presents and 62 percent said they sign their dog's names to letters and cards (Goodavage 1996).[8]

---

[8]In 1987, the United States had more than 10,000 pet shops, 19,000 dog food vendors, 11,000 grooming shops, 7,000 kennels, and 300 pet cemeteries, as well as 200 products for dogs, ranging from feeding dishes to raincoats and sunglasses (Rosenfeld 1987). No doubt these figures have increased over the past decade.

U.S. in Perspective

The Importance of Corn

Corn is believed to have originated in the area of the world known today as Mexico and Central America. It is now grown all over the world. The United States, the largest corn producer in the world, produces half of the world's supply, exporting 40 percent of all that it grows. China is the second-largest producer of corn. Other countries that grow significant amounts of corn include Egypt, Thailand, Indonesia, Russia, Iran, and many African countries (Perry 1993). Corn is more than a source of nourishment. There is also the world of "hidden corn" in which corn is the equivalent of "industrial gold" (Shapiro 1992).

The driving wheel of the supermarket is not always visible: it is not the business of a driving wheel to be ostentatious. But it is there—everywhere. It is American corn, or maize. You cannot buy anything at all in a North American supermarket which has been untouched by corn, with the occasional and single exception of fresh fish—and even that has almost certainly been delivered to the store in cartons or wrappings which are partially created out of corn. So is milk: American livestock and poultry are fed and fattened on corn and cornstalks. Frozen meat and fish [have] a light corn starch coating on [them] to prevent excessive drying. The brown and golden colouring which constitutes the visual appeal of many soft drinks and puddings comes from corn. All canned foods are bathed in liquid containing corn. Every carton, every wrapping, every plastic container depends on corn products—indeed all modern paper and cardboard, with the exception of newspaper and tissues, [are] coated in corn.

One primary product of the maize plant is corn oil, which is not only a cooking fat but is important in margarine *(butter, remember, is also corn). Corn oil is an essential ingredient in soap, in insecticides (all vegetables and fruits in a supermarket have been treated with insecticides), and of course in such factory-made products as mayonnaise and salad dressings. The taste-bud sensitizer, monosodium glutanate or MSG, is commonly made of corn protein.*

Corn syrup—viscous, cheap, not too sweet—is the very basis of candy, ketchup, and commercial ice cream. It is used in processed meats, condensed milk, soft drinks, many modern beers, gin, and vodka. It even goes into the purple marks stamped on meat and other foods. Corn syrup provides body where "body" is lacking, in sauces and soups, for instance (the trade says it adds "mouth-feel"). It prevents crystallization and discolouring; it makes foods hold their shape, prevents ingredients from separating, and stabilizes moisture content. It is extremely useful when long shelf-life is the goal.

Corn starch is to be found in baby foods, jams, pickles, vinegar, yeast. It serves as a carrier for the bubbling agents in baking powder; is mixed in with table salt, sugar (especially icing sugar), and many instant coffees in order to promote easy pouring. It is essential in anything dehydrated, such as milk (already corn, of course) or instant potato flakes. Corn starch is white, odourless, tasteless, and easily moulded. It is the invisible coating and the active neutral carrier for the active ingredients in thousands of products, from headache tablets, toothpastes, and cosmetics to detergents, dog food, match heads, and charcoal briquettes.

All textiles, all leathers are covered in corn. Corn is used when making things stick (adhesives contain corn)—and also whenever it is necessary that things should not stick: candy is dusted or coated with corn, all kinds of metal and plastic moulds use corn. North Americans eat only one-tenth of the corn their countries produce, but that tenth amounts to one and a third kilograms (3 lb.) of corn—in milk, poultry, cheese, meat, butter, and the rest—per person per day.*

The supermarket does not by any means represent all the uses of corn in our culture. If you live in North America—and even very possibly if you do not—the house you live in and the furniture in it, the car you drive, even the road you drive on, all depend for their very existence on corn. Modern corn production "grew up" with the industrial and technological revolutions, and the makers of those revolutions were often North American. They turned their problem-solving attention to the most readily available raw materials and made whatever they wanted to make—antibiotics or deep-drilling oil-well mud or ceramic spark plug installators or embalming fluids—out of the material at hand. And that material was the hardy and obliging fruit of the grass which the Indians called maïs.

In English, the word corn *denotes the staple grain of a country. Wheat is "corn" to the people of a country where wheaten bread is the staple. Oats is "corn" to people who eat oats; rye is "corn" if the staple is rye. When Europeans arrived in America they saw that, for the Indians, maize was the basic food, so the English-speaking newcomers called it "Indian corn." We continue in North America to recognize the primacy of maize in our culture by calling it "corn."*

Source: Pages 22–23 in *Much Depends on Dinner*, by Margaret Visser. Copyright © 1988 by Grove Press, New York. Reprinted with permission.

Among other factors, these differences in attitudes toward dogs are rooted in historical and environmental factors. Whereas the United States uses an abundance of fertile, flat land for grazing cattle, many Asian countries such as Korea with limited space employ available land to grow crops, not to graze cattle. The few existing cattle are important to the agricultural system as a source of labor—to pull plows. Even today, cattle are more efficient than tractors in tilling steep inclines. The agricultural importance of cattle, combined with the lack of land to support a cattle industry, discourages the widespread practice of eating beef and encourages the consumption of dogs and snakes as alternative food sources in Korea.[9]

Another interesting cultural difference is that rice is the staple of the Korean diet, whereas corn is the staple of the U.S. diet. Most of the American diet is affected by corn, though few Americans realize its pervasiveness. In addition to being a vegetable, corn (in one form or another) appears in soft drinks, canned foods, candy, condensed milk, baby food, jams, instant coffee, instant potatoes, and soup, among other things (see "The Importance of Corn").

Corn is a "gift" from Native Americans to all people who settled in the United States. The Native Americans recognized the significance of corn to life by referring to it as "our mother," "our life," or "she who sustains us" (Visser 1988). Interestingly, among the many colors of corn, only the yellow and the white varieties are defined as edible by the dominant U.S. culture; the more exotic colors (blue, green, orange, black, and red) are considered fit only for decoration at Thanksgiving time (although blue corn chips and other blue corn products have become available in gourmet food stores and trendy restaurants). One might speculate that this arbitrary preference for some colors of corn over others reflects the early American immigrants' rejection of the "exotic" elements of Native American cultures.

Koreans have no such lack of awareness of rice as a staple. They recognize and appreciate the significance of rice in their lives and eat it at all meals—breakfast, lunch, and dinner. In fact, in the Korean language *rice* can be synonymous with food.[10] Most Koreans are aware of the many uses of rice and the by-products of rice plants: to feed livestock; to make soap, margarine, beer, wine, cosmetics, paper, and laundry starch; to warm houses; to provide inexpensive fuel for steam engines; to make bricks, plaster, hats, sandals, and raincoats; and to use as packing material to prevent items from breaking in shipping.

Cultural Formulas for Social Emotions

Culture also influences the expression of emotion, just as it influences people's responses to food needs. **Social emotions** are internal bodily sensations that we experience in relationships with other people. Empathy, grief, love, guilt, jealousy, and embarrassment are a few examples of social emotions. Grief, for instance, is felt at the loss of a relationship; love reflects the strong attachment that one person feels for another person; jealousy can arise from fear of losing the affection of another (Gordon 1981). People do not simply express social emotions directly, however. Rather, they also interpret, evaluate, and modify their internal bodily sensations upon considering "feeling rules" (Hochschild 1976, 1979).

Feeling rules are norms that specify appropriate ways to express the internal sensations. They define sensations that one should feel toward another person. In the dominant culture of the United States, for example, same-sex friends are supposed to like one another but not feel anything resembling romantic love. Consequently it is generally unacceptable for same-sex people to hold one another or to "celebrate" their friendship by holding hands in public.[11] The process by which we come to learn feeling rules is complex; it evolves through observing others' actions and in interactions with others.

In her novel *Rubyfruit Jungle*, Rita Mae Brown (1988) describes a situation in which feeling rules shape the way the central character, Molly, evaluates an encounter between her father, Carl, and his friend, Ep. Ep's wife has just died and Carl is comforting his friend. In this passage Molly reflects on the feeling rules that apply to men:

> I was planning to hotfoot it out on the porch and watch the stars but I never made it because Ep and Carl were in the living room and Carl was holding Ep. He had both arms around him and every now and then he'd smooth down Ep's hair or put his cheek next to his head. Ep was crying just like Leroy. I couldn't make out what they were saying to each other. A couple of times I could hear Carl

[9]Even in the United States, cattle are relatively new entrants as a major source of meat. The widespread use of beef as food began only after the westward expansion of the 1870s and the development of the tractor. Before then, the population was largely confined to the densely wooded eastern states, where pigs rather than cows thrived. Thus pork was the major source of meat. The settlement of the West opened up an abundance of rich flatland for grazing and food production (Tuleja 1987).

[10]In the Korean language, a more precise word for food is *umsik.*

[11]The Army publishes a list of "Must Know Items" about South Korea for American soldiers stationed there. One item says, "Don't be surprised if you see two Korean women or men walking arm in arm. They are just good friends and there is nothing sexual implied" (U.S. Army 1998).

Social emotions Internal bodily sensations that we experience in relationships with other people.

Feeling rules Norms that specify appropriate ways to express the internal sensations.

U.S. in Perspective
Feeling Rules in Korea and the United States

There are metaphors galore on both sides. In the West, the squeaky wheel gets the grease. But across the Pacific, the nail that sticks out gets pounded down, and trees that stand tallest catch the most wind. Here [in the United States], one has to toot one's horn to be heard. In the East, only a half-baked fool would boast, because people in positions of authority are presumed to recognize a person's qualities. In the West, a man who speaks well of his wife outside the home is considered a good spouse. But in the East, nothing could breach social etiquette as much as bragging about one's spouse beyond one's family. In the West, particularly in the United States, people communicate by explaining their views. Among Koreans, important communication is often nonverbal, and sometimes what you say may be just the opposite of what you mean. A Korean of sound mind could not imagine holding up a placard saying, "Hi, Mom," and waving into a TV camera—especially on an occasion as auspicious as the Seoul Olympics. Yet, to his American counterpart, no occasion is so dignified that he cannot send a message to his mother. Even on the most august occasions, Americans take the time to kiss and hug, and Europeans have made such gestures an art form. This is the way they show their feelings. Koreans transmit their feelings silently. They will send a son off to war, and not touch him one last time.

I have kept in mind always, when comparing Americans and Koreans, the story of the famous Korean General Yu-Shin Kim. The general was leading his troops as they passed through his native village. In front of him stood the house in which he had been born. If he had only knocked at the gate, he could have seen the face of his beloved mother, whom he had yearned so long to see. Instead, General Kim had his soldiers go to his house and bring back some hot bean-paste sauce. He remained on his horse and tasted it. When he confirmed that the bean paste was as he remembered it, he knew that his mother was well, and the general left without a word. Whenever I think of the story, I can imagine the general's mother, tears streaming down her wrinkled face, wondering whether she would live long enough to see him again, but bound to our timeless cultural code. I know these Korean mothers. Koreans love this story, but Americans would probably think how silly he was to have been at his mother's doorstep without going inside. In Hollywood Westerns, the cowboy rescues his love and they ride off into the sunset. In Korean movies, mismatched lovers part because they must yield to society's demands.

Source: Pages 296–297 in *Home Was the Land of Morning Calm: A Saga of a Korean-American Family*, by K. Connie Kang. Copyright © 1995 by K. Connie Kang. Reprinted by permission.

---

telling Ep he had to hang on, that's all anybody can do is hang on. I was afraid they were going to get up and see me so I hurried back to my room. I'd never seen men hold each other. I thought the only things they were allowed to do was shake hands or fight. But if Carl was holding Ep maybe it wasn't against the rules. Since I wasn't sure, I thought I'd keep it to myself and never tell. I was glad they could touch each other. Maybe all men did that after everyone went to bed so no one would know the toughness was for show. Or maybe they only did it when someone died. I wasn't sure at all and it bothered me. (p. 28)

This example shows that people learn norms that specify how, when, where, and to whom to display emotions. Somehow Molly learned that men do not hold one another—perhaps because she had never encountered such images. That is, her culture provided her with no images of men comforting one another in the way Carl was comforting Ep.

Feeling rules apply to male-female relationships as well as to same-sex friends. Different norms, for example, govern the body language that men and women can use to show affection toward one another when in public. A Korean husband and wife almost never touch, hug, or kiss in public; instead, they express affection inwardly. Indeed, Koreans tend to value the concealment of such emotions. When they see American couples express love overtly by touching, caressing, and kissing in public, they regard that behavior as a sign of insecurity. Korean couples have no need to express love overtly; they know that they love each other (Park 1979).

Sociologist Choong Soon Kim (1989) found that these feeling rules applied even to Korean couples separated from one another for 30 years or more. Kim observed the "Family Reunion" program that took place in South Korea in June 1983. This television campaign was designed to reunite relatives living throughout South Korea who had been separated from one another as a result of the Korean War (Jun and Dayan 1986). Kim noted that no husband and wife kissed when reunited, and most couples did not hug. No Korean onlookers commented that they found this behavior unusual. In the United States, by contrast, if reunited couples did not display affection, onlookers would wonder about the quality of the relationship. (See "Feeling Rules in Korea and the United States.")

Another example of the role that culture plays in channeling expressions of emotion has to do with laughter. Laughter is *not* something that happens only when people are amused. Instead, laughter can be an expression of emotions such as anxiety, sadness, nervousness, happiness, or despair.

No matter which emotion laughter releases or defuses, however, it occurs when something has gone awry, when behavior in a specific circumstance differs from what is expected, or when something takes place that caricatures everyday behavior. Because culture provides the guidelines for expected behavior in a specific set of circumstances, this observation about laughter implies that the situations that make someone laugh in one culture may not seem funny in another. In fact, communication specialists generally agree that jokes do not translate well. The following anecdote in which a Japanese translator decides not to translate a foreign speaker's joke because the audience will not understand the cultural context that makes the joke funny represents one way of handling this dilemma:

> I began my speech with a joke that took me about two minutes to tell. Then my interpreter translated my story, and about thirty seconds later the Japanese audience laughed loudly. I continued with my talk, which seemed well received but at the end, just to make sure, I asked the interpreter, "How did you translate my joke so quickly?" The interpreter replied: "Oh, I didn't translate your story at all. I didn't understand it. I simply said, 'Our foreign speaker has just told a joke, so would you all please laugh.'" (Moran 1987:74)

To this point, we have discussed a number of principles about culture. We have emphasized that material and nonmaterial culture are tools that people employ to meet the challenges of living. The discussion may have led you to believe that material and nonmaterial items represent two separate categories of culture. In fact, the material and nonmaterial components of culture are interrelated: the nonmaterial shapes the material, and the material shapes the nonmaterial.

The Relationship Between Material and Nonmaterial Culture

Principle 6: It is difficult to separate the effects of non-material and material cultures on behavior. To see how nonmaterial components of culture shape the material, consider the flags of the United States and South Korea. (The South Korean flag was the flag of Korea before the country was divided at the 38th parallel into North and South Korea.) A flag can be designed in any number of ways, so why did Korean and American designers settle on these particular designs? One could argue that designers drew on symbols that reflected important themes in each country's history.

The South Korean flag has a circle in the middle that is divided into two comma-shaped halves, the red half for yin (feminine) and the blue half for yang (masculine). The yin/yang design represents contributing, yet opposing, forces in the universe and depicts balance and harmony as a solution to contradictory or conflicting forces. The four trigrams (☰) around the circle represent heaven (☰) opposite earth (☷) and water (☵) opposite fire (☲). The message conveyed in the design is that humans have the "responsibility to balance these forces for optimum social harmony and human progress" (Reid 1988:171). The use of the yin/yang symbol and the trigrams, taken from the *I Ching,* or *Book of Changes,* an ancient Chinese book (1122 B.C.), acknowledges the value of the past and China's influence on Korean culture. The flag's design also reflects a belief in the

The text describes how South Korea's flag reflects a belief in the interdependence among natural forces. How would you interpret the cultural significance of some other flag (besides the U.S. flag) with which you are familiar?

interdependence between forces rather than the dominance of one force over another. Finally, it represents an important norm by which Koreans are guided: People have the responsibility to balance opposing or contradictory forces to ensure human progress (Yoo 1987).

Similarly, the flag of the United States reflects its nonmaterial culture. Each of the 13 alternating red and white stripes represents one of the 13 original colonies. Each of the 50 stars on the blue field represents a state. Although the United States began as a British colony, the flag does not acknowledge that influence in an obvious way (except perhaps in the red, white, and blue colors). Nor do the stars and stripes overlap. This separateness reflects the value placed on independence or freedom from a strong central power. The absence of any symbol that represents Great Britain or Europe also reflects the value placed on shedding the influence of the past to make a clean start. It matches the norm that rejects the role of the past in defining a person.

We can see another influence of nonmaterial culture on material culture in the different burial customs of Koreans and Americans. Almost 80 percent of Americans, especially those from Christian backgrounds, have the deceased placed in a wooden coffin and then in a concrete-lined metal vault for burial (Powell 1995). The double enclosure keeps the dead body separate from the earth as it decays. This method ensures that the deceased will not become "confused" with the earth; it reflects the belief that humans and their environment are independent (see "Cremation Statistics—United States and Canada"). In contrast, the Korean practice of burying the dead in wooden coffins that readily decay corresponds with the belief that humans and their environment are interdependent and that decay represents part of that interdependence.

Just as nonmaterial components of culture shape flags and coffins, material components shape the nonmaterial ones. For example, the microwave oven is an important material influence on American values and norms—specifically, values that emphasize the individual over the group and norms that govern when the members of a family should eat.

Approximately 70 percent of American households have microwave ovens, which have eliminated one incentive for families to eat together. Before the microwave became available, it was more efficient for one member of the household to cook for everyone because of the time—whether calculated in person time or in oven time—required to cook a meal. Now that meals can take only minutes to prepare, it is no longer considered inefficient for family members to cook and eat separately. The microwave also enhances independence, in that a person is no longer tied to a particular menu, meal schedule, or even family group:

> The old dining room table required each individual to give up some personal autonomy and bow to the dictates of the group and the social system. If we stop eating together, we shall save time for ourselves and achieve mealtime self-sufficiency. . . . The communal meal is our primary ritual for encouraging the family to gather together every day. If it is lost to us, we shall have to invent new ways to be a family. It is worth considering whether the shared joy that food can provide is worth giving up. (Visser 1989:42)

The examples discussed in this subsection show that inventions often have consequences that go beyond their intended purpose. The microwave oven not only changed the time and effort required to cook a meal, but also changed norms regarding how family members relate to one another at mealtime. The examples also illustrate how nonmaterial culture can shape material culture. Nonmaterial culture shaped the design of U.S. and Korean flags and the two countries' burial practices, for instance.

Most people tend to think that the material and nonmaterial culture that surrounds them is "homegrown"—that it originated in their society. Journalist Nicholas D. Kristof (1998) offers an example:

> I once asked my 5-year-old son, who has grown up largely in Tokyo, about his favorite Japanese foods. Gregory thought for a moment and decided on rice balls and McDonald's. It makes perfect sense to think of Big Macs as Japanese food, since McDonald's is a much more important part of his life in Japan than it is when we are on vacation in America. In particular, given the cramped homes in which most Japanese live, when his Japanese friends have birthdays the most common place to hold parties is McDonald's. (p.18)

The point is that most people tend to underestimate, ignore, or distort the extent to which familiar ideas, materials, products, and other inventions are connected in some way to outside sources or are borrowed outright from those sources (Liu 1994).

U.S. in Perspective

Cremation Statistics—United States and Canada

In 1997, 21 percent of people in the U.S. who died were cremated compared with 36 percent in Canada. Cremation rates vary by state and province. Which states have the highest cremation rates? The lowest? Which Canadian province has the highest cremation rate? The lowest? What factors might account for these differences?

Source: Cremation Association of North America (1998).

Alaska 54%

British Columbia 65%

Alberta 37%

Sask. 26%

Manitoba 43%

Ontario 34%

Quebec 31%

Newfoundland 1%

New Brunswick 13%

Nova Scotia 27%

Wash. 53%

Oregon 42%

Idaho 34%

Montana 45%

N. Dakota

Minn. 23%

S. Dakota

Wis. 18%

Mich. 22%

Me 35%

Vt. 43%

N.H. 32%

Mass. 21%

R.I. 29%

Conn. 22%

N.J. 15%

Del. 17%

Md. 17%

W. Va. 6%

Nevada 50%

Wyoming

N.Y. 15%

Penn. 14%

Calif. 42%

Utah 14%

Colorado 42%

Nebraska 13%

Iowa 13%

Ill. 19%

Ind. 11%

Ohio 17%

Va. 15%

Arizona 40%

New Mexico 27%

Kansas 8%

Mo. 9%

Ky. 5%

N.C. 12%

Oklahoma 7%

Ark. 12%

Tenn. 6%

S.C. 10%

Texas 11%

La. 7%

Miss. 3%

Ala. 4%

Ga. 10%

Fla 40%

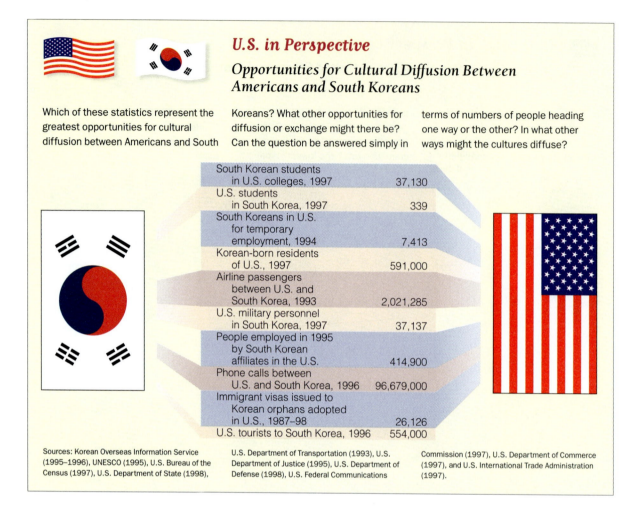

U.S. in Perspective

Opportunities for Cultural Diffusion Between Americans and South Koreans

Which of these statistics represent the greatest opportunities for cultural diffusion between Americans and South Koreans? What other opportunities for diffusion or exchange might there be? Can the question be answered simply in terms of numbers of people heading one way or the other? In what other ways might the cultures diffuse?

South Korean students in U.S. colleges, 1997	37,130
U.S. students in South Korea, 1997	339
South Koreans in U.S. for temporary employment, 1994	7,413
Korean-born residents of U.S., 1997	591,000
Airline passengers between U.S. and South Korea, 1993	2,021,285
U.S. military personnel in South Korea, 1997	37,137
People employed in 1995 by South Korean affiliates in the U.S.	414,900
Phone calls between U.S. and South Korea, 1996	96,679,000
Immigrant visas issued to Korean orphans adopted in U.S., 1987–98	26,126
U.S. tourists to South Korea, 1996	554,000

Sources: Korean Overseas Information Service (1995–1996), UNESCO (1995), U.S. Bureau of the Census (1997), U.S. Department of State (1998), U.S. Department of Transportation (1993), U.S. Department of Justice (1995), U.S. Department of Defense (1998), U.S. Federal Communications Commission (1997), U.S. Department of Commerce (1997), and U.S. International Trade Administration (1997).

Cultural Diffusion

Principle 7: People borrow ideas, materials, products, and other inventions from other societies. The process by which an idea, an invention, or some other cultural item is borrowed from a foreign source is called **diffusion**. In this definition, the term *borrow* is used in the broadest sense; it can mean to usurp, pirate, steal, imitate, plagiarize, purchase, or copy. The opportunity to borrow occurs whenever two people who have different cultural traits make contact, whether face to face, by phone or fax, or via the Internet.

Basketball, a U.S. invention, has been borrowed by people in 75 countries, including those in South Korea, where 21 clubs are registered with the *Federation Internationale de Basketball*. Baseball, another U.S. invention, has been borrowed by people in more than 90 countries, including Korea, which won the twenty-seventh world baseball championships in 1982 after it upset the United States and Japan (An 1997; *World Monitor* 1992, 1993).[12]

Instances of cultural diffusion are endless and can easily be found by skimming the newspaper headlines. Consider the following examples:

> "School for Japan's Executives in the United States"

Diffusion The process by which an idea, an invention, or some other cultural item is borrowed from a foreign source.

[12]There are at least 48,167,750 members of the International Baseball Association (International Baseball Association 1993).

U.S. in Perspective

Culture and the Pacific Rim

Consider the position of the United States in relation to 12 countries and city-states of the Pacific Rim. What aspects of material and nonmaterial culture are you already aware of being shared between the United States and these countries? What aspects of culture would you expect to be increasingly shared in the twenty-first century? Trade is naturally a major force in the exchange of culture. What other forces might promote cultural exchange across the Pacific Rim? What forces might hold it back?

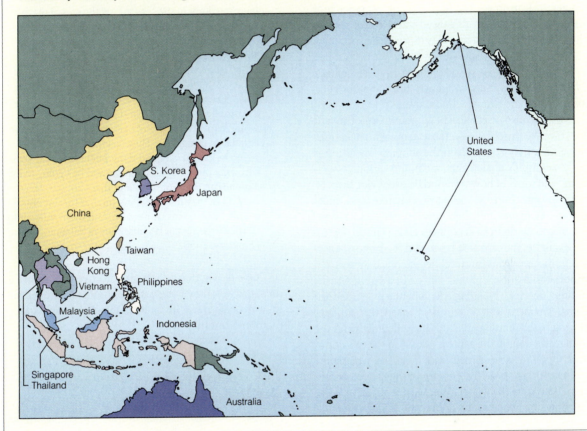

"In China, Beauty Is a Big Western Nose"

"Thai Publisher Plans to Expand Empire in U.S."

"Invasion of the Discounters: American-Style Bargain Shopping Comes to the United Kingdom"

"Global Goliath: Coke Conquers the World"

"Japan's Favorite Import from America: English"

"Skilled Asians Leaving U.S. for High-Tech Jobs at Home"

Opportunities for cultural diffusion between the United States and South Korea exist as a result of the 50-year presence of U.S. troops, international trade, and the departure and return of many Koreans who have attended colleges and universities in the United States. More than 37,000 South Koreans enrolled in colleges and universities in the United States in 1996–1997 (Institute of International Education 1997; see "Opportunities for Cultural Diffusion Between Americans and South Koreans" and "Culture and the Pacific Rim").

Likewise, the presence of the American military in South Korea since 1945 has encouraged diffusion in a number of ways. First, the media service, American Forces in Korea Network (AFKN), exposes Koreans to U.S. television and music and (by extension) to its dominant values and norms. Second, many jobs on U.S. bases are filled by Koreans, and a service industry (bars, gift shops, brothels, tailor shops, laundries) exists to meet the soldiers' needs.[13] Third, U.S. soldiers, in collaboration with Korean black marketeers, leak PX goods into Korean society. At one point, 60 percent of the available PX materials flowed into Korean society. This situation stimulated a desire for these products among the Korean people (Bok 1987).

People of one society do not borrow ideas, materials, or inventions indiscriminately from another society. Instead, borrowing is almost always selective. Even if people in one society accept a foreign idea or invention, they are nevertheless choosy about which features of the item they adopt. Even the simplest invention is really a complex of elements, including various associations and ideas of how it should be used. Not surprisingly, people borrow the most concrete and most tangible elements and then develop new associations and shape the item to serve new ends (Linton 1936). For example, the Japanese borrowed the game of baseball from Americans but modified it considerably to fit important Japanese cultural values. In a Public Broadcasting Service *Frontline* documentary, "American Game, Japanese Rules," reporters asked U.S. athletes playing baseball in Japan about the differences and the similarities in how the game is played in the two countries. One American athlete said, "Well, they play nine innings. That's about the only thing they have in common [with us in the game]. After we put on the uniform I'm not sure what we're doing out there" (*Frontline* 1988:2).

During this documentary, American athletes described a number of striking differences:

> I think that the biggest difference is, they play for ties. You know, that's unheard of back home. You'll play all day and all night to break a tie. But over here, you play for ties and you have a time limit on a game that is three hours and twenty minutes.

Culture shock The strain that people from one culture experience when they must reorient themselves to the ways of a new culture.

A tie is a wonderful thing in Japan. In the U.S., being a professional athlete when I played it, a tie is like kissing your sister. I mean, I'd just as soon not have that happen, I'd just as soon not play the game. In Japan they're ecstatic when you have a tie because everybody came out, everybody had the big fight, they had the big confrontation. They all fought together but nobody lost—no face was lost. So that is a perfect day.

When they win the championship they only win by a couple of games. [Say] they're up by ten games [at some point in the season], by the end of the year maybe they'll be up by two. That looks OK; it doesn't look like one team blew out the other team and embarrassed them.

What they feel is that, for instance, if a foreigner comes over and he's just hitting a lot of home runs, a great deal of home runs, the umpire—I wouldn't say the umpire—the league and people themselves would expect the umpire to expand the strike zone to balance things out. They feel that this foreigner has to be very strong and if the pitchers would keep throwing the ball over the plate and he's going to hit a home run every time up and that's not fair. So what they do is, they'll start calling pitches this far outside a strike, that far inside a strike. And their philosophy is, they're being fair because they're balancing it out. (*Frontline* 1988:2–3, 8)

One reason that people are selective with regard to the items that they borrow from another culture is that they evaluate those items in terms of the standards of their home culture.

The Home Culture as the Standard

Principle 8: The home culture is usually the standard that people use to make judgments about the material and nonmaterial cultures of another society. Most people come to learn and accept the ways of their culture as natural. When they encounter foreign cultures, they can experience mental and physical strain. Sociologists use the term **culture shock** to describe the strain that people from one culture experience when they must reorient themselves to the ways of a new culture. In particular, they must adjust to a new language and to the idea that the behaviors and responses they learned in their home

---

[13]Companies such as Hyundai and Daelin got their start by filling contracts for the U.S. military.

culture and now take for granted do not apply in the foreign setting. The intensity of culture shock depends on several factors: (1) the extent to which the home and foreign cultures differ, (2) the level of the person's preparation or knowledge about the new culture, and (3) the circumstances (vacation, job transfer, or war) surrounding the encounter. Some cases of culture shock are so intense and so unsettling that people become ill. Among the symptoms are "obsessive concern with cleanliness, depression, compulsive eating and drinking, excessive sleeping, irritability, lack of self-confidence, fits of weeping, nausea" (Lamb 1987:270).

In his book *Communication Styles in Two Different Cultures,* Myung-Seok Park, a professor of communication, describes some examples of the kinds of stress encountered when someone enters a foreign society. These experiences are typical of the adjustments that Koreans must make when they come to the United States. Taken as separate incidents, the two encounters are not especially stressful. Rather, the cumulative effect of a series of such encounters causes culture shock.

> When I studied at the University of Hawaii, my academic advisor was an old, retired English professor, Dr. Elizabeth Carr. All of the participants were struck by her enthusiasm, deep devotion and her unfailing health. So at the end of the fall semester I said to her, "I would like to extend my sincere thanks to you for the enormous help and enlightening guidance you gave us in spite of your great age." Suddenly she put on a serious look, and I saw a portion of her mouth twisting. I had an inkling that she seemed unhappy about the way I expressed my thanks to her. Understandably enough, I was not a little embarrassed. A few hours later she told me that my remark "in spite of your great age" had reminded her suddenly that she was very old. I felt as if I had committed a big crime. I restrained myself from commenting on age any more. (Park 1979:66)

> One Saturday afternoon I was drinking in an American drinking establishment with some of my new American friends. What surprised me at that moment was that whenever the bar girl brought some bottles of beer, she made change and took away a certain amount out of the money in front of each drinker. It was only I who did not put money on the table. (We Korean people pay the total amount for drinks when we leave.) When the money placed in front of each person ran out, the girl proceeded to take another person's money. What was still more surprising to me was that each one filled his own glass and drank without passing the glass to his friend and without asking him to

drink. I was somewhat bewildered because I had never poured my own glass before. . . . I realized that there was something cold and unfriendly in the American way of drinking. (pp. 38–39)

Some accounts from a Japanese physician visiting the United States for the first time provide additional illustrations of culture shock:

> Another thing that made me nervous was the custom whereby the American host will ask a guest, before the meal, whether he would prefer a strong or a soft drink. Then, if the guest asks for liquor, he will ask him whether, for example, he prefers scotch or bourbon. When the guest has made this decision, he next has to give instructions as to how much he wishes to drink, and how he wants it served. With the main meal, fortunately, one has only to eat what one is served, but once it is over one has to choose whether to take coffee or tea, and—in even greater detail—whether one wants it with sugar, and milk, and so on. I soon realized that this was only the American's way of showing politeness to his guest, but in my own mind I had a strong feeling that I couldn't care less. What a lot of trivial choices they were obliging one to make—I sometimes felt—almost as though they were doing it to reassure themselves of their own freedom. My perplexity, of course, undoubtedly came from my unfamiliarity with American social customs, and I would perhaps have done better to accept it as it stood, as an American custom. (Doi 1986:12)

> The "please help yourself" that Americans use so often had a rather unpleasant ring in my ears before I became used to English conversation. The meaning, of course, is simply "please take what you want without hesitation," but literally translated it has somehow the flavor of "nobody else will help you," and I could not see how it came to be an expression of good will. The Japanese sensibility would demand that, in entertaining, a host should show sensitivity in detecting what was required and should himself "help" his guests. To leave a guest unfamiliar with the house to "help himself" would seem excessively lacking in consideration. This increased still further my feeling that Americans were a people who did not show the same consideration and sensitivity towards others as the Japanese. As a result, my early days in America, which would have been lonely at any rate, so far from home, were made lonelier still. (p. 13)

Americans in Korea or Japan would be equally confused and bewildered by the Korean habits of pouring and drinking from each other's glasses, by their constant references to age (especially mature age), and by Japanese norms about hospitality.

Do not assume that culture shock is limited to experiences with foreign cultures. People can also experience **reentry shock,** or culture shock in reverse, upon returning home after living in another culture (Koehler 1986). In fact, some researchers have discovered that many people find it surprisingly difficult to readjust to the return "home" after spending a significant amount of time elsewhere. As in the experience of culture shock, they face a situation in which differences jump to the forefront.

As with culture shock, the intensity of reentry shock depends on an array of factors, including the length of time lived in the host culture and the extent of the returnee's immersion in the everyday lives of people in that culture. Symptoms of reentry shock are essentially the mirror image of those associated with culture shock. They include panic attacks ("I thought I was going crazy"), glorification of the host culture, nostalgia for the foreign ways, panic, a sense of isolation or estrangement, and a feeling of being misunderstood by people in the home culture. These comments by Americans returning from abroad illustrate these reactions:

> People pushed and shoved you in New York subways; they treated you as if you simply don't exist. I hated everyone and everything I saw here and had to tell myself over and over again: "Whoa, this is your country; it is what you are part of." (Werkman 1986:5)
>
> America was a smorgasbord. But within two weeks, I had indigestion. Then things began to make me angry. Why did Americans have such big gas-guzzling cars? Why were all the commercials telling me I had to buy this product in order to be liked? Material possessions and dressing for success were not top priorities in the highlands. And American TV? I missed the BBC. (Sobie 1986:96)

Although many people expect to have problems in adjusting to a stay in a foreign culture and even prepare

for such difficulties, most do not expect comparable trouble upon returning home. Because reentry shock is unexpected, many people become anxious and confused and feel guilty about having problems with readjustment ("How could I possibly think the American way was anything but the biggest and the best?"). In addition, they become apprehensive about how their family, friends, and other acquaintances might react to their critical views of the home culture; for example, they are afraid that others might view them as unpatriotic.

The experience of reentry shock points to the transforming effect of an encounter with another culture (Sobie 1986). The fact that the returnees go through reentry shock means that they have experienced up close another way of life and have been exposed to new norms, values, and beliefs. Consequently, when they come home, they see things in a new light.

One reason people experience cultural shock upon visiting a foreign culture or returning home after becoming immersed in a foreign culture is that they hold the viewpoint of **ethnocentrism.** That is, they use one culture as the standard for judging the worth of foreign ways. From this viewpoint, "one's group is the center of everything, and all others are scaled and rated with reference to it" (Sumner 1907:13). Thus, other cultures are seen as "strange" or, worse, as "inferior."

Ethnocentrism

Several different levels and consequences of ethnocentrism exist. The most harmless is simply defining foreign ways as peculiar, as did some Americans who attended the 1988 Summer Olympic Games in Seoul. Learning that some Koreans eat dog meat, some visitors made jokes about it. People speculated about the consequences of asking for a doggy bag, and they made puns about dog-oriented dishes—Great Danish, fettuccine Alfido, and Greyhound as the favorite fast food (Henry 1988).[14] (See "Projecting a Clean Image.")

The most extreme and most destructive form of ethnocentrism is **cultural genocide,** in which the people of one society define the culture of another society

---

Reentry shock Culture shock in reverse; experienced upon returning home after living in another culture.

Ethnocentrism A viewpoint that uses one culture as the standard for judging the worth of foreign ways.

Cultural genocide A form of ethnocentrism in which the people of one society define the culture of another society not as merely offensive, but as so intolerable that they attempt to destroy it.

---

[14]After Seoul was awarded the 1988 Summer Olympic Games, the South Korean government took steps to rid the city of snake and dog shops. Several factors prompted this move, including Korean sensitivity to foreign opinion and the power of global mass media. The history of foreign involvement in Korean affairs has made the Koreans self-conscious about what people from other countries think about their culture. Thus, to avoid ridicule and embarrassment, the South Korean government moved to close the dog and snake shops so they would not be present for the cameras.

Projecting a Clean Image

With the Seoul Olympics just a month away, Olympic countdown fever began to grip the city. Radio stations aired English lessons several times a day for Seoulites to learn how to communicate to visitors expected during the Olympics. Taxi companies and department stores held English classes for their employees. Public service announcements over radio and television urged Koreans to "put their best foot forward" and to behave as a people worthy of an Olympic-host country.

To project a "clean" image of Seoul, thousands of street vendors, who scraped out a living by selling everything from magnifying glasses to barbecued beef, were ordered to leave town during the games, scheduled for September 15 to October 2. The vendors protested. Thousands of them staged a rally and demanded that the city "get off the backs of the poor."

"Which is more important—the Olympics or one's right to make a living?" shouted one vendor at the rally I attended. "We may be poor, but we want to live on what we earn." The vendor's protests inspired newspaper columnists on influential papers to come to their defense. One columnist suggested that the government capitalize on the vendors' street stalls by turning them into tourist attractions. But the vendors lost, and were banned.

Others lost out, too. Restaurants offering canine cuisine were ordered closed throughout the Olympics as a result of pressure from animal rights groups in Great Britain and the United States. In July and August foreign correspondents wrote so many stories about canine cuisine that it appeared from abroad that every Korean ate dog meat stew three times a day. The controversy generated by these stories was substantial. Unfortunately, it also was one of the few stories my newspaper specifically requested me to do. I kept putting my editors off because the idea did not appeal to me. For one thing, it was another grating example of Western media zooming in on what to them was sensational exotica. I never found stories making fun of other people's cultures appealing. But stories about dog meat began appearing in major papers, including the *Chicago Tribune*. My editors faxed me these stories—a nudging reminder. Finally, I gave in, reasoning that at least I could write about the subject with a little more understanding than these Western reporters.

A couple of weeks before the start of the games, I went to the House of Happiness in an old section of downtown, famous among a select group of middle-aged Korean males I knew for its extraordinary canine cuisine. I asked a middle-aged male friend to go with me as I felt I had to witness the serving of a canine dish in order to do the story. My friend said he occasionally enjoyed a bowl of "Vitality Soup" after work when he felt run-down. He said he could not understand why so many foreigners were upset. "Koreans don't eat their pet dogs," he said, explaining that a special breed of large, tan-colored dogs were raised especially for canine cuisine. "Besides, people who eat cows and pigs and lamb are in no position to poke fun at people who eat dog meat," he said. I agreed with him.

On the following day, I went to a big market where canine meat was sold. I also interviewed the owners of several restaurants where health stew was served, and consumers who told me why they relished it. I learned that for centuries some Koreans have considered dog meat to have special curative powers, providing stamina and strength. I filed a long story, complete with the fascinating history of how dog meat became a delicacy. But I still cringe whenever I think about the story, because it was an example of American journalism at its worst (not unlike the mass insanity of editors unduly paying attention and allocating resources to cover the O. J. Simpson double-murder case in 1994 and 1995).

Government officials failed also in another hide-and-seek game with foreign correspondents. The officials had decided it would be a good idea to build a wall to hide a stretch of hovels along the route of the Olympic marathon in the southern part of the country. A British reporter discovered this and wrote a story poking fun at the government. Soon, a legion of foreign reporters were on the scene. They took pictures not only of the wall but of the unsightly houses the government had wanted to hide.

This, coming after the controversy involving dog meat, angered a lot of Koreans. "I don't understand what motivates foreign reporters," said a Korean reporter with a major Seoul daily. "What is the purpose of those people going out of their way to uncover things Koreans don't particularly want to show to the outside world?"

Dog meat and shantytowns notwithstanding, the twenty-fourth Olympiad, an event that had been South Korea's official obsession for six years, got under way at 11:00 A.M. on September 17. For South Korea, whose image to the Western world was that of war orphans and peasants, as conveyed in the M*A*S*H television series, the Olympics was an extraordinary undertaking that required extraordinary sacrifices.

Forced Culture Change of Native Americans by Bureau of Indian Affairs Schools

The Eastern band of Cherokee Indians lives on a federal reservation in the mountains of North Carolina and today numbers more than 8,000 people. In the fall of 1990, the North Carolina Cherokees took direct control of their school system. For almost a century, however, dating back to 1892, Cherokee schools, like schools for most other Native Americans, were operated directly by the Bureau of Indian Affairs (BIA) and financed by federal taxes.

For most of the twentieth century, formal education was a tool to wipe out Indian culture and transform Indians into "red-skinned whites." The process of acculturation was accomplished over the objections of parents and tribal leaders and often at the expense of the physical and emotional well-being of Indian students. For most Native Americans, the most oppressive phase of BIA education occurred during the first three decades of the twentieth century. Even today, however, there are lingering effects.

This Cherokee school was one of the special schools established by the U.S. Bureau of Indian Affairs to "Americanize" the native people of the United States. To further this goal, the school required pupils to speak English and to wear uniforms.
Courtesy Museum of the Cherokee Indian

The hallmark of BIA education as it evolved around the turn of the century was the boarding school. In 1920, the BIA superintendent on the Cherokee reservation wrote about the importance of "taking the most promising young men and women away from reservation influences" (Eastern Cherokee Agency File 1920). The "bad influences" of the reservation were not crime or other social problems but parents who spoke Cherokee or wore moccasins.

All over the United States, Indian students were shuffled around to maximize the distance between themselves and their families. Thus, although a boarding school existed on the Cherokee reservation, many of the students

not as merely offensive, but as so intolerable that they attempt to destroy it. There is overwhelming evidence, for example, that the Japanese tried to exterminate Korean culture between 1910 and 1945. After Japan annexed Korea in 1910, Japanese became the official language, Koreans were given Japanese names, Korean children were taught by Japanese teachers, Korean literature and history were abandoned, ancient temples—important symbols of Korean heritage—were razed, the Korean national anthem was banned, and the Korean flag could not be flown. Even Korean flowers were banned as Japanese officials forced Koreans to dig up their national flowers and plant cherry trees (Kang 1995).[15] The Japanese brutally suppressed all resistance on the part of the Korean people. When Koreans tried to declare their right to self-determina-

tion in March 1919, thousands of people were injured or killed in clashes with the Japanese military.[16] Unfortunately, American history is filled with similar

[15]The national flower of Korea is the rose of Sharon.

[16]The Japanese maintained that their intent in destroying Korean culture was to modernize Korea. Even today, many Japanese argue that under their occupation, Korean agricultural and industrial productivity improved and that highways, railroads, ports, communication systems, and industrial plants were built. The Koreans' earnings and output, however, went to support Japanese society and left most Koreans impoverished.

As Japan became more deeply involved in World War II, Koreans were forced to serve in the military or to fill the factory positions of Japanese workers fighting in the war. During this 35-year period, the Koreans were subservient to the Japanese in every way. In essence, the Japanese attempted to "Japanize" the Koreans by taking away their language, ideas, and material culture.

were non-Cherokee Indians from distant reservations; in turn, many Cherokees, even as young as age six, were shipped to Chilocco Boarding School in Oklahoma, to Haskell in Kansas, to Carlisle in Pennsylvania, or to Hampton in Virginia. Even where day schools were available for elementary school-age students, the situation often changed with junior high or high school, where boarding schools became the norm.

In 1918, the Cherokee BIA superintendent commented on one small boy whose whole family "have been known to stand him [the superintendent] off with drawn knives" (Commissioner of Indian Affairs 1918). For many Native Americans, whether Navajo, Cheyenne, Cherokee, or others, the months of August and September were spent trying to hide their children from the truant officers. Captured children were taken crying off to the out-of-state boarding schools. If they were lucky, they got to return home for a visit the following summer. In many cases, however, it was years before children saw their parents again. It was not uncommon for a weeping six-year-old to be taken away and return to his parents only at age sixteen when he could legally drop out of school.

To make things even worse, the boarding school was often an alien environment where only English could be spoken and uniforms had to be worn. The schools stressed vocational subjects: boys learned to farm—ironic for Cherokee boys who were removed from their family farms—and girls were trained to be servants. Girls at the Cherokee Boarding Schools were taught to do laundry and to cook: "the demand for them as house servants in Asheville and other neighboring towns could not be supplied" (Young 1894:172).

Children caught speaking their native language had their mouths washed out with soap or were beaten if they were repeat offenders. At the Cherokee Boarding School, children who attempted to run away were chained to their dormitory beds at night.

Perhaps the best description of the BIA boarding school era came from the anthropologist John Collier who, as Commissioner of Indian Affairs under Franklin Roosevelt's New Deal, attempted to repeal many of the oppressive policies. In 1941, Collier, writing in a magazine called *Indians at Work*, reflected on a night in the 1920s

when he was allowed to sleep over at the boys' dormitory at the Cherokee Boarding School. Collier had gotten lost hiking in the mountains and asked to stay over the night until he could get transportation off the reservation: "The little boy inmates could not or would not talk English to each other, and they dared not talk Cherokee. . . . This was an Oliver Twist place but with every light and shade of imagination disbarred. . . . Horror itself gave up the fight" (Collier 1941:3).

References
Collier, John. 1941, September 1–8. Editorial. *Indians at Work.*
Commissioner of Indian Affairs. 1918. *Annual Report of the Commissioner of Indian Affairs, Interior Department Report.* Vol. 2. Indian Affairs and Territories, 65th Congress, 3rd Session, House Document 1455.
Eastern Cherokee Agency File. 1892–1958. Eastern Cherokee Agency correspondence. Federal Archives and Records Center, East Point, GA.
Young, Virginia. 1894. "A Sketch of the Cherokee People on the Indian Reservation of North Carolina." *Woman's Progress,* pp. 171–172.

Source: Sharlotte Neely, Northern Kentucky University (1991).

instances of this type of ethnocentrism, as when the U.S. Bureau of Indian Affairs forced Native Americans to attend boarding schools (see "Forced Culture Change of Native Americans by Bureau of Indian Affairs Schools").

Sociologist Everett Hughes (1984) identifies yet another type of ethnocentrism:

One can think so exclusively in terms of his own social world that he simply has no set of concepts for comparing one social world with another. He can believe so deeply in the ways and the ideas of his own world that he has no point of reference for discussing those of other peoples, times, and places. Or he can be so engrossed in his own world that he lacks curiosity about any other; others simply do not concern him. (p. 474)

Since the end of the Korean War, millions of Americans—military personnel, technicians, social workers, and educators—have spent time in South Korea as advisers but not as learners. For this reason, "surprisingly few Americans have come away from their Korean experience with much understanding of the country, its people, language, or culture" (Hurst 1984:1). Few of the U.S. servicewomen and servicemen stationed in South Korea venture beyond their bases and the "tourist" towns surrounding them (Cherni 1998).

A conversation between author Simon Winchester (1988) and the Korean owner of a bar frequented by American service personnel shows that from the Korean viewpoint, many American military people display the kind of ethnocentrism that Everett Hughes describes:

Just then two burly and unshaven airmen walked past. One wore a patch on his jacket that said "Munitions Storage—We tell you where to stick it!" The other had a T-shirt with the words: "Kill 'Em All: Let God Sort the Bastards Out." Both were sporting newly stitched shoulder patches showing what appeared to be a small plane—the fuselage looking remarkably phallic—beneath the rubric "One Hundred Successful Missions to A-Town." Mr. Kwong shook his head with distaste.

"That's what I can't take. Don't they ever learn? We need to be respected here, and they're not respecting us. They treat us like we're some backward Third World country, and you know we're not. We're proud, we've got good reason to be. But this. . . ." He gestured with despair.

I said I hadn't found anything very offensive about the two passing airmen. "Maybe not, maybe I react too much," he said. "I've worked for twenty-five years trying to bring the two communities together. I organize them to go out to meet families. I try to persuade them to learn a bit of Korean, to eat some of the food, to understand why they're here. But they don't want to know. And it's the way some of them treat our women, and our men too. Some of them just have no respect for us. The way they see it, they're top of the pile, and everyone else is nothing. It makes me mad." (p. 144)

These examples may lead you to conclude that ethnocentric thinking is an American characteristic, but unfortunately it exists in every society. In Korea, for example, a dissident poet recently released from prison displayed an ethnocentric perspective when he argued that "the Korean race" is destined to take over the world. As another example, the ultimate feel-good movie for Koreans is one in which North and South Koreans team up to defeat Japan (Kim and Kirby 1996).

Another relatively unknown type of ethnocentrism is **reverse ethnocentrism**, in which the home culture is regarded as inferior to a foreign culture. People who engage in this kind of thinking often idealize other cultures as utopias. For example, the former Soviet Union is labeled as the model of equality, Japanese culture as the model of human connectedness, the United States as the model of self-actualization, India as the model of other-worldliness, Israel as the model of pioneering spirit, Nigeria as the model of family values, or Native Americans as the original environmentalists (Hannerz 1992).

People who engage in reverse ethnocentrism not only idealize other cultures, but also reject any information disproving that view. Journalist K. Connie Kang (1995) offers an excellent example of reverse ethnocentrism among the Korean students she taught while a visiting professor between 1967 and 1970 at Hankuk University of Foreign Studies:

> [G]oing to the United States was a preoccupation with Koreans. America was akin to paradise, Koreans thought, and the image of America they carried in their heads was exaggerated. They really believed everyone was rich and lived in big houses with winding staircases. I did not know how to begin to make my students at the university understand that what they believed about America was very different from what America really was. They were motivated to study English because they saw going to the United States as the best way to improve their lot. . . . Fielding their endless queries, I sometimes felt like a one-person U.S. Information Service. I tried to dispel their notions of what America was by saying that most people worked in uninteresting jobs to pay their bills and put their kids through school, but I do not think I succeeded. People want to hear only what they want to hear; it was hard to compete with the images of American life created by Hollywood. (p. 205)

Cultural Relativism

A perspective that runs counter to ethnocentrism is cultural relativism. **Cultural relativism** means two things: (1) that a foreign culture should not be judged by the standards of a home culture and (2) that a behavior or way of thinking must be examined in its cultural context—that is, in terms of the society's values, norms, beliefs, environmental challenges, and history. Cultural relativism is a perspective that aims to understand—not condone or discredit—foreign behavior and thinking.

For example, the Korean practice of defining infants as one year of age at birth cannot be understood if it is evaluated according to dominant U.S. values, which emphasize the future over the past. U.S. values support the idea that the birth of a baby represents a fresh start with unlimited possibilities, no matter what the social

Reverse ethnocentrism A type of ethnocentrism in which the home culture is regarded as inferior to a foreign culture.

Cultural relativism The perspective that a foreign culture should not be judged by the standards of a home culture and that a behavior or way of thinking must be examined in its cultural context.

class, ethnicity, or educational level of the baby's parents. From this point of view, it makes sense to define age at birth as zero. The Korean practice, however, must be considered in light of Korean values. Defining age in this way corresponds with Korean values about the importance of the past and beliefs about the relationship of the individual to the group. The message is that generations are interdependent and that past events (even those occuring prior to conception) are important to a person's life. Thus the Korean way of defining age ties together past, present, and future generations.

Similarly, whereas most Americans cannot understand why some Koreans eat dog meat, most Koreans are equally appalled that Americans often let dogs live in their homes, allow them to lick their faces, and spend so much money on them when the U.S. population includes many poor and homeless people. When we consider the historical and environmental challenges surrounding the Korean decision to eat dog meat, this practice might not seem so unreasonable.

Cultural relativism, however, does not entail an "anything goes" position. Such a position allows every cultural trait—even some of the most harmful and violent ones (for example, infanticide, human sacrifice, foot binding, death threats against those holding controversial ideas, witch hunts, scientific experimentation on naive human subjects)—to escape judgment or criticism. Unfortunately, uninformed critics often interpret cultural relativism as a search to justify any behavior and as support for one or more of the following attitudes: "Whatever they do is fine"; "That's just the way that culture is"; "It's none of my business what others do"; or "Everything is relative. What's good in one culture is bad in another, and vice versa."

Anthropologist Clifford Geertz (1995) maintains that if one is serious about addressing a problematic cultural trait, the first step is to ask the persons affected (supporters, critics, victims, beneficiaries) about the practice. Investigators should not assume that they know the truth of the situation and proceed as if they have the solution. As a rule of thumb, one should avoid making judgments and recommendations that affect another culture when these decisions are based on superficial knowledge. With regard to education, for example, Iranian education critic, reformer, and scholar Samad Behrangi maintains that

unless we have seen the school's environment and surrounding community, unless we have lived among the people, unless we have been friends with the people, we have not heard their voice and have not known their desires, it is not even proper to have sympathy for the environment, or impose unrelated educational policies, or even to write stories or textbooks for them. (Behrangi 1994:28)

Understanding a cultural trait or taking a position of cultural relativism does not mean that one accepts the trait unconditionally, nor does it mean that one has no values of one's own. As Geertz (1995) argues, the challenge of taking this position is finding "a way to keep one's values and identity while living with other values—values you can neither destroy or approve" (Berreby 1995:47). In Geertz's words:

> I hold democratic values, but I have to recognize that a lot of other people don't hold them. So it doesn't help much to say, "This is the truth." That doesn't mean I don't believe anything. . . . You can't assert yourself in the world as if nobody else was there. Because this isn't [just] a clash of ideas. There are people attached to those ideas. If you want to live without violence, you have to realize that other people are as real as you are. (p. 47)

Cultural relativism is an especially useful perspective when people must make a decision supporting one cultural trait over another or must contemplate taking action to change, modify, or even eliminate a cultural trait that "others" share. In all cases, cultural relativism facilitates decision making and/or negotiation by making the cultural elements in question intelligible to everyone involved. Certainly a decision is more appropriately made by people who understand the traits than by those who do not.

Studying and learning about other cultures teach us that morality is both relative and universal (Redfield 1962). Morality is relative in that norms, values, and beliefs about rightness and wrongness vary across time and place. It is also universal, however, in that every culture has its own conceptions of morality. Although mores do exist that can make virtually any idea or behavior seem right or wrong, "some mores have a harder time making some things right than others" (Redfield 1962:451).

Subcultures

Principle 9: In every society there are groups that possess distinctive traits that set them apart from the main culture. Groups that share in some parts of the dominant culture but have their own distinctive values, norms, language, or material culture are called **subcultures.**

Often we think we can identify subcultures on the basis of physical traits, ethnicity, religious background, geographic region, age, gender, socioeconomic or occupational status, dress, or behavior defined as deviant by society. Determining which people constitute a subculture, however, is actually a complex task that requires careful thought; it must go beyond simply including everyone who shares a particular trait. For example, using broad ethnic or racial categories as a criterion for identifying the various subcultures within the United States makes little sense. The broad racial category "Native American," for example, ignores the fact that the early residents of North America "practiced a multiplicity of customs and lifestyles, held an enormous variety of values and beliefs, spoke numerous languages mutually unintelligible to the many speakers, and did not conceive of themselves as a single people—if they knew about each other at all" (Berkhofer 1978:3).

Sociologists determine whether a group of people constitutes a subculture by learning whether they share a language, values, norms, or a territory and whether they interact with one another more than with people outside the group. One characteristic central to all subcultures is that their members are separated or cut off in some way from other people in the society. This separation may be total or it may be limited to selected aspects of life such as work, school, recreation and leisure, dating and marriage, friendships, religion, medical care, or housing. It may be voluntary, result from an accident of geography, or be imposed consciously or unconsciously by a dominant group. It could also result from a combination of these three factors.

Subcultures within the United States experience separation in different ways or to different degrees. Some integrate themselves into various areas of mainstream culture when possible but remain excluded from other areas of life. In general, African Americans who work or attend school primarily with whites are often excluded or feel excluded from personal and social relationships with them. This exclusion forces them to form their own fraternities, study groups, support groups, and other organizations. Other subcultures are **institutionally complete** (Breton 1967); their members do not interact with anyone outside their subculture to shop for food, attend school, receive medical care, or find companionship because the subculture satisfies these needs. Often we find a clear association between institutional completeness and language differences. Persons who cannot speak the language of the dominant culture are very likely to live in institutionally complete ethnic communities (for example, Little Italy, Chinatown, Koreatown, Mexican barrios). Of the 750,000 Koreans living in the United States, for example, approximately 300,000 live in southern California. A large portion of these California residents live in an institutionally complete Koreatown west of downtown Los Angeles. Nevertheless, the Korean experience in the United States cannot be described by a few generalizations. For one thing, "Korean immigrants to the United States are not always from Korea. Significantly large Korean communities

Subcultures Groups that share in some parts of the dominant culture but have their own distinctive values, norms, language, or material culture.

Institutionally complete Subcultures whose members do not interact with anyone outside their subculture to shop for food, attend school, receive medical care, or find companionship because the subculture satisfies these needs.

For more than half a century, the demilitarized zone has divided Korea into what are now a capitalist economy in the south and a halting socialist economy in the north. If the two countries were to reunite, how would the lack of exchange between the two societies for so long make it difficult to reunite?

© T. Matsumoto/Sygma

exist outside Korea in Siberia, Canada, Japan, and even Brazil" (Lee 1994:39).

Despite the ethnic similarity among the people, Korea has been divided since 1945 into two institutionally complete subcultures: North Korea and South Korea. The people of the two Koreas have no relationship with one another (including no correspondence by mail, phone, or travel) and technically remain at war with one another. The line that separates the two people at the 38th parallel is the most heavily militarized region of the world; each side is poised to stop the other from invading. Both sides recognize their common language, ethnicity, history, and culture, and both believe in unification. Yet, despite these similarities, neither side has been able to compromise on economic or government structure, the elective process, or the appointive process for government offices.[17]

Summary and Implications

We began this chapter by identifying three challenges that sociologists face when they study culture: (1) describing a culture, (2) determining who belongs to a designated culture, and (3) identifying characteristics that set one culture apart from another. Sociologist Immanuel Wallerstein (1990) explains the challenges of describing a culture and identifying its members and its distinguishing characteristics:

> [I]t is surely true that people in different parts of the world, or different epochs, or in different religious or linguistic communities do indeed behave differently from each other, and in certain ways that can be specified and fairly easily observed. For example, anyone who travels from Norway to Spain will note that the hour at which restaurants are most crowded for the "evening meal" is quite different in the two countries. Anyone who travels from France to the U.S. will observe that the frequency with which foreign strangers are invited to homes is quite different. The length of women's skirts in Brazil and Iran is surely strikingly different . . . on the one hand, differences are obvious . . . and yet the degree to which groups are in fact uniform in their behavior is distressingly difficult to maintain. (p. 34)

Several factors account for these challenges. First, although people are products of cultural experiences, people from the same culture are not replicas of one another. Because people possess the ability to interpret, select, manipulate, revive, and create culture, they are not passive agents who absorb or transmit to others one version of a culture.

Second, people borrow ideas, materials, products, and other inventions from those in other societies. Cultural diffusion occurs through many kinds of relationships including everyday mingling (the routine talking, looking, and listening people do as they live their lives). For example, as the 37,000 exchange students from South Korea and the 37,000 U.S. troops stationed in Korea go about their lives, they are exposed to the others' culture (See "Foreign Influences on Professional Women's Basketball in the United States"). Another kind of relationship is the marketplace, the arena in which transactions between sellers and buyers occur. The processes by which goods and services are exchanged between people from two cultures (producing, advertising, shipping, selling) also involve interaction between the parties. Moreover, the goods and services exchanged have the potential of transforming each culture.

Cultural diffusion also occurs when "outsiders" make organized, deliberate efforts to transform, reform, or replace some cultural characteristic. The work of Wycliff Bible Translators represents one example. The organization's aim is to translate the Bible into all the languages of the world. The existence of the Bible transforms those who read it *and* the language into which it is translated transforms the Bible in subtle, and sometimes dramatic, ways.

The facts of cultural diffusion and individual autonomy may leave you wondering whether such a thing as Korean or American culture even exists. Ulf Hannerz (1993) maintains that *culture* "is still the most useful keyword we have to summarize that peculiar capacity of human beings for creating and maintaining

---

[17]In view of the unification of Germany, we might speculate that the two Koreas might reunite soon. South Korean officials predict, however, that it may take 20 years before they eventually reunite. One important difference is that West Germany, in comparison with South Korea, was far wealthier and thus better able to absorb the staggering economic costs ($250 billion) of reunification. Recently, South Korea's assistant minister for reunification commented: "After seeing what happened in Germany, especially in the economic area, it has changed attitudes toward unification in Korea. The biggest lesson we learned is that we need to prepare steadily, and we need to have realistic expectations. We cannot just be ruled by sentiment" (Protzman 1991; Sterngold 1991:C1).

U.S. in Perspective

Foreign Influences on Professional Women's Basketball in the United States

This map shows the countries from which the Women's National Basketball Association draws its international players. The circled numbers indicate the number of athletes from each foreign country. How might the presence of foreign-born players affect U.S.-born players? What kinds of things might members of each group learn from each other?

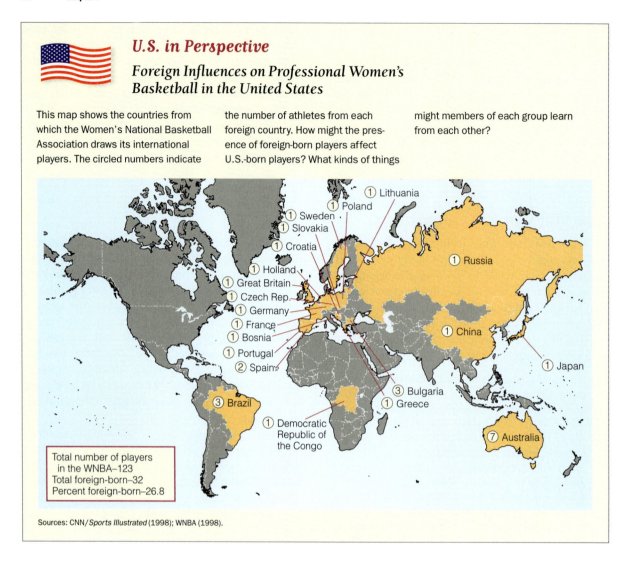

Total number of players in the WNBA–123
Total foreign-born–32
Percent foreign-born–26.8

Sources: CNN/*Sports Illustrated* (1998); WNBA (1998).

their own lives together" (p. 109). He suggests that the problem lies not so much in the idea of culture itself, but in our narrow application of the concept. The problem is that we use the term "culture" (1) in conjunction with specific places and categories of people, (2) in ways that emphasize differences (cultural gaps), (3) as if cultures have clear, identifiable boundaries, and (4) in ways that suggest people from one culture are replicas of one another.

The facts of cultural diffusion and individual autonomy demand that we expand our conception of culture. An expanded view prepares us to see overlap, influences, and exchanges between cultures. Thus culture is viewed as more than a barrier to interactions and a source of differences. Expanding our view of culture to think of it as a force that shapes individuality (see principle 4), a force that borrows (albeit selectively) from other cultures, and a force that people manipulate, revive, and re-create prepares us to approach people as more than simply representatives of a particular culture and to treat cultures as parts of an intercultural tapestry.

Hannerz (1992) offers the following metaphor that applies to this expanded vision of a culture:

> When you see a river from afar, it may look like a blue (or green, or brown) line across a landscape; something of awesome permanence. But at the same time, "you cannot step into the same river twice," for it is always moving, and only in this way does it achieve its durability. The same way with culture—even as you perceive structure, it is entirely dependent on ongoing process. (p. 4)

Key Concepts

Use this outline to organize your review of the key chapter ideas.

Culture
    Material culture
    Nonmaterial culture
        Beliefs
        Values
        Norms
            Folkways
            Mores
Social emotion

Diffusion
Culture shock
Reentry shock
Ethnocentrism
    Cultural genocide
    Reverse ethnocentrism
Cultural relativism
Subculture
    Institutionally complete

internet assignment

Find examples of Internet sites that promote and describe some kind of interaction between people in the United States and South Korea. That interaction may involve learning the Korean language, tourist attractions, sister cities and other cultural exchanges, or friendships.

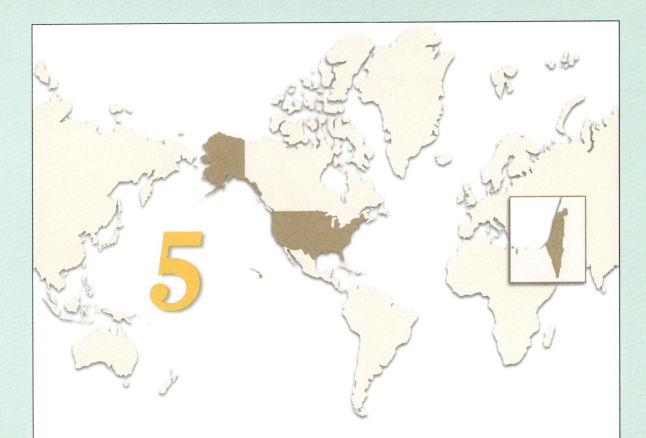

5

Socialization

With Emphasis on Israel, the West Bank, and Gaza

Jerusalem remains the great unresolved issue between Israel and the Palestinians. The Dome of the Rock dominates a view of the walled Old City. (Will Yuman/Gamma Liaison.)

Source: Chaliand and Rageau (1995).

Palestine and Israel

How does a person learn to identify with a specific family, history, land, or nation, even when that individual or his or her ancestors have not lived on that land for decades, centuries, or even several millennia? This chapter explores the importance of socialization in creating and maintaining a sense of who one is and the groups to which one does or does not "belong."

[**I**n May 1998, the people of Israel] marked the 50th anniversary of the founding of the state with celebrations and displays of military power. Large crowds watched air force fighters in formation and gunboats off the coast. For other Israelis it was a day of picnics in parks and beach barbecues. The highlight of the festivities was an evening gala concert in a Jerusalem stadium attended by American Vice-President Al Gore. Mr. Gore addressed the crowd in Hebrew as "the people of Israel." He noted that the United States was the first country to recognise Israel at its founding in 1948. America has been Israel's staunchest supporter since its creation when founding father David Ben-Gurion declared independence. Prime Minister Binyamin Netanyahu drew applause from the crowd, including Mr. Gore, when he said of Jerusalem: "We have united this city, never to be divided again." (From BBC 1998a)

* * *

As Jewish Israelis celebrate 50 years since the creation of Israel, for the Palestinians, the anniversary is a day of mourning. They call Israel's creation *"al naqba"*—the catastrophe—and blame the Jewish state for usurping their land. Most Palestinians have also been barred from the festivities although there are few who wish to celebrate. The Gaza Strip and the West Bank have both been sealed off, and the people who live there have been told they are not allowed to travel into Israel until the celebrations are over. (From BBC News 1998b)

Why Focus on Israel, the West Bank, and Gaza?

Our emphasis in this chapter is on the fierce century-long conflict between Jews and Palestinians. This conflict is one of an estimated 60 internal conflicts going on in the world between people who share a territory but who differ from one another in ethnicity, race, language, or religion. Most of these conflicts have long histories, which means that the conflict has been passed on from one generation to the next. We concentrate on Israel and the West Bank and Gaza because the United States has taken on the role of "peace broker" in this region. According to President Bill Clinton (1996), "The United States has often played a pivotal role in bringing Arabs and Israelis together to work out their differences in peace. It is our responsibility to do whatever we can to protect the peace process and to help move it forward." Achieving peace will not be easy, however, because everyone involved has been affected in some way by the conflict. For most, the conflict has been a part of their lives since they were born. In this chapter, we draw on socialization concepts and theories to help us understand how the conflict has been "passed down" from one generation to the next.

Keep in mind that socialization is just one factor that helps us understand why the Israeli-Palestinian conflict has lasted at least a century. Although we concentrate on two groups (Palestinians and Israelis) in this chapter, keep in mind that the issues raised here are relevant to the strategies people everywhere use to teach newcomers how to participate in the society in which they are born. The newcomers, however, do not become carbon copies of their teachers. They learn about the environment they inherit and then come to terms with it in unique ways.

By the time children are two years old, most are biologically ready to show concern for what adults regard as "rules of life." They are bothered when things do not match their expectations: paint peeling from a table, broken toys, small holes in clothing, and persons in distress raise troubling questions. From a young child's point of view, when something is "broken," then somebody somewhere has done something very wrong (Kagan 1988, 1989).[1]

To show this kind of concern with standards, however, two-year-olds must first be exposed to information that leads them to expect behavior, people, and objects to be a certain way (Kagan 1989). They develop these expectations as they interact with others in their world. For example, children go to adults with their questions and needs. Adults respond in different ways—perhaps by offering explanations, expressing concern, trying to help, showing no concern, or paying no attention. Through many such simple exchanges, children learn how to think about objects and people. They learn about the social group to which they primarily belong and about other groups to which they do not belong. In addition, these exchanges enable children to acquire basic skills, such as the ability to talk, walk upright, and reason. This learning from others about the social world is part of a complex lifelong process called **socialization.**

Socialization begins immediately after birth and continues throughout life. It is a process by which newcomers develop their human capacities and acquire a unique personality and identity. This process also facilitates the process of **internalization,** in which people take as their own and accept as binding the norms, values, beliefs, and language needed to participate in the larger community. Socialization also enables culture to be passed on from generation to generation. In this chapter, we explore the socialization process and its significance for both societies and individuals.

An Israeli soldier accompanies a preschool outing in Galilee near the border of Lebanon. He is there because armed conflicts are common in that area. How might socialization in the United States be different if soldiers routinely had to accompany school outings? How might it be different if military service were required of every U.S. citizen?

© Annie Griffiths Belt/Aurora

Coming to Terms: The Palestinians and Israelis

The following excerpts are voices of Israeli soldiers describing what it was like to serve in the West Bank and Gaza during the years of the *intifada,* the Palestinian grassroots uprising against Israeli military occupation:

- The way I see myself in all this . . . is I have a job to do and I have to carry it out. That is . . . as a person it's disturbing to me, too. It upsets me to beat up a child, it's upsetting to see children throwing (stones) . . . and to see myself chasing after them. It disturbs me to shoot rubber bullets at boys. It disturbs me also to see soldiers . . . shoot with other things . . . But on the other hand, this is what we have got to do . . . I have no choice . . . I would be glad not to have to go there.

Socialization A process by which people develop their human capacities and acquire a unique personality and identity and by which culture is passed from generation to generation.

Internalization The process in which people take as their own and accept as binding the norms, values, beliefs, and language needed to participate in the larger community.

[1]Kagan (1989) reports the results of a study in which 14- and 19-month-old children were given a set of 22 toys to play with. Of the 22 toys, 10 were unflawed, 10 were flawed (for example, a boat with holes in the bottom, a doll with black streaks on its face), and 2 were meaningless wooden forms. No 14-month-old child showed concern with the flawed toys, but 60 percent of the 19-month-old children showed clear concern with the flawed toys: they brought the toys to their mothers and said things like "broke," "bad," or "yucky." According to Kagan, this "moral sense is one of the most profound accomplishments. We should view it just as we view singing and speaking and walking. It is a maturational milestone that will be acquired by every child as long as they live in a world of human beings" (Kagan 1988:62).

- Someone tried to kill me. He stood above me with a rock and if this rock had hit me. . . . He was on the roof and I pointed my gun at him and told him in Arabic that I would . . . shoot him. He said: "Shoot me, I don't care." He aimed and threw the rock at me. . . . I jumped to one side . . . escaping the stone.

- You see all sorts of things . . . strange things . . . a toothless old woman screams . . . she curses at me. . . . You see their hatred . . . the fact that they are throwing stones at you . . . suddenly, you feel scared (Liebes and Blum-Kulka 1994:54,56,57).

- And as you walk through a village, and out of every window somebody is shouting, you begin to be afraid. An incredible fear takes hold of you, and then comes the moment when you can't stand it any longer and you hit out at the next person you see. For a fleeting moment you simply have to feel you're still stronger and able to defend yourself or you'll go crazy. Most of us . . . just hit out, without looking, and you see these terrible explosions of rage. In the evening or on weekends they sit around and talk about what's happened (Sichrovsky 1991:61).

The situation that these Israeli soldiers found themselves in is connected to a century-long dispute between Palestinian Arabs and Israeli Jews over the land between the Jordan River and the Mediterranean Sea. Both sides call this land "home." The Palestinians, descendants of Canaanites, Muslims, and Christians, had lived on this land for more than 2,000 years. The Hebrews (Israelites), exiled from Egypt, arrived in the area around 1200 B.C. and established a kingdom (Eretz Israel) with Jerusalem as its capital. The Romans conquered this land around 70 B.C. They treated the Jews harshly and suppressed all expression of Jewish culture, including religion and language. The Jews actively resisted, rebelling against Roman rule. The Romans responded by expelling most of the Jews. Only a very small number of determined Jews remained through successive occupations by Persians, Arabs, European Crusaders, Turks, and, ultimately, the British. The dispersed Jews, victims of discrimination abroad, never forgot their "home" in Palestine (see "The Jewish Population of the World").

In the late nineteenth century, a growing climate of anti-Semitism arose throughout Europe and Russia. In response, Theodor Herzl founded the modern Zionist movement.[2] Herzl, a Jewish émigré living in England,

Haunted by the Holocaust and widespread persecution throughout Europe, Jewish refugees fled to the territory they called Israel and the Palestinian people called Palestine.

© Robert Capa/Magnum

believed that the only way to combat European anti-Semitism was to establish a Jewish state. He formulated a plan to return dispersed Jews to Palestine, which he thought belonged rightfully to Jews everywhere. Young European and Russian Jews followed Herzl's lead and emigrated to Palestine to buy land and build settlements.

Shortly after the Jewish return movement began, World War I came to an end. The British, who had defeated the Turks, were given control over Palestine by the Allies. Although the British set limits on the number of returning Jews, the Nazi Holocaust during World War II increased the flow of Jewish refugees. It gave a decisive push and a desperate urgency to the Jewish return movement. Haunted by the Holocaust, which had claimed more than 6 million lives (approximately one-third of all European Jews), the Jews were determined to establish their own state and their own army to protect themselves from future aggression and to reduce their dependency on other countries for food, shelter, jobs, and passports when they assumed refugee status.[3]

---

[2]Zionism is the plan and the movement to establish a homeland in Palestine for Jews scattered across the globe. After the establishment of Israel, "Zionism was widened to include material and moral support for Israel" (Patai 1971:1262).

[3]One of the last messages received by the outside world from the Warsaw ghetto as it was being crushed by the Nazis was the following: "The world is silent. The world *knows* (it is inconceivable that it should not) and stays silent. God's vicar in the Vatican is silent; there is silence in London and Washington; the American Jews are silent. This silence is astonishing and horrifying" (Steiner 1967:160).

The Jewish Population of the World

After the Romans expelled the Jews in 70 B.C., they relocated in various locations around the world. According to the World Jewish Congress, an estimated 13 million Jews now live in 120 countries. The largest number of Jews live in the United States (5.8 million), followed by Israel (4.6 million). The map shows the estimated number of Jews living in each country. Keep in mind that the European *pogroms* (government-instigated massacres and persecution) and severe economic crises between 1880 and 1914 also pushed Jews from Europe to many countries around the world, but especially the United States.

In addition, the Nazi Holocaust during World War II increased the flow of Jewish refugees out of Europe to other countries, but especially to Israel. (Circled numbers on the map correspond with the numbers listed with the countries in the table on page 123.)

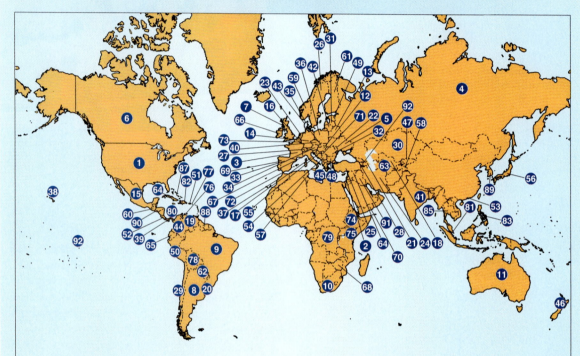

Cities with the Largest Jewish Population

City	Population
New York, USA	1,750,000
Miami, USA	535,000
Los Angeles, USA	490,000
Paris, France	350,000
Philadelphia, USA	254,000
Chicago, USA	248,000
San Francisco, USA	210,000
Boston, USA	208,000
London, UK	200,000
Moscow, Russia	200,000
Buenos Aires, Argentina	180,000
Toronto, Canada	175,000
Washington DC, USA	165,000
Kiev, Ukraine	110,000
Montreal, Canada	100,000
St. Petersburg, Russia	100,000

Communities with Fewer Than 100 Jewish Persons

Afghanistan	Haiti
Albania	Honduras
Algeria	Indonesia
Bahrain	Macedonia
Barbados	Malta
Bermuda	Martinique
Botswana	Mozambique
Cayman Islands	Myanmar (Burma)
China	Namibia
Cyprus	New Caledonia
Egypt	Nicaragua
Fiji Islands	Slovenia
French Guyana	Taiwan
Guadeloupe	Zambia

Jewish Population Worldwide

Country	Jewish Population	Country	Jewish Population
1 United States	5,800,000	47 Kyrgyzstan	4,500
2 Israel	4,600,000	48 Bulgaria	3,000
3 France	600,000	49 Estonia	3,000
4 Russia	550,000	50 Peru	3,000
5 Ukraine	400,000	51 Puerto Rico	3,000
6 Canada	360,000	52 Costa Rica	2,500
7 United Kingdom	300,000	53 Hong Kong	2,500
8 Argentina	250,000	54 Yugoslavia	2,500
9 Brazil	130,000	55 Croatia	2,000
10 South Africa	106,000	56 Japan	2,000
11 Australia	100,000	57 Tunisia	2,000
12 Hungary	80,000	58 Tajikistan	1,800
13 Belarus	60,000	59 Norway	1,500
14 Germany	60,000	60 Guatemala	1,200
15 Mexico	40,700	61 Finland	1,200
16 Belgium	40,000	62 Paraguay	1,200
17 Italy	35,000	63 Turkmenistan	1,200
18 Uzbekistan	35,000	64 Cuba	1,000
19 Venezuela	35,000	65 Ecuador	1,000
20 Uruguay	32,500	66 Ireland	1,000
21 Azerbaijan	30,000	67 Monaco	1,000
22 Moldova	30,000	68 Zimbabwe	925
23 Netherlands	30,000	69 Portugal	900
24 Iran	25,000	70 Yemen	800
25 Turkey	25,000	71 Bosnia-Herzegovina	600
26 Sweden	18,000	72 Gibraltar	600
27 Switzerland	18,000	73 Luxembourg	600
28 Georgia	17,000	74 Ethiopia	500
29 Chile	15,000	75 Kenya	400
30 Kazakhstan	15,000	76 Netherlands Antilles	400
31 Latvia	15,000	77 US Virgin Islands	400
32 Romania	14,000	78 Bolivia	380
33 Spain	14,000	79 Zaire	320
34 Austria	10,000	80 Jamaica	300
35 Denmark	8,000	81 Singapore	300
36 Poland	8,000	82 Dominican Republic	250
37 Morocco	7,500	83 Philippines	250
38 Hawaii, USA	7,000	84 Syria	250
39 Panama	7,000	85 Thailand	250
40 Czech Republic	6,000	86 Armenia	200
41 India	6,000	87 Bahamas	200
42 Lithuania	6,000	88 Suriname	200
43 Slovakia	6,000	89 South Korea	150
44 Colombia	5,650	90 El Salvador	120
45 Greece	5,000	91 Iraq	120
46 New Zealand	5,000	92 Tahiti	120

Source: *Jewish Communities of the World*, lists compiled by Institute of the World Jewish Congress. Copyright © by Virtual Communities, Inc., 1996–1998.

With the defeat of Arab armies in 1948, the country of Palestine ceased to exist, and about 100,000 Palestinians became refugees.

© UPI/Corbis-Bettmann

The steady stream of Jewish colonists notwithstanding, Palestine was then home to approximately 1.2 million Palestinian Arabs (Smooha 1980).[4] From the time that Jews began to return, Palestinians and Jews fought local battles over the land. Members of each group burned crops, destroyed trees, stole animals, sabotaged irrigation systems, and destroyed agricultural equipment belonging to members of the other group. As British forces prepared to withdraw from the region following World War II, they asked the United Nations (UN) to act as an outside mediator in the growing dispute. On November 29, 1947, the UN voted to partition Palestine into two independent states, one Jewish and the other Palestinian.

The Palestinians could not tolerate this arrangement. From their point of view, it was inconceivable that representatives from other countries could vote to divide and give away their land. As the British gradu-

ally withdrew troops and military equipment, the struggle between Palestinians and Israelis over the now-evacuated territory escalated.

On May 14, 1948, Jewish leaders declared Israel to be an independent state. The next day, Palestinian and other Arab armies from Egypt, Syria, Jordan, Lebanon, and Iraq attacked from all sides. When the Jews defeated the Arab armies, the country of Palestine ceased to exist, and about 100,000[5] Palestinian Arabs fled or were driven out to refugee camps controlled by Jordan, Syria, Egypt, and the UN. Other Palestinians remained in the newly declared state of Israel (Palestinian National Authority 1998). A second large Palestinian displacement occurred after Israel defeated armies from Jordan, Syria, Iraq, and Egypt in the Six Day War in 1967. Between 190,000 and 350,000 Palestinians fled to neighboring Arab countries when Israeli troops took control of the West Bank of the Jordan River, East Jerusalem, Gaza, and Golan Heights seized from Syria (see "Palestinian Migration Out of Palestine"). Arab governments responded by expelling their Jewish citizens, many of whom found asylum in Israel. Today, approximately 3.4 million Palestinian refugees and their descendants are registered with the UN's Reliefs and Works Agency for Palestinian Refugees in the Near East [UNRWA] (Palestinian National Authority 1998).

Today, approximately 1 million Palestinians (20 percent of Israel's population), descendants of those who remained in the country in 1948, live in Israel proper. The West Bank and Gaza are home to some 3.7 million Palestinians and 275,000 Israelis who have built settlements there (Palestinian Central Bureau of Statistics 1997; see Figure 5.1).

The conflict between Jews and Palestinians has lasted more than a century, and its meaning has passed from generation to generation. Only the form of the conflict has changed. Before 1948, both sides engaged in terrorist activities as well as land and property destruction. Since Israel declared statehood in 1948, Israelis, Palestinians, and Arabs have fought six major wars (1948, 1956, 1967, 1968–1971, 1973, and 1982); hardly a day has passed without acts of retaliation by both sides. For almost two decades the Palestinians lived under military occupation, waiting impatiently for a resolution to the "Palestinian question." Then, beginning in 1987, a new generation

---

[4]In his biography of Theodor Herzl, Ernst Pawel (1989) writes that Herzl never acknowledged the presence of large Arab villages, contrary to the findings of Leo Motzkin, the man he commissioned to survey the land. Herzl formulated his plan on the basis of two misconceptions: (1) that the land was unpopulated and (2) that any people who lived there would welcome the modernization brought by the Jews.

[5]In doing research on the number of displaced Palestinians, the author found numbers ranging from 100,000 to one million. This estimate of 100,000 was a UN estimate reported by Dr. Manuel Hassassian (1998).

U.S. in Perspective

Palestinian Migration Out of Palestine

The creation of the state of Israel and the subsequent 1948–1949 Arab-Israeli War resulted in the Palestinian diaspora. While approximately 100,000 Palestinians emigrated to the United States, several million others emigrated elsewhere. The table shows the diaspora as of 1998.

Considering the moment in history and cultural factors, what response might you suggest for the fact that about 100,000 Palestinians came to the United States? Is the number larger or smaller than you might have guessed?

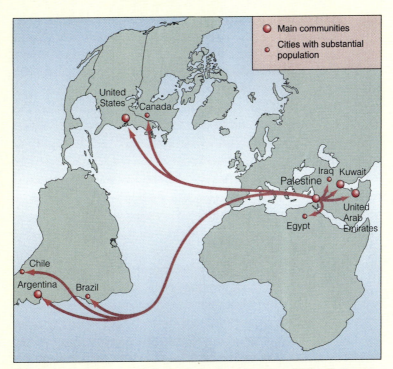

Source: "The Palestinians in the Modern World." Map by Catherine Petit. P. 179 in *The Penguin Atlas of Diasporas* by Gerard Chaliand and Jean-Pierre Rageau. Copyright ©1995 by Gerard Chaliand and Jean-Pierre Rageau. Reprinted by permission of Penguin Books.

Palestinians in Historic Palestine and Diaspora, 1996–1997 Estimates	
United States	100,000
Jordan	2,225,000
Gulf countries (Kuwait, Saudi Arabia, United Arab Emirates, Qatar)	3,711,000
Iraq	450,000
Libya	40,000
South America	25,000
Europe	50,000
Lebanon	350,000
Egypt	100,000
Syria	340,000
Other countries	150,000

Source: Passia (1998).

Figure 5.1   Israeli Settlements in Gaza, the West Bank, and East Jerusalem and the Question of Safe Passage

Israeli settlements in the West Bank, Gaza, and East Jerusalem represent one of the most difficult issues of the peace process. Critics argue that the settlements are attempts to establish a significant Jewish presence in Palestinian territory so that a permanent solution with regard to land cannot be possible. Another difficult issue revolves around "safe passage." Note that Palestinian territories will eventually consist of two disconnected lands—Gaza and the West Bank. If the two lands are to be eventually regarded as one state, how can they be geographically linked? Who will control the access roads from one territory to the other—Palestinians? Israelis? or a joint force? If Israelis control access roads, then Palestinians are subjected to Israeli control over access to their land. If Palestinians control access roads, then Israelis face the prospect of their state being divided into north-south segments (Klieman 1998). If Palestinians and Israelis are free to travel through each other's territories, how will the two governments protect their people from terrorist factions on both sides?

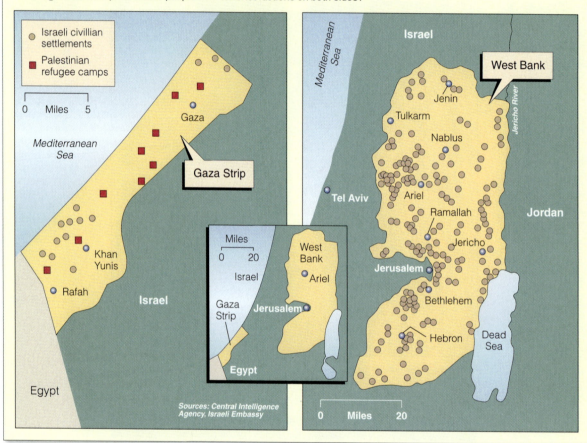

launched the *intifada*,[6] fighting the occupation with knives, Molotov cocktails, flags, rocks, graffiti, strikes, boycotts, and barricades. The Israelis responded with

gunfire, deportations, imprisonments, curfews, and school closures.

This long sequence of historical events, in conjunction with the end of the Cold War, pushed Palestinians and Israelis to participate in peace talks[7] jointly sponsored by Russia and the United States. On September 13, 1993, the prime minister of Israel, Yitzhak Rabin, and the chairman of the Palestinian Liberation

[6]Translated literally, the word *intifada* means a "tremor, shudder, or shiver." It is derived from the Arabic *nafada*, which means "to shake, to shake off, shake out, dust off, to shake off one's laziness, to have reached the end of, be finished with, to rid oneself of something, to refuse to have anything to do with . . . someone" (Friedman 1989:375). The *intifada* was waged most visibly by the "children of the stones"—young Palestinians who threw rocks and bottles at Israeli settlers and soldiers. Although Palestinian youth played a critical and very confrontational role in the *intifada*, a complex structure of committees and leadership directed the uprising. The *intifada* was supported through boycotts, business closures, and strikes.

[7]This cooperation between Moscow and Washington would not have been possible before the end of the Cold War, when foreign policy in the Middle East was dominated by U.S. and Soviet efforts to contain the spread of one another's economic and political systems.

Israelis and Palestinians marked the 50-year anniversary of the founding of the state of Israel in very different ways.

© Antoine Gyori/Sygma (left); © Nadav Neuhaus/Sygma (right)

Organization (PLO), Yasir Arafat, after several years of negotiations (including 18 months of secret negotiations held in and mediated by leaders in Oslo, Norway) met on the U.S. White House lawn to sign a peace accord known as the Declaration of Principles. For the first time, each side acknowledged the other's right to exist and took the first steps toward a permanent settlement of their conflict. In 1994, Israel withdrew its armed forces from most of the Gaza Strip and Jericho in the West Bank. In 1995, Israeli troops withdrew from another six West Bank cities. Many unresolved issues persist, however, including (1) Jewish settlements in the West Bank, Gaza, and East Jerusalem, (2) security arrangements, (3) the return of Palestinian refugees, and (4) Palestinian statehood, including "safe passage" between Gaza and the West Bank (Schmemann 1996; Jensen 1998; see Figure 5.1). The most complicated issue revolves around the status of Jerusalem, which Israel claims as its capital. The Palestinians, on the other hand, claim East Jerusalem as the capital of their hoped-for state (Greenberg 1997). To complicate matters, various factions on each side have tried to influence the course of the peace settlement to match their own vision of the future. Furthermore, factions on both sides con-

tinue to spur on violent confrontations between Palestinians and Israelis with the hope that such actions will end the possible peace settlement and future talks. As of July 1998, the peace process between the Palestinian and Israeli governments remained stalled.

In studying this conflict, sociologists ask two questions: (1) How do members of a new generation learn about and come to terms with the environment they have inherited, and (2) how is conflict passed down from one generation to another? For sociologists, part of the answer lies with socialization, a process that involves nature and nurture.

Nature and Nurture

No discussion of socialization can ignore the importance of how nature and nurture affect physical, intellectual, social, and personality development. **Nature** comprises one's human genetic makeup or biological inheritance. **Nurture** refers to the environment or the interaction experiences that make up every individual's life. Some scientists debate the relative importance of genes and environment, arguing that one is ultimately more important than the other to all phases of human

Nature Human genetic makeup or biological inheritance.

Nurture The environment or the interaction experiences that make up every individual's life.

development. Such a debate is futile, because it is impossible to separate the influence of the two factors or to say that one is more forceful. Both are essential to socialization. Trying to distinguish between the separate contributions of nature and nurture is analogous to examining a tape player and a cassette tape separately to determine what is recorded on the tape rather than studying how the two work together to produce the sound (Ornstein and Thompson 1984).

The development of the human brain illustrates rather dramatically the inseparable qualities of nature and nurture. As part of our human genetic makeup, we possess a cerebral cortex[8]—the thinking part of the brain—which allows us to organize, remember, communicate, understand, and create. The cortex consists of at least 100 billion neurons or nerve cells (Montgomery 1989); the fibers of each cell form thousands of synapses, which make connections with other cells. The number of interconnections is nearly infinite; it would take 32 million years, counting one synapse per second, to establish the number (Hellerstein 1988; Montgomery 1989).

Perhaps the most outstanding feature of the human brain is its flexibility. Scientists believe that the brain may be "set up" to learn any of the more than 6,000 to 9,000 known human languages. In the first months of life, babies are able to babble the sounds needed to speak all of these languages; as children grow, this enormous potential is reduced by the language (or languages) that each baby hears and eventually learns. Evidence suggests that the brain's language flexibility begins to diminish when a child reaches one year of age (Ornstein and Thompson 1984; Restak 1988). The larger implication is that genetic makeup provides essential raw materials but that these materials can be shaped by the environment in many different ways.

The human genetic makeup is flexible enough to enable a person to learn the values, beliefs, norms, behavior, and language of any culture. A multitude of experiences must combine with genetic makeup to create a Palestinian who desires a homeland, believes that he or she should not "pay" for the Holocaust, and fights

to establish a homeland. Likewise, nature and nurture combine to create an Israeli who values a homeland, believes the Jews have a legitimate right to land they were forced to leave 2,000 years ago, and fights to preserve the boundaries of his or her country. These ideas, however, are learned through interactions with others. If no such contact occurs, a person cannot ever become a normally functioning human being, let alone learn to become part of society.

The Importance of Social Contact

Cases of children raised in extreme isolation or in restrictive and unstimulating environments show the importance of social contact (nurture) to normal development. Some of the earliest and most systematic work in this area was done by sociologist Kingsley Davis, psychiatrists Anna Freud and Rene Spitz, and sociologist Peter Townsend. Their work demonstrates how neglect and lack of socialization influence emotional, mental, and even physical development.

Cases of Extreme Isolation

In two classic articles, "Extreme Isolation of a Child" and "Final Note on a Case of Extreme Isolation," sociologist Kingsley Davis (1940, 1947) documented and compared the separate yet similar lives of two girls in the United States, Anna and Isabelle. Each girl had received a minimum of human care during the first six years of her life. Both were illegitimate children, and for that reason both were rejected and forced into seclusion. When authorities discovered the girls, each was living in a dark, attic-like room, shut off from the rest of the family and from daily activities. Although both girls were six years old when authorities intervened, they exhibited behavior comparable to that of six-month-old children. Anna "had no glimmering of speech, absolutely no ability to walk, no sense of gesture, not the least capacity to feed herself even when food was put in front of her, and no comprehension of cleanliness. She was so apathetic that it was hard to tell whether or not she could hear" (Davis 1947:434).

Like Anna, Isabelle had not developed speech; she communicated with gestures and croaks. Because of a lack of sunshine and a poor diet, she had developed rickets: "Her legs in particular were affected; they 'were so bowed that as she stood erect the soles of her shoes came nearly flat together, and she got about with a skittering gait'" (Davis 1947:436). Isabelle also exhibited extreme fear of and hostility toward strangers.

---

[8]Ornstein and Thompson (1984) provide an excellent metaphor for visualizing the cortex:

> Form your hands into fists. Each is about the size of one of the brain's hemispheres, and when both fists are joined at the heel of the hand they describe not only the approximate size and shape of the entire brain but also its symmetrical structure. Next, put on a pair of thick gloves, preferably light gray. These represent the cortex (Latin for "bark")—the newest part of the brain and the area whose functioning results in the most characteristically human creations, such as language and art. (pp. 21–22)

Anna was placed in a private home for retarded children until she died four years later. At the time of her death, she behaved and thought at the level of a two-year-old child. Isabelle, on the other hand, entered into an intensive and systematic program designed to help her master speech, reading, and other important skills. After two years in the program, Isabelle had achieved a level of thought and behavior normal for someone her age.

Because of Isabelle's achievements, Davis concluded that extreme "isolation up to the age of six, with the failure to acquire any form of speech and hence failure to grasp nearly the whole world of cultural meaning, does not preclude the subsequent acquisition of these" (Davis 1947:437).

Davis speculated on the question of why Isabelle did so much better than Anna. He offered two possible explanations: (1) Anna may have inherited a less hardy physical and mental constitution than did Isabelle, and (2) Anna did not receive the same intensive and systematic therapy as Isabelle did. The case history comparisons are inconclusive, though. Because Anna died at age ten, researchers will never know whether she eventually could have achieved a state of normal development.

In drawing these conclusions, Davis overlooked one important factor. Although Isabelle had spent her early childhood years in a dark room, shut off from the rest of her mother's family, she spent most of this time with her deaf-mute mother. Davis and the medical staff who treated Isabelle seemed to equate deafness and muteness with feeble-mindedness (not an unusual association in the 1940s). In addition, they apparently assumed that being in a room with a deaf-mute is equivalent to a state of isolation. A considerable body of evidence, however, now indicates that deaf-mutes have rich symbolic capacities (see Sacks 1989). The fact that Isabelle could communicate through gestures and croaks suggests that she had established an important and meaningful bond with another human being. Although the bond was less than ideal, it gave her an advantage over Anna. In view of this possibility, we must question Davis's conclusion that children may be able to overcome the effects of extreme isolation during the first six years of life, provided they have a "good enough" constitution and systematic training. We use the term "good enough" because researchers do not know the exact profile of a person or training regime that would allow someone to overcome such effects.

Children of the Holocaust

Anna Freud and Sophie Dann (1958) studied six German-Jewish children whose parents had been killed in the gas chambers of Nazi Germany. The children were shuttled from one foster home to another for a year before being sent to the ward for motherless children at the Tereszin concentration camp. The ward was staffed by malnourished and overworked nurses, themselves concentration camp inmates. After the war, the six children were housed in three different institution-like environments. Eventually, they were sent to a country cottage where they received intensive social and emotional care.

During their short lives, these six children had been deprived of stable emotional ties and relationships with caring adults. Freud and Dann found that the children were ignorant of the meaning of family and grew excessively upset when they were separated from one another, even for a few seconds. In addition:

> they showed no pleasure in the arrangements which had been made for them and behaved in a wild, restless, and uncontrollably noisy manner. During the first days after arrival they destroyed all the toys and damaged much of the furniture. Toward the staff they behaved either with cold indifference or with active hostility, making no exception for the young assistant Maureen who had accompanied them from Windermere and was their only link with the immediate past. At times they ignored the adults so completely that they would not look up when one of them entered the room. They would turn to an adult when in some immediate need, but treat the same person as nonexistent once more when the need was fulfilled. In anger, they would hit the adults, bite or spit. (Freud and Dann 1958:130)

Less Extreme Cases of Extreme Isolation

Other evidence of the importance of social contact comes from less extreme cases of neglect. Rene Spitz (1951) studied 91 infants who were raised by their parents during their first three to four months of life but who were later placed in orphanages. At the time of their entry into these institutions, the infants were physically and emotionally normal. At the orphanages they received adequate care with regard to bodily needs—good food, clothing, diaper changes, clean nurseries—but little personal attention. Because only one nurse was available for every 8 to 12 children, the children were starved emotionally. The emotional starvation caused by the lack of social contact resulted in such rapid physical and developmental deterioration that a significant number of the children died. Others became completely passive, lying on their backs in their cots. Many were unable to stand, walk, or talk (Spitz 1951).

Such cases teach us that children need close contact with and stimulation from others to develop normally. Adequate stimulation means the existence of strong ties with a caring adult. The ties must be characterized by a bond of mutual expectation between caregiver and baby. In other words, there must be at least one person who knows the baby well enough to understand his or her needs and feelings and who will act to satisfy them. Under such conditions, the child learns that certain actions on his or her part elicit predictable responses: getting excited may cause the child's father to become equally excited; crying may prompt the mother to soothe the child. When researchers set up experimental situations in which a parent failed to respond to his or her infant in expected[9] ways (even for a few moments), they found that the baby suffers considerable tension and distress (*Nova* 1986).

Meaningful social contact with and stimulation from others are important at any age. Strong social ties with caring people are linked to overall social, psychological, and physical well-being. British sociologist Peter Townsend (1962) studied the effects of minimal interaction that can characterize life for the elderly in nursing homes. The consequences for the institutionalized elderly are strikingly similar to those described by Spitz in his studies of institutionalized children:

> In the institution people live communally with a minimum of privacy, and yet their relationships with each other are slender. Many subsist in a kind of defensive shell of isolation. Their mobility is restricted, and they have little access to general society. Their social experiences are limited, and the staff leads a rather separate existence from them. They are subtly oriented toward a system in which they submit to orderly routine and lack creative occupation, and cannot exercise much self-determination. They are deprived of intimate family relationships. . . . The result for the individual seems to be a gradual process of depersonalization. He may become resigned and depressed and may display no interest in the future or things not immediately personal. He sometimes becomes apathetic, talks little, and lacks initiative. His personal and toilet habits may deteriorate. (Townsend 1962:146–147)

The work of Kingsley Davis, Rene Spitz, Anna Freud, and Peter Townsend supports the idea that a person's overall well-being depends on meaningful interaction experiences with others. On a more fundamental level, social interaction is essential to a developing sense of self. Sociologists, psychologists, and biologists agree that "it is impossible to conceive of a self arising outside of social experience" (Mead 1934:135). Yet, if the biological mechanisms involved in remembering or learning and recalling names, faces, words, and the meaning of significant symbols were not present, people could not interact with one another in meaningful ways: "You have to begin to lose your memory, if only in bits and pieces, to realize that memory is what makes our lives. Life without memory is no life at all. . . . Our memory is our coherence, our reason, our feeling, even our action. Without it we are nothing" (Bunuel 1985:22).

Individual and Collective Memory

Memory, the capacity to retain and recall past experiences, is easily overlooked in exploring socialization. Yet without memory individuals and even whole societies would be cut off from the past. On the individual level, memory allows people to retain their experiences; on the societal level, memory preserves the cultural past.

How memory works remains largely a mystery. The latest neurological evidence suggests that some physical trace remains in the brain after new learning takes place, stored in an anatomical entity called an engram. **Engrams,** or *memory traces* as they are sometimes called, are formed by chemicals produced in the brain. They store in physical form the recollections of experiences—a mass of information, impressions, and images unique to each person.

> It may have been a time of listening to music, a time of looking in at the door of a dance hall, a time of imagining the action of robbers from a comic strip, a time of waking from a vivid dream, a time of laughing conversation with friends, a time of listening to a little son to make sure he was safe, a time of watching illuminated signs, a time of lying in the delivery room at childbirth, a time of being frightened by a menacing man, a time of watching people enter the room with snow on their clothes. (Penfield and Perot 1963:687)

Scientists do not believe that engrams store actual records of past events, like films stored on videocassettes. More likely, engrams store edited or consolidated versions of experiences and events, which are edited further each time they are recalled.

Engrams Chemically formed entities in the brain that store in physical form a person's recollections of experiences.

[9]The key word in this sentence is "expected," meaning that the child becomes upset when parents fail ro respond in ways the child has come to expect.

As we noted previously, memory has more than an individual quality; it is strongly social. First, no one can participate in society without an ability to remember and recall such things as names, faces, places, words, symbols, and norms. Second, most "newcomers" easily learn the language, norms, values, and beliefs of the surrounding culture. We take it for granted that people have this information stored in memory. Third, people born at approximately the same time and place have likely lived through many of the same events. These experiences, each uniquely personal and yet similar to one another, remain in memory long after the event has passed. We will use the term **collective memory** to describe the experiences shared and recalled by significant numbers of people (Coser 1992; Halbwachs 1980). Such memories are revived, preserved, shared, passed on, and recast in many forms, such as stories, holidays, rituals, and monuments.[10]

Writer David Grossman (1998) explains that Israel is alive with memories and reminders of the past:

> There are nine hundred memorials to the war dead in this small country. (In Israel, there is, on the average, one memorial for every twenty-two dead soldiers; in Europe, the ratio is one memorial for every ten thousand dead.) There is no week on the Israeli calendar in which there is not a memorial day of some sort for a traumatic event.
>
> A person walking through downtown Tel Aviv can, in the space of five minutes, set out from Dizengoff Street (where a suicide bomber murdered thirteen people two years ago), glance apprehensively at the No. 5 bus (on which another suicide bomber killed twenty-two civilians four years ago), and try to regain his composure at the Apropo Cafe (where a year ago yet another suicide bomber killed three women; none of us can forget the images of a blood-spattered, newly orphaned baby). When my daughter took her first field trip with her kindergarten classmates, they went to Yitzhak Rabin's grave and to the adjacent military cemetery (p.56).

Palestinians displaced by the 1948 and 1967 wars retain memories of their former homeland, and, like the Jews, they pass them down to those who lack personal memories. Some name their children after the cities and towns in which they lived before the 1948

Passover seder is a socialization ritual for the children of the family that promotes their identification with the Jewish historic tradition of oppression and with Israel as an ideal or nation.
© Bill Aron/Photo Researchers Inc.

war (Al-Batrawi and Rabbani 1991). Most tell their children about the places where they used to live; they teach their offspring to call those places home and even to respond with the name of that land if they are asked where they come from. When author David Grossman (1988) asked a group of Palestinian children in a West Bank refugee camp to tell him their birthplace, each replied with the name of a former Arab town:

> Everyone I spoke to in the camp is trained—almost from birth—to live this double life; they sit here, very much here . . . but they are also there. . . . I ask a five-year-old boy where he is from, and he immediately answers, "Jaffa," which is today part of Tel Aviv.
> "Have you ever seen Jaffa?"
> "No, but my grandfather saw it." His father, apparently, was born here, but his grandfather came from Jaffa.
> "And is it beautiful, Jaffa?"
> "Yes. It has orchards and vineyards and the sea."
> And farther down . . . I meet a young girl sitting on a cement wall, reading an illustrated magazine. . . . She is from Lod, not far from Ben-Gurion International Airport, forty years ago an Arab town. She is sixteen. She tells me, giggling, of the beauty of Lod. Of its houses, which were big as palaces. "And in every room a hand-painted carpet. And the land was wonderful, and the sky was always blue. . . .
> "And the tomatoes there were red and big, and everything came to us from the earth, and the earth gave us and gave us more."
> "Have you visited there, Lod?"
> "Of course not."

---

[10]A case in point is the U.S. Holocaust Memorial Museum in Washington, D.C. This museum was founded with the realization that most surviving witnesses to the Holocaust will be dead in about 10 to 20 years. It will help preserve Jewish collective memory of this event by housing "object survivors"—letters, diaries, identity papers, armbands, clothes—that document life in the concentration camps and ghettos (Goldman 1989).

Collective memory The experiences shared and recalled by significant numbers of people.

"Aren't you curious to see it now?"

"Only when we return." (Grossman 1988:6–7)

The Jews who went to Palestine to escape persecution and who fought to establish the state of Israel hold memories that are both parallel to but different from Palestinian memories. Although members of both groups experienced many of the same historical events, their memories differ because they witnessed these events from different viewpoints. Because Israel has participated in six wars with neighboring countries, has occupied the West Bank and Gaza for more than 30 years, and serves as a refuge for persecuted Jews from more than 100 countries around the world, virtually everyone in Israel has memories of war and persecution. These memories may concern personal involvement in war, waiting for a loved one to return, or flight from places where they were deemed unfit to exist.

The point is that socialization is not possible without memory. Memory is the mechanism by which group expectations become internalized in individuals and, by extension, in the whole society; it is also the mechanism by which the past remains an integral, living part of the present. The picture that any one individual holds cannot provide a complete record of the past. Yet memories include the perceived reasons that things are the way they are.

Memory of past experiences allows individuals to participate in society and shapes their viewpoint. In his essay "The Problem of Generations," sociologist Karl Mannheim (1952) maintained that first impressions or early childhood experiences are fundamental to a person's view of the world. In fact, Mannheim believed that an event has more biographical significance if it is experienced early in life than if it is experienced later in life. Many of our most significant early experiences take place in groups, some of which leave a powerful and lasting impression.

The Role of Groups

In the most general sense, a **group** is two or more persons who do the following:

1. Share a distinct identity (the ability to speak a specific language; the biological children of a specific couple; team members; or soldiers);

2. Feel a sense of belonging; and

3. Interact directly or indirectly with one another.

Groups vary according to a whole host of characteristics, including size, cohesion, purpose, and duration. Sociologists identify primary groups, ingroups, and outgroups as particularly important groups.

Primary Groups

Primary groups, such as a family or a high school sports team, are characterized by face-to-face contact and strong ties among members. Primary groups are not always united by harmony and love, however; they can be united by hatred for another group. In either case, the ties are emotional. The members of a primary group strive to achieve "some desired place in the thoughts of [the] others" and feel allegiance to the other members (Cooley 1909:24). Although a person may never achieve the desired place, he or she may nevertheless remain preoccupied with that goal. In this sense, primary groups are "fundamental in forming the social nature and ideals of the individual" (Cooley 1909:23). The family is an important primary group because it gives the individual his or her deepest and earliest experiences with relationships and because it gives newcomers their first exposure to the "rules of life." In addition, the family can serve to buffer its members against the effects of stressful events or negative circumstances; alternatively, it can exacerbate these effects.

Sociologists Amith Ben-David and Yoav Lavee (1992) interviewed 64 Israelis to learn how members of their families behaved toward one another during the 1990 SCUD missile attacks, which were launched by Iraq on Israel during the Persian Gulf War. During these attacks, families gathered in sealed rooms and put on gas masks. The researchers found that families varied in their responses to this life-threatening situation. Some respondents reported that interaction was positive and supportive: "We laughed and we took pictures of each other with the gas masks on" or "We talked about different things, about the war, we told jokes, we heard the announcements on the radio" (p. 39).

Other respondents reported that interaction was minimal but that a feeling of togetherness prevailed: "I was quiet, immersed in my thoughts. We were all around the radio . . . nobody talked much. We all sat there and we were trying to listen to what was happening outside" (p. 40).

Group Two or more people who share a distinct identity, feel a sense of belonging, and interact directly or indirectly with one another.

Primary groups Social groups characterized by face-to-face contact and strong emotional ties among members.

Finally, some respondents reported that interaction among family members was tense: "We fought with the kids about putting on their masks, and also between us about whether the kids should put on their masks. There was much shouting and noise" (p. 39). The point is that, even under extremely stressful circumstances such as war, the family can respond in ways that increase or decrease that stress. Clearly, children in families that make constructive responses to stressful events are more advantaged than children whose parents respond in destructive ways.

The kibbutz system (see "The Kibbutz") of child-rearing in Israel represents an alternative primary group to the family as the only emotional center in children's lives, offering a second emotional center to them (Aviezer et al. 1994). The underlying logic is that the kibbutz protects children from inconsistent and unreliable parenting.

Like the family and kibbutz, a military unit is also a primary group. A unit's success in battle depends on strong ties among its members. Soldiers in this primary group become so close that they fight for one another, rather than for victory per se, in the heat of battle (Dyer 1985). In Israel, the military represents a place where immigrants from almost 100 national and cultural backgrounds become "Israelis," bonding with each other and with the native-born Israelis (Rowley 1998b). Military units train their recruits always to think of the group before themselves. In fact, the paramount goal of military training is to make individuals feel inseparable from their unit. (See "How Does Military Training Turn a Group of Individuals into a Cohesive Unit?") Some common strategies to achieve this goal include ordering recruits to wear uniforms, shave their heads, march in unison, sleep and eat together, live in isolation from the larger society, and perform tasks that require the successful participation of all unit members: if one member fails, the entire unit fails. Another key strategy is to focus the unit's attention on fighting together against a common enemy. An external enemy gives a group a singular direction and thus increases its internal cohesiveness.

Almost every Israeli can claim membership in this type of primary group because virtually every Israeli citizen, male and female, serves in the military (three years for men and two years for women). Men must serve on active duty for at least one month every year until they are 51 years old (Rodgers 1998). Military training is similarly an important experience for many Palestinians, although it takes place on a less formal basis. Palestinian youths, especially those living in Syrian, Lebanese, Egyptian, and Jordanian refugee camps

Primary groups such as military units provide meaningful and emotional interactions with others. In the process, primary groups have an immense influence on forming an individual's sense of self and his or her view of the world.

© Esaias Baitel/Gamma-Liaison

(which are outside Israeli control), join youth clubs and train to protect the camps from attack. The focus on a common enemy helps establish and maintain the boundaries of the military unit. All types of primary groups, however, have boundaries—a sense of who is in the group and who is outside the group.

Ingroups and Outgroups

Sociologists use the term **ingroup** to describe a group with which people identify and to which they feel closely attached, particularly when that attachment is founded on hatred from or opposition toward another group known as an outgroup. Ingroups exert their influence on our social identity in conjunction with outgroups. An **outgroup** is a group of individuals toward which members of an ingroup feel a sense of separateness, opposition, or even hatred. Obviously, one person's ingroup is another's outgroup.

Ingroup A group with which people identify and to which they feel closely attached, particularly when that attachment is founded on hatred from or opposition toward another group.

Outgroup A group toward which members of an ingroup feel a sense of separateness, opposition, or even hatred.

The Kibbutz

The *kibbutz* (the Hebrew word for "communal settlement") is a unique rural community; a society dedicated to mutual aid and social justice; a socioeconomic system based on the principle of joint ownership of property, equality, and cooperation of production, consumption, and education; the fulfillment of the idea "from each according to his ability, to each according to his needs"; and a home for those who have chosen it.

The Dream of an Egalitarian Society Became a Reality

The first kibbutzim (plural of *kibbutz*) were founded some 40 years before the establishment of the State of Israel (1948). Their founders were young Jewish pioneers, mainly from Eastern Europe, who came not only to reclaim the soil of their ancient homeland but also to forge a new way of life. Their path was not easy. A hostile environment, inexperience with physical labor, a lack of agricultural know-how, desolate land neglected for centuries, scarcity of water, and a shortage of funds were among the difficulties confronting them. Overcoming many hardships, they succeeded in developing thriving communities that have played a dominant role in the establishment and building of the state.

Today some 270 kibbutzim, with memberships ranging from 40 to over 1,000, are scattered throughout the country. Most of them have between 300 and 400 adult members and a population of 500–600. The number of people living in kibbutzim totals approximately 130,000, about 2.5 percent of the country's population. Most kibbutzim belong to one of three national kibbutz movements, each identified with a particular ideology.

People of All Ages Live Together in a Rural Setting

Most kibbutzim are laid out according to a similar plan. The residential area encompasses carefully tended members' homes and gardens, children's houses and playgrounds for every age group, and communal facilities such as a dining hall, auditorium, library, swimming pool, tennis court, medical clinic, laundry, grocery, and so on. Adjacent to the living quarters are sheds for dairy cattle and modern chicken coops, as well as one or more industrial plants. Agricultural fields, orchards, and fish ponds are located around the perimeter, only a short tractor ride from the center.

The Decision-Making Process Is Democracy in Action

The kibbutz functions as a direct democracy. The general assembly of all its members formulates policy, elects officers, authorizes the kibbutz budget and approves new members. It serves not only as a decision-making body but also as a forum where members may express their opinions and views.

Day-to-day affairs are handled by elected committees, which deal with areas such as housing, finance, production planning, health, and culture. The chairpersons of some of these committees, together with the secretary (who holds the top position in the kibbutz), form the kibbutz executive.

Kibbutz Ideology Made Farmers Out of City Dwellers

For the founders, tilling the soil of their ancient homeland and transforming city dwellers into farmers was an ideology, not just a way to earn a livelihood. Over the years kibbutz farmers made barren lands bloom. Field crops, orchards, poultry, dairy and fish farming, and—more recently—organic agriculture became mainstays of their economy. Through a combination of hard work and advanced farming methods, they achieved remarkable results, which accounts for a large percentage of Israel's agricultural output to this day.

The Kibbutz Economy Has Diversified

Production activities of the kibbutzim are organized in several autonomous branches. Although most of them are still in agriculture, today virtually all kibbutzim have also expanded into various kinds of industry. The majority of kibbutz industry is concentrated in three main branches: metal work, plastics, and processed foods. Most industrial facilities are rather small, with less than a hundred workers. They also manufacture a wide range of products, from fashion clothing to irrigation systems.

In many areas, kibbutzim have pooled their resources, establishing regional enterprises such as cotton gins and poultry-packing plants, as well as providing a gamut of services ranging from computer data compilation to joint purchasing and marketing. The contribution of the kibbutzim to the country's production, both in agriculture (33 percent of farm produce) and in industry (6.3 percent of manufactured goods) is far greater than their share of the population (2.5 percent). In recent years, increasing numbers of kibbutzim have become centers for tourism, with recreational facilities such as guest houses, swimming pools, horseback riding, tennis courts, museums, exotic animal farms, and water parks for Israelis and foreign visitors.

As Israel's population grew and urban centers expanded, some kibbutzim found themselves virtually suburbs of cities. Due to this proximity, many of them now offer services to the public such as commercial laundries, catering, factory outlet stores, and child care, including summer camps.

Commitment to Work Is an Integral Part of Kibbutz Ideology

Work is a value in and of itself, the concept of the dignity of labor elevating the most menial job, with no special status, material or otherwise, attached to any task. Members are assigned to positions for varying lengths of time, while routine functions such as kitchen and dining hall duty are performed on a rotation basis. Each economic branch is headed by an elected administrator who is replaced every 2–3 years. An economic coordinator is responsible for organizing the work of the different branches and for implementing production and investment plans.

Although management positions are increasingly professionalized, the kibbutzim have adopted various methods of administration and organization to adapt their economic structure to the needs of the times without losing a sense of mutual responsibility and equality of work.

Women are equal participants in the labor force, with jobs in all parts of the kibbutz open to them. In contrast to kibbutz women two generations ago who sought to prove their worth by doing "men's work," the majority of women today are reluctant to become involved in agriculture and industry, preferring jobs in education, health care, and other services. Older members receive suitable work assignments according to their health and stamina.

Most members work in the kibbutz itself. Some, however, are employed in regional kibbutz enterprises, a few are sent by the kibbutz to perform educational and political functions under the aegis of its national movement, and others pursue their own special talent or profession outside the kibbutz framework. The income of these outside workers is turned over to the kibbutz.

The occasional lack of personnel for factories, agricultural tasks, tourism services, and other jobs necessitates hiring paid workers, although this practice is contrary to the kibbutz principle of self-reliance in labor. Many kibbutzim host young volunteers from Israel and abroad for periods of one month or longer in exchange for work, thus partially solving the dilemma of obtaining outside labor.

Children Grow Up with Their Peer Group

Unlike former times when they lived in communal children's houses, children in the majority of kibbutzim today sleep at their parents' home until they reach high school age. Most of their waking hours, however, are still spent with their peers in facilities adapted specifically for each age group. At the same time, parents are becoming increasingly involved in their children's activities, and the family unit is gaining more importance in the structure of the kibbutz community. Thus the granddaughters of women who 75 years ago insisted on being released from domestic chores are now the leading force within the kibbutz for more parental involvement in the upbringing of young children and for allocating women more time at home with their families. Children grow up knowing the value and importance of work and that everyone must do their share. From kindergarten, the educational system emphasizes cooperation in daily life and, from the early school grades, youngsters are assigned duties and take decisions with regard to their peer group. Young children perform regular age-appropriate tasks, older children assume certain jobs in the kibbutz and, at high school level, they devote one full day each week to work in a branch of the kibbutz economy.

Elementary schools are usually on the kibbutz premises, while older children attend a regional kibbutz high school serving several area kibbutzim, in order to experience a broader range of academic subjects and social contacts. At all age levels, accommodations are available for youngsters with special talents or needs.

Some 40 percent of all kibbutz children return to settle on their kibbutz after army service. The majority of kibbutz members today grew up in the kibbutz and decided to build their life there.

Kibbutz Society Strives to Meet the Aspirations of the Individual

Based on the voluntary participation of its members, the kibbutz is a communal society that assumes responsibility for its members' needs throughout their lives. It is a society that strives to allow individuals to develop to their fullest potential, while demanding responsibility and commitment from each person to contribute to the welfare of the community. For some, the feelings of security and satisfaction engendered by belonging to a small, closed community are among the advantages of kibbutz living, whereas others might find communal life very confining.

At first kibbutz society as a whole took precedence over the family unit. In time, this priority shifted, as the community became increasingly family-centered. Today, in the context of a normal society of grandparents, mothers and fathers, aunts and uncles, sons and daughters, the kibbutz still offers a level of cooperation that provides a social framework and personal economic security.

Compared to the past, kibbutzim today offer their members a much wider range of individual choices. Members have more latitude in all aspects of their lives, from the selection of clothing and home furnishings to where and how to spend their vacations. More opportunities are available for participation in higher education, and the special needs of artists and writers are recognized, with time given them to pursue their own projects. Although money doesn't actually change hands, members allot themselves a predetermined amount of credit each year to spend as they wish.

The Kibbutzim's Contribution to the State Exceeds Their Share of the Population

The kibbutz is not only a form of settlement and a lifestyle, it is also an integral part of Israeli society. Before the establishment of the State of Israel and in the first years of statehood, the kibbutz assumed central functions in settlement, immigration, defense, and agricultural development. When these functions were transferred to the government, the interaction between the kibbutz and the society at large decreased, though it never stopped completely. Besides active involvement in the country's political life, the kibbutz has also carried out various national tasks over the years.

A considerable number of kibbutzim run five-month study courses for new immigrants, which combine intensive Hebrew language instruction, in-depth tours of the country, and lectures on various aspects of Israeli life with periods of work on the kibbutz. Participants who decide to stay in the kibbutz may apply for membership. Some kibbutzim take part in a project in which they accept youth from disadvantaged families for their high school years. Some of these young people choose to continue living on the kibbutz and become members.

Source: The State of Israel © 1999. Reprinted by permission.

How Does Military Training Turn a Group of Individuals into a Cohesive Unit?

How does the Israeli military take people from as many as 100 national origins, speaking 80 different languages, and turn them into a unit that identifies with one country—Israel? They do it the same way militaries everywhere do—according to a set of time-tested principles and strategies that includes giving recruits a valued goal to fight for, such as freedom, peace, or land; stripping recruits of everything that makes them individuals; and exposing recruits to hardships and pressures that ultimately bring them together into a motivated and team-oriented unit.

In this essay I will give concrete examples of these time-tested principles by sharing my experiences at a two-week camp that is part of Officer Candidate School (OCS) in the U.S. Army National Guard. I was a part of a group of 16 officer candidates (OC) (most of whom had never met one another) who were selected and then thrown together for the camp. OCS is somewhat like Basic Training, as Teach, Assess, and Counsel (TAC) Officers are constantly in the OCs' faces, yelling and dropping them for pushups, flutter-kicks, and other exercise activities.

During the two-week camp, we were required to do everything together, in unison and in synchronized fashion, even moving up and down together as we did pushups and sit-ups. We also had to dress alike in every way, down to the most minute detail. If an OC did not shine his or her boots or shined them in a half-hearted way, he or she stood out from the others. Therefore we had to help shine each other's boots so as to

look exactly alike. Also, everyone's ID card was to be worn inside the exact same pocket and our watches had to be black and worn on the left wrist. Everything had to be exactly alike so that no one stood out as an individual. From an outsider's perspective and often even to the soldiers, this requirement seems ridiculous. However, the sameness spurred a sense of togetherness and provided the foundation of a very important bond.

Of course, we practiced other soldiering skills, such as marching (all heads and eyes looking forward only and moving from the waist down; no "bee-bopping" or strutting). Because OCS training is purposely designed to be stressful, we usually ran everywhere. Everyone had to run the same way: dress right,[1] eyes forward, and singing the cadence as loudly as possible. We constantly practiced what the military calls "movements," such as removing helmets from our heads, removing map cases from our shoulder, and taking off Load Carrying Equipment (LCE) (basically a belt that holds the soldier's canteens, ammunition pouches, flashlights, and so on). Everyone in the entire company (about 50 soldiers) had to take their ballistic helmet ("kevlar") off in the same way, at the same time, and place it on the ground in synchronized fashion. Everyone's movements had to be exactly synchronized or we repeated the movements over and over again until we got it right. (This requirement was especially frustrating when we had to remove the equipment to eat meals; the more we practiced getting it right,

the longer we had to wait to eat.) Always pressed for time, we had to do movements rather rapidly. Therefore our timing was off much of the time. But it was a great—even indescribable—feeling when we finally did it all together.

OCS is stressful because there is always a "mission" to be accomplished. It might be getting ready for an inspection, moving to a field site, or simply ensuring that all of the soldiers get to eat within a certain time frame. The TACs give the soldiers about half the time needed to get the job done. To increase the stress, TACs assign a second mission, and sometimes a third, before the soldiers have had time to complete the first one. Keep in mind that the missions are endless; there is always another mission to be accomplished. Furthermore, recruits are not given time to calmly plan out a strategy for accomplishing missions. Decisions must be made on the spot and usually with a TAC Officer yelling, "What are you doing? Where are you going? You are going to kill your soldiers! Why are you doing that?"

At first OCs approach a mission as if they should do only their share of the work. Ultimately this mindset cannot work because at any given time some OCs in the unit may be sent off on special missions (details) and unable to take part in a platoon mission. Therefore if someone doesn't step up and assume the missing OC's "share" of the work, the unit will fail.

After about a week into the two-week camp, most of us learned that we could be successful only if we worked

The very existence of an outgroup heightens loyalty among ingroup members and magnifies characteristics that distinguish the ingroup from the outgroup. Recognition of an outgroup can unify an ingroup even when the ingroup members are extremely different from one another. For example, one could argue that the presence of Palestinians provides a

thread of unification among Israelis, who are themselves culturally, linguistically, religiously, and politically diverse. One reason for Israel's diverse population is that, since 1948, Jews from 102 different countries, speaking 80 different languages have settled there (Peres 1998; see Figure 5.2). To ease communication problems caused by diversity, Israeli law requires that

as a team. On the sixth night of training, we had another inspection. (Inspections were constantly taking place and passing one is difficult because TACs can always find something wrong or dirty.) This time it was like a tornado had struck: our bunks were turned over, sheets were pulled off the beds, and boots and clothes were scattered and thrown all over the barracks. We were given 15 minutes to clean the barracks, shower, and get ready for bed. Working as individuals, it would be impossible to accomplish the mission on time. But, for the first time, we broke the "code" and figured out that to be successful we needed to pull together. Instead of everyone searching for personal belongings we decided on the spot that we would get everything back in place and then sort everything out later, after lights were out.[2] While we were cleaning up, we managed to get everyone showered in the three shower stalls, into PT uniforms,[3] and ready for bed on time. We finally accomplished something in the allotted time frame and we all knew it was only because we had worked together as a team. From that point on we began to work as a team, accomplishing tasks that could never be accomplished by individuals. The euphoric feeling that accompanies a mission accomplishment still gives me goosebumps when I think about it.

Another thing that made the training experience even more challenging was that TACs kept us sleep-deprived, allowing us about four or five hours of sleep each night, yet expecting us to stay alert, while we sat in boring classes,

learning things many of us already knew such as map reading. We were also deprived of basic privileges such as soft drinks, chewing gum, and access to phones. The lack of privileges had the effect of bringing us together, as we all drank water, blessed the orange juice we received at breakfast, and dreamed about drinking a cool refreshing soda!

It would be misleading to think the soldiers become a cohesive unit because everyone likes each other. Rather, the stress of training makes people realize that they are stuck with each other and that to survive the training, differences and conflicts must be worked out and compromises made. You can't say, "O.K., we don't get along—guess we'd better change teams." Instead, everyone learns that "Real teams aren't about 'Do we like each other?' They're about 'Can we respect the diversity—do we have the skill sets covered at all the bases within the team?' Anything can knock you down—a new member comes in, you get a leader who communicates differently. What the team's got to be good at is getting through it"[4] (Ellin, 1998:9).

I remember thinking, over the course of my military career, that I am enduring these hardships for the freedom of my country and sometimes even the world. In some ways, it seems almost humorous to think that we are fighting for freedom when soldiers have almost none of the freedoms that ordinary citizens do. For the United States, the idea of freedom is a very strong motivating idea, strong enough to make people endure the training, put

their lives on the line, and fight to maintain the quality of the American way of life. I imagine that Israeli soldiers respond to the idea of national security and that Palestinian soldiers respond to visions of autonomy. Regardless, every military uses strategies like these to pull together people who would never have the opportunities to interact with each other in a civilian society. Like basic training, OCS emphasizes team goals over individual goals while subjecting them to relentless stress, overwhelming soldiers to the point that they simply do not have time to be distracted by differences in national origin, race, or language.

[1]Dressing right means that if you look down a row or a column of soldiers in formation, you should see only one soldier, as the others would fall directly behind or beside the first soldier in the rank.

[2]For example, we made sure that a pair of boots was under each bed, according to Standard Operating Procedure (SOP), which specifies how everything should be done, from placement of shoes to eating in the dining facility. It did not matter whose boots were under the bed, as long as there were boots there.

[3]Once again, everyone had to be alike so we all wore Physical Training (PT) clothes to sleep in. This also made us uniform for the morning formation to do PT.

[4]We probably learned more about one another's capabilities under stress than do the OCs' families and closest friends. Yet, we really did not know much about each other's personal lives because, in the broad scheme of things, it did not matter.

Source: Leigh Cherni, student Northern Kentucky University.

everyone learn Hebrew. In addition to a common language, unifying threads for Israelis include the desire for a homeland free of persecution and the ongoing conflict with an outgroup—Palestinians and other Arabs in surrounding states. Similarly, the presence of Israelis acts to unite an equally diverse Palestinian society, which includes West Bank Palestinians, Gaza Pales-

tinians, and Israeli Palestinians. Palestinians also come from different ethnic and religious groups, clans, and political orientations.

Loyalty to an ingroup and opposition to an outgroup are accompanied by an us-versus-them consciousness. To complicate matters even further, consciousness can be traced to the fact that the two

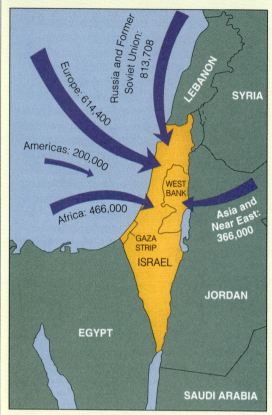

Figure 5.2   Countries of Origin of Immigrants to Israel Since 1948

Russia and Former Soviet Union: 813,708

Europe: 614,400

Americas: 200,000

Africa: 466,000

Asia and Near East: 366,000

WEST BANK

GAZA STRIP

ISRAEL

LEBANON

SYRIA

JORDAN

EGYPT

SAUDI ARABIA

Sources: Israeli Ministry of Foreign Affairs (1998b); Jewish Student Online Research Center (1998).

Because little interaction occurs between ingroup and outgroup members, they know little about one another. This lack of firsthand experience deepens and reinforces misrepresentations, mistrust, and misunderstandings between members of the two groups. Members of one group tend to view members of the other in the most stereotypical of terms.

Dr. Yorum Bilu at Hebrew University of Jerusalem designed and conducted a particularly creative study to examine the consequences of ingroup-outgroup relations on Israel's West Bank. Dr. Bilu and two of his students asked youths aged 11 to 13 from Palestinian refugee camps and Israeli settlements on the West Bank to keep journals of their dreams over a specified period. Seventeen percent of Israeli children wrote that they dreamed about encounters with Arabs; 30 percent of the Palestinian children dreamed about meeting Jews:

> Among thirty-two dreams of meetings (Jews and Arabs) there is not one character identified by name. There is not a single figure defined by a personal, individual appearance. All the descriptions, without exception, are completely stereotyped; the characters defined only by their ethnic identification (Jew, Arab, Zionist, etc.) or by value-laden terms with negative connotations (the terrorists, the oppressors, etc.). . . .
>
> The majority of the interactions in the dreams indicate a hard and threatening reality, a fragile world with no defense. . . .
>
> An Arab child dreams: "The Zionist Army surrounds our house and breaks in. My big brother is taken to prison and is tortured there. The soldiers continue to search the house. They throw everything around, but do not find the person they want [the dreamer himself]. They leave the house, but return, helped by a treacherous neighbor. This time they find me, and my relatives, after we have all hidden in the closet in fright."
>
> A Jewish child dreams: ". . . suddenly someone grabs me, and I see that it is happening in my house, but my family went away, and Arab children are walking through our rooms, and their father holds me, he has a *kaffiyeh* and his face is cruel, and I am not surprised that it is happening, that these Arabs now live in my house." (Grossman 1988:30, 32–33)

Often an ingroup and an outgroup clash over symbols—objects or gestures that are clearly associated with and valued by one group. These objects can be defined by members of the other group as so threatening that they seek to eliminate them: destroying the objects becomes a way of destroying the group. During the *intifada,* Israeli reporter Danny Rubinstein (1988) witnessed hundreds of clashes between Palestinian

groups live segregated lives. In the case of the Palestinians and Israelis, they have little in common. For the most part they do not share a language,[11] religion, schools, residence, or military service. Although they do share an economic relationship,[12] Palestinians are concentrated in low-status jobs. Palestinians, many of whom have graduated from technical colleges or universities, can obtain only manual labor jobs and low-status service jobs.

[11]Arabic is taught in Jewish schools but is not mandatory (Abu-Rabin 1998). A large segment of the Jewish population are Sephardic, meaning that they come from Arab countries and their native language is Arabic.

[12]In 1998, 19.4 percent of all Palestinian workers in the West Bank commuted to jobs in Israel or the settlements. Almost 12 percent of Palestinian workers in Gaza commuted to jobs in Israel or the settlements (Palestinian Central Bureau of Statistics 1998). In the Palestinian territories of the West Bank and Gaza, which are no longer under Israeli control, 17.5 percent of Palestinians work at jobs in Israel or the settlements.

youths and Israeli soldiers over symbols, especially the Palestinian flag, which was outlawed at that time. Palestinian Arabs

> hoist the flag (which is very much like the Jordanian flag), while Israeli soldiers bring it down and attempt to catch and punish the perpetrators. At times the situation takes a ridiculous turn. Some time ago, in Bethlehem, I heard an Israeli officer issue an order to close down for a week all shops on a certain street where a Palestinian flag had been hoisted on the corner utility pole the night before. I saw schoolgirls in Hebron knitting satchels modeled on the Palestinian flag and clothing stores with window displays arranged to fit its color and pattern. (p. 24)

Although ingroup and outgroup clashes center around symbols, the symbol alone is not the cause. Rather, it is the meaning attached to it that functions as a rallying point for people on both sides. For example, in December 1997 census takers from the Palestinian Central Bureau of Statistics clashed with Israeli officials when they attempted to survey Palestinian neighborhoods in East Jerusalem. For the Palestinians, the self-census symbolized an important step toward laying the foundations for independent statehood. For many Israelis, the same census symbolized a threat to their identity and security. The Israeli government issued warnings over the radio that census-takers would be jailed for conducting a survey of East Jerusalem. Israeli spokesmen made it clear that "Jerusalem is Israel's capital" and "the State of Israel will not permit the division of Jerusalem. It will not allow *them* a foothold in East Jerusalem" (Greenberg 1997:A3).

It is important to know that ingroups and outgroups also exist within Palestinian and Israeli society. Among Israelis, for example, clear divisions exist among Sephardic Jews with North African and Middle Eastern roots and Ashkenazi Jews with European roots. An estimated 50 percent of Israelis do not actively practice a religion, 30 percent consider themselves practicing Jews, and 18 percent are ultra-orthodox. The ultra-orthodox are exempt from military service, which creates resentment among those Jews who must serve in the military and then the reserves until they are 51 years old (BBC 1998c).

While we have applied the concepts "ingroup" and "outgroup" to a very dramatic situation—the Israeli-Palestinian conflict—keep in mind that ingroup-outgroup experiences help shape people's identity and sense of belonging. Three students in my 1998 Introduction to Sociology class offered some less dramatic examples:

Often members of ingroups and outgroups clash over symbols. Before the 1993 peace accord, it was illegal in Israel to display the Palestinian flag. The flag thus became the focus of many conflicts between Israelis and Palestinians.

© Stephane Compoint/Sygma

- The section on "ingroups" and "outgroups" reminded me of a neighborhood rivalry. When I was about eight, we moved to a new subdivision where many kids around that age lived. For some unknown reason, two groups formed according to street. Members of each group pulled pranks on the other, such as ringing doorbells and then running away, and throwing eggs and rocks at the other's houses. Today I am good friends with two of the boys in the "outgroup" I used to torment years ago. None of us knows why we ever disliked each other.

- When reading about ingroups and outgroups I instantly thought about my experiences in high school. When I was in high school, there were different groups such as "preps," "hoods," and "nerds." It was easy to tell who belonged to each group simply by looking at their dress and general physical appearance. People who belonged to one group didn't have much to do with those in the other two groups.

- In reading the information on ingroups and outgroups, I thought of my niece. Until she was about 20 years old, her family was her primary group and ingroup. Her thoughts, time, and energies focused on her family, and she saw them as a source of support. But then she began hanging out with people her age with "distinct" tastes in clothes, music, and interests. She modeled her appearance and her demeanor after these friends. Her family became her "outgroup" and her newfound friends became her new "ingroup." She now avoids family events because she fears their

Symbolic gestures, such as the arm and hand movements directed at airplane pilots, help convey meaning from one person to another.

© Corbis

reprimand or ridicule and because she believes they will make her feel guilty for the choices she has made. I believe that she will, some day, think of her family as her ingroup again.

To this point, we have examined how socialization is a product of nature and nurture. We have discussed how genetic makeup provides each individual with potentials that are developed to the extent made possible by the environment. We also have considered the importance of stimulation from caregivers in developing our genetic potential and the connection between group membership and self-awareness. Even groups to which we do not belong can exert powerful influences on our sense of self. For example, an outgroup makes us clearly aware of who "they" are, which in turn reminds us of who "we" are. Next we will examine the theories of sociologists George Herbert Mead and Charles Horton Cooley, regarding some specific ways in which the self develops and information is transmitted to newcomers.

Reflexive thinking Stepping outside the self and observing and evaluating it from another's viewpoint.

Significant symbol A word, gesture, or other learned sign used to convey a meaning from one person to another.

Symbolic gestures Nonverbal cues, such as tone of voice and other body movements, that convey meaning from one person to another.

Symbolic Interactionism and Self-Development

Humans are not born with a sense of self; rather, this consciousness evolves through regular interaction with others. The emergence of a sense of self depends on our physiological capacity for **reflexive thinking**—stepping outside the self and observing and evaluating it from another's viewpoint. Reflexive thinking allows individuals to learn how they come across to others and to adjust and direct their behavior in ways that meet others' expectations. In essence, self-awareness emerges hand in hand with awareness of others and of their evaluations of one's behavior and appearance.

The Emergence of Self-Awareness

According to George Herbert Mead, significant symbols and gestures are the mechanisms that allow an individual to interact with others and in the process to learn about the self. A **significant symbol** is a word, gesture, or other learned sign that is used to "convey a meaning . . . and that has the same meaning for the person transmitting it as for the person receiving it" (Theodorson and Theodorson 1979:430). Language is a particularly important significant symbol because the shared meanings attached to words allow us to communicate with others. Significant symbols also include **symbolic gestures** or signs are nonverbal cues, such as tone of voice, inflection, facial expression, posture, and other body movements or positions that convey meaning from one person to another.

As people learn significant symbols, they also acquire the ability to carry out reflexive thinking and to adjust their presentation of self to meet other people's expectations. Mead believed, however, that humans do not adhere mechanically to others' expectations. Instead a dialogue goes on continuously between two aspects of the self—the *me* and the *I*.

The *me* is Mead's term for the social self—that part of the self that has learned and internalized society's expectations about what constitutes appropriate behavior and appearances. In other words, the *me* is the part of the self that knows the norms that govern behavior in specific situations. Before an individual acts or speaks, the *me* takes into account or anticipates how others will respond. The individual then proceeds to act or speak as planned or alters behavior accordingly.

The *I* is the spontaneous, autonomous, creative self capable of rejecting expectations and acting in unconventional, inappropriate, or unexpected ways. The *I* takes chances and violates expectations. Some-

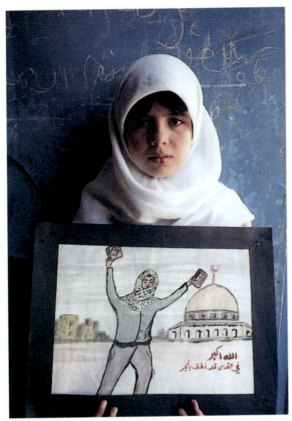

Through socialization this Palestinian girl has learned to dream of living someday in a glorious Jerusalem—a city she has never seen.

© Joanna B. Pinneo/Aurora

times taking a chance pays off and people define the individual as unique, exceptional, or one of a kind. At other times, taking a chance backfires and the individual is punished or ostracized.

Although Mead does not specify how the *I* emerges, we know that a spontaneous, creative self must exist; otherwise, human life would never change and would stagnate. Mead is more specific about how the *me* develops: through imitation, play, and games, all of which allow the child to practice role taking.

Role Taking

Mead assumed that the self is a product of interaction experiences. He maintained that children acquire a sense of self when they become objects to themselves. Children become objects to themselves when they are able to imagine the effect of their words and actions on other people. According to Mead, a person can see himself or herself as an object after learning to role-take. **Role taking** involves stepping outside the self and

imagining how others view its appearance and behavior imaginatively from an outsider's perspective.

Researchers have devised an ingenious method for determining when a child is developmentally capable of role taking. A researcher puts a spot of rouge on the child's nose and then places the child in front of a mirror. If the child ignores the rouge, he or she presumably has not yet acquired a set of standards about how he or she ought to look; that is, the child cannot role-take or see himself or herself from another person's viewpoint. If the child shows concern over the rouge, however, then he or she presumably has formed some notion of self-appearance and therefore can role-take (Kagan 1989).

Mead hypothesized that children learn to take the role of others through three stages: (1) imitations, (2) play, and (3) games. Each of these stages involves a progressively sophisticated level of role taking.

THE PREPARATORY STAGE   In this stage, children have not yet developed the mental capabilities that allow them to role-take. They mimic or imitate people in their environment but have almost no understanding of the behaviors that they are imitating. Children may imitate spontaneously (by mimicking a parent writing, cooking, reading, and so on), or they may repeat things that adults encourage them to say and reward them for saying. In the process of imitating, children learn to function symbolically; that is, they learn that particular actions and words arouse predictable responses from others. For example, Israeli children may be taught early on that "Jerusalem is the capital of Israel." Similarly, Palestinian parents teach their children where they come from, even before the children learn notions of geography and understand the historical circumstances of their living arrangements. A typical exchange between a Palestinian parent and child would go something like this:

PARENT: What is Israel?

CHILD: The real name for Israel is Palestine.

Both Palestinian and Israeli children, like children in nearly every culture, learn to sing patriotic songs and say prayers before they can understand the meaning of the words. Jenny Bourne, a political activist and a member of a delegation that visited the West Bank and Gaza during the *intifada*, was struck by the fact that as soon as some two-year-old Palestinian children "saw the cameras

Role taking Stepping outside the self and imagining how others view its appearance and behavior imaginatively from an outsider's perspective.

Palestinian children in Nablus, a West Bank city, flash the peace sign while holding weapons that were illegal under Israeli occupation. The men with them are Palestinian Authority police who replaced the Israeli military when Nablus came into self-rule in December 1995.

© Miki Kratsman/Corbis

come out, they were up and alert, hands outstretched as taut fingers made in unison the victory sign for our photos. No [Palestinian] child we met anywhere wanted to be photographed without that sign" (Bourne 1990:70).

THE PLAY STAGE   Mead saw children's play as the mechanism by which they practice role taking. **Play** is a voluntary and often spontaneous activity, with few or no formal rules, that is not subject to constraints of time (for example, 20-minute halves, 15-minute quarters) or place (for example, a gymnasium, a regulation-size field). Children, in particular, play whenever and wherever the urge strikes. If rules exist, they are developed by the children on their own and not imposed on participants by higher authorities (for example, rule

Play Voluntary and often spontaneous activity, with few or no formal rules, that is not subject to constraints of time or place.

Significant others People or characters who are important in an individual's life, in that they have considerable influence on that person's self-evaluation and encourage him or her to behave in a particular manner.

books, officials). Participants undertake play for their amusement, entertainment, or relaxation. These characteristics make play less socially complicated than organized games such as Little League Baseball (Corsaro 1985; Figler and Whitaker 1991).

In the play stage, children pretend to be **significant others**—people or the characters (for example, cartoon characters, the family pet) who are important in their lives, in that they have considerable influence on a child's self-evaluation and/or encourage the child to behave in a particular manner. Children recognize behavior patterns characteristic of these significant persons and incorporate them into their play. For example, when a little girl plays with a doll and pretends she is the doll's mother, she talks and acts toward the doll in the same way that her mother talks and acts toward her. By pretending to be the mother, she gains a sense of the mother's expectations and perspective and learns to see herself as an object. Similarly, two children playing doctor and patient are learning to see the world from viewpoints other than their own and to understand how a patient acts in relation to a doctor, and vice versa.[13]

In the play stage, children's role taking comes from what they see and hear going on around them. For the most part, Palestinian children in the West Bank and Gaza have never seen an adult male Israeli without a gun. Palestinian children's play reflects their experiences: The children pretend to be Israeli soldiers arresting and beating other Palestinian children who are pretending to be stone throwers. They use sticks and cola cans as if they were guns and tear-gas canisters (Usher 1991). One evening ABC News featured a segment on Palestinian children engaged in this type of play. When asked by the reporter which they preferred to be, soldiers or stone throwers, the children replied, "Soldiers, because they have more power and can kill."

Israeli children have had little experience with Palestinians except as manual laborers or "terrorists." Thus, it is hardly surprising that some Israeli kindergartners pretend that Israelis are Smurfs (good guys) and Palestinians portray Gargamel (a bad guy in a TV program). Israeli children, like Palestinian children, pretend to be soldiers because both men and women must serve in the Israeli military beginning at age 18.

[13]One student in my Introduction to Sociology class wrote about her two-year-old cousin who for a time acted like the family dog. "My aunt would come into the kitchen and find her crawling on the floor drinking out of the dog's water dish. The point is my niece has a great imagination and likes to play at being different things."

THE GAME STAGE   In Mead's theory, the play stage is followed by the game stage. **Games** are structured, organized activities that almost always involve more than one person. They are characterized by a number of constraints, including one or more of the following: established roles and rules and an outcome toward which all activity is directed. Through games children learn to (1) follow established rules, (2) take simultaneously the role of all participants, and (3) see how their position fits in relation to all other positions.

When children first take part in games such as organized sports, their efforts seem chaotic. Instead of making an organized response to a ball hit to the infield, for example, everyone tries to retrieve the ball, leaving nobody at the base to catch the throw needed to put the runner out. This chaos exists because children have not developed to the point at which they can see how their role fits with the roles of everyone else in the game. Without such knowledge, a game cannot have order. Through playing games, children learn to organize their behavior around the **generalized other**—that is, around a system of expected behaviors, meanings, and viewpoints that transcend those of the people participating. "The attitude of the generalized other is the attitude of the whole community. Thus, for example, in the case of such a social group as a baseball team, the team is the generalized other insofar as it enters—as an organized process or activity—into the experience of [those participating]" (Mead 1934:119). In other words, when children play organized sports such as baseball, they practice fitting their behavior into an already established behavior system that governs the game.

In view of this information, not surprisingly, games are the tools used in programs designed to break down barriers between Palestinian and Jewish children and adolescents. The games involve activities like

> throwing an orange into the air, calling a person's name to catch it, throwing it again with another's name, and again and again as the whoops of laughter fill the room. Then they all crowd together, take each other's hands, and turn around until they are enmeshed in a tangle of arms. Intertwined with each other, they try to unravel themselves without letting go. They talk to each other, giving advice, crouching so another can step over an arm, stooping so others can swing arms over heads, spinning around, trying to turn the snarled mess of Arab and Jewish bodies into a clean circle. (Shipler 1986:537)

Although these games seem merely fun, sociologists contend that participants are learning to see things from another perspective and to play their parts successfully in a shared activity. The participants cannot be effective unless they understand their own roles in relation to everyone else's. Although the children trying to untangle themselves may not be fully aware of it, they are learning that a Palestinian (or an Israeli) can be in positions like their own. If anyone can become untangled, participants must be able to understand everyone else's situation.

As we have learned, George Herbert Mead assumed that the self develops through interaction with others. Mead identified the interaction that occurs in play and games as important to children's self-development. When children participate in play and games, they practice at seeing the world from the viewpoint of others and gain a sense of how others expect them to behave. Sociologist Charles Horton Cooley offered a more general theory about how the self develops.

The Looking-Glass Self

Like Mead, Charles Horton Cooley assumed that the self is a product of interaction experiences. Cooley coined the term **looking-glass self** to describe the way in which a sense of self develops: People act as mirrors for one another. We see ourselves reflected in others' reactions to our appearance and behaviors. We acquire a sense of self by being sensitive to the appraisals of ourselves that we perceive others to have: "Each to each a looking glass, / Reflects the other that [does] pass" (Cooley 1961:824). As we interact, we visualize how we appear to others, we imagine a judgment of that appearance, and we develop a feeling somewhere between pride and shame: "The thing that moves us to pride or shame is not the mere mechanical reflection of ourselves but . . . the imagined effect of this reflection upon another's mind" (Cooley 1961:824). Cooley went so far as to argue that "the solid facts of social life are the facts of the imagination."

Because Cooley defined the imagining or interpreting of others' reactions as critical to self-awareness, he believed that people are affected deeply even when

Games   Structured, organized activities that usually involve more than one person and a number of constraints concerning roles, rules, time, place, and outcome.

Generalized other   A system of expected behaviors, meanings, and viewpoints that transcend those of the people participating.

Looking-glass self   A way in which a sense of self develops in which people see themselves reflected in others' reactions to their appearance and behaviors.

the image they see reflected is exaggerated or distorted. One responds to the perceived reaction rather than to the actual reaction.

On the other hand, we cannot overlook the fact that more often than not our imaginations of how other people will react and behave rest on past experiences with others. In the case of Palestinians and Israelis, for example, each group aims a number of powerful images at the other. For example, many Palestinians call the Israelis "Nazis" and equate the Israeli presence in West Bank and Gaza and the accompanying system of identification cards, checks, and imprisonments with the concentration camps of World War II. These labels are quite painful to Israelis, who see little similarity between the Holocaust and the Palestinian situation. Many Israelis, on the other hand, react by defining the Palestinians as culturally primitive and incapable of managing their own affairs. They tell the Palestinians that the Israelis are responsible for turning the worthless desert land occupied previously by a backward Palestinian people into a modern, high-technology state.

Both Mead's and Cooley's theories suggest that self-awareness derives from an ability to think reflexively—to step outside oneself and view the self from another's perspective. Although Cooley and Mead describe the mechanisms (imitation, play, games) by which people learn about themselves, neither theorist addressed how a person acquires this level of cognitive sophistication. To answer this question, we must turn to the work of Swiss psychologist Jean Piaget.

Cognitive Development

Piaget is the author of many influential and provocative books about how children think, reason, and learn. The titles of some of his many books—*The Language and Thought of the Child* (1923), *The Child's Conception of the World* (1929), *The Moral Judgment of the Child* (1932), *The Child's Conception of Time* (1946), and *On the Development of Memory and Identity* (1967)—give some clues about the many categories of childhood thinking that Piaget investigated.

Piaget's influence reaches across many disciplines: biology, education, sociology, psychiatry, psychology, and philosophy. His ideas about how children develop increasingly sophisticated levels of reasoning stem from his study of water snails (*Limnaea stagnalis*),

which spend their early life in stagnant waters. When transferred to tidal water, these lazy snails engage in motor activity that develops the size and shape of the shell to help them remain on the rocks and avoid being swept away (Satterly 1987).

Building on this observation, Piaget arrived at the concept of **active adaptation,** a biologically based tendency to adjust to and resolve environmental challenges.[14] The theme of active adaptation runs through almost all of Piaget's writings. He believed that learning and reasoning form an important adaptive tool that helps people to meet and resolve environmental challenges. Logical thought emerges according to a gradually unfolding genetic timetable. This unfolding must be accompanied by direct experiences with persons and objects; otherwise, a child will not realize his or her potential ability. Children construct and reconstruct their conceptions of the world as they experience the realities of living.

Piaget's model of cognitive development includes four broad stages—sensorimotor, preoperational, concrete operational, and formal operational—each characterized by a progressively more sophisticated reasoning level. A child cannot proceed from one stage to another until he or she masters the reasoning challenges of earlier stages. Piaget maintained that reasoning abilities cannot be hurried; a more sophisticated level of understanding will not show itself until the brain is ready.

- *Sensorimotor stage (from birth to about age two).* In this stage, children explore the world with their senses (taste, touch, sight, hearing, and smell). The cognitive accomplishments of this stage include an understanding of the self as separate from other persons and the realization that objects and persons exist even when they are out of sight. Before this notion takes hold, children under seven months act as if an object does not exist when they can no longer see it (Kotre 1995).

- *Preoperational stage (from about ages two to seven).* Children in this stage typically demonstrate three

---

[14]One of my students offered an excellent example to illustrate the general principle of active adaptation: "Reading about active adaptation reminded me of the 'bodily adjustments' I made when I moved from Kentucky to New Mexico for about six months. In Kentucky I grew up in the hot humid summers wearing as little clothing as possible. When I moved to New Mexico the climate was much hotter but drier. I was surprised at the dress of the Native residents. Most people dressed in heavy flannel shirts with tee-shirts underneath and jeans. I soon realized that there was as much as a 60-degree difference between day- and night-time temperatures. I was very surprised that in about a month my body adapted to wearing clothes I had always associated with cold weather and would never think of wearing on 100-degree days, and I really didn't feel that hot."

Active adaptation A biologically based tendency to adjust to and resolve environmental challenges.

characteristic types of thinking. They think *anthropomorphically*—that is, they assign human feelings to inanimate objects. Thus they believe that objects such as the sun, the moon, nails, marbles, trees, and clouds have motives, feelings, and intentions (for example, dark clouds are angry; a nail that sinks to the bottom of a glass filled with water is tired). Second, they think *nonconservatively*, a term Piaget used to signify an inability to appreciate that matter can change form but still remain the same in quantity. Third, they think *egocentrically* in that they cannot conceive how the world looks from another point of view. Thus, if a child facing a roomful of people (all of whom are looking in his or her direction) is asked to draw a picture of how a person in the back of the room sees the people, the child will draw the picture as he or she sees the people. Related to egocentric thinking is *centration*, a tendency to center attention on one detail of an event. As a result of this tendency, the child fails to process other features of a situation (see Figure 5.3).

• *Concrete operational stage (from about ages seven to twelve)*. By the time children enter this stage, they have mastered preoperational tasks but have difficulty in thinking hypothetically or abstractly without reference to a concrete event or image. For example, a child in this stage has difficulty envisioning a life without him or her in it. One 12-year-old struggling to grasp this idea said to me, "I am the beginning and the end; the world begins with and ends with me."

• *Formal operational stage (from the onset of adolescence onward)*. At this point, a person is able to think abstractly. For example, a person can conceptualize his or her existence as a part of a much larger historical continuum and a larger context.

This progression by stages toward increasingly sophisticated levels of reasoning is apparently universal, though the content of people's thinking varies across cultures. For example, all Palestinian and Israeli children learn the following rule: "If you ever see an unattended package or bag on a street or bus, don't touch it. Notify an adult immediately." Knowing this rule is a matter of safety because the package might contain a bomb. American children typically are not exposed to the same dangers and hence have no need to consider situations in which this rule can be relaxed.

New York Times reporter David Shipler, in his book *Arab and Jew*, describes his frustrations in explaining to his young children that they did not have to follow this rule when they returned to the United States. When he set some bags of newspapers on the curb to be picked up, his children reported suspicious packages outside:

Figure 5.3   Centration in a Child's Drawing

As they draw, young children can focus on only one detail at a time. For example, they draw the passengers outside the automobile. When they think about the passengers in a car, they fail to consider that they should draw the passenger's bodies inside the car and show only a portion of their bodies.

Source: From "Children's Drawing of Human Figures," by Norman H. Freeman. Page 138 in *The Oxford Companion to the Mind*. Copyright © 1987. Oxford University Press. Reprinted by permission.

Michael [age 7] ran in another day to report a plastic cup of some sort in the street. I had seen it and asked him to throw it away. He adamantly refused to go near it, and he remained solidly unmoved by my extravagant assurances that we didn't have to worry about bombs on a quiet, tree-lined suburban street in America. (Shipler 1986:83)

From the perspective of Piaget's theory, Michael centered all of his attention on one detail (the rule) and could not respond to other aspects of the situation that made the rule irrelevant, such as geographic location.

The theories of Mead, Cooley, and Piaget all suggest that the process of social development is multifaceted and continues over time. It is important to realize that socialization is a lifelong process in which people make any number of transitions over a lifetime: from single to married, from married to divorced or widowed, from childlessness to parenthood, from healthy to disabled,

Piaget's theory of cognitive development suggests that the emergence of social concern in many young people may be explained in part by the development of their ability to think abstractly and, consequently, to identify with issues and causes.

© Tony Freeman/PhotoEdit

from one career to another, from civilian status to military status, from employed to retired, and so on. In making such transitions, people undergo resocialization.

Resocialization

Resocialization is the process of becoming socialized over again. In particular, this process involves discarding values and behaviors unsuited to new circumstances and replacing them with new, more appropriate values and norms (standards of appearance and behavior). A considerable amount of resocialization happens naturally over a lifetime and involves no formal training; people simply learn as they go. For example, people marry, change jobs, become parents, change religions, and retire without formal preparation or training. However, some resocialization requires that,

---

Resocialization The process of discarding values and behaviors unsuited to new circumstances and replacing them with new, more appropriate values and norms.

Total institutions Institutions in which people surrender control of their lives, voluntarily or involuntarily, to an administrative staff and carry out daily activities with others required to do the same thing.

---

to occupy new positions, people must undergo formal, systematic training and demonstrate that they have internalized appropriate knowledge, suitable values, and correct codes of conduct.

Such systematic resocialization can be voluntary or imposed (Rose, Glazer, and Glazer 1979). Voluntary resocialization occurs when people choose to participate in a process or program designed to "remake" them. Examples of voluntary resocialization are wide-ranging—the unemployed youth who enlists in the army to acquire a technical skill, the college graduate who pursues medical education, the drug addict who seeks treatment, the alcoholic who joins Alcoholics Anonymous (AA).

Imposed resocializaion occurs when people are forced to undergo a program designed to train them, rehabilitate them, or correct some supposed deficiency in their earlier socialization. Military boot camp (when a draft exists), prisons, mental institutions, and schools (when the law forces citizens to attend school for a specified length of time) are examples of environments that are designed to resocialize individuals but that people also enter involuntarily.

In *Asylums: Essays on the Social Situation of Mental Patients and Other Inmates,* sociologist Erving Goffman writes about a setting—total institutions (with particular focus on mental institutions)—where people undergo systematic socialization. In **total institutions,** people surrender control of their lives, voluntarily or involuntarily, to an administrative staff and, as inmates, carry out daily activities (eating, sleeping, recreation) in the "immediate company of a large batch of others, all of whom are [theoretically] treated alike and required to do the same thing together" (Goffman 1961:6). Total institutions include homes for the blind, the elderly, the orphaned, and the indigent; mental hospitals; jails and penitentiaries; prisoner-of-war camps and concentration camps; army barracks; boarding schools; and monasteries and convents. Their total character is symbolized by barriers to social interaction, "such as locked doors, high walls, barbed wire, cliffs, water, forests, or moors" (p. 4).

Goffman identified wide-ranging general and standard mechanisms that the staffs of all total institutions employ to resocialize "inmates." When the inmates arrive, the staff strips them of their possessions and their usual appearances (and the equipment and services by which their appearances are maintained). In addition, the staff sharply limits interactions with people outside the institution to establish a "deep initial break with past roles" (p. 14).

We very generally find staff employing what are called admission procedures, such as taking a life history, photographing, weighing, finger-printing, assigning numbers, searching, listing personal possessions for storage, undressing, bathing, disinfecting, haircutting, issuing institutional clothing, instructing as to rules, and assigning to quarters. The new arrival allows himself to be shaped and coded into an object that can be fed into the administrative machinery of the establishment. (p. 16)

Goffman maintained that the admission procedures function to prepare inmates to shed past roles and assume new ones by participating in the various enforced activities that staff members have designed so as to fulfill the official aims of the total institution—whether to care for the incapable, to keep inmates out of the community, or to teach people new roles (for example, to be a soldier, priest, or nun).

In general, it is easier to resocialize people when they want to be resocialized than when they are forced to abandon old values and behaviors. Furthermore, resocialization is likely to occur more readily if acquiring new values and behaviors requires competence rather than subservience (Rose et al. 1979). Consider the resocialization that takes place in medical school. Theoretically, medical students learn (among other things) to be emotionally detached in their attitudes toward patients, not prefer one patient over another (that is, patients of a particular ethnicity, gender, or age, or even level of cooperation), and provide medical care whenever it is required (Merton 1976). These attitudes are necessary for proper diagnosis and treatment.

Michael Gorkin (1986) believes that many problems can develop between Israeli psychiatrists and Palestinian patients during therapy. (The same could be said of Palestinian psychiatrists and Israeli patients, but very few Palestinian psychiatrists practice in Israel. Consequently, most Palestinians who go to psychiatrists consult Israeli ones.) If Israeli psychiatrists do not learn to manage their stereotypes and prejudices with regard to Palestinians and to familiarize themselves with Arab culture, they may treat the client in counterproductive ways. In addition, the patient must trust the psychiatrist if therapy is to be successful.

Establishing trust is complicated by the fact that almost all Israeli psychiatrists serve in army reserve units and must cancel therapy hours several times a year for several weeks. Many are reluctant to tell their Palestinian clients the reason for their absence; most hope that their patients will not inquire. Palestinians, of course, are aware of this commitment. Gorkin maintains that the Israeli psychiatrist must address these strains constructively with the Palestinian client in the initial sessions. He does not believe that the discussion will resolve these differences but thinks that it "sets the stage for openness in the therapeutic interaction and conveys the message that this crucial issue is not taboo" (Griffith 1977:38). The Israeli psychiatrist is likely to come to terms with prejudices and stereotypes because he or she has chosen to become a physician and because coming to terms with these beliefs demonstrates professional competence achieved through resocialization during the medical training period.

Therein lies the dilemma in finding a resolution to the Palestinian-Israeli conflict. Factions on both sides attempt to use resocialization measures to force the other side to change its position about land rights. Israelis deport, imprison, impose curfews, close schools, level houses, and kill. Palestinians throw stones, strike, boycott Israeli products, and kill. A small, but significant percentage of Palestinians and Israelis seem to believe that if they make life miserable enough for the other group, each will gain a homeland. The problem is that if one side wins through intimidation, the other side by definition assumes a subservient position. On the other hand, if the peace is achieved, people who have developed identities around the "group in conflict" must then redefine themselves to fit in a new environment (Elbedour, Bastien, and Center 1997).

Summary and Implications

Humans are born with a genetic endowment. By way of their genes, parents transmit to their offspring a biological heritage common to all human beings. On the one hand, this heritage causes us to depend on others for a relatively long period of time. On the other hand, it presents us with a great capacity for learning—to speak (or sign) innumerable languages, perform countless movements, retain and recall past experiences, and devise and use a seemingly unlimited number of objects. The biological process of learning, inasmuch as it is tied to the central nervous system, the cerebral cortex, and other physiological equipment, is similar, if not virtually identical, for everyone.

The fact that we are born with this great learning capacity suggests that we come into the world "unfinished," lacking the information and skills needed to meet the challenges of living. To fill this gap, the unfinished person must participate in social life (Hannerz 1992). Thus we can say that among the most significant of human biological needs is a need for social contact. Without such contact, a person cannot become a normally functioning human being. The rare instances of children raised in extreme isolation or in very restric-tive and unstimulating environments, such as the cases of Anna and Isabelle, clearly demonstrate the importance of social contact to normal development. Our strong dependence on interaction supports the view that for the individual, social contact "is a reality from which everything that matters to us flows" (Durkheim 1964:252). The need for social contact is perhaps the most important universal trait possessed by all humans.

The need for social contact indicates that we cannot speak of the individual as if he or she existed apart from others. In *Human Nature and the Social Order,* Charles Horton Cooley (1964) argues that "a separate individual is an abstraction unknown to experience" (p. 36). The few "completely" separate individuals we know about, such as Anna and Isabelle, did not possess the qualities that make people human when they were found.

It is obvious, then, that an individual is a product of two major sources: heredity and social interaction. The genes that parents transmit to their offspring reach back over an indefinite period of time through four biological grandparents, and eight biological great-grandparents, and beyond these individuals to a common ancestor. As a result, each of us has a unique biological heritage, yet this heritage is common to all humans. In addition, through contact with a unique combination of various others (parents, grandparents, baby-sitters, peers, and so on), individuals realize their human capacities.

Socialization goes beyond the needs of creating an individual with "human qualities." It functions to link people to one another in orderly and predictable ways. Without the benefits of social interaction, newcomers fail to thrive physically and to learn the skills needed to achieve meaningful connections with others. In addition, without meaningful contact between the generations, culture (solutions to the problems of living) cannot be passed from one generation to the next. Socialization can also be a process by which newcomers learn to think and behave in ways that reflect the interests of their teachers and to accept and fit into a system that benefits some groups more than others.

Our knowledge of the socialization processes suggests that life does not have to unfold in a seemingly predictable fashion. Human genetic and social makeup contain considerable potential for change.

First, people are not born with preconceived notions about standards of appearance and behavior. To develop standards, people must be exposed to information that leads them to expect people, behavior, and objects to be a certain way. Second, the cerebral cortex allows people to think reflexively—to step outside the self and observe and evaluate it from another viewpoint. Third, the mechanisms that teach prejudice and hatred for another group—such as imitation, play, and games—also teach respect and understanding. The problem is that these mechanisms often are used to teach children respect and understanding only after the children already have learned prejudice and hatred through the same pathways. Finally, people can be resocialized to abandon one way of thinking and behaving for another. Preferably, the socialization process should allow people to choose to abandon old habits and, in doing so, gain a feeling of competence and personal empowerment that comes with choice.

Each generation learns about the environment it inherits and comes to terms with it in unique ways. Table 5.1 gives some idea of the challenges facing Israel/Palestine in this regard. Although the older generations may share their personally acquired memories with later generations, these memories cannot affect the behavior of the younger generations in the same way. The continuous emergence of new generations "serves the necessary purpose of enabling us to forget" or at least brings a fresh perspective to a situation (Mannheim 1952:294).

As a case in point, consider Israel's Ministry of Education sponsoring of the project "My Israel," which encouraged fourth through eighth graders to draw their vision of Israel. One purpose of the project was to determine whether children's images of Israel as reflected in their drawings have changed since 1948, the year the state was founded. With the exception of Israeli children who live along the Israeli-Lebanon border, where armed conflict continues, the 1998 drawings included many fewer guns, tanks, and machine guns. These images have been replaced by Israel's flag, the peace dove, handshaking among Jews and Palestinians, and words of peace like *shalom* (Rowley 1998a).

This example, however, should not lead you to believe that members of older generations cannot change the way they think about and approach their environment. Assassinated Israeli Prime Minister Yitzhak Rabin and Palestinian President Yasir Arafat, lifelong bitter enemies, eventually recognized the other's right to exist. They did so in an exchange of letters before the peace accord ceremony in 1993:

Table 5.1	Future Generations' Population Age and Growth (estimated for 1995)

Perpetuating cultural and political identities depends on the size and socialization of future genera-tions. Suppose you saw your interests primarily in terms of the future of Israel as a Jewish homeland. What might these numbers suggest to you? Suppose you saw your interests primarily in terms of an Arab-Palestinian homeland. What might the numbers then suggest? Suppose you saw your interests in terms of long-term peace. What might the numbers suggest then?

	Gaza	West Bank	Israel
Fertility rate (children born per woman)	7.4	5.4	2.74
Infant mortality rate (deaths per 1,000 live births)	33.0	27.0	8.3
Population growth rate	6%	6%	2.01%
Life expectancy at birth (years)	71.8	71.8	78.2
Profile of ages			
0–14 years	51%	46%	28%
15–64 years	46%	50%	62%
65 and over	3%	4%	10%
Percentage in college	5%	4%	34%
Unemployement	31.6%	18.2%	6.5%

Sources: Palestine Central Bureau of Statistics (1997); Israeli Ministry of Foreign Affairs (1998a).

Mr. Prime Minister
The PLO recognizes the right of the State of Israel to exist in peace and security . . . [and] renounces the use of terrorism and other acts of violence. . . . (Sin-cerely, Yasir Arafat, Chairman, The Palestinian Liber-ation Organization)

Mr. Chairman . . .
The Government of Israel has decided to recognize the PLO as the representative of the Palestinian people. . . . (Yitzhak Rabin, Prime Minister of Israel)

Key Concepts

Use this outline to organize your review of the key chapter ideas.

Socialization
　Nature
　Nurture
Collective memory
　Engrams
Groups
　Primary
　Ingroups
　Outgroups
Reflexive thinking
Significant symbol

Symbolic gestures
Role taking
　Play
　　Significant others
　Games
　　Generalized others
Looking-glass self
Active adaptation
Resocialization
　Total institutions

internet assignment

In this chapter, we studied the concept "collective memory" to describe the experiences shared and recalled by significant numbers of people and considered how versions of past expe-riences are transmitted through stories, holidays, rituals, museums, and monuments to those who were not there. Use a search engine to find three or four Web sites related to muse-ums, monuments, or other memorials. What part of past expe-rience does the museum record? What messages are conveyed directly or indirectly about the past?

Social Interaction and the Social Construction of Reality

With Emphasis on the Democratic Republic of the Congo

A French hospital at the Croix Rouge in the Congo. (Patrick Robert.)

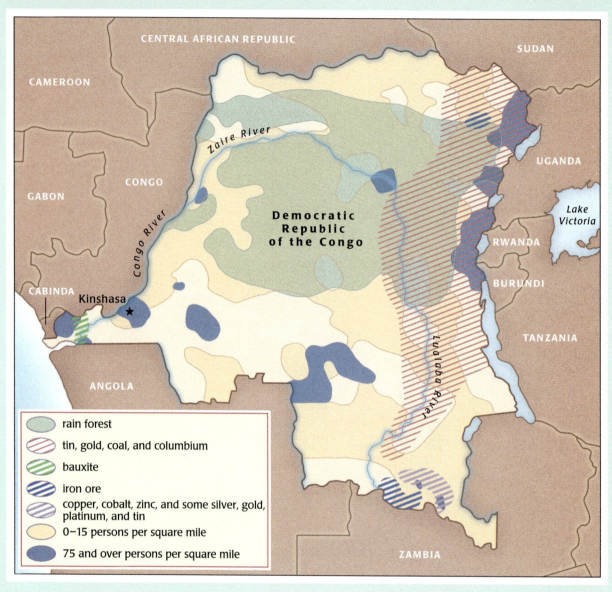

CENTRAL AFRICAN REPUBLIC

SUDAN

CAMEROON

CONGO

GABON

Zaire River

Congo River

UGANDA

Democratic
Republic
of the Congo

Lake
Victoria

CABINDA

Kinshasa

RWANDA

BURUNDI

Lualaba River

TANZANIA

ANGOLA

ZAMBIA

rain forest

tin, gold, coal, and columbium

bauxite

iron ore

copper, cobalt, zinc, and some silver, gold,
platinum, and tin

0–15 persons per square mile

75 and over persons per square mile

Source: Kurian (1992).

Roots of Interaction

This map of the Democratic Republic of the Congo helps explain why European countries sought to control Africa in the nineteenth and twentieth centuries. The Belgian effort to tap resources and centralize power in what is now known as the Congo led to 90,000 miles of road and significant river transport by the 1960s. By the mid-1990s, ruled by a dictator supported by the United States and other industrialized countries, the Congo had about 1,738 miles of passable roads. These facts provide a context for understanding broad patterns of social interaction within the Congo and between the Congo and other countries.

Margrethe Rask, a Danish surgeon, was exposed to the virus now known as human immunodeficiency virus, or HIV, while working in a small village clinic in an African country then known as Zaire[1] in 1977. The excerpt that follows highlights some of the final events in Dr. Rask's life, her last interactions with colleagues and close friends.

Grethe Rask gasped her short, sparse breaths from an oxygen bottle. . . . "I'd better go home to die," Grethe had told [her friend] Ib Bygbjerg matter-of-factly. The only thing her doctors could agree on was the woman's terminal prognosis. All else was mystery. Also newly returned from Africa, Bygbjerg pondered the compounding mysteries of Grethe's health. None of it made sense. In early 1977, it appeared that she might be getting better; at least the swelling in her lymph nodes had gone down, even as she became more fatigued. But she had continued working, finally taking a brief vacation in South Africa in early July.

Suddenly, she could not breathe. Terrified, Grethe flew to Copenhagen, sustained on the flight by bottled oxygen. For months now, the top medical specialists of Denmark had tested and studied the surgeon. None, however, could fathom why the woman should, for no apparent

reason, be dying. There was also the curious array of health problems that suddenly appeared. Her mouth became covered with yeast infections. Staph infections spread in her blood. Serum tests showed that something had gone awry in her immune system; her body lacked T-cells, the [essential parts of] the body's defensive line against disease. But biopsies showed she was not suffering from a lymph cancer that might explain not only the T-cell deficiency but her body's apparent inability to stave off infection. The doctors could only gravely tell her that she was suffering from progressive lung disease of unknown cause. And, yes, in answer to her blunt questions, she would die.

Finally, tired of the poking and endless testing by the Copenhagen doctors, Grethe Rask retreated to her cottage near Thisted. A local doctor fitted out her bedroom with oxygen bottles. Grethe's longtime female companion, who was a nurse in a nearby hospital, tended her. Grethe lay in the lonely white-washed farmhouse and remembered her years in Africa while the North Sea winds piled the first winter snows across Jutland.

In Copenhagen, Ib Bygbjerg, now at the State University Hospital, fretted continually about his friend. Certainly, there must be an

answer to the mysteries of her medical charts. Maybe if they ran more tests. . . . It could be some common tropical culprit they had overlooked, he argued. She would be cured, and they would all chuckle over how easily the problem had been solved when they sipped wine and ate goose on the Feast of the Hearts. Bygbjerg pleaded with the doctors, and the doctors pleaded with Grethe Rask, and reluctantly the wan surgeon returned to the old Rigshospitalet in Copenhagen for one last chance. On December 12, 1977, just twelve days before the Feast of the Hearts, Margrethe P. Rask died. She was forty-seven years old. (Shilts 1987:6–7)

Why Focus on the Democratic Republic of the Congo?

In this chapter, as we explore the sociological theories and concepts that sociologists use to analyze any social interaction in terms of context and content, we give particular focus to social interaction as it relates to the transmission of HIV and the treatment of AIDS. In doing so, we look closely at the central African country of the Democratic Republic of the Congo for two important reasons.

First, focusing on the Congo and its relationship to other countries helps us connect the transmission of HIV to a complex set of intercontinental, international, and intrasocietal interactions. Specifically, these interactions involve unprecedented levels of international and intercontinental air travel of the privileged for pleasure and business, as well as legal and illegal migrations of the underprivileged from villages to cities and from country to country (Sontag 1989).

Second, focusing on the Congo highlights evidence that HIV existed as early as 1959. This evidence takes the form of an unidentified blood sample frozen in that year and stored in a Congo blood bank. To date, it marks the earliest confirmed case of HIV infection (Balter 1998). Although this sample hardly proves that HIV originated in the central African country, this hypothesis has received considerable scientific and popular support in Western countries.

Whether the Congo is actually the country of origin of HIV is irrelevant to our purpose. Far more important is the idea that reality is a social construction. That is, people assign meaning to what is going on around them. The meaning assigned almost always emphasizes some aspect of reality and ignores others. For example, to say that HIV traveled from Africa to the United States ignores the possibility that it traveled from the United States to Africa. When we compare Western values and beliefs related to the origin of the virus and the treatment of AIDS with African values and beliefs, we realize that the Western framework is only one way of viewing the AIDS phenomenon. Moreover, by comparing the two frameworks, we can see that the meanings people give to an event have enormous consequences for the individuals involved (medical personnel, infected persons and those close to them, and noninfected persons). These meanings influence how people interact with one another and what decisions they make and actions they take to deal with HIV and AIDS.

[1]On May 18, 1997 the territory once known as Zaire became the Democratic Republic of the Congo (DRC). Before that time, this territory was called the Belgian Congo. Throughout this chapter we will refer to the DRC as the Congo. A second country bordering on the Congo bears a very similar name—the Republic of the Congo.

We can visualize some of the interactions between Grethe Rask and her friends and colleagues, presented in the opening quote. For example, we can visualize Dr. Rask telling her friend Ib Bygbjerg, "I'd better go home to die," or asking the Copenhagen doctors whether she would die after they tell her that she is "suffering from progressive lung disease of unknown cause." Finally, when Dr. Rask decides to leave the hospital and die at home, we can imagine Ib Bygbjerg pleading, "Please come back to the hospital for more tests; maybe there is still hope."

Sociologists looking at this situation would agree that Dr. Rask's illness is the focus of these **social interactions**—everyday events in which at least two people communicate and respond through language and symbolic gestures to affect one another's behavior and thinking. In the process, the parties involved define, interpret, and attach meaning to the encounter. Sociologists also assume that any social interaction reflects forces beyond any one individual's control. Hence, they strive to locate the interaction according to time (history) and place (culture).

When sociologists study social interaction, they seek to understand and explain the forces of context and content. **Context** consists of the larger historical circumstances that bring people together. **Content** includes the cultural frameworks (norms, values, beliefs, language, material culture) that guide behavior, dialogue, and interpretations of events. In the case of Dr. Rask, sociologists would place her interactions with others in context by asking what historical events brought Dr. Rask to Africa in the first place and what further events put her in direct contact with a deadly virus. To understand the content of her interactions, sociologists would ask how the parties involved are influenced

Social interactions Everyday events in which at least two people communicate and respond through language and symbolic gestures to affect one another's behavior and thinking.

Context The larger historical circumstances that bring people together.

Content The cultural frameworks (norms, values, beliefs, material culture) that guide behavior, dialogue, and interpretations of events.

What Is the Difference Between AIDS and HIV?

Acquired immunodeficiency syndrome (AIDS) is a fatal disease that severely compromises the human body's ability to fight infections and is caused by the human immunodeficiency virus (HIV). HIV infection can be transmitted by sexual intercourse between men and between men and women; by exposure to contaminated blood or blood products; by sharing or reusing contaminated needles; and during pregnancy, childbirth, and possibly breastfeeding, from mother to child. No evidence indicates that HIV infection is transmitted through casual contact, water, air, or insects.

Although persons infected with HIV may not show any clinical symptoms of AIDS for months or years, they may never become free of the virus and may infect others without realizing it. An individual is considered to have AIDS if a blood test indicates the presence of antibodies to the (HIV) virus and if he or she has one or more debilitating and potentially fatal cancers, neurological disorders, or bacterial, protozoal, or fungal infections that are characteristic of the syndrome.

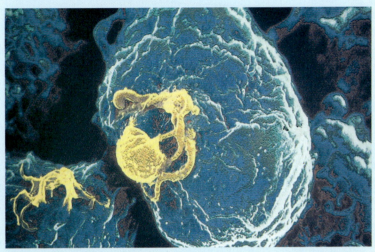

AIDS virus budding from a human lymphocyte, one of the white blood cells that form in lymphoid tissues.

© Bill Longcore/Science Source/Photo Researchers

Source: U.S. General Accounting Office (1987), p. 7.

---

by cultural frameworks as they strive to define, interpret, and respond meaningfully to her condition.

Continue to think about Dr. Rask as you read this chapter. As we explore issues of context and content, we will see that an individual's seemingly unique and personal interactions are affected by history and culture. We begin by exploring the context of Dr. Rask's social interactions—the historical events that originally brought her to Africa. If we can understand these larger, seemingly impersonal forces, we can understand more about the transmission of viruses in general and the transmission of HIV in particular (see "What Is the Difference Between AIDS and HIV?"). We will learn that Dr. Rask was caught up in the larger historical forces that resulted in the unprecedented mixing of the world's peoples and the emergence of worldwide economic interdependence.

The Context of Social Interaction

Emile Durkheim was among the first sociologists to provide insights into the social forces that contributed to the rise of a "global village." In *The Division of Labor in Society* ([1933] 1964), Durkheim gives us a general framework for understanding global interdependence. More specifically, his ideas provide a framework for understanding how the Congo was transformed, in less than 200 years, from a land with no overarching political structure governing 450 distinct yet interdependent nations, to a participant in the world economy. (A **nation** is a geographical area occupied by people who share a culture and a history; a **country** is a political entity, recognized by foreign governments, with a civilian and military bureaucracy to enforce its rules.)

According to Durkheim, an increase in population size and density increases the demand for resources (such as food, clothing, and shelter). This demand, in

Nation A geographical area occupied by people who share a culture and a history.

Country A political entity, recognized by foreign governments, with a civilian and military bureaucracy to enforce its rules.

turn, stimulates people to develop more efficient methods for producing goods and services. As population size and density increase, society "advances steadily towards powerful machines, towards great concentrations of [labor] forces and capital, and consequently to the extreme division of labor" (Durkheim [1933] 1964:39). As Durkheim describes it, **division of labor** refers to work that is broken down into specialized tasks, with each task being performed by a different set of persons trained to do that task. Not only are the tasks themselves specialized, but the parts and materials needed to manufacture products also come from many geographical regions of the world.

Faced with a growing demand for resources, Western European governments vigorously colonized much of Asia, Africa, and the Pacific in the late nineteenth and early twentieth centuries. Colonizers forced local populations to cultivate and harvest crops and extract minerals and ores for export. The Belgian government claimed territory in central Africa, named it the Belgian Congo, and forced the people living there to acquire ivory, extract rubber, and mine copper. As industrialization proceeded in Europe, so did the demand for various raw materials. Over time, the world grew to depend on the Congo as a source of copper, cobalt (needed to manufacture jet engines), industrial diamonds, zinc, silver, gold, manganese (needed to make steel and aluminum dry-cell batteries), and uranium (needed to generate atomic energy and fuel the atomic bomb). The worldwide division of labor now included the indigenous people of the Congo, who mined the raw materials needed for products in distant parts of the world.

Durkheim notes that as the division of labor becomes more specialized and as the sources of materials for products become more geographically diverse, a new kind of solidarity emerges. He uses the term **solidarity** to describe the ties that bind people to one another in a society. Durkheim refers to the solidarity that characterizes preindustrial society as *mechanical*

Division of labor Work that is broken down into specialized tasks, with each task being performed by a different set of persons.

Solidarity The ties that bind people to one another in a society.

Mechanical solidarity Social order and cohesion based on a common conscience or uniform thinking and behavior.

and the solidarity that characterizes industrial societies as *organic*.

Mechanical Solidarity

Mechanical solidarity consists of social order and cohesion based on a common conscience or uniform thinking and behavior. In this situation everyone views the world in much the same way. A person's "first duty is to resemble everybody else"—that is, "not to have anything personal about one's [core] beliefs and actions" (Durkheim [1933] 1964:396). Such uniformity derives from the simple division of labor and the corresponding lack of specialization.[2] A simple division of labor means that people are more alike than different. People are bound together because similarity in how they maintain their livelihood gives rise to common experiences, similar skills, and core beliefs, attitudes, and thoughts. In societies characterized by mechanical solidarity, the ties that bind individuals to one another are based primarily on kinship, religion, and a shared way of life.

As one of approximately 450 nations or ethnic groups in the Congo (French 1997a), each of which has a distinct language and belief system, the Mbuti pygmies, a hunting-and-gathering people who live in the Ituri Forest (an equatorial rain forest) of northeastern Congo, exhibit this type of solidarity. Their society represents one way of life that many people were forced to abandon after colonization began.

The Mbuti share a forest-oriented value system. Their common conscience derives from the fact that the forest gives them food, firewood, and materials for shelter and clothing. Anthropologist Colin Turnbull has written extensively about the Mbuti and their core value system in three books, *The Forest People* (1961), *Wayward Servants* (1965), and *The Human Cycle* (1983). Excerpts from these books show the extent to which forest-centered values permeate the lives of the Mbuti:

> For them the forest is sacred, it is the very source of their existence, of all goodness. . . . Young or old, male or female . . . the Mbuti talk, shout, whisper, and sing to the forest, addressing it as mother or father or both. (1983:30)

---

[2]A simple division of labor does not mean that no division of labor occurs. Perhaps the best way to illustrate the meaning of a simple division of labor is to contrast it with the division of labor in the United States today, which is considered complex. As one indicator of its complexity, consider that the U.S. Bureau of Labor Statistics (1998) lists 323 occupations in its 1998–1999 *Occupational Outlook Handbook,* ranging from account executives to zookeepers.

The Mbuti, a hunting-and-gathering people, exemplify Durkheim's concept of mechanical solidarity. The core of their "common conscience" is a value system centered on the forest, which provides them with all the necessities of life.

© Sarah Errington/Hutchison Library

It is not surprising that the Mbuti recognize their dependence upon the forest and refer to it as "Father" or "Mother" because as they say it gives them food, warmth, shelter, and clothing just like their parents. What is perhaps surprising is that the Mbuti say that the forest also, like their parents, gives them affection. . . . The forest is more than mere environment to the Mbuti. It is a living, conscious thing, both natural and supernatural, something that has to be depended upon, respected, trusted, obeyed, and loved. The love demanded of the Mbuti is no romanticism, and perhaps it might be better included under "respect." It is their world, and in return for their affection and trust it supplies them with all their needs. (1965:19)

Turnbull provides several examples of the intimacy between the Mbuti and the forest. In one instance, he came upon a youth dancing and singing by himself in the forest under the moonlight: "He was adorned with a forest flower in his hair and with for-

est leaves in his belt of vines and his loin cloth of forest bark. Alone with his inner world he danced and sang in evident ecstasy" (1983:32). When questioned as to why he was dancing alone, he answered, "'I am not dancing alone, I am dancing with the forest'" (1965:253).

In a second instance, Turnbull asked a Mbuti pygmy whether he would like to see a part of the world outside the forest. The pygmy hesitated a long time before asking how far they would venture beyond the forest. Not more than a day's drive from the last of the trees, replied Turnbull, to which the Mbuti pygmy responded with disbelief, "No trees? No trees at all?" He was highly disturbed about this prospect and asked whether this site was a good country. From the Mbuti perspective, people living without trees must be very bad to deserve that punishment. In the end, he agreed to go if they took enough food to last them until they returned to the forest. "He was going to have nothing to do with 'savages' who lived in a land without trees" (1961:248).

Finally, Turnbull notes that the pygmies are aware of the ongoing destruction of the rain forest by companies that push them farther into the forest's interior. By consensus, the pygmies do not wish to leave the forest and become part of the modern world: "The forest is our home; when we leave the forest, or when the forest dies, we shall die. We are the people of the forest" (1961:260).

Organic Solidarity

A society with a complex division of labor is characterized by **organic solidarity**—social order based on interdependence and cooperation among people performing a wide range of diverse and specialized tasks. A complex division of labor increases differences among people, in turn leading to a decrease in common conscience. Nevertheless, Durkheim argues, the ties that bind people to one another can be very strong. In societies characterized by a complex division of labor, these ties are no longer based on similarity and common conscience but rather on differences and interdependence. When the division of labor is complex and when the materials for products are geographically scattered, few individuals possess

Organic solidarity Social order based on interdependence and cooperation among people performing a wide range of diverse and specialized tasks.

Sociologist Emile Durkheim was particularly concerned about the kind of events that break down the ability of individuals to connect with one another in meaningful ways through their labor. Those events included war and massive layoffs. The war in Rwanda resulted in millions of people fleeing to the Congo. Massive layoffs in the United States increase the numbers of people in unemployment lines.

© B. Press/Panos Pictures (left); © R. Crandall/The Image Works (right)

the knowledge, skills, and materials to permit self-sufficiency. Consequently, people find that they must depend on others. Social ties remain strong because people need each other to survive.

Specialization and interdependence mean that every individual contributes a small part in creating a product or delivering a service. Because of specialization, relationships among people take on a transitory, limited, impersonal, and abstract character. That is, we relate to one another in terms of our specialized roles. We buy tires from a dealer; we interact with a sales clerk by telephone, computer, and fax; we fly from city to city in a matter of hours and are served by flight attendants; we pay a supermarket cashier for coffee; and we deal with a lab technician for only a few minutes when we give blood. We do not need to know these people personally to interact with them. Like-wise, we do not need to know that the rubber in the tires, the cobalt from which the jet engine is built, and the coffee we purchase come from the Congo. Similarly, we do not need to know whether the blood we give remains in the United States or is exported elsewhere.

When we interact in this manner, we can ignore personal differences and treat those who perform the same tasks as interchangeable. Yet, members of society "are united by ties which extend deeper and far beyond the short moments during which the exchange is made. Each of the functions that they exercise is, in a fixed way, dependent upon others. . . . [W]e are involved in a complex of obligations from which we have no right to free ourselves" (Durkheim [1933] 1964:227). In other

words, because everyone is dependent on everyone else, each individual has a stake in preserving the system as a whole.

A curious feature of organic solidarity is that although people live in a state of interdependence with others, they maintain little awareness of it, possibly because of the fleeting and impersonal nature of the relationships created. Because the ties with most of the people with whom we come in contact during a given day are largely instrumental (we interact with them for a specific reason) rather than emotional, we seem to live independently of one another.

Durkheim hypothesized that societies become more vulnerable as the division of labor becomes more complex and more specialized. He was particularly concerned with the kinds of events that break down individuals' ability to connect with one another in meaningful ways through their labor, a process we take for granted until something disrupts those connections. Such events include (1) industrial and commercial crises caused by such occurrences as plant closings, massive layoffs, crop failures, technological revolutions, and war; (2) workers' strikes; (3) job specialization, insofar as workers are so isolated that few people grasp the workings and consequences of the overall enterprise; (4) forced division of labor to such an extent that occupations are based on inherited traits (race, sex) rather than on ability, in which case the "lower" groups aspire to the positions that are closed to them and seek to dispossess those who occupy such positions; and (5) inefficient management and development of workers' talents and abilities, so that work for them is nonexistent, irregular, intermittent, or subject to high turnover.[3]

As one example of how disruptions to the division of labor interfere with individuals' ability to connect with one another in meaningful ways, consider the life of Joel Goddard after he is laid off from Ford, as profiled in David Halberstam's (1986) *The Reckoning*. Goddard is married and has two children. Halberstam chronicles the changes in Goddard's life that affect his ability to sustain meaningful connections in his life. For example, after Goddard loses his job, his daily contacts shift from colleagues at work to contacts with those at the unemployment office. Moreover, Goddard loses the structure to his life that comes with the routine of his job. Instead, he watches TV, fishes, or reads want ads. His ties are further disrupted when some of his friends from work move to Texas to find employment. Goddard eventually takes a job selling insurance but finds himself selling to his acquaintances. Eventually he quits, and then his wife decides to go to work. However, her success at work strains their marriage as Joel is reminded of his failures.

The Congo in Transition

From 1883 to the present, at least one of the five disruptive situations that Durkheim postulated has existed in the Congo. A brief summary of the Congo's history in the past 100 years shows the extent to which these disruptions to the division of labor have created a social order that severs the connections that people have to one another.

Belgian Imperialism (1883–1960)

Before 1883, inhabitants of what is now the Congo lived in villages characterized by common conscience and a simple division of labor. In 1883, however, King Leopold II of Belgium claimed the land as his private property, and millions of people were forced from their villages to build roads, work the land, and mine raw materials (forced division of labor). Leopold's personal hold over the land was formally legitimized in 1885 by leaders of 14 European countries attending the Berlin West Africa Conference. The purpose of this conference was to carve Africa into colonies and divide the continent's natural resources among competing colo-

King Leopold II of Belgium, and then the Belgian government, claimed the territory now known as the Democratic Republic of the Congo, exploiting the people and resources.

© Eric A. Wessman/Stock, Boston

nial powers (Witte 1992). The continent was divided without regard to preexisting national boundaries, so that friendly nations were split apart and hostile nations were thrown together (see Figure 6.1).

For 23 years, Leopold capitalized on the world's growing demand for rubber. His reign over the Congo was the "vilest scramble for loot that ever disfigured the history of human conscience and geographical location" (Conrad 1971:118). The methods he used to extract rubber for his own personal gain involved atrocities so ghastly that in 1908 international outrage forced the Belgian government to assume administration of the Belgian Congo (see "The Essay That Mark Twain Could Not Get Published").

The Belgian government operated more humanely than Leopold, but it too forced the indigenous peoples to build roads so that minerals and crops could be transported from mines and fields across the country for export. Africans were forced to leave their villages to work the mines, cultivate and harvest the crops, and live alongside the roads and maintain them. Keep in mind that

> [t]he roads were built by Africans. The roads were not built for Africans. Their purpose was to provide the Europeans with easy access into the interior in order to maintain order, increase trade, and get raw materials out. Africans were not supposed to travel except on official business. They had to have a permit to be on the road, an explanation of where they

---

[3]As an example, a country might not develop enough workers (teachers, scientists, nurses) or too many workers (athletes and entertainers) for available positions, or it might fail to retrain people whose positions are vulnerable to layoff or obsolescence.

Figure 6.1   Colonial Africa in 1913

The national heritage today of many African states is an outward appearance of administrative unity for the nation as a whole, contrasted with complex internal interactions based on many competing ethnic identities and differing histories, languages, and customs. According to this map, how much of Africa was under colonial claim in 1913? What meanings do you think the colonial boundaries had for the European powers that drew them on the maps? What meanings might they have had for the people actually living in these regions?

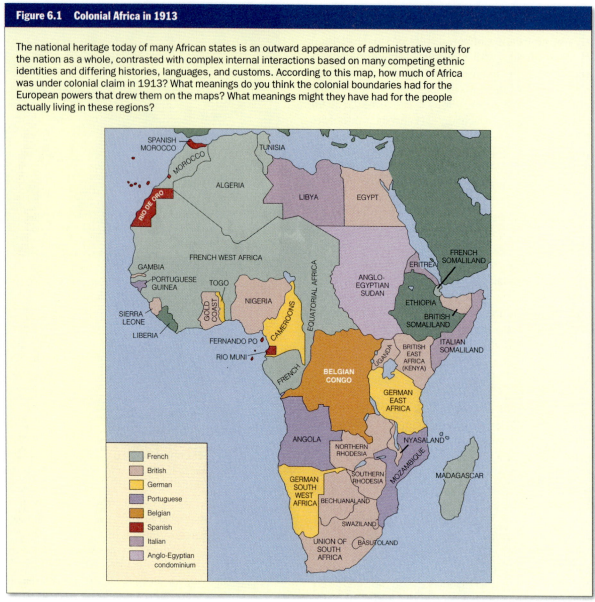

Source: *Times Atlas of World History* (1984).

were going and why, written by a white man. The roads were for Europeans. (Mark 1995:220)

The Belgians introduced a cash economy, imported goods from Europe that eventually became essential to native life, established a government, and built schools, hospitals, and roads. Under this system, the African people could acquire cash in one of two ways: growing cash crops or selling their labor. They were no longer allowed to be self-sufficient, as they had been prior to European colonization.[4] In addition, under European colonization the African people were denied access to most

professional-level and high-skilled occupations (forced division of labor; inefficient management and development of worker's talents).

[4]In June 1997, Nicholas Kristof, a reporter for *The New York Times*, described how even today the pygmies are at a severe disadvantage as most are not part of the Congo's cash economy. As the world closes in on them (for example, timber companies, poachers with guns who deplete the game), the pygmies' hunting-and-gathering way of life is disappearing. At the same time, it is difficult for them to integrate into the societies outside the forest. For one, they cannot attend Congo schools because they cannot pay the fees as the hunting-and-gathering way of life earns them no cash income.

U.S. in Perspective

The Essay That Mark Twain Could Not Get Published

Most people associate Mark Twain with the novels *Huckleberry Finn* and *Tom Sawyer*. They do not think of him as an avid critic of American and European imperialism who wrote essays and pamphlets expressing his outrage at the "Great Powers for the way they exercised their 'unwilling' missions in South Africa, China, and the Philippines" (Meltzer 1960:256). When the Congo Reform Association approached Twain in early 1905 to "lend his voice 'for the cause of the Congo natives'" (p. 257), Twain responded by writing "King Leopold's Soliloquy." In the essay, he presents a report filed by the Reverend H. E. Scrivener, a British missionary, regarding the plight of the people of the Belgian Congo under King Leopold's rule. Since no magazine editor in the United States would agree to publish this essay, Congo reform groups issued it as a pamphlet and sold it for 25 cents, with proceeds going toward the relief of the Congo people. The following is excerpted from that report.

Soon we began talking, and without any encouragement on my part the natives began the tales I had become so accustomed to. They were living in peace and quietness when the white men came in from the lake with all sorts of requests to do this and that, and they thought it meant slavery. So they attempted to keep the white men out of their country but without avail. The rifles were too much for them. So they submitted and made up their minds to do the best they could under the altered circumstances.

First came the command to build houses for the soldiers, and this was done without a murmur. Then they had to feed the soldiers and all the men and women—

hangers on—who accompanied them. Then they were told to bring in rubber. This was quite a new thing for them to do. There was rubber in the forest several days away from their home, but that it was worth anything was news to them. A small reward was offered and a rush was made for the rubber. "What strange white men, to give us cloth and beads for the sap of a wild vine." They rejoiced in what they thought their good fortune. But soon the reward was reduced until at last they were told to bring in the rubber for nothing. To this they tried to demur; but to their great surprise several were shot by the soldiers, and the rest were told, with many curses and blows, to go at once or more would be killed. Terrified, they began to prepare their food for the fortnight's absence from the village which the collection of rubber entailed. The soldiers discovered them sitting about. "What, not gone yet?" Bang! bang! bang! and down fell one and another, dead, in the midst of wives and companions. There is a terrible wail and an attempt made to prepare the dead for burial, but this is not allowed. All must go at once to the forest. Without food? Yes, without food. And off the poor wretches had to go without even their tinder boxes to make fires. Many died in the forests of hunger and exposure, and still more from the rifles of the ferocious soldiers in charge of the post. In spite of all their efforts the amount fell off and more and more were killed. I was shown around the place, and the sites of former big chiefs' settlements were pointed out. A careful estimate made the population of, say, seven years ago, to be 2,000 people in and about the post, within a radius of, say, a quarter of a mile. All told, they would not muster 200 now, and there is so much sadness and gloom about them that they are fast decreasing.

We stayed there all day on Monday and had many talks with the people. On the Sunday some of the boys had told me of some bones which they had seen, so on the Monday I asked to be shown these bones. Lying about on the grass, within a few yards of the house I was occupying, were numbers of human skulls, bones, in some cases complete skeletons. I counted thirty-six skulls, and saw many sets of bones from which the skulls were missing. I called one of the men and asked the meaning of it. "When the rubber palaver began," said he, "the soldiers shot so many we grew tired of burying, and very often we were not allowed to bury; and so just dragged the bodies out into the grass and left them. There are hundreds all around if you would like to see them." But I had seen more than enough, and was sickened by the stories that came from men and women alike of the awful time they had passed through. The Bulgarian atrocities might be considered as mildness itself when compared with what was done here. How the people submitted I don't know, and even now I wonder as I think of their patience. That some of them managed to run away is some cause for thankfulness. I stayed there two days and the one thing that impressed itself upon me was the collection of rubber. I saw long files of men come in, as at Bongo, with their little baskets under their arms; saw them paid their milk tin full of salt, and the two yards of calico flung to the headmen; saw their trembling timidity, and in fact a great deal that all went to prove the state of terrorism that exists and the virtual slavery in which the people are held.

Source: From "King Leopold's Soliloquy on the Belgian Congo," in *Mark Twain and the Three R's*, by Mark Twain, pp. 47–48. Copyright © 1973 by Maxwell Geismar. Reprinted with permission of International Publishers.

The introduction of European goods and a cash economy pulled the people who inhabited the Belgian Congo into a worldwide division of labor and created a migrant labor system within the country. Since colonization, people have moved continuously from the villages to the cities, mining camps, and plantations (Watson 1970). In addition, family members have endured prolonged separations as a result of the migrant labor system.

> The migrant labor system affected Africans' lives in many fundamental ways. . . . [M]ale workers were typically recruited from designated labour supply areas great distances from the centers of economic activity. This entailed prolonged family separations which had serious physical and psychological repercussions for all concerned. The populations of African towns "recruited by migration" were characterized by a heavy preponderance of men living in intolerably insecure and depressing conditions and lacking the benefits of family life or other customary supports. (Doyal 1981:114)

The Belgians did not anticipate the Africans' anger about the exploitation of their land, minerals, and people and were not prepared for the violent confrontations that took place in the late 1950s. The Belgians in power termed the revolutions "savage" and claimed that the Africans did not appreciate the "benefits" of colonialism. When the Belgians pulled out suddenly in 1960, the Belgian Congo became an independent country without trained military officers, businesspeople, teachers, doctors, or civil servants. In fact, only 120 medical doctors served a country of 33 million people (Fox 1988).

Independence of Zaire (1960–Present)

In the vacuum left by the Belgians, the various African ethnic groups that had been forced together to form the Belgian Congo now fought one another to obtain power (industrial and commercial crises). Civil wars raged until 1965, when Sese Mobutu took power with the support of the U.S. Central Intelligence Agency (CIA) and white mercenaries and renamed the country Zaire (Gourevitch 1998). After Mobutu seized control, several power struggles occurred between various nations within Zaire, especially in the late 1970s. To stop the rebellions, Mobutu called on mercenary forces from Morocco, Belgium, and France. To make up for the lack of skilled personnel, Mobutu invited French, Danes, Haitians, Portuguese, Greeks, Arabs, Lebanese, Pakistanis, and Indians to work as civil servants, teachers, doctors, traders, businesspeople, and researchers.

It was through this invitation that Dr. Rask arrived in Zaire.

In addition to problems posed by civil war, several other major problems made life difficult for the people of Zaire after independence. First, many cash crops were priced out of competition in a growing world economy, and a technological revolution in synthetic products such as plastics reduced the demand for African raw materials (industrial and commercial crises). Second, civil wars raged in neighboring countries and have caused hundreds of thousands of refugees from Sudan, Rwanda, Angola, Uganda, the Congo Republic, and Burundi to flee to the Congo. Meanwhile, the Congolese suffering from their own civil wars and economic problems sought refuge in those same countries (Brooke 1988; U.S. Bureau for Refugee Programs 1988; U.S. Central Intelligence Agency 1995). Finally, during the 32 years Mobutu remained in power, he diverted much of the Congo's wealth to European banks and invested it in property outside Africa. Some people estimate that Mobutu's personal fortune is worth between $5 billion and $8 billion.

On May 18, 1997, rebel leader Laurent Kabila declared himself president and renamed Zaire the Democratic Republic of the Congo. His rise to power was supported by the Alliances of Democratic Forces for the Liberation of the Congo, a rebel militia supported primarily by Congolese Tutsi and by the governments of Rwanda, Uganda, and Angola (French 1997b, 1997c). The Congolese Tutsi and Rwandian army, which was also composed of Tutsi troops, supported Kabola because helping him seize power allowed them to fight Rwandan Hutu guerrillas, who were based in refugee camps in the Congo (McKinley 1997).[5]

The local populations have suffered greatly and are still suffering from the ongoing massive upheaval in the Congo. These events disrupted the division of labor, and people lost important social connections to one another. Such "change of existence, whether it be sudden or prepared, always brings forth a painful crisis" (Durkheim [1933] 1964:241). Out of economic necessity, a desire for a higher standard of living, and a need to escape war, many people left their villages for the cities and for industrial, plantation, and mining sites.

---

[5]In 1994, ethnic conflicts between the Hutu and Tutsi in Rwanda resulted in the massacre of 500,000 Tutsi by Hutu hardliners. Tutsi troops retaliated and pursued innocent and guilty Hutus fleeing to refugee camps in the Congo (French 1998).

Women, children, and the elderly left behind in the villages have had little choice but to change to higher-yield and less labor-intensive crops to survive. Unfortunately, the new crops, such as cassava, are low in protein and high in carbohydrates. (Low-protein diets compromise the human immune system, making people more vulnerable to infection.) To further complicate matters, significant numbers of single women with no means of supporting themselves in the villages and rural areas migrated to labor sites in search of employment. Because few employment opportunities existed for women at the labor sites, many survived only by entering into prostitution (inefficient management and development of workers' talent). In the meantime, when the men and women who had migrated out of the villages to find employment became sick and could no longer work, they returned home to their villages and infected an already vulnerable population with whatever diseases they carried (Hunt 1989).

Major dislocations in the social structure, such as those described here, clearly put stress on the ecological system and alter the equilibrium between people and microbes. They can therefore lead to plagues and epidemics (Krause 1993:xii).

The magnitude of these migrations is reflected in the population increase of Kinshasa, the capital city, which grew from 390,000 in 1950 to 5.0 million in 1993. A large portion of its population (almost 60 percent) lives in squatter slums, the largest of which is named the Cite. Here the "streets [are] stuffed with children and families living under cardboard roofs held down by rocks. Kinshasa [is] a wasteland of flooded streets and cracked sidewalks, smoldering garbage and bars catering to whores and lonely white men. There [is] an end-of-civilization atmosphere, with survivors finding shelter in the rubble" (Clarke 1988:175, 178).

As a result of this upheaval and mismanagement, Zaire fell from its position as one of the wealthiest colonies in Africa to become the poorest independent country and one of the 12 poorest countries in the world. Its gross national product per capita is an estimated $400 (U.S. Central Intelligence Agency 1998). As another indicator of how dire the situation has become, in 1960 Zaire had 90,000 miles of passable highway; by 1988 that number had dwindled to only 6,000 miles. According to the U.S. Central Intelligence Agency (1995), 1,738 miles of paved roads are found in Zaire today. Obviously this poor road system impedes the transportation of medical supplies, food, and fertilizer to villages that have become dependent on these commodities (Noble 1992).

These historical events are the contextual forces that brought Dr. Margrethe Rask to a small village clinic in what was then Zaire. Amid such chaos and poverty, Dr. Rask had to perform operations on less than a shoe-string budget, with only minimal supplies. "Even a favored clinic would never have such basics as sterile rubber gloves or disposable needles. You just used needles again and again until they wore out; once gloves had worn through you risked dipping your hands in your patient's blood because that was what needed to be done" (Shilts 1987:4).

The importance of considering the Congo's history and global connections so as to understand the origins of AIDS is supported by the fact that disease patterns historically are affected by changes in population density and transportation patterns (McNeill 1976). Leading AIDS researchers believe that the transmission of HIV is indeed linked to changes in population density and transportation. Interestingly, "the medical condition which was later to be called AIDS began to be noticed in the late 1970s and early 1980s in several widely separated locations, including Belgium, France, Haiti, the United States, Zaire, and Zambia" (Panos Institute 1989:72). Recall that all of these countries have some historical connections with the Congo (see "Regional Statistics: Adults and Children Living with HIV/AIDS").

In view of this information about the global context of interaction, it is difficult to say who is responsible for triggering and transmitting the virus that causes AIDS. Clearly, "the foreigners introduced hitherto unknown diseases and probably aggravated some previously endemic diseases to epidemic proportions by the facilitation of transportation, forced migration of rural populations to work sites, and the creation of congested cities" (Lasker 1977:280). For example, the destruction of the Ituri rain forest in Zaire by companies from around the world brings developers into contact with previously isolated populations[6] (such as the Mbuti) and forces many people to migrate to the city because they have lost their homes. In addition to disrupting people's lives, the commercial activities in the rain forest cause climatic changes (such as the greenhouse effect) that can alter the structure of

---

[6]Before HIV was discovered, it may have survived in a dormant state in an isolated population with a tolerance to the virus but then became activated when this population came in contact with another population with no tolerance. Alternatively, two harmless retroviruses, each existing in a previously isolated population, may have interacted to produce a third, lethal virus.

Regional Statistics: Adults and Children Living with HIV/AIDS

The number of AIDS cases worldwide by region are shown in the map. The accompanying chart shows when the epidemic started and the main modes of transmission. Why do you think "MSM" is not considered a main mode of transmission in some regions of the world?

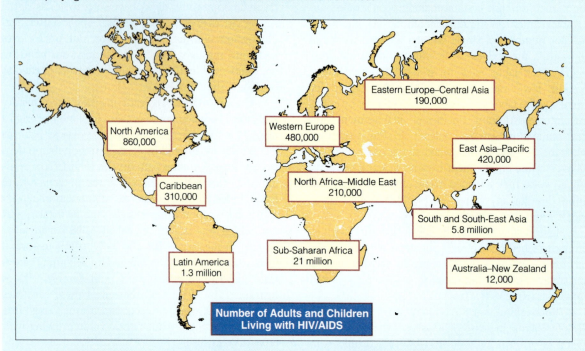

Region	Epidemic Started	Main Mode(s) of Transmission for Those Living with HIV/AIDS
Sub-Saharan Africa	Late 1970s–early 1980s	Heterosexual contact
South and South-East Asia	Late 1980s	Heterosexual contact; injecting drug use
Latin America	Late 1970s–early 1980s	MSM*; injecting drug use; heterosexual contact
North America	Late 1970s–early 1980s	MSM*; injecting drug use; heterosexual contact
Western Europe	Late 1970s–early 1980s	MSM*; injecting drug use; heterosexual contact
Australia–New Zealand	Late 1970s–early 1980s	MSM*; injecting drug use; heterosexual contact
Caribbean	Late 1970s–early 1980s	Heterosexual contact
Eastern Europe–Central Asia	Early 1990s	Injecting drug use; MSM*
East Asia–Pacific	Late 1980s	Injecting drug use; heterosexual contact; MSM*
North Africa–Middle East	Late 1980s	Injecting drug use; heterosexual contact

*Instead of the term "homosexual," the United Nations and other agencies use "MSM," meaning "men who have sex with men," because not all men who have ever had sex with men identify themselves as homosexual.

Source: UNAIDS/ World Health Organization (1998b).

viruses. What is important is not to determine who started the transmission (because it is impossible to ascertain), but to recognize the extent to which the world's people are interacting with one another and to become aware that the actions of one group can affect other groups.

Placing Dr. Rask's interactions in this global context helps us see how historical events bring people into interaction with one another. The opportunity for Dr. Rask to practice medicine in Zaire, the circumstances that placed her in direct contact with patients' blood, and her subsequent illness arose from a unique sequence of historical events. In other words, Dr. Rask was merely one individual caught up in the unprecedented mixing of people from all over the world. The import and export of blood (which brings people into contact with one another in indirect ways),[7,8] the development of wide-body jet aircraft, and large-scale migrations are cited as events that increased opportunities for large numbers of people from different countries and from different regions of the same country to interact with one another and to transmit the HIV infection through unprotected sexual intercourse, needle sharing, and other activities that involve blood and blood products (De Cock and McCormick 1988).

To this point in the chapter we have used the case of Dr. Rask to consider how context or larger historical circumstances shape social interaction. Dr. Rask and the patients in African villages were brought together because of Belgium's historical connection to the Congo. She contracted HIV because her clinic lacked the money to buy the most basic medical supplies to protect her from direct contact with her patients' blood and other bodily fluids. The severe budgetary constraints are clearly the result of the chaos and upheaval associated with colonization and subsequent independence. In the second part of this chapter we consider the forces shaping the content of Dr. Rask's interactions as she interacted with physicians and friends and came to terms with her illness.

The Content of Social Interaction

When people interact, they identify the social status of the people with whom they interact. Once they determine another person's status in relation to their own, people proceed to interact on the basis of role expectations.

Durkheim argued that, as the division of labor becomes more specialized and the sources of labor and

When we think about how HIV/AIDS has become a global problem, we must consider the forces that bring people from different geographical locations together. As one example, this photo shows Congolese soldiers—mostly Hutus from Rwanda—being trained in hand-to-hand combat by military instructors from Serbia. In 1997, President Mobutu relied on Serbs to prepare his soldiers to defend his government against rebel attacks.

© Grossman/Sygma

raw materials become more geographically diverse, the ties that connect people to one another shift from mechanical to organic; the number of interactions people have with strangers also increases. How can people interact smoothly with individuals about whom they know nothing? They eliminate "strangeness" by identifying the social positions or social status of these strangers. Knowing a person's social status gives us some idea of the behaviors we can expect from him or

Zaire is considered to be one of the 12 poorest countries in the world. The poverty is exacerbated by a transportation system incapable of delivering medical supplies and goods to those in need.

© Robert Caputo/Aurora

[7]An Associate Medical Director of an Iowa blood bank noted, "Blood transfusion is the most promiscuous thing that human beings do. It's even more promiscuous than having sex with people we haven't selected well."

[8]As one indicator of the amount of "indirect" interaction between blood donors and receivers, consider that the blood industry collects 26 million blood donations each year (U.S. General Accounting Office 1997b). Consider also that the American Red Cross, the largest blood supplier in the United States, collects approximately 5.7 million blood donations annually. The Red Cross also maintains "the world's largest registry of rare blood donors and maintains a frozen supply of rare blood available for immediate shipment across the globe" (American National Red Cross 1996). In addition, the Red Cross provides the human tissue used in approximately 25 percent of transplantation surgeries.

her. It also affects how we will interact with that person. To grasp this principle, think about what happens when you meet someone for the first time. How do you start the interaction? You ask the stranger a question to determine his or her social status. You might ask, "What do you do?" That question sets the interaction into motion.

Social Status

In everyday language, people use the term *social status* to mean rank or prestige. To sociologists, **social status** refers to a position in a social structure. A **social structure** consists of two or more people interacting and interrelating in specific expected ways, regardless of the unique personalities involved. For example, a social structure can consist of the two statuses of doctor and patient and their relationship. Other familiar examples of statuses in a two-person social structure include husband and wife, professor and student, sister and brother, and employer and employee.

Examples of multiple-status social structures are a family, an athletic team, a school, a large corporation, and a government. Again, the common characteristic of all social structures is that it is possible to generalize about the behavior of people in each of the statuses, no matter who occupies them. Just as the behavior of a person occupying the status of a football quarterback is broadly predictable, so too is the behavior of a person occupying the status of nurse, secretary, mechanic, patient, or physician. Once we know a person's status, we have enough information to interact with him or her.

Every person occupies a number of statuses. For example, Dr. Rask was middle-aged, a female, a physician, a patient, and Danish. A status has meaning, however, only in relation to other statuses. For instance, the

status of a physician takes on quite different meanings depending on whether the interaction occurs with someone who occupies the same status or with a person of a different status, such as patient, spouse, or nurse. That is, a physician's role varies according to the status of the person with whom he or she interacts.

Social Roles

Sociologists use the term **role** to describe the behavior expected of a status in relationship to another status (for example, professor to student). The distinction between role and status is subtle but noteworthy: people *occupy* statuses and *enact* roles.

Associated with every status is a **role set,** or an array of roles. For example, the status of physician entails, among other roles, the role of physician in relationship to patients, nurses, other doctors, and patients' family members. The sociological significance of statuses and roles is that they make it possible for us to interact with other people without knowing them. Once we determine another person's status in relation to our own, we interact on the basis of role expectations attached to that status relationship.

Role expectations are socially prescribed and include both rights and obligations. The **rights** associated with a role define what a person assuming that role can demand or expect from others. For example, professors have the right to demand and expect that students be prepared for class. Students have the right to demand and expect professors to follow their syllabi. The **obligations** associated with a role define the appropriate relationship and behavior expected of a person assuming that role. For instance, professors have an obligation to their students to come to class prepared and to not keep students past the end of a class period. Students have an obligation to not pack up books and notebooks before class officially ends (Appleby 1990). As these examples make clear, professors' rights correspond with students' obligations, and vice versa.

Although roles set general limits on how we think and act, we should not expect behavior to be totally predictable. Sometimes people do not meet their role obligations, as when professors come to class unprepared, students sleep during class, or patients do not cooperate with their physicians' treatment plans. By definition, when people fail to meet their role obligations, other people are not accorded their role rights. When professors are unprepared, students' rights are violated; when patients do not follow treatment plans, physicians' rights are violated. Moreover, the idea of role does not imply that all people occupying the same

Social status A position in a social structure.

Social structure Two or more people interacting and interrelating in specific expected ways, regardless of the unique personalities involved.

Role The behavior expected of a status in relationship to another status.

Role set An array of roles.

Rights The behaviors that a person assuming a role can demand or expect from others.

Obligations The relationship and behavior that the person enacting a role must assume toward others in a particular status.

status enact the roles of that status in exactly the same way. Roles are enacted differently because of the existence of individual personalities and diverse interpretations of how the role should be carried out. Finally, roles are enacted differently because people resolve role strain and role conflict differently.

Role strain is a predicament in which contradictory or conflicting expectations are associated with a single role that a person is enacting. For example, military doctors, as physicians, have an obligation to preserve life and to "first do no harm." At the same time, they are employed to care for soldiers who are deliberately placed in situations that threaten their health and lives. **Role conflict** is a predicament in which the expectations associated with *two or more roles* in a role set contradict one another. For example, a person in a sick role has an obligation to want to get better and comply with treatment plans. This obligation, however, can interfere with a second role that the person holds, as in the case of a woman who finds that the side effects of a prescribed drug (such as drowsiness) prevent her from carrying out her obligations in her role as mother if she is not alert enough to properly care for her children.

Socially prescribed rights and obligations notwithstanding, room always exists to exercise improvisation and personal style. Yet, despite variations in how people enact roles, role is still a useful concept because, for the most part, a predictability generally exists, "sufficient to enable most of the people, most of the time to go about [the] business of social life without having to improvise judgments anew in each newly confronted situation" (Merton 1957:370). In other words, these variations usually fall within "a certain range of culturally acceptable behavior—if the performance of a role deviates very much from the expected range of behavior, the individual will be negatively sanctioned" (p. 370).

CULTURAL VARIATIONS IN ROLES: THE PATIENT-PHYSICIAN INTERACTION   The behaviors expected of one status in relation to another status vary across cultures. Role expectations are intertwined with norms, values, beliefs, and nonmaterial culture. In the United States, for example, the major objective of the patient-physician interaction is to determine the exact physiological malfunction and to use the available material culture and technology (tests, equipment, machines, drugs, surgery, transfusions) to treat it.

This objective is shaped by a profound cultural belief in the ability of science to solve problems. Practitioners of Western medicine are expected to use all

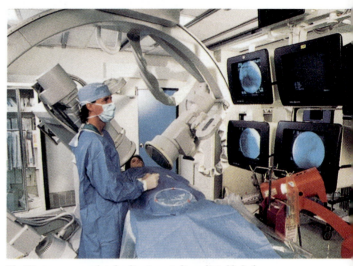

The Western approach to healing is rooted in cultural beliefs about the nature of disease and the power of science and technology to combat it.

© Bob Daemmrich/Stock, Boston

available tools of science to establish the cause, combat the disease for as long as possible, and return the body to a healthy state. When physicians violate this norm, their actions are almost always accompanied by intense public debate.

In view of this cultural orientation, it is not surprising that U.S. physicians and their patients rely heavily on technological elements such as X-rays and CAT scans to diagnose conditions and on vaccines, drugs, and surgery to cure them. This reliance is reflected in the fact that the United States, with a population of 2–5 percent of the world's population, consumes an estimated 23 percent of the world's pharmaceutical supply (Peretz 1984). Given this emphasis, it is also not surprising that tremendous effort is devoted to finding a technological solution to the AIDS problem. The technological emphasis on treatment is reflected in the number of antiretroviral agents and medications for prophylaxis and symptom management consumed by HIV patients. In one study, U.S. Centers for Disease Control and Prevention (CDC)

Role strain A predicament in which contradictory or conflicting expectations are associated with a person's role.

Role conflict A predicament in which the expectations associated with two or more roles in a role set contradict one another.

Alternative Medicine in the United States

Type of Therapy	Percentage of People Who Used Therapy in the Last 12 Months (%)
Relaxation techniques	13
Chiropractic	10
Massage	7
Imagery	4
Spiritual healing	4
Commercial weight-loss program	4
Lifestyle diets (for example, macrobiotics)	4
Herbal medicine	3
Megavitamin therapy	2
Self-help groups	2
Energy healing	1
Biofeedback	1
Hypnosis	1
Homeopathy	less than 1
Acupuncture	less than 1
Total using some form of unconventional therapy (figure does not include respondents who used exercise and prayer)	**34**

Source: Eisenberg et al. (1993).

researchers studied 474 patients taking medications for HIV and found that "the average number of pills prescribed per day was 13.4 with a range of 2 pills (4 patients) to 60 (1 patient)" (Von Bargen, Moorman, and Holmberg 1998).

We can contrast the U.S. physician-patient relationship with the traditional African[9] healer-patient relationship. Although the Congo has modern health care facilities, the majority of people prefer to consult traditional healers.[10] The social interaction between the healer and patient is very different from the U.S. physician-patient social interaction.

Just as Western physicians do, traditional healers recognize and treat[11] the organic and physical aspects of disease. But they also attach considerable importance to other factors—supernatural causes, social relationships (hostilities, stress, family strain), and psychological distress. This holistic perspective allows for a more personal relationship between healer and patient. Another significant difference from Western medicine is that healers concentrate on providing symptom relief instead of searching for a total cure. African healers, for example, focus on treating the debilitating symptoms of AIDS such as diarrhea, headaches, fevers, and weight loss, and they employ remedies with few side effects so as not to make people feel sicker (Hilts 1988). When traditional methods fail, Africans may make a trip to a hospital or clinic to consult with a doctor of Western medicine.

Obviously, each system has its attractions and its weaknesses, as evidenced by the fact that Westerners

[9]Unfortunately, many Americans associate the phrase "traditional African healer" with witch doctors. An interesting interview with U.S.-educated Nigerian psychiatrist Thomas Adeoye Lambo appeared in *Omni* magazine in 1990. The interviewer, Thomas Bass (1992), asked Lambo, "What's your opinion of the term 'witch doctor'?" Lambo replied, "It is a derogatory term coined by missionaries. When I went to mission school, every Sunday we were sent into the villages to collect all the idols and carved objects that now fetch millions of dollars at Christie's auction house. We'd pile them in the middle of the village and burn them. This was part of our mission to convert the savage to Christianity. But just as there's no one single religion, there's no one single way to practice medicine. I could call quite a number of modern physicians 'witch doctors,' such as the ones who do 'exploratory' surgery so they can hand you a heavy bill" (p. 72).

[10]In the Congo, the chief "physicians" are traditional healers. Only a small proportion are like the witch doctors of legend. (In fact, many wear business suits.) Traditional healers operate from a variety of beliefs and practices, but most approach health care from a holistic perspective.

[11]U.S.-educated Nigerian psychiatrist Thomas Adeoye Lambo (quoted in Bass 1992) reminds us that "the Masai were suturing blood vessels, removing appendixes and practicing other sophisticated surgical techniques long before the British" (p. 96).

suffering from incurable diseases sometimes turn to alternative treatments and medicines.[12] In one random national survey of 1,539 U.S. adults, researchers found that 34 percent of respondents had used some form of alternative medicine in the past 12 months (Olmos 1993; see "Alternative Medicine in the United States").[13]

Sociologist Ruth Kornfield observed Western-trained physicians working in the Congo's urban hospitals and found that success in treating patients was linked to the foreign physician's ability to tolerate and respect other models of illness and include them in a treatment plan. For example, among some ethnic groups in the Congo, when a person becomes ill, the patient's kin form a therapy management group and make decisions about administering treatments. Because many people in the Congo believe that illnesses are caused by disturbances in social relationships, the cure must involve a "reorganization of the social relations of the sick person that [is] satisfactory for those involved" (Kornfield 1986:369; also see Kaptchuk and Croucher 1986:106–108).

The point of this discussion is not to evaluate the quality or outcome of either culture's patient-practitioner interaction. Rather, by comparing the two cultures, we can see more clearly that people think and behave in largely automatic ways because they are influenced by norms, values, beliefs, and nonmaterial culture. Those involved with Dr. Rask automatically assumed a scientific framework to define the origin and treatment of her condition. Such a conceptual framework defines illness as "a state of disease and dysfunction 'impersonally' caused by microorganisms, inborn metabolic disturbances, or physical or psychic stress" (Fox 1988:505). On this basis, we can begin to understand why Dr. Rask, other physicians, and her close friends all defined her illness as a condition contracted in Africa, requiring hospitalization, having biological origins, and being related to direct contact with a patient's blood. The interactions and dialogue among Dr. Rask, other medical personnel, and friends would

have been quite different if they had assumed a "non-scientific" model of illness.

The Dramaturgical Model of Social Interaction

A number of sociologists have compared roles with the dramatic roles played by actors. Erving Goffman is a sociologist associated with the **dramaturgical model** of social interaction. In this model, social interaction is viewed as though it were theater, people as though they were actors, and roles as though they were performances presented before an audience in a particular setting. People in social situations resemble actors in that they must be convincing to others and must demonstrate who they are and what they want through verbal and nonverbal cues. In social situations, as on a stage, people manage the setting, their dress, their words, and their gestures to correspond to the impression they are trying to make or the image they are trying to project. This process is called **impression management.**

Impression Management

On the surface, the process of impression management may strike us as manipulative and deceitful. Most of the time, however, people are not even aware that they are engaged in impression management because they are simply behaving in ways they regard as natural. Women engage in impression management when they remove hair from their faces, legs, armpits, and other areas of their bodies and present themselves as hairless in these areas. From Goffman's perspective, even if people are aware that they are engaged in impression management, it can be both a constructive and a normal feature of social interaction because smooth interactions depend on everyone's behaving in socially expected and appropriate ways. If people spoke and behaved entirely as they pleased (as in the movie "Liar, Liar" starring Jim Carrey), civilization would break down. Goffman (1959) also recognized the dark side of

---

[12]Many Western doctors are now showing considerable interest in traditional African medicine. There seems to be a movement toward incorporating elements of both scientific and traditional medicine into health care and learning more about the curative properties of the herbs and remedies used by healers before the plants that supply them become extinct (because of the destruction of the rain forests). Apparently, traditional medicines work faster and better for some diseases (such as hepatitis) than prescribed pharmaceuticals do (Lamb 1987).

[13]In 1992, the U.S. Congress created the Office of Alternative Medicine to encourage research investigating and evaluating unorthodox medical ideas (Global Childnet 1995).

Dramaturgical model A model in which interaction is viewed as though it were theater, people as though they were actors, and roles as though they were performances presented before an audience in a particular setting.

Impression management The process by which people in social situations manage the setting and their dress, words, and gestures to correspond to the impressions they are trying to make or the image they are trying to project.

impression management, which occurs when people manipulate their audience in deliberately deceitful and hurtful ways.

Impression management often presents us with a dilemma. If we reveal inappropriate, unacceptable, or unpleasant information, we risk offending or losing our audience. Yet, if we conceal this information, we may feel that we are being deceitful, insincere, or dishonest. According to Goffman, in most social interactions people weigh the costs of losing their audience against the costs of losing their integrity. If keeping our audience is important, concealment of some information becomes necessary; if being honest is perceived as more important, we may risk losing the audience.

In the United States, people who test positive for HIV antibodies face the dilemma of impression management. If they disclose the test results, they risk discrimination, including loss of their jobs, insurance coverage, friends, and family. This risk explains why many HIV-infected people fail to disclose their HIV status even to sexual partners or when giving blood. In a study of 203 HIV-infected patients treated at Boston City Hospital and Rhode Island Hospital, Michael D. Stein (1998) and his colleagues found that more than half (129) were sexually active. Of these 129 patients, 40 percent had not disclosed their HIV-positive status to partners with whom they had had sex in the past six months. In addition, only 42 percent reported that they always used a condom. In another study of 304 HIV-positive blood donors, almost one-third indicated they had donated blood because their colleagues had pressured them to do so (U.S. General Accounting Office 1997a). Other research studies have found that between 14 and 30 percent of blood donors believe that blood bank screening areas provide inadequate privacy, enabling others to hear answers to the blood screening questions technicians ask donors. Twenty percent of donors claim they would have answered screening questions differently if a more private setting had been provided (U.S. General Accounting Office 1997a).

The dramaturgical model and the idea of impression management represent useful concepts for understanding the dilemma that one partner may face if he or she suggests using a condom as a precautionary condition of sexual intercourse. This dilemma is quite different in the United States than in the Congo. In both countries, health officials recommend condom use as a way to reduce substantially the risk of HIV infection.

In both the Congo and the United States, the subject of condom use is a sensitive one because of the message that the condom conveys to potential sexual partners. In the Congo, condom use is associated with birth control. The taboo against birth control is strong because many Africans measure their spiritual and material wealth by the number of offspring. If children survive,[14] they become economic assets to the family as well as links to ancestors (Whitaker 1988). In view of the strong pressures to have children, many believe that condom use virtually deprives them of the approval of their families and ancestors.

Researcher Kathleen Irwin (1991) and 15 colleagues interviewed healthy factory workers and their wives from Kinshasa, Zaire. They found that, although many respondents had heard of condoms, few actually used them. The researchers also discovered that, among these respondents, the men decide whether to use and purchase condoms.

Whereas the high value placed on many children lowers incentives to use condoms in the Congo, in the United States, the widespread availability and use of female contraceptives lowers the incentives (Urban Institute 1996). In the United States, if a person suggests using a condom during sex, he or she is implying that the partner's sexual orientation or sexual history is suspect. Advertisers try to package and market condoms in ways that counteract this message by claiming that condoms enhance sexual pleasure. Condoms come in all colors, are designed and manufactured in forms that stimulate greater sexual sensation, and show sexually appealing scenes on outer packages.

In both the Congo and the United States, "sexual behavior is based on long-standing cultural traditions and social values and may be very difficult to change" (U.S. General Accounting Office 1987). This fact makes it difficult for any person to manipulate the meaning of a condom without offending his or her sexual partner. Therefore, many people resist using condoms, even in high-risk situations (Giese 1987).

The dramaturgical model of social interaction helps us see how the need to convey the right impressions can discourage people from behaving in ways that are not in their best (or others' best) interest. Goffman uses another theater analogy—staging behavior—to identify situations in which people are most likely to engage in impression management.

---

[14]In the Congo, the infant mortality rate is 130 per 1,000; one-half of all children in the Congo die before they reach age five.

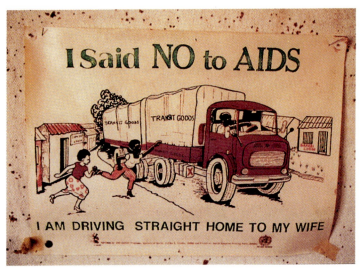

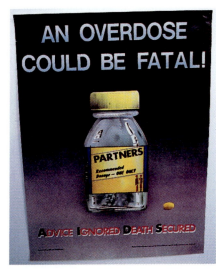

Almost every country in the world produces AIDS prevention messages tailored to the values and beliefs of people in that country. These posters are from Uganda (left) and Barbados (right).
© T. Cambre Pierce/Saba (left); © Jeremy Hartley/Panos Pictures (right)

Staging Behavior

Just as the theater has a front stage and a back stage, so too does everyday life. The **front stage** is the area visible to the audience, where people take care to create and maintain expected images and behavior. The **back stage** is the area out of the audience's sight, where individuals can "let their hair down" and do things that would be inappropriate or unexpected on the front stage. Because back-stage behavior frequently contradicts front-stage behavior, we take great care to conceal it from the audience. Goffman uses the restaurant as an example. Restaurant employees do things in the kitchen and pantry (back stage) that they would not do in the dining areas (front stage), such as eating from customers' plates, dropping food on the floor and putting it back on a plate, and yelling at one another. Once they enter the dining area, however, such behavior stops. How often have you, as a customer, seen a server eat a scallop or a french fry from a customer's plate while in the dining area? If you have ever worked in a restaurant, however, you know that such actions are fairly common back-stage behavior.

The division between front stage and back stage is hardly unique to restaurants, but is found in nearly every social setting. In relation to the AIDS crisis, we can name a host of environments that have a front stage and a back stage, including hospitals, doctors' offices, and blood banks. Much as restaurant personnel shield diners from behavior in the kitchen, medical personnel shield patients from backstage behavior. For example, most people know little about the blood bank industry beyond what they see when they donate, sell, or receive blood. Although the public can research such industries and learn about their inner, "back-stage" workings, most people do not have the time to study every industry or do not know what questions to ask about industries that provide them with goods and services.

For example, most people probably give little thought to the donors from whom the blood industry collects blood and would never think to ask questions about the blood's origins. They would be surprised to find out that many Mexicans cross the border to give blood at plasma centers located in the United States. In 1980, a few years before HIV was discovered, *Newsweek* magazine ran a story on the thousands of poor Mexicans who earned $10 by undergoing a procedure called plasmapheresis (in which technicians take blood, separate the red cells from the plasma, and reinject the donor with his or her own cells minus the

Front stage The region where people take care to create and maintain expected images and behavior.

Back stage The region out of sight where individuals can do things that would be inappropriate or unexpected on the front stage.

The Blood Services Industry

The blood services industry has a volunteer and a commercial sector. Voluntary donors are unpaid and usually donate whole blood. Commercial facilities collect plasma from paid donors for manufacturing various derivatives. Table 1.1 outlines the different types of blood collection services and the amount of blood they collect annually.

Table 1.1	U.S. Blood Collection Facilities and the Blood Units They Collect		
	Volunteer Sector*		Commercial Sector
Type of Facility	**Licensed**	**Unlicensed**	**Plasma center †**
Number of facilities	308	2,274	463
Number of units collected (millions)	12.6	1.4	12.0

*Licensed facilities ship blood across state lines.
†All plasma centers are licensed.

The three types of facilities in the volunteer sector are (1) regional and community blood centers, which usually collect and distribute blood and blood components to hospitals within circumscribed geographical areas; (2) hospital blood facilities, which collect and transfuse whole blood and blood components; and (3) hospitals, which primarily store and transfuse blood but do not collect it. Regional and community blood centers provide a full range of blood services to a surrounding geographical area. They generally collect, test, and label blood, as well as distribute blood and blood products to hospitals, physicians, and hemophilia care centers. Hospital blood facilities usually provide a smaller range of services, limited to collecting and storing whole blood and its components. Some hospitals conduct their own viral testing; others send blood and blood products to outside laboratories for viral testing.

The volunteer sector is represented by three organizations: the American Association of Blood Banks (AABB), the American Red Cross (ARC), and America's Blood Centers (ABC), formerly known as the Council of Community Blood Centers (CCBC). ABC member centers collect approximately 45 percent of all blood, ARC collects another 45 percent, and independent facilities collect the remaining 10 percent. The members of the AABB include both ARC and the majority of ABC member centers.

AABB is the professional society of blood facilities and transfusion services; it also includes individual members such as physicians, scientists, nurses, and administrators, among others. ABC is a council of community-based blood-collection facilities. ARC is a single corporation consisting of all ARC blood centers. Until 1994, ARC served as an organizational framework for its centers, each operating somewhat independently and self-sufficiently. In an organizational change that was completed in 1995, ARC centralized and standardized its operations, reducing the number of regions and limiting testing to a few centralized laboratories.

The commercial sector, which is generally called the "source plasma sector" and receives plasma from paid donors, has three main components: (1) collectors, or plasmapheresis centers; (2) fractionators; and (3) brokers.

(Brokers do not collect source plasma.) The plasmapheresis centers collect plasma that they either sell to U.S. fractionators (which manufacture derivatives, such as albumin, from it) or export to fractionators in Europe, Japan, and South America. Some fractionators also operate their own source plasma collection centers. Plasma brokers purchase and market recovered plasma from whole-blood facilities (that is, the volunteer sector) and sell these products directly to fractionators. Plasma is "recovered" after components have been removed from whole blood or after whole blood has become outdated.

The commercial sector is represented by the American Blood Resources Association (ABRA), a nonprofit trade association that represents the interests of businesses that collect certain biological products (in particular, plasma) for further manufacturing. This sector is also represented by the International Plasma Products Industry Association (IPPIA), which represents the commercial processors of plasma-based therapies in the United States.

Source: U.S. General Accounting Office (1997a).

plasma). Because red cells are returned to donors, they may "donate" blood up to twice a week. This article highlighted one element of the international nature of blood collection (Clark and McGuire 1980).

THE BACK STAGE OF BLOOD BANKS   Blood bank officials found themselves in a quandary in 1981, when officials at the CDC made known their suspicion that HIV was being transmitted through blood products.

U.S. in Perspective

Imports and Exports of Blood and Blood Products

To truly understand the global AIDS epidemic, we need to consider the role of exported blood products in the transmission of HIV. Figure 6.2 graphs the amount of blood products exported from the United States to selected other countries as a percentage of the receiving country's total need. The data are from 1981,

the year that HIV was "discovered" and that scientists first learned that it was in the blood supply. Notice that at the time, Japan imported 98 percent of its blood products from the United States—46 million units of concentrated blood products and 3.14 million liters of blood plasma (Yasuda 1994).

Figure 6.3 maps the United States' pattern of export and import of human blood plasma around 1996.

Sources: Figure 6.2: International Federation of Pharmaceutical Manufacturers Associations (1981); Figure 6.3: International Trade Administration 1997a, 1997b.

Figure 6.2

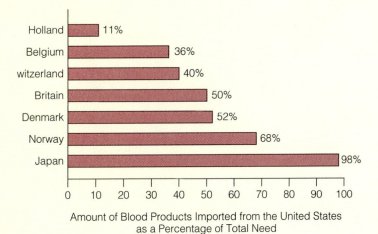

Amount of Blood Products Imported from the United States
as a Percentage of Total Need

- Holland 11%
- Belgium 36%
- witzerland 40%
- Britain 50%
- Denmark 52%
- Norway 68%
- Japan 98%

Figure 6.3

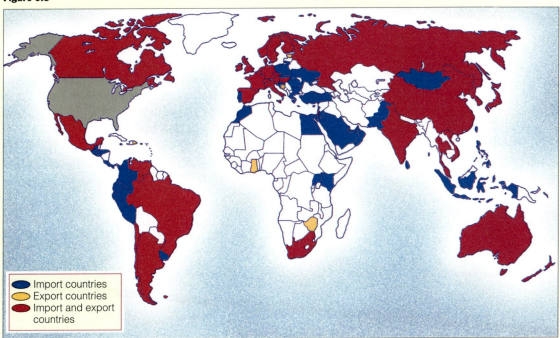

- ● Import countries
- ● Export countries
- ● Import and export countries

This revelation meant that not only the United States' blood supply, but also the world's blood supply, was contaminated, because the corporations in the United States supply about 60 percent of the world's plasma and blood products (see "Imports and Exports of Blood and Blood Products" and "The Blood Services Industry").

Until spring 1985, U.S. blood bank officials continued to affirm publicly their faith in the safety of the country's blood supply, insisting that screening donors was unnecessary. Yet, these officials never revealed to the public the many shortcomings in production methods that could jeopardize the safety of blood. (By *shortcomings*, we mean deficiencies in medical knowledge and the level of technology rather than negligence.) Blood bank officials later argued that they practiced this concealment to prevent a worldwide panic. (In Goffman's terminology, they did not want to lose their audience.) Such a panic would have brought chaos to the medical system, which depends on blood products.[15] This delay in implementing screening ultimately exposed many people to infection, especially hemophiliacs.

Hemophiliacs were especially at risk because their plasma lacks Factor VIII, a substance that aids in clotting, or because their plasma contains an excess of anticlotting material (U.S. Department of Health and Human Services 1990). In fact, we now know that 50 percent of hemophiliacs became infected with HIV from Factor VIII treatments before the first case of AIDS appeared in this group (*Frontline* 1993).[16]

Even after blood bank officials agreed to start screening blood for HIV infection in spring 1985 (after years of debate over testing), they still pronounced the blood supply safe. They did not, however, announce the shortcomings of the screening tests: (1) the antibodies measured by the test may not appear in the blood for 11 months after infection with the virus, and (2) a small but unknown percentage of HIV carriers never develop detectable antibodies (Kolata 1989).

In hindsight, it is easy to criticize blood bank officials' response to the situation. Yet, in fairness to the blood industry, we must acknowledge the legitimacy of their wish not to induce worldwide panic, especially very early on when no tests were available to screen blood for the infection. As the head of the New York Blood Center argued, "You shouldn't yell fire in a crowded theater, even if there is a fire, because the resulting panic can cause more deaths than the threat" (Grmek 1990:162). This consideration is especially important in light of the fact that "blood is typically prescribed for patients who have very serious trauma or disease" (U.S. Government Accounting Office 1997b:14). In addition, we must recognize that it is impossible to eliminate every element of risk. Even so, one troubling fact suggests that the decision made by blood bank officials may have been motivated by profit. Although U.S. companies tested new blood for the domestic market, they did not test blood already stored in their inventories, and they continued to supply untested blood to foreign countries for at least six months after the tests became available (*Frontline* 1993; Hiatt 1988; Johnson and Murray 1988).

Today, blood bank officials regularly reassure the public that the risk of HIV infection from blood products is virtually nonexistent. Knowledge of the backstage collection, production, and distribution of blood products, however, leaves one with no doubt that the risks are actually higher. First, consider that the CDC has no data on the numbers of people who are HIV-infected as a result of blood transfusions and products. It keeps statistics only on the number of people known to have developed AIDS. Second, blood banks "do not always notify recipients of blood donations that they might have received tainted blood," nor are they required to report "mistakes" connected to testing and collection (Newman and Podolsky 1994).

To this point, we have examined a number of concepts that help us analyze the content of any social interaction. Specifically, when examining content sociologists identify the statuses of those people involved. Once statuses have been identified, they focus on the roles expected of each status in relation to the other, placing special emphasis on rights and obligations. Sociologists use a dramaturgical model to think about how people

---

[15]Many medical treatments depend on blood products:

Blood transfusions save the lives of premature infants and are crucial for children with Cooley's anemia and other hereditary blood disorders. A vaccine against hepatitis B is derived from blood. Injections of gamma globulin, prepared from blood, are effective in helping prevent hepatitis A, chicken pox, rabies, and other ailments. Platelet transfusions are key to some cancer treatments. (Altman 1986:A1)

Here lies a potential connection to the Congo. Malaria is a common disease in the Congo, especially among children. The disease leaves its victims vulnerable to severe anemia, and blood transfusions are used to treat the anemia. As mentioned previously, the United States is a large exporter of blood products. Many countries that import this blood reexport it, and the Red Cross delivers blood products to countries in need. Therefore, some of the blood used in the Congo is likely to have originated in the United States.

[16]Hemophiliacs were highly vulnerable to HIV infection because the clotting agent of blood used to treat their condition—Factor VIII—is manufactured from batches of plasma pooled from hundreds of donors. One HIV-infected person can contaminate an entire batch and expose anyone who receives Factor VIII from that batch to infection (Rock 1986).

enact roles. Thus they focus on impression management (how people manage the setting, their dress, their words, and their gestures) to correspond to role expectations. Impression management is most important when people are front stage as opposed to back stage.

Knowing about statuses, roles, impression management, and front/back stages helps sociologists predict much of the content of social interaction. When we interact, however, we do more than identify the statuses of the people involved and behave in ways consistent with role expectations. We also try to assign causes to our own and others' behaviors. That is, we posit explanations for behavior, and we may act differently depending on what explanations we develop. A theoretical approach that helps us understand how we arrive at our everyday explanations of behavior is attribution theory.

Attribution Theory

Social life is complex. As we have seen, people need a great deal of historical, cultural, and biographical information if they are to understand the causes of even the most routine behaviors. Unfortunately, it is nearly impossible for people to have this information at hand every time they want to explain why a behavior occurs. For one thing, "the real environment is altogether too big, too complex, and too fleeting for direct acquaintance" (Lippmann 1976:178).

Yet, despite our limitations in understanding causes of behavior, most people do attempt to determine a cause, even if they rarely stop to examine critically the accuracy of their explanations. As most of us know very well, ill-defined, incorrect, and inaccurate perceptions of cause do not keep people from forming opinions and taking action. Such perceptions, however, result in actions that have real consequences. Sociologists William and Dorothy Thomas described this process very simply: "If [people] define situations as real they are real in their consequences" (Thomas and Thomas [1928] 1970:572).

Attribution theory relies on the assumption that people assign (or attribute) a cause to behavior in an effort to make sense of it. People usually attribute cause to either dispositional traits or situational factors. **Dispositional traits** include personal or group traits, such as personality traits, motivation level, mood, or innate ability. **Situational factors** include forces outside an individual's control, such as environmental conditions or bad luck. When evaluating the causes of their own behavior, people tend to favor situational factors. When evaluating the causes of

another person's behavior, however, they tend to point to dispositional traits. That is, people often use different criteria when they attribute cause to their own behavior than when they attribute cause to another's behavior. On the one hand, with regard to other people's failures or shortcomings, we tend to overestimate the extent to which their failures flow from dispositional factors. With regard to success, we tend to exaggerate the extent to which others' successes are caused by situational factors (a lucky break; they were in the right place at the right time). On the other hand, we tend to explain our own successes as resulting from dispositional factors such as personal effort and sacrifice (if you try hard "like me," you can succeed at anything). We explain our failures by pointing to situational factors (I failed the test because the teacher can't explain the subject and asked tricky questions).

Right or wrong, the attributions that people make shape the content of social interaction. In other words, the way in which people construct reality affects how they treat and deal with different individuals and groups. For example, in searching for an explanation for why HIV is widespread in Africa, dispositional thinkers from Western countries point to rituals involving blood, such as one described by anthropologist Colin Turnbull. The Mbuti pygmies made Turnbull a member of their camp by making tiny tattoo marks above his eyes and in the center of his forehead and then mixing a small amount of Turnbull's blood with the blood of another pygmy (Mark 1995:199). At the same time, these dispositional thinkers would ignore high-risk "rituals" Western people have engaged in while in Africa. It is well known, for example, that anthropologists and other scientists drew blood from various African ethnic groups in an attempt to learn whether "it might be possible to distinguish races by the prevalence of certain blood types" (Mark 1995:6). On just one research trip, anthropologist Patrick Tracy Lowell Putnam carried 500 blood tubes to be used for this purpose (Mark 1995).

Explaining the Origin of HIV

Throughout history, whenever medical professionals have lacked the knowledge or technology to combat a

Dispositional traits Personal or group traits, such as motivation level, mood, and inherent ability.

Situational factors Forces outside an individual's control, such as environmental conditions or bad luck.

disease, especially one of epidemic proportions, the general population has tended to hold some groups of people within or outside of the society responsible for causing the disease (Swenson 1988). In the sixteenth and seventeenth centuries, for example, the English called syphilis "the French disease," and the French called it "the German disease." In 1918, the worldwide influenza epidemic, which infected more than 1 billion people and killed more than 25 million, was called "the Spanish flu" even though no evidence supported the idea that it originated in Spain. American history holds many examples of groups blamed for bringing disease to the country, including the Irish (cholera), Italians (polio), Chinese (bubonic plague), Jews (tuberculosis), and Haitians (AIDS) (Kraut 1994).

In a similar vein, U.S. medical researchers trying to map the geographic origin and spread of AIDS have hypothesized various interaction scenarios between specific groups of people who are inferred to behave in careless, irresponsible, or immoral ways to explain how AIDS has spread transcontinentally. The hypotheses usually assume that the disease originated in the Congo or another Central African country and then spread to the United States via Europe, Haiti, or Cuba. Yet, the possibility exists that it could have spread transcontinentally from the United States. Two of the most prevalent hypotheses related to the origin of HIV are as follows:

- The virus traveled from the Congo to Haiti to the United States. In the mid-1960s, a large number of Haitians went to the Congo to fill middle management positions in the newly independent state. In the 1970s, Mobutu sent them home. American homosexuals vacationing in Port-au-Prince later brought the virus back to New York and San Francisco (Cohen 1997).

- The virus traveled from the Congo to Cuba to the United States. Cubans brought it back from Angola, which shares a long border with the Congo. In the late 1970s, the Cuban government purged the army of undesirables, including some homosexual veterans who had served in Angola. Many of these Cubans then migrated to Miami.

Attributing cause to dispositional factors seems to reduce uncertainty about the source and spread of the disease. The rules are clear: If we do not interact with members of that group or behave like members of that group, then we can avoid the disease. Such rules provide us with the secure feeling that the disease cannot affect *us* because it affects *them* (Grover 1987).

In the United States, many people believe that

HIV and AIDS are confined to members of a few high-risk groups, notably hemophiliacs, male homosexuals, and intravenous (IV) drug users. Dispositional explanations for the high risk among members of these groups imply that these individuals "earned" their disease as a penalty for perverse, indulgent, and illegal behaviors (Sontag 1989). As late as 1985, medical and government officials referred to AIDS as the "gay plague," even though overwhelming evidence indicated that HIV infection was also transmitted through heterosexual intercourse, needle sharing, and other exchanges of blood and blood products or some other bodily fluids. In his book *And the Band Played On,* journalist Randy Shilts (1987) argues that the public health and medical research response to AIDS was delayed several years because policy makers believed that casualties were limited to unpopular and socially powerless groups such as homosexuals, IV drug users, and Haitians.

Similarly, dispositional thinkers might explain the high incidence of HIV infection and AIDS among both males and females in the Congo (or other African countries) in terms of excessive sexual promiscuity or polygynous marriages[17] or in terms of rituals involving monkeys, female circumcision, and scarification.

Whereas Americans tend to regard HIV infection as originating in Africa and as being transmitted through bizarre and indulgent behaviors, the Congolese tend to view AIDS as a foreign disease:

> Westerners had brought AIDS to Africa with their "weird sexual propositions"—a view echoed by *La Gazette* in July 1987, which referred to Westerners coming to Africa with their "sexual perversions." "Many of the venereal diseases now found in Kenya," said an editorial in the *Kenyan Standard,* "were brought into the country by the same foreigners who are now waging a smear campaign against us." (Panos Institute 1989:75)

In the Congo and other African countries, people often refer to the United States as the "United States of AIDS." Condoms are called "American socks," and aid in the form of condoms is known as "foreign AIDS" (Brooke 1987; Hilts 1988). People commonly believe that (1) AIDS arrived via rich American sports fans who

---

[17]African officials become outraged when the Western media equate polygynous marriage arrangements with promiscuity. One villager responded as follows to this charge: "We have several wives, and we are faithful to them all, and we care for all their children until we die. You people cannot even be faithful to one wife, and your children are such a nuisance to you that you send them away from home as soon as they can walk" (Turnbull 1962:28).

came to Kinshasa to watch the Ali-Foreman boxing match in 1977, (2) the virus was manufactured in an American laboratory for military germ warfare and was unleashed deliberately on Africans,[18] (3) the disease came from American canned goods sent as foreign aid, or (4) the AIDS epidemic can be traced to the way in which the polio vaccine was manufactured and administered to Africans 30 years ago.

In each dispositional scenario, one can infer that those giving the disease to Africans are morally suspect, evil plotters, profit-driven, or careless; these are all dispositional characteristics. Such dispositional theories, whether American or African, are alike in that they define a clear culprit or scapegoat. A **scapegoat** is a person or a group that is assigned blame for conditions that cannot be controlled, threaten a community's sense of well-being, or shake the foundations of a trusted institution. Often the scapegoat belongs to a group whose members are already vulnerable, hated, powerless, or viewed as different.

The public identification of scapegoats gives the appearance that something is being done to protect the so-called general public; at the same time, it diverts public attention from those who have the power to assign labels. In the United States, the early identification of AIDS as the "gay plague" diverted attention from the blood banks and the risks associated with medical treatments involving blood products. Blood bank officials maintained that the supply was safe as long as homosexuals abstained from giving blood. In the Congo, identifying AIDS as an American disease diverted attention from corrupt officials who were funneling money out of the Congo, leaving its people poor and malnourished and (by extension) vulnerable to HIV infection.

From a sociological perspective, dispositional explanations that point to a group—or characteristics supposedly inherent in members of that group—are simplistic and potentially destructive not only to the group but also to the search for solutions. When a specific group and that group's behavior emerge as a target, the solution is framed in terms of controlling that

group. In the meantime, the problem can spread to members of other groups who believe that they are not at risk because they do not share the problematic attribute of the supposedly high-risk groups. This kind of misguided thinking about risk applies even to physicians and medical researchers.

Medical sociologist Michael Bloor (1991) and colleagues argue that the official statistics on the modes of HIV transmission are influenced by researchers' beliefs about the relative riskiness of behaviors. For example, an HIV-positive male who has received a blood transfusion and who has had sexual relations with another male is placed in the transmission category "men who have sex with men" rather than "blood recipient" (Centers for Disease Control and Prevention 1998). This approach to classification ignores the real possibility that a person who has engaged in a homosexual act can become infected with HIV through a blood transfusion or other means (see "Research Decisions About Labeling and Categorizing HIV Risk").

Similarly, until 1993 the official definition of AIDS did not include HIV-related gynecological disorders, such as cervical cancer, as conditions that constituted a diagnosis of AIDS. (Also not included under the official definition of AIDS until 1993 were HIV-related pulmonary tuberculosis and recurrent pneumonia.) Under the old definition, HIV-positive persons were said to have AIDS only if they developed one of 23 illnesses, many of which were unique to the gay population with AIDS. Under the new, revised definition, HIV-positive women with cervical cancer are officially diagnosed as having AIDS. Before this change, physicians did not advise women with cervical cancer to be tested for HIV and did not give HIV-positive women with this condition the opportunity to be treated for AIDS (Barr 1990; Stolberg 1996). In a similar vein, "HIV-positive patients older than 50 years tend to be more immunosuppressed and more often have AIDS at first evaluation than their younger counterparts" (Reuters Health Information 1999:1). These examples show that attributions about who "should" have AIDS affect the way in which the condition of AIDS is defined and influence the statistics about who has AIDS. They also illustrate that such attributions affect the content of the physician-patient interaction. In

---

[18]This theory has some basis in fact. In 1969, the U.S. Defense Department discussed the theoretical possibility of the following development:

> Within the next five to ten years it would probably be possible to make a new infective micro-organism which could differ in certain important respects from any known disease-causing organisms. Most important of these is that it might be refractory to the immunological and therapeutic processes upon which we depend to maintain our relative freedom from infectious disease. (Harris and Paxman 1982:241)

Scapegoat A person or a group that is assigned blame for conditions that cannot be controlled, threaten a community's sense of well-being, or shake the foundations of a trusted institution.

Research Decisions About Labeling and Categorizing HIV Risk

Exposure Category	AIDS cases Number	%
Single mode of exposure		
Men who have sex with men	296,483	47
Injecting drug use	129,990	21
Hemophilia/coagulation disorder	3,784	1
Heterosexual contact	57,360	9
Receipt of transfusion	8,201	1
Receipt of transplant of tissues, organs, or artificial insemination	13	0
Other	114	0
Single mode of exposure subtotal	**495,945**	**78**
Multiple modes of exposure		
Men who have sex with men; injecting drug use	34,845	6
Men who have sex with men; hemophilia/coagulation disorder	155	0
Men who have sex with men; heterosexual contact	8,966	1
Men who have sex with men; receipt of transfusion/transplant	3,315	1
Injecting drug use; hemophilia/coagulation disorder	183	0
Injecting drug use; heterosexual contact	29,051	5
Injecting drug use; receipt of transfusion/transplant	1,581	0
Hemophilia/coagulation disorder; heterosexual contact	86	0
Hemophilia/coagulation disorder; receipt of transfusion/transplant	785	0
Heterosexual contact; receipt of transfusion/transplant	1,524	0
Men who have sex with men; injecting drug use; hemophilia/coagulation disorder	44	0
Men who have sex with men; injecting drug use; heterosexual contact	4,868	1
Men who have sex with men; injecting drug use; receipt of transfusion/transplant	587	0
Men who have sex with men; hemophilia/coagulation disorder; heterosexual contact	22	0
Men who have sex with men; hemophilia/coagulation disorder; receipt of transfusion/transplant	35	0
Men who have sex with men; heterosexual contact; receipt of transfusion/transplant	266	0
Injecting drug use; hemophilia/coagulation disorder; heterosexual contact	68	0
Injecting drug use; hemophilia/coagulation disorder; receipt of transfusion/transplant	38	0
Injecting drug use; heterosexual contact; receipt of transfusion/transplant	938	0
Hemophilia/coagulation disorder; heterosexual contact; receipt of transfusion/transplant	34	0
Men who have sex with men; injecting drug use; hemophilia/coagulation disorder; heterosexual contact	11	0

other words, if physicians do not believe a patient could be HIV-positive because he or she does not possess well-publicized attributes associated with risk, they will not recommend testing (Henderson 1998).

In addition to acknowledging the shortcomings related to AIDS classification and diagnosis, we must acknowledge that we simply do not know how many people are HIV-infected in the United States or worldwide. In the United States, physicians in all 50 states are required to report AIDS cases to public health officials. Only 28 states, however, require physicians to report HIV infection cases (Richardson 1998).

Determining Who Is HIV-Infected

To obtain information on who is actually infected with HIV, every country in the world would have to administer blood tests to a random sample of its population. Unfortunately (but perhaps not surprisingly), people resist being tested.

A planned random sampling of the U.S. population sponsored by the CDC was abandoned after 31 percent of the people in the pilot study refused to participate, despite assurances of confidentiality (Johnson

Exposure Category	AIDS cases	
	Number	**%**
Multiple modes of exposure		
Men who have sex with men; injecting drug use; hemophilia/coagulation disorder; receipt of transfusion/transplant	14	0
Men who have sex with men; injecting drug use; heterosexual contact; receipt of transfusion/transplant	160	0
Men who have sex with men; hemophilia/coagulation disorder; heterosexual contact; receipt of transfusion/transplant	5	0
Injecting drug use; hemophilia/coagulation disorder; heterosexual contact; receipt of transfusion/transplant	23	0
Men who have sex with men; injecting drug use; hemophilia/coagulation disorder; heterosexual contact; receipt of transfusion/transplant	5	0
Multiple modes of exposure subtotal	**87,609**	**14**
Risk not reported or identified	**49,446**	**8**
Total	**633,000**	**100**

Source: Centers for Disease Control and Prevention (1998).

According to the Centers for Disease Control and Prevention (1998) report *HIV/AIDS: Surveillance Report,* "HIV infection cases and AIDS cases are counted only once in a hierarchy of exposure categories. [This means that] persons with more than one reported mode of exposure to HIV are [ultimately] classified in the exposure category listed [first] (p. 41). Notice that "men who have sex with men" is always listed first in the hierarchy of risks. Also notice that for 8 percent of AIDS cases, "risk was not reported or identified." It is also significant that researchers use the general-risk category "men who have sex with men" and "heterosexual sex," which implies that the sex act must be different between people in the two categories. It is possible that "heterosexual contact" is "anal" and "oral," not just vaginal. According to the researchers who conducted the most comprehensive study on sexual behavior in the United States, between 13.2 and 37.5 percent of people said they have engaged in anal sex at some time. Between 55.6 and 83.6 percent said they have had oral sex in the past year (Michael et al. 1994). Would we learn more if we classified AIDS cases by sex act rather than identifying the people having sex (for example, two males versus one male and one female)? Should researchers separate "hemophilia," "receipt of transfusion," and "receipt of transplant . . . " into three separate categories since all involve "blood"? Why is there no "females who have sex with females" category when "studies have shown that the prevalence of female-to-female sex among HIV-infected or at-risk women is not insignificant" (Kennedy et al. 1998:29)?

and Murray 1988). Two researchers involved with this project concluded that "it does not seem likely that studies using data on HIV risk or infection status, even with complete protection of individual identity, will be practical until the stigma of AIDS diminishes" (Hurley and Pinder 1992:625).

The United States is not the only country whose people do not want to be tested. This resistance seems to be universal. For example, at one time Congolese officials were reluctant to disclose the number of AIDS cases and infection rates to United Nations officials or to allow foreign medical researchers to test their citizens because of sensitivity to the unsubstantiated but widely held belief that the Congo was the cradle of AIDS (Noble 1989).

Random sampling, however, is the most dependable method we have of determining the number of HIV-infected persons. Until we have such information, we cannot know what factors cause a person with HIV infection to develop AIDS. Random blood samples would permit comparisons between the lifestyles of infected but symptom-free people and infected people who have developed AIDS or AIDS-related complex (ARC).[19] From such comparisons, we could learn which cofactors cause a healthy carrier to develop ARC or AIDS. Such cofactors might include diet, exposure to hazardous materials or pesticides (*New York Times* 1996), or prolonged exposure to the sun or tanning booth rays—anything that might compromise the immune system in such a way as to activate a dormant infection. For example, why do some people remain HIV-infected for years "free of serious symptoms and complications" (Wood, Whittet, and Bradbeer 1997)? Why do other people develop AIDS shortly after exposure to HIV (Altman 1986, 1995) and about 13 percent survive for 25 years or longer without progressing to AIDS (Reuters Health Information 1998)? Why is HIV absent from the bodies of some patients with AIDS-like symptoms (Liversidge 1993)? Why do some "high-risk" individuals, such as hemophiliacs and others who received tainted blood products or gay men who have had unprotected sex with HIV-infected partners, never become infected (U.S. General Accounting Office 1997b; Radetsky 1997; Cohen 1998)? Finally, how is it that at one time HIV may have been a harmless virus? How did the virus change to become the causative pathogen of AIDS? (Some scientists speculate that HIV has been present in an inactive state for at least a century.)[20]

The point is that a person may develop AIDS as a result of other factors besides the behavior that causes an individual to contract HIV infection. As long as no systematic and random sampling of populations is performed, people will continue to speculate about these factors, either overestimating or underestimating the prevalence of infection and the projected numbers of AIDS cases worldwide.

This situation leaves people with a dilemma in dealing with a complex health problem such as AIDS when so many unanswered questions persist. Most people do not have the time to inform themselves about all of the contextual forces underlying HIV and AIDS. Yet, this constraint does not stop people from

acting or attributing cause on the basis of limited information. Even if people do not have the time to fully research this issue, they do have the option of at least being critical of their information sources; for most people, that source is television.

Public health officials believe that the media need to deliver at least one important message with regard to HIV: "It is not who you are; it is how you live and what you do" (Kramer 1988:43). For the most part, however, this message has not been transmitted to the American public. Because television news and information shows serve as an important source of information for most Americans (98 percent of U.S. households have television sets), we will examine how television producers present information in general and data about the AIDS phenomenon in particular.

Television: A Special Case of Reality Construction

When sociologists say that reality is constructed, they mean that people assign meaning to interaction or to some other event. When people assign meaning, they almost always emphasize some aspects of that event and ignore others. In this chapter, we have looked at some of the strategies that people use to construct reality. Consider the following points:

1. When people assign meaning, they tend to ignore the larger context.

2. When people interact with others, they assign meaning by first determining their own social status in relation to the other parties, then drawing upon learned expectations of how people in some social statuses behave.

3. People create reality when they engage in impression management; that is, they manage the setting, their dress, their words, and their gestures to correspond to impressions they are trying to make.

4. People control access to the back stage so that outsiders to the back stage form opinions on the basis of the front stage.

5. People attribute cause to dispositional traits when evaluating others' behavior, and they attribute cause to situational factors when evaluating their own behavior.

People also assign meaning to events based on firsthand experiences. Television gives people access

---

[19]AIDS-related complex (ARC) is a term applied to HIV-infected persons whose symptoms do not meet the definition of AIDS set forth by the CDC. ARC encompasses a wide range of symptoms, including fever, rash, and bacterial or viral infections, most of which point to a diminished ability of the body's immune system.

[20]Actually, the HIV-AIDS connection is not a clear-cut one. In recognition of this fact, more than 100 biologists from around the world have formed an organization (Group for Scientific Reappraisal of the HIV/AIDS Hypothesis) with the purpose of rethinking this connection. This group also issues a newsletter titled *Rethinking AIDS* (Liversidge 1993).

to events that they would never have the chance to experience if left to their own resources. Thus our analysis of how a person constructs his or her reality would be incomplete if we ignored the format that television, especially television news, uses to present information about the world.

Television conquers time and space: it allows us to see world events as soon as they happen. Consider, however, the following features of national and local news—television at its most serious and most informative:

- The average length of a camera shot is 3.5 seconds.

- Every three or four news items, no matter how serious, are followed by three or four commercials.

- The news of the day is often presented as a series of sensationalized images.

- Approximately 15 news items are presented in a 30-minute news program with approximately 12 different commercials.

- Most news coverage of events focuses on the moment; each item is usually presented without a context.

In *Amusing Ourselves to Death,* Neil Postman (1985) examines the format of news programs and asks how it affects the way people think about the world. Overall, he believes, this format gives viewers the impression that the world is unmanageable and that events just seem to happen. More to the point, it gives the public only the most superficial facts about world events. For example, most people in the United States know basic facts about how HIV is spread and about how to reduce risk of transmission. A large percentage (71 percent), however, also admit that they do not know a lot about AIDS (U.S. Department of Health and Human Services 1992).

What characteristics of the news format produce this consequence? The brevity of camera shots is probably one of the main problems. Postman argues that with the average length of a shot being just 3.5 seconds, facts are pushed into and out of consciousness in rapid succession so that viewers do not have sufficient time to reflect on what they have seen and to evaluate it properly. Furthermore, commercials defuse the effects of any news event; they give the following message: I cannot do anything about what is happening in the world, but I can do something to feel good about myself if I eat the right cereal, own the right car, and use the right hair spray.

The story of Ryan White, a hemophiliac who contracted HIV, was widely publicized in the news media. How does this kind of selective coverage change people's view of a phenomenon like AIDS?

© Mary Ann Carter/Sipa Press

Television is image-oriented; a picture is a moment in time and, by definition, is removed from any context. This quality often causes television news to be sensationalistic. The highly publicized case of Ryan White (who died in April 1990 at age 18 from AIDS-related respiratory failure) illustrates just how sensationalistic news reports can be. As a child, White had received HIV-infected Factor VIII while being treated for hemophilia and later was diagnosed with AIDS. White was barred from attending Western Middle School in Kokomo, Indiana, in August 1985 after school officials learned that he had AIDS (Kerr 1990). When White returned to school in April 1986, reporters and camera crews covered the event in what the school principal termed a sensationalistic manner:

> I understand that the media has a job to do, but I think there is a fine line between informing the public and creating controversy in order for a story to keep continuing.
>
> It seems like the problems were brought out by those who jumped on the sensational. These were published and displayed on television and it created an excited atmosphere in what was really a pretty calm school situation. . . .
>
> When Ryan did come back, there were thirty kids whose parents took them out of school. That's what made the news, but there were 365 other children who stayed in the school. (Colby 1986:19)

An image-oriented format tends to ignore those historical, social, cultural, or political contexts that would make the event more understandable. Viewers are left with vivid, sensationalistic images of enraged parents, an emaciated gay AIDS patient, a skid row drug addict, a prostitute, a family home destroyed by fearful neighbors, Africans walking to an AIDS clinic, or gays protesting discrimination. Rarely, however, do television producers present AIDS as a chronic but often manageable condition or portray those with AIDS as leading responsible lives. Imagine what people's initial attitudes toward AIDS might have been if the first discussion of the disease had centered around the life of someone like Dr. Rask instead of a small group of homosexual males. Trying to understand how Dr. Rask contracted HIV would certainly have told us more about the complex social origins of the disease than did focusing on homosexual practices.

This is not to deny that television is an important tool for informing large audiences. Despite the sensationalistic coverage, most adults in the United States know that AIDS exists and that HIV is transmitted from one person to another in only certain ways (Henderson 1997).

To illustrate how little context television provides, you might list the things you have learned about AIDS from this chapter that you did not learn from the media. One might argue that there is not enough time to learn about AIDS from television because so many other important events simultaneously compete for attention. Still, this point does not eliminate the need for context in understanding events.

Another way to see how little context television provides is to count how many events (other than sports and weather items) you remember from last night's news. If you cannot remember many events, then the information may have been presented so quickly that you could not reflect long enough to absorb it. To remedy this problem, newscasters might reduce the number of stories, increase the time given to context, and cover stories in less reactive and more reflective ways. In the case of AIDS, for example, producers could show segments explaining how HIV differs from AIDS or how the television image of AIDS differs from the experience of AIDS. With regard to this last suggestion, television producers are quite good at covering and discussing presidential campaigns as media events. They certainly could cover AIDS as a media event as well.

Summary and Implications

In this chapter, we learned that sociological concepts and theories contribute to our understanding of HIV and AIDS and, by extension, other infectious diseases. AIDS (and other infectious diseases) cannot be viewed simply as a biological event. This condition is a social phenomenon if only because certain kinds of "high-risk" interactions are associated with the transmission of HIV. Sociology gives us a framework for conceptualizing the role of social interaction in HIV transmission and the content of interactions that occur in reaction to HIV.

The importance of considering interaction on a global scale is supported by the fact that disease patterns historically are affected by changes in population density and changes in transportation. Physician Mary

E. Wilson (1996) describes the relationship between travel and the emergence of infectious diseases:

> Travel is a potent force in the emergence of disease. Migration of humans has been the pathway for disseminating infectious diseases throughout recorded history and will continue to shape the emergence, frequency, and spread of infections in geographic areas and populations. The current volume, speed, and reach of travel are unprecedented. The consequences of travel extend beyond the traveler to the population visited and the ecosystem. When they travel, humans carry their genetic makeup, immunologic sequelae of past infections, cultural preferences, customs, and behavioral patterns. Microbes, animals, and other biologic life also

Disease patterns historically are affected by changes in population density and transportation patterns.

© Crispin Hughes/Panos Pictures

accompany them. Today's massive movement of humans and materials sets the stage for mixing diverse genetic pools at rates and in combinations previously unknown. Concomitant changes in the environment, climate, technology, land use, human behavior, and demographics converge to favor the emergence of infectious diseases caused by a broad range of organisms in humans, as well as in plants and animals.

As one indicator of the massive movement of human beings, consider that 595 million people took an international trip in 1996 (Crossette 1998). In that year, approximately 46.3 million foreign visitors entered the United States (excluding same-day visitors), and 47.3 million U.S. residents traveled to a foreign country (Crossette 1998).

We focused on the Congo and on Dr. Rask to show that a complex set of interactions underlies the transmission of HIV infection. Dr. Rask represents one person with AIDS who was caught up in the historical events that led to worldwide economic interdependence, which in turn brought people from all over the world into contact with one another. The complexity of the interaction that brought Dr. Rask to the Congo and in contact with HIV makes it impossible to state conclusively that HIV originated in that country. Even if a previously isolated group in the Congo, such as the Mbuti, was identified as the group that harbored the virus in a dormant state, could we in good conscience define those people as the source? What about the many forces that brought them out of isolation into contact with groups having no immunity to the virus?

The historical context of interaction is only one dimension of the interaction process that sociologists examine. They also seek to understand the *content* of social interaction. Sociology offers a rich conceptual vocabulary to analyze the dialogue, actions, and reactions that take place when people interact, as well as to analyze how the parties involved in social interaction strive to define, interpret, and attach meaning to the encounter. People associated with Dr. Rask, for example, adopted a scientific framework to define the origin, significance, and treatment of her condition. As a result, Dr. Rask experienced not only a physical state of sickness but a social state in which her behavior and the behavior of the involved parties reflected the assumption that her physiological malfunction could be understood and corrected with the available medical technologies (tests, drugs, machines, surgery, and so on).

Key sociological concepts that help us analyze interaction include social status, social roles, impression management, front stage, back stage, dispositional traits, and situational factors. These concepts enable us to identify a number of mechanisms that impede progress and prevent effective action in dealing with HIV and AIDS. For example:

- Expectations associated with the physician's role focus his or her attention on finding a cure for the condition without regard to side effects and to the neglect of symptom relief and patient comfort.

- Back-stage behavior plays a significant role in determining the eventual course of the disease. (How many people around the world ultimately contracted HIV because of the way blood and blood products were processed? What role did "profit" play in exposing some people to HIV who might not have otherwise been exposed?)

- Dispositional explanations that point to a group or characteristics supposedly inherent in members of that group are simplistic and potentially destructive. When the focus is the group, the solutions are framed in terms of controlling that group rather than understanding the problem and finding solutions.

- Official statistics on modes of transmission and official definitions of AIDS are influenced by researchers' beliefs about the relative riskiness of behavior and by attributions about who "should" have AIDS.

- The public identification of scapegoats diverts attention away from a consideration of other factors that may be causing the problem. (Initially identifying AIDS as the "gay plague," for example, diverted attention away from blood banks and the risks associated with medical treatments involving blood products.)

- Social meanings associated with preventive measures such as condoms can cause people to resist using those measures, even in clearly high-risk situations.

The global transmission of HIV infection illustrates a point about interdependence on a global scale: When something goes wrong in one part of the world, other parts of the world are affected as well. Our discussions of global interdependence in relation to colonization and of many countries' reliance on blood

products from the United States illustrate this point; so do many other phenomena, such as global warming and illegal drug trade. Interdependence is not necessarily a negative situation, however. It can also lead to greater efforts to solve problems and keep the system running. As futurist John Naisbitt (1984) suggests, "If we get sufficiently interlaced economically, we will probably *not* bomb each other off the face of the planet" (p. 79). Such interdependence also means that "AIDS cannot be stopped in any country unless it is stopped in all countries" (Mahler, quoted in Sontag 1989:91); "it cannot be mastered in the West unless it is overcome everywhere" (Rozenbaum, quoted in Sontag 1989:91). These comments suggest that effective AIDS policies must necessarily involve worldwide effort.

Such an effort is now under way. The World Health Organization (WHO) is sponsoring, directing, and coordinating a global strategy to prevent HIV and control its transmission. WHO's Global Programme on AIDS supports national AIDS programs in 150 countries that include health education, prevention information, blood transfusion services, and cross-cultural research on human behavior and effective communication. WHO officials maintain that this program offers health benefits beyond the prevention of AIDS: "The global response to AIDS offers a great opportunity to accelerate the strengthening of our health care infrastructures" in general (World Health Organization 1988:15).

Key Concepts

Use this outline to organize your review of the key chapter ideas.

Social interaction
  Context
    Division of labor
    Solidarity
      Mechanical solidarity
      Organic solidarity
  Content
    Social structure
    Social status
      Ascribed status
      Achieved status
    Stigma
    Role
      Role set
      Rights

      Obligations
      Sick role
      Role strain
      Role conflict
  Dramaturgical model
    Impression management
    Front stage
    Back stage
  Attribution theory
    Dispositional traits
    Situational factors

internet assignment

Use a search engine to find two U.S. General Accounting Office (GAO) reports: *Blood Supply: Transfusion-Associated Risks* (February 1997) and *Blood Supply: FDA Oversight and Remaining Issues of Safety* (February 1997).

The American Association of Blood Banks (1998) claims that the risk of contracting HIV from a single blood transfusion is 1 in 676,000. Based on the information presented in the two GAO reports, do you believe this estimate is accurate? Explain.

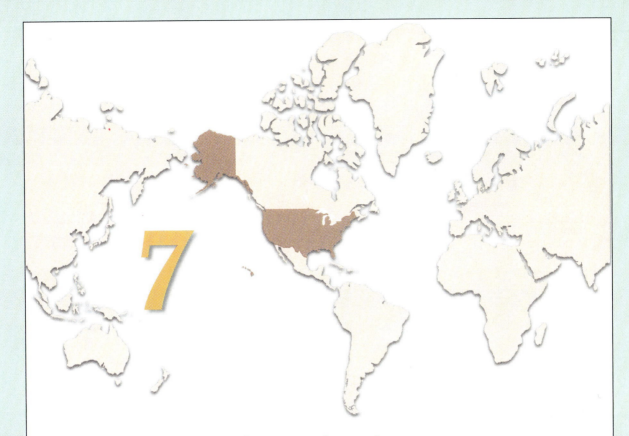

Social Organizations

With Emphasis on McDonald's, a U.S.-Based Multinational Corporation with Operations in 111 Countries

McDonald's golden arches in Xiamen, China. (Michael Justice/Gamma Liaison.)

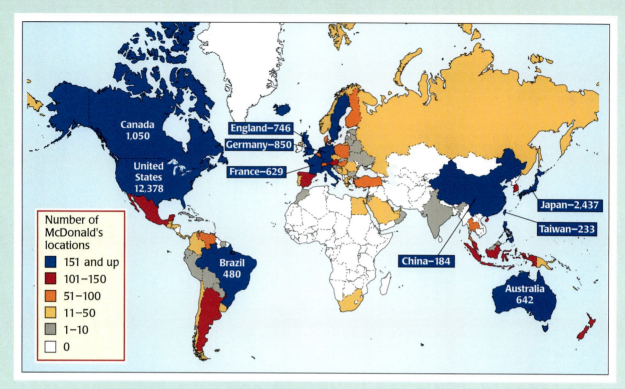

Source: McDonald's Corporation (1998).

McDonald's: A Multinational Corporation Headquartered in the United States

How many locations does McDonald's have in the United States? Which countries have the greatest number of locations? Can you tell from the map which parts of the world do not have McDonald's-affiliated operations? Why do you think this is the case?

In 1994, the U.S.-based McDonald's Corporation sued British vegetarian activists Helen Steel and Dave Morris. The multinational corporation accused the two of libel based on information contained in pamphlets they were distributing outside British McDonald's fast-food restaurants. Among other things, the pamphlets titled "What's Wrong with McDonald's? Everything They Don't Want You to Know" accused the company of (1) starving people in Third World countries, (2) destroying the rain forests, (3) falsely promoting the company as environmentally friendly, (4) hiding evidence of a connection between fast-food diets and specific medical conditions (in particular, heart disease, breast cancer, and cancer of the bowel), (5) supporting cruelty to animals (associated with rearing and slaughtering), and (6) practicing unfair labor practices (anti-union politics, poor working conditions) (London Greenpeace Group 1986).

Almost three years later, Judge Rodger Bell issued an 805-page ruling stating that McDonald's was "wrongly defamed" on many core allegations outlined in the pamphlet (Bloyd-Pashkin 1997). Bell ordered the activists to pay McDonald's $98,000 in damages. The judge did find, however, that McDonald's was cruel to some animals, "that the company ran ads aimed at getting children to 'pester' their parents into going to McDonald's" (Koenig 1997:20), and that the company paid low wages and sometimes sent teenage employees home when business was slow, further depressing their wages (Beveridge 1997).

In the end, both sides declared victory. A McDonald's spokesperson commented on the ruling: "We are, as you can imagine, broadly satisfied with the judgment . . . we wanted to show these allegations to be false and I am pleased" (Preston 1997:24). Supporters of Steel and Morris celebrated their victory over a corporation that spent $16 million to silence its critics and lauded the fact that the court case validated many of their claims against the fast-food giant, even claims that the judge ruled as "false." For example, the judge ruled that it was "unjust to blame McDonald's for the poor diets of many of its customers" (Koenig 1997:21). The judge did not rule that McDonald's served healthy food.

Why Focus on McDonald's?

In this chapter, we focus on one kind of organization—the multinational corporation. Multinational corporations, especially the 300 or so largest companies, are a major force in the globalization of the world economy and the prime agents behind the economic transformations taking place over the past 25 years (Barnet and Cavanagh 1994; McNeely 1996). In particular, we focus on a multinational corporation headquartered in the United States—McDonald's. We choose McDonald's because it is the number one fast-food chain in the United States and in the world. In addition, it is the world's most successful franchising operation (see Table 7.1); in fact, it has achieved global dominance through franchising. McDonald's is "the company that made fast food fast, that pioneered everything from drive-through windows to breakfast sandwiches" (Strother 1998). At the time of this writing, McDonald's has 14,000-plus franchises in the United States and more than 9,000 units in 110 other countries (McDonald's Corporation 1998).[1] On any given day, an estimated 38 million people eat at a McDonald's (Lowe and Nicholas 1997) and the fast-food chain estimates that it handles 14 billion customer visits per year (McDonald's Corporation 1998). Sixty percent of McDonald's sales and profits are derived from its international units. As one measure of its global presence, 45 percent of all globally branded fast-food restaurants outside the United States are McDonald's franchises (McDonald's Corporation 1998).

[1]McDonald's closest global competitor is Burger King with 7,539 restaurants in the United States and 2,060 international restaurants (Burger King 1998).

McDonald's is the focus of this chapter for reasons other than its size. First, the fast-food giant aggressively markets its products to children all over the world. Most of the company's promotions are aimed at children (Watson 1997). For example, Happy Meals are child-sized orders, priced at about $1.99 in the United States, that come with free toys such as Teenie Beanie Babies[2] or Disney movie characters such as 101 Dalmatians and

Mulan.[3] The toys are offered as sets; each toy in the set is distributed at different times during a promotion, encouraging several visits to collect the entire set.

A second reason for focusing on McDonald's relates to the terms under which the fast-food giant enters a foreign market. McDonald's, like all franchisers, licenses owners not just to sell its products, but also to use its name, business format, and operating methods (Hollander 1991). "The

franchisee is both an employee and an independent business person. They operate their own stores, but generally must adhere to stringent rules and regulations. . . . Rigorous attention to rule books is imperative to maintain 'shared identity'" (Waters 1998:2). For example, Chicken McNuggets in Croatia taste and look exactly like Mc-Nuggets in Kuwait City, Beijing, or Bison, Wyoming. One way the corporation builds its "shared identity" across 23,000-plus units is that it requires all franchise holders to attend a two-week course on quality control and management at one of its four Hamburger

[2]In 1997, McDonald's distributed more than 100 million Teenie Beanie Babies in about a week's time (Steen-huysen 1997).

[3]The "Mulan" toy is the first Happy Meal promotion carried out on a global scale (Hoffman 1998).

Table 7.1	Top 50 U.S. Restaurant Concepts, 1997		
Company	Location of Headquarters	Sales ($ millions)	Total Units
1. McDonald's	Oak Brook, IL	33,638	23,132
2. Burger King	Miami, FL	9,800	9,400
3. KFC	Louisville, KY	8,200	10,237
4. Pizza Hut	Dallas, TX	7,300	14,400
5. Wendy's	Dublin, OH	5,226	5,207
6. Taco Bell	Irvine, CA	4,900	6,941
7. Hardee's	Rocky Mount, NC	3,526	3,056
8. Subway	Milford, CT	3,300	13,016
9. Domino's Pizza	Ann Arbor, MI	3,200	5,950
10. Dairy Queen	Minneapolis, MN	2,540	5,792
11. Dunkin' Donuts	Randolph, MA	2,229	4,876
12. Little Caesars	Detroit, MI	2,100	4,300
13. Arby's	Ft. Lauderdale, FL	2,060	3,091
14. Denny's	Spartanburg, SC	1,900	1,652
15. Red Lobster	Orlando, FL	1,900	682
16. Applebee's	Overland Park, KS	1,818	960
17. The Olive Garden	Orlando, FL	1,400	466
18. Jack in the Box	San Diego, CA	1,300	1,323
19. T.G.I. Friday's	Dallas, TX	1,291	444
20. Chili's Grill & Bar	Dallas, TX	1,259	559
21. Outback Steakhouse	Tampa, FL	1,246	459
22. Boston Market	Golden, CO	1,200	1,166
23. Sonic Drive-Ins	Oklahoma City, OK	1,191	1,717
24. Shoney's	Nashville, TN	1,100	770
25. IHOP	Glendale, CA	903	787

Table 7.1 shows the top 50 U.S. restaurant concepts, 1997 sales, and the total units in 1997. If you were directing operations for one of these corporations, what policies might you want to establish to avoid problems, improve service, or enhance the way your company was regarded in other countries?

Universities or other international training centers (Watson 1997).[4]

In light of its stringent rules and regulations governing its operations, it is interesting that McDonald's prides itself on becoming as much a part of the local culture as possible. It leases its franchises to local partners and cultivates business relationships with local suppliers as well, coaching them on matters including raising, slaughtering, storing, deboning, and processing

[4]With the help of translators and electronic equipment, professors at McDonald's international training centers can communicate in 14 languages at once (McDonald's Corporation 1998).

the chickens and beef (Waters 1998; McGurn 1997). Although the basic menu is the same all over the world, McDonald's does allow its franchises to market food items that appeal to local clientele, such as the Shogun Burger in Japan, McSpaghetti (with chunks of tomato and hot dog) in the Philippines, McLaks (fresh grilled salmon sandwich) in Norway, and the McHuevo (a hamburger topped with a poached egg) in Uruguay (McDonald's Corporation 1998; Gilo and Welsh 1997).

Our emphasis of McDonald's operations does not mean that McDonald's is better or worse than other fast-

food corporations or multinational corporations. In fact, in 1997 a sample of 12,000 *Fortune* 1000 senior executives ranked McDonald's fourth among the largest 500 U.S. corporations on "corporate responsibility," suggesting the fast-food giant is considered a model (*Fortune* 1998b). Focusing on this well-known corporation, however, allows us to apply concepts sociologists use to analyze organizations to a specific case and allows us to consider a range of social and environmental issues connected with any organization, but especially with multinational corporations (see Table 7.2).

Company	Location of Headquarters	Sales ($ millions)	Total Units
26. Papa John's	Louisville, KY	867	1,517
27. Cracker Barrel	Lebanon, TN	863	307
28. Popeyes Chicken	Atlanta, GA	847	1,132
29. Long John Silver's	Lexington, KY	846	1,397
30. Baskin-Robbins	Glendale, CA	800	4,400
31. Tim Hortons	Dublin, OH	772	1,578
32. Golden Corral	Raleigh, NC	770	454
33. Coco's	Irvine, CA	730	493
34. Church's Chicken	Atlanta, GA	720	1,366
35. Perkins	Memphis, TN	711	473
36. Carl's Jr.	Anaheim, CA	703	708
37. Big Boy	Warren, MI	697	550
38. Friendly's	Wilbraham, MA	682	696
39. Sizzler	Culver City, CA	677	350
40. Chick-Fil-A	Atlanta, GA	672	763
41. Ruby Tuesday	Mobile, AL	650	369
42. Ryan's Steakhouse	Greer, SC	635	295
43. Ponderosa	Plano, TX	630	540
44. Waffle House	Norcross, GA	612	1,150
45. Starbucks Coffee	Seattle, WA	596	1,381
46. Bob Evans	Columbus, OH	594	398
47. Lone Star	Wichita, KS	585	308
48. Old Country Buffet	Eden Prairie, MN	537	245
49. Whataburger	Corpus Christie, TX	530	588
50. Luby's Cafeteria	San Antonio, TX	495	229

Source: From the "Top 400 Restaurant Concepts," *Restaurants and Institutions.* Copyright © 1997–1999 Cahners Business Information, a division of Reed Elsevier, Inc. Reprinted by permission.

Organizations

In this chapter, we examine concepts that sociologists use to analyze an **organization,** defined as a coordinating mechanism created by people to achieve stated objectives. Organizations are coordinating mechanisms because they bring together people, resources, and technology in such a way as to achieve a stated goal. Those objectives may be to maintain order (for example, a police department), to challenge an established order (for example, the McSpotlight Web site), to keep track of people (a census bureau), to grow, harvest, or process food (PepsiCo), to produce goods (Sony), to make pesticides (Union Carbide), or to provide a service (a hospital) (Aldrich and Marsden 1988).

From a sociological perspective, organizations can be studied apart from the people who form them. Indeed, organizations have a life that extends beyond the people who constitute them. This idea is supported by the simple fact that organizations continue to exist even as their members die, quit, or retire or get fired, promoted, or transferred. Organizations are a taken-for-granted aspect of life:

> Consider, however, that much of an individual's biography could be written in terms of encounters with [them]: born in a hospital, educated in a school system, licensed to drive by a state agency, loaned money by a financial institution, employed by a corporation, cared for by a hospital and/or nursing home, and at death served by as many as five organizations—a law firm, a probate court, a religious organization, a mortician, and a florist. (Aldrich and Marsden 1988:362)

Because organizations are so much a part of our lives, we rarely consider how they operate, how much power they wield, and how much social responsibility they assume. The concepts that sociologists use to study organizations help us understand how they can be powerful coordinating mechanisms that channel individual effort into achieving goals that benefit the lives of many people. At the same time, we can use these concepts to help us see how these coordinating mechanisms contribute to a wide range of social and environmental problems as well as the failure to take responsibility for these problems.

Organization A coordinating mechanism created by people to achieve stated objectives.

Multinational corporations Enterprises that own or control production and service facilities in countries other than the one in which their headquarters are located.

Table 7.2   Benefits and Drawbacks of Multinational Corporations

Frequently Heard Multinational Corporation Claims of Benefits for Host Nations

Provide new products

Introduce and develop new technical skills

Introduce new managerial and organizational techniques

Promote higher employment

Yield higher productivity

Provide greater access to international markets

Provide for greater accumulation of foreign exchange

Supplement foreign aid objectives and programs of home countries directed toward the host

Serve as a point of contact for host-country business-people and officials in the home country

Encourage the development of new ancillary or spin-off industries

Assume investment risks that might not have been undertaken by others

Mobilize capital for productive purposes that might have gone to other, less fruitful uses

Frequently Heard Criticisms of Multinational Corporations by Host Nations

Lead to a loss of cultural identity and traditions with the creation of new consumer tastes and demands

Be used as channels for foreign (especially U.S.) political influence

Possess a competitive advantage over local industries

Create inflationary pressures

Misapply host-country resources

Exploit host-country wealth for the primary benefit of the citizens of other nations

Lead to loss of control by hosts over their own economies

Possess neither sufficient understanding nor concern for the local economy, labor conditions, and national security requirements

Dominate key industries

Divert local savings from investment by nationals

Restrict access to modern technology by centralizing research and development facilities in the home country and by employing home-country nationals in key management positions

Source: Adapted from U.S. General Accounting Office (1978).

The Multinational Corporation: Agent of Colonialism or Progress?

Multinational corporations (or just "multinationals") are enterprises that own, control, or license production and service facilities in countries other than the one in which their headquarters are located. The United

U.S. in Perspective
The World's Largest Global Corporations, 1997

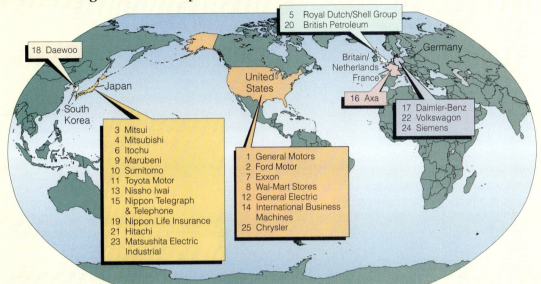

Source: *Fortune* (1998c).

This map shows the headquarters of the 25 largest industrial corporations in the world in 1997. Should it be any cause for concern to the rest of the world that 20 of the 25 have their headquarters in just three countries? What factors might cause companies based in the United States, Germany, and Japan to lose touch with the needs and interests of the people in the many other countries in which they hold great economic power? What sorts of "local" concerns might they tend to overlook in those other countries?

If you were directing operations for one of these corporations, what policies might you want to establish to avoid problems, improve service, or enhance the way your company was regarded in other countries?

The World's Largest Global Corporations, 1997

Company	Profits ($ millions)	Number of Employees
1. General Motors	6,698.0	608,000
2. Ford Motor	6,920.0	363,892
3. Mitsui	268.7	40,000
4. Mitsubishi	388.1	36,000
5. Royal Dutch/Shell Group	7,758.2	105,000
6. Itochu	(773.9)	6,675
7. Exxon	8,460.0	80,000
8. Wal-Mart Stores	3,526.0	825,000
9. Marubeni	140.4	64,000
10. Sumitomo	209.8	29,500
11. Toyota Motor	3,701.3	159,035
12. General Electric	8,203.0	276,000
13. Nissho Iwai	24.7	18,158
14. International Business Machines	6,093.0	269,465
15. Nippon Telegraph & Telephone	2,361.3	226,000
16. Axa	1,357.0	80,613
17. Daimler-Benz	4,639.2	300,068
18. Daewoo	526.9	265,044
19. Nippon Life Insurance	2,118.3	75,851
20. British Petroleum	4,046.2	56,450
21. Hitachi	28.3	331,494
22. Volkswagen	772.4	279,892
23. Matsushita Electric Industrial	762.5	275,962
24. Siemens	1,427.4	386,000
25. Chrysler	2,805.0	121,000

Source: *Fortune* (1998a).

Nations estimates that at least 35,000 multinationals operate worldwide with 150,000 foreign affiliates (Clark 1993).[5] Multinationals are headquartered disproportionately in the United States, Japan, and Western Europe (see "The World's Largest Global Corporations, 1997"). They compete against rival corporations for global market share, and they plan, produce, and sell on a multicountry and even a global scale. In addition, they recruit employees, extract resources, acquire capital, and borrow technology on a multicountry or worldwide scale (Kennedy 1993; Khan 1986; U.S. General Accounting Office 1978) (See "How Did McDonald's Become a Global Corporation?").

Multinationals establish operations in foreign countries for many reasons, including to obtain raw materials or use an inexpensive labor force (for example, a *maquila* assembly plant in Mexico, a lumber company in Brazil, a mining company in South Africa). They also establish subsidiary companies in foreign countries and employ their citizens to manufacture goods, provide services, or sell products that are marketed to customers in those countries (for example, IBM Japan).

Critics maintain that multinational corporations are engines of destruction. That is, they exploit people and resources to manufacture products inexpensively. They take advantage of cheap and desperately poor labor forces, lenient environmental regulations, and sometimes nonexistent worker safety standards. According to these critics, multinational corporations represent another kind of colonialism. Supporters of multinational corporations, on the other hand, maintain that these companies are agents of progress. They praise the multinationals' ability to raise standards of living, increase employment opportunities,[6] transcend political hostilities, transfer technology, and promote cultural understanding.

In this regard, the president of McDonald's international operations noted that his corporation does not "force" itself on foreign countries, rather, governments around the world actively recruit the company:

> I feel these countries want McDonald's as a symbol of something—an economic maturity and that they are open to foreign investments. I don't think there is

a country out there we haven't gotten inquiries from. I have a parade of ambassadors and trade representatives in here regularly to tell us about their country and why McDonald's would be good for the country. (Freidman 1996:E15)

In reality, we cannot make a simple evaluation that would apply to all multinationals (see Table 7.2). Obviously, at some level they "do spread goods, capital, and technology around the globe. They do contribute to a rise in overall economic activity. They do employ hundreds of thousands of workers around the world, often paying more than the prevailing wage" (Barnet and Müller 1974:151). George Keller (1996), chairman of the board and chief executive officer of Chevron Corporation, explains:

> To conduct our operations, we had to help create the necessary environment. We drilled water wells, built roads, and developed electrical power. As we made progress, the local communities grew and developed into prosperous cities. Schools and hospitals were built, and local industries emerged. (pp. 125–126)

Even so, the means that multinational companies employ to achieve the maximum profit for owners and stockholders (the valued goal) do not necessarily alleviate a host country's problems of poverty, hunger, mass unemployment, and gross inequality (see "The Prawn Aquaculture Case"). Critics argue that, if anything, multinationals aggravate these problems because their pursuit of profit is closely related to social inequalities and ecological imbalances that are visible to anyone visiting a country such as India:

> What a curious contradiction of rags and riches. One out of every 10,000 persons lives in a palace with high walls and gardens and a Cadillac in the driveway. A few blocks away hundreds are sleeping in the streets, which they share with beggars, chewing gum hawkers, prostitutes, and shoeshine boys. Around the corner tens of thousands are jammed in huts without electricity or plumbing. . . . The stock market is booming, but babies die and children with distended bellies and spindly legs are everywhere. (Barnet and Müller 1974:133–134)

How are multinationals connected with this kind of social imbalance, when it seems that "without the technologies and the capital that multinationals help to introduce, such countries would have little hope of eradicating poverty and hunger" (*Union Carbide Annual Report* 1984:107)?

As one example of how such imbalances arise, consider that in 1995 the government of India agreed

---

[5]According to the U.S. Internal Revenue Service (1998), in 1995—the last year for which data are available—60,157 corporations in the United States were foreign-controlled.

[6]In its 1997 annual report, McDonald's notes that by the turn of the century the company will create as many as 200,000 new jobs— "many of them in countries where generating new employment opportunities is a critical issue" (McDonald's Corporation 1998).

How Did McDonald's Become a Global Corporation?

One answer to how corporations such as McDonald's grew from one restaurant (in Illinois in 1955) to a global giant lies with the means the company employed to increase profits. McDonald's used the following strategies to grow the company.

Lower production costs. This is achieved by hiring employees who will work for lower wages, by introducing labor-saving technologies, by reducing the number of employees,[1] and by moving production facilities out of high-wage zones. One example of a major labor-saving technology occurred in the mid-1960s, when McDonald's replaced in-store potato peeling and slicing appliances with a system of flash-freezing half-cooked potatoes. This change not only lowered the labor cost associated with producing french fries, but also allowed the company to distribute uniform and consistently prepared french fries (Johnson 1997). A second strategy to lower production costs is to identify new sources of low-wage labor. McDonald's boasts that it is the largest employer of teenagers and that it actively recruits minorities, senior citizens, and the disabled (Sheridan 1998). While McDonald's is to be congratulated for its policy of inclusiveness, we must also acknowledge that the company is identifying and finding new sources from which to draw low-wage workers.

Create new products that consumers need to buy. Obviously, every new menu item represents a "new product." McDonald's and other fast-food chains have been particularly successful in marketing their products to children. The company offers Happy Meals that come with toys that many children (and some adults) "need to have," such as Teenie Beanie Babies and toys made popular by Disney movies such as "Mulan," "Babe," "The Jungle Book," and "Pocahontas."[2]

Improve on existing products and make previous versions obsolete. This strategy includes making existing food items bigger and better. Fast-food

companies compete, for example, to make french fries better (hotter or crispier) than their competitors do. Jack-in-the-Box has developed a transparent potato starch to coat its fries. The coating keeps fries hotter and crispier for as much as 15 minutes longer than uncoated fries (Johnson 1997).

Expand the outer boundaries of the world economy and create new markets. The opening map to this chapter shows the countries in which McDonald's has fast-food restaurants. McDonald's began its expansion outside the United States in 1967, when it opened a unit in Canada. It now operates units in 110 foreign countries. In its annual report to stockholders, McDonald's distinguishes among established, well-penetrated markets, such as Australia, Brazil, Japan, and the United States, and emerging markets that "have been in operation for more than five years, and have relatively low penetration and significant near-term growth potential," such as Argentina, China, Italy, and Spain. The company has also identified more than three dozen countries where it is introducing the fast-food industry, including Bolivia, Egypt, India, and the Ukraine. Major competitors to McDonald's also have units in foreign countries (see Table 7.3). Wendy's boasts restaurants in 34 countries and, according to the company's 1997 shareholder report, it plans "to add units in markets where we have a meaningful presence or recognize an opportunity to establish a base in countries with strong or emerging middle classes" (p. 12).

In their drive to create new markets, executives of fast-food restaurants have come to realize that customers make choices about where to eat on the spur of the moment and based on convenience (the closer the restaurant, the better the chance that customers will choose to eat there). As a result, corporations have located restaurants in Wal-Mart and Home Depot stores, gas stations, malls,

airports, college campuses, hospitals, and military bases (Mannex 1996; Crecca 1997).

Identify ways that enable or encourage people to purchase more products and services. Suggestive selling is one way to increase sales. Order takers ask the customers if they would like to "Supersize" their value meal, meaning that, for a small charge (between 39 and 50 cents), customers can receive a large fry and drink instead of a medium fry and drink with their meal.

Market researchers and financial analysts have also found that when consumers defer payments until a later date they are likely to spend more at the time of purchase. Although credit cards are not widely used to purchase fast food, we are beginning to see a trend in this direction (Stone 1998). McDonald's, for example, has located restaurants in approximately 700 Wal-Mart stores. When Wal-Mart customers check out after shopping in the store, they have an opportunity to order food. Their order is relayed from the register to the McDonald's kitchen and the food is brought to the customer as he or she leaves. As many people use credit cards to purchase Wal-Mart items, those short of cash may decide to order dinner on credit as well (Frank 1997).

Recall that McDonald's corporate headquarters coordinates the activities of 1 million employees in 23,000 restaurants located in 111 countries to feed 38 million people each day. The concept of bureaucracy helps us to understand the organizational mechanisms underlying this monumental task.

---

[1] For example, Burger King reduced the number of employees per store to an average of 30 (down from an average of 45) (Stone 1997).

[2] As one measure of the success of this marketing strategy, consider that the book *Fast Food Figures* features hundreds of color photographs dating back to 1987 for collectors of such toys (Beech 1998).

Table 7.3 The Global Reach of McDonald's and Its Major Competitors

Restaurant	Founded	Number of Units/Countries	Domestic/Foreign	Number of People Fed
McDonald's	1955	23,000+/111	13,000/10,000	38 million/day
Burger King	1954	9,644/56	7,539/2,060	13.2 million/day
KFC	1952	10,237/79	5,120/5,117	7 million/day
Pizza Hut	1959	12,434/86	8,698/3,836	——
Wendy's	1969	5,207/34	4,575/632	——
Taco Bell	1962	6,941/16	6,768/173	7.9 million/day
Hardee's	1961	3,023/11	——	——
Subway	1974	13,000/64	11,000/3,000	——
Dominos	1960	5,952/62	4,431/1,521	325 million pizzas/year
Dairy Queen	1940	4,790/25	——	

Sources: Burger King (1998); Dairy Queen (1998); Dominos (1998); Hardees Food Systems (1998); McDonald's (1998); Subway (1998); Triconglobal (1998).

U.S. in Perspective

The Prawn Aquaculture Case

In an interview with a reporter for *Multinational Monitor* (1995), environmental lawyer M. C. Mehta describes his work on the shrimp aquaculture case before the Supreme Court of India.

MM: Could you describe the shrimp aquaculture case which you have brought in the Supreme Court?

MEHTA: The prawn aquaculture case, filed in 1994, challenges the operation of prawn farms in the coastal areas, especially in Tamil Nadu, where they are destroying the local ecology. After four or five years, the lands where the prawn farms are located will become useless; it is very difficult to reclaim the land once it has been flooded and made into a prawn farm. First, the land becomes unfertile. Second, the land becomes saline. Third, the mangroves and other vegetation cover along the coastline are destroyed.

Community residents have undertaken a grassroots fight

This prawn farm, located in Sri Lanka, exports prawns raised there to the United States, Japan, and other countries.

© Dominic Sansoni/Panos Pictures (left); © Anne Dowie (right)

to allow Pepsi Foods to open 30 Pizza Hut and 30 Kentucky Fried Chicken restaurants. Livestock farms[7] in India supply the meat needed to make fast-food products for the domestic market and to export to other fast-food restaurants abroad. In light of the fact that 40 percent of India's population remains too poor to afford an adequate diet (U.S. Department of Commerce Guides 1997), can India afford to divert grains such as corn[8] and soybeans to feed the animals slaughtered for fast-food products that are then purchased by middle-

[7]In India, Hindus consider cows to be sacred and Muslims avoid pork. Consequently, livestock farms raise chickens and sheep for McDonald's restaurants in India (Karp 1996).

[8]To get some idea of the amount of corn consumed by cattle, consider that according to National Cattlemen's Beef Association (1998) U.S. ranchers fed 2.9 billion bushels of corn to 99.5 million cattle in 1997.

class consumers? Should the grain be distributed instead to the masses of malnourished and undernourished people? Because it is more profitable, farmers would rather grow grain to sell to the poultry and other meat industries rather than to people in village markets (Gandhi 1995) (see Figure 7.1). Of course, Pepsi Foods and other fast-food chains such as McDonald's cannot shoulder the entire blame for the gross inequalities and ecological imbalances in India and other countries with large populations living in poverty. After all, this example illustrates only one business practice among hundreds that exacerbate such inequalities.

One can also argue that multinationals are not responsible for the behavior of the people who purchase or supply their products. Nevertheless, many people question whether the multinationals and other corporations should have the right to ignore the larger

against the prawn companies. For the last three years, the people suffering from this activity have agitated, held nonviolent protests, sat in satyagrahas, and many activists have been arrested on a number of occasions.

These people came to me and asked for assistance. I went to Tamil Nadu, and collected facts. On the basis of all those facts, we filed a petition in the Supreme Court of India challenging the violations committed by the prawn companies. The Supreme Court directed scientists from the National Environment Engineering Research Institute (NEERI), a prestigious institute, to go and see the situation and report to the Court.

The NEERI scientists went there and examined the situation and concluded that the prawn farms are seriously degrading the environment. Their report concludes that this activity is more harmful than beneficial.

After NEERI submitted the report, the Supreme Court issued an order saying no more agricultural land or salt pans can be converted to prawn farming.

But the mangrove forests are still being destroyed all along the coastline. The trees are being cut, and the salinity is increasing in those waters, and the people have even lost their drinking water sources.

The big multinational companies, in conjunction with Indian businessmen, have set up these prawn farms without taking any precautionary measures to control pollution. They have destroyed thousands of acres of land in this area, which is all along the nine coastal states of India.

The prawns are exported to the United States, Japan and other countries. The argument is being made that India needs foreign exchange, and that the prawn farming benefits India because it brings in foreign exchange. But I think that U.S. citizens should boycott such things, particularly from India. The prawns may be a delicacy for U.S. citizens who are eating them, but

this delicacy is depriving the livelihood and bringing untold suffering to thousands of fishermen and other people who have become victims of serious pollution and environmental degradation caused by the prawn industries. The people who have lived for centuries in these areas are being displaced.

MM: What has been the reaction to the Court order?

MEHTA: The Supreme Court order is an interim one, with the final hearing on the matter set for August, 1995. Because the multinationals have a lot of money, they feel they can do anything. The rich multinationals and big businessmen housed in India have started a campaign to portray prawn farming as a very good activity. They are hiring consultants and scientists and are trying to say that prawn farming will generate employment in the country and help poor people.

Source: Interview with M. C. Mehta for *Multinational Monitor.* Copyright © 1995 Essential Information. Reprinted by permission of Multinational Monitor, PO Box 19405, Washington, DC 20036.

Figure 7.1   Livestock Numbers (in millions)

This chart shows the number of livestock by region in 1984 and 1994. Think about the amount of grain and the facilities needed to support these numbers. Do you think people's access to grain is affected by the need to feed livestock?

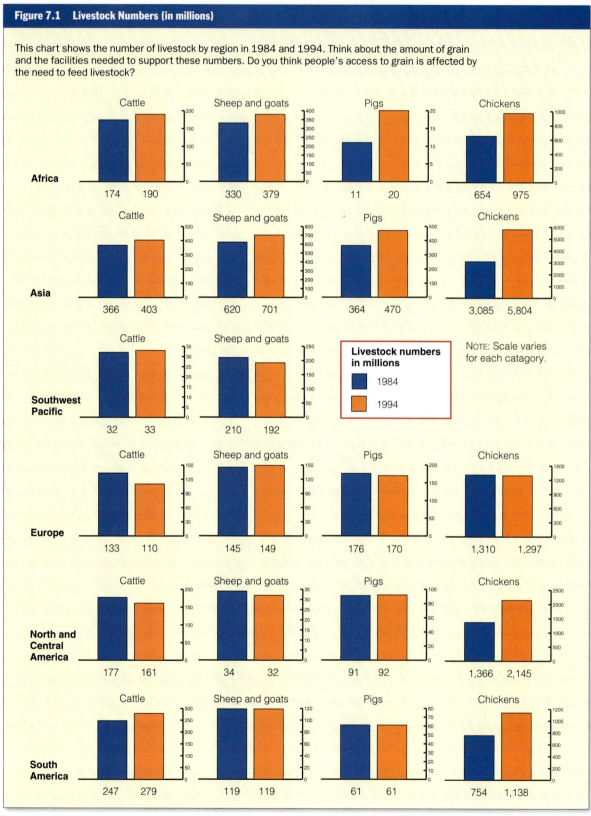

Source: Food and Agriculture Organization of the United Nations (1996).

long-term effects of their products and business practices on people and the environment, even as they respond to consumer demand.

Consider the statement of Keith Richardson, the public affairs chief of B.A.T Industries,[9] the world's largest manufacturer of cigarettes. When asked whether the company felt some obligation to put health warnings on its cigarette packs sold in the Third World similar to the kind of health warnings required in Great Britain, the United States, and other countries, he replied, "These are sovereign countries, and they are unenthusiastic about being told what to do by pressure groups in the U.K. and the United States. We fit in with what the government wants in each country. We let the marketplace decide. We do not try to impose" (Moskowitz 1987:40–41).

The fast-food industry also responds to consumer tastes. Virtually all of the major fast-food companies have introduced low-fat foods on their menus, and most have proved unpopular with consumers. For example, McDonald's introduced the reduced-fat McLean hamburger and Hardee's responded with the Lean One. Taco Bell even gave away 8 million Border Lights products (low-fat tacos and burritos) and spent $20 million on advertising in 1995 to launch lower-fat menu items. In the end, Taco Bell lost customers who mistakenly perceived the company as shifting its menu from high-fat to low-fat items. Stockholders pressured the chain to abandon the campaign. Despite marketing campaigns promoting healthier choices, it is clear that consumers prefer menu items higher in calories and fat (such as McDonald's Arch Deluxe and the Hardee's Monster Burger) (Stone 1997a).

These examples show that the most profitable product for a corporation may not be profitable for a society, because the society in question pays tremendous **externality costs**—costs that are not figured into the price of a product but that are nevertheless eventually paid by the consumer when using or creating a product (Lepkowski 1985). Examples of such costs include the cost of restoring contaminated and barren environments, restoring health, and so on. Regardless of the position we take toward corporate responsibility, our views will be more realistic and well informed if we understand how organizations in general operate. We turn first to the systematic procedures that they follow to produce and distribute goods and services in the most efficient (especially the most cost-efficient) manner. A key concept to understanding these workings is that of the bureaucracy.

[9]B.A.T Industries does not use a period after the letter "T" (Moskowitz 1987:37).

This cigarette advertisement in Tokyo does not carry the kind of health warning that is legally required in the United States. When profit-making organizations are not operating under legal constraints, do they have any obligations to protect consumers from the hazards associated with their products?

© Robert Wallis

Features of Modern Organizations

Sociologist Max Weber gives us one framework for understanding the two faces of organizations: organizations are capable of (1) efficiently managing people, information, goods, and services on a worldwide scale and (2) promoting inefficient, irresponsible, and destructive actions that can affect the well-being of the entire planet. Weber's ideas about organizations rely on an understanding of rationalization and its significance in modern life.

Rationalization as a Tool in Modern Organizations

In Chapter 1, we learned that ever since the onset of the Industrial Revolution, people's thoughts and actions were less likely to be guided by tradition or emotion and more likely to be value-rational. Weber defined

Externality costs Costs that are not figured into the price of a product but that are nevertheless eventually paid by the consumer when using or creating a product.

rationalization as a process whereby thought and action rooted in emotion (love, hatred, revenge, joy), superstition, respect for mysterious forces, and tradition become replaced by thought and action grounded in the logical assessment of the most efficient means to achieve a particular valued goal or end (Freund 1968).

One way to show how the thought and action guided by tradition, superstition, and emotion differ from the thought and action guided by value-rational action is to compare two distinctly different meanings applied to trees. One meaning is held by the Bonda, a small tribe that lives in India's Orissa Mountains; the other is held by modern science and industry.

The Bonda culture is rooted in emotion, superstition, and respect for mysterious forces. The Bonda believe that spirits inhabit the earth, the sky, and the water; they believe that sickness, death, and poor harvests are caused by evil spirits. They are particularly respectful of spirits who live in trees and plants. Bonda priests specify which trees can and cannot be cut down. Members of the tribe do not touch those trees considered to be the homes of gods and genies. Some trees are left standing if the priests believe that their removal would displease phantoms or demons and would cause them to send poor harvests, sickness, and deaths to avenge crimes against trees (Chenevière 1987). In essence, the belief that trees, plants, and animals possess souls leads people to feel reverence and respect for nature and to behave accordingly toward it.

Science and technology, on the other hand, have enabled us to break down trees and plants into various parts and assign them a precise function. In contemporary society, for example, trees are envisioned in value-rational terms—as a means to an end or valued goal. Trees are viewed as a source of food, wood, rubber, quinine (a drug used to combat malaria), turpentine (an ingredient of paint thinner and solvents), cellulose (used to produce paper, textiles, and explosives), and resins (used in lacquers, varnishes, inks, adhesives, plastics, and pharmaceuticals). From a value-rational point of view, nature is something to use: "Rivers are something to dam; swamps are something to drain; oaks are something to cut; mountains are something to sell; and lakes are sewers to use for corporate waste"

Rationalization A process whereby thought and action rooted in emotion, superstition, respect for mysterious forces, and tradition are replaced by thought and action grounded in the logical assessment of cause and effect or the means to achieve a particular end.

(Young 1975:29). It is not that people in "rational" environments do not value nature on some level; it is just that they place greater value on the goals of profit, employment, convenience, and global competition.

As another example of "means to an end" thinking, consider that Pizza Hut (1997) views cows as the source of milk for pizza cheese. According to the company Web site, "that cheese production requires a herd of 250,000 dairy cows producing at full capacity 365 days a year." Likewise, the National Cattlemen's Beef Association (1998) views cows as a source of meat and its members have succeeded in "creating" a cow that, in 1998, yielded an average of 586 pounds of meat, up from 449 pounds in 1980.

Weber made several important qualifications regarding value-rational thought and action. First, he used the term *rationalization* to refer to the way in which daily life is organized socially to accommodate large numbers of people, and not necessarily the way that individuals actually think (Freund 1968). With regard to satisfying hunger, for example, fast food is marketed as the most efficient means of achieving a state of feeling full:

> In a highly mobile society in which people are rushing, usually by car, from one spot to another, the efficiency of a fast-food meal, perhaps without leaving one's car while passing by the drive-through window, often proves impossible to resist. The fast-food model offers us, or at least appears to offer us, an efficient method for satisfying many of our needs. (Ritzer 1993:9)[10]

Second, rationalization does not assume better understanding or greater knowledge. People who live in a value-rational environment typically know little about their surroundings (nature, technology, the economy). "The consumer buys any number of products in the grocery without knowing how they are made or of what substances they are made. By contrast, 'primitive' man in the bush knows infinitely more about the conditions under which he lives, the tools he uses and the food he consumes" (Freund 1968:20).

Chemical corporations, for example, manufacture pesticides and advertise them as a rational means for the quick and efficient killing of bugs in the house, in the garden, or on pets. Yet, the millions of people who buy and use these products have no idea how the chemicals

---

[10]Sixty percent of Burger King's sales are from drive-through business; 55 percent of Wendy's International's sales are from drive-through business (Lindeman 1998).

work, where they come from, or what consequences they bring except that they kill bugs. Thus, on a personal level, people deal with pesticides as if they were magic. In a similar vein, most people who eat fast-food products on a regular basis have no concept as to how these foods are produced or the food's effect on the body—in other words, they have no understanding of the processes by which the digestive system "ingests, absorbs, transports, utilizes and excretes food substances" (National Library of Medicine 1998). Most people are not troubled by such ignorance but are content to let specialists or experts make corrections when something goes wrong.

Finally, when people are determining a valued goal and deciding on the means (actions) to achieve it, they seldom consider less profitable or slower ways to reach the stated goal. For example, people often turn to technology as the means of solving problems that they define as important. The problem may be something as seemingly simple as growing potatoes that can be processed at a low cost into a uniform-looking and -tasting french fry. In making the potato, we fail to consider that the demand for uniformity limits the varietal range of potatoes to those that are high-yielding, high in dry matter, low in sugar content, long and oval in shape, and uniform in color and flavor (International Potato Center 1998). With the help of science, such potatoes can be produced—but they require heavy doses of chemicals and threaten the longer-term viability of domestic potato production[11] to meet commercial processing requirements (International Potato Center 1998).

When creating the "perfect" potato or any other product, rarely does anyone ask questions such as the four listed here to evaluate its consequences for the overall quality of life on the planet (adapted from Standke 1986, p. 66):

1. Is this technology directed toward helping achieve the highest possible human goals?

2. Does this technology use mineral and energy resources efficiently and preserve or enhance the environment?

3. Does this technology preserve or enhance "good work" for the maximum number of human beings?

4. Is this technology founded on the very best scientific and technical information in combination with the wisdom and highest values of the culture?

More often than not, people set valued goals without first evaluating the possible disruptive or

Cars drive through a foreign McDonald's.
© Sue McDermott

Uniformity is evident in mass-produced french fries.
© Peter Menzel/Stock Boston

[11]Keep in mind that hundreds of varieties of potatoes exist in all shapes, colors, and sizes.

destructive social consequences of the means or strategies used to reach them. This failure to consider such consequences lies at the heart of the "destructive side" of organizations in general and of multinational corporations in particular.

The Concept of Bureaucracy

Max Weber defined a **bureaucracy,** in theory, as a completely rational organization—one that uses the most efficient means to achieve a valued goal, whether that goal is feeding people, making money, recruiting soldiers, counting people, or collecting taxes. The following are some major characteristics of a bureaucracy that allow it to coordinate people so all of their actions center on achieving the goals of the organization. After identifying each characteristic, we provide one example in parentheses that shows how it applies to McDonald's organizational structure.

- A clear-cut division of labor exists: each office or position is assigned a specific task toward accomplishing the organizational goals. (One of McDonald's organizational goals is to deliver a meal 90 seconds after it is ordered. When a customer places an order, such as a Big Mac, it appears on a video screen in the kitchen. The employee known as the initiator reads the screen and sends the order off to be prepared. The next step is to warm the bun. The third step is assembly, which includes adding condiments and other toppings. The fourth step involves wrapping the order, at which point a hot hamburger patty is added.) (*CNN Newstand Fortune* 1998)

- Authority is hierarchical: each lower office or position is under the control and supervision of a higher one. (Individual McDonald's franchises are under the control of the corporate office.)

- Written rules and regulations specify the exact nature of relationships among personnel and describe the way in which tasks should be carried out. (McDonald's issues a 600-page *Operations and Training Manual* that specifies everything from where sauces should be placed on

buns to how thick pickle slices should be.) (Watson 1997)

- Positions are filled on the basis of qualifications determined by objective criteria (academic degree, seniority, merit points, or test results) and not on the basis of emotional considerations such as family ties or friendship. (All franchisers must attend a two-week training program at Hamburger University.)

- Administrative decisions, rules, regulations, procedures, and activities are recorded in a standardized format and preserved in permanent files (such as McDonald's *Operations and Training Manual*).

- Authority belongs to the position and not to the particular person who fills the position or office. The implication is that one person can have authority over another on the job because he or she holds a higher position, but those in higher positions can have no authority over another worker's personal life away from the job. (A McDonald's manager has authority over employees only when they are on the time clock.)

- Organizational personnel treat clients as "cases" and "without hatred or passion, and hence without affection or enthusiasm" (Weber 1947:340). This approach is believed necessary because emotion and special circumstances can interfere with the efficient delivery of goods and services. (According to standard operating procedures, every customer should be greeted with the words "Welcome to McDonald's. May I take your order?")

Taken together, these characteristics describe a bureaucracy as an **ideal type**—ideal not in the sense of being desirable but as a standard against which real cases can be compared. An ideal type is a deliberate simplification or caricature in that the characteristics emphasized exaggerate certain aspects of a bureaucracy, which highlights them as important and essential features (Sadri 1996). In other words, real cases can be compared against the ideal. Anyone involved with an organization realizes that actual behavior departs from this ideal. Thus, one might ask, why should we list essential characteristics if no organization exemplifies them? This list is a useful tool because it identifies important organizational features. (Note, however, that having these traits does not guarantee that things will run perfectly; in fact, the rules and policies themselves can cause problems.) By comparing the actual operation with the ideal, one possesses a "checklist" of characteristics to

Bureaucracy An organization that uses the most efficient means to achieve a valued goal.

Ideal type A deliberate simplification or caricature in that the characteristics emphasized exaggerate certain aspects of a bureaucracy, which makes them the objects of comparison.

evaluate the extent to which an organization departs from them or adheres to them too rigidly. As we will learn, most organizational problems result from following too closely or varying too dramatically from these essential characteristics.

Factors That Influence Behavior in Organizations

On paper, the job descriptions, relationships among personnel, and procedures for performing work-related tasks are well defined and predictable. The actual workings of organizations are not as predictable, however, because the people involved with organizations vary in the extent to which they adhere to rules and regulations. Three factors influence how people act: (1) informal actions that depart from formal policy, (2) the way employees are trained to do their jobs, and (3) the way workers' performances are evaluated.

Formal Versus Informal Dimensions of Organizations

Sociologists distinguish between formal and informal aspects of organizations. The **formal dimension** consists of the official, written guidelines, rules, regulations, and policies that define the goals of the organization and its relationships to other organizations and to integral parties (for example, suppliers, government, or stockholders).

The **informal dimension** includes employee-generated norms that evade, bypass, do not correspond with, or are not systematically stated in official policies, rules, and regulations. In the most general sense, this term applies to behavior that does not correspond to written plans (Sekulic 1978). Examples include unwritten rules governing interactions among employees, customers, and management. A boss who expects employees to work off the clock so he or she can meet goals related to labor costs is an example of the informal dimension of organization. An employee who gives free food and soft drinks when the manager is not looking to his or her friends or who spits in a rude customer's drink represents another example.

Another aspect of the informal organization is worker-generated norms that govern output or physical effort. These include informal norms against working too hard (workers who do so are often called "rate busters"), working too slowly, or slacking off, as well as norms about the number and length of coffee breaks and the length of the lunch break. According to the results of the Second Annual Survey of Restaurants and Fast Food Employees, worker-generated norms may govern how workers respond to managers in the industry who treat them unfairly. On average, entry-level fast-food service employees claim they steal $238.72 in cash and merchandise over a year's time and they are more likely to do so when they believe they have been treated unfairly (Lowe 1997).

In the British McDonald's court case, many former employees testified about employees and managers who routinely violated formal corporate policies so as to meet profit-related goals, such as squeezing fry boxes before filling to make them look fuller; watering down soft drinks, syrups, and shake mix; and failing to throw away food that had dropped on the floor (Coton 1995; Beech 1994; Brett 1993).

Two other factors affect the way that people behave in organizations—specifically, whether they behave in flexible or rigid ways. One factor is how people are trained to do their jobs. The other is how the organization evaluates worker performance.

Trained Incapacity

If an organization is to operate in a safe, creditable, predictable, and efficient manner, its members need to follow rules, guidelines, regulations, and procedures. Organizations train workers to perform their jobs a certain way and reward them for good performances. When workers are trained to respond mechanically or mindlessly to the dictates of the job, however, they risk developing what economist and social critic Thorstein Veblen (1933) called **trained incapacity,** the inability to respond to new and unusual circumstances or to recognize when official rules and procedures are outmoded or no longer applicable. In other words, workers are trained to do their jobs only under normal circumstances and in a certain way; they are not trained to respond in imaginative and creative ways or to

---

Formal dimension The official, written guidelines, rules, regulations, and policies that define the goals and roles of the organization and its relationship to other organizations and with integral parties.

Informal dimension Owner- or employee-generated norms that evade, bypass, do not correspond with, or are not systematically stated in official policies, rules, and regulations.

Trained incapacity The inability to respond to new and unusual circumstances or to recognize when official rules and procedures are outmoded or no longer applicable.

Because of advances in computer technology, many organizations now run daily production and manufacturing operations from computer terminals and control panels.

© John Coletti/Stock, Boston

anticipate what-if scenarios so that they can perform under a variety of changing circumstances.

In her 1988 book, In *the Age of the Smart Machine,* social psychologist Shoshana Zuboff[12] distinguishes between work environments that promote trained incapacity and those that promote empowering behavior. In her research, Zuboff found that management can choose to use computers as automating tools or as informating tools. *To automate* means to use the computer to increase workers' speed and consistency or as a source of surveillance (for example, by checking up on workers or keeping precise records on the number of keystrokes per minute). "Smart" equipment such as automatic timers, cash registers that calculate change, and time clocks monitoring speed by which orders are filled—anything that does the thinking for employees, "watches" them, or pushes them to produce—represents examples of computers as automating tools.

On the other hand, management can choose to use computers as informating tools. *To informate* means to

empower workers with decision-making tools, such as employee-scheduling software that ensures that a sufficient number of employees are scheduled for the busiest times and shifts. Other examples of decision-making tools include software to keep track of payroll, sales, inventory, and purchasing. On the surface it may seem as if the software is doing the work for managers. Keep in mind that managers must interpret the results and use this information to make decisions. Workers who use the computer as an informating tool experience work very differently than those who use the computer as an automating tool.

So far we have looked at informal relationships between people in organizations and at the ways in which people are trained to do their jobs to understand the factors that determine how closely workers adhere to organizational rules and regulations. Now we turn to a third factor: how organizations evaluate job performances.

Statistical Records of Performance

In large organizations, executives and other managers often compile statistics on profits, losses, market share, customer satisfaction, total sales, production quotas, and employee turnover as a way to measure individual, departmental, and overall organizational performance. Such measures can be convenient and useful management tools for two reasons: they are considered to be objective and precise, and they permit systematic comparison of individuals across time and departments. On the basis of numbers, management can reward good performances through salary increases, profit-sharing, and promotions and can take action to correct poor performances. Sociologist Peter Blau examined the problems that can occur when managers use statistical measures without taking their shortcomings into consideration.

One problem with statistical measures of performance is that a chosen measure may not be a valid indicator of what it is intended to measure or may measure performance by a too-narrow criterion. For example, occupational safety is often measured by the number of accidents that occur on the job. On the basis of this indicator, the chemical industry has one of the lowest accident rates among all industries. This measure, however, has been criticized as too narrow and lacking validity: chemical workers are more likely to suffer from exposure-related illnesses with symptoms that may take years to develop and that cannot be directly connected to their work in the same way that heat burns from cooking hamburgers can.

[12]Zuboff's conclusions result from more than a decade of field research in various work environments, including pulp mills, a telecommunications company, a dental insurance claims office, a large pharmaceutical company, and the Brazilian offices of a global bank. All of these workplaces had one trait in common: the workers were learning to use computers.

A second problem with statistical measures of performance is that they encourage employees to concentrate on achieving good scores and to ignore problems generated by their drive to score well. In other words, people tend to pay attention only to those areas that are being measured and to overlook those for which no measures exist. "Sales increases" are a common statistical measure of performance. Employees are asked to meet hourly, weekly, monthly, or annual sales goals. In addition, sales goals often increase from one evaluation period to the next. Even after achieving record sales, employees are expected to achieve even better levels in the future. This attitude — "we can do better; there is no limit to profit increases" — is reflected in McDonald's 1997 *Annual Report to Stockholders,* where the chairman and CEO wrote, "I'm happy to tell you that McDonald's 1997 financial results were again record-setting, yet 1997 was a disappointing year — our financial performance wasn't what we wanted and our stock price lagged the market" (Quinlan 1998:2).

If sales increases and profits are the main criteria by which employees, especially managers, are evaluated, then problems are inevitable. In an industry such as fast-food service, where profits depend on being cost-conscious about every item used, statistical measures of performance exist for everything,

> whether it be the amount of milk-shakes sold per gallon of shake mix used, or the amount of cola drinks per litre of cola syrup, the amount of burgers sold per box of burgers used, the number or portions of chips per kilo used, the monetary amount of cleaning materials used as a percentage of the taking, even small things like the amount of sauces used per portion of Chicken McNuggets sold, or the amount of ketchup used per burger sold, and so on. (Gibney 1993)

Managers under pressure to make and increase profits may force employees to work unpaid overtime, as in the case of 62 Seattle-area Taco Bells. A jury found that the managers of these restaurants had illegally required at least 12,000 workers to "prepare food, pick up trash, or do other chores" while off the time clock (Rousseau 1997).[13]

Keep in mind that corporations are not the only kind of organizations that uses statistical measures of performance. Watchdog groups also use such criteria to monitor corporate responsibility. For example, the authors of the pamphlet "What's Wrong with McDonalds?" offer readers a statistical measure of performance in bold headlines — "50 Acres of Rain Forest Every Minute" each year adding up to an area "the size of Britain is cut down or defoliated and burnt." The authors mention that 100,000 beef ranches operate in the Amazon region. This account of the problem is followed by a question in bold headlines "Why Is It Wrong for McDonald's to Destroy Rain Forests?" According to UN data, however, McDonald's cannot possibly be responsible for Amazon ranches because that "Amazon area never produced more than 5 percent of the total beef supply of Brazil and that production was not exported" (UN Food and Agricultural Organization 1998).[14] Although McDonald's does have 300 franchises in Brazil that could use beef from the Amazon, the pamphlet does not clearly explain this matter.

As we have seen, many potential problems are associated with statistical measures of performance. To ensure that important conditions, such as occupational safety, are monitored, it is advisable to develop thoughtful and accurate indicators to measure them, and then tie those measures to the performance of specific franchises and to persons occupying a definite position. When no such system exists, responsibility never rests squarely with specific people. When managers are concerned with meeting the goal of serving customers within 90 seconds of the time their orders were placed and keeping labor costs low, worker safety may suffer. One possible reason that the food service industry rates first in total recordable injuries and illnesses, with heat burns representing the largest injury category, could be related to pressures placed on employees to fill orders quickly and to chronic understaffing (Personick 1991).[15]

---

[13]These problems associated with statistical measures of performance do not mean that none should ever be used. Some performance indicators can certainly improve service and product quality. For example, fast-food restaurants are evaluated on any number of indicators, ranging from the time it takes to fill orders, the accuracy with which orders are filled, and even speaker clarity at drive-through windows. Some kinds of measures must be in place to ensure customer satisfaction.

[14]When activists are concerned with alerting the public to a major global problem such as deforestation, they may exaggerate one corporation's responsibility in the matter and/or present misleading information. If both sides use inaccurate, inappropriate, or misleading measures of performance, then the consumer is left with the well-founded belief that no one can be trusted and the only choice is to proceed as usual.

[15]The chronic understaffing is due to a shortage of workers and to management's desire to keep labor costs low.

Obstacles to Good Decision Making

In his writings about bureaucracy, Weber emphasizes that power is not located in the person but in the position that a person occupies in the division of labor. The kind of power Weber described is clear-cut and familiar: a superior gives orders to subordinates, who are required to carry out those orders. The superior's power is supported by the threat of sanctions, such as demotions, layoffs, or firings. Sociologists Peter Blau and Richard Schoenherr (1973) recognize the importance of this form of power but identify a second, more ambiguous type—expert power—that they believe is "more dangerous than seems evident for democracy and . . . is not readily identifiable as power" (p. 19).

Expert Knowledge and Responsibility

According to Blau and Schoenherr (1973), expert power is connected to the fact that organizations are becoming increasingly professionalized. **Professionalization** is a trend in which organizations "hire" experts (such as chemists, physicists, accountants, lawyers, engineers, psychologists, or sociologists) as consultants or full-time employees. Experts have formal training in a particular subject or activity that is essential to achiev-ing organizational goals. Experts are not trained by the organization, however; they receive their training in colleges and universities. Theoretically, they are self-directed and not subject to narrow job descriptions or direct supervision. Experts use the frameworks of their chosen profession to analyze situations, solve problems, or invent new technologies. From the experts' viewpoints, the information, service, or innovation they provide to the organization is technical and neutral. They do not have, nor do they seek, control over the application of that information, service, or invention.

Consider the testimony of an expert witness for the McDonald's Corporation in its 1994 court case against activists Steel and Morris—Dr. Neville George Gregory (1994), an expert on the "science of pre-slaughter stunning of animals and its application to the improvement of farm animal welfare" (see "Chicken Preslaughter Handling and Slaughter"). As you read his account, notice how Gregory presents information in a technical and neutral way and how he uses his knowledge of industry standards as a basis on which to draw conclusions about a particular slaughterhouse.

Blau and Schoenherr regard this arrangement between experts and organizations as problematic because it leaves no one accountable for the actions of powerful corporations and because it complicates attempts to find individuals "whose judgments [are] the ultimate source of a given action" (pp. 20–21). This situation is complicated because the recommendations and judgments of experts rely on their specialized knowledge and training. The experts may understand principles of animal preslaughter handling, accounting, physics, biology, chemistry, or sociology, but their training for the most part is compartmentalized; that is, they know one subject very well, but they have less knowledge of other subjects. For example, experts such as Gregory have been trained to understand the technology governing animal slaughter and to focus on the animals. They are not trained, however, to consider the strengths and limitations of the people who work for animal slaughter corporations. In his report to the court, for example, Gregory does not consider the possibility that the people who carry out the stunning and slaughtering of animals may deviate from formal operating procedures. If McDonald's employees find ways to meet profit-related goals by directly violating corporate policies, could workers in the meat slaughtering industry likewise find ways to sidestep official rules and regulations as they strive to meet production

Chickens are going to slaughter.
© Guy Mansfield/Panos Pictures

Professionalization A trend in which organizations hire experts with formal training in a particular subject or activity that is essential to achieving organizational goals.

Chicken Preslaughter Handling and Slaughter

There was not an opportunity to inspect the birds being caught at the farm, but some general comments can be made. The system used by the company for transporting the birds is that known as the Anglia Autoflow system. Modules containing plastic drawers are unloaded from a lorry and placed inside or near to the entrance of the shed. A team of catchers [picks] up the birds and [takes] them to the modules where they are loaded into the drawers. If the birds are handled roughly they can be damaged and if there is sufficient hemorrhaging they can die. These birds will appear as "dead on arrival" at the processing plant.

In 1990/1992 a survey of causes of trauma in dead on arrival birds was conducted in England. The company in question was included in that survey, and its results are shown as plant F in the enclosed paper (Gregory and Austin 1989).* The most common cause of fatal trauma in the survey was dislocated hips, but for plant F this damage was very low. The overall level of dead on arrival birds at plant F was close to the average for the survey. Head trauma was the principal unusual feature for this plant, and it is suspected that it would be common to other plants that use the Anglia Autoflow system. This system is gaining in popularity and it is thought that it is likely to replace most of the alternative systems in the future.

On the day I visited the plant for the purpose of this report (19/4/93) I asked to see the dead on arrival bin. It contained 44 birds, 6 of which had died from a crushed neck but this was not certain as it could have had its neck dislocated manually for euthanasia

purposes. The prevalence of dislocated hips (14%) was higher than that recorded in the survey, but it must be noted that it was based on fewer birds.

When the live birds arrive at the processing plant they are hung upside down on an overhead shackle which conveys them to a waterbath stunner. On 19/4/93 the company was operating two such killing lines, one for male and the other for female birds. Two weeks previously there had been only one killing line which operated at twice the speed and carried both male and female birds. On 19/4/93 measurements were made on only one of the lines—usually the male line.

The line speed on 19/4/93 was 87 birds per minute. The lighting at the hanging-on position was subdued; a green light was used to allow the shackling staff to see. A normal reaction of the birds to being taken from transport crates and being suspended on the shackles is to flap their wings (Gregory and Bell 1987).* Shackling was closely observed for 26 birds. Only one of them displayed any wing flapping and it lasted for about half a second. One hundred shackled birds were palpated for broken wing bones, and none was found. It was concluded that the disturbance of hanging the birds on the line was as low as can be hoped for. Between the hanging-on point and the waterbath stunner there was a continuous breast comforter; a device designed to quieten the birds and prevent them from flapping their wings. Over a 3-minute recording period, no birds flapped their wings between these points in the line. At other processing plants resurgences of flapping during this period are said to be associated

with disturbance of birds because of unevenness in the line, temporary loss of visual contact between adjacent birds when they go round a corner, and sudden exposure to bright lights (e.g., sunshine). These were not a cause for concern at this plant.

The birds reached the stunner at 1 minute 16 seconds after hanging-on. At this point they should be drawn into the water, which is electrically live, and immediately stunned. In some situations birds receive an electric shock before they are stunned. This causes them to recoil before being conveyed into the water and before being stunned. This situation can be common in poultry processing plants, and it was observed during a previous visit to this particular plant (Gregory, 18/2/93 report, appended).* The interval between the initial electric shock and the stun can be very brief (about 0.5 second). Owing to the difficult access to the entrance of the waterbath stunners it was difficult to observe the birds closely and estimate the extent of this problem. However, it was thought to have occurred in 13.5% of 200 birds examined. One percent of the birds were thought to have missed the water of the waterbath stunner, and hence were not stunned. These birds also missed the automatic cutter because their necks were not extended, and instead had their heads cut off by the manual back-up neck cutter. This is normal practice for poultry lines, although it is more usual to just cut one side of the neck instead of taking the whole head off.

*Source: Excerpted from Dr. Neville George Gregory (1994), witness statement for the prosecution. For the complete transcript see http://www.mcspotlight.org/people/witnesses/animals/gregory.html

goals? Keep in mind that Blau and Schoenherr emphasize that the men and women who give expert advice are decent people, but that their position, training, and point of view make them unable to anticipate, plan, or control unintended consequences.

In addition, decision making in large organizations is complex because no single person provides all of the input that goes into a decision. Rather, a decision is a joint product of information and judgments by a variety of experts. Often the decision maker does not understand the principles underlying an expert's recommendations and judgments. For example, Gregory notes in general comments to his signed witness statements that "it is assumed that the reader is reasonably familiar with farming practices." The problem with the specialization of knowledge is that when something goes wrong, the experts claim that they provided only the patent, information, suggestions, or recommendations and that they cannot control the ultimate implementation. Members of management, on the other hand, claim that they relied on the expert's advice to make decisions about a process they really do not understand.

The Problems with Oligarchy

Oligarchy is rule by the few, or the concentration of decision-making power in the hands of a few persons who hold the top positions in a hierarchy.

> One of the most bizarre features of any advanced industrial society in our time is that the cardinal choices have to be made by a handful of men . . . who cannot have firsthand knowledge of what those choices depend upon or what their results may be. . . . [And by] "cardinal choices," I mean those which determine in the crudest sense whether we live or die. For instance, the choice in England and the United States in 1940 and 1941, to go ahead with work on the fission bomb: the choice in 1945 to use that bomb when it was made. (Snow 1961:1)

Political analyst Robert Michels believed that large formal organizations tended inevitably to become oligarchical, for the following reasons. First, democratic participation is virtually impossible in large organizations. Size alone makes it "impossible for the collectivity to undertake the direct settlement of all the controversies that may arise" (Michels 1962:66). For example, McDonald's employs more than 1 million people in its corporate offices and has franchises located in more than 110 foreign countries (see "McDonald's at a Glance"). "It is obvious that such a gigantic number of persons belonging to a unitary organization cannot do any practical work upon a system of direct discussion" (Michels 1962:65).

Second, as the world becomes more interdependent and technology grows increasingly complex, many

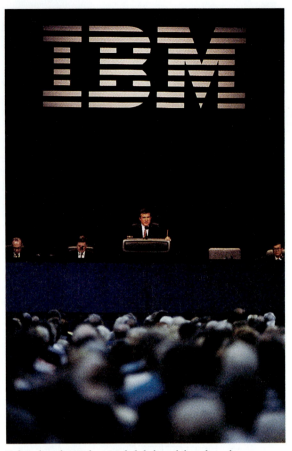

Political analyst Robert Michels believed that oligarchy, or rule by the few, is the inevitable tendency of large organizations. Size alone makes it impossible to get everyone's input into organizational decisions.

© Brad Markel/Gamma-Liaison

organizational features inevitably become incomprehensible to workers. As a result, many employees work toward achieving organizational goals that they did not define, cannot control, may not share, or may not understand. This lack of knowledge prevents workers from participating in or evaluating decisions made by executives.

A danger of oligarchy is that those who make decisions may not have the necessary background to understand the full implications of those choices. In addition, decision makers may not consider the greater good and may instead become preoccupied with preserving their own leadership. Guarding against these effects of oligarchy requires that the average worker and the general public be interested, attentive, and informed.

Clearly, an uninformed, inattentive public that consumes unthinkingly also bears responsibility for poor decision making. An informed and attentive public is

Oligarchy Rule by the few, or the concentration of decision-making power in the hands of a few persons who hold the top positions in a hierarchy.

McDonald's at a Glance

Headquarters	Oak Brook, Illinois
First restaurant	Des Plaines, Illinois, 1955
First foreign franchise	Canada, 1967
Latest foreign franchise	Republic of Georgia
Number of customer visits to McDonald's worldwide	38 million/day; 14 billion/year
Share of U.S. fast-food burger market	42%
Percentage of corporation profits derived from foreign sales	60%
Major sponsorship	NBA; Winter Olympics, Nagano Japan; NASCAR; World Cup Soccer
Business alliances/partnerships	Disney, Wal-Mart, Home Depot, Amoco, Chevron, Coca-Cola
Total number of employees worldwide	1,150,000
Total number of employees in the United States	650,000
Estimated percentage of labor force whose first work experience was McDonald's	7%
Most well-known charitable activities	Ronald McDonald House
Projected franchises added by 2000	2000-plus
U.S. *Fortune* 500 rank	135
Global *Fortune* 500 rank	397

Sources: Berstein (1997); *Fortune* (1998a, 1998b); Leohnhardt (1998); Marquesee (1994); McDonald's Corporation (1998).

important because consumer demand drives corporate decision making. Moreover, alert and active consumers are capable of shaping market trends in more constructive directions.

To this point we have discussed a number of important concepts that help us understand how some characteristics of organizations make them coordinating mechanisms with the potential for both constructive and destructive consequences. Now we turn to Karl Marx and his concept of alienation to understand how workers may be dominated so strongly by the forces of production that they remain uninformed or uncritical about their role in the production process.

Alienation of Rank-and-File Workers

Human control over nature increased with the development of increasingly sophisticated instruments and tools and the growth of bureaucracies to coordinate the efforts of humans and machines. Machines and bureaucratic organizations combined to extract raw materials from the earth more quickly and more efficiently and to increase the speed with which necessities such as food, clothing, and shelter could be produced and distributed.

Karl Marx believed that increased control over nature is accompanied by **alienation,** a state in which human life is dominated by the forces of human inventions. Chemical substances represent one such invention; they have reduced the physical demands and risk of crop failure involved in producing goods such as potatoes. Fertilizers, herbicides, pesticides, and chemically treated seeds give people control over nature because they eliminate the need to fight weeds with hoes, they prevent pests from destroying crops, and they help people produce unprecedented amounts of standard and uniform-looking fruits and vegetables.

These gains also have a dark side, however. In the long run, people are dominated by the effects of this invention. For example, heavy reliance on chemical technologies causes the soil to erode and become less productive; it also prompts insects and disease-causing agents to develop resistance to the chemicals. In addition, chemical technologies have altered the ways in which farmers plant crops: planting patterns have changed from many varieties to a single standard and uniform cash crop, planted in rows. As a

Alienation A state in which human life is dominated by the forces of human inventions.

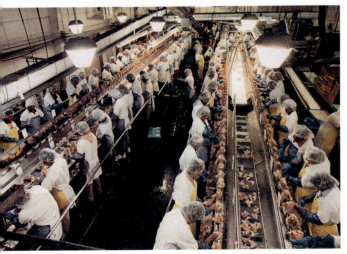

Karl Marx believed that alienation occurs when the production process is divided up so that workers are treated like parts of a machine rather than as active, creative, social beings.

© Michael Schwarz/The Image Works

result of these changes, some farmers have lost knowledge of how to control insects and diseases without chemicals by interplanting a variety of flowers, herbs, and vegetables. Many farmers are now economically dependent on a single crop and the chemical industry.

Although Marx discussed alienation in general, he wrote more specifically about alienation in the workplace. Marx maintained that workers are alienated on four levels: (1) from the process of production, (2) from the product, (3) from the family and the community of fellow workers, and (4) from the self. Workers are alienated from the process because they produce not for themselves or for known consumers but for an abstract, impersonal market. In addition, they do not own the tools of production.

Workers are alienated from the product because their roles are rote and limited—each fast-food worker, for instance, performs a specialized task such as warming buns, adding condiments, or wrapping the order. Many workers are treated as being replaceable or as being interchangeable as machine parts. That is, they are treated as economic components rather than as active, creative social beings (Young 1975). Marx believed that the conditions of work usually are such that they impair an individual's "capacity to become a multidimensional, authentic being with human qualities of compassion, reflection, judgment, and action" (Young 1975:27).

Workers are alienated from the community of fellow workers and their families because households

and work environments remain separate from one another. In the fast-food industry, workers can become alienated from their families when they work shifts (that is, late at night, early in the morning, or on weekends) that keep them from participating in the other family members' lives. Workers are alienated from the community of fellow workers because they compete for jobs, business, advancement, and awards. As they compete, they fail to consider how they might unite as a force and control their working conditions. Often workers are alienated from more than the families and community of fellow workers—they are alienated from other communities as well, such as the school. In the case of McDonald's and other fast-food restaurants, aggressive recruiting of high-school-aged employees helps to supply 65 percent of the labor force (Waters 1998).[16] While McDonald's considers homework schedules and other school demands, we must also question the extent to which after-school jobs shortchange involvement in academic and social activities (Crispell 1995).[17]

Finally, workers are alienated from themselves because "one's genius, one's skills, one's talent is used or disused at the convenience of management in the quest of private profit. If private profit requires skill, then skill is permitted. If private profit requires subdivision of labor and elimination of craftsmanship, then skill is sacrificed" (Young 1975: 28). When Karl Marx developed these ideas in 1888, he was describing "alienation from self" as it relates to industrial society. More recently, sociologist Robin Leidner (1993) has described the "alienation from self" that can occur in service industries when management standardizes or routinizes virtually every aspect of the service-recipient relationship such that neither party feels authentic, autonomous, or sincere:

> Employers may try to specify exactly how workers look, exactly what they say, their demeanors, their gestures, even their thoughts. The means available for standardizing interactions include scripting; uniforms or detailed dress codes; rules and guidelines for dealing with service-recipients and sometimes with co-workers . . . surveillance and a range of incentives and disincentives can be used to encourage or enforce compliance. (pp. 8–9)

---

[16]A subminimum wage provision in the Fair Labor Standards Act allows employers to pay employees under the age of 20 a minimum wage of $4.25 per hour during the first 90 days of employment (Camp 1998).

[17]While many students have to work, a large segment work to pay for cars, clothes, and entertainment.

Summary and Implications

Organizations are powerful coordinating mechanisms that permit goods and services to be produced and delivered efficiently to millions of people. Although they manage people, information, goods, and services efficiently, organizations also can promote inefficient, irresponsible actions (see Table 7.4). This chapter introduced concepts that sociologists use to understand the two sides of organizations with special emphasis on a major multinational corporation, McDonald's.

An analysis of the organizational issues facing McDonald's reveals that many factors within organizations can cause problems. These factors include (1) rational decision making that emphasizes the quickest and most cost-efficient means to achieve a goal without considering the merits of other methods and goals, (2) informal departures from rules and regulations, (3) excessive adherence to rules and regulations, (4) reliance on experts with compartmentalized training, (5) use of inappropriate statistical measures of performance, (6) the development of an oligarchy, and (7) an alienated work force.

We also have learned that a complex relationship exists between an organization and the environment, especially when the organization is a multinational corporation. The case of McDonald's shows that an organization not only draws labor and raw materials from the environment, but also affects the environment through the products it manufactures, the services it provides, and the waste it disposes of.

The case of McDonald's represents a larger organizational trend that sociologist George Ritzer (1993) has called the **McDonaldization of society**—"the process by which the principles of the fast-food restaurant are coming to dominate more and more sectors of American society as well as the rest of the world" (p. 1). What are these principles? Ritzer maintains that they include (1) efficiency, (2) quantification and calculation, (3) predictability, and (4) control. **Efficiency** means that a corporation or organization offers the "best" products and services that allow consumers to get quickly from one state to another (that is, from hungry to full, from fat to thin, from uneducated to educated, and so on). **Quantification and calculation** mean that customers can easily evaluate the product and service with numerical indicators (that is, delivery within 30 minutes, a combination meal that is less expensive than the same items purchased separately, an order of chicken McNuggets with exactly six pieces, a college degree in 19 months). **Predictability** means

that, for all intents and purposes, the service and product will be the same no matter where or when it is purchased. **Control**, "especially through the substitution of nonhuman for human technology," means that the process of producing and acquiring the service or product is planned out in detail (by assigning a limited task to each worker, filling soft drinks from dispensers that automatically shut off, having customers stand in line, and so on).

While this model may allow McDonald's to serve 38 million customers worldwide each year (an amazing organizational task), it also has its drawbacks. Sociologist Max Weber used the phrase **iron cage of rationality** to describe the irrationalities that rational systems generate. "Ultimately, we must ask whether the creations of these rationalized systems create an even greater number of irrationalities" (Ritzer 1993). For example, it may seem efficient to divide up the production task in such a way that "even a moron could learn to do the job." At the same time, a work setting that requires so little skill generates high employee turnover—in the fast-food industry the turnover can be as high as 100 percent (Ritzer 1993). Likewise, it may seem rational to create and produce uniform potatoes, but feeding middle-class consumers at the expense of the poor is irrational.

McDonaldization of society "The process by which the principles of the fast-food restaurant are coming to dominate more and more sectors of American society as well as the rest of the world" (Ritzer 1993:1).

Efficiency A corporation or organization offering the "best" products and services that allow consumers to get quickly from one state to another (that is, from hungry to full, from fat to thin, from uneducated to educated, and so on).

Quantification and calculation The ability of customers to easily evaluate a product or service with numerical indicators.

Predictability For all intents and purposes, a state in which the service or product will be the same no matter where or when it is purchased.

Control When the process of producing and acquiring the service or product is planned out in detail.

Iron cage of rationality The irrationalities that rational systems generate.

Table 7.4   Organizations: What the Critics Say Versus What the Defenders Say

What the Critics Say

Organizations force workers to do simplified, meaningless tasks, so robbing them of initiative, creativity, and independence. Look at the assembly line.

Organizations do not serve society or consumers well. Automobiles are unsafe, factories pollute the air, and the sheer size of organizations makes them dangerous. The multinationals dominate life in company towns and even overthrow governments.

Organizations don't even work well. Cars break down before you get them home from the showroom, prisons don't rehabilitate, and schools turn out illiterates. We need alternative institutions.

Organizations are out of control. An arrogant power elite of interlocking directors controls them, and an army of ever-increasing bureaucrats administers them. The client or citizen is powerless against organizations.

What the Defenders Say

We can enrich jobs by deliberately building in autonomy and discretion. In Sweden, the assembly line has been modified to do just this. In any case, very few people actually work on assembly lines anymore. We are becoming a service society.

We can control pollution and make better products with quality-control circles and other advances pioneered by Japan and other countries. Large size produces economies of scale. Besides, only very large organizations can afford to do research on new products.

You go too far. Cars give trouble because you want so much from them—trouble-free driving at high speeds with minimal maintenance. If we produced a serviceable car with minimum features, you wouldn't buy it. Prisons could rehabilitate if you would pay the price for vocational counseling and training. The schools do a great job—name another country in which over a third of all high-school students go on to higher education.

The evidence for any monolithic power elite is exaggerated or so biased as to be invalid. Client and citizen power are increasing. Look at the detail on contents now provided on labels at the supermarket. Look at the growth of consumer action groups.

Source: From *Organizations in Society,* by Edward Gross and Amitai Etzioni, p. 4. Copyright ©1985 by Prentice-Hall, Inc. Reprinted by permission of Prentice-Hall, Upper Saddle River, NJ.

This discussion may leave you with the feeling that we have created a **social trap,** a situation where people and the organizations they have created lead them into some direction or some set of relationships that later proves to be not only unpleasant or lethal, but also such that they see "no easy way to back out of or to avoid" (Platt 1973:641). Yet everything is not hopeless. As a case in point, note that in the 1994 McDonald's court case the defense called upon witnesses who described their work as impossible:

- Due to there being so few staff, we were constantly "encouraged" to use the "hustle" system. "Hustle" in McDonald's does not mean move efficiently and quickly, as their handbooks say; it does, in fact, mean run and scramble about slightly faster than is humanly possible while being shouted at and frequently burning oneself and colleagues. Work as fast as you possibly can and you are still shouted at to work faster, be-

Social trap A situation where people and the organizations they have created lead them into some direction or some set of relationships that later proves to be not only unpleasant or lethal, but also such that they see no easy way to avoid it.

cause you are doing the work of four or five due to understaffing and can't possibly keep up. (Beech 1994)

- I remember that frozen food was to be cooked from frozen. However, I recall numerous occasions where, due to various reasons—e.g., breakdown of freezers, meat or fish was accidentally defrosted. To reduce waste, in most cases this defrosted meat or fish was cooked and served to the customers. This happened with the knowledge of the manager. (Alimi 1995)

McDonald's, on the other hand, called upon seven witnesses to counter such claims. Interestingly, many of McDonald's witnesses acknowledged claims made by defense witnesses as valid but insisted these problems could be traced to incompetent managers rather than to problems of organizational structure.

- When I arrived at the Heathrow restaurant it was immediately obvious that the basic problem with the restaurant was one of understaffing. There were only about 60 crew members on the payroll, which was far too few for a store of that size and volume. It was also clear that there was a lack of discipline both amongst the managers and the crew members and that the atmosphere in the restaurant was not a happy one. In my view, the store had clearly been suffering from poor management by my predecessor. . . . As an emergency interim measure I arranged for the transfer of a number of experienced crew members from other restaurants to tide me over while I recruited and trained new crew members. In the first month we took on 40 new crew. I also set about improving the discipline within the restaurant of both managers and crew by making it clear what standards I expected in the store. (Roberts 1996)

- As I am a breakfast manager I am responsible for checking the fridge in the morning. If I find salads from the previous day I always throw them away and in my experience this is what everyone else does. Salads are freshly made each day. I regularly see the waste bin filled with discarded products, including salads. (Perrett 1995)

While McDonald's might have us believe that managers as individuals are ultimately responsible for each restaurant's problems and successes, we must also recognize the ways in which organizational structure contributes to poor decision making on the part of managers. Perhaps in the end the best managers are those who understand that organizations are coordinating mechanisms capable of generating irrational consequences.

Key Concepts

Use this outline to organize your review of the key chapter ideas.

Organizations
Multinational corporations
Rationalization
   Iron cage of rationality
   Externality costs
   Social trap
   Bureaucracy
     Ideal type
McDonaldization of society
   Efficiency
   Quantification and calculation
   Predictability
   Control

Formal dimensions of organizations
Informal dimensions of organizations
Trained incapacity
   Informate
   Automate
Oligarchy

internet assignment

Identify industries that operate as major suppliers to the fast-food industry (for example, toymakers, bun makers, meat processors, potato growers and processors, ketchup and other condiment makers). Try to find information on the Internet about these industries and relate what you find to some concepts covered in this chapter.

8

Deviance, Conformity, and Social Control

With Emphasis on the People's Republic of China

The Zhengzhou Rail station in Beijing, where thousands of migrant workers are waiting for trains to take them home for the New Year holiday. (Jeffrey Aaronson/Network Aspen.)

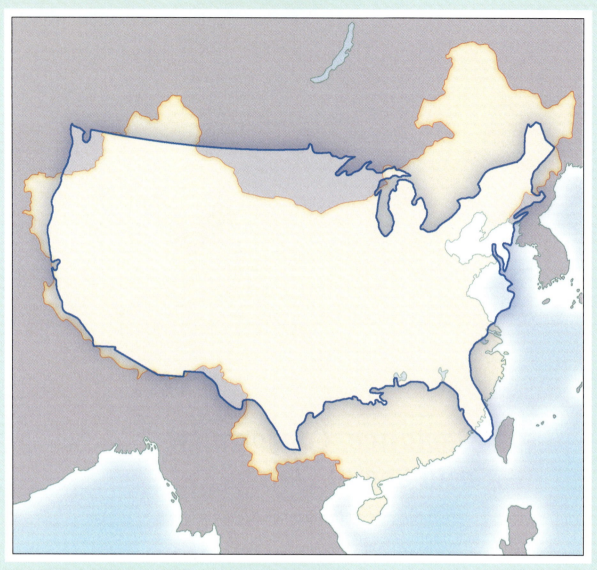

Source: U.S. Central Intelligence Agency (1995).

The Problems of Governing More Than 1.2 Billion People

China's system of social control is much more rigid than our own. One major reason for this difference may be the size of China's population relative to its available resources. More than 1.26 billion Chinese live in a space roughly the size of the United States. The United States, with 268 million people, has an abundant 11.7 percent of the world's arable land and 4.5 percent of the world's population. In contrast, China has only 6.3 percent of the world's arable land and 21 percent of the world's population.

Another reason may relate to the difficulties inherent in managing a huge and centralized society. To get an idea of the scale of Chinese society, consider these facts: As of 1994, the United States had only 1 city with more than 5 million people (New York), and 8 million people (New York), and 8 cities with more than 1 million residents. China has 96 cities with more than 1 million residents each and 5 cities with more than 5 million people (UN 1998).

Or look at it this way: The central authorities of China are attempting to govern more than 1 billion people. In Europe and North and South America, it takes about 50 separate countries to accomplish the same task.

When the Cultural Revolution began . . . my [family's house] was one of the first in the city to be ransacked.

I later found out that it was my mother's ignorance that started the ransacking. Both my grandfather and father were working in banks. They were well-known capitalists. At that time all the funds capitalists had in the banks were frozen. You couldn't withdraw anything. It was called "money made by exploitation." All the names of capitalists were listed just outside the banks. My mother didn't know about that. She went to get some money out. The bank clerks immediately called the Red Guards. They showed up in no time at my home and started to search and ransack our apartment.

[When I finally got there,] I glanced over the rooms from the corridor. Red Guards were standing everywhere, searching for things. No sign of my family. Lots of things were in shreds, smashed and torn.

My parents moved in with my mother's family. Their home had been searched and sealed, too. My parents just blocked off a tiny area with a piece of cloth in the corridor. They found some wooden boards to use as a bed. When I went to see my mother, her hair had been cut and shaved by the Red Guards. (Feng Jicai 1991:58–60)

This event took place during the ten years of the Cultural Revolution (1966–1976), a period in which more than a half-million Chinese were imprisoned, tortured, humiliated, and executed. During this time, any person who held a position of authority, worked to earn a profit, showed the slightest leaning toward foreign ways, or had academic interests was subject to interrogation, arrest, and punishment. Included in this group were scientists, teachers, athletes, performers, artists, writers, private business owners, and people who had relatives living outside China, wore glasses, wore makeup, spoke a foreign language, owned a camera or a radio, or had traveled abroad (Mathews and Mathews 1983).

Contrast the events of the Cultural Revolution with Asian scholar Robert Oxman's description of China today:

All over China people are jumping into the sea. That's the new Chinese expression for going into business, making money privately in a country that used to forbid any form of capitalism. It means saving a few dollars for the simple tools to fix bicycles at a curbside shop, or running a sidewalk shoe repair stand because new soles are still cheaper than new shoes, or cutting hair at 5:30 in the morning before going to a regular job. In the country, a wife grows produce to make extra

income to buy consumer goods for her family. Unleashing the natural business instincts of the Chinese people, plus the addition of huge foreign investments, have fueled one of the most explosive economic take-offs in history. Around major cities, buildings, housing projects, and roads seem to emerge almost overnight. It's even been suggested that China's national bird should be the crane. Every urban horizon is filled with cranes. In Shanghai's frenzied harbor, ships from around the world compete with tiny river barges as they haul goods and move people. Once the people of China were caught in economic slow motion by a state [that] prized ideology above all else. Now the Chinese are rushing in fast forward in a national quest for prosperity. (Oxman 1993a:9)

Why Focus on the People's Republic of China?

In this chapter, we pay special attention to the People's Republic of China for one major reason. In 15 to 20 years, China has undergone a lifetime of economic reform that is startling when contrasted with its economic policies of the Cultural Revolution of the 1960s and 1970s. Those policies made profit-making activities criminal and closed the country off to foreign investment and influences. During the Cultural Revolution, if a family planted small, but extra, crops to sell after the autumn harvest for money to buy food during the winter, they were criticized for taking the capitalist road. At that time, the pressure against acquiring even small amounts of extra wealth drew such severe criticism that peasants accepted living in poverty rather than being equated with the so-called bourgeoisie (Bernstein 1983).

In contrast, an estimated 220,000 foreign-invested enterprises are currently registered in China (U.S. Department of State 1996), up from 70,000 in 1993. Among the foreign-invested firms are the U.S.-based shoe giants Nike, Converse, and Reebok; the fast-food giants PepsiCo (KFC, Pizza Hut) and McDonald's; and Procter & Gamble. Beijing alone boasts at least 100 luxury hotels, up from 9 in 1981, where foreigners and wealthy Chinese can stay (Montalbano 1993).

Because of this vast amount of economic change and freedom, which has been described as one of the largest economic experiments the world has ever known (Broadfoot 1993), many investors view China as potentially the world's largest market and a "gigantic reservoir of cheap labor" (Carrel and Hornik 1994:A15)—a potential gold mine. Although China has been defined as a market that must be penetrated, however, the country has also received considerable international criticism and pressure directed at improving its human rights record. The U.S. government, for example, regularly warns the Chinese leadership about its (1) unacceptably high trade surplus with the United States (growing at a faster rate than any other U.S. trading partner's trade surplus); (2) infringements against copyrights, trademarks, and patents reflected in the pirating of U.S. movies, computer software, music, and books; and (3) human rights violations against political dissidents and, most recently, orphans.

China represents an interesting case for studying issues of deviance, conformity, and social control, because many of the behaviors that constituted deviance during the Cultural Revolution are no longer judged in the same way today. In addition, the Chinese government's ideas about what constitutes deviance and its methods of social control are often the object of international criticism.

The topics of this chapter are deviance, conformity, and social control—three of the most complex issues in sociology because, as we will learn, almost any behavior or appearance can qualify as deviant or conformist under the right circumstances. **Deviance** is any behavior or physical appearance that is socially challenged and/or condemned because it departs from the norms and expectations of a group. **Conformity,** on the other hand, comprises behavior and appearances that follow and maintain the standards of a group. All groups employ mechanisms of **social control**—methods used to teach, persuade, or force their members, and even nonmembers, to comply with and not deviate from norms and expectations.[1]

As shown by the vignettes at the start of this chapter, depending on the cultural circumstances, a deviant act or appearance in China can be something as seemingly minor as wearing eyeglasses or withdrawing money from a bank. We can illustrate the same point within U.S. society. When sociologist J. L. Simmons asked 180 men and women in the United States from a variety of age, educational, occupational, and religious groups to "list those things or types of persons whom you regard as deviant," they identified a total of 1,154 items.

Deviance Any behavior or physical appearance that is socially challenged and/or condemned because it departs from the norms and expectations of a group.

Conformity Behavior and appearances that follow and maintain the standards of a group. Also, the acceptance of the cultural goals and the pursuit of these goals through legitimate means.

Social control Methods used to teach, persuade, or force a group's members, and even nonmembers, to comply with and not deviate from norms and expectations.

---

[1]The controlling agent may be parents, peers, coaches, elected officials, governing bodies, or the organizational mechanisms of governments, justice systems, or international bodies.

Even with a certain amount of grouping and collapsing, these included no less than 252 different acts and persons as "deviant." The sheer range of responses included such expected items as homosexuals, prostitutes, drug addicts, beatniks, and murderers; it also included liars, democrats, reckless drivers, atheists, self-pitiers, the retired, career women, divorcées, movie stars, perpetual bridge players, prudes, pacifists, psychiatrists, priests, liberals, conservatives, junior executives, girls who wear make up, and know-it-all professors. (Simmons 1965:223–224)

Although this study was conducted more than 30 years ago, Simmons's conclusions remain relevant today: almost any behavior or appearance can qualify as deviant under the right circumstances. The only characteristic common to all forms of deviance is "the fact that some social audience regards them and treats them as deviant" (p. 225). In light of this fact, it is difficult to generate a precise list of deviant behaviors and appearances, because something that some people consider deviant may not be considered deviant by others. Likewise, something that is considered deviant at one time and place may not be considered deviant at another. For example, wearing makeup is no longer considered a deviant behavior in China,[2] as evidenced by the fact that Avon, a U.S. cosmetics company, has recruited 18,000 salespeople to sell its products in China (WuDunn 1993). As another example, cocaine and other now-illegal drugs once were legal substances in the United States. In fact, "We unwittingly acknowledge the previous legality of cocaine every time we ask for the world's most popular cola by brand name" (Gould 1990:74). Originally Coca-Cola was marketed as a medicine that could cure various ailments. One ingredient used to make the drink came from the coca leaf, which is also used to produce cocaine (Henriques 1993). Finally, considering how cigarette smokers are treated today, it is interesting to note that smoking was not merely accepted but considered glamorous in the United States just a few decades ago.

Such wide variations in responses to questions about "what is deviant" alert us to the fact that deviance exists only in relation to norms in effect at a particular time and place. The sociological contribution to understanding deviant behavior lies in going beyond studying the individual to instead emphasizing the context

A society's definitions of deviance change over time. Considering how smokers are viewed today, it is interesting to note that cigarette smoking was not merely accepted but glamorized in the United States just a few decades ago.

Cover drawing by Barry Blitt; © The New Yorker Magazine, Inc.

under which deviant behavior occurs. Thus sociologists are interested in at least two general but fundamental questions about the nature of deviance. First, how is it that almost any behavior or appearance can qualify as deviant under the "right" circumstances? Second, who defines what is deviant? That is, who decides that a particular group, behavior, appearance, or person is deviant? The fact that an activity or an appearance can be deviant at one time and place and not deviant under other circumstances suggests that it must be defined as deviant by some particular process. In the case of China, who determined that wearing eyeglasses, speaking a foreign language, and withdrawing money constituted deviant behavior and warranted such severe punishments?

In this chapter, we explore the concepts and theories that sociologists use to understand deviance, and we use them to answer these fundamental but complex questions about the social nature of deviance.

[2]Whenever the term *China* is used in this chapter, it denotes the People's Republic of China, whose capital is Beijing. The use of *China* does not include Hong Kong (a former British colony that reverted to the People's Republic of China in 1997) or Macao (currently a Portuguese colony but expected to revert to the People's Republic of China in 1999).

The Role of Context in Defining Deviance

It is impossible to fully understand how the Cultural Revolution disrupted Chinese lives without having experienced it (Bernstein 1982). Chinese author Wang Shuo (1997) described the Cultural Revolution, which began in 1966 and lasted approximately a decade, as a time when "we were out of control. Everything was turned upside down. The teachers who used to do the educating were sent away to be reeducated. Children were allowed to correct their parents" (p. 51).

During the Cultural Revolution, many artifacts from China's long history (tombstones, relics, manuscripts, art objects, books, scrolls of poetry) and any other objects that suggested special status or the accumulation of worldly possessions were destroyed. Any person in a position of authority or with the slightest leaning toward foreign ways, including farmers who planted extra crops, were considered suspect. If someone simply remarked that a foreign-made product such as a can opener was better than its Chinese counterpart, or if someone wrapped some food or garbage in a piece of newspaper with then Communist Party Chairman Mao Zedong's picture, he or she was regarded with distrust (Mathews and Mathews 1983).

"We believed deeply in Chairman Mao": The decade-long Cultural Revolution was an attempt to rekindle revolutionary spirits in China and enforce conformity to Mao's doctrine.

© Max Scheler/Black Star

As Jung Chang writes in *Wild Swans: Three Daughters of China* (1991), such conditions reduced many people "to a state where they did not dare even to think, in case their thoughts came out involuntarily" (1992:6). The slightest misstep could make one a target of intense criticism:

> Targets might be required to stand on a platform, heads bowed respectfully to the masses, while acknowledging and repeating their ideological crimes. Typically they had to "airplane," stretching their arms out behind them like the wings of a jet. In the audience tears of sympathy might be in a friend's eyes, but from his mouth would come only curses and derisive jeering, especially if the victim after an hour or two fell over from muscular collapse. . . .

To Chinese people, who were especially sensitive to peer-group esteem, to be beaten and humiliated in public before a jeering crowd including colleagues and old friends was like having one's skin taken off (Fairbank 1987:336).

During the Cultural Revolution, why were people considered deviant and punished so severely for such seemingly harmless acts?[3] The forces behind this movement help us to understand more clearly. The Cultural Revolution was Chairman Mao Zedong's response to the failure of an important national plan—the Great Leap Forward.

The Great Leap Forward was Mao's plan to mobilize the masses and transform China from a country of poverty into a land of agricultural abundance in five short years. Under this plan, Mao mobilized hundreds of thousands of people for projects ranging from killing insects to building giant dams with shovels and wheelbarrows (Butterfield 1976). The Great Leap Forward was an ill-conceived, hastily planned, sweeping reorganization of Chinese society that created economic disruption on a massive scale. The nature of this disruption is illustrated by the failure of a plan to increase the availability of fertilizers by cutting down trees, burning them, and spreading their ashes over the fields. This plan left peasants without fuel for cooking and heating. In one region, peasants stopped harvesting crops and dug tunnels in their fields in an effort to find coal, which Communist Party officials believed was plentiful there. No coal was found, however, and the crops rotted. As a result

[3]The circumstances leading to this revolution are not easy to summarize; one would have to go back at least 185 years to trace its origin.

of the Great Leap Forward, 30 to 50 million Chinese died from human-induced famine (Leys 1990; Liu Binyan 1993).[4] In addition, the plan caused considerable environmental destruction. Desperate to grow food and obtain fuel, Chinese peasants destroyed more than half of their country's grasslands and one-third of its forests (Leys 1990).

In light of these catastrophic problems, Mao was particularly vulnerable to political attack. He blamed the failure of his plan on entrenched authority, which he loosely defined as the "Four Olds": old ideas, old culture, old customs, and old habits. Mao also blamed the failure of the Great Leap Forward on the abandonment of revolutionary spirit[5] and on the "evils of special status and special accumulation of worldly goods" (Fairbank 1987:319). He used the Cultural Revolution as an attempt to eliminate anyone in the Communist Party and in the masses who opposed his policies.

Mao initially assigned the Red Guards (his name for the youths of China between the ages of 9 and 18),

and eventually the People's Liberation Army,[6] the task of finding and purging those responsible. The reflections of a former Red Guard, some 15 years after the event, show the intensity with which people were hunted down and persecuted:

> "I was very young when the Cultural Revolution began. . . . My schoolmates and I were among the first in Peking to become Red Guards; we believed deeply in Chairman Mao. I could recite the entire book of the Chairman's quotations backward and forward, we spent hours just shouting the slogans at our teachers."

> Hong remembered in particular a winter day, with the temperature below freezing, when she and her faction of Red Guards put on their red armbands and made three of the teachers from their high school kneel on the ground outside without coats or gloves. "We had gone to their houses to conduct an investigation, to search them, and we found some English-language books. They were probably old textbooks, but to us it was proof they were worshipping foreign things and were slaves to the foreigners. We held a bonfire and burned everything we had found."

> After that, she recalled, the leader of her group, a tall, charismatic eighteen-year-old boy, the son of an army general, whose nickname was "Old Dog," ordered them to beat the teachers. He produced some wooden boards, and the students started hitting the teachers on their bodies. "We kept on till one of the teachers start[ed] coughing blood," Hong said. . . . "We felt very proud of ourselves. It seemed very revolutionary." (Butterfield 1982:183)

In view of everything that happened during the Cultural Revolution, why is making money acceptable in China today? Again, this question must be examined in the context of the earlier era.

After Mao's death in 1976, the Cultural Revolution ended, and China's new leaders faced a great many problems. The revolution had taken place at the expense of China's economic, technological, scientific, cultural, and agricultural development and drained the Chinese physically and mentally. It also created a 10-year gap in trained workers. "'I can't honestly let any of the young doctors in my hospital operate on a patient,'

[4]In *Hungry Ghosts: Mao's Secret Famine*, Jasper Becker (1996) interviews a survivor of the famine more than 35 years later. "Mrs. Liu Xiaohua, now aged 65, still vividly remembers the events of thirty-six years ago. . . . On the muddy path leading from her village, dozens of corpses lay unburied. In the barren fields there were others; and amongst the dead, the survivors crawled slowly on their hands and knees searching for wild grass seeds to eat. In the ponds and ditches people squatted in the mud hunting for frogs and trying to gather weeds. . . . She remembered, too, the unnatural silence. The village oxen had died, the dogs had been eaten and the chickens and ducks had long ago been confiscated by the Communist party in lieu of grain taxes. There were no birds left in the trees, and the trees themselves had been stripped of their leaves and bark. At night there was no longer even the scratching of rats and mice, for they too had been eaten or had starved to death. . . . Most of all she missed the cries of young babies, for no one had been able to give birth for some time. The youngest children had all perished, the girls first. Mrs. Liu had lost a daughter. The milk in her breasts had dried up and she had been forced to watch her child die. Her aunt, her mother and two brothers had also died" (pp. 1–2).

[5]The phrase "the abandonment of revolutionary spirit" refers to the 1949 revolution that resulted in the establishment of the People's Republic of China. Before this revolution, China was very poor. Between 1849 and 1949, the Chinese people were victims of every imaginable sort of exploitation by foreign countries. To survive starvation, many Chinese ate the leaves off trees and the grass off the ground, and a large portion of the population became addicted to opium. "A peasant party had come out of the hills to put an end to corruption, invasion, and humiliation; they wore straw sandals and told the truth" (Wang Ruowang 1989:40). Mao led this party; he instilled a revolutionary spirit into the masses that motivated the Chinese people to stand up to these immense problems (Strebeigh 1989). Mao was a remarkable leader and hero to his followers. The Chinese intellectual Liu Zaifu noted that "in the 1950s 'we did not believe in ourselves, but only in the all-wise, all powerful Mao'" (1989:40).

[6]Within three years, the Red Guards had disrupted and divided the country so deeply that Mao denounced them, claiming that they had failed to understand and implement his strategy. He banished 17 million youths to the countryside to live among the peasants and to be reeducated, and he reassigned the task of monitoring the population to the People's Liberation Army.

said a leading surgeon. 'They went to medical school. But they studied Mao's thought, planting rice or making tractor parts. They never had to take exams and a lot of them don't know basic anatomy'" (Butterfield 1980:32). Similar shortcomings affected more than the medical profession; during the Cultural Revolution, millions of people secured positions because they were loyal to Mao—not because they were qualified (Broaded 1991).

To solve these problems, the Communist Party, under the leadership of Deng Xiaoping, needed the support of those teachers, technicians, artists, and scientists who had been struck down during the Cultural Revolution. To make up for "Ten Lost Years," the Chinese leaders sent thousands of students overseas to study in the capitalist West. They allowed farmers and factory workers to keep profits from surplus crops after meeting government quotas, and they established five Special Economic Zones (SEZs—designated areas within the People's Republic of China that enjoy capitalist privileges).[7] In reversing the policies of the Cultural Revolution, Deng Xiaoping employed a rhetoric that defined the new policies as being in the best interest of the country, not as promoting individual self-interest. With regard to SEZs, Deng argued, "A few regions that have the right conditions will develop first. . . . The developed regions will then carry the developing regions until they finally reach common prosperity" (Deng 1995:A5).

As one measure of the success of the new policies, consider that foreign investments to China amounted to $33.8 billion in 1994 (U.S. Department of State 1996) and China has run a trade surplus with the United States since the beginning of 1995, creating trade tensions between the two countries (Hong Kong Trade Development Council 1996). As another measure of economic success, personal income increased an average of 329 percent in the cities and 355 percent in the countryside during the 1980s. Between 1990 and 1995, personal income rose an average of 226 percent in the cities and 196 percent in the countryside (Liu and Link 1998).

This increase in average personal income has been very uneven. The gap between the rich and poor has widened so that the wealthiest fifth of the Chinese pop-

A woman sells Mao memorabilia in contemporary China. Ironically, it is now acceptable to make money from objects celebrating the man who forbade any form of capitalism.

© Murray White/Sipa Press

ulation now own 50.2 percent of the wealth, compared with the poorest fifth, who own only 4.3 percent of the wealth (Liu and Link 1998). This state of economic affairs is truly remarkable when we remember that as recently as 1976 one could suffer imprisonment and hard labor simply for knowing some foreign words. On the other hand, we should not interpret these economic changes to mean that China is moving toward establishing a capitalist economy (see "The Importance of Ideological Commitment"). While Chinese leaders have been trying to transform a sluggish Soviet-style, centrally planned system into a more dynamic and flexible economy with market elements, the new economy remains under Communist control (U.S. Central Intelligence Agency 1998). The Chinese Embassy (1998) predicts that by this century's end "the state-owned sector will generate a third of the country's economic output, and their assets will account for two-thirds of the country's total." Also keep in mind that, in the cities, about 85 percent of the work force still works for the state (Liu and Link 1998).

---

[7]Special Economic Zones are designed to attract foreign investment and capital; foreign investors are lured there by the cheap labor force and by the potential market of more than 1.2 billion Chinese consumers. The five SEZs are Hainan Island, Chantou, Shenzhen, Xiamen, and Zhuhai.

The dramatic changes in China since 1978 make it an ideal case for illustrating two major assumptions that guide sociological thinking about the nature of deviance:

1. Almost any behavior or appearance can qualify as deviant under the right circumstances.

2. Conceptions of what is deviant vary over time and place.

In Chapter 4, we examined norms, a sociological concept that underscores these two assumptions. As you recall, norms are written and unwritten rules specifying appropriate and inappropriate behaviors. The important concept of norms cannot be overlooked when discussing deviance, because it is the violation of norms that constitutes deviance.

Deviance: The Violation of Norms

In Chapter 4, we learned that some norms are considered more important than others. In that chapter, we highlighted two kinds of norms—folkways and mores—distinguished by sociologist William Graham Sumner. We learned that **folkways** are customary ways of doing things that apply to the details of life or routine matters—how one should look, eat, greet another person, or express affection toward same-sex and opposite-sex persons.

Mores, on the other hand, are norms that people define as essential to the well-being of their group or nation. People who violate mores usually are punished severely; they are ostracized, institutionalized in prisons or mental hospitals, and sometimes executed. In comparison with folkways, people consider mores to be "the only way" and "the truth." Consequently, they consider mores to be final and unchangeable. The United States has a large number of mores that protect individual privacy, property, rights, and freedoms. For example, an individual has the right to marry, have children, choose a career, and change jobs and residences without

The Chagan shopping center, Beijing, would be very much at home in the United States. This scene is quite remarkable when you consider that during the Cultural Revolution people were criticized for possessing foreign-made goods or appearing to have extra wealth or "bourgeoisie" status.

© Adrian Bradshaw/Saba

appealing to a higher authority for permission. It does not matter that the country as a whole has a shortage of scientists, teachers, and nurses or that no physicians practice in some geographic areas and a glut of physicians exists in other regions. What matters is an individual's right to choose his or her occupation. It is unthinkable that Americans could not change jobs or residences whenever they wanted, for whatever reason.

Throughout China's long history, its mores have reflected the traditional values of conformity, collectivism, and obedience to authority.[8] In China it is unthinkable that an individual could marry, have a baby, or obtain housing without first obtaining approval from the Communist Party–controlled work unit or neighborhood committee (Oxman 1993b). Until recently, a person could not select an occupation, purchase a train ticket, bicycle, or television, secure a hotel room, obtain employment, or buy food without written approval from a unit or a committee.

[8]In fact, the Communist Party, which was formed in the 1920s and eventually came to power under Mao's leadership in 1948, sought to maintain these traditional values but to shift them away from the family to the party and party-led institutions. At the same time, the party wanted to discard many old mores that were based on a disdain for physical labor, allegiance to the family, religious beliefs, and reliance on personal networks and to substitute mores that reflected new values, including a love for physical labor, loyalty to the party and its leader, and atheism. The Communists aimed to create a new style of person ("socialist man"), and they built upon some old mores and values and introduced a new set of values, mores, and sanctions to discourage deviance and enforce their own rule (personal correspondence 1993).

Folkways Customary ways of doing things that apply to the details of life or routine matters.

Mores Norms that people define as essential to the well-being of their group or nation.

The Importance of Ideological Commitment

China seems to be home to two groups of people: those who are ideologically sound and support the Communist Party, and those who lack ideological commitment and conspire against the party. Today, the ideologically sound must support four major political principles: (1) China is a socialist country, and hence the creation of not a capitalist system but a socialist market economy; (2) power and leadership reside with the Chinese Communist Party; (3) a combination of Marxism, Leninism, and Mao Zedong thought is the guiding ideology; and (4) the state is a dictatorship by the proletariat. The principles have been left intentionally vague so that the party or individual leaders can claim violations against these principles when it suits their purposes (personal correspondence 1993). Any behavior that disrupts the progress or the smooth operation of the workplace, the neighborhood, or the country is conceived as an offense against the people or the country, which the Communist Party oversees. The most widespread formal sanction used in China is rehabilitation through reeducation and labor.

A single man risks his life to face down a line of tanks sent to suppress the 1989 demonstration in Tiananmen Square.
© AP/Wide World Photos

Obviously, Chinese who have robbed, murdered, assaulted, or raped someone have disrupted the social fabric. In addition, those who commit adultery, disobey work assignments, lose their jobs, have no honest occupation, are vagrants, disrupt a group's discipline, are expelled or drop out of school, or engage in behaviors that endanger state security are considered disruptive and in need of rehabilitation (Faison 1997). One activity defined as a political crime was the series of 1989 demonstrations in Tiananmen Square

---

Usually people abide by established folkways and mores because they accept them as "good and proper, appropriate and worthy" (Sumner 1907:60). For the great majority of people, "the rule to do as all do suffices." Recall from Chapter 5 that socialization is the process by which most people come to learn and accept as natural the ways of their culture. Because socialization begins as soon as a person enters the world, one has little opportunity to avoid exposure to the culture's folkways and mores.

If we compare the ways in which life is structured for four-year-olds in Chinese preschools and American preschools, we can see that different but important cultural lessons are incorporated into their daily activities. Even though it is impossible to generalize about preschools in countries as large and as diverse as the United States and China, we can identify some broad, outstanding differences. For the most part, Chinese preschoolers are taught to suppress individual impulses, play cooperatively with other children, and attune themselves to group enterprises. In contrast, American preschoolers are taught to cultivate individual interests and compete with other children for success and recognition by the teacher.

Socialization as a Means of Social Control: Preschool in China and in the United States

Professors Joseph Tobin and Dana Davidson and researcher David Wu filmed daily life in Chinese and U.S. preschools to learn how teachers in each system socialize children to participate effectively in their respective societies.[9] (It is worth noting that the great majority of preschool children whom they filmed in

[9]Actually, the authors studied preschools in three countries: China, the United States, and Japan.

by Chinese students in support of democracy,[1] which was eventually suppressed violently in June of that year. The students were accused of "bourgeois liberalization,"

a term that has never been clearly defined but, based on its usage by different sources in the media, it could perhaps be interpreted as wanton expression of individual freedom (individualism) that poses a threat to the stability and unity of the country. Such tendencies had to be curbed. In the government's view, students participated in the protests because they had led a sheltered life and were ignorant of the complexities of the reform process. Their youthful outburst did not take into consideration larger collective interests and concerns. (Kwong 1988:983–984)

The government undertook a number of measures to persuade the students to become more knowledgeable about the complexities of life and less responsive to subversive ideas. These measures included sending them "to rural areas to teach them to endure hardship, work hard and appreciate the daily difficulties faced by China's mostly rural population" (Kristof

1989:Y1). The rationale is that proper ideological commitment can be instilled through contact with the masses and through manual labor.[2] Other measures included placing limits on the number of students entering the humanities and social sciences. In the year following the Tiananmen Square incident, almost no students were admitted to study academic areas that government officials considered to be "ideologically suspect" by nature, such as history, political science, sociology, and international studies (Goldman 1989).

Perhaps one of the most intriguing things about China is that, since 1949, the leaders in power have adhered to the following code:

In any circumstance and at any cost, political power must be retained in its totality. This rule is absolute, it tolerates no exception and must take precedence over any other consideration. The bankruptcy of the entire country, the ruin of its credit abroad, the destruction of national prestige, the annihilation of all efforts toward overture and modernization—none of these could ever enter into consideration once the Party's authority was at stake. (Leys 1989:17)

This code still applies. Even today, as China undergoes massive economic growth and change, "its leaders have made it abundantly clear that they intend to preserve an unchallengeable Communist Party dictatorship, even as they pull back from trying to dominate every detail of economic and social life" (Holley 1993:H15).

[1]The movement toward democracy has been suppressed many times. In 1979, it was suppressed on the grounds that it was "unstabilizing" to China. In 1983, it was labeled a case of "spiritual pollution" (Link 1989).

[2]"A Chinese professor who at the end of the Cultural Revolution was sent to Anhui province, in central China . . . described what it was like to live there. Everything was made from mud, he recalled: the floors, the walls, even the beds and the stools the peasants sat on were constructed from pounded earth. The villagers had no wood for fuel—all the trees in the region had long ago been chopped down—so the women and children gathered grass and wheat stalks, depriving the earth of valuable natural nutrients. During the year the professor was there, he ate no meat, and the family he was quartered with had no matches, no soap, and most of the time no cooking oil, an essential part of the Chinese diet" (Butterfield 1982:16).

both countries seemed happy and productive.) The researchers found that, in comparison with U.S. preschools, Chinese preschools are highly structured and socially minded: Chinese teachers discipline their four-year-old students "by stopping them from misbehaving before they even know they are about to misbehave" (Tobin, Wu, and Davidson 1989:94), and they promote loyalty to the group. The bathroom scene described here is an example of the extent to which Chinese children are taught to follow instructions and attune themselves to group enterprises:

It is now 10:00, time for children to go to the bathroom. Following Ms. Wang, the twenty-six children walk in single file across the courtyard to a small cement building toward the back of the school grounds. Inside there is only a long ditch running along three walls. Under Ms. Wang's direction and, in a few cases, with her assistance, all twenty-six children pull down their pants and squat over the

ditch, boys on one side of the room, girls on the other. After five minutes Ms. Wang distributes toilet paper, and the children wipe themselves. Leaving the toilet, again in single file, the children line up in front of a pump, where two daily monitors are kept busy filling and refilling a bucket with water that the children use to wash their hands. (pp. 78–79)

When the researchers showed U.S. parents and teachers the film portraying daily life in Chinese preschools, Americans were particularly disturbed by the bathroom scene and asked why children were forced to go to the bathroom in this manner.[10] To this question, one Chinese educator replied:

[10]The Chinese system seems harsh from an American point of view, but it ensures that children wipe themselves properly and that their hands are washed. This method has some practical benefits, because many diseases are spread through contact with bodily substances, including saliva, mucus, and feces.

Quite different mores govern behavior in American and Chinese preschools. Whereas Chinese preschool teachers emphasize structured activities and social-mindedness, teachers in the United States are more likely to teach pupils self-direction, freedom of choice, and individuality.

© Forrest Anderson/Gamma-Liaison (left); © Paul Conklin/Monkmeyer Press (right)

Why not? Why have small children go to the bathroom separately? It is much easier to have everyone go at the same time. Of course, if a child cannot wait, he is allowed to go to the bathroom when he needs to. But, as a matter of routine, it's good for children to learn to regulate their bodies and attune their rhythms to those of their classmates. (p. 105)

In the United States, preschoolers are also taught discipline, but they are more likely to be corrected after they do something wrong or get out of hand. In comparison with their Chinese counterparts, U.S. teachers encourage self-direction, freedom of choice, independence, and individuality—qualities that often depend on a supply of material items. For example, American children typically use as much paper as they want; they start a drawing, decide they do not like it, and crumple up the paper. When they play house, store, or firefighter, they use costumes, plastic dishes, children's versions of household appliances, plastic food items, and so on. American teachers also encourage children to choose from a number of activities. A typical exchange between a preschool teacher and his or her students follows. It is difficult to imagine this exchange taking place in a Chinese classroom:

Who would like to paint? Michelle. Mayumi. Nicole. Okay, you three get your smocks from your cubbies and you can paint. [The teacher holds up a wooden block.] Who wants to do this? Mike? Okay, that's one. Stu, that makes two. Billy is three. . . . Here's a puzzle piece. You want to start on the puzzles? Okay? [The teacher holds up a toy frying pan.] Who wants to start in the house? Lisa, Rose, Derek. Go ahead to the housekeeping corner. Kerry, what do you want to do? The Legos? You're going to work on the radio, Carl? That's fine. Who is going to come over to the book corner to read this book? It's called *Stone Soup*. Okay, come on with me. (p. 130)

Reaction to Socialization of Another Culture

Both the Chinese and the Americans who watched the films were disturbed by their counterparts' system of handling preschool. Comments by Chinese viewers showed their clear preference for their own way of structuring early education. They maintained that their form of discipline expresses care and concern, and they regarded U.S. preschools as chaotic, undisciplined, and promoting self-centeredness. As one Chinese viewer remarked, "There are so many toys in the classroom that children must get spoiled. When they have so much, children don't appreciate what they have" (p. 88). On the other side, the U.S. viewers criticized the Chinese preschools for being rigid, totalitarian, too group-oriented, and overrestrictive, "making children drab, colorless, and robot-like" (p. 138). The people in each country were uncomfortable with the other's system because the lessons being taught in that country clashed with the prevailing mores or ideas about which behaviors are essential to their own country's well-being.

Such early socialization experiences are intended to help children to fit within the existing system. Each society tries to prepare its people to mesh with and accept their respective environments. In the case of preschool socialization, most Chinese preschoolers begin kindergarten after having learned a great deal about the need for group cooperation. Most American preschoolers graduate knowing that they will continue to be evaluated and rewarded based on their individual performance. Even so, primary socialization experiences such as those that take place during preschool are uneven at best. Not all preschools are alike, and some children do not attend preschools. Even among those who attend, some children do not internalize (take as their own and accept as binding) the values, norms, and expectations to which they are exposed. Therefore, all societies establish other mechanisms of social control to ensure conformity.

Ideally, conformity should be voluntary. That is, people should be internally motivated to maintain group standards and to feel guilty if they deviate from them. As noted earlier, during the Cultural Revolution it was considered deviant to wear glasses, use makeup, speak a foreign language, or break or destroy items that displayed Mao Zedong's picture. Many Chinese conformed to these rules on their own and punished themselves for violations, even if they broke the rules by accident. The memories of one Chinese man illustrate this point:

> As a boy, I did not know what a god looked like, but I knew that Mao was the god of our lives. When I was six, I accidentally broke a large porcelain Mao badge. Fear gripped me. In my life until that moment, the breaking of the badge seemed the worst thing I had ever done. Desperate to hide my crime, I took the pieces and threw them down a public toilet. For months I felt guilty. (Author X 1992:22)

In this case, the author's own guilt was a sign of voluntary conformity. If conformity cannot be achieved voluntarily, however, people may employ various means to teach, persuade, or force others to conform.

Mechanisms of Social Control

Ideally, socialization brings about conformity and conformity is voluntary. When conformity cannot be achieved voluntarily, however, other mechanisms of social control may be used to convey and enforce

norms and expectations. Such mechanisms are known as **sanctions**—reactions of approval and disapproval to others' behavior and appearances. Sanctions can be positive or negative, formal or informal.

A **positive sanction** is an expression of approval and a reward for compliance; such a sanction may take the form of applause, a smile, or a pat on the back. In contrast, a **negative sanction** is an expression of disapproval for noncompliance; the punishment may be withdrawal of affection, ridicule, ostracism, banishment, physical harm, imprisonment, solitary confinement, or even death.

Informal sanctions are spontaneous and unofficial expressions of approval or disapproval; they are not backed by the force of law. **Formal sanctions,** on the other hand, are definite and systematic laws, rules, regulations, and policies that specify (usually in writing) the conditions under which people should be rewarded or punished and that define the procedures for allocating rewards and imposing punishments. Examples of formal positive sanctions include medals, cash bonuses, and diplomas. Formal negative sanctions may take the form of fines, prison sentences, the death penalty, corporal punishment, or the application of tear gas to disperse demonstrators. Sociologists use the word **crime** to refer to deviance that breaks the laws of society and is punished by formal sanctions. People in every society have different views of what constitutes crime, what causes people to commit crimes, and how to handle offenders (see "The Extent of Social Control in China").

Sanctions Reactions of approval and disapproval to others' behavior and appearances.

Positive sanction An expression of approval and a reward for compliance.

Negative sanction An expression of disapproval for noncompliance.

Informal sanctions Spontaneous, unofficial expressions of approval or disapproval that are not backed by the force of law.

Formal sanctions Definite and systematic laws, rules, regulations, and policies that specify (usually in writing) the conditions under which people should be rewarded or punished and that define the procedures for allocating rewards and imposing punishments.

Crime Deviance that breaks the laws of society and is punished by formal sanctions.

The Extent of Social Control in China

In China, "social control is everywhere and involves everyone" (Clark and Clark 1985:109), including relatives, friends, colleagues, employers, and everyone with whom a person comes in contact (Xu 1995). Thus "each person has a social duty to participate in group activities and to 'help' others in the collective living arrangement. The mandate to 'help' means assuming responsibility for others and correcting their faults" (Rojek 1985:119).

Theoretically, every Chinese person belongs to a work unit and a neighborhood committee headed by Communist Party members. The work unit and neighborhood committee are "a basic social cell that receives and executes various policy programs of the communist party and government" (Shaw 1996:xi). Until recently, almost every important area of life was supervised by the unit or the committee. The work unit issued job assignments, determined salary, distributed ration coupons and other goods (light bulbs, contraceptives, bicycles, television

This billboard in Beijing is part of the Chinese effort to control population growth through a policy of one child per family.

© Forrest Anderson/Gamma-Liaison

sets), and granted permission to travel, change jobs, and change residences. The neighborhood committee is part of the "chain" connecting its members to local leaders and the

police, and up the various levels of leadership between it and the central authority structure in Beijing (Farley 1995). It still scrutinizes requests to marry and have children, and it determines whether a worker or his or her children may take college entrance examinations. The work unit maintains a confidential file[1] on every person, which contains information on education, work, class background as far back as three generations, the party's evaluation of the person (as an activist or as a counterrevolutionary), and any political charges made against him or her by informants. Members of the work unit are encouraged to report wrongdoings committed by other members. Whereas Americans are hesitant to report their suspicions of misconduct such as child abuse, spousal abuse, or drug use because they believe in a person's right to live without interference, the Chinese typically are afraid *not* to voice such suspicions because, by remaining silent, they are viewed as accomplices to such acts.

Defining Deviance: The Functionalist Perspective

According to sociologist Randall Collins (1982), Emile Durkheim presented one of the most sophisticated sociological theories of deviance. Durkheim ([1901] 1982) argued that, although deviance does not take the same form everywhere, it is present in all societies. He defined deviance as those acts that offend collective norms and expectations. The fact that, always and everywhere, some people will exist who offend collective sentiments led him to conclude that deviance is normal as long as it is not excessive and that "it is completely impossible for any society entirely free of it to exist" (p. 99). According to Durkheim, deviance will be present even in a "community of saints in an exemplary and perfect monastery" (p. 100). Even in seemingly perfect societies, acts that most persons would view as minor may offend, create a sense of scandal, or be treated as crimes.

Durkheim drew an analogy between a society populated with exemplary individuals and the "perfect and

upright" person. Just as such a person judges his or her smallest failings with a severity that others reserve for the most serious offenses, so too do societies that supposedly contain the most exemplary people. Even among such exemplary individuals, some act or appearance will offend, simply because "it is impossible for everyone to be alike if only because each of us cannot stand in the same spot. It is also inevitable that among these deviations some assume a criminal character" (p. 100). What makes an act or appearance criminal is not so much the character or the consequences of that act or appearance, but the fact that the group has defined it as something dangerous or threatening to its well-being. Wearing eyeglasses, for example, clearly does not harm others. As we have learned, however, this behavior was identified as a clear indicator of other threatening behaviors, such as the crimes of "special status" or abandonment of revolutionary spirit, that were not so easily observable.

According to Durkheim, deviance has an important function for society for at least two reasons. First,

The neighborhood committee monitors life outside the workplace. It enforces birth control policies, monitors contacts between its members and outsiders, settles domestic problems, scrutinizes each household's activities, investigates disputes among neighbors, deals with petty crime, and educates members about new policy and law. The committee has the right to search living quarters without the occupants' consent. As in the work unit, members are encouraged to inform on other members' wrongdoings. "When society is so tightly organized, where can one run to [when one unintentionally or intentionally violates the rules]?" (Wu Han 1981:39). Recently, multinational corporations have employed neighborhood committees to pass out product samples to its members and to survey them about product preferences (Farley 1995).

Among the aspects of Chinese life that are controlled by work units and neighborhood committees, perhaps the most widely publicized outside of China is procreation. Because of its huge population, China has tried to impose a limit of one child per couple in hopes of slowing population growth so that it will stabilize at 1.26 billion by the year 2000 (Tien 1990). Each province and each city is assigned a yearly quota with regard to how many children can be born, and the neighborhood committee or work unit determines which couples will be included in that quota. In essence, the work unit and the neighborhood committee decide which married couples can have a baby, determine when they can start trying, oversee contraceptive use, and even record women's menstrual cycles. Female officials check to see that IUDs are in place. If a woman misses several periods and has not been authorized to try to conceive, she is persuaded to have an abortion (Ignatius 1988). As one Chinese official told an American journalist, "We don't force her. . . . We talk to her again and again until she agrees" (Hareven 1987:73).[3] Couples who have only one child and sign an agreement to have no more children are rewarded with positive sanctions—salary bonuses and other financial incentives, educational opportunities, housing priority, and extra living space. If couples request permission to have a second child, they are asked to wait at least four years; if they have two children, they are persuaded not to have any more. Various kinds of economic sanctions or penalties are imposed on couples who have more than one child.[2]

[1]In China, information that people living in the United States would consider in the public domain, such as demographic statistics, is considered classified information. Chinese students and professors interested in such information must make a formal request. Such requests are entered into their permanent file (Rorty 1982).

[2]The policy of one child per couple has been more successful in urban areas than in rural areas. Chinese peasants have resisted government efforts because they believe that more children bring more happiness, that sons are more effective laborers in the fields than daughters, and that sons offer security to parents in old age.

[3]There is some evidence that China is gradually moving toward less forceful approaches to controlling reproduction.

the ritual of identifying and exposing the wrongdoing, determining a punishment, and carrying it out is an emotional experience that binds together the members of a group and establishes a sense of community. Durkheim argued that a group that went too long without noticing crime or doing something about it would lose its identity as a group. In evaluating Durkheim's argument, consider your reaction when you learn that your government has criticized the Chinese system (for example, "I am glad I don't live in China"). On the other hand, consider how people in China might react when its officials criticize something about the U.S. system or another system. The merits of the respective criticisms aside, the act of exposing "wrongdoing" functions to generate patriotic feelings in many people, as the indirect message is "Be glad you live here and not there!"

Second, deviance is functional because it is useful in making necessary changes and preparing people for change. It is the first step toward the future. Nothing would change if someone did not step forward and introduce a new perspective or new ways of doing things. Almost every invention or behavior is rejected by some group when it first comes into existence.

Durkheim's theory offers an intriguing explanation for why almost anything can be defined as deviant. Yet, Durkheim did not address an important question: Who decides that a particular activity or appearance is deviant? Labeling theory provides one answer to this question.

Labeling Theory

In *Outsiders: Studies in the Sociology of Deviance*, sociologist Howard Becker[11] states the central thesis of labeling theory: "All social groups make rules and attempt, at some times and under some circumstances, to enforce them. When a rule is enforced, the person who

[11]A number of sociologists—including Frank Tannenbaum (1938), Edwin Lemert (1951), John Kitsuse (1962), Kai Erikson (1966), and Howard Becker (1963)—are linked to the development of what is conventionally called labeling theory. The scholar most frequently associated with labeling theory, however, is Howard Becker.

Are people who use marijuana for medical purposes criminals because their behavior leads to harmful consequences or because some powerful segments of society have defined the behavior as illegal?

© Mark Richards/PhotoEdit

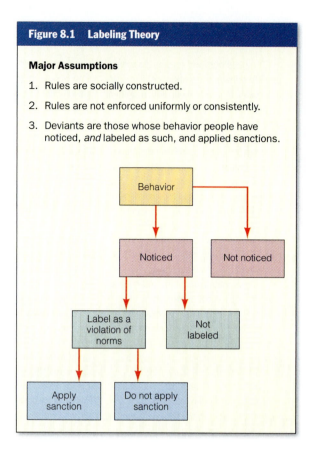

Figure 8.1   Labeling Theory

Major Assumptions

1. Rules are socially constructed.
2. Rules are not enforced uniformly or consistently.
3. Deviants are those whose behavior people have noticed, *and* labeled as such, and applied sanctions.

is supposed to have broken it may be seen as a special kind of person, one who cannot be trusted to live by the rules agreed on by the group. He is regarded as an outsider" (1963:1).

As Becker's statement suggests, labeling theorists operate under two assumptions: (1) that rules are socially constructed and (2) that these rules are not enforced uniformly or consistently. Support for the first assumption comes from the fact that definitions of deviant behavior vary across time and place. People must therefore decide what is deviant. The second assumption is supported by the fact that some people break rules and escape detection, whereas others are treated as offenders even though they have broken no rules. Labeling theorists maintain that whether an act is deviant depends on whether people notice it and, if they do notice, on whether they label it as a violation of a rule and subsequently apply sanctions. Such contin-

gencies suggest that violating a rule does not automatically make a person deviant. That is, from a sociological point of view, a rule breaker is not deviant (in the strict sense of the word) unless someone notices the violation and decides to take corrective action (see Figure 8.1). "The critical variable in the study of deviance, then is the social audience rather than the individual actor, since it is the social audience [that] eventually determines whether or not any episode or behavior . . . is labeled deviant" (Erikson 1966:11).

Labeling theorists suggest that for every rule a social group creates, four categories of people exist: conformists, pure deviants, secret deviants, and the falsely accused. The category to which one belongs depends on whether a rule has been violated and on whether sanctions are applied.

Conformists are people who have not violated the rules of a group and are treated accordingly. **Pure deviants,** on the other hand, are people who have broken the rules and are caught, punished, and labeled as outsiders. As a result, these rule breakers take on the **master status of deviant,** an identification that "proves to be more important than most others. One will be identified as a deviant first, before other identi-

Conformists People who have not violated the rules of a group and are treated accordingly.

Pure deviants People who have broken the rules and are caught, punished, and labeled as outsiders.

Master status of deviant An identification marking a rule breaker first as a deviant and then as having any other identification.

Table 8.1	Percentage Distribution of Victimizations and Crimes Reported to the Police	
Sector and Type of Crime	**Number of Victimizations**	**Percentage of Victimizations Reported to the Police**
All crimes	39,582,880	36.5
All personal crimes	10,268,280	43.0
Crimes of violence	9,866,200	43.2
Crimes of theft	22,883,060	27.3
Property crimes	10,035,060	34.2

Source: U.S. Bureau of Justice Statistics (1996a, 1996b).

fications are made" (Becker 1963:33). We must remember that, although pure deviants undeniably violate rules, rule enforcers "select" the people whom they apprehend and punish.

Consider how highway patrol officers assigned to Maryland Interstate 95 choose, from among all of the cars speeding along the highway, which drivers to pull over and search for drugs. A study of vehicles' and drivers' characteristics on Maryland Interstate 95 showed that 98 percent of all drivers were violating traffic laws. Of those violating traffic laws, 17.5 percent were "black."[12]

Seventy-two percent of the 823 drivers Maryland State Police stopped and searched on that stretch of highway between January 1995 and September 1996 were classified as "black" (Lamberth 1996). Upon stopping and searching the vehicle, police found some reason to arrest 29.9 percent of drivers. Significantly, this arrest rate is the same for both blacks and whites. In other words, approximately one-third of all whites and one-third of all blacks stopped are arrested. Because police stop and search three times as many "black" drivers as whites, however, the higher number of blacks arrested reflects the police's practice of targeting black drivers (Lamberth 1996). The higher numbers do not mean blacks are more "deviant" than whites; rather, more whites assume the status of "secret deviant."

Secret deviants are people who have broken the rules but whose violation goes unnoticed or, if it is noticed, prompts no one to enforce the law. Becker maintains that "no one really knows how much of this phenomenon exists," but he is convinced that the "amount is very sizable, much more so than we are apt to think" (1963:20). For example, a 1994 U.S. Bureau of Justice survey of crime victims documented that

42.4 million crimes were committed against U.S. residents 12 years of age and older. Of these 42.4 million crimes, only 38 percent of the victims reported the crime to police (U.S. Bureau of Justice Statistics 1996a, 1996b; see Table 8.1).

The **falsely accused** are people who have not broken the rules but who are treated as if they have done so. The ranks of the falsely accused include victims of such things as eyewitness errors or police cover-ups; they also include innocent suspects who make false confessions under the pressure of interrogation. In their book, *In Spite of Innocence,* sociologist Michael L. Radelet and philosopher Adam Bedau reviewed more than 400 cases of innocent people convicted of capital crimes and found that 56 had made false confessions. Some innocent suspects apparently admitted guilt, even regarding heinous crimes, to escape the stress of interrogation (Jerome 1995). As with the phenomenon of secret deviants, no one knows how often people are falsely accused, but it probably happens more often than we think. In any case, the status of accused often lingers even if the person is cleared of all charges. Such cases lead us to ask a larger question: Under which circumstances are people most likely to be falsely accused?

The Circumstances of the Falsely Accused

Sociologist Kai Erikson (1966) identifies a particular situation in which people are likely to be falsely accused of a crime: when the well-being of a country or a group is threatened. The threat can take the form of

---

[12]This violation rate corresponds to the percentage of black drivers on I-95, which is 17 percent.

Secret deviants People who have broken the rules but whose violation goes unnoticed, or, if it is noticed, prompts no one to enforce the law.

Falsely accused People who have not broken the rules but who are treated as if they have done so.

Dale E. Mahan and his brother Ronnie spent 14 years in prison after being falsely accused of raping a woman in 1983. DNA tests proved their innocence.

© Alan S. Weiner/NYT Pictures

an economic crisis (an economic depression or recession), a moral crisis (family breakdown, for example), a health crisis (such as AIDS), or a national security crisis (such as war). At these times, people need to identify a seemingly clear source of the threat. Whenever a catastrophe occurs, it is common practice to blame someone for it. Identifying the threat gives an illusion of control. The person blamed is likely to be someone who is at best indirectly responsible, someone in the wrong place at the wrong time, and/or someone who is viewed as different.

This defining activity can take the form of a **witch-hunt,** a campaign to identify, investigate, and correct behavior that undermines a group or a country. In actuality, a witch-hunt rarely accomplishes this goal, because the real cause of a problem is often complex and lies far beyond the behavior of a targeted person or group. Often the people who are defined as the problem are not, in fact, the cause of the threat but are simply used to make the cause of a complicated situation appear manageable. For example, as happened during the Cultural Revolution, sometimes the seemingly most insignificant acts—such as wearing makeup or eyeglasses—were classified as crimes against the country. Of course, such behaviors could not possibly be responsible for the failure of the Great Leap Forward. Nevertheless, targeting such behaviors diverted the

public's attention from the shortcomings of those in power, united the public behind a cause, and took people's attention away from this disruptive event.[13]

The internment of more than 110,000 people of Japanese descent (80 percent of whom were American citizens) living on the West Coast of the United States during World War II is another example of a situation in which a group was targeted in conjunction with a crisis. Japanese Americans were forced from their homes and taken to desert prisons surrounded by barbed wire and guarded with machine guns (Kometani 1987). The Japanese Americans had not pursued any anti-American activity. Yet, the wartime hysteria, combined with long-standing prejudices, led to the deportment of men, women, and children to concentration camps.

The existence of the falsely accused underscores the fact that the study of deviance must look beyond people identified or labeled as rule breakers. After all, the falsely accused are labeled as deviant by those with the power to do so.

Rule Makers and Rule Enforcers

Sociologist Howard Becker (1973) recommends that researchers pay particular attention to who the rule makers and rule enforcers are and to how they achieve power and then use that power to define how others "will be regarded, understood, and treated" (p. 204). This topic, of course, interests not only labeling theorists but also conflict theorists. According to conflict theorists, members of a society with the most wealth, power, and authority have the power to create laws and crime-stopping and -monitoring institutions. Consequently, we should not be surprised to learn that law enforcement efforts tend to focus disproportionately on crimes committed by the poor and other powerless groups rather than on those committed by the wealthy and politically powerful. This uneven focus gives the widespread impression that the poor, the uneducated, and minority group members are more prone to criminal behavior than are people in the middle and upper classes, the educated, and majority group members. In fact, crime exists in all social classes, but the type of crime, the extent to which the laws are enforced, access to legal aid, and the power to shape laws to one's advan-

---

Witch-hunt A campaign to identify, investigate, and correct behavior that undermines a group or a country.

[13]Another example involves the timing of the 1993 debate over whether gays should be allowed to serve in the U.S. military. Sociologists might ask, Is it a coincidence that gays became the focus of media attention just when the Clinton administration and Congress were planning such large cuts in the defense budget?

Conflict theorists emphasize that not all rules and rule breakers are treated equally. A sharp financial dealer like Michael Milken (shown here doing community service as part of his punishment) may inflict far more harm on people than street thieves (right). Yet the white-collar felon is often treated far less harshly by the criminal justice system.

© Bart Bartholomew/Black Star (left); © Rick Friedman/Black Star (right)

tage vary across class lines (Chambliss 1974). In the United States, for example, police efforts are largely directed at controlling crimes against individual life and property (crimes such as drug offenses, robbery, assault, homicide, and rape) rather than against white-collar and corporate crime.

White-collar crime consists of "crimes committed by persons of respectability and high social status in the course of their occupations" (Sutherland and Cressey 1978:44). **Corporate crime** is crime committed by a corporation as it competes with other companies for market share and profits. Usually white-collar and corporate crimes—such as the manufacturing and marketing of unsafe products, unlawful disposal of hazardous waste, tax evasion, and money laundering—are handled not by the police but by regulatory agencies, such as the Environmental Protection Agency and Food and Drug Administration, that have minimal staff to monitor compliance. Escaping punishment is easier for perpetrators of white-collar and corporate crimes. In the case of white-collar crime, offenders are part of the system: they occupy positions in the organization that permit them to carry out illegal activities discreetly. In the case of corporate crime, the crimes are carried out by everyone in the organization during the course of doing their jobs. In addition, both white-collar and corporate crime are "directed against impersonal—and often vaguely defined—entities such as the tax system, the physical environment, competitive conditions in the

market economy, etc." (National Council for Crime Prevention in Sweden 1985:13). These crimes are without victims in the usual sense, because they are "seldom directed against a particular person who goes to the police and reports an offense" (p. 13). In 1998, .07 percent (659 persons) of all people sentenced to U.S. federal prisons were classified as white-collar criminals as compared with 58.9 percent (56,291 persons) classified as drug offenders (see "White-Collar and Corporate Crime"; Federal Bureau of Prisons 1998).

Deviance is a consequence not of a particular behavior or appearance, but rather of the application of rules and sanctions to a so-called offender by others (see "The Presumption of Innocence"). For these reasons, sociologists are concerned less with rule violators than with those persons who make the rules and those individuals who support and enforce these rules. Any effort to explain deviance is complicated by the fact that the powerful play an important role in defining what is deviant and in establishing the sanctions that should be

White-collar crime "Crimes committed by persons of respectability and high social status in the course of their occupations" (Sutherland and Cressey 1978:44).

Corporate crime Crime committed by a corporation as it competes with other companies for market share and profits.

White-Collar and Corporate Crimes

Two U.S. Department of Justice press releases illustrate how corporate and white-collar crimes are carried out in the context of a larger organization. Read the press releases and decide whether each describes corporate or white-collar crime.

Tuesday, July 21, 1998

COMPANY PRESIDENT PLEADS GUILTY TO MAIL FRAUD

Washington, D.C.—The president of Tucker Environmental Consultants (TEC), Paducah, Ky., has pleaded guilty to mail fraud for mailing false laboratory analyses regarding contamination from an underground storage tank (UST) site to the owner of the site and to the Kentucky Department of Environmental Protection (KDEP), the Department of Justice announced today.

Janice Tucker, who was initially indicted in the case on March 5, [1998,] pleaded guilty on Monday in U.S. District Court for the Western District of Kentucky. According to the indictment, Tucker had submitted false lab analyses for a number of UST sites in the Paducah area.

As part of her plea, Tucker admitted that in April 1993, she submitted false lab analyses from the clean-up of USTs located at VMV Enterprises, Inc., a railroad yard in Paducah. Tucker charged VMV Enterprises about $6,000 for the false lab analyses of contamination at the site and mailed the false analyses to both VMV Enterprises and KDEP.

"Those whose business is monitoring, removing and cleaning up contaminated sites owe a duty to the surrounding community to tell the whole truth," said Lois Schiffer, Assistant Attorney General for the Environment and Natural Resources Division.

"In the case at hand, Janice Tucker did not tell the truth. In fact, she lied about test data from one of the contaminated sites," Schiffer said.

Tucker faces up to five years in federal prison, without parole, and a fine of up to $250,000. She is scheduled to be sentenced on October 16. The case was investigated by the Federal Bureau of Investigation with assistance from KDEP.

Thursday, August 6, 1998

USX SETTLES FEDERAL, STATE ENVIRONMENTAL CLAIMS

Nation's Largest Steel Maker Plans $30 Million Clean-Up of Grand Calumet River

Washington, DC—USX Corporation, the nation's largest steel maker, has agreed to settle federal and state environmental claims for illegally discharging wastewater from its Gary, Indiana, plant into the Grand Calumet River, announced the U.S. Departments of Justice, the Environmental Protection Agency, the U.S. Department of Interior and the State of Indiana today. Among other things, the company will pay $30 million to clean up the River, $22 million on future pollution abatement and a $2.9 million civil penalty. The Gary plant is the largest steel making plant in the nation.

Two consent agreements were lodged today in U.S. District Court in Hammond, Indiana. Under the first, USX will spend approximately $30 million to remove and dispose of nearly 700,000 cubic yards of contaminated sediments from five miles of the River. USX must remove the sediments over the next five years.

"These agreements will help put the 'Grand' back in the Calumet River," said Lois J. Schiffer, Assistant Attorney General for Environment and Natural Resources. "USX will clean up its past pollution and reduce any future pollution. This means a cleaner river for fishing and swimming and a healthier environment for the citizens of Northwest Indiana."

Violating both the Clean Water Act and a 1990 federal consent decree since 1992, USX's Gary facility illegally discharged wastewater contaminated with PCBs, heavy metals (such as iron, lead, zinc, cadmium and chromium), oil and grease, benzene, polycyclic aromatic hydrocarbons and cyanide. The contaminated sediments kill fish, injure migratory birds and release pollutants into the River and Lake Michigan.

"The clean-up of this stretch of the Grand Calumet River will be a crucial step in the revitalization of the river corridor," said David A. Ullrich, Acting EPA Regional Administrator. "It will result in cleaner water, healthier fish

and a visibly improved environment for the people of Gary and all those who enjoy Lake Michigan."

As part of the clean-up, USX must restore 32 acres of wetlands next to the river and construct a disposal facility for contaminated sediments, including PCBs, from the river. In addition, USX must improve the Gary facility to eliminate spills and other illegal discharges. The cost of these capital improvements is estimated at $22 million.

The first consent agreement also requires USX to pay a $2.9 million civil penalty. . . . The second consent agreement addresses natural resource damages. Under Superfund authority, USX will buy 139 acres of property for the National Park Service in order to expand the rare dune and swale wildlife habitat along the southern shore of Lake Michigan. A separate 77 acre acquisition for the State of Indiana will enhance wildlife habitat nearer to the River and will increase access for fishermen for recreational use.

"Fish and wildlife habitat along this important waterway will be restored, and the removal of PCB-contaminated sediment along a five mile stretch of the Grand Calumet River will improve the water quality of Lake Michigan," said Larry Macklin, Director of the Indiana Department of Natural Resources. "Also, 77 acres have been acquired along Salt Creek, about 10 miles east of Grand Calumet. The land will provide people with a new fishing site that should be excellent for trout and salmon, wildlife habitat will be protected, and the new site will help compensate the public for the lost use of an important natural resource."

In addition to the two consent agreements, USX has entered into an agreement with EPA, known as an administrative order, to identify releases of hazardous materials at its Gary plant and clean up the waste that poses a public health or environmental threat. Entered into under the authority of the Resource Conservation and Recovery Act, the order will help protect the River from recontamination.

Source: U.S. Department of Justice 1998a, 1998b

U.S. in Perspective
The Presumption of Innocence

"China is oriented toward crime control not due process" (Xu 1995:84). In China a defendant is presumed to be guilty, not innocent until proved guilty as in the U.S. system. Before appearing in court, the defendant will have been scrutinized already by a work unit or neighborhood committee. If the work unit or committee members cannot correct the defendant's outlook or behavior, they will refer the case to the control committee, which can either assign a punishment or refer the case to the courts. If a case is referred to the courts, it is presumed that the offender is guilty and that some punishment is warranted. The court appearance is likely to be just a sentencing hearing, as evidenced by the fact that approximately 99 percent of persons whose cases are heard by the Supreme People's Court are convicted. Theoretically, a Chinese offender can appeal a decision, but the court system allows only one appeal (Chiu 1988).

In theory, a defendant in the United States is assumed innocent until proved guilty; the burden of proving guilt rests with the prosecution. Furthermore, sentences and punishments can be reduced through plea bargaining, shock probation (sudden and unexpected release from prison), or plea of insanity or self-defense—or simply because prisons are overcrowded. In 1990, 300,000 criminal cases were processed in New York City alone. Only 10 percent, or 30,000 cases, were tried

A man accused of counterrevolutionary activity is brought to trial before being executed. By the time his case reached the trial stage, he had already been presumed guilty.
© Agence Vu

to completion. The remaining 270,000 cases ended in dismissals, reduction in charges, or plea bargaining (Shipp, Baquet, and Gottlieb 1991). Finally, defendants who can afford the fees are able to purchase the best legal defense and thus have the greatest chance of being found innocent or receiving lighter sentences.

The role of lawyers in each society dramatically reflects the basic differences between Chinese and American views on how defendants should be handled. In China, lawyers are paid by

the state and must remit any fees to the state. Although the Chinese lawyer is trained to protect the legitimate rights of clients, those rights are secondary to the rights and interests of the state and the people. In addition, Chinese lawyers cannot use the impression management strategies that American lawyers employ to convince a jury of a client's innocence when in fact the client is guilty (Lubman 1983). If Chinese lawyers followed such techniques, they would be accused of conspiring with an enemy of the state.

applied to correct such behavior. We must therefore question how the powerful influence the public to accept such definitions and apply the recommended sanctions. The work of social psychologist Stanley Milgram gives us one answer to this question.

Obedience to Authority

When Stanley Milgram (1974) conducted his research for *Obedience to Authority,* he was interested in learning about how people in positions of authority manage to get

people to accept their definitions of deviance and to conform to orders about how to treat people classified as deviant. His study gives us insights into how events like the Holocaust and the Cultural Revolution could have taken place. The fact that such atrocities required the cooperation of a large number of people raises important questions about people's capacity to obey authority.

> The person who, with inner conviction, loathes stealing, killing, and assault may find himself per-

forming these acts with relative ease when commanded by authority. Behavior that is unthinkable in an individual who is acting on his own may be executed without hesitation when carried out under orders. (p. xi)

Milgram designed an experiment to see how far people would go before they would refuse to conform to an authority's orders. The findings of his experiment have considerable relevance for understanding the conditions under which rules handed down by authorities are enforced by the masses.

The participants for this experiment were volunteers who answered an ad placed by Milgram in a local paper. When each participant arrived at the study site, he or she was greeted by a man in a laboratory jacket who explained to the participant and another apparent volunteer the purpose of the study: to test whether the use of punishment improves the ability to learn. (Unknown to the subject, the other "volunteer" was actually a **confederate**—someone who works in cooperation with the person conducting the research study.) The participant and the confederate drew lots to determine who would be the teacher and who would be the learner. The draw was fixed, however, so that the confederate was always the learner and the real volunteer was always the teacher.

The learner was strapped to a chair, and electrodes were placed on his or her wrists. The teacher, who could not see the learner, was placed in front of an instrument panel containing a line of shock-generating switches. The switches ranged from 15 to 450 volts and were labeled accordingly, from "slight shock" to "danger, severe shock." The experimenter explained that when the learner made a first mistake, the teacher was to administer a 15-volt shock; the teacher would then increase the voltage with each subsequent mistake. In each case, as the strength of the shock increased, the learner expressed greater discomfort. One learner even said that his heart was bothering him; that statement was followed by complete silence. When the volunteers expressed concern about the learner's safety, the experimenter firmly told them to continue administering the shock. Although many of the volunteers

protested against administering such severe shocks, a substantial number obeyed and continued to administer shocks, "no matter how vehement the pleading of the person being shocked, no matter how painful the shocks seemed to be, and no matter how much the victim pleaded to be let out" (Milgram 1987:567).

The results of Milgram's experiments are especially significant when one considers that the participants received no penalty if they refused to administer a shock. Obedience in this situation was founded simply on the firm command of a person with a status that gave minimal authority over the subject. If this level of obedience is possible under the circumstances of Milgram's experiments, one can imagine the extent to which it is possible in a situation in which disobedience brings severe penalties or negative consequences.

In view of this need to understand more about the role that rule makers play in molding deviance, we turn to the constructionist approach, which concentrates on a particular type of rule maker—the claims maker.

The Constructionist Approach

The **constructionist approach** focuses on the process by which specific groups (for example, illegal immigrants or homosexuals), activities (such as child abuse), conditions (teenage pregnancy, infertility, pollution), or artifacts (song lyrics, guns, art, eyeglasses) become defined as problems. In particular, constructionists examine the claims-making activities that underlie this process.[14] Claims-making activities include "demanding services, filling out forms, lodging complaints, filing lawsuits, calling press conferences, writing letters of protest, passing resolutions, publishing exposés, placing ads in newspapers, supporting or opposing some governmental practice or policy, setting up picket lines or boycotts" (Spector and Kitsuse 1977:79). Claims makers are people who articulate and promote claims and who tend to gain in some way if the targeted audience accepts their claims as true. Examples of claims makers include government offi-

---

Confederate Someone who works in cooperation with a person conducting a research study.

Constructionist approach A sociological approach that focuses on the process by which specific groups, activities, conditions, or artifacts become defined as problems.

[14]Sociologist Joel Best (1989) assembled 13 articles expressing a constructionist viewpoint, written by various sociologists, in a single volume called *Images of Issues: Typifying Contemporary Social Problems.* The titles include "Horror Stories and the Construction of Child Abuse," "Dark Figures and Child Victims: Statistical Claims about Missing Children," "AIDS and the Press: The Creation and Transformation of a Social Problem," and "The Surprising Resurgence of the Smoking Problem." In spite of the diversity of topics, all 13 authors take a constructionist approach, which has strong roots in symbolic interaction but also appeals to conflict theorists.

cials, marketers, scientists, professors, and other interest groups.

Claims Makers

The success of a claims-making campaign depends on a number of factors, including access to the media, available resources, and the claims maker's status and skills at fund raising, promotion, and organization (Best 1989). During the 1989 Tiananmen Square incident in Beijing,[15] for example, government officials engaged in claims-making activities by controlling the information that the troops received about the demonstration. Essentially, the soldiers were denied access to newspapers and television for roughly one week. When soldiers and students became friendly with one another, the government sent in replacements, some of whom came from northern China and were not fluent in standard Chinese (Calhoun 1989). Thus the replacement soldiers could not communicate with the demonstrators. Obviously, the government officials, because of their positions, were better able than the students to control the soldiers' understanding of the situation.

According to sociologist Joel Best, when constructionists study the process by which a group or behavior is defined as a problem to the society, they focus on who makes the claims, whose claims are heard, and how audiences respond. Constructionists are guided by one or more of the following questions: What kinds of claims are made about the problem? Who makes the claims? Which claims are heard? Why is the claim made at a particular time? What are the responses to the claim? Is there evidence that the claims maker has misrepresented or inaccurately characterized the situation? In answering this last question, constructionists examine how claims makers characterize a condition. Specifically, they pay attention to any labels that claims makers attach to a condition, the examples they use to illustrate the nature of the problem, and their orientation toward the problem (describing it as a medical, moral, genetic, educational, or character problem).

Labels, examples, and orientation are important because they tend to evoke a particular cause and a particular solution to a problem (Best 1989). For example,

to label AIDS as being a moral problem is to locate its cause in the goodness or badness of human action and to suggest that the solution depends upon the victims changing their evil ways. To call it a medical problem is to locate its cause in the biological workings of the body or mind and to suggest that the solution rests with a drug, a vaccine, or surgery. To label homosexuality as being a legitimate form of sexual expression is to suggest that homosexuality is not a sin or something shameful. Similarly, a claims maker who uses the example of a promiscuous homosexual male to illustrate the nature of the AIDS problem sends a much different message about AIDS than does a claims maker who uses hemophiliacs or HIV-infected children as examples (see "Two Claims-making Organizations: The Family Research Council and Human Rights Campaign" and "Grounds for Divorce or Annulment").

To this point, we have examined several sociological concepts—socialization, norms (folkways and mores), and mechanisms of social control—and discussed how they relate to deviance. In addition, we have examined Durkheim's theory of deviance to gain insights into how any behavior or appearance can come to be defined as deviant. We have looked at labeling theory and the constructionist approach to gain insights into the role that rule makers play in shaping deviance. Now we turn to the theory of structural strain, which helps us to answer the following question: Under what conditions do people engage in behavior defined as deviant?

Structural Strain Theory

Robert K. Merton's theory of structural strain takes into account three factors: (1) culturally valued goals defined as legitimate for all members of society, (2) norms that specify the legitimate means of achieving these goals, and (3) the actual number of legitimate opportunities available to people to achieve the culturally valued goals. According to Merton, **structural strain** is any situation in which (1) the valued goals have unclear limits (that is, people are unsure

---

[15]The Tiananmen Square incident is the most well-known event in a series of demonstrations that took place in China between April 15 and June 4, 1989. In all, 84 cities and as many as 3 million students joined by workers, teachers, and even police and soldiers were involved in demonstrations against those in power. The image that remains with us is that of a young Chinese man facing down a tank. Although no official figures are available, hundreds are estimated to have been killed and thousands injured.

Structural strain Any situation in which (1) the valued goals have unclear limits, (2) people are unsure whether the legitimate means that society provides will allow them to achieve the valued goals, and (3) legitimate opportunities for meeting the goals remain closed to a significant portion of the population.

Two Claims-making Organizations: The Family Research Council and Human Rights Campaign

Two national organizations—the Family Research Council and the Human Rights Campaign—ran advertisements in *USA Today* proclaiming their views on homosexuality. Which perspective do you find most compelling? Why?

Anne Paulk – wife, mother, former lesbian

I'm living proof that Truth can set you free.

"Recently, several prominent people like Trent Lott, Reggie White, and Angie and Debbie Winans have spoken out on homosexuality...calling it a sin. When I was living as a lesbian I didn't like hearing words like that...until I realized that God's love was truly meant for me."

One boy's sin and the making of a lesbian.

"I was four years old when a teenage boy molested me. When he warned me not to say a thing, I went silent. But as I grew, the pain wouldn't stay quiet. It would shout when I made male friends. It laughed when I tried to feel pretty. It told me I was unlovable. But saddest of all, when I wanted to turn to my parents to make everything all right – it wouldn't let me tell them why."

Being a woman became a mystery.

"By the time I hit my teens I was rough...my heart cold. I believed being *'feminine'* meant being weak and vulnerable...so looking and dressing hard felt right. I had so thoroughly rejected my own femininity that, even though I had a lot of male friends, I just wasn't attracted to men sexually. I became drawn to other women who had what I felt was missing in me. But the pain inside kept yelling."

There's a God-shaped hole in everyone's heart.

"My sexual attraction to women blossomed in college, and after a gay counselor affirmed my feelings I joined the campus gay/lesbian group. But it was in the course of those group meetings that I knew something was still missing. While I longed for a female life-partner, I knew it just wouldn't work. That's when I went home and prayed, *"God, please show me who You are, and fill the void in my heart."*

Knock and He'll answer. But the next step is still yours.

"Change didn't come overnight. Within six months I'd made a firm decision to forsake homosexuality, but I still had sexual desire for women. Even though I filled my days with Christian activity, I fell back into a relationship with someone who quickly

Thousands of ex-gays like these have walked away from their homosexual identities. While the paths each took into homosexuality may vary, their stories of hope and healing through the transforming love of Jesus Christ are the same.

Ex-gay ministries throughout the U.S. work daily with homosexuals seeking change, and many provide outreach programs to their families and loved ones.

became my priority in life, over work, over family and friends...over God. By now the pain inside was throbbing, and my delight eroded into conviction, deception and emotional instability."

Once God answers He never hangs up.

"I knew I was running from God, and one day just put it to him: *"Lord, You know that I really enjoy this lifestyle, but I want You to be my first love. I need Your help. I need You to change my heart."* Shortly after that prayer, I met a Christian woman, *a former lesbian*, who listened patiently to my story and led me to a ministry helping people overcome homosexuality. Because they loved me without judgement, I was able to finally give all my relationships to God, and begin the real road to healing."

Changing hearts. Changing lives.

"Leaving homosexuality was the hardest thing I've ever had to do. I finally saw the patterns of my same-sex attraction and came to understand the underlying needs that had sparked my longings. As I grew in my relationship with God, I knew He had changed me forever. Gone was the hardness. Gone was the hurt. And gone was the shrill cry inside, replaced with God's still, small voice. In His gentle love I found forgiveness, and the acceptance I sought so hard on my own."

There is another way out.

Please, if you, or someone you know or love, is struggling with homosexuality, show them this story. If you truly love someone, you'll tell them the truth. *And the truth that God loves them could just be the truth that sets them free.*

If you really love someone, you'll tell them the truth.

Advertisement from the Family Research Council.

Courtesy of the Family Research Council

I'm living proof that lesbian and gay Americans can be whole and happy and are worthy of all that this great country promises.

Dreams of Equality

Hello. My name is Stampp. I'm a computer executive. I work hard. I live in the heartland of America. I love to swim and run. And, oh yes, I happen to be gay. For too long, gay and lesbian Americans of all backgrounds have been singled out for discrimination. Lately, some people have spent tens of thousands of dollars on a national ad campaign trying to convince the American public that gay people are not fully human. They want all of us to believe gay people are incapable of whole, happy and healthy lives. To be blunt, these are lies. There have always been those who want to single out one group or another for discrimination. This is contrary to the values of fairness and nondiscrimination — values to which we aspire as a nation.

When I was growing up, my parents encouraged me to take advantage of every opportunity. They could not always protect me from racial discrimination. But they tried to create a world in which I could flourish. This is what all children need. My parents knew the best thing they could do for me, straight or gay, is love me with all their heart and soul. That's exactly what they did. For some parents, it can be a great struggle. When I finally told my parents in my late teens that I was gay, there really wasn't the trauma you often hear about. They were relieved I could finally tell the truth. I hope every parent can find this simple message of love and support for their child.

A Mother's Support

About two years ago, I went to my mother and told her the time had come for me to become more deeply involved in the work of achieving basic equal rights for gay and lesbian Americans. I explained that you could be fired in most places in this country simply because you are gay. I explained there are no basic protections under the law. I told my mother I would be more visible in my advocacy, and that, as a result, some of our relatives or her friends or neighbors might see me on a brochure or even on television. I asked for her blessing.

It was an emotional moment and I remember tears came to my mother's eyes. I was dismayed and assured her this was a happy thing — an opportunity to tell the truth about my life and the lives of other gay people. She said they were not tears of anguish, but tears of joy. She explained that if she had not participated in the black civil rights movement in the 1960s, she would have regretted it for the rest of her life. She was glad that in the 1990s, I had found my voice and the courage to speak out for the equality gay people deserve.

The Church Was at the Center of Our Family Life

I grew up in the church. In fact, the church was at the center of our family life. Spirituality remains very important to me. Not everyone in this nation is a Christian, and I guess I'm old-fashioned. I think one's religious journey is a very private matter. I would never try to tell another person their personal spiritual path is wrong, whatever their faith. When I was young, Christian ministries primarily focused on caring for the sick, the poor, the elderly and children in need. The so-called Christian ministries sponsoring this expensive ad campaign are doing nothing more than standing in judgment of others.

You know, it is interesting — Jesus never said one word about homosexuality. Some people think they need to explain what Jesus *really* thought. What Jesus did address was the importance of love and acceptance. You would think that as Christians, they would listen to what Jesus *actually* said, rather than what people think he *should* have said. His central message is about love and unity, not about judging, attacking, discriminating and diminishing — and you don't have to be Christian to understand these values. Like my mom says, God doesn't make mistakes. She says her boys are the perfect reflection of God's best work. *All* her boys.

An Important Message to Gay and Lesbian Americans, Young and Old

Right-wing organizations in this country are currently spending thousands upon thousands of dollars to try to convince you that you are flawed in some way. They want you to believe that you are less than whole and that you need to be "converted" into something different. They want you to feel bad about yourself. Don't listen to them! Being gay or lesbian is as natural as the sun. There have always been gay people; there will always be gay people. More than ever, there are millions of gay and lesbian Americans who are happy, fulfilled and have terrific, loving relationships.

Of course, many are still paralyzed by the bigotry and discrimination demonstrated in this extreme right ad series. These right-wing groups want to foster feelings of inadequacy and doubt. Over time, things will continue to change for the better. You are absolutely perfect as you are. Accepting yourself for who you are, and feeling good and whole, will help bring about important change. For more information on coming out, call the National Coming Out Project, a program of the Human Rights Campaign Foundation at 1-800-866-6263.

If you really love someone, you'll fight for their right to live with dignity and fairness.

Advertisement from the Human Rights Campaign.

Photograph by Don Flood at the Human Rights Campaign 1996.
Used with the permission from the Human Rights Campaign.

U.S. in Perspective
Grounds for Divorce or Annulment

This box provides a way to look at how U.S. society varies in its attitude toward alcohol, other drugs, and adultery, another behavior often considered "deviant." The map shows which states define alcoholism, drug addiction, and adultery as definite grounds for divorce or annulment. Making such behaviors grounds for divorce is one way of con-trolling the behaviors without making them strictly illegal. What patterns do you see in the data?

Practically speaking, the illegality of drugs like marijuana and cocaine is currently maintained based on federal law. Suppose someone proposed that grounds for divorce should also be standardized at the national level rather than being controlled by the states. What arguments would you make for or against such a proposal?

Suppose someone proposed that a spouse's addiction to prescription drugs should be grounds for divorce. Could a case be made for such a law?

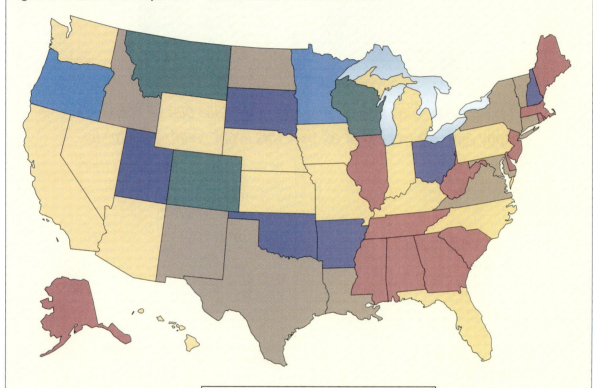

	Grounds for Divorce		
	Adultery	Alcoholism	Drug Addiction
	Yes	Yes	Yes
	Yes	Yes	No
	No	Yes	Yes
	Yes	No	No
	No	No	Yes
	No	No	No

Source: *The World Almanac and Book of Facts 1996* (p. 729)

whether they have achieved them), (2) people are unsure whether the legitimate means that society provides will allow them to achieve the valued goals, and (3) legitimate opportunities for meeting the goals remain closed to a significant portion of the population. The rate of deviance is likely to be high under any one of these situations. Merton uses the United States, a country where all three conditions exist, to show the relationship between structural strain and deviance.

Structural Strain in the United States

In the United States, most people place a high value on the culturally valued goal of economic affluence and social mobility.[16] Americans tend to believe that anyone can achieve success—that is, that all people, regardless of the circumstances in which they are born, can achieve affluence and stature. Such a viewpoint suggests that persons who fail have only themselves to blame. Merton argues that "Americans are bombarded on every side" with the message that this goal is achievable, "even in the face of repeated frustration" (1957:137).

Even among those who do achieve monetary success, there is no point at which those people can say they feel secure about having achieved their ultimate goal. "At each income level . . . Americans want just about 25 percent more (but of course this 'just a bit more' continues to operate once it is obtained)" (p. 136).

Merton also maintains that structural strain exists in the United States because the legitimate means of achieving affluence do not always lead to this goal. The individual's task is to choose a path that leads to success. That path might involve education, hard work, or the development of a talent. The problem is that school, hard work, and talent do not guarantee success. With regard to education, for example, many Americans believe that a diploma (and especially a college diploma) in itself entitles them to a high-paying job. In reality, however, a diploma is only one of many components needed to achieve success.

Finally, structural strain exists in the United States because too few legitimate opportunities are available to achieve the desired goals. Although Americans are supposed to seek financial success, they cannot all expect to achieve it legitimately; the opportunities for achieving success remain closed to many people, especially those in the lower classes. For example, many young African American men living in poverty believe that one seemingly sure way to achieve success is through sports. The opportunities dwindle rapidly, however, as an athlete advances: fewer than 2 percent of the athletes who play college basketball, for instance, even have a chance at joining the professional ranks.

Merton believed that people respond in identifiable ways to structural strain and that their response involves some combination of acceptance and rejection of the valued goals and means (see Figure 8.2). He identified the following five responses—only one of which is not a deviant response:

- **Conformity** is the acceptance of the cultural goals and the pursuit of these goals through legitimate means.

Figure 8.2 Merton's Typology of Responses to Structural Strain

Merton believed that people respond to structural strain and that their responses involve some combination of acceptance and rejection of valued goals and means.

Mode of Adaptation	Goals	Means
Conformity	+	+
Innovation	+	−
Ritualism	−	+
Retreatism	−	−
Rebellion	+/−	+/−

+ Acceptance/achievement of valued goals or means

− Rejection/failure to achieve valued goals or means

Source: Adapted from Merton (1957:140), "A Typology of Modes of Individual Adaptations."

Conformity Behavior and appearances that follow and maintain the standards of a group. Also, the acceptance of the cultural goals and the pursuit of these goals through legitimate means.

---

[16]Merton does not believe that the same proportion of people in all social classes accept the cultural goal of monetary success, but he does think that a significant number of people across all classes do so.

- **Innovation** involves the acceptance of the cultural goals but the rejection of legitimate means to obtain these goals. For the innovator, success is equated with winning the game rather than playing by the rules. After all, money may be used to purchase the same goods and services whether it was acquired legally or illegally. According to Merton, when the life circumstances of the middle and upper classes are compared with those of the lower classes, the lower classes clearly face the greatest pressure to innovate, though no evidence suggests that they do so (for example, white-collar and corporate crime are forms of innovation).

- **Ritualism** involves the rejection of the cultural goals but a rigid adherence to the legitimate means of attaining those goals. This response is the opposite of innovation; the game is played according to the rules despite defeat. Merton maintains that this response can be a reaction to the status anxiety that accompanies the ceaseless competitive struggle to stay on top or to get ahead. Ritualism finds expression in clichés such as "Don't aim high and you won't be disappointed" and "I'm not sticking my neck out." It can also be the response of people who have few employment opportunities open to them. If one wonders how a ritualist might be defined as deviant, consider the case of a college graduate who can find only a job bagging groceries at minimum wage. Most people would react as if this person is a failure, even though he or she may be working full-time.

- **Retreatism** involves the rejection of both cultural goals and the means of achieving these goals. The people who respond in this way have not succeeded by either legitimate or illegitimate means and thus have resigned from society. According to Merton, retreatists are the true aliens or the socially disinherited—the outcasts, vagrants,

vagabonds, tramps, drunks, and addicts. They are "in the society but not of it."

- **Rebellion** involves the full or partial rejection of both the goals and the means of attaining those goals and the introduction of a new set of goals and means. When this response is confined to a small segment of society, it provides the potential for the emergence of subgroups as diverse as street gangs and the Old Order Amish. When rebellion is the response of a large number of people who wish to reshape the entire structure of society, a great potential for a revolution exists.

Structural Strain in China

In China, one source of structural strain can be traced to the number of legitimate opportunities open to married couples to achieve the culturally valued goal of producing children, but especially a son. As pointed out in "The Extent of Social Control in China," since 1979 China has tried to impose a limit of one child per couple in urban areas and a limit of two children per couple in rural areas in hopes of slowing population growth (Chinese Embassy 1998). This family-planning policy poses a problem, because a cultural preference for boys exists in China, especially among the people living in the countryside. One reason that sons are valued more than daughters is that sons and their families are expected to care for parents in old age. Thus couples who produce a son can be confident that someone will care for them later in life. The legitimate means of starting a family that is open to couples in China includes the following steps: obtain permission to have a baby, accept the sex of the child (that is, do not abort female fetuses or kill a daughter), report the birth and sex of the child to appropriate agencies, and practice birth control to avoid conceiving other children.

We can apply Merton's typology of responses to structural strain to describe the reactions of Chinese couples (Figure 8.3). Those most likely to be conformists are those whose first child is a healthy son. Conformists would also include those couples who have no preference as to the sex of their child and those who are firmly committed to upholding the laws related to birth control because they see them as critical to China's quality of life (Bolido 1993:6). The majority of people in urban China can be classified as conformists.

Innovators accept the culturally valued goal of one child per couple but reject the package of legitimate means to obtain this goal. Upon learning that she is expecting a child, the woman may undergo ultrasound

Innovation The acceptance of the cultural goals but the rejection of legitimate means to obtain these goals.

Ritualism The rejection of cultural goals but a rigid adherence to the legitimate means of attaining those goals.

Retreatism The rejection of both cultural goals and the means of achieving those goals.

Rebellion The full or partial rejection of both the cultural goals and the means of attaining those goals and the introduction of a new set of goals and means.

Figure 8.3   Merton's Typology Applied to China's One-Child Policy

Culturally valued goal: population control
Culturally valued means: one-child limit
Sources of structural strain: preference for boys

Response	Goal: Population Control	Means: One-Child Policy	Examples
Conformists	+	+	Couples with no preference "Ideologically sound" couples
Innovators	+	−	Couples who abort female or unhealthy babies Couples who abandon babies
Ritualists	−	+	Couples who follow the rules but reject policies
Retreatists	−	−	Couples who reject the idea of population control and hide the "extra" babies
Rebels	+/−	+/−	Couples who replace official goals and means with new goals and means

+ Acceptance/achievement of valued goals or means

− Rejection/failure to achieve valued goals or means

to learn the sex of the fetus. If it is a girl, the couple may decide to abort the baby. Alternatively, upon the birth of a girl baby, the parents may kill, or arrange to have a midwife kill, the infant. Such practices are blamed for the so-called missing girls problem; the 1990 census in China showed that for every 114 boys born, there are 100 girls (Oxman 1993c). A second category of innovators includes couples who abandon an unhealthy and/or female child or abandon a second child born in violation of family-planning regulations. These children fill the country's 67 state-run orphanages (Tyler 1996b).[17]

Ritualists reject the one-child-per-couple goal of population control, but they adhere to the rules. They do not agree with the government policies, but they are afraid they will be punished if they do not follow them (Remez 1991).

Retreatists, on the other hand, reject the one-child goal and reject the legitimate means open to them. This category of deviants may include those couples who continue having children until a son is born and who hide the birth of the baby girls from party officials. In fact, Sterling Scruggs of the United Nations Population Fund argues that many of the so-called missing girls are actually unreported children (Oxman 1993c). In a sense, these parents are in society but, because of their secret, "not of it," as Merton would say.

Rebels reject the cultural goal of population control as well as the legitimate means; instead, they introduce new goals and new means. One could argue that this response applies to couples who are part of China's 55 ethnic minority populations that make up 9 percent (91.2 million) of the total population. The majority of these ethnic groups are exempt from the one-child policy. The government permits them to have two children, and some groups are permitted to have three and, in special cases, four children. For these groups, the cultural goal is to increase the size of the minority

[17]The annual death rates in some orphanages are higher than 20 percent, and in some institutions the death rate is between 57.2 and 72.5 percent (Schell 1996). Human rights groups argue that the high mortality rate reflects a deliberate policy to minimize the number of abandoned children. Chinese officials argue, however, that the high orphan mortality rate reflects the poor health of the abandoned children and the shortage of electricity needed to keep buildings warm (Tyler 1996a, 1996c).

populations, and the means of achieving that goal is exemption from the one-child policy. This option has prompted many couples to claim ethnic minority status (Tien et al. 1992).

Differential Association Theory

Sociologists Edwin H. Sutherland and Donald R. Cressey advanced a theory of socialization called **differential association** to explain how deviant behavior, and especially delinquent behavior, is learned. This theory refers to the idea that "when persons become criminal, they do so because of contacts with criminal patterns and also because of isolation from anticriminal patterns" (1978:78). These contacts take place within **deviant subcultures**—groups that are part of the larger society but whose members adhere to norms and values that favor violation of the larger society's laws. That is, people learn techniques of committing crime from close association and interaction with those who engage in and approve of criminal behaviors.

Sutherland and Cressey maintain that impersonal forms of communication, such as television, movies, and newspapers, play only a relatively small role in the genesis of criminal behavior. If we accept the premise that criminal behavior is learned, then criminals constitute a special type of conformist. That is, they conform to the norms of the group with which they associate. Furthermore, the theory of differential association does not explain how a person makes the initial contact with a deviant subculture (unless, of course, the person is born into such a subculture). Once contact is made, however, the individual learns the subculture's rules for behavior in the same way that all behavior is learned.

Sociologist Terry Williams studied a group of teenagers, some as young as age 14, who sold cocaine in the Washington Heights section of New York City. These youths were recruited by major drug suppliers

Differential association A theory of socialization that explains how delinquent behavior is learned. It refers to the idea that "when persons become criminal, they do so because of contacts with criminal patterns and also because of isolation from anticriminal patterns" (Sutherland and Cressey 1978:78).

Deviant subcultures Groups that are part of the larger society but whose members adhere to norms and values that favor violation of the larger society's laws.

because, as minors, they could not be sent to prison. Williams (1989) argues that the teenagers were susceptible to recruitment for two reasons: they saw little chance of finding well-paying jobs, and they perceived drug dealing as a way to earn money that would enable them to pursue a new life.

To become a successful drug dealer, a youth must learn a number of skills. Williams (1989) describes some of these skills:

> Each week, Max is fronted [consigned] three to five kilos (in 1985, this would have a street value of $180,000 to $350,000) to distribute to the kids. The quantity he receives varies according to the quantity he has previously sold, how much he has on hand, and how much he is committed to deliver both to the crew and directly to other customers.
>
> The quality, variety and amount of cocaine each crew member receives, then, is determined by Max and by his suppliers. The kids might ask for more but whether or not they get more depends on how much they sold from their last consignment. Personal factors are also involved. (p. 34)

Williams's findings suggest that once teenagers become involved in drug networks, they learn the skills to perform their jobs in the same way that everyone learns to do a job. Indeed, success in a "deviant" job is measured in much the same way that success is measured in mainstream jobs: pleasing the boss, meeting goals, and getting along with associates.

In China, notions of differential association are the basis of the philosophy underlying rehabilitation: a deviant individual, whether a thief or a revisionist, becomes deviant because of "bad" education or associations with "bad" influences (see "Is There a Rationale for Control?"). To correct these influences, society must reeducate deviant individuals politically, inspire them to support the Communist Party, and teach them a love of labor. Particular attention is paid to the 5 million or so university or advanced technical institute students, who represent an elite few (3 percent of middle or high school students). As one Beijing University professor commented in an interview with Asian scholar Robert Oxman (1994b), "It is very challenging job here on the campus how to learn those advanced technology and also very interesting ideas, value systems from the outside [and at] the same time to keep our own traditions" (p. 7). In other words, the Chinese government must prepare new generations of young people for coping with global interdependence and for running the country's economy and other affairs. In the course of their education, these students may associate with people

U.S. in Perspective

Is There a Rationale for Control?

The Chinese believe in the malleability of human beings and emphasize social reasons and factors that contribute to deviant behavior (Zhou 1995:84). As a result, they support early intervention and have a definite rehabilitation program for offenders, consisting of thought reform but especially reeducation through labor:

The Chinese believe that it is important to intervene early, before deviance has become extreme or caused too much damage. To refrain from correcting minor expressions of deviance on the grounds that such correction interferes with the individual's civil rights strikes the Chinese as analogous to withholding early treatment of a disease on the grounds that the disease has a right to develop until it proves to be lethal. This is particularly true since society (through bad education) is at least partially to blame for the deviance; it thus has a corresponding duty to correct it. (Bracey 1985:142)

In *New Ghosts, Old Ghosts: Prisoners and Labor Reform,* James D. Seymour and Richard Anderson (1998), "two respected experts on human rights in China" (Berstein 1998:B6), argue that, relative to other countries, China's rate of imprisonment—166 per 100,000 population—is only moderately above the world's average, which is estimated at 105 per 100,000. The U.S. rate is 440 per 100,000 and, if the jail population is included, it is 614 per 100,000. Seymour and Anderson (1998) believe, however, "that the prisoner-to-crime ratio is much higher in China than in almost all other countries. That is, for each serious crime committed, someone is much more likely to be arrested and imprisoned than elsewhere" (p. 220).

The most widespread prison regime is labor reform, but there are other forms of detention. Offenders are placed in one of six types of correctional facilities, including reeducation through labor, shelter and investigation, and home surveillance. The official purpose of incarceration is to rehabilitate or "rehumanize" offenders. Rehabilitation programs are designed to teach offenders a productive skill and to turn them into socially responsible citizens. Offenders labor in mines, farms, or factories, make bricks, do agricultural work, or work on large infrastructure projects such as road building and irrigation.

Many critics in the United States believe that China should be denied Most Favored Nation status (a trading status that has been given to all but about 12 countries in the world) until it improves its human rights record, especially with regard to political prisoners. For example, the human rights watch group Asia Watch maintains that 1,300 people arrested during Tiananmen Square remain in prison. Chinese officials contend that the figure is 70.

Critics also claim that China uses prisoners in labor camps to manufacture products for export. According to the U.S.–China Business Council, however, prisoner-made exports cannot possibly exceed 1 percent of all exports. China has responded to the charge by opening to inspection five labor camps believed to be sites of such manufacturing activities (Oxman 1994a). Because China gives no monitoring group such as the International Committee of the Red Cross access to its prisons, accounts of human rights abuses such as torture, solitary confinement, and other physical and mental abuses are based primarily on anecdotal evidence. Chinese officials deny that such abuses take place and further state that such practices are against the law. Seymour and Anderson (1998) maintain that it is difficult to verify such claims because the Chinese government has been secretive and intentionally evasive about its prison system and the number of prisoners held there. They believe that the fundamental issues revolve not around the size of China's penal system but "the substantial number of prisoners who do not belong there, and the harshness of the conditions in the less well-managed institutions" (p. 223).

In the United States, on the other hand, no clear rationale, such as reeducation, underlies the formal sanctions. In other words, there is little agreement about the purpose of corrections: Should it be rehabilitation? Restitution? Revenge? Officially, the Federal Bureau of Prisons' mission is "to protect society by confining offenders in the controlled environments of prisons and community-based facilities that are safe, humane, and appropriately secure, and that provide work and other self-improvement opportunities to assist offenders in becoming law-abiding citizens" (Federal Bureau of Prisons 1998).

Moreover, no coherent philosophy guides the operation of the 1,375 state and 125 federal prisons housing more than 1 million prisoners (Federal Bureau of Prisons 1998). American politicians and criminal justice personnel debate continually not only the purpose of prisons but also the effectiveness of the various methods, which may include an uneven array of educational, vocational, and therapeutic programs. Currently, the overwhelming emphasis is placed on ensuring that criminals serve time by establishing mandatory punishment schedules free of judicial discretion (Xu 1995). This emphasis explains in part why state prisons house 15 to 24 percent, and federal prisons hold 19 percent, more inmates than they were built to accommodate.

and ideas that challenge Chinese ways of doing things. To counteract such exposures, Communist ideology is officially espoused at the universities. Also, Chinese students are required to take a course in Marxism, and campus security remains tight (Oxman 1994b).

We close this chapter by discussing the factors that influence the Chinese and U.S. systems of social control. They include population size, the age of the two countries, and the degree of internal turmoil.

Factors That Shape U.S. and Chinese Systems of Social Control

The size of the Chinese population is one major reason for the country's rigid system of social control. More than 1.26 billion Chinese—one of every five people alive in the world—live in a space roughly the same size as the United States. After subtracting deserts and uninhabitable mountain ranges, China's habitable land area is about half that of the United States. Although almost 21 percent of the world's population lives in China, this country has 6.3 percent of the world's agricultural land (U.S. Central Intelligence Agency 1998). Even though China manages to feed most of its population well, approximately 65 million people live in a state of absolute poverty; that is, they lack the resources to satisfy the basic needs of food and shelter. Another 290 million live in a state of substantial deprivation. World Bank poverty specialist Alan Piazza (1996) maintains that "most Americans could not begin to comprehend what 65 million Chinese have to endure and what that level of deprivation is all about" (p. 47). On the average, 75,616 births and 22,027 deaths occur every day in China. If these patterns persist, the total population could increase by another 168 million over the next 10 years. To complicate the situation, the population has expanded rapidly in the past 45 years, growing from approximately 500 million in 1949 to 1.26 billion in 1998 (U.S. Central Intelligence Agency 1999).

Such rapid growth has strained the country's ability to house, clothe, educate, employ, and feed its people and has proved capable of overshadowing industrial and agricultural advancements. When one considers that approximately 1 billion people live under approximately 50 independent governments in Europe, North America, and South America (Fairbank 1989), the magnitude of China's population problem appears even more overwhelming. China's large population explains in part why jobs and housing are assigned and why until recently nearly everything was

rationed. Rice and other grains, cooking oil, soap, light bulbs, bicycles, and television sets all have been rationed. For example, Chinese people had to purchase rice with ration coupons. Even if a bowl of rice was consumed in a restaurant, it counted as part of a person's monthly allotted ration. In light of the rapid economic growth and the accompanying abundance of goods and food in recent years, this rationing system has become obsolete (Oxman 1993b).

The Chinese government must create work for many people, especially the large number of people awaiting employment. (Theoretically, there are no unemployed people in China—only those awaiting employment.)[18] Currently, 583.6 million people (160 million in urban areas and 420 million in rural areas) constitute the Chinese work force. Approximately 120 million people in the rural work force (which is growing by 20 million people per year), however, are surplus workers (Chinese Embassy 1996). As many as 100 million surplus rural workers known as the "floating population" have migrated to the cities in search of work but could find only part-time, low-paying jobs. Consequently, a considerable amount of "make-work" occurs. For example, "every park employs dozens of ticket takers, some to collect the ten cents for a ticket, some to collect the ticket, and some to collect the stub" (Mathews and Mathews 1983:34).

Living space is another problem. Only 10 percent of housing in China is owned privately. Usually housing is allocated by the government or work units. Space is tight. Family members sleep in the same room, and most households share bathroom and kitchen facilities with other households (Butterfield 1982).

A second reason for the difference between Chinese and U.S. social control systems is rooted in history. Most Americans are immigrants or descendants of voluntary or involuntary immigrants from many nations and cultures who settled in the United States over the past 200 years. Geographic separation from their native lands enabled these newcomers to break with tradition and either retain or discard elements of their native cultures (Fairbank 1989). In addition, the United States is a land with abundant resources and an impressive people-to-resources ratio. The material abundance, combined with geographic separation from native cultures, permitted immigrants to the United States to create a society in which, in theory, citizens live wherever they can afford and where they can manage their own lives.

---

[18]The Chinese government estimates unemployment to be 3 percent. The U.S. Department of State estimates unemployment to be between 10 and 15 percent (U.S. Department of State 1999).

In contrast, the Chinese not only are members of the most populous country in the world, but also belong to the longest-continuing civilization, which dates back some 3,700 years. This long and complex history has been marked by at least four strong and persistent traits:

1. The Confucian[19] system of ethics—which emphasizes order, justice, harmony, personal virtue and obligation, devotion to the family (including the spirits of one's ancestors), and respect for tradition, age, and authority—assigns everyone an unalterable role and place in society.

2. The system of family responsibility makes each member responsible for the conduct of other family members.

3. An imperial tradition gives rulers supreme authority over the lives of the people—historically, those in power have been considered above the law—and their power has not been constrained by institutionalized checks and balances.

4. No regime in China has ever relinquished its power without first resorting to bloodshed (Fairbank 1987, 1989).

A third issue that explains the rigid system of social control in China also relates to history. Since the beginning of the twentieth century, the Chinese people have suffered through several wars of foreign aggression and four revolutionary civil wars.[20] The most recent civil war, the Cultural Revolution (1966–1976), ended less than 20 years ago; in contrast, the last civil war in the United States took place more than 125 years ago. As a people, Americans are unfamiliar with the division, chaos, and destruction that war and revolution can bring. Often, strict control is the only way to restore order and to bring the stability needed to rebuild a society whose citizens have opposed each other violently.[21]

Summary and Implications

Deviance is a complex concept for several reasons: (1) definitions of deviance change over time and place; (2) not everyone who commits a deviant act is caught, and not everyone who is punished actually committed the crime of which he or she is accused; and (3) rule makers, rule enforcers, and the larger social audience affect how some behaviors or appearances come to be defined as deviant and others do not.

For these reasons, sociologists maintain that deviance does not result from a particular behavior or appearance but rather arises as a consequence of time, place, and the application of rules and sanctions by others to an "offender." The case of China's dramatic shift in attitude toward profit-making activities illustrates that examining a specific behavior or the personality of someone defined as deviant cannot give us a true understanding about the nature of deviance. Examining the social and historical context surrounding a behavior or appearance brings significant insights into the causes of deviant behavior.

The concepts of "secret deviants" and "falsely accused" also have tremendous and important implications for the study of deviance. Their existence helps us build the case that the causes of deviant behavior go beyond the personal characteristics of the officially accused (pure deviants and the falsely accused) and must include the behavior of rule makers and rule enforcers. Researchers who compare prison populations with a population that has never been imprisoned to discover traits that distinguish the two will learn little about what prompts people to break rules, because such studies do not identify the characteristics that create criminals but only those that cause a person to end up in prison. As the U.S. Bureau of Justice Statistics (1996a, 1996b) victimization studies show, the so-called never-imprisoned population actually includes a large number of people who have committed the same crimes as prisoners but who have escaped detection or conviction in part because victims fail to report crimes. To complicate matters even further, the prison

---

[19]Confucius lived between 551 and 479 B.C., a time of constant war, destruction, and chaos. It is not surprising, therefore, that his teachings emphasize rules that support order and harmony.

[20]The four revolutionary civil wars include the Republican Revolution of 1911, the Nationalist Revolution of 1925–1928, the Kuomintang-Communist Revolution of 1945–1949, and the Great Proletarian Cultural Revolution of 1966–1976.

[21]Mao Zedong, the charismatic Chinese leader who mobilized the masses in two of these civil wars, stated, "A revolution is not a dinner party, or writing an essay, or painting a picture, or doing embroidery; it cannot be so refined, so leisurely and gentle, so temperate, kind, courteous, restrained and magnanimous. A revolution is an insurrection, an act of violence by which one class overthrows another" (Mao Zedong 1965:28).

Political purges directed by Communists are not unique to the Cultural Revolution. Whenever China has faced serious economic problems or the authority of Chinese officials has been questioned, campaigns have been launched to hunt down and purge the purported opposition.

Theories of Deviance and Conformity

Theory	Central Question	Answer
Functionalist	How does deviance contribute to order and stability?	Deviance—especially the ritual of identifying and exposing wrongdoing, determining a punishment, and carrying it out—is an emotional experience that binds together members of groups and establishes a sense of community.
Labeling Theory	What is deviance?	Deviance depends on whether people notice it and, if they do notice it, on whether they label it as such and subsequently apply sanctions/punishment.
Obedience to Authority	How do rule makers get people to accept their definition of deviance and to act on those definitions?	Behavior that is unthinkable in an individual acting on his or her own may be executed without hesitation when authority figures command such behavior.
Constructionist	How do specific groups, activities, conditions, or artifacts come to be defined as problems?	Claims makers with the ability to define something as a problem, to have those claims heard, and to shape public responses play important roles in defining deviance and responses to it.
Structural Strain	Under what conditions do people engage in behavior defined as deviant?	People engage in behavior defined as deviant when valued goals have no limits or clear boundaries, when legitimate means do not guarantee valued goals, and when the number of legitimate opportunities are in short supply.
Differential Association Theory	How is behavior defined as deviant learned?	People learn deviant behavior through close associations and interactions with those who engage in and approve of criminal behavior.

population can contain innocent people. The contribution of sociology to the study of deviance is that it reminds us to consider factors other than the individual defined as deviant. By considering the larger context, we will be better able to understand the true nature of deviance and to design policies that change not just the individual but the larger society (see "Theories of Deviance and Conformity").

Key Concepts

Use this outline to organize your review of key chapter ideas.

Deviance
  Pure deviants
  Master status of deviant
  Secret deviants
  Falsely accused
    Witch-hunts
Conformity
Social control
  Sanctions
    Positive sanctions
    Negative sanctions
    Formal sanctions
    Informal sanctions

White-collar crime
Corporate crime
Constructionist approach
  Claims makers
Structural strain
  Conformity
  Innovation
  Ritualism
  Retreatism
  Rebellion
Differential association

internet assignment

In this chapter we considered how two claims-making organizations, the Family Research Council and the Human Rights Campaign, presented views on homosexuality. Use the Internet to find an organization that is using the Internet as a forum for claims making.

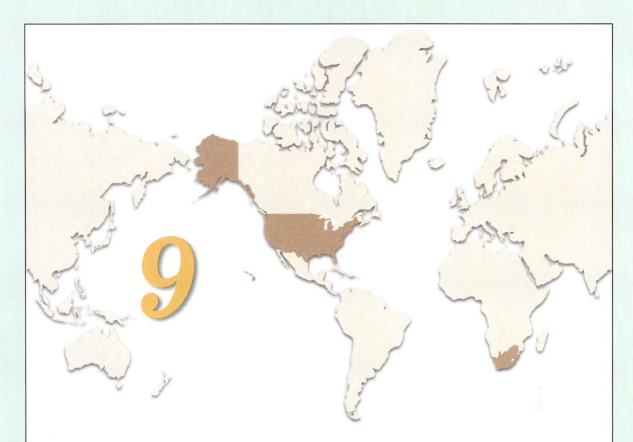

Social Stratification

With Emphasis on South Africa

Capetown, South Africa. (© Milner/Sygma.)

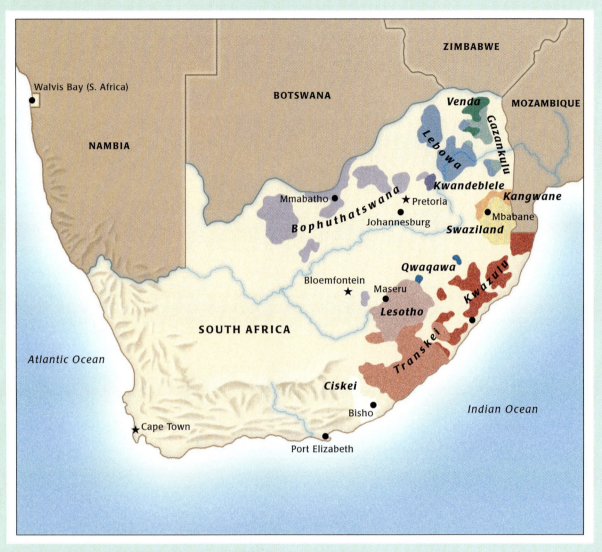

Source: Central Intelligence Agency Maps.

The Geography of Apartheid

South African apartheid was an attempt to create a society in which a person's race determined his or her rights and privileges in life. It divided the country into land for whites and land for blacks. The Land Acts of 1913 and 1936 mandated that 10 "homelands" be set aside, one for each of the major African ethnic groups. As the map shows, a single homeland could consist of several dozen pieces of land. The irregular shapes reflect the all-white government's decision to give the best land to the whites.

have no confidence in myself, I am sometimes suicidal. I do not know whether I can carry myself alone. I am messed up because of what I went through during my high school years. I would not like to focus on myself, but I would like to focus on the story of my forgotten Comrades. Those students who sacrificed their youth, their lives for the sake of our freedom. Those Comrades who fought for liberation, who never knew what it was like to enjoy life as teenagers. They did not have time to develop their relationships or to participate in sport. Those students whose days [were comprised] of meetings, protest marches, facing rubber bullets and often live ammunition, but this was part of the daily struggle against an unjust enemy. Our primary objective was to make the country ungovernable so that our leaders could return and lead us into a true democracy. . . .

Through the pressure and sacrifices of all involved in the liberation struggle, we inevitably got our freedom we fought for so hard. We reached our objectives at a price. However, many never realized the kind of psychological stress trauma we have been subjected to. Many Comrades had to go into hiding, others were in exile away from their loved ones and friends,

but the majority of us remained behind to continue the struggle for liberation. For those Comrades who were sought by security police it was the start of a nightmare, a nightmare that was going to be so horrendous, it was going to destroy our lives. Never could we have imagined that at the age of 15, 16, 17 and 18, we would have been running away from police and going to safe houses only to find police coming to look for us there. We were going crazy with all the thoughts of being captured, tortured, maimed and even killed at the hands of the security police. We were never sure whether we would see our parents, brothers and sisters again. One thing was certain, we were not teenagers any longer. We have aged 10 to 15 years in a matter of months and we had to be prepared to face whatever came our way whether we were prepared for it or not.

Many of us were caught by security police. In detention we were interrogated, tortured, maimed, violated and continuously harassed. After this horrid experience we suffered from nervous tension and were all nervous wrecks. We could not continue school like normal pupils. Our lives were destroyed. Some of us managed to complete matric under

extremely difficult circumstances. Others could not take the pressure; they left school to later become drug addicts and gangsters. If only there were people to help us through our trauma. We so much [wanted] the people to be there for us, because we could not carry on alone. Too many brilliant students never got the chance to reach their full potential as they were school drop-outs. They are still searching for themselves amongst all the turmoil in their lives. To tell you the truth, many of us ask the price that we had to pay for freedom, whether it was worth it in the end.

Riefaat Hattas
Youth Hearings
Truth and Reconciliation
    Commission
May 22, 1997

Why Focus on South Africa?

In this chapter, we focus on South Africa, a country that is attempting to dismantle its system of social stratification known as apartheid.

Under apartheid 11 percent of the population (the whites) benefited at the expense of the remaining 89 percent.[1] Political violence such as that described in the opening quotation pushed the transistion from apartheid to a multiracial democracy. On October 21, 1994, South Africa's first postapartheid government voted to establish a Commission on Truth and Reconciliation "to determine who suffered and/or caused harm to others during the political struggle of the past" and to bring the atrocities of apartheid, long hidden from view, to light (Jones 1998). The Commission, chaired by Archbishop Desmond Tutu, was charged with (1) determining the "nature, causes, and the extent of the gross violations of human rights that occurred between March 1, 1960, and December 31, 1998"; (2) granting amnesty to those who committed such gross violations provided the perpetrators were completely truthful, were motivated by political objectives, or had carried out the acts in a nonsadistic manner; (3) offering victims a public forum to tell their stories and, in that process, to restore their human dignity; (4) verifying victims' claims and determining whether they were eligible for reparations; and (5) reporting to the nation and making recommendations by October 31, 1998, for preventing such human rights violations in the future (Ignatieff 1997; Amnesty Committee 1998). More than 20,000 victims filed petitions to have their cases heard. "Victims were asked to put aside their claims for punishment and vengeance in return for getting at the truth" (Ignatieff 1997:84). More than 7,000 people have petitioned the commission for amnesty (Amnesty Commission 1998). Those guilty parties who ignore the amnesty offer will be prosecuted. The Commission began its hearings in December 1995 and is expected to continue to hear cases through 1999.

The 7,000 applications for amnesty came primarily from police and security forces working for the South African apartheid government. In addition, some petitions were filed by anti-apartheid activists and guerrillas, as well as by *askavis,* former anti-apartheid guerillas turned informers (Ignatieff 1997).

Among the thousands who received amnesty were the following:

- Those whites who invented and produced the chemical and biological tools, gadgetry, and weapons that the South African security forces used to brutalize anti-apartheid activists. Examples include poison-tipped umbrellas, soap boxes packed with explosives, walking sticks capable of firing poisonous bullets, substances to reduce fertility and virility, anthrax-laced chocolates and cigarettes, beer bottles laced with botulism, and sugar laced with salmonella (Daley 1998).
- The killers of American student Amy Beihl, who, while on a Fulbright scholarship to South Africa, was mistakenly targeted as a member of the South African white community. As many as seven youths stoned and stabbed Beihl after attending an anti-apartheid rally where they had been urged to help end apartheid by making the country ungovernable.
- ANC activists who had killed men and women known to be or believed to be active supporters of the apartheid regime.

When Nelson Mandela became president of South Africa in 1994, he inherited a country characterized by enormous economic disparities, created over 300 years, between whites and nonwhites. Mandela promised that his government would work to undo apartheid's legacy by building more than 1 million houses for low-income families, creating jobs, investing heavily in health and education, bringing electricity to 20 million people, purchasing and redistributing farmland and residential property, and taking control of the mining industry so that wealth beneath the land would go to the people of South Africa (Keller 1993). Most importantly, funding for these projects would come not from redistribution of the existing wealth, but from new wealth generated from foreign and domestic investment.

We focus on South Africa because it allows us to consider the process by which an extreme form of social stratification is dismantled. Does a clear understanding of past historical forces facilitate the dismantling process? Is it realistic to believe that equality can be achieved without redistributing the existing wealth?

[1]These figures are based on data from the 1996 census released on October 20, 1998. Depending on the year and the source, the percentage of the South African population that is white is estimated at between 10 and 15 percent.

This South African man testified before the Truth and Reconciliation Commission. South African police shot him in the face, leaving him blind.

© Rodger Bosch/Impact Visuals

Social stratification is the systematic process by which people are classified into categories that are ranked on a scale of social worth. The categories into which people are placed affect their **life chances**, a critical set of potential social advantages including "everything from the chance to stay alive during the first year after birth to the chance to view fine art, the chance to remain healthy and grow tall, and if sick to get well again quickly, the chance to avoid becoming a juvenile delinquent—and very crucially, the chance to complete an intermediary or higher educational grade" (Gerth and Mills 1954:313).

This chapter examines several important and interrelated features of social stratification, including the kinds of social categories into which people are classified, the symbols that designate a person's category (for example, skin color, type of car owned, occupation), types of stratification systems, reasons for stratification systems, and the effects of stratification on life chances.

Social Categories

Every society in the world classifies its people into categories—whether formally by the government and other institutions or informally in the course of social interaction. Almost any criterion can be used (and at one time or another has been used) to categorize people: hair color and texture, eye color, physical attractiveness, weight, height, occupation, sexual preferences, age, grades in school, test scores, and many others. Two major kinds of criteria are used as the basis for classification: ascribed and achieved characteristics.

Ascribed characteristics are attributes that people have at birth (such as skin color, sex, or hair color), develop over time (such as baldness, gray hair, wrinkles, retirement, or reproductive capacity), or possess through no effort or fault of their own (national origin or religious affiliation that was "inherited" from parents).

Achieved characteristics, on the other hand, are acquired through some combination of choice, effort, and ability. In other words, people must act in some way to acquire these attributes. Examples of achieved characteristics include occupation, marital status, level of education, and income.

Ascribed and achieved characteristics may seem clearly distinguishable, but that is not always the case. For example, how many of the estimated 14.1 million people under the age of 18 living below poverty in the United States earned that status? On the one hand, we might argue that these children inherited their status from their parents. On the other hand, we might argue that their parents "achieved" this status for them.

Sociologists are especially interested in those ascribed and achieved characteristics that take on social significance and **status value:** When that occurs, persons who possess one feature of a characteristic (white skin versus brown skin, blond hair versus dark hair, low income versus high income, professional athlete versus high school teacher) are regarded and treated as more valuable or more worthy than persons who possess features from other categories (Ridgeway

---

Social stratification The systematic process by which people are divided into categories that are ranked on a scale of social worth.

Life chances A critical set of potential social advantages.

Ascribed characteristics Attributes that people have at birth, develop over time, or possess through no effort or fault of their own.

Achieved characteristics Attributes acquired through some combination of choice, effort, and ability.

Status value The value that some characteristics impart to those who possess them, causing that individual to be regarded and treated as more valuable or more worthy than persons who possess features from other categories.

Teacher Jane Elliot's now-classic experiment showed how easy it is for people to assign social worth, explain behavior, and build a reward system around a seemingly insignificant physical attribute—eye color.

© Anne Dowie

1991).[2] Sociologists are particularly interested in studying different social values attached to ascribed characteristics, because people have no control over these attributes.

A classic demonstration of how ascribed characteristics can affect people's access to valued resources involved a third-grade class in Riceville, Iowa. In 1970, teacher Jane Elliot conducted an experiment in which she placed her students in either of two groups based on a physical attribute—eye color—and rewarded them accordingly. She conducted this experiment to show her class how easy it is for people to (1) assign social worth, (2) explain behavior in terms of an ascribed characteristic such as eye color, and (3) build a reward system around this seemingly insignificant physical attribute. The following excerpt from the transcript of the program "A Class Divided" (*Frontline* 1985) shows how Elliot established the ground rules for the classroom experiment:

ELLIOT: It might be interesting to judge people today by the color of their eyes . . . would you like to try this?

CHILDREN: Yeah!

ELLIOT: Sounds like fun, doesn't it? Since I'm the teacher and I have blue eyes, I think maybe the blue-eyed people should be on top the first day. . . . I mean the blue-eyed people are the better people in this room. . . . Oh yes they are, the blue-eyed people are smarter than brown-eyed people.

BRIAN: My dad isn't that . . . stupid.

ELLIOT: Is your dad brown-eyed?

BRIAN: Yeah.

ELLIOT: One day you came to school and you told us that he kicked you.

BRIAN: He did.

ELLIOT: Do you think a blue-eyed father would kick his son? My dad's blue-eyed, he's never kicked me. Ray's dad is blue-eyed, he's never kicked him. Rex's dad is blue-eyed, he's never kicked him. This is a fact. Blue-eyed people are better than brown-eyed people. Are you brown-eyed or blue-eyed?

BRIAN: Blue.

ELLIOT: Why are you shaking your head?

BRIAN: I don't know.

ELLIOT: Are you sure that you're right? Why? What makes you sure that you're right?

BRIAN: I don't know.

ELLIOT: The blue-eyed people get five extra minutes of recess, while the brown-eyed people have to stay in. . . . The brown-eyed people do not get to use the drinking fountain. You'll have to use the paper cups. You brown-eyed people are not to play with the blue-eyed people on the playground, because you are not as good as blue-eyed people. The brown-eyed people in this room today are going to wear collars. So that we can tell from a distance what color your eyes are. [Now], on page 127—one hundred twenty-seven. Is everyone ready? Everyone but Laurie. Ready, Laurie?

CHILD: She's a brown-eye.

ELLIOT: She's a brown-eye. You'll begin to notice today that we spend a great deal of time waiting for brown-eyed people. (*Frontline* 1985:3–5)

Once Elliot set the rules, the blue-eyed children eagerly accepted and enforced them. During recess, the children took to calling each other by their eye colors, and some brown-eyed children got into fights with blue-eyed children who called them "brown-eye." The teacher observed that these "marvelous, cooperative, wonderful, thoughtful children" turned into "nasty,

---

[2]South Africa's constitution names several statuses that cannot take on status value. They are race, gender, pregnancy, marital status, ethnic or social origin, color, sexual orientation, age, disability, religion, conscience, culture, language, and birth.

vicious, discriminating little third-graders in a space of fifteen minutes" (p. 7).

This experiment illustrates on a small scale how categories and their status value are reflected in the distribution of valued resources. The category to which one is assigned determines status and is related to life chances.

Sociologists are particularly interested in any classification scheme that incorporates the belief that certain important abilities (such as athletic talent or intelligence), social traits (such as criminal tendencies or aggressiveness), and cultural practices (dress, language) are passed on genetically or the belief that some categories of people are inferior to others by virtue of genetic traits and therefore should receive less wealth, income, and other socially valued items. They study classification schemes because these systems have enormous consequences for society and the relationships among people.

For sociologists, one important dimension of stratification systems is the extent to which people "are treated as members of a category, irrespective of their individual merits" (Wirth [1945] 1985:310). In this vein, they examine how "open" or "closed" a stratification system is.

"Open" and "Closed" Stratification Systems

Real-world stratification systems fall somewhere on a continuum between two extremes: a **caste system** (or "closed" system, in which people are ranked on the basis of traits over which they have no control) and a **class system** (or "open" system, in which people are ranked on the basis of merit, talent, ability, or past performance). No form of stratification exists in any pure form. Three characteristics most clearly distinguish a caste from a class system of stratification: (1) the rigidity of the system, or how difficult it is for people to change their category, (2) the relative importance of ascribed and achieved characteristics in determining life chances, and (3) the extent to which restrictions are placed on social interaction between people in different categories. Caste systems are considered closed because of their extreme rigidity—they are rigid in the sense that ascribed characteristics determine life chances and that social interaction (such as marriage) among people

in different categories is restricted. In comparison, class systems are open in the sense that achieved characteristics determine life chances and that no barriers exist to social interaction among people in different categories.[3]

Caste Systems

When people hear the term *caste,* what usually comes to mind is India and its caste system, especially as it existed before World War II. India's caste system, now outlawed in its constitution, consisted of a strict classification of people into four basic categories with 1,000 subdivisions. To see the strict nature of this caste system, consider some of the rules that specified the amount of physical distances that were to be maintained between people of different castes: "a Nayar must keep 7 feet (2.13 m) from a Nambudiri Brahmin, an Iravan must keep 32 feet (9.75 m), a Cheruman 64 feet (19.5 m), and a Nyadi from 74 to 124 feet (22.6 to 37.8 m)" (Eiseley 1990:896).

Most sociologists use the term *caste* not to refer to one specific system but rather to describe any form of social stratification in which people are ranked on the basis of characteristics over which they have no control and that they usually cannot change. Whenever people are ranked and rewarded on the basis of such traits, they are part of a caste system of stratification.

A direct connection exists between caste rank and the opportunities available to members of that caste to wield power and acquire wealth and other valued items. People in lower castes are seen as innately inferior in intelligence, personality, morality, capability, ambition, and many other traits. Conversely, people in higher castes consider themselves to be superior in such traits. Moreover, caste distinctions are treated as if they are absolute; that is, their significance is never doubted or questioned (especially by those who occupy higher castes), and they are viewed as unalterable and clear-cut (as if everyone fits neatly into a category). Finally, heavy restrictions constrain social interactions among people in higher and lower castes. For example, marriage between people of different castes is forbidden. South Africa's system of apartheid, which was abolished on paper in 1994, is a classic example of a caste system of social stratification.

[3]Class and caste systems are ideal types. We learned in Chapter 7 that an ideal type is ideal not in the sense of having desirable characteristics, but rather as a standard against which "real" cases can be compared. Actual stratification systems depart in some way from the ideal types.

Caste system A system in which people are ranked on the basis of traits over which they have no control.

Class system A system in which people are ranked on the basis of merit, talent, ability, or past performance.

Table 9.1   Life Chances and Racial Classification in South Africa

Which group in South Africa has the "best chance" of getting to work by car?, having water piped into the home?, being employed? Note that these statistics apply two years after apartheid ended. How long do you think it will take to make life chances equal across the four major groups?

	White	Indian	Mixed or Colored	African
Percentage of the total population	11.0	7.0	9.0	77.0
Infant mortality rate	12/1,000	18/1,000	52/1,000	110/1,000
Percentage of babies born underweight:				
Urban (%)	16.0	35.0	49.0	28.0
Rural (%)	—	—	—	43.0
Number of pupils per teacher	18	25	29	42
Government expenditures per pupil	$1,700	$1,100	$600	$220
Mean years of schooling	11.6	8.8	6.9	5.3
Literacy rate	99/100	69/100	62/100	50/100
Life expectancy in years	70	65	58	59
Access to:				
Piped water in home (%)	96.9	96.8	72.2	32.7
Electricity (%)	99.8	100.0	86.2	36.5
Flush toilet (%)	99.8	97.8	99.5	41.5
Car to drive to work (%)	82.0	67.7	25.7	8.1
Occupation				
Male				
Professional/managerial (%)	41.7	32.0	9.7	7.2
Laborer/semi-skilled (%)	0.8	17.3	41.4	31.2
Female				
Professional/managerial (%)	43.8	23.4	13.3	15.3
Laborer/semi-skilled (%)	0.8	22.2	52.0	31.1
Unemployed (%)				
Male	5.5	9.9	17.8	28.9
Female	3.7	19.9	27.8	46.9

Sources: U.S. CIA World Factbook (1997), South Africa Labour Research Unit (1994), African National Congress (1998), Statistics South Africa (1998a).

Apartheid: A Caste System of Stratification

Apartheid was a system of laws in which everyone in South Africa was classified into a racial category and issued an identity card denoting his or her race. The system was one in which skin color overwhelmingly determined a person's life chances (see Table 9.1). Under apartheid, a person was not just born a baby; a person was born a black, colored, Indian, or white baby—and that label profoundly affected every aspect

> **Apartheid** A system of laws in which everyone in South Africa was put into a racial category and issued an identity card denoting his or her race.

of a person's life (Mabuza 1990), determining where and with whom he or she could "live, work, eat, travel, play, learn, sleep, and be buried" (Roberts 1994:54).

Apartheid (the Afrikaans word for "apartness") was practiced in South Africa for hundreds of years. It did not actually become official policy until 1948, however, when the conservative white Nationalist Party, led by D. F. Malan, won control of the government. Once in power, the Nationalists passed hundreds of laws and acts mandating racial separation in almost every area of life and severely restricting blacks' rights and opportunities. The ultimate result was legalized political, social, and economic domination by whites (14 percent of the population) over nonwhites (86 percent of the population).

It is virtually impossible to give an overview of the effects of apartheid, a system that has inflicted enduring misery, economic damage, and daily indignities on South Africa's nonwhite population (Wilson and Ramphele 1989). Although apartheid has been officially dismantled, the real challenge lies in dismantling its effects on life chances for nonwhites. To fully grasp how apartheid laws determined life chances, we turn to an overview of these policies. As you read this overview, do not try to memorize names, dates, and specific purposes of the acts. Instead, think about how every aspect of life in South Africa was regulated based on a person's skin color. Also, consider the inevitable challenges that must accompany the dismantling of the apartheid system.

APARTHEID POLICIES Apartheid policies centered on one aim: maintaining separate black (that is, mixed-race, Asian, and African) and white areas. The Transvaal Province Law of 1922 (paragraph 267) stated that a nonwhite "should only be allowed to enter urban areas, which are essentially the white man's creation, when he is willing to enter and to minister to the needs of the white man and should depart therefrom when he ceases so to minister" (Wilson and Ramphele 1989:192). Other legislation—the Separate Amenities Act, the Group Areas Act, the Land Acts, and the Population Registration Act—presented additional ways in which a white-ruled government worked to achieve this aim.

The Separate Amenities Act of 1953, which was abolished on October 16, 1990,[4] authorized the creation of separate and unequal (in quantity and quality) public facilities—parks, trains, swimming pools, libraries, hotels, restaurants, hospitals, waiting rooms, pay telephones, beaches, cemeteries, and so on—for whites and nonwhites. In areas officially designated for "whites only," local governments constructed only the most basic and necessary public facilities (such as toilets and bus stations) for blacks "ministering to the needs of whites" and prohibited nonwhites from using all other public facilities (for example, parks and libraries).

The Land Acts of 1913 and 1936, which were repealed on June 5, 1991, reserved 85 percent of the land for South Africa's 5 million whites (14 percent of the population) and set aside the remaining 15 percent

During the same period in which the 1953 Separate Amenities Act was passed, the United States had its own version—the "Jim Crow" laws—in a number of southern states that legalized racial segregation of all public facilities.

© Hulton Getty/Liaison Agency

of the land for the rest of the population. These acts mandated the creation of ten homelands in the eastern half of the republic, one for each of the ten major African ethnic groups (as defined by white South Africans). These homelands consisted of land that was arid, underdeveloped, and lacking in natural resources. Needless to say, the Africans had no influence in determining the location or type of land that would constitute their homelands. For example, a homeland might be composed of as many as two dozen pieces of land, separated by large, white-owned plantations, farms, and industries (Liebenow 1986; see the map on page 252).

Regardless of where he or she was born or had lived most of his or her life, each African was declared a citizen, not of South Africa, but of one of the ten homelands. Four of these homelands were declared independent (Bophuthatswana, Transkei, Ciskei, and Venda), and the government hoped that the other six eventually would accept independent status. No other country in the world recognized the independence of any of these homelands, however, because they were

---

[4]Despite the 1990 legislation, many in power have managed to avoid the legal obligation to integrate public facilities. Residents of white areas, for example, can treat nonwhites as "nonresidents" and charge them fees as high as $200 per year to use public facilities such as the library (Wren 1991).

Apartheid policies Policies centered on maintaining separate black and white areas.

established only to keep black South Africans from living in white areas and to maintain a low-wage employment force to serve white South Africans.

The logic underlying the Land Acts was as follows: If all Africans had citizenship in a homeland, they had no claim to economic and political rights when they were outside their homeland. Thus the South African government could deny them a vote and treat them as guest workers and noncitizens:

> The homelands have a more sinister side. On the one hand, they are a labor pool for industry; on the other, they are a rubbish heap for the people industry has no use for—the old, the sick, the very young. By law, if you cannot work you must go to your homeland, and there you are dumped. Sometimes there are "resettlement camps" of tents or huts for groups who are moved [by the government] to a countryside that they do not know. Sometimes people are just left on the land, and told to survive off relatives. Schools are bad or nonexistent, medical facilities the same. Some people survive. Many die. The homelands policy is a policy of polite extermination. Gas chambers are not needed when people can simply starve to death. (Seidman 1978:111)[5]

The Group Areas Act of 1950, which was repealed on June 5, 1991, designated separate living areas for whites and nonwhites. Under this act, Africans were permitted to work in white areas, but they had to be housed in adjacent townships or in single-sex (usually male) closed compounds called hostels. Buses brought blacks from the townships or compounds into white areas to work and then took them home after their day was done. Most blacks spent three or more hours a day traveling to and from their jobs:

> Well, let's take a common scenario of a middle-aged black man living in a township 30 miles from his job. He lives there because that's where he's been put. About 4 A.M. he gets up so he can make the two-hour trek into the city each day. He rides a bus or train that makes our worst rush-hour scene look peaceful by comparison.
>
> When the workday is over, the man must leave the city by a certain time because of his curfew and take his two-hour trek back home again.
>
> Or let's say the man happens to be a miner. He'll live in a crowded dormitory and make about $5000 a year, while white miners live in rent-free homes

and make around $20,000 a year for often doing the same work. It's against the law for the black miner to bring his wife and children to live with or near him or for them to visit him. Except for his two short visits home a year, he stays at the mine and works at a very hazardous job.

> Now what about his wife? Well, she can't live with her husband if he works at the mines, and she can't live with her family if she's a domestic. If she works for a white family, they often will insist that she live in housing quarters behind their home. If so, her kids have to be left back in the homeland with grandma or auntie. Occasionally, she too will be permitted to travel back home to visit her family. (Lambert 1988:29–30)[6]

Under the Group Areas Act, every square foot of land was officially classified as reserved for whites, coloreds, Asians, or Africans. Anyone found living in the wrong area could be resettled or forced out. In addition, the government had the right to change the classification of an area. Between 1960 and 1983, the South African government "resettled" an estimated 3.5 million nonwhites to places where they did not choose to go (Wilson and Ramphele 1989).

In addition, South Africa's pass laws restricted the free movement of blacks. Under these laws, which were abolished in 1986, blacks were required to carry passbooks that contained information about their racial, residential, and employment status. Without their passbooks, blacks were not permitted to be present in areas designated as white (85 percent of the land). Between 1916 (when statistics on pass-law violations first were recorded) and 1986, 17 million Africans were prosecuted for pass law violations. One white employer explained:

> We protested the pass laws, too. They were dreadful. If our garden man left our garden to water some plants on the other side of the hedge and if he didn't have his pass in his working shirt (not wanting to get it dirty and spoiled because without it he was a lost soul), he could have been carted off by any policeman who happened to walk by. We wouldn't know where he was. We wouldn't know what had happened. We'd ring up the police stations, and if we happened to strike somebody a bit simpatico, we might learn his whereabouts. This used to happen all the time in the suburbs. It was driving us mad. Every evening you used to see lines of these people

---

[5]Many of the people sent to the homelands were born in the cities and had no strong family ties in the homelands. There were no public facilities to care for them, and they were not wanted (*Weekend World* 1977).

[6]A South African government official explained the purpose of this labor system: "The African labor force must not be burdened with superfluous appendages such as wives, children, and dependents who could not provide service" (African National Congress 1996).

handcuffed to one another being dragged to the nearest police station. We succeeded in getting that done away with. Of course, the passes still exist, but there is usually less harassment.

When I began thinking, I was distressed to have to write a piece of paper. This was before they issued passbooks. I had an African man servant whom we absolutely adored. He had worked for us for over twenty-five years. We had built very beautiful servants' quarters for him and our other help. We all loved him. And this man was really noble. He was old enough to be my father, and yet he had to come to me if he wanted to go out. I would have to scribble, "Please pass Amos until six o'clock—or eight o'clock—tonight." (Crapanzano 1985:131–132)

The Population Registration Act of 1950, which was repealed on June 18, 1991, required that everyone in South Africa be classified by race and issued an identification card denoting his or her race. Classification boards composed of white social workers ruled on questionable race identity cases and heard appeals from those who wished to change their racial classification. It is not known how many people applied for such a change:

Deliberations take months, sometimes years, with the board paying careful attention to skin tint, facial features, and hair texture. In one typical twelve month period, 150 coloreds were reclassified as white; ten whites became colored; six Indians became Malay; two Malay became Indians; two

The Population Registration Act required that everyone in South Africa be classified by race and that they carry an identification card denoting their race.

© Reuters/Corbis-Bettmann

coloreds became Chinese; ten Indians became coloreds; one Indian became white; one white became Malay; four blacks became Indians; three whites became Chinese.

The Chinese are officially classified as a white subgroup. The Japanese, most of whom are visiting businessmen, are given the status of "honorary whites." Absurd? Yes, if all this were part of an anti-utopian novel. But in real life it is a chilling part of the most complex system of human control in the world. (Lamb 1987:320–321)

In light of this history, it is not surprising that South Africa's new constitution seeks to protect rights denied to nonwhites under apartheid (see "The South African Bill of Rights").

Apartheid is a classic example of a caste system of stratification: People are categorized according to physical traits such as skin color, ascribed characteristics determine life chances, and many barriers limit social interaction among people who belong to different racial categories. The legacy of apartheid is evident in Tables 9.1 and 9.2.

Class Systems

In class systems of stratification, "people rise and fall on the strength of their abilities" (Yeutter 1992:A13). Class systems of stratification contain economic and occupational inequality, but that inequality is not systematic. By "*not* systematic," we mean that a person's sex, race, age, or other ascribed characteristics are not connected

Table 9.2	Average Hourly Wages of South African Employees by Race and Sex	
Race/Sex		**Rands/Hour***
African		
Female		7.67
Male		8.61
Colored		
Female		7.07
Male		8.53
Indian		
Female		11.99
Male		16.19
White		
Female		17.56
Male		29.13

*1 rand is valued at approximately 0.173 U.S. dollar.
Sources: Statistics South Africa (1998b), Federal Reserve Bank of New York (1998).

U.S. in Perspective

The South African Bill of Rights

The following is an excerpt from the 1996 Constitution of South Africa, Bill of Rights, as adopted May 8, 1996. The document was adopted six years after the then-South African President F. W. de Klerk traveled to the United States to convince its government and people that South Africa was beginning a sincere and irreversible effort to dismantle apartheid. During that visit, he emphasized that the principles of the U.S. political system (such as the Bill of Rights) deserved to be emulated.

As you read the South African Bill of Rights, consider the ways in which it is similar to and goes beyond the U.S. Bill of Rights. The South African Bill of Rights also alerts us to the various areas of life in which life chances are affected by a country's system of social stratification. What are some of those areas? Which Bill of Rights would you rather live under? Why?

Everyone is equal before the law and has the right to equal protection and benefit of the law.

Everyone has inherent dignity and the right to have their dignity respected and protected.

Everyone has the right to life.

Everyone has the right to freedom and security of the person, which includes the right—

a. not to be deprived of freedom arbitrarily or without just cause;
b. not to be detained without trial;
c. to be free from all forms of violence from both public and private sources;
d. not to be tortured in any way; and
e. not to be treated or punished in a cruel, inhumane, or degrading way.

Everyone has the right to bodily and psychological integrity, which includes the right—
a. to make decisions concerning reproduction;
b. to security in and control over their body; and
c. not to be subjected to medical or scientific experiments without their informed consent.

No one may be subjected to slavery, servitude, or forced labor.

Everyone has the right to privacy, which includes the right not to have—
a. their person or home searched;
b. their property searched;
c. their possessions seized; or
d. the privacy of their communications infringed.

Everyone has the right to freedom of conscience, religion, thought, belief, or opinion.

Everyone has the right to freedom of expression, which includes—
a. freedom of the press and other media;
b. freedom to receive and impart information and ideas;

c. freedom of artistic creativity; and
d. academic freedom and freedom of scientific research.

Everyone has the right, peacefully and unarmed, to assemble, to demonstrate, to picket, and to present petitions.

Everyone has the right to freedom of association.

Every citizen is free to make political choices.

No citizen may be deprived of citizenship.

Everyone has the right to freedom of movement.

Every citizen has the right to choose their trade, occupation, or profession freely.

Everyone has the right to fair labor practices.

Everyone has the right—
a. to an environment that is not harmful to their health or well-being; and
b. to have the environment protected, for the benefit of present and future generations, through reasonable legislative and other measures that—
i. prevent pollution and ecological degradation;
ii. promote conservation; and
iii. secure ecologically sustainable development and use of natural resources while promoting justifiable

---

to the inequality that exists. A class system differs from a caste system in that the inequality found in a society can be explained by differences in talent, ability, and past performance and not by attributes over which people have no control, such as skin color or sex. In class systems, people assume that they can achieve a desired level of education, income, and standard of living through hard work. Furthermore, people can raise their class position during their own lifetimes, and their children's class position can differ from (and ideally be higher than) their own.

Movement from one class to another is termed **social mobility.** Many kinds of social mobility exist in class systems. **Vertical mobility** involves a change in class status that corresponds to a gain or loss in rank, as when a medical student moves up in rank and becomes a physician or when a wage earner loses a job, goes on unemployment, and moves down in rank. A loss of rank

Social mobility   Movement from one class to another.

Vertical mobility   Movement that occurs when a change in class status corresponds to a gain or loss in rank.

economic and social development.

No one may be deprived of property except in terms of law of general application, and no law may permit arbitrary deprivation of property.

Everyone has the right to have access to adequate housing.

Everyone has the right to have access to—
a. health care services, including reproductive health care;
b. sufficient food and water; and
c. social security, including, if they are unable to support themselves and their dependents, appropriate social assistance.

Every child has the right—
a. to a name and a nationality from birth;
b. to family care, parental care, or appropriate alternative care when removed from the family environment;
c. to basic nutrition, shelter, basic health care services, and social services;
d. to be protected from maltreatment, neglect, abuse, or degradation;
e. to be protected from exploitative labor practices;
f. not to be required or permitted to perform work or provide services that—
   i. are inappropriate for a person of that child's age; or
   ii. place at risk the child's well-being, education, physical or mental health, or spiritual, moral, or social development.

Everyone has the right—
a. to a basic education, including adult basic education; and
b. to further education, which the state must take reasonable measures to make progressively available and accessible.

Everyone has the right to use the language and to participate in the cultural life of their choice, but no one exercising these rights may do so in a manner inconsistent with any provision of the Bill of Rights.

Persons belonging to a cultural, religious, or linguistic community may not be denied the right, with other members of their community, to—
a. enjoy their culture, practice their religion, and use their language; and
b. form, join, and maintain cultural, religious, and linguistic associations and other organs of society.

Everyone has the right of access to—
a. any information held by the states; and
b. any information that is held by another person and that is required for the exercise or protection of any rights.

Everyone has the right to administrative action that is lawful, reasonable, and procedurally fair.

Everyone has the right to have any dispute that can be resolved by the application of law decided in a fair public hearing in a court or, where appropriate, another independent and impartial forum.

Everyone who is arrested for allegedly committing an offense has the right—
a. to remain silent;
b. to be informed promptly—
   i. of the right to remain silent; and
   ii. of the consequences of not remaining silent;
c. not to be compelled to make any confession or admission that could be used in evidence against that person;
d. to be brought before a court as soon as reasonably possible, but not later than 48 hours after the arrest, but if that period expires outside ordinary court hours, to be brought before a court on the first court day after the end of that period;
e. at the first court appearance after being arrested, to be charged or to be informed of the reason for the detention to continue, or to be released; and
f. to be released from detention if the interests of justice permit, subject to reasonable conditions.

Note: The South African Bill of Rights also includes a detailed list of rights that apply to all phases of justice after arrest. The complete document can be accessed via the following URL address: gopher://gopher.anc.org.za/00/anc/misc/sacon96.txt.

---

is **downward mobility,** whereas a gain in rank is **upward mobility. Intragenerational mobility** is movement (upward or downward) during an individual's lifetime. **Intergenerational mobility** is a change up or down in rank over two or more generations, as when a son or daughter goes into an occupation that is higher or lower in rank and prestige than a parent's occupation.

Unlike the clear-cut distinctions in caste systems, distinctions between classes are not always readily apparent. In principle, people can move from one class to another. In fact, it is often difficult to identify a person's class through observation alone. That is, a person may drive a BMW and wear designer clothes—symbols of solid middle-class status—but may work at

Downward mobility A loss of rank or status.

Upward mobility A gain in rank or status.

Intragenerational mobility Movement (upward or downward) during an individual's lifetime.

Intergenerational mobility A change (upward or downward) in rank over two or more generations.

This painting by John Trumbull shows the first U.S. Congress in 1776. Who do you not see in this picture? About how many years would pass before you could expect to see those missing groups in a picture of U.S. legislators?

A detail from John Trumbull, *The Declaration of Independence, July 4, 1776.* Yale University Art Gallery. Trumbull Collection.

Table 9.3	Money Income by Household, Percentage of 102.5 Million U.S. Households in Each Income Category, 1997	
Income Category		**Percentage**
Under $10,000		11.0
$10,000–14,999		8.1
$15,000–24,999		14.9
$25,000–34,999		13.3
$35,000–49,999		16.3
$50,000–74,999		18.0
$75,000–99,999		9.0
$100,000+		9.4
Median income		$37,005

Source: U.S. Bureau of the Census (1998b).

a low-paying job, live in a low-rent apartment, and have exceeded his or her line of credit. Or a person may change class position through marriage, graduation, inheritance, or job promotions but may not assume the lifestyle of that class immediately.

Does the United States Have a Class System?

Before we answer this question, let us recall that in a true class system, ascribed characteristics do not determine life chances. That is not to say that a class system of stratification provides economic equality for everyone. Instead, it simply means that no connection exists between the economic inequality and ascribed characteristics such as race, gender, age, disability, national origin, and so on. On paper, the United States has a class system. The Declaration of Independence, the Constitution, and the Bill of Rights prohibit discrimination on the basis of ascribed characteristics. We must, however, distinguish between the ideal and reality.

Although Americans may want to believe that they live in a true class system, evidence indicates otherwise. Life chances are connected to forces over which people have little control, such as the family into which one is born, skin color, and massive restructuring of the economy. To put it bluntly, a clear

pattern of inequality can be observed in the United States. Although this pattern of inequality is not nearly as systematic as that which existed under apartheid and that persists in present-day South Africa, it is striking nonetheless.

INCOME INEQUALITY IN THE UNITED STATES   The U.S. Bureau of the Census (1998a) classifies households according to income categories. Table 9.3 shows how income is distributed across the 102.5 million households in the United States. As can be seen in the table, 11.0 percent of all households are in the lowest income category, whereas 9.4 percent are in the highest income category. Nevertheless, the presence of inequality as shown in Table 9.3 by itself is not enough evidence to support a claim that the United States is not a class system.

When we examine income profiles for households classified as white, black, and Hispanic, however, we see that 9.5 percent of white households, 21.4 percent of black households, and 16.8 percent of Hispanic households fall in the lowest income categories (see Table 9.4 and "Income Inequality in the United States"). Such differences suggest that the United States is not a class system in the true sense of the word. If it were a true class system, the percentages of households in each income category would be the same across all three groups.

The systematic character of inequality in the United States becomes even more obvious when we compare median incomes of various types of families classified as white, black, and Hispanic households. Table 9.5 shows that, of these household types, female-

Table 9.4 Money Income by Household, Percentage Distribution of U.S. Households Classified as White, Black, and Hispanic in Each Income Category, 1997

What percentage of households classified as white make over $100,000? How does that compare with households classified as black and Hispanic? Which household type is most likely to be concentrated in the under-$10,000 income category?

Income Category	White	Black	Hispanic
Under $10,000	9.5	21.4	16.8
$10,000–14,999	7.8	10.6	10.7
$15,000–24,999	14.6	17.9	19.7
$25,000–34,999	13.2	14.2	15.0
$35,000–49,999	16.5	14.9	16.6
$50,000–74,999	18.8	13.1	12.2
$75,000–99,999	9.5	4.6	5.0
$100,000+	10.2	3.3	4.0
Median income	$38,972	$25,050	$26,628

Source: U.S Bureau of the Census (1998b).

headed households in general have the lowest median income ($23,040). When we look at female-headed families classified as black and Hispanic, we see that the median incomes are $17,962 and $16,393, respectively. In view of these findings, we should not be surprised to learn that some groups are concentrated more heavily than others in the lowest-paying and least prestigious occupational categories.

OCCUPATIONAL INEQUALITY IN THE UNITED STATES

Occupations are considered to be segregated according to ascribed characteristics when some occupations (librarian, secretary, physician, lawyer, chief executive officer) are filled primarily by people of a particular race, ethnicity, sex, or age. The occupational categories shown in Tables 9.6 and 9.7 are examples of those with the highest concentrations by race and by sex.

Occupational segregation is a problem when certain groups become concentrated in low-paying, low-ranking jobs. When we compare the annual income of nonfamily, single-person households further classified as black, white, and Hispanic, the effect of occupational segregation United States becomes evident (see Figure 9.1). When we consider the information in Tables 9.3 through 9.7 and in Figure 9.1 and recognize the fact that income and occupation are connected to sex, race, and ethnicity (an ascribed characteristic), we have no choice but to conclude that the United States is at best a mixture of class and caste systems of stratification.

Table 9.5 Median Income of U.S. Families Classified as White, Black, and Hispanic, 1998 (in dollars)

Note that families classified as "white two-earners" have the highest median income of all categories. Which type of household ranks second and third? Which type ranks last?

Type of Family	White	Black	Hispanic	All
Married couple	$52,199	$45,372	$34,317	$51,681
Male householder, wife absent	$38,511	$28,593	$28,249	$36,634
Female householder, husband absent	$25,670	$17,962	$16,393	$23,040
Two earners	$55,474	$44,728	$37,106	$54,192

Source: U.S. Bureau of the Census (1998b).

U.S. in Perspective

Income Inequality in the United States

Top Ten Counties in the United States: Highest Percentage of Families Below the Poverty Line

How many of the 10 poorest counties are predominately white? What do you make of the relatively low percentage of single-mother households in counties with predominately Hispanic populations?

County Name	Largest Racial Group and Percentage of the Population	Percentage Below the Poverty Line	Percentage of Households Classified as Single-Mother
Shannon, SD	Native American, 94.1	56.9	54.0
Starr, TX	Hispanic, 97.5	56.8	16.3
East Carroll, LA	Black, 64.8	50.8	42.8
Tunica, MS	Black, 75.4	50.5	30.8
Owsley, KY	White, 99.3	46.8	9.0
Todd, SD	Native American, 82.4	46.6	49.8
Zavala, TX	Hispanic, 89.3	46.1	22.2
Maverick, TX	Hispanic, 93.5	45.6	18.3
Holmes, MS	Black, 75.8	45.5	44.2
Dimmit, TX	Hispanic, 83.9	43.5	18.8

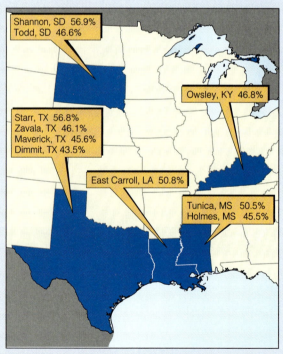

Shannon, SD 56.9%
Todd, SD 46.6%

Owsley, KY 46.8%

Starr, TX 56.8%
Zavala, TX 46.1%
Maverick, TX 45.6%
Dimmit, TX 43.5%

East Carroll, LA 50.8%

Tunica, MS 50.5%
Holmes, MS 45.5%

The percentages indicate the percentage of families living below the poverty line.

Richest Counties in the United States: Highest Percentage of Families with Incomes of $75,000 or More

Why do you think North Slope, Alaska, ranks among the richest counties? How does the percentage of households classified as single-mother compare with the ten poorest?

County Name	Largest Racial Group and Percentage of the Population	Percentage with Income of $75,000 or More	Percentage of Households Classified as Single-Mother
Fairfax, VA	White, 81.0	33.7	6.6
Morris, NJ	White, 91.9	32.2	4.6
Nassau, NY	White, 86.8	30.9	5.3
Montgomery, MD	White, 76.8	30.5	7.3
Hunterdon, NJ	White, 96.4	30.2	3.5
Somerset, NJ	White, 88.0	30.1	5.1
Rockland, NY	White, 84.0	29.2	6.8
Fairfield, CT	White, 84.8	28.9	9.4
Westchester, NY	White, 79.5	28.9	8.3
North Slope, AK	Native American, Eskimo, or Aleut, 73.0	28.6	19.7

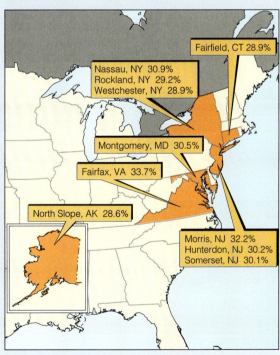

The percentages indicate the percentage of families with incomes of $75,000 or more.

Source: 1990 U.S. Census data.

Table 9.6   Occupations in Which People Who Declare Themselves to Be Black and Hispanic Are Disproportionately Under- and Overrepresented, 1996

According to the 1990 census, blacks constitute 12.1 percent, persons of Hispanic origin constitute 9.0 percent, and whites constitute 80.3 percent of the U.S. population. If race or ethnicity has no bearing on occupation, we would expect the percentages of people in each category to match these percentages. In which category are blacks most overrepresented? Underrepresented? How about Hispanics?

Underrepresented Occupation	Percentage of All Employed		Overrepresented Occupation	Percentage of All Employed	
	Black	Hispanic Origin		Black	Hispanic Origin
Dental hygienists	0	2.9	Social workers	22.6	7.7*
Managerial and professional	7.4	4.5	Maids and housemen	29.6	21.1
Scientists	3.3	1.9	Nurses' aides, orderlies, and attendants	33.2	8.1*
Physicians	4.5	5.1			
Dentists	1.2	0.9	Cleaners and servants (private households)	18.5	32.4
Pharmacists	6.8	1.3			
College professors	6.5	4.1	Postal clerks (except mail carriers)	28.3	7.9*
Lawyers and judges	3.4	2.9			
Authors	5.4	0.9	Pressing machine operators	22.7	38.4
Technical writers	7.4	2.6	Barbers	30.3	8.0*
Designers	2.1	5.9	Textile sewing machine operators	16.7	28.1
Speech therapists	3.8	4.3			
Airplane pilots and navigators	1.4	3.5	Dieticians	29.3	7.4*
Sales occupations	7.9	7.0			
Waiters and waitresses	4.8	9.0			
Bartenders	2.4	7.0	*Hispanic not overrepresented.		

Source: U.S. Bureau of the Census, Statistical Abstracts of the United States (1997).

Figure 9.1   Median Income in the United States Among Single-Person Households Classified as White, Black, and Hispanic

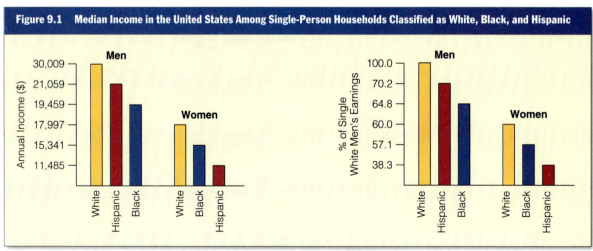

Source: U.S. Census Bureau (1998b).

Table 9.7 Examples of Occupations in Which Women Are Disproportionately Under- and Overrepresented, 1996

According to the Department of Labor (1998), females constitute 46.4 percent of the total workforce in the United States. If gender has no bearing on a person's occupation, we would expect 46.4 percent of people in each occupational category to be women. In what occupation are women most overrepresented? In what occupation are they most underrepresented?

Underrepresented Occupation	Percentage of All Employed	Overrepresented Occupation	Percentage of All Employed
Engineers	8.5	Registered nurses	93.3
Dentists	13.7	Speech therapists	93.3
Clergy	12.3	Prekindergarten and kindergarten teachers	98.1
Firefighting and fire prevention	2.1	Dental hygienists*	98.2
Mechanics and repairers	4.1	Licensed practical nurses	95.3
Construction trades	2.5	Secretaries, stenographers, typists	97.8
Transportation and material-moving occupations	9.5	Receptionists	96.9
Forestry and logging operations	3.5	Financial records processing (bookkeepers)	91.1
Airplane pilots and navigators	1.4	Eligibility clerks, social welfare	90.5
		Data entry keyers	90.1
		Teachers' aides	92.1
		Child care (private household)	97.1
		Cleaners and servants (private)	93.6
		Dental assistants	99.1
		Family childcare providers	98.5

*1991 figures.
Source: Adapted from the U.S. Bureau of the Census (1995).

Mixed Systems: Class and Caste

Systems of stratification usually possess elements of both class and caste characteristics. In the United States, virtually every occupation contains members of different ethnic, racial, age, and sex groups. At the same time, however, some groups such as women and blacks are severely over- or underrepresented in some occupations (Table 9.7). Moreover, in the United States, a person's ascribed characteristics can overshadow his or her achievements in such a way that "no amount of class mobility will exempt a person from the crucial implications of . . . birth" (Berreman 1972:399). For example, even though a person holds a high-ranking occupation, others may question or overlook that person's talent, merit, and accomplishments if the person has dark skin or female reproductive organs. The interrelations between caste and class come to mind when considering the advantages and disadvantages associated with persons who are judges, garbage collectors, airline pilots, factory workers, professors, farmers, nurses, CEOs,

unemployed, or homeless. Now imagine encountering "two Americans, for example, in each of the above mentioned occupations, one of whom is white and one of whom is black, or encountering one who is male and the other female. This quite literally changes the complexion of the matter. . . . Obviously something significant has been added to the picture of stratification in these examples that is entirely missing in the first instance—something over which the individual generally has no control, which is determined at birth, which cannot be changed, which is shared by all those of like birth, which is crucial to social identity, and which vitally affects one's opportunities, rewards, and social roles. The new element is race (color), caste, ethnicity (religion, language, national origin), or sex" (Berreman 1972:385–386).

Two examples demonstrate how class and caste stratification can operate together in the United States. Consider how positions are filled and how players relate to one another on a coed softball team. Usually, females are assigned to the least central positions (those requiring the least amount of involvement in completing a play). Also, when a player hits the ball toward a female player, at least one male comes over to "help out." Rarely does a female charge over to aid a male teammate. Finally, when a female player comes to the plate to bat, the four outfielders move in, nearer to the infield; when a male player bats, they stay in the outfield. (A female might be a good singles hitter, but it is difficult to hit a single through nine infielders.)

We could make the case that women play the least central positions and are helped by men in adjacent positions because they lack the skills and experience to field, throw, and hit the ball adequately. When talent and skill are the only criteria for assigning positions to men and women, a class system of stratification is at work. When a predictable relationship exists between an ascribed characteristic such as sex and an achievement such as shortstop, however, it is a signal for sociologists to investigate the source of this pattern. Often the reasons for the association can be traced to the fact that an undetermined, yet significant amount of the male-female difference in talent and ability is imposed externally; that is, these characteristics are by-products of the ways in which males and females are socialized.

Starting in infancy, parents tend to elicit more active and more physical behavior from sons than from daughters. In many cases, they also channel their children toward sex-appropriate sports. Girls are guided into predominantly noncontact sports—and often individual sports—that require grace, flowing movements, flexibility, and aesthetics, such as gymnastics, tennis, and swimming. By contrast, boys are encouraged to participate in team sports that involve contact, lifting, throwing, catching, and running. Because boys and girls participate in different kinds of athletic activities, they develop different skills and styles. Finally, the number of teams organized for each sex (from T-ball to professional) shows that more human and monetary resources are devoted to male than to female athletic development. Keep in mind that sociologists do not single out socialization forces as the only explanation for such discrepancies. They would acknowledge that innate physical differences may be responsible as well. Nevertheless, they focus their attention on how social practices artificially widen male-female differences in physical ability.

In the case of coed softball, a class system of stratification is operating on one level, because players who have the most ability and experience are assigned to the most central positions. Yet, a caste system is also at work on another level, because social practices con-

Are you inclined to guess that one or the other of these players has more "God-given talent"? Do you expect that the other has more of some other quality? In what ways may your beliefs about race and athletic ability affect how you interpret what you see?

© Dennis MacDonald/PhotoEdit

tribute to differences between males' and females' talent and ability.

Caste and class systems of stratification operate together in professional sports as well. Sociologists have long noted that black athletes tend to be concentrated in positions that require strength, speed, and agility, whereas whites are concentrated in central leadership, "thinking," and play-making positions (Loy and Elvogue 1971; Medoff 1977). In professional baseball, for instance, whites tend to play the infield positions (including pitcher and catcher), whereas blacks tend to play in the outfield. Because no on-field position in professional baseball excludes black athletes completely, some element of class stratification must be at work. At the same time, however, coaches, who are predominantly white, obviously assign blacks to noncentral positions and whites to leadership positions. This hypothesis is supported by the fact that most elementary and high school athletes play on predominantly white or black teams, which means that black athletes have experience in playing all positions. Sports sociologists would argue that black athletes are removed systematically from positions of leadership and assigned to the less central positions as they advance from the elementary to high school to college to the professional levels. This practice continues after their on-field professional careers end: relative to the number of white athletes in the three major professional sports, a smaller percentage of black athletes become head coaches or general managers. For example, 68 percent of the players in the National Football League (NFL) are classified as black, yet only 6.7 percent of head coaches and 22.1 percent of assistant coaches are "black." In the NBA, 79 percent of the players are classified as black, with approximately 18.5 percent of the head coaches and 42.3 percent of assistant coaches classified as black. If "being white" offered no advantage, the percentages of head and assistant coaches who are white would be 32 percent for the NFL and 21 percent for the NBA.

From a sociological point of view, any person who explains these differences as being due only to biological differences between men and women and across race and ethnic groups is, in the succinct words of social psychologist E. A. Ross ([1908] 1929), "too lazy" to trace these differences to the social environment or historical conditions.

Clearly, social stratification is an important feature of a society, and one with significant consequences for the life chances of the advantaged and disadvantaged alike. In the remainder of this chapter, we examine various theories that seek to explain why stratification occurs and the forms it takes.

Theories of Stratification

The theory of functionalism seeks to explain why resources are distributed unequally in society. A second set of theories deals with the identification of various strata within society.

A Functionalist View of Stratification

Sociologists Kingsley Davis and Wilbert Moore (1945) are associated with the functionalist perspective of stratification. They claim that social inequality—the unequal distribution of social rewards—is the device by which societies ensure that the most functionally important occupations are filled by the best-qualified people. How do we determine an occupation's functional importance? Davis and Moore offer two indicators: (1) the degree to which the occupation is functionally unique (that is, whether few other people can perform the same function adequately) and (2) the degree to which other occupations depend on the one in question. In view of these indicators, garbage collectors, although functionally important to sanitation, need not be rewarded highly because little training and talent are required to perform that job. Even though literally every occupation and area of life depend on garbage collectors to maintain sanitary environments, many people possess the skills needed to do the work.[7]

Davis and Moore argue that society must offer extra incentives to induce the most talented individuals to undergo the long and arduous training needed to fill the most functionally important occupations. They specify that the incentives must be great enough to prevent the best-qualified and most capable people from finding less functionally important occupations as attractive as the most important occupations.

Davis and Moore concede that the efficiency of a stratification system in attracting the best-qualified people is weakened when (1) capable individuals are overlooked or not granted access to training, (2) elite groups control the avenues of training (as through admissions quotas), and (3) parents' influence and wealth (rather than the ability of their offspring) determine the status that their children attain. Nevertheless, these sociologists maintain that the system adjusts to such inefficiencies. For example, when there are shortages of personnel to fill functionally important occupations, the society must increase individuals' opportunities to enter those occupations, that is, allowing in those who were previously denied entry. If this step is not taken, the society as a whole will suffer and will be unable to compete with other societies.

A functionalist might argue that white South Africans' recent moves to repeal the system of job reservation, which restricted each race to certain types of jobs and prohibited the advance of a nonwhite over a white in the same occupation, reflects such an "adjustment." To compete effectively in a world economy, South Africa needs a skilled work force, which cannot be maintained by the white population alone. Even though the long-standing job restrictions placed on nonwhite workers have now been repealed, it will nevertheless take many years to train those who were denied the educational opportunities to obtain such

---

[7]Imagine what would happen if the people who pick up and haul garbage to landfills stopped performing this important task. The spaces in which we work, live, and drive would overflow with garbage, not to mention the smell, which would be unbearable.

positions. Thus the case of South Africa casts doubt on the functionalist assumption that society will adjust and that the situation will right itself, and it alerts us to a critical moral question: How do the systematically disadvantaged react while they are waiting for society to make the necessary adjustments? It also introduces another more general question: Is social inequality necessary to ensure that the most important occupations are filled by the most qualified people?

Critique of the Functionalist Perspective

As you might imagine, many people would challenge the fundamental assumption underlying the Davis and Moore theory—that social inequality is a necessary and universal device used by societies to ensure that the most important occupations attract the best qualified people. Two especially insightful critiques are Melvin M. Tumin's (1953) "Some Principles of Stratification: A Critical Analysis" and Richard L. Simpson's (1956) "A Modification of the Functional Theory of Social Stratification."

Neither Tumin nor Simpson believes that a position commands great social rewards simply because it is functionally important or because few possess the talent to perform the job. Some positions command large salaries and bring other valued rewards even though their contribution to the society is questionable. Consider the salaries of athletes, for example. For the 1997–1998 season, 55 percent of the 384 athletes who played in the National Basketball Association (NBA) teams were paid $1 million or more (DuPree 1998). We might argue that professional athletes deserve this enormous pay because they generate income for owners, cities, advertisers, and media giants. This argument becomes less convincing, however, when we consider that elementary and secondary teachers are paid an average of $39,580 per year (Lewis, 1996), despite a teacher shortage and the suspicion that a significant percentage of teachers are unqualified. This difference in pay raises several questions: Are professional athletes and entertainers more essential to society than teachers? If there is a shortage of qualified teachers, why do we not raise salaries to attract the most qualified, just as we raise salaries to attract the most qualified athletes?

Critics of functionalism also question why a worker should receive a lower salary for the same job just because the person is of a certain race, age, sex, or national origin. After all, the workers are performing the same job, so functional importance is not the issue. This question relates to issues connected with pay equity. For example, why do females working as registered nurses in the United States earn on average of $0.98 for every dollar that a male registered nurse makes? On the other hand, why do female mechanics in the United States earn an average of $1.06 for every dollar that their male counterparts make?

In addition, critics of the functionalist perspective on social stratification point to the "comparable worth" debate. Advocates of comparable pay for comparable work ask whether women who work in predominantly female occupations (such as registered nurse, secretary, day-care worker) should receive salaries comparable to those earned by men who work in predominantly male occupations that are judged to be of roughly comparable worth (such as housepainter, carpenter, automotive mechanic). For example, assuming comparable worth, why should workers at a child day-care center (a traditionally female occupation) receive an average weekly salary of $234.86 while a person working in an auto supply store (a traditionally male occupation) earns $389.71 (U.S. Department of Labor 1998; see Figure 9.2 and Table 9.8)?

Tumin and Simpson argued that it is very difficult to determine the functional importance of an occupation, especially in societies characterized by a complex division of labor. The specialization and interdependence that accompany a complex division of labor imply that every individual contributes to the overall operation. In light of this interdependence, one could argue that every individual makes an essential contribution. "Thus to judge that the engineers in a factory are functionally more important to the factory than the unskilled workmen involves a notion regarding the dispensability of the unskilled workmen, or their replaceability, relative to that of the engineers" (Tumin 1953:388). Even if engineers, supervisors, and CEOs have the more functionally important positions, how much inequality in salary is necessary to ensure that people choose these positions over unskilled ones? In the United States, for example, the average salary of the CEOs of U.S. industrial corporations in 1998 was $1.1 million—32 times the average salary of a factory worker (Bryant 1999).[8] Are such high salaries really necessary to make sure that someone chooses the job of CEO over the job of factory worker? Probably not. Nevertheless, such high salaries have been justified as necessary to recruit the most able people to run a corporation in the context of a global economy. It is

---

[8]This salary is figured for the CEO of industrial corporations with revenues between $250 million and $500 million.

Figure 9.2   Two Approaches to Fairness

Pay Equity
When men and women work in the same firm in the same occupation, they must not be paid differently.

Comparable Worth
When occupational categories are agreed to be *equivalently valuable within a firm*, then compensation must be equivalent across those categories *at that firm.*

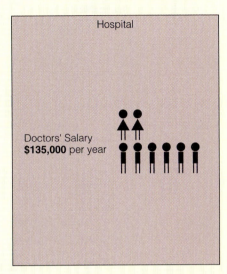

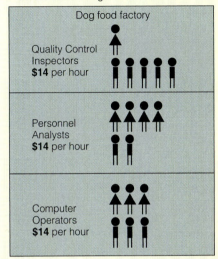

What happens (that is, what interpretation can we attach to the situation?) when pay differentials show up between equivalent firms whose chief difference is the gender composition of the workers?

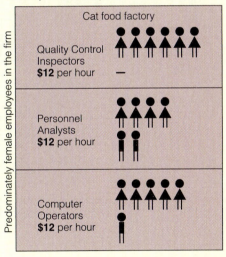

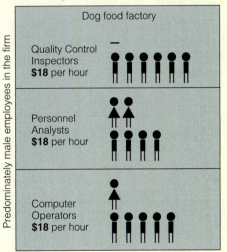

Alternatively. . .

What can we interpret from data that show that certain occupations or industries contain disproportionate shares of either male or female employees, and there is a pay differential? Does the pay differential go with the occupation or industry or with the gender of the worker?

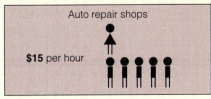

Source: U.S. Bureau of the Census (1995).

Table 9.8   Average Weekly Earnings for Employees Working in Predominantly Male and Predominantly Female Occupations

Which field of work employs the greatest percentage of females? What is the average weekly earnings of women working in that field? Which field employs the greatest percentage of males? What is the average earnings of men in that field?

Field of Work	Male (%)	Average Weekly Earnings
Coal mining	94.3	$809.59
Heavy construction equipment rental	85.6	693.88
Automotive repair shops	84.4	479.13
General building contractors	84.2	609.49
Trucking and warehousing	84.0	549.04
Sanitary services	83.6	671.67
Auto and home supply stores	83.0	389.71
New and used car dealers	82.4	575.05
Surveying services	79.1	528.64
Automotive dealers	79.0	495.47
Electric, gas, and sanitary services	77.6	836.83
Disinfecting and pest-control services	77.0	429.73
Automotive dealers and service stations	75.0	439.06
Local and suburban transportation	73.2	458.64
Agricultural services	61.4	354.33

Field of Work	Female (%)	Average Weekly Earnings
Child day-care services	93.1	$234.86
Home health care services	92.4	335.52
Beauty shops	90.6	266.72
Offices and clinics of dentists	88.4	398.18
Intermediate-care facilities	88.0	298.56
Nursing and personal-care facilities	86.2	321.43
Health services	81.9	453.47
Offices and clinics of medical doctors	80.1	460.85
Hospitals	79.9	539.21
Social services	77.9	286.08
Veterinary services	76.9	291.76
Apparel and accessory stores	75.0	229.06
Variety stores	73.0	221.37
Physical fitness facilities	70.5	171.86
Miscellaneous personal services	69.6	250.61

Source: U.S. Department of Labor (1995, 1998).

unclear whether such salaries accurately reflect the CEO's contribution to society relative to that of the factory worker.[9] Even though unskilled workers might be replaced more easily than engineers or CEOs, an industrialized society depends on motivated and qualified people in all positions. Moreover, should low-skill workers be denied a livable wage simply because they can be easily replaced? This question becomes espe-

[9]The average salary of the CEOs of comparable-sized foreign-headquartered industrial corporations is substantially lower than the U.S. average. The average CEO salary for Brazil is $701.2 thousand, for Hong Kong $681 thousand, for Britain $645.5 thousand, for France $520.4 thousand, for Canada $498.2 thousand, for Mexico $456.9 thousand, for Japan $420.9 thousand, for Germany $398.4 thousand, and for South Korea $151 thousand (Bryant 1999).

How do we explain the salary differences between an auto parts store clerk (left) and a day-care worker (right)? Is one job more functionally important than the other?

© Jon Levy/Gamma Liaison (left); © Jeff Greenberg/Photo Edit (right)

cially critical when we consider that 4.5 million families with at least one person working live below the poverty line (Rockefeller Foundation 1996c).

Finally, both Tumin and Simpson challenge the functionalist claim that a system of stratification evolves so as to meet the overall needs of the society. In evaluating such a claim, one must identify whose needs are being met by the system. In the case of apartheid, the needs of whites unquestionably were met at the expense of the needs of blacks. As an example, consider the way in which electricity is still distributed to the people of South Africa (Passell 1994). South Africa produces 60 percent of all of the electricity used throughout the African continent, but 66 percent of all South Africans (mostly nonwhites) do not have electricity in their homes. Those without electricity face enormous difficulties in obtaining the fuel to cook food and to warm and light their houses. As a result, people without electricity[10] must resort to gathering wood:

> In the high grassland areas of KwaZulu [one of the ten African homelands], for example, the average distance walked in collecting one headload [of fuel] was a little over 5 miles (8.3 km) and the average time taken in collecting the one load was 4.5 hours.
> Not only is the collecting of firewood exhausting, time-consuming, and dangerous but it has serious

ecological consequences. Each household uses between three and four tons of wood a year. . . . Over the relatively brief span of the past 50 years, 200 of the 250 forests in KwaZulu have disappeared. (Wilson and Ramphele 1989:44)

This example presents a clear challenge to the idea that stratification functions to meet the needs of a society. In light of the environmental consequences resulting from the inequality in distribution of electricity, it is clear that only one segment of South Africa's society benefits—the whites.

Analyses of Social Class

Although sociologists use the term *class* to refer to one form of stratification, they also use **class** to denote a category that designates a person's overall status in society. Sociologists consider social class to be an important factor in determining life chances. They are, however, preoccupied with two questions: How many social classes are there, and what constitutes a social class? For some answers to these questions, we turn to the works of Karl Marx and Max Weber.

Karl Marx and Social Class
Marx's focus on class and the class divisions in society underscores his belief that the most important engine

---

[10]Since Nelson Mandela came to power in April 1994, 525,000 homes (almost all in impoverished black townships) have been wired for electricity. The plan is to wire 300,000 homes each year until 2000. Even at that pace, more than 17 million people will be without electricity in South Africa at the start of the new millennium (Drogin 1996).

Class A category that designates a person's overall status in society.

of change is class struggle. Marx gave several answers to the question, How many social classes are there? In *The Communist Manifesto* ([1848] 1996), he named two: the bourgeoisie and the proletariat. In *Capital: A Critique of Political Economy* (1909), he identified three social classes: wage laborers, capitalists, and landlords. In *The Class Struggles in France 1848–1850* ([1895] 1976), he named at least six: the finance aristocracy, the bourgeoisie, the petty bourgeoisie, the proletariat, landlords, and peasants. A careful reading of Marx's writings suggests that he believed "that the number of classes to be defined depends on the reason why we want to define them" (Boudon and Bourricand 1989: 341). A brief overview of Marx's three works clarifies this point.

In *The Communist Manifesto,* written with Friedrich Engels in 1848, Marx described how class conflict between two distinct classes propels society from one historical epoch to another. Throughout human history, free men and slaves, nobles and commoners, barons and serfs, and guildmasters and journeymen have confronted each other. Marx observed that the rise of factories and mechanization as a means of production created two fundamental classes: the property owners or bourgeoisie (the owners of the means of production) and the propertyless workers or proletariat (those individuals who must sell their labor to the bourgeoisie).

In *Capital: A Critique of Political Economy,* Marx (1909) named three classes: wage laborers, capitalists, and landlords. Each class is composed of people whose revenues or income "flow from the same common sources" (p. 1032). For wage laborers, the source is wages; for capitalists, profit; for landowners, ground rent. Marx acknowledged that the boundaries separating landowners from capitalists are not clear-cut. In this three-category classification scheme, for example, Henry Ford, the founder of Ford Motor Company, is both a landlord (because he owned a rubber plantation in Brazil) and a capitalist (because he owned the factories and the machines and purchased the labor).[11] In this case, Marx was interested in categorizing people based on their sources of income, so a broad three-category social class scheme made sense.

The Class Struggles in France 1848–1850 ([1895] 1976) is a historical study of an event in progress—

the failed 1848 revolutions against several European governments (Germany, Austria, France, Italy, and Belgium), with special emphasis on France. In this book, Marx sought to describe and explain "a concrete situation in its complexity" (Boudon and Bourricaud 1989:341). To accomplish this task Marx named at least six classes. He described the 1848 revolution as a struggle for the necessities of life and as "a fight for the preservation or annihilation of the bourgeois order" (Marx [1895] 1976:56). The "concrete situation" involved a class of people he labeled as the finance aristocracy, which lived in obvious luxury among masses of the starving, low-paid, unemployed urban proletariat class.

The widespread discontent that fueled the 1848 revolutions resulted from the potato blight and the bad harvests of 1845 and 1846, which further raised the already high level of frustration among the people. The resulting rise in the cost of living spurred bloody conflict both in France and on the rest of the European continent. A second incendiary event was a general commercial and industrial crisis that resulted in an economic depression and a collapse of international credit. The resulting revolutions were centered in the cities, where the Industrial Revolution had created a proletariat class from persons who had migrated there in search of work. Generally, the workers were paid very low wages, lived in squalor, and lacked the most basic necessities.

Although the faces of those who ruled the French government changed as a result of the 1848 revolution, the exploitive structure remained. In the end, the workers were put down by "unheard of brutality" (Marx [1895] 1976:57). Marx believed that the uprising failed because, even though the workers displayed unprecedented bravery and talent, they were "without chiefs, without a common plan, without means and for the most part, lacking weapons" (p. 56). Also, the revolution failed because the "other" classes did not support the proletariat when they moved against the finance aristocracy.

It is difficult to apply Marx's ideas about social class in their totality because, as he made clear in *The Class Struggles in France 1848–1850,* the reality of class is very complex. Marx left us, however, with some useful ideas with which to approach the study of social class. First is the idea that conflict between two distinct classes propels us from one historical epoch to another. South Africa clearly contained two distinct classes: those designated as white and those designated as something else. In the United States, one division that

---

[11]Marx also acknowledged that each of the three classes can be subdivided further. The category of landowners, for example, can be divided into owners of vineyards, farms, forests, mines, fisheries, and so on.

has received considerable attention is the "class" division between lenders and debtors.

A second important idea left to us by Marx is the concept of viewing social class in terms of the sources of income. Approaching social class in this way carries our understanding of social class beyond the simple notion of occupation, the most obvious source of earnings. Other legal sources include inheritance, rent, credit, capital gains, and so on.[12]

Finally, Marx's ideas remind us that the conditions that lead to a successful revolt by an exploited class against the exploiting class are multifaceted and complex. He recognized that unbearable conditions can trigger uprisings, but observed that other factors such as a well-thought-out plan, effective leadership, the support of other classes, and access to weapons ultimately determine the success or failure of the revolt.

To illustrate this point, consider that two of the best-known demonstrations against apartheid occurred in 1960 and 1976. In 1960, the residents of Sharpville, an African township, marched to the police station without their passes to protest the pass laws (U.S. Department of State 1996). The police fired into the crowd, killing 69 persons and wounding 180. Although the police response initiated riots throughout South Africa, the white government eventually won and declared all anti-apartheid organizations illegal. In 1976, 20,000 Soweto schoolchildren marched in protest against the use of Afrikaans as the language of classroom instruction. The police opened fire on the protesters; hundreds of children were killed and thousands were wounded. This brutal response set off riots throughout the country. The point is that nonwhite South Africans have always protested apartheid, but only since the mid-1980s has the other class—whites—taken significant steps to end the system.

What factors other than protests by the exploited class contributed to the movement to end apartheid? First, the United Nations stripped South Africa of its member status in 1974. Then, in 1976, the United States and Europe applied new and more forceful economic sanctions while resistance from within increased. Also, the collapse of communism and the end of the Cold War between the United States and the Soviet Union meant that the African continent and South Africa lost their strategic importance. As a result, the United States no longer needed to support the white South African government so as to counter communist influences within South Africa and surrounding countries.

Max Weber and Social Class

Although Karl Marx did not consistently specify an exact number of social classes in society, he clearly stated that a person's social class was based on his or her relationship to the means of production and sources of income. Like Marx, Max Weber did not specify how many social classes exist. For Weber, though, the basis for a social class was not merely one's relationship to the means of production or one's source of income.

According to Weber ([1947] 1985), people's class standing depends on their marketable abilities (work experience and qualifications), their access to consumer goods and services, their control over the means of production, and their ability to invest in property and other sources of income. Persons completely unskilled, lacking property, and dependent on seasonal or sporadic employment constitute the very bottom of the class system. They form the "**negatively privileged" property class.** Individuals at the very top—the **"positively privileged" social class**—monopolize the purchase of the highest-priced consumer goods, have access to the most socially advantageous kinds of education, control the highest executive positions, own the means of production, and live on income from property and other investments. Between the top and the bottom of the ladder is a continuum of rungs.

Weber states that class ranking is complicated by the status groups and political parties to which people belong. He defines a **status group** as an amorphous group of persons held together by virtue of a lifestyle that has come to be "expected of all those who wish to belong to the circle" (Weber 1948: 187) and "by the level of social esteem and honor accorded

"Negatively privileged" property class Persons completely unskilled, lacking property, and dependent on seasonal or sporadic employment who constitute the very bottom of the class system.

"Positively privileged" social class Those individuals at the very top of the class system.

Status group An amorphous group of persons held together by virtue of a lifestyle that has come to be expected of all those who wish to belong to the circle.

---

[12]There are illegal and informal sources of income as well.

African National Congress leader, anti-apartheid activist, and now president of South Africa, Nelson Mandela was sent to prison as a young man in 1964 (left) and finally released 27 years later (right).

© Selon Mention/Sipa Press (left); © Jacques Witt/Sipa Press (right)

to them by others" (Coser 1977:229). The lifestyle encompasses education, family background, or occupation. This definition suggests that wealth, income, and position are not the only factors that determine an individual's status group. That is, some people may possess equivalent amounts of wealth but yet be of different status due to their upbringing and education. The episodes of the television situation comedy "The Beverly Hillbillies," for example, focus on this issue of lifestyle and wealth. In South Africa, English-speaking whites form a status group distinct from Afrikaans-speaking white Afrikaners. The two groups practice what sociologist Diana Russell calls a voluntary apartheid—"speaking different languages, attending different schools, living in different areas, and voting for different political parties" (Russell 1989:4).

Political parties are organizations "oriented toward the planned acquisition of social power [and] toward influencing social action no matter what its content may be" (Weber 1982:68). Parties are organized to represent people of a certain class, status, and

other interests; they exist at all levels (within an organization, a city, a country). The means employed to obtain power can include violence, vote canvassing, bribery, donations, the force of speech, suggestion, and fraud. Recall that the president of South Africa—Nelson Mandela—a member of the African National Congress (ANC) party, was convicted of treason and sentenced to life in prison in 1964 for acts of sabotage (that is, damaging property so as to disrupt the functioning of society) directed against apartheid. At his trial, Mandela (1990) explained why his political party believed sabotage was necessary:

> The initial plan was based on careful analysis of the political and economic situation of our country. We believed that South Africa depended to a large extent on foreign capital and foreign trade. We felt that planned destruction of power plants and interference with rail and telephone communications would tend to scare away capital from the country, make it more difficult for goods from the industrial areas to reach the seaports on schedule, and would in the long run be a heavy drain on the economic life of the country, thus compelling the [white] voters of the country to reconsider their position. (pp. 26–27)

A second example of a South African political party oriented toward the acquisition of power is the Third Force, security forces who worked to sabotage

Political parties Organizations oriented toward the planned acquisition of social power and toward influencing social action no matter what its content may be.

the transition from apartheid to democracy. Their acts of sabotage included infecting Johannesburg prostitutes with AIDS and supplying the bitter rivals of the ANC party, the Inkatha Freedom party, with assault weapons to help escalate the fighting among black factions. The Third Force also employed death squads, terrorist bombs, and torture to control anti-apartheid sympathizers and to destabilize the country (Drogin 1995).

Examples of political parties are endless. Some political parties that have received media coverage in the United States include the National Organization of Women, Promise Keepers, the National Association for the Advancement of Colored People, the AFL-CIO, United Auto Workers, the National Rifle Association, the American Association of Retired Persons, and so on.

Weber's conception of social class enriches that put forth by Marx. As noted earlier, Weber views class as a continuum of rungs on a social ladder, with the top and the bottom rungs being the positively privileged class and the negatively privileged property class. He argues that a "uniform class situation prevails only when completely unskilled and propertyless persons are dependent on irregular employment" (1982:69). We cannot speak of a uniform situation with regard to the other classes, because class standing is complicated by such elements as occupation, education, income, the status and groups to which people belong, differences in property, consumption patterns, and so on.

Weber's idea of top and bottom rungs, with everyone else somewhere between, inspires us to compare the situation of the wealthiest person with that of the poorest. In "Measuring Income Inequality Within the Fifty States and the District of Columbia," the average income of the bottom and top fifth of families with children are compared. For example, in New York, the wealthiest fifth have income almost 20 times that of the poorest fifth.

Weber's ideas about social class also draw our attention to the negatively privileged classes. The proportion of negatively privileged persons in a society tells us something important about the extremes of inequality found there. For example, although some people of all racial groups in South Africa have high incomes, income distribution is clearly related to race. Table 9.9 shows the average annual per capita income by race and the percentage of households in each racial group in various income categories. Sixty-one percent of white households have incomes exceeding 50,000 rand, as compared with 6.3% of households classified

as colored, and 1.9% classified as African. Even more extreme inequality is apparent in the former homelands, where 80 percent of households are in a state of dire poverty (Wilson and Ramphele 1989). The plight of the negatively privileged in South Africa can be traced directly to apartheid.

The existence of a negatively privileged property class in the United States can also be traced to structural factors—in particular, to changes in the occupational structure. This finding may come as a surprise to some Americans who attribute mobility, whether upward or downward, to individual effort and do not consider changes in the occupational structure as the cause. Many Americans may not recognize that some groups are affected by changes in the occupational structure more strongly than others. In *The Truly Disadvantaged* (1987) and other related articles and books (Wilson 1983, 1991, 1994), sociologist William Julius Wilson describes how structural changes in the U.S. economy helped create what he termed, in his 1990 presidential address to the American Sociological Association, the "ghetto poor." A number of economic transformations have taken place, including the restructuring of the American economy from a manufacturing-based economy to a service- and information-based economy; the rise of a labor surplus since the 1970s, marked by the entry of women and the large "baby boom" segment of the population into the labor market; a massive exodus of jobs from the cities to the suburbs; and the transfer of low-skilled manufacturing jobs out of the United States to offshore locations (see Chapter 2). These changes are major forces behind the emergence of the ghetto poor or **urban underclass,** a "heterogeneous grouping of families and individuals in the inner city that are outside the mainstream of the American occupational system and that consequently represent the very bottom of the economic hierarchy" (Wilson 1983:80).

Wilson (in collaboration with sociologist Loic J. D. Wacquant; Wacquant and Wilson 1989) has studied Chicago to illustrate this point. (Actually, the point applies to every large city in the United States—Los Angeles, New York, Detroit, and so on.) In 1954,

Urban underclass The group of families and individuals in the inner city who are outside the mainstream of the American occupational system and who consequently represent the very bottom of the economic hierarchy.

U.S. in Perspective

Measuring Income Inequality Within the Fifty States and the District of Columbia

The chart below compares the average income of the poorest 20 percent of families with children. The District of Columbia has the most inequality, with the average income of the wealthiest families being 28.2 times that of the poorest families. Utah has the least inequality. Is great income inequality desirable? Why do you think that some states have greater inequality than others?

Source: Center on Budget and Policy Priorities (1997).

Top to Bottom Income Ratios

State	Average Income of Bottom Fifth of Families with Children	Average Income of Top Fifth of Families with Children	Ratio
1. New York	$6,787	$132,390	19.5
2. Louisiana	$6,430	$102,339	15.9
3. New Mexico	$6,408	$91,741	14.3
4. Arizona	$7,273	$103,392	14.2
5. Connecticut	$10,415	$147,594	14.2
6. California	$9,033	$127,719	14.1
7. Florida	$7,705	$107,811	14.0
8. Kentucky	$7,364	$99,210	13.5
9. Alabama	$7,531	$99,062	13.2
10. West Virginia	$6,439	$84,479	13.1
11. Tennessee	$8,156	$106,966	13.1
12. Texas	$8,642	$113,149	13.1
13. Mississippi	$6,257	$80,980	12.9
14. Michigan	$9,257	$117,107	12.7
15. Oklahoma	$7,483	$94,380	12.6
16. Massachusetts	$10,694	$132,962	12.4
17. Georgia	$9,978	$123,837	12.4
18. Illinois	$10,002	$123,233	12.3
19. Ohio	$9,346	$111,894	12.0
20. South Carolina	$8,146	$96,712	11.9
21. Pennsylvania	$10,512	$124,537	11.8
22. North Carolina	$9,363	$107,490	11.5
23. Rhode Island	$9,914	$111,015	11.2
24. Washington	$10,116	$112,501	11.1
25. Maryland	$13,346	$147,971	11.1
26. Virginia	$10,816	$116,202	10.7
27. Kansas	$10,790	$110,341	10.2

Chicago was at the height of its industrial power. Between 1954 and 1982, however, the number of manufacturing establishments within the city limits dropped from more than 10,000 to 5,000, and the number of jobs declined from 616,000 to 277,000. This reduction, in conjunction with the outmigration of stably employed working-class and middle-class black families, which was fueled by new access to housing opportunities outside the inner city, had a profound impact on the daily life of people left behind. The exodus of the stably employed resulted in the closing of hundreds of local businesses, service establishments, and stores. According to Wacquant (1989), the single most significant consequence of these historical and economic events was the "disruption of the networks of occupational contacts that are so crucial in moving individuals into and up job chains . . . [because] ghetto residents lack parents, friends, and

State	Average Income of Bottom Fifth of Families with Children	Average Income of Top Fifth of Families with Children	Ratio
28. Oregon	$9,672	$97,589	10.1
29. New Jersey	$14,211	$143,010	10.1
30. Indiana	$11,115	$110,876	10.0
31. Montana	$9,051	$89,902	9.9
32. South Dakota	$9,474	$93,822	9.9
33. Idaho	$10,721	$104,725	9.8
34. Delaware	$12,041	$116,965	9.7
35. Arkansas	$8,995	$83,434	9.3
36. Colorado	$14,326	$131,368	9.2
37. Hawaii	$12,735	$116,060	9.1
38. Missouri	$11,090	$100,837	9.1
39. Alaska	$14,868	$129,025	8.7
40. Wyoming	$11,174	$94,845	8.5
41. Minnesota	$14,655	$120,344	8.2
42. Nebraska	$12,546	$102,992	8.2
43. Maine	$11,275	$92,457	8.2
44. New Hampshire	$14,299	$116,018	8.1
45. Nevada	$12,276	$98,693	8.0
46. Iowa	$13,148	$104,253	7.9
47. Wisconsin	$13,398	$103,551	7.7
48. Vermont	$13,107	$97,898	7.5
49. North Dakota	$12,424	$91,041	7.3
50. Utah	$15,709	$110,938	7.1
District of Columbia	$5,293	$149,508	28.2
Total U.S.	**$9,254**	**$117,499**	**12.7**

acquaintances who are stably employed and can therefore function as diverse ties to firms . . . by telling them about a possible opening and assisting them in

[13]In numerical terms, people classified as white represent the largest category of people living in poverty. The percentage of "whites" living in poverty in 1994, however, was 11.7 percent compared with 30.6 percent of "blacks" and 30.7 percent of "Hispanics" (O'Hare 1996).

applying [for] and retaining a job" (Wacquant 1989:515–516).

Because the ghetto poor are the most visible and most publicized underclass in the United States, many Americans associate poverty with minority groups. In fact, approximately 50 percent of the population classified as living in poverty is white.[13] In addition, demographers William P. O'Hare and Brenda Curry-White (1992) estimate that approximately 736,000

Table 9.9   Annual Household Income, by Race (in rands)

This table shows the average income for South Africans in each racial category. It also shows the percentages of people in each income category by race. Note that 61 percent of South Africans classified as white earn $50,000 or more per year compared with 1.9 percent of people classified as African or black.

		Percentage of People in Each Income Bracket (in Rands)			
	Average*	Less than 4,999	5,000–14,999	15,000–49,999	50,000+
White	$25,344	9.3	5.8	31.9	61.0
Indian	$11,112	3.6	13.5	49.7	33.1
Colored	$5,196	8.9	30.1	53.7	6.3
African	$2,520	30.2	45.0	22.7	1.9
Urban	$3,192	—	—	—	—
Rural	$1,920	—	—	—	—

*1 rand is equivalent to $.173 U.S.
Source: South Africa Labour Development Research Unit (1994).

rural residents can be classified as constituting an underclass. More than two-thirds (70 percent) of the less visible rural poor are white (Rockefeller Foundation 1996b). Like their urban counterparts, members of the rural underclass are concentrated in geographic areas with high poverty rates. They, too, have been affected by economic restructuring, including the decline of farming, mining, and timber industries and the transfer of routine manufacturing out of the United States.

The rural and urban underclass represent two distinct segments of the U.S. population that live below the poverty line, which was set at approximately $16,400 for a family of four in 1997. On the basis of this definition, almost 35.6 million Americans (or 13.3 percent of the population) live below the poverty line. Included in this number are 12 million people whose income is less than half the amount officially defined as the poverty level (U.S. Bureau of the Census 1998). The definition of poverty also encompasses diverse groups of people, some of whom might not be considered really poor (such as graduate students and retired people who have assets but who live on a low fixed income). Nevertheless, it is important to point out that one out of every three households[14] headed by women are below the poverty line (U.S. Bureau of the Census 1998b). For many of these women "their only 'behavioral deviancy' is that their husbands or boyfriends left them" (Jencks 1990:42); in the case of many older women, their husbands have died. For the most part, two reasons account for women's poverty: the economic burden of children and women's disadvantaged position in the labor market. These issues will be explored further in Chapter 11.

Summary and Implications

In this chapter, we have examined the workings of social stratification (the systematic division of people into categories). More importantly, we have learned that classification schemes have a profound impact on people's life chances. Some forms of social stratification affect life chances more strongly than others. Caste systems such as apartheid have a decided impact, because they ensure that life chances are determined by characteristics over which people have no control. Class systems, although not models of equality, stratify people on the basis of individual effort. Class and caste systems can be considered the two extremes on a continuum.

The current South African society still approaches a caste system, because in that country people's life

[14]A Rockefeller Foundation (1996a) survey found that the average person in the United States estimated the cutoff for poverty-level income for a family of four to be approximately $3,000 higher than the federal measure.

chances and access to scarce and valued resources are clearly connected to race. As much as we would like to believe that the United States represents a pure class system, the evidence tells us otherwise—it is more castelike than we care to acknowledge. At the same time, the United States is classlike in the sense that every occupation contains people of different ethnic, racial, age, and sex groups. On the other hand, some ethnic, racial, age, and sex groups are clearly concentrated in the low-status occupations.

The most intriguing and most problematic feature of stratification systems in general involves the criteria used to rank people, especially when ascribed characteristics are the important criteria. How can ranking systems exist in which people who belong to one category of an ascribed characteristic (such as white skin or blue eyes) are treated as more valuable or worthy than people who belong to other categories? Jane Elliot, the third-grade teacher who separated her students by eye color and rewarded them accordingly, gives one answer: "This is not something I can do alone" (*Frontline* 1985:20). Elliot meant that the experiment could not have worked without the cooperation of the people on top. Her observation suggests that people work together to maintain systems of stratification. But why do people cooperate? One answer comes from a blue-eyed member of Jane Elliot's class, now an adult, as he recalled the experience. "Yeah, I felt like I was—like a king, like I ruled them brown-eyes, like I was better than them, happy" (p. 13). The feeling of being "better" translated into an unexpected result. Elliot explains:

> The second year I did this exercise I gave little spelling tests, math tests, reading tests two weeks before the exercise, each day of the exercise and two weeks later and, almost without exception, the students' scores go up on the day they're on the top, down the days they're on the bottom and then maintain a higher level for the rest of the year, after they've been through the exercise. We sent some of those tests to Stanford University to the Psychology Department and they did sort of an informal review of them, and they said that what's happening here is kids' academic ability is being changed in a 24-hour period. And it isn't possible but it's happening. Something very strange is happening to these children because suddenly they're finding out how really great they are and they are responding to what they know now they are able to do. And it's happened consistently with third graders. (p. 17)

One clear answer to the question of why the people on top cooperate to maintain a system of stratification is that they benefit (whether they know it or not) from the system of stratification and the way in which rewards are distributed. In the third-grade class, the blue-eyed children benefited from the system that distributed rewards on the basis of eye color. In South Africa, whites clearly have benefited from a system that rewards people on the basis of race.

In comparison with South Africa, it is more difficult to see the inequality that exists in the United States. Perhaps this difficulty arises because we believe that the United States is a model of equal opportunity because (1) we can find examples of people from all racial, ethnic, sex, and age groups who do achieve rewards, (2) no

Simply dismantling the apartheid laws will not reverse the enormous disparities between black and white South Africans that have developed over 300 years.

© A. Tannenbaum/Sygma

Although individual effort is one important variable in upward and downward mobility, we must also acknowledge structure factors that significantly affect people's ability to find and keep well-paying jobs. The dismantling of apartheid opened to blacks some jobs previously reserved for white South Africans.

© Eric Miller/Impact Visuals

obvious laws govern ascribed characteristics and life chances, and (3) we believe that anyone can transcend his or her environment through hard work.

The case of South Africa reminds us how difficult it is for people to acknowledge and give up their privileges and to put into practice a new ranking and reward system. "Privilege is often invisible to those who enjoy it, because to them it is simply the normal state of affairs" (Howard and Hollander 1997:23). One goal of the Truth and Reconciliation Commission (TRC) is to show the extreme lengths to which some white South Africans went to preserve the apartheid system for all whites, most of whom could not have imagined how the system was being held together. The work of the TRC makes it impossible for whites to say they "didn't know." Dr. Alex Boraine, Deputy Chairperson of the TRC, explained the importance of knowing and coming to terms with the truth:

> The process will not be completed until all South Africans who benefited from apartheid confront the reality of the past, accept the uncomfortable truth of complicity, give practical expression to remorse and a commitment to a way of life which accepts and offers the dignity of humanness. Closure is decidedly messy and agonizing, but it is absolutely necessary. Dealing with the past is essential but our focus must be on building a future. Our journey together as a Commission has been demanding, traumatic and often painful, but I know that I would not have missed being a small part of it for all the world. Closure is a little bit like dying to the old and has to take place in order for new life to begin. Commissioners and staff alike move towards these new opportunities and challenges deeply enriched by our common search for truth and healing. My very special thanks to all my colleagues for a special time in all our lives. (Boraine 1998)

The sociological perspective is valuable because it enables us to see how social stratification systems connect with life chances. When we know what is going on, we have an obligation to work to change things. People are rarely so clear-sighted, however. In the case of South Africa, it took pressure from outside (in the form of economic sanctions and cultural isolation) and from within (in the form of mass demonstrations, strikes, and bloodshed) to push South African whites to take the first steps toward dismantling apartheid and creating a multiracial democracy. The success of South Africa's attempts to eliminate apartheid will hinge on whether it can end the legacy of this policy—the profound social and economic inequalities that now keep South Africans apart. Everyone on the outside looking in at South Africa can see that those who benefited from apartheid must give up something. Many whites, however, cannot seem to see how their advantages over the nonwhite population are connected to apartheid (Mandela 1997).

Key Concepts

Use this outline to organize your review of key chapter ideas.

Life chances

Social stratification

    Caste

       Apartheid

          Ascribed characteristics

        Class

           Achieved characteristics

Status value

Social mobility
 Vertical mobility
  Downward mobility
  Upward mobility
 Horizontal mobility
 Intergenerational mobility
 Intragenerational mobility

Social class
 "Negatively privileged" property class
 "Positively privileged" property class
 Status group
 Political parties
 Urban underclass

internet assignment

Find the home page for the U.S. Bureau of the Census. Browse the Web site and look for reports showing some connection between an ascribed characteristic and life chances.

10

Race and Ethnicity

With Emphasis on Germany

Brandenburg Gate, Berlin. (© Dave Bartruff/Artistry International.)

Source: Adapted from *The Times Atlas of World History* (1984).

Race, Ethnicity, and the Berlin Wall

At the end of World War II, Germany was divided into Soviet-occupied East Germany and British-, French-, and U.S.-occupied West Germany. The city of Berlin was also divided. Between 1945 and 1961, 3 million East Germans and 9 million people from former German territories migrated to West Germany, providing labor for West Germany's rapid postwar reconstruction. In 1961, when 3,000 migrants per day were joining the East–West flow, the Soviets sealed off East Germany and East Berlin from West Germany and West Berlin with more than 970 miles of wall and fortified fences.

As a result of this barrier, West Germany faced a labor shortage. In response, it began to recruit millions of "guest workers" from other countries. Although West Germany depended on this foreign labor, it was not willing to offer German citizenship to the guest workers or even to their offspring who were born and raised in Germany.

On November 9, 1989, the Berlin Wall was torn down and the "Iron Curtain" that divided Eastern and Western Europe was removed. The subsequent reunification of East Germany (with a population of 17 million) with West Germany (with a population of 63 million) followed less than one year later on October 3, 1990. In addition, after the fall of the Wall, 2.2 million ethnic Germans who had been living in Eastern European countries and in the former Soviet Union settled in Germany.

The entry of 17 million East Germans and 2.2 million ethnic Germans into West German society was not easy. It soon became clear that German "blood" did not make someone German. In fact, the guest workers—many of whom had lived and worked in the same German town for 30 or more years and their offspring who were born in Germany and knew only the German language—were in many ways more "German" than the ethnic Germans (Thränhardt 1989).

Ender Bsaran is twenty-five now. He looks no different from most young Germans on his street. He has the latest German haircut— long in back, and short and spiky on top. . . . [W]ith . . . his baggy jeans and the pack of Marlboros sticking out of his shirt pocket, it's hard to think of him as "foreign." He likes to say that his family is "more of a European family," although his father goes to the mosque every day and sits with the men after prayers and listens to the gossip. . . . He wants you to know that his mother is nothing at all like the squat, ruddy Turkish women you see on the streets of Hemshof, wrapped in the bulky coats and gabardine head shawls that they brought from Turkey. He likes Turkey, but he doesn't think that either he or his mother belongs in Turkey. (Kramer 1993:56)

- Emmi Schleicher arrived in Germany from her native Kazakstan six years ago and thought she had finally come home to the land her ancestors left over two centuries ago. Since then, she has learned a much more painful lesson. As an ethnic German in Kazakstan, she said, she faced nationalist hostility among Kazaks who called her a foreigner. Now she confronts another kind of animosity from Germans who do not accept her as a German and who say she and others like her receive social benefits they do not deserve. "In Kazakstan they said we were Germans and should go back to Germany. . . . Now we are in Germany and people here curse us as 'the Russians.'"

- Alena's father, a white American, met her mother in Thailand when he was stationed there while in the military. Alena has many of her mother's physical characteristics: dark skin, dark hair, big eyes, and full lips. Many people try to classify her as Hawaiian or Filipino. When

people ask, "Where are you from?" or "What are you?" she usually answers, "American." It is obvious from her that she has been raised in America. She dresses and talks like an American. When she was a child, her paternal grandmother took care of her a lot, and all of her friends are white. She knows very little Thai and is Catholic (her mother is Buddhist). (Northern Kentucky University student 1996)

- Three years ago my good friends Cathy and Sam found out they were going to have their first baby. They spent the next nine months preparing to make everything perfect for the new arrival. Cathy eventually gave birth to a blue-eyed, blond-haired baby boy whom they named Michael. The hospital nurse who prepared the birth certificate told the parents that Michael would be classified according to the race of his father, black. Cathy was outraged by this idea, afraid that her son would be an object of discrimination. How will people react when Michael checks "black" as his race on the various forms and applications that will come his way in the future? Will he be viewed as an impostor if he claims his black heritage? (Northern Kentucky University student 1996)

- I can't be anything but what my skin color tells people I am. I am black because I look black. It does not matter that my family has a complicated biological heritage. One of my great-great-grandmothers looked white but she was of French and African-American descent. Another great-great-grandmother looked Indian but she was three-fourths Cherokee and one-quarter black. My great-grandfather looked white but his sister was so black she looked purple. My coloring is a middle shade of brown, but I have picked up a lot of red tones in my hair from my Indian heritage. My family

is a good example of how classifying people according to skin color is ridiculous. (Northern Kentucky University student 1996)

Why Focus on Germany?

In this chapter, we focus on Germany because it shares something in common with the United States: Each country's government classifies the people within its borders according to race and/or ethnicity. We will examine the U.S. system of racial and ethnic classification and consider a recent change made to that system. On October 30, 1997, the U.S. Office of Management and Budget (OMB) declared that for the first time in the history of the United States, people could identify themselves on the census and other federal forms as belonging to more than one racial category. The official racial categories with which they can identify are limited to only five choices, however: (1) black or African American; (2) white; (3) American Indian or Alaskan native; (4) Native Hawaiian or other Pacific Islander; and (5) Asian

(Holmes 1997).[1] The OMB has yet to decide how it will count people who identify with more than one race. One thing is clear, it will not classify them as multiracial. This recent shift in policy leads us to ask several questions, including the following: Why did this change in racial categories occur? And how did the U.S. government account for people who identified with more than one race before October 30, 1997?

The German government also is concerned with classifying people within its borders. According to German law, people who can prove that they are ethnic Germans are entitled to citizenship regardless of their country of birth. It does not matter whether they cannot speak German or whether they know nothing about German culture; the criterion for automatic German citizenship is biological (i.e., blood). This law excludes from citizenship people who have lived in Germany all of their lives but who are not biologically German. Under this law, nonethnic Germans who have lived and worked in the same German town for 30 or more years or those who were born in Germany and know only the German language are classified as foreigners (Thränhardt 1989). In 1998 a proposal granting automatic citizenship to third-generation foreigners and to nonethnic German-born babies, provided that at least one of their parents has lived in Germany since age 14 was voted down. The change in law would have allowed 3 million foreigners to become German citizens (Cohen 1998).

Studying the systems of racial and ethnic classification used in Germany and the United States and the proposed changes in those systems helps us to understand the concepts of "race" and "ethnicity." Among other things, comparing the two systems helps us to see that governments differ in the ways in which they classify people within their borders. In addition, we will learn that the systems of racial and ethnic classification have real consequences, in that the categories to which people are assigned are not neutral. People in some categories are treated as more worthy than people in other categories.

[1]The United States also classifies people according to ethnicity and officially recognizes two broad categories: Hispanic and non-Hispanic.

Classifying People by Race in the United States

Most people in the United States equate race with physical features. In their minds, the term *race* refers to a group of people who possess certain distinctive physical traits related to skin color, hair texture, and the shape of their eyes, nose, and lips. Officially, the U.S. government recognizes five racial categories, each of which evokes an imaged appearance in most peoples' minds. As mentioned earlier, for the first time in U.S. history, the people living throughout the 50 states are now permitted to identify themselves with more than one of the five official racial categories. Previously, people could claim to be members of only one race, and if they could not do so, the government decided their race.

As sociologist Sharon Lee (1993) points out, although the official racial categories have changed since 1790, the year of the first census, the government continues to maintain three dominant themes:

Benetton, a clothing company, runs ads that feature people whose physical appearances reinforce the idea that people can be classified into clear-cut racial categories.

© Benetton/Sipa Press

Figure 10.1   Assigning an Official Race Category

Study the physical traits of the people in the photographs and assign each to one of the five official racial categories (described on page 292) used in the United States.

© Lawrence Migdale/Stock Boston

© Jeff Greenberg/PhotoEdit

© Brent Jones/Stock Boston

© Barbara Stitzer/PhotoEdit

© Richard Pasley/Stock Boston

© Tamara Reynolds/Tony Stone Images

Top row, left to right: Alaskan Native, Native American, Black
Bottom row, left to right: Native Hawaiian, Black, Caucasian

1. A belief in racial purity. The U.S. government has operated under the false assumption that all people can be sorted and classified into racial categories and that it is *not* possible for someone to belong to more than one category. So the people pictured in Figure 10.1 were categorized as belonging to one racial category, no matter what the facts are with regard to their ancestry. While "mixed race" people have always existed, the U.S. government has asked us to accept the idea that one parent contributes a disproportionate amount of genetic material to the child—so large a genetic contribution that it negates the genetic contribution of the other parent. Even in light of the 1998 historic decision allowing people to identify with more than one race, the OMB rejected a general "multiracial" category (Holmes 1997).

2. A pattern of transforming many ethnic groups into one racial group. While the official number of racial categories has changed over time, the U.S. government has persisted with the practice of using broad racial categories as umbrella terms or supercategories to encompass peoples who vary widely according to

What "race" is Gregory Howard Williams? The left photo shows Williams' grandmother with his father around 1920. The photo above shows Williams, far left, with his immediate family in 1992. How meaningful or valid is it for a government to routinely categorize people—or for people to categorize themselves—by racial constructs that ignore the extent and biological realities of racial blending? If distinctions of race are drawn solely on the basis of outward appearances, what real purpose do they serve?

Courtesy of Dr. Gregory H. Williams, author of *Life on the Color Line: The True Story of a White Boy Who Discovered He Was Black.*

nationality, ethnicity, language, culture, and time of arrival in the United States. The following definitions explain the five supercategories in place today:

- *American Indian or Alaskan Native*—any person having origins in any of the original peoples of North America, which by some estimates includes more than 2,000 distinct groups

- *Asian*—any person having origins in any of the original peoples of the Far East, Southeast Asia, or the Indian subcontinent

- *Black*—any person having origins in any of the black racial groups of Africa

- *White*—any person having origins in any of the original peoples of Europe, North America, or the Middle East

- *Native Hawaiian or other Pacific Islander*—any person having origins in any of the original peoples of Hawaii or the Pacific Islands

3. Confusing and contradictory guidelines for determining race. The guidelines or rules for placing people into distinct racial categories have been vague, contradictory, unevenly applied, and subject to change. The lack of clarity and changeability can be illustrated by the following examples. For the 1990 census, coders were instructed to classify as white people who self-classified as "white-black" and to classify as "black" individuals who classified themselves as "black-white" (U.S. Bureau of Census 1994). Before 1989, a newborn child was designated as "white" if both parents were "white"; if only one parent was "white," the child was classified according to the race of the nonwhite parent; if the parents were of different nonwhite races, the child was assigned to the race of the father. If the race of one parent was unknown, the infant was assigned to the race of the parent whose race was known. After 1989, the rules for classifying newborns changed. Now the race of the infant is the same as that of the mother (Loch 1993), as if identifying the mother's race presents no challenges. The examples of the arbitrary nature of classification in the United States are endless. Perhaps the most severe rule was the "one-drop rule," whereby any amount of "black" blood in a person's ancestry was considered sufficient to count him or her as "black."

The system of racial classification in the United States helps us to see that **race** is not an easily observed characteristic immediately evident on the basis of a person's physical characteristics, but rather a category defined and maintained by people through a complex array of formal and informal mechanisms. One such formal mechanism comprises the rules that the U.S. government uses to ensure that people fit into its official racial categories. An informal mechanism operates when we use visible clues, such as skin color and hair texture, to determine race rather than taking the time to learn the facts of that person's ancestry.

Classifying People by Ethnicity in the United States

In addition to classifying people by race, U.S. residents are also classified by ethnicity. The only official ethnic categories that the United States recognizes are Hispanic/Spanish and non-Hispanic origin. Contrary to popular belief, "Hispanic" is not a race. People classified as Hispanic can be of any race. How can this be the case? The history of Latin America (and all of the Americas, for that matter) is intertwined with that of Asia, Europe, the Middle East, and Africa. As a result of this interconnected history, the countries of Latin America are populated *not* by a homogeneous group popularly known as Hispanics, but by persons from every conceivable ancestry and combinations of those ancestries (see "The Complexity of North American and 'Hispanic' Origins").

The United States' official definition of Hispanic ignores this complex history and intermixing of peo-

ples, however, and declares an Hispanic to be "a person of Mexican, Puerto Rican, Cuban, Central or South American, or other Spanish culture or origin, regardless of race." According to this official definition, baseball legend Ted Williams (whose mother is of Mexican ancestry) and Walt Disney (who was born in Spain) should be classified as Hispanic, as should Vanna White, Raquel Welch, and Linda Carter (Toro 1995). The Hispanic category even includes someone born in Costa Rica of Chinese ancestry[2] who later immigrated to the United States.

The variety of people included in the ethnic category "Hispanic" reminds us that ethnicity is a complex concept. To illustrate, when sociologist Raymond Breton and his colleagues studied ethnicity in Toronto, they asked respondents 167 questions to determine their ethnicity (Breton et al. 1990), including questions about language spoken at home, the importance of ethnic background, the ethnic composition of the neighborhood in which they lived as children and as adults, whether they subscribed to ethnic magazines, and their parents' ethnic background.

If race and ethnicity are such vague categories, perhaps the most appropriate definition of a **racial** or **ethnic group** is that its members believe (or outsiders believe) that they share a common national origin, cultural traits, or distinctive physical features. It does not matter whether this belief is based on reality. Even though racial and ethnic classification schemes are obviously problematic, many people argue that they know a "white," "black," "Asian," or "Indian" person when they see one.[3] If they meet someone who does not fit the image, they say, "But you don't look 'black,' 'white,'" and so on. For these people, their vision of human variety is limited by the images of people portrayed on airline posters, magazine ads, and television situation comedies (Houston 1991). The United States, however, is not the only country in the world that classifies, even forces, its population into racial and ethnic categories that are not clear-cut. Next, we consider Germany's classification system, which revolves around ethnicity.

---

[2]In the article "Distribution of the Overseas Chinese in the Contemporary World," Poston and Yu (1990) estimate that approximately 220,000 Chinese live in the Americas, excluding Canada and the United States.

[3]One student in my class noted that we may even believe we know someone's race by listening to his or her voice. The student told me that she didn't even question rap singer Jon B's race, although she had never seen a photo of him. She just assumed he was "black" until she purchased his CD and learned he was "white." The same student also believed that the lead singer in the band Hootie and the Blowfish was "white" because he sounded "white." Later, she learned he was "black."

Race A term that refers to a group of people who possess certain distinctive physical characteristics.

Racial or **ethnic group** A group whose members believe (or outsiders believe) that they share a common national origin, cultural traits, or distinctive physical features.

U.S. in Perspective
The Complexity of North American and "Hispanic" Origins

The U.S. government recognizes two ethnic categories and points out that Hispanics can be of any race. The table shows the racial categories into which Hispanics have been further classified. How many Hispanics in the United States are classified as "white"? Does this number surprise you? Why or why not?

Hispanic Origin

White	11,402,291
Black	645,928
American Indian, Eskimo, or Aleut	148,336
Asian or Pacific Islander	232,684
Other race	94,708

The map and notes show estimates of the numbers of people migrating to the Americas from other regions of the world. They document the multiracial origins of the immigrant American peoples. With regard to Latin America, the map and tables suggest a wider range of ethnic and linguistic origins than the word *Hispanic* implies, in addition to many indigenous peoples.

Based on these data, what are some possible ethnic and linguistic origins of a person who has immigrated to the United States from South or Central America?

Modern emigration to the Americas (in millions)

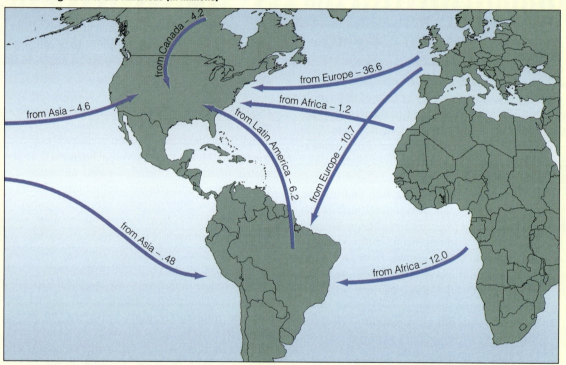

Sources: Segal (1993), Cheliand and Rageau (1995); *Encyclopedia of Latin American History and Culture* (1996); Stalker (1994).

The German System of Classification

Officially, Germany considers itself a "nonimmigration" country (German Government 1998; Jopke 1996). Its constitution does "not seek to increase its national population through naturalization of foreigners" (Holzner 1982:67). Thus all nonethnic Germans who live in Germany, even those who have lived and worked in the same town for 30 years and those who were born there and know only the German language, are considered foreigners.[4]

According to German citizenship and naturalization laws,

- A person's nationality is determined by the nationality of the parents, not the place of birth.

- A child is a German citizen if one parent is a German citizen. This rule applies whether the child is born in Germany or in another country.

- Inheriting the German nationality of one's parents is the only way to become a German citizen automatically or by right.

- Descendants of German farmers and craftspeople who settled in Russia and Romania and other parts of Eastern Europe in the eighteenth century are entitled to citizenship. These ethnic Germans were resettled within their countries by Stalin after Hitler attacked the Soviet Union in World War II. After the war, some of these people fled to West Germany, but the Cold War prevented most of them from leaving.

- The German constitution extends the right to German citizenship to persons who, between January 30, 1933, and May 8, 1945, were deprived of their citizenship on political, racial, or religious grounds.

Taken together the citizenship law established two broad categories of people in Germany: (1) biologically German individuals and (2) foreigners. One major group of people who fall under the category "foreigner" are the 2.6 million guest workers who were recruited to work in what was then West Germany between 1961 and 1973.

"Foreigners" in Germany

When the Soviet Communists erected the Berlin Wall in 1961, West German employers lost a major labor pool.[5]

---

[4]According to Faist and Häubermann (1996), most of the guest workers of the 1960s have since acquired an "unlimited residence permit." Some "temporary labor migrants" can apply for citizenship after 10 to 15 years of residence, provided they have "proven their willingness to assimilate to life, language, and culture of the German demos" (Kurthen 1995:930).

[5]Between 1945, when World War II ended, and 1961, 3 million refugees from East Germany and 9 million from former German territories migrated to West Germany. These migrants provided the labor needed to offset the labor loss due to war casualties and to fuel postwar economic reconstruction. In 1961, when the East–West flow of migrants reached an average of 3,000 per day, the Soviet Communists erected the Berlin Wall—103 miles of heavily guarded, 10-foot-high, steel-fortified concrete—to seal off East Berlin from West Berlin and to stop the flow of people from East to West (McFadden 1989). Also, the Soviets helped construct fortified fences and walls along the additional 860-mile border that separated East Germany from West Germany.

To make up for the shortage, the German government established labor recruitment offices in Turkey, Yugoslavia, Italy, Greece, Portugal, and Spain. From these offices, officials screened male and female job applicants. Those who possessed the needed occupational skills, had no police records, and enjoyed good health were admitted to West Germany as guest workers to live in employer-provided, work-site housing. These people were admitted to West Germany on the assumption that they were temporary workers and would eventually return to their home countries (Castles 1986; Herbert 1995). Officials from labor-exporting countries agreed to this arrangement for three reasons: it helped lower their high unemployment rates; it generated revenue as migrant workers sent home a portion of their earnings; and it gave workers an opportunity to acquire skills and training that would benefit their countries on their return (Sayari 1986). The recruits accepted guest worker status because they believed they could work in West Germany for a few years, save some money, and return home with enough funds to start a business or buy land (Castles 1986). In time, however, workers found that they could not save enough to return home in better financial condition than when they had left. Eventually, many temporary stays became permanent as German employers lobbied to extend guest worker permits. They argued that to be competitive, businesses needed a stable work force, not a temporary one.

After the so-called world oil crisis in 1973, the German government stopped recruiting guest workers and initiated policies, including offers of financial incentives, to encourage long-time guest workers to return home. The number of people classified as foreigners in West Germany continued to increase, however, because workers produced offspring and family members joined workers already there. With the end of the guest worker program, the only legal way for nonethnic Germans without family members already in Germany to enter the country was to seek asylum. Although Germany receives the most asylum requests of any other European Union country, only a small percentage of asylum seekers eventually are granted asylum. Nevertheless, all such applicants receive free housing and meals at state-operated shelters, plus a small stipend to cover other living costs, until their cases are heard (often years later). In the end, the German government has allowed many immigrants to remain indefinitely for humanitarian reasons (for example, until a war ends in their home country), even though their applications for asylum have been rejected.

In 1961, Communist authorities decided to stop the flow of people from the East to the West by erecting the Berlin Wall (left). Nearly 30 years later, demonstrators were eager to have a hand in taking down the infamous Wall while East German border guards looked on placidly (right).

© UPI/Corbis-Bettmann (left); © Reuters/Corbis-Bettmann (right)

Until the fall of the Berlin Wall in 1989 and the subsequent reunification, East Germans, in contrast to their West German counterparts, had little or no interaction with foreigners. The only foreigners who lived in East Germany were approximately 500,000 Soviet troops (including the soldiers' families), students from other Communist countries, and about 190,000 conscript workers who worked in heavy industry and in jobs that East Germans found unattractive (Ireland 1997). The workers lived in hostels on or near work sites and were subject to strict rules: Pregnant workers were deported, only single individuals were permitted to be workers, and they were allowed to visit their home country every five years (Heilig, Büttner, and Lutz 1990).

On November 9, 1989, the Berlin Wall was torn down.[6] The subsequent reunification of East Germany with West Germany followed less than one year later. After the fall of the Wall, 2.2 million ethnic Germans who had been living in Eastern European countries and in the former Soviet Union settled in Germany. The absorption of 17 million East Germans and 2.2 million

ethnic Germans into the West German system was not easy. It became evident that German "blood" did not make someone "German." In fact, many of the foreigners who had lived in Germany for their entire lives were in many ways more "German" than the East Germans and other ethnic Germans.

Today, united Germany has a population of 81.8 million. This figure includes 7.2 million foreigners, most of whom live in large cities in the western part of the country. The most numerous, poorest, and most visibly different foreigners are the 2 million Turks who make up about 28 percent of the foreign-born

[6]In 1988, when Mikhail Gorbachev was named president of the Soviet Union, he instituted policies of glasnost ("openness"), perestroika ("economic restructuring"), and noninterference in Eastern Europe. Gorbachev also announced plans to withdraw a half-million troops and 10,000 tanks from Eastern Europe over two years. These policies led to the overthrow of hard-line Communist regimes in Poland, East Germany, Czechoslovakia, Bulgaria, and Romania. (In addition, they eventually prompted the breakup of the Soviet Union into separate and independent countries.) On November 9, 1989, the Berlin Wall was dismantled.

Table 10.1 Attitudes Toward Foreigners of Various Ethnic Groups Who Live in Germany

Category	Nationality of Foreigner	Attitude Toward Foreigner
Noble foreigners	British, French, Americans, Swedes	Positive
Foreigners	Spaniards, Yugoslavs, Greeks	Neutral
Strange foreigners	Portuguese, Italians, Vietnamese	Neutral, with tendency toward negative
Rejected foreigners	North Africans, Black Africans, Pakistani, Persians, Turks	Rejected by substantial parts of the population

Source: Adapted from Thränhardt (1989:13).

population (German Government 1998; Jones and Pope 1993; Martin and Miller 1990). The Turks, along with North Africans, sub-Saharan Africans, Pakistanis, and Persians (virtually all of whom are Muslims), are considered "rejected foreigners" and are targets of prejudice and discrimination (Safran 1986; see Table 10.1 and Figure 10.2). The irony is that many people from these groups have more in common with the West Germans than the West Germans have in common with the East Germans (see "East Meets West").

Although foreigners who have settled in Germany permanently and who are integrated into German society can become German citizens if they wish to do so, the government sends the clear message that naturalization should be the last stage of social integration, not the first. The rigorous citizenship process discourages foreigners from applying. As a result, Germany has one of the lowest naturalization rates in the world. Germany attempted to change its citizenship laws in 1998 by granting automatic citizenship to any third-generation "foreigner" born in Germany and to any child whose parents have lived in Germany since the age of 14 (Cohen 1998). But the legislation failed to pass.

The problems with racial and ethnic classification go beyond just placing people into official categories. The real problem lies with the status values associated with particular categories. In other words, classification becomes problematic when people in some categories are treated as more worthy than people assigned to other categories. We turn to the concept of minority groups to gain insight into this matter.

Minority Groups

Minority groups are subgroups within a society that can be distinguished from members of the dominant groups by visible and identifying characteristics, including physical and cultural attributes. Members of such subgroups are regarded and treated as inherently different from those in dominant groups. For these reasons they are systematically excluded—whether consciously or unconsciously—from full participation in society and denied equal access to positions of power, prestige, and wealth. Thus members of minority groups tend to be concentrated in inferior political and economic positions and to be isolated socially and spatially from members of the dominant groups.[7]

Sociologist Louis Wirth (1945) made a classic statement on minority groups, identifying a number of essential traits that are characteristic of all minority groups. First, membership is involuntary: As long as people are free to join or leave a group, no matter how unpopular the group, they do not by virtue of that membership constitute a minority. This first trait is quite controversial, because the meaning of "free to join or leave" is unclear. For example, if a very light-skinned person of African and German descent can pass as German, is he or she "free" to leave the African connections in his or her life? Second, minority status is not necessarily based on numbers; that is, a minority may be the numerical majority in a society. The key to minority status, then, is not size but access to and control over valued resources. South Africa is an obvious example: Roughly 11 percent of the South African population, even under the post-apartheid multiracial government,

[7]On the basis of these characteristics, many groups can be classified as minorities, including some racial, ethnic, and religious groups, women, the very old, the very young, and the physically different (for example, visually impaired people or overweight people). Although we focus on ethnic and racial minorities in this chapter, the concepts described here can be applied to any minority.

Minority groups Subgroups within a society that can be distinguished from members of the dominant groups by visible and identifying characteristics, including physical and cultural attributes.

Figure 10.2   Foreign Migrations into Germany (in thousands)

This map shows major flows of immigration into Germany by region of the world. The accompanying chart shows the number of ethnic Germans and foreigners living in Germany.

Ethnic Germans		Other Foreigners	
West Germans		Turks	2,014,000
East Germans	17 million	Yugoslavs	798,000
Repatriated ethnic Germans	2.2 million (since 1989)	Bosnia-Herzegovinians	316,000
European Union Foreigners		Croatians	185,000
Italians	586,000	Romanians	109,000
Greeks	360,000	Moroccans	82,000
Austrians	184,000	Tunisians	26,000
Spaniards	132,000	Ghanans	22,000
Portuguese	125,000	Brazilians	17,000
British	116,000	Afghanistanis	59,000
Dutch	113,000	Chinese	33,000
French	99,000	Indians	35,000
U.S. Foreigners	108,000	Iranians	107,000
Eastern European and former Soviet Union Foreigners		Lebanese	55,000
Poles	277,000	Pakistanis	37,000
Soviets	58,000	Sri Lankans	55,000
Hungarians	57,000	Vietnamese	96,000
Jews	40,000		

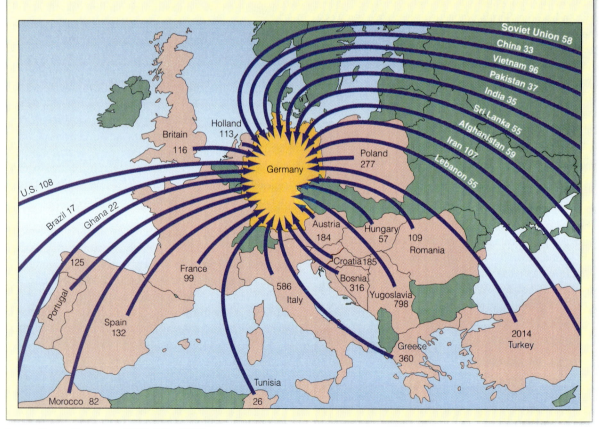

Sources: German Government (1998); Sallen (1995).

East Meets West

The following excerpt from the *New York Times Magazine,* written on the eve of reunification, pointed out the inevitable strains between East and West Germans that arise from a 45-year separation. Five years later, opinion polls indicated that 67 percent of East Germans agree with the statement that "even though the Wall is gone, the Wall in people's minds keeps growing," and East Germans as a group rated the following areas of life better under Communism: security from crime (88%), women's equality (87%), social security (92%), schooling (64%), vocational training (70%), health services (57%), and housing (53%) (Kienbaum and Grote 1997).

It has been forty years. I think it is time to stand back from this social experiment that we call the divided Germany and assess the results.

True, an experiment was the last thing the allies had in mind when they agreed to that border running through the middle of Europe. Call it an involuntary experiment, then, born of the pressures of victory, in which the allies functioned as principal researchers and the Germans as extraordinarily cooperative white mice.

Perhaps it would clarify matters if I mention a related scientific area, one in which I have some expertise as an amateur; I am speaking of research on twins.

Let us assume we are dealing with two hell-raising twins who share a criminal past. Through allied efforts, they are finally forcibly separated and sent to two extremely different boarding schools. One twin grows up in the bracing climate of Western values; first with difficulty and then with growing enthusiasm, he learns to appreciate as basic values democracy, capitalism and individual freedom; and he develops great respect for the Western principal researcher.

The other twin has quite a different fate. He is often beaten and brutalized, and finally learns just as assiduously the basic values of Eastern culture: "solidarity," "social commitment," "passion for socialism," and of course "eternal friendship" for the Eastern principal researcher.

Let us further assume that a wall is constructed between the twins and an odd system of visitation rights is established. The Western twin can move in any direction he likes, including east; he can visit his brother on the other side, chat with him and compare experiences, bring him presents, then return to the Western half to have dinner in a French restaurant.

The Eastern twin, on the other hand, has some freedom of movement north and south, and to the east has access to an almost unlimited recreational area (which, however, has only recently come to be considered a place worth visiting). But access to the West is blocked. There is the famous forbidden door, a good 1,400 kilometers wide. It still opens legally only in exceptional cases, and after a wait of at least two years; anyone who does not wish to wait that long must take his life in his hands and jump, or dig.

Let us further assume that the twin in the West, thanks to assistance such as the Marshall Plan and the Western market economy, gradually gets rich. His twin brother, meanwhile, not only has to pay war debts to the far poorer principal researcher in the East; he also has to adopt the researcher's inefficient economic system.

At least one result we can safely predict—the twin in the East will fall victim to a psychological law. Every wall in the world, whether German or Chinese, begs to be overcome. Also, because he finds himself in the awkward situation of having to wait for his Western sibling to come to him, a certain reproachful attitude forms. "That guy over there really could come to see me more often," thinks the Eastern twin. "At the very least, he could write or phone regularly. And he could be a bit more generous, because it's turned out that he has the better deal. Not that he has done anything to deserve it, by the way; it was pure luck that he happened to be living on the right side of the Elbe at the right time.

"But now his success has gone to his head. Instead of sharing, he acts as though he had all the talent and claims he works harder. He can't fool me; I know him, we started out under the same roof, and he's just as industrious and just as lazy as I am. He's gotten pretty

arrogant, even self-righteous. Actually, he's still living off the misery of the world's poor, whom he exploits, but he doesn't want to hear that.

"Well, if he doesn't have a conscience, at least he could show a little family feeling, a little interest in his relatives in the East, at least listen to us—is that asking too much? He's gained an awful lot of weight, by the way; even if he is rich, that doesn't look like happiness to me."

Meanwhile, the Western twin is concocting a monologue of his own. He feels pressured by what he calls his relative's "eternal posture of expectation."

"I can see that the poor guy behind his wall doesn't have an easy time of it," thinks the Western twin. "But these demands, these unspoken reproaches really cast a pall over our relationship. God knows I'm happy to give him things, but it's no fun bringing a present when the other person always expects you to. Those people over there seem to think that cars and color televisions grow on trees. But we're not born with a Mercedes; you have to earn it; you have debts, interest payments—concepts my brother knows only by hearsay. . . .

"I'd like to explain this, but he won't listen; he just talks and talks. Of course, it's not his fault that he always has to stand in line to buy oranges; but at least he should admit that he bet on the wrong horse, that the socialist economic system is a disaster—no one's criticizing him personally. The problem is—and it's so German of him—he takes any kind of criticism personally. Instead of agreeing with me, he tries to spoil my success. He claims to be an idealist—I'm glad to let him call himself that, because the poor fellow has a lot to compensate for. . . . But he accuses me of being a conspicuous consumer and a 'conformist'—a compliment I'm glad to return by pointing out that his so-called 'socialist' or even 'revolutionary' virtues are all a pose; I know him, after all.

"Sometimes I feel downright relieved when the visit is over; there's an unpleasant tension between us that we really should talk about some day. Next time."

And what happens when the Eastern twin comes west? Every immigrant is expected to declare

continued

still controls access to valued resources (see Chapter 9). A third characteristic, and the most important, is nonparticipation by the minority group in the life of the larger society. That is, minorities do not enjoy the freedom or the privilege to move within the society in the same way as members of the dominant group do. Sociologist Peggy McIntosh (1992) identifies a number of privileges that most members of the dominant group take for granted, including the following:

- I can, if I wish, arrange to be in the company of people of my race [or ethnic group] most of the time.

- If I should need to move, I can be pretty sure of renting or purchasing housing in an area which I can afford and in which I would want to live.

- I can go shopping alone most of the time, fairly well assured that I will not be followed or harassed by store detectives.

- I can be late to a meeting without having the lateness reflect on my race [or ethnicity].

- Whether I use checks, credit cards, or cash, I can count on my skin color not to work against the appearance that I am financially reliable. (pp. 73–75)

The final and most troublesome characteristic of a minority group is that people who belong to such a group are "treated as members of a category, irrespective of their individual merits" (Wirth 1945:349) and often irrespective of context. In other words, people outside the minority group focus on the visible characteristics that identify someone as belonging to a minor-

ity. These visible characteristics become the focus of interaction, as in the scene described here, in which a woman can focus only on the physical features of the people she encounters to the neglect of what is going on around her:

"It obsesses everybody," declaimed my impassioned friend, "even those who think they are not obsessed. My wife was driving down the street in a black neighborhood. The people at the corners were all gesticulating at her. She was very frightened, turned up the windows, and drove determinedly. She discovered, after several blocks, she was going the wrong way on a one-way street and they were trying to help her. Her assumption was they were blacks and were out to get her. Mind you, she's a very enlightened person. You'd never associate her with racism, yet her first reaction was that they were dangerous." (Terkel 1992:3)

The characteristics that Wirth identifies as associated with minority group status indicate that minorities stand apart from the dominant culture. Some people argue that minorities stand apart because they do not wish to assimilate into mainstream culture. To assess this claim, we turn to the work of sociologist Milton M. Gordon, who has written extensively on assimilation.

Perspectives on Assimilation

Assimilation is a process by which ethnic and racial distinctions between groups disappear. Two main types of assimilation exist: absorption assimilation and melting pot assimilation.

Absorption Assimilation

In absorption assimilation, members of a minority ethnic or racial group adapt to the ways of the dominant group, which sets the standards to which they must

Assimilation A process by which ethnic and racial distinctions between groups disappear.

adjust (Gordon 1978). According to Gordon, absorption assimilation has at least seven levels. That is, an ethnic or racial group is completely "absorbed" into the dominant group when it goes through all of the following levels:

1. The group abandons its culture (language, dress, food, religion, and so on) for that of the dominant group (an action known as acculturation).

2. The group enters into the dominant group's social networks and institutions (structural assimilation).

3. The group intermarries and procreates with members of the dominant group (marital assimilation).

4. The group identifies with the dominant group (identification assimilation).

5. The group encounters no widespread prejudice from members of the dominant group (attitude receptional assimilation).

6. The group encounters no widespread discrimination from members of the dominant group (behavior receptional assimilation).

7. The group has no value conflicts with members of the dominant group (civic assimilation). (See "Liberal and Illiberal Citizens.")

Gordon advances a number of hypotheses to explain how the various levels of assimilation relate to one another. First, he maintains that acculturation is likely to take place before the other six levels of assimilation are achieved. Gordon also states, however, that even if acculturation is total, it does not always lead to the other levels of assimilation.

In addition, Gordon proposes that a clear connection exists between the structural and marital levels of assimilation. That is, if the dominant group permits people from ethnic and racial minority groups to join its social cliques, clubs, and institutions on a large enough scale, a substantial number of interracial or interethnic marriages are bound to occur:

> If children of different ethnic backgrounds belong to the same play group, later the same adolescent cliques, and at college the same fraternities and sororities; if the parents belong to the same country club and invite each other to their homes for dinner; it is completely unrealistic not to expect these children, now grown, to love and to marry each other, blithely oblivious to previous ethnic extraction. (pp. 177–178)

Of the seven levels of assimilation, Gordon believes that the structural level is the most important—if it occurs, the other levels of assimilation inevitably will follow. Yet, the structural level is very difficult to achieve in practice. Why? Because members of ethnic or racial minorities are denied easy and comfortable access to the dominant group's networks and institutions. In fact, all of the important and meaningful primary relationships that are close to the core of personality and selfhood are confined largely to people of the same racial or ethnic group:

> From the cradle in the sectarian hospital to the child's play group, the social clique in high school, the fraternity and religious center in college, the dating group within which he [or she] searches for a spouse, the marriage partner, the neighborhood of his [or her] residence, the church affiliation and the church clubs, the men's and the women's social and service organizations, the adult clique of "marrieds," the vacation resort, and then, as the age cycle nears completion, the rest home for the elderly and, finally, the sectarian cemetery—in all these activities and relationships which are close to the core of personality and selfhood—the member of the ethnic group may if he wishes follow a path which never takes him across the boundaries of his [or her] ethnic structural network. (p. 204)

This scenario especially characterizes the primary group relations of **involuntary minorities,** ethnic and racial groups that did not choose to be a part of a country. Rather, they were forced to become part of a country by slavery, conquest, or colonization. Native Americans, African Americans, Mexican Americans, and native Hawaiians are examples of involuntary minorities in the United States. Unlike **voluntary minorities,** whose members come to a country expecting to improve their way of life, members of involuntary minorities have no such expectations. Their forced incorporation involves a loss of freedom and status (Ogbu 1990).

A goal of absorption assimilation describes the mindset by which West Germany approached unification with East Germany. In fact, it can easily be argued that West Germany "took over" East Germany rather than merging with it (Kienbaum and Grote 1997). Put another way, the smaller and poorer East was swallowed by the larger and richer West, whose experts and

Involuntary minorities Ethnic and racial groups that were forced to become part of a country by slavery, conquest, or colonization.

Voluntary minorities Members of racial or ethnic groups that came to a country expecting to improve their way of life.

Liberal and Illiberal Citizens

Since unification, West German social scientists (often with the financial support of government) have carried out numerous studies of the political culture of East Germans. During the Seventies and Eighties, West German social scientists (often with the financial support of government) carried out numerous studies of the political culture of foreigners. These studies share unmistakable similarities and produce virtually identical images of their subjects. First, foreigners and East Germans are persons who have been socialized in authoritarian political cultures outside the Federal Republic of Germany (FRG). Second, they are persons experiencing rapid social change. Third, they suffer from a number of social and psychological pathologies, one of which is a tendency toward anti-democratic political extremism. And fourth, they represent a threat to the "basic free democratic order" of the FRG and must be re-socialized to respect the liberal democratic values of West German political culture before they can be fully trusted.

West German studies invariably note the major differences between the predominant norms and values in the FRG and in the German Democratic Republic (GDR) and emphasize the great difficulties East Germans face in adapting to their new society. "There exists a significant discrepancy," concluded a study commissioned by the Federal Agency for All-German Efforts, "between the behavioral repertoire of the erstwhile GDR citizens and the new challenges. The mental demands which pluralistic democracy and social market economy place on social actors are totally different." Another study implored readers to keep in mind "that 16 million GDR citizens have with the regime and system transformation lost their identity overnight." Both studies conclude that East Germans suffer from various social and psychological problems such as a "collapse of a holy, ordered world," "anomie," "culture shock," and "identity vacuum" which stem from feeling like "a foreigner in one's own land." They exhibit such symptoms as "depression," "GDR nostalgia," "self-doubt," "suicide," "excessive consumption," and "criminality."

But the most disturbing deviance is a tendency to support political extremism, far right or left. Thus countless studies interpret the rise in neo-Nazi activities and organizations in East Germany as a reaction against rapid social change combined with deficient appreciation of liberal democratic norms and values. For example, Hans-Gerd Jaschke writes: "the reasons for the increase in the number of right-wing extremists' seem to be . . . the feeling of social discrimination, the lack of other traditionally established political orientations, the lack of experience in dealing with strangers and practical everyday problems." "The sum total of all the frustrations," concludes another study, "becomes vented in slogans like 'Germany for the Germans—Foreigners out!' Violent hooligans and applauding observers do not want to accept that indigent refugees, victims of political persecution or religious minorities are being integrated into the commonwealth. Instead, the alleged uniform, homogeneous society is preferred." Others see persistent support for the Party of Democratic Socialism (PDS) as evidence that the "authoritarian-repressive structures of the society have not yet really been overcome." The same holds true for "GDR nostalgia." "This can be interpreted as a 'new myth of the social welfare in the former GDR' and hints at a political instability in the eastern part of Germany, comparable to the one in the FRG in the 1960s." The conclusion to be drawn from these various studies is that East Germans have yet to become genuine liberal democrats.

officials dictated the social, economic, and cultural terms of reunification:

> In order to catch up, eastern Germany had to utilize western know-how, which meant importing large armies of experts. Not always with the greatest of tact, these experts would then articulate how deplorable conditions in the east were, typically adding a healthy dose of self-righteousness and

condescension. The fact that government personnel assigned to the east receives hardship pay has not helped, either. All of this has led to the feeling among eastern Germans that they have been colonized and dismissed as inferior. (Kienbaum and Grote 1997:3)

Melting Pot Assimilation

Assimilation need not be a one-sided process in which a minority group disappears, or is absorbed, into the dominant group. Ethnic and racial distinctions can also disappear through **melting pot assimilation** (Gordon 1978). In this process, the groups involved accept many new behaviors and values from one another. This

Melting pot assimilation A process of cultural blending in which the groups involved accept many new behaviors and values from one another.

This image of East Germans mirrors the one of foreigners produced by social scientists during the Seventies and Eighties. The many scholars who moved into so-called "foreigner research" in those decades diagnosed foreigners with virtually the same sociopsychological illnesses and pathologies discussed above. Moreover, the root cause was understood to be the rapid transformation from traditional agricultural societies to the modern industrial society of the FRG. Thus Ursula Neumann in a prominent study of migrant families argued that their tendency to cling to tradition in Germany:

is not to be understood as a natural continuation of the lifestyle in the homeland, rather as a defense against the changed environment. The confrontation with the divergent ways of the surrounding world creates in every case a sense of uncertainty, a strain on the personality. [This leads] to signs of retreat and compensation, such as exaggeration of traditional norms and values, idealization of the homeland, avoidance of contact with the German environment . . .

The re-socialization of East Germans can also be expected to take a long time, West Germans insist. Hans-Joachim Maaz writes in his study of East German political culture: "Democracy cannot . . . be put on like a coat, rather it must take root in the minds and hearts. . . . Democracy can only prevail . . . if it begins in the souls of men." Katharina Belwe argues that "it will take a longer individual and collective learning process before the East German citizens can be made equal to the citizens of the old federal states in competition for jobs but also political and social positions."

Such rationalizations of postponed equality for East Germans sound unmistakably similar to those resident aliens have heard for years. Delaying full sovereignty to newcomers seems natural to West Germans. After World War II, they were welcomed into the western alliance of liberal democratic nation-states, but not as fully sovereign members. As the Allies built the Federal Republic and integrated it into the western alliance, they only gradually relinquished sovereignty to West Germany. They first had to be satisfied not only that liberal democratic structures and institutions were in place, but that a liberal democratic political culture was widely internalized by the citizenry. West Germans are simply doing to East Germans and foreigners what was once done to themselves. And since the latter is generally understood as a grand success story, there seems no good reason not to repeat (actually extend) the process of converting more newcomers to the nation-state to its core liberal democratic values. This mode of thinking receives clear expression in the following passage of a major study on East German political culture commissioned by the Federal Ministries for Technology and Research and for Labor and Social Order:

The development of a stable democracy needs time. So it was after the Second World War in West Germany. Democratization is a long-term process. This is true in particular for the development of a democratic political culture. . . . The population of the former GDR initially oriented itself mainly around the living standards of the old Federal Republic . . . and not very much around the democratic values. The development of a democratic value system will demand considerable time. These developments thoroughly correspond to the experiences in the old Federal Republic following the war.

Source: "Liberal and Illiberal Citizens," excerpted from *Germany's New Aliens: The East Germans* by Peter O'Brien. Pages 554–555 in *Eastern European Quarterly* 30(4):449–559. Copyright © 1996 by East European Quarterly.

exchange produces a new cultural system, which is a blend of the previous separate systems. Melting pot assimilation becomes total when significant numbers of people from each ethnic and racial group take on cultural patterns of the other, enter each other's social network, intermarry and procreate, and identify with the blended culture.

The melting pot concept can be applied to the various African ethnic groups imported to the United States as slaves. They were "not one but many peoples" (Cornell 1990:376), who spoke many languages and came from many cultures. Slave traders capitalized on this diversity: "Advertisements of new slave cargoes frequently referred to ethnic origins, while slaveowners often purchased slaves on the basis of national identities and the characteristics they supposedly indicated" (Cornell 1990:376; see also Rawley 1981). Although slave owners and traders acknowledged ethnic differences among Africans, they treated Africans from various ethnic groups as belonging to the same category of people—slaves. Because slave traders sold and slave owners purchased individual human beings, not ethnic groups, this treatment had the effect of breaking down ethnic concentrations. In addition, slave owners tended to mix together slaves of different ethnic origins so as to decrease the likelihood of the slaves' plotting a rebellion. To communicate with each other, the slaves

Although many Turkish people have been living in Germany for 40 years, they may not have been assimilated into German society.

© Regis Bossu/Sygma

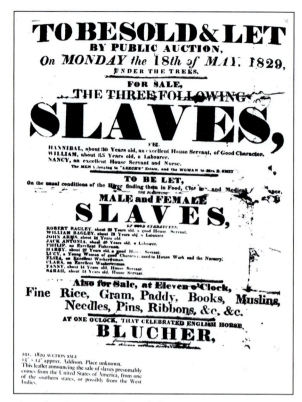

Throughout history, members of ethnic or racial groups have become involuntary minorities through conquest, annexation, or slavery.

© Corbis/Bettmann

invented pidgin and Creole languages. In addition to developing a new language, the slaves created a common and distinctive culture based on kinship, religion, food, songs, stories, and other features. The harsh conditions of slavery, in combination with the mixing together of people from many ethnic groups, encouraged slaves to borrow aspects of each other's cultures and to create a new, blended culture.

In Germany (which, as noted earlier, called itself a nonimmigration country until October 1998) and the United States (which calls itself a melting pot culture), assimilation has been mainly a one-sided process— that is, newcomers are expected to adapt to the dominant culture. In the case of Germany, the categories or labels that the German people use to refer to Turks and other foreigners help us to see why assimilation is difficult. Names assigned to nonethnic Germans have included "foreign workers," "guest workers," "foreigners," "foreign employees," "migrants," and "foreign co-citizens" (Mandel 1989; White 1997). All of these terms reinforce the belief that "foreigners," no matter

how many years they have lived in the country, remain outsiders.

When most people in majority groups think of assimilation, they do not envision it as a process of mutual exchange in which members of the dominant group identify with a blended culture (Opitz 1992a). Rather, they think of it as a one-way process whereby a minority group comes to fit into the dominant culture without disturbing it. As one Turkish student in Germany described it:

> What [Germans] want is that there be no more women with head scarves. What they want is that, when you are invited [to dinner], you don't first ask, "Is there any pork?" [or say] "I'm sorry, I don't eat [pork]." (White 1997:759)

We do not mean to say that the dominant culture has not been shaped and influenced by minority cultures; rather, members of the dominant group more often than not fail to acknowledge others' contributions to the society in which they live or they come to believe that nonminority accomplishments/contributions are not relevant to their lives.[8] Stratification theory is one major approach to understanding forces that work against assimilation (both the absorption and melting pot forms) between dominant and minority groups.

Stratification Theory and Assimilation

Stratification theory is guided by the assumption that racial and ethnic groups compete with one another for scarce and valued resources. As a whole, the dominant group retains an advantage over competing groups, because its members are in a position to preserve the system that gives them these advantages. Stratification theorists focus on the mechanisms that people in the dominant group employ to preserve this inequality (Alba 1992). These mechanisms include racist ideologies, prejudice and stereotyping, discrimination, and institutional discrimination.

Racist Ideologies

An **ideology** is a set of beliefs that are not challenged or subjected to scrutiny by the people who hold them. Thus ideologies are taken to be accurate accounts and explanations of why things are as they are. On closer analysis, however, ideologies are at best half-truths, based on misleading arguments, incomplete analysis, unsupported assertions, and implausible premises. They "cast a veil over clear thinking and allow inequalities to persist" (Carver 1987:89–90). One such ideology is racism. People who embrace this ideology believe that something in the biological makeup of an ethnic or racial group explains and justifies its subordinate or superior status.

Racist ideologies are structured around three notions: (1) People can be classified into categories on the basis of physical characteristics; (2) a close correspondence exists between physical traits and characteristics such as language, dress, personality, intelligence, and athletic capabilities; and (3) physical attributes such as skin color are so significant that they explain and determine behavior and inequalities. Any racial or ethnic group may use racist ideologies to explain their own or another group's behavior. (See "Is There Such a Thing as Black Racism?") One example of a racist ideology is the hypothesis offered by former Los Angeles police chief Daryl Gates to explain why so many blacks have died from restraining chokeholds: their "veins and arteries do not open up as fast as they do on normal people" (Dunne 1991:26). The fact behind these statistics is that many police officers, black and white, tend to handle black suspects more harshly than they do their white counterparts. Another example of how racist ideology shapes thinking comes from students. Many students in the author's class work as waiters and waitresses (or waitpersons) and claim that "black" and "white" customers tip differently. In particular, they note that "black" customers do not tip as well as "white" customers do. As one student wrote, "We talk about it at work all the time; it's that big an issue. One co-worker believes it's 'black people's way of paying us back for enslaving them in historical times. They must get off seeing us wait on them and they rub it in by stiffing us.'" To the author's knowledge, no study has ever examined "black/white" tipping—but there is a possible explanation for differences in tipping, if they exist. Think about this question: "Who are the best tippers?" When members of the author's class were asked this question, practically in unison they answered, "former or current waiters and waitresses." According to U.S. Bureau of Labor Statistics data, only 4.8% of waiters and waitresses are classified as "black" and only 2.4% of bartenders are classified as such. A more likely explanation for differences in tipping is that customers classified as "black" have not learned the "norms" about what constitutes a good tip[9] because, until very recently, they have not been allowed to work in these kinds of restaurant jobs.

For their part, German Nazis rationalized the extermination and persecution of Jews, Gypsies, Poles,

---

[8]One "white" student in my class wrote that she felt she could not really share in the Martin Luther King Day celebration. As you read her thoughts, consider the following question: "How did she come to feel this way?"

> I remember a time I went with my Dad to a Martin Luther King, Jr., Day candlelight vigil/march. I remember being one of only a very few white people in attendance. I was pretty young, but seeing the looks on these peoples' faces and the feelings they were showing, I realized the strength of this man. Some of the people were crying, some were singing, and some, like me, were just thinking and walking. I began to think how much this man accomplished. I remember I wanted to cry. I also remember that I could not cry. I could not cry because I felt that I was not allowed because I was white. I felt that since I was not black I could not "fully" participate in this day. I felt that Dr. King could never mean for me what he means to "black" people.

[9]As a former waitress, I know that I am always telling relatives who have never worked as waitpersons to leave bigger tips.

> **Ideology** A set of beliefs that are not challenged or subjected to scrutiny by the people who hold them.

U.S. in Perspective

Is There Such a Thing as Black Racism?

Black racism. For some blacks it is a laughable oxymoron. ("How can the victims of racism be racist?") For some whites it is an excuse for doing little to reduce the inequalities that still plague the country. ("See, blacks are just as bad as they say we are.")

The recent diatribe against Catholics, Jews and homosexuals (among others) by Khalid Abdual Muhammad, a member of the hierarchy of the Nation of Islam, has focused attention on the disturbing question of how much racism permeates black America. Indeed, the public fascination with the Nation of Islam—a fringe group with negligible political and economic power—is partly explained by white fear that its views are shared by other blacks. Looking at the cleaning women, the bank tellers, partners in their law firm, some whites silently ask, "Do you agree with Louis Farrakhan?"

Until recently, the question of how blacks view whites has been largely unexplored territory for scholars and sociologists. Why have sociologists not probed the black psyche to see how much racial animus resides there? Perhaps because they are afraid of what they might find; perhaps because they feel that since whites are the dominant group, their views are more important; perhaps because they feel that blacks have no opinions that whites are bound to respect.

But recent work by pollsters, sociologists and political scientists has uncovered evidence of specific negative attitudes that blacks may harbor toward whites in general and Jews in particular. Seeking responses to a series of assertions (Jews tend to stick together more than other Americans; Jews wield too much power on Wall Street; Jews are more willing than others to use shady practices to get what they want), a 1992 survey by the Anti-Defamation League of B'nai B'rith found that blacks are more than twice as likely as non-Jewish whites to hold stereotypical views of Jews.

Preliminary findings in a study to be published soon by the National Conference of Christians and Jews found that blacks feel that whites think themselves superior, and that they do not want to share power and wealth with nonwhites. The poll found that black antipathy toward whites is far greater than that of other minorities, such as Latinos and Asian-Americans.

"Of course, there's black racism," said Roger Wilkins, a professor of history at George Mason University, who is black. "I think that any time that you say that a whole group of people is this or that, fill in the blanks, and the blanks are all negative adjectives, and then you go on to prove your conclusions by telling stories that support your use of those adjectives, that's racism."

But, as another black writer, Ellis Cose, points out in his 1993 book, *The Rage of a Privileged Class* (Harper-Collins), the harboring of negative views by blacks does not tell the whole story. Surveys of blacks also find that many give American whites, both Jew and gentile, high marks for intelligence and for creating a democratic society. Viewed in this light, the results of the surveys may be less an affirmation of racism than an expression of envy and a plea to share power.

Racism among blacks, whatever its extent, has been like magma bubbling under black nationalism. Like other nationalist movements—Zionists, the Quebecois in Canada, Serbs seeking a "Greater Serbia"—those embraced by some blacks have used ethnic solidarity as a political organizing tool.

For some mainstream black politicians, black anger toward whites presents both opportunity and temptation. Just as white politicians often take advantage, some discreetly and others more cynically, of racist attitudes among whites, so black politicians are aware of hostility among *their* constituents. And when they tap into it, they confront a question of conscience: When do they cross the line into demagoguery?

A call for unity is by implication a call for exclusivity. The desire to maintain that solidarity makes it difficult for many blacks to denounce black racism. But the censure of Mr. Muhammad by people like the Rev. Jesse Jackson, Benjamin Chavis, executive director of the National Association for the Advancement of Colored People, and Representative Kweisi Mfume of Maryland, chairman of the Congressional Black Caucus, shows that difficult does not mean impossible. Indeed, black officials have shown more willingness to move beyond solidarity than have members of the United States Senate who met with silence the remarks of Senator Ernest F. Hollings when he jokingly implied in December, while commenting on trade talks in Switzerland, that African heads of state are cannibals. A Hollings aide later said the Senator meant no offense.

There are those who argue that even if blacks espouse racist views, it makes little difference since they lack the power to turn those views into action. Such a rationale was articulated by Mr. Farrakhan at his recent press conference: "Really, racism has to be coupled with a sense of real power."

Mr. Farrakhan may have a point. For all the racial bombast by the Nation of Islam and other strident black nationalists, America has had no episodes of large-scale coordinated attacks by blacks against whites. But individual acts like the beating of the trucker Reginald O. Denny during the Los Angeles riots demonstrate that black bigotry can maim people and poison that atmosphere.

continued

U.S. in Perspective

Is There Such a Thing as Black Racism? *continued*

There is no denying that black bigotry is fed by social and economic conditions. Even discounting, as some whites do, the country's history of slavery, segregation and discrimination, there are trends that ensure black resentment of whites will not diminish soon, and that black demagogues will find allies among the disaffected.

Black poverty rates remain stubbornly high. Progress toward closing the income gap between the races stalled in 1975, according to census data, and since then the gap has widened. (For instance, male black college graduates who entered the job market in 1971 earned on average 2 percent more than their white counterparts, but by 1989 the same blacks were earning 25 percent less on average than white men who graduated in 1971.)

Blacks remain the most segregated group, according to the census. Whether it is the result of whites making blacks feel unwelcome, or of declining faith in integration among some blacks, black Americans remain huddled in their corners, afraid that any attempt to reach out will tar them with labels: bleeding heart, Uncle Tom. "We don't live next door to each other,' said Lani Guinier, the law school professor whose nomination as Assistant Attorney General for civil rights was withdrawn in a controversy over her voting-rights theories. "No one talks to each other. Racial stereotypes fester in isolation."

Source: "Behind a Dark Mirror: Traditional Victims Give Vent to Racism," by Steven A. Holmes, p. 4E in *The New York Times* (February 15). Copyright © 1994 by The New York Times Company. Reprinted by permission.

Russians, blacks, and others "deemed unfit to breed" (including disabled, homosexual, or mentally ill persons) with the racist ideology that the Aryan race was superior to other races. According to this ideology, the Jews were responsible for Germany's defeat in World War I and did not even belong to a race of people; they were "nonhuman; blood suckers, lice, parasites, fleas . . . organisms to be squashed or exterminated by chemical means" (Fein 1978:283).

The premise of racial superiority also lies at the heart of other rationalizations used by one group to dominate another. Sociologist Larry T. Reynolds (1992) observes that race as a concept for classifying humans is a product of the 1700s, a time of widespread European exploration, conquest, and colonization that did not begin to subside until the end of World War II. Racist ideology also supported Japan's annexation and domination of Korea, Taiwan, Karafuto (the southern half of the former Soviet island Sakhalin), and the Pacific Islands prior to World War II. Both the Japanese and Europeans used racial schemes to classify the people they encountered; the idea of racial differences became the "cornerstone of self-righteous ideology," justifying their right by virtue of racial superiority to exploit, dominate, and even annihilate conquered peoples and their cultures (Lieberman 1968).

Prejudice and Stereotyping

A **prejudice** is a rigid and usually unfavorable judgment about an outgroup (see Chapter 5) that does not

Among the victims of the Nazi ideology of a "pure race" were Gypsies like those being held at this concentration camp in Belzec, Poland. Historians estimate that between 20 and 50 percent of European Gypsies died in the "Gypsy Holocaust."

Archives of Mechanical Documentation, Warsaw, Poland. Courtesy of the United States Holocaust Memorial Museum.

Prejudice A rigid and usually unfavorable judgment about an outgroup that does not change in the face of contradictory evidence and that applies to anyone who shares the distinguishing characteristics of that group.

| Table 10.2   Prejudice in the United States, 1997 |

This table shows findings from the Gallup Poll Social Audit on Black/White Relations in the United States in which 3,036 adults (1,269 self-identified as black and 1,680 identified as white) were asked questions about feelings of prejudice. What percentage of whites rated their prejudice toward blacks as being a 5 or higher? What percentage of whites do black respondents believe feel prejudice toward blacks that is a 5 or higher? How do white respondents assess "other whites'" prejudice against blacks?

	Mean: 0 to 10 Scale	% 0 (No Prejudice)	% 1 or 2 (Low Prejudice)	% 5 or Higher (High Prejudice)
Blacks (prejudice against whites)	2.2	41	21	22
Whites (prejudice against blacks)	2.1	27	36	14
Blacks (white prejudice against blacks)	3.9	21	13	44
Whites (other whites' prejudice against blacks)	4.2	7	13	44

Note: On the 10-point scale, 0 means not at all, and 10 means extremely prejudiced.
Source: The Gallup Poll Organization (1997).

change in the face of contradictory evidence and that applies to anyone who shares the distinguishing characteristics of that group (see Table 10.2). Prejudices are based on **stereotypes**—exaggerated and inaccurate generalizations about people who are members of an outgroup—and they are applied to everyone who possesses that distinguishing characteristic on sight. For instance, many Germans stereotype Turks as backward people who come from villages where "their houses are huts, the streets are unpaved, . . . and where there are goats, sheep, and chickens, and fields of corn, wheat, and melons" (Teraoka 1989:111).

Stereotypes "give the illusion that one [group] knows the other" and "confirms the picture one has of oneself" (Crapanzano 1985:271–272). Germans who claim that Turks are backward are saying that they, by definition, are not. In addition, Germans who hold such stereotypes about Turks fail to see Turks in Germany as a highly heterogeneous group with regard to social origin, cultural norms, and length of residence in Germany (Teraoka 1989).[10]

Stereotypes are supported and reinforced in a number of ways. In **selective perception**, prejudiced persons notice only those behaviors or events that support their stereotypes about an outgroup. In other words, people experience "these beliefs, not as prejudices, not as prejudgments, but as irresistible products of their own observations. The facts of the case permit them no other conclusion" (Merton 1957:424).

In a poignant editorial, columnist Clarence Page (1990) reflects on the fact that stereotypes about black teenagers and violence can cause people to fear contact with any black teenager:

As much as everyone tells us how cute and handsome our little boy is now (and they're right!), I also know that someday he will be a teen-ager and many adults have an outright fear of black male teen-agers.
It's getting to be an old story. See a group of small black children coming toward you on the street and you want to scoop the little darlings up and hug them. See the same kids approaching a few years later as teen-agers in sneakers and fashionably sculpted hair-cuts and, depending on who you are, you very well might want to cross the street, thinking to yourself about how you can't be too careful.
I dread the experiences of a family friend who is trying to help his seventh-grade son cope with his transformation from the cuddly cuteness of childhood

Stereotypes Exaggerated and inaccurate generalizations about people who are members of an outgroup.

Selective perception The process in which prejudiced persons notice only those behaviors or events that support their stereotypes about an outgroup.

[10]For example, if we know that someone comes from Turkey, it does not follow that he or she is a Turk in an ethnic, linguistic, or religious sense. Turkey's population includes 3 to 10 million Kurds, Arabs (who migrated from Egypt and North Africa in the 1930s), Armenians, Greeks, and Jews. Even the so-called Turkish population contains at least three distinct regional groups.

Young Turkish women in Germany, one in the modern dress of a devout Muslim (right), the other "integrated" in dress, though she probably would not eat pork (left). The young Turks desire German acceptance of the full range of their dress and behavior styles.

© Ergun Cagatay/Gamma Liaison

to the imposing presence of a full-sized African American teen-ager. Formerly greeted with delight, he now is sometimes greeted with apprehension. Positive thinker that I am, I cannot help but think negatively about the damage done by the negative signals of fear we send young black males, simply because some have sinned. (p. A10)

Stereotypes also persist in another way: When a prejudiced person encounters a minority person who contradicts the stereotype, he or she sees that person as an exception. The encounter with a minority group member who is "different" only serves to reinforce the prejudice person's stereotypes. In addition, prejudiced people use facts to support their stereotypes. A prejudiced person can point to the small number of black or other minority quarterbacks, pitchers, and baseball managers as evidence that they do not possess the leadership qualities to be successful at such positions. It does not occur to the prejudiced person that prejudice and discrimination by owners and general managers may actually explain the small numbers. In a similar vein, a prejudiced person may point to the fact that "black" athletes dominate sports such as basketball and track, claiming it as evidence of natural leaping ability and quickness. At the same time, this person does not use the same kind of logic to explain why "white" athletes dominate sports like gymnastics, golf, and hockey.

Finally, prejudiced individuals keep stereotypes alive when they evaluate the same behavior differently at different times, depending on the person who exhibits that behavior (Merton 1957). For example, incompetent behavior on the part of racial and ethnic minority mem-

bers often is attributed to innate flaws in their biological makeup; in contrast, incompetence exhibited by someone from the dominant group is almost always treated as an individual issue. Similarly, prejudiced people treat certain ideas, when expressed by members of a minority group, as more threatening than when those ideas are expressed by members of a dominant group. Spike Lee drew considerable criticism for ending his film *Do the Right Thing* with a quote by Malcolm X: "I don't even call it violence; when it's self-defense, I call it intelligence." From Lee's point of view, such criticism meant only that "we're not allowed to do what everyone else can. The idea of self-defense is supposed to be what America is based on. But when black people talk about self-defense, they're militant. When whites talk about it, they're freedom fighters" (McDowell 1989:92).

Discrimination

In contrast to prejudice, **discrimination** is not an attitude but a behavior (see Table 10.3). It includes intentional or unintentional unequal treatment of individuals or groups based on attributes unrelated to merit, ability, or past performance. Discrimination is behavior aimed at denying people equal opportunities to achieve valued goals (education, employment, health care, long life) and/or blocking their access to valued goods and services.

Robert K. Merton has explored the relationship between prejudice (the attitude) and discrimination (the behavior). He distinguishes between two types of individuals: the nonprejudiced (who believe in equal opportunity) and the prejudiced (who do not). Merton asserts that people's beliefs about equal opportunity are not necessarily related to their conduct, as he makes clear in his four-part typology (see Figure 10.3).

Nonprejudiced nondiscriminators (all-weather liberals) accept the creed of equal opportunity, and their conduct conforms to that creed. They represent a "reservoir of culturally legitimized goodwill" (Merton 1976:193). In Germany, for example, some religious, industrial, and political groups have formed a coalition and have launched Action Courage, a national

Discrimination Intentional or unintentional unequal treatment of individuals or groups on the basis of attributes unrelated to merit, ability, or past performance.

Nonprejudiced nondiscriminators (all-weather liberals) Persons who accept the creed of equal opportunity, and their conduct conforms to that creed.

Table 10.3   Percentage of Blacks Experiencing Discrimination Within the Last 30 Days, 1997

The 1997 Gallup Poll described in Table 10.2 asked the 1,269 black respondents if they had experienced discrimination within the past 30 days. In which areas are blacks most likely to experience discrimination? Least likely? Which age group experiences the most discrimination?

	All	18–34 Men	18–34 Women	35+ Men	35+ Women
Shopping	30	45	28	25	26
Dining out	21	32	24	19	15
At work	21	23	10	19	14
With police	15	34	8	17	9
Public transportation	6	12	2	8	3

Source: The Gallup Poll Organization (1997).

campaign encouraging Germans to challenge anti-foreign and anti-ethnic remarks, to intervene and protect foreigners from discrimination, and to wear a button bearing the word "courage" to symbolize their commitment to these goals (Marshall 1992). People who act to realize these goals would qualify as nonprejudiced nondiscriminators in Merton's typology.

Unprejudiced discriminators (fair-weather liberals) believe in equal opportunity but engage in discriminatory behaviors because they hold prejudices against members of a certain group (because it is to their advantage to do so) or because they fail to consider the discriminatory consequences of some of their actions. In one example, unprejudiced persons decide to move out of their neighborhood after a black family moves in—not because they don't like blacks, but because they are afraid that property values might start to decline and they want to "get out" while time remains. In another

instance, a white personnel officer tells friends and neighbors (who also are likely to be white) about a job opening. This word-of-mouth method of recruiting reduces the chances that a minority candidate will learn of the opening. Examples of unprejudiced discriminators specific to Germany include people who stand by and fail to challenge young sports fans for yelling "Jew to Auschwitz" at referees who make "bad" calls and individuals who join in singing songs whose rhyming words note that the Christmas season is the time "to burn a Turk" (Marshall 1992:A13).

One of the author's students admitted that he fell into this category. "Whenever someone would break out the inevitable 'black joke,' I would be somewhat angered at what they were saying, but on the other hand, I didn't want to jeopardize my own social standing by speaking up. I am ashamed to say that when a group of friends went out to paint racist comments on the town overpass, I went along as the lookout."

Prejudiced nondiscriminators (timid bigots) do not accept the creed of equal opportunity but refrain from discriminatory actions primarily because they fear possible sanctions if they are caught. Timid bigots rarely express their true opinions about racial and ethnic groups; instead, they use code words. In *Race: How Blacks and Whites Think and Feel About the American Obsession,* Studs Terkel (1992) cites several examples of speaking in code, including this one: During a black mayor's first campaign in Chicago, his white opponent's slogan was "Before It's Too Late."

Prejudiced discriminators (active bigots) reject the notion of equal opportunity and profess a right, even a duty, to discriminate. They derive significant

Unprejudiced discriminators (fair-weather liberals) Persons who believe in equal opportunity but engage in discriminatory behaviors because it is to their advantage to do so or because they fail to consider the discriminatory consequences of some of their actions.

Prejudiced nondiscriminators (timid bigots) Persons who do not accept the creed of equal opportunity but refrain from discriminatory actions primarily because they fear the sanctions they may encounter if they are caught.

Prejudiced discriminators (active bigots) Persons who reject the notion of equal opportunity and profess a right, even a duty, to discriminate.

Figure 10.3   A Typology of Ethnic Prejudice and Discrimination

	Attitude Dimension: Prejudice and Nonprejudice	Behavior Dimension: Discrimination and Nondiscrimination
Type I: Unprejudiced nondiscriminator	+	+
Type II: Unprejudiced discriminator	+	−
Type III: Prejudiced nondiscriminator	−	+
Type IV: Prejudiced discriminator	−	−

* + Attitude/behavior supports equal opportunity

− Attitude/behavior rejects equal opportunity

Source: Adapted from Merton (1976:192).

social and psychological gains from the conviction that anyone from the ingroup (including the village idiot) is superior to any members of the outgroup (Merton 1976:198). Active bigots are most likely to believe that they "have the moral right" to destroy the people whom they see as threatening their values and way of life. Police officers who believe "blacks" and other minorities are problem citizens and who make decisions based on that belief fall into this category of discrimination. One student in the author's class wrote about an encounter he had with campus police that illustrates this point:

> My friend (who is "black") and I decided to meet on campus and go together to a basketball game across town. On our way to the game, it began to get dark and I had to turn my headlights on. While waiting at a stoplight, my friend noticed that one of the headlights was out. He commented that we would probably get pulled over [by the police], especially if they saw a "black" man in the car. I told him he was stupid for thinking that.
>
> After the game ended we drove back to campus. As we drove to my friend's car, I noticed two campus police officers in their cars talking. I dropped my friend at his car, waited until he started it, and then left. I looked in my rearview mirror and noticed that one of the officers was following my friend and the second was following me with his flashers on. To make a long story short, they asked me what I was doing on campus and a lot of questions about my friend and if I had purchased drugs from him. When

I got home, I called my friend, who told me the police wanted to know if he had sold me drugs and had warned him that people like him should not be out so late at night. I asked him if he wanted to press charges, and he said that kind of stuff happens all the time. He also said, "Now you know what it feels like to be 'black.'"

Of the four categories in Merton's typology, prejudiced discriminators are the most likely to commit hate crimes, actions aimed at humiliating minority-group people and destroying their property or lives (see Table 10.4). The government of Saxony (one of 15 states in Germany) published a profile of 1,244 persons suspected of committing hate crimes between 1991 and 1992. Ninety-six percent were males; two-thirds were age 18 or younger; 50 percent had been drinking before the incident; 90 percent of the hate crimes occurred near the suspects' homes (within 12 miles); more than 98 percent of the suspects had less than a tenth-grade education. Surprisingly, only 20 percent of the suspects were unemployed or school dropouts. Eighty percent were first-time offenders. Approximately 30 percent justified their violent action with firm right-wing ideological views (Marshall 1993). A second systematic investigation of 1,398 police files related to anti-foreigner offenses committed between January 1991 and April 1992 produced similar findings (Krell, Nicklas, and Ostermann 1996). One study of foreigners in the eastern part of Germany found that 70 percent

Ku Klux Klan members in the United States (left) and neo-Nazis in Germany (right) exemplify active bigotry based on a racist ideology.

© Bob Daemmrich/Sygma (left); © Adenis/Sipa Press (right)

Table 10.4   Hate Crimes in the United States

The data below show that people of all racial categories commit hate crimes, sometimes even against members of their own racial categories. How many hate crimes were classified as anti-white? Why do you think people commit hate crimes against someone of their own race? What do you make of the "race unknown" category? Under what circumstances would we not know the offender's race?

				Suspected Offender's Race			
	Total Offenses	White	Black	American Indian/Alaskan Native	Asian/Pacific Islander	Multiracial Group	Unknown
Total Single-bias incidents	10,706	4,892	1,258	50	106	177	4,223
Race:	**6,767**	**3,334**	**995**	**40**	**71**	**116**	**2.211**
Anti-white	1,383	219	818	21	25	42	259
Anti-black	4,469	2,647	103	13	32	42	1,632
Anti-American Indian/Alaskan Native	69	46	8	1	1	0	13
Anti-Asian/Pacific Islander	527	291	44	5	13	9	165
Anti-multi-racial group	318	131	22	0	0	23	142
Ethnicity/ National Origin	**1,163**	**625**	**100**	**7**	**17**	**14**	**400**
Anti-Hispanic	710	425	71	5	12	5	192
Anti-other ethnicity/ national origin	453	200	29	2	5	9	208
Religion	**1,500**	**260**	**24**	**1**	**9**	**4**	**1,202**
Sexual Orientation	**1,256**	**659**	**139**	**2**	**9**	**39**	**408**

Source: U.S. Department of Justice (1996c).

claimed they had been insulted by a German citizen; 40 percent had suffered discrimination in business establishments; 20 percent had been attacked physically in some way (Wilpert 1991).

Institutionalized Discrimination

Sociologists distinguish between individual and institutionalized discrimination. **Individual discrimination** is any overt action on the part of an individual that depreciates someone from the outgroup, denies outgroup opportunities to participate, or does violence to lives and property. The U.S. Commission on Civil Rights (1981) describes some examples of how individuals can take discriminatory actions against each other:

- A personnel officer who holds stereotyped beliefs about members of outgroups may justify hiring them for low-level and low-paying jobs exclusively, regardless of their potential experience or qualifications for higher-level jobs.

- A teacher may interpret linguistic and cultural differences as indications of low potential or lack of academic interest.

- A guidance counselor and teacher may hold low expectations that lead them to steer members of outgroups away from "hard" subjects, such as mathematics and science, toward subjects that do not prepare them for higher-paying jobs (p.10).

- An owner of an apartment complex may refuse to rent to members of a minority by falsely informing them that no apartments are available (U.S. Department of Justice 1996b).

Institutionalized discrimination, on the other hand, is the established and customary way of doing things in society—the unchallenged rules, policies, and day-to-day practices established by dominant groups that impede or limit minority members' achievements and keep them in a subordinate and disadvantaged position. It is the "systematic discrimination through the regular operations of societal institutions" (Davis 1978:30). Laws and practices designed with the clear intention of keeping members of outgroups in subordinate positions, such as the 1831 Act Prohibiting the Teaching of Slaves to Read, are examples of institutionalized discrimination.

Two examples of institutionalized discrimination in Germany are laws that limit the proportion of foreign children in German classrooms to 20 to 30 percent and laws that limit the percentage of foreigners living in an area to 9 percent (Wilpert 1991). Another example of institutionalized discrimination, found in both the United States and Germany, is the selective law enforcement practice in which one racial or ethnic group is more likely than another to be arrested or to receive stiffer penalties for breaking the law (see the Interstate 95 study described in Chapter 8).

Institutionalized discrimination is more difficult to identify, condemn, hold in check, and punish than individual discrimination, because it can exist in a society even if the society's members are not prejudiced. Institutionalized discrimination cannot be traced to the motives and actions of particular individuals; instead, discriminatory actions result from simply following established practices that seem on the surface to be impersonal and fair or part of the standard operating procedures. Some examples of institutional discrimination include the following cases, which draw the attention of the U.S. Department of Justice:

- The First National Bank (of Dona Ana County in New Mexico) unfairly denied loans to Hispanic applicants by applying more stringent standards to the Hispanic applicants than to similarly situated Anglo applicants (U.S. Department of Justice 1997).

- The city of Garland, Texas, discriminated against African Americans and Hispanics by "knowingly" administering "written examinations for jobs in the police and fire departments that do not predict how well an applicant will perform on the job and that disproportionately exclude qualified African American and Hispanic applicants" (U.S. Department of Justice 1998).

- The city of Waukegan, Illinois, enacted and enforced a housing code restricting "the number of persons related by blood or marriage who could live together in the same home." Specifically, a husband, wife, and children and no more than

Individual discrimination Any overt action on the part of an individual that depreciates someone from the outgroup, denies outgroup opportunities to participate, or does violence to lives and property.

Institutionalized discrimination The established and customary way of doing things in society—the unchallenged rules, policies, and day-to-day practices that impede or limit minority members' achievements and keep them in a subordinate and disadvantaged position.

two additional relatives could live together in a house or apartment regardless of its size. The intent of the code was to limit the number of Hispanic relatives living in the same dwelling (U.S. Department of Justice 1996b).

To this point, we have looked at barriers that exist in society to prevent racial and ethnic minorities from adapting and/or assimilating. These barriers include racist ideology, prejudice, and discrimination (individual and institutional). Although we have viewed these barriers in a general way, we have not examined how they operate in everyday interaction. In *Stigma: Notes on the Management of Spoiled Identity,* sociologist Erving Goffman (1963) gives us such a framework.

Social Identity and Stigma

When people encounter a stranger, they make many assumptions about what that person ought to be.[11] Basing their judgments on an array of clues such as physical appearance, mannerisms, posture, behavior, and accent, people anticipate the stranger's social identity—the category (for example, male or female, black or white, professional or blue-collar, under 40 or over 40, American-born or foreign-born) to which he or she belongs and the qualities that they believe, rightly or wrongly, to be "ordinary and natural" (Goffman 1963:2) for a member of that category.

Goffman was particularly interested in social encounters in which one of the parties possesses a stigma. A **stigma** is an attribute that is defined as deeply discrediting in that person's society. A stigma is considered discrediting because it draws people's attention away from other qualities that a person might possess. To illustrate this point, Goffman uses the case of a 16-year-old girl born without a nose. Although she is a good student, has a good figure, and is an excellent dancer, no one she meets can really "see" these qualities because they cannot get past the fact that this girl has no nose.

Goffman classified stigmas into three broad categories: (1) physical deformities; (2) attributes socially defined as character blemishes, such as homosexuality or a history of mental hospitalization or imprisonment;

Stigma An attribute defined as deeply discrediting because it overshadows all other attributes that a person might possess.

and (3) stigmas based on ethnicity, race, nationality, or religion. The important quality underlying all stigmas is that those who possess them are not seen by others as multidimensional, complex persons but rather as one-dimensional beings. When many Germans encounter a Turkish woman wearing a head scarf, for example, they do not see the scarf in all its complexities—as "signifying the relation between the wearer and any number of things, such as her male relatives, her personal religious/political views, her financial means, or her region

---

[11]People come to learn these assumptions by a complex process. The learning takes place in very subtle ways. Two of my students shared interesting memories with me.

- Growing up in an urban town, I really never had much experience with interracial interactions, although I have had, throughout my schooling (in kindergarten and high school), friends that were labeled "black." Starting out in kindergarten, I made friends with three "black" boys for the first time. Although these boys were not in my class, I rode a private bus home with them every day. Curiosity struck my mind after the first few times I rode home with them; the bus driver always dropped off the other boys after me, so I never knew where they lived. I had such a good time on the bus every day—just the four of us—I wanted to know where they lived so I could have my mother take me to visit. I told the bus driver one day that she should drop the other boys off first. After finding out where they lived, I informed my mother, and she told me that I could not go over there because it was a bad apartment complex and that it would not be safe for me. I know now that it was an apartment complex inhabited mainly by "blacks." Looking back, this was the first time I was introduced to how communities organized along racial lines.

- It never occurred to me to question why I grew up in an all-white community. I wonder if my parents ever asked themselves the same question. It is certainly not a new convention that so-called black and so-called white people live in separate neighborhoods. I grew up in an all-white neighborhood, went to all-white schools, an all-white grocery store, all-white church, all-white YMCA, all-white EVERYTHING. I never got the impression that my family disliked black people; they were just never mentioned. I can't recall my first experience having contact with a black person, but I can remember being about fourteen or fifteen and riding a TANK bus where there were some black passengers. I felt pretty uncomfortable. I remember not wanting to stare so that I wouldn't offend anyone, but I was curious about the people on the bus that I had no knowledge about—except for the social prejudices that existed. At that time, I didn't ask myself or anyone else WHY "black" people and "white" people were so separated from one another, nor did I ask WHY these preconceived notions of "black" people existed.

  My first recollection of prejudice against black people was when I was much younger than fourteen or fifteen. I was riding with my mom to the eye doctor's office located in Covington. She made no mention of black people whatsoever, but she insisted that I lock my car door. I could see that we were passing "black" people. She said that it was a bad neighborhood. I think that is when I was influenced for the first time to have a prejudgment of "black" people. Why is our society this way? Who taught my mom to be afraid of black people, and who taught the person who taught her?

of origin" (Mandel 1989:30). Instead, they see only a foreigner, pointing to the scarf as proof that Turks cannot integrate into German society. In other words, this single attribute overshadows any other attribute that the woman might possess.

The same dynamic applies to how many Germans view people of African and German ancestry. Many Germans see only one attribute—dark skin—and fail to acknowledge other attributes that Afro-German people possess, such as the ability to speak German. An example shows how the dark skin symbolizes—African and not German—ancestry and dominates the course of interaction:

> I have dark skin, too, but I am a German. No one believes that, without some further explanation. I used to say that I was from the Ivory Coast, in order to avoid further questions. I don't know that country, but to me it sounded so nice and far away. And after this answer, I didn't get any more questions either. Germans are that ignorant. I could tell people any story I wanted to, the main thing was that it sounded foreign and exotic. But no one ever believes that I am German. When I respond to the remark, "Oh, you speak German so well," by saying, "So do you," people's mouths drop open. (Emde 1992:109–110)

The situation of Afro-Germans in Germany is similar to the situation of African Americans in the United States, most of whom have ancestries in their genetic makeup other than "African." Pro golfer Tiger Woods is perhaps one of the most visible examples. Most people would label Tiger Woods as "black" because of his skin color. But Woods has protested that he is a unique mixture of Thai, Chinese, white, black, and Native American (Page 1996:285). Nevertheless, the stigma of "skin color" and "hair texture" keeps people from acknowledging that he is related to people from five different national origins.[12]

According to Goffman, the discrediting trait itself does not constitute the stigma; rather, the stigma consists of the set of beliefs held by others regarding the trait. Furthermore, from a sociological point of view, the person who possesses the trait is not the problem; the problem is how others react to the trait. Thus Goffman maintained that sociologists should not focus on the attribute that is defined as a stigma. Instead, the focus should be on interaction—specifically, interac-

Sociologist Erving Goffman wrote about a pattern that characterizes "mixed contacts." Each party is unsure how the other views him or her, making both parties self-conscious about what they say and do in each other's presence.

© Bob Daemmrich/The Image Works

tion between the stigmatized and normals. Goffman did not use the term *normal* in the literal sense of "well adjusted" or "healthy." Instead, he used it to refer to those people who are in the majority or who possess no discrediting attributes. Goffman's choice of this word is unfortunate, because some readers ignore Goffman's intentions.

Mixed Contact Between the Stigmatized and the Dominant Population

In keeping with this focus, Goffman (1963) wrote about **mixed contacts**, "the moments when stigmatized and normals are in the same 'social situation,' that is, in one another's immediate physical presence, whether in a conversation-like encounter or in the mere co-presence of an unfocused gathering" (p. 12). According to Goffman, when normals and the stigmatized interact, the stigma comes to dominate the course of interaction in at least five ways. First, the very anticipation of contact can cause normals and stigmatized individuals to try to avoid one another. One student in my class pointed out that he almost dropped (that is,

---

[12]"If all Tiger Woods' heritages were acknowledged, Tiger would be just a 'golfer.' Instead, we 'force' Tiger to be a 'black' golfer, even though he has revealed to us his true background" (Cherni 1998).

Mixed contacts "The moments when stigmatized normals are in the same 'social situation,' that is, in one another's immediate physical presence, whether in a conversation-like encounter or in the mere co-presence of an unfocused gathering" (Goffman 1962:12).

avoided taking) a history class upon learning the professor was "black":

> The professor walked into the room and he was well dressed in African-style clothes. My first thought was, "Great, a class about how bad the slaves were treated two hundred years ago for the next fifteen weeks of class." I had no evidence to back up this reasoning except that the color of the skin of the professor immediately conditioned me to believe that the class would revolve around race.[13]

Sometimes the stigmatized and the normals avoid one another simply to escape each other's and outside scrutiny. Persons of the same race may prefer to interact with each other so as to avoid the discomfort, rejections, and suspicions they encounter from people of another racial or ethnic group. ("I have a safe space. I don't have to defend myself or hide anything, and I'm not judged on my physical appearance" [Atkins 1991:B8].) Social psychologist Claude Steele (1995) explains:

> Imagine a black and a white man meeting for the first time. Because the black person knows the stereotypes of his group, he attempts to deflect those negative traits, finding ways of trying to communicate, in effect, "Don't think of me as incompetent." The white, for his part, is busy deflecting the stereotypes of his group: "Don't think of me as a racist." Every action becomes loaded with the potential of confirming the stereotype, and you end up with these phantoms they're only half aware of. The discomfort and tension [are] often mistaken for racial animosity."

Another reason that stigmatized persons and normals make conscious efforts to avoid one another is because they believe that widespread social disapproval will undermine any relationship. One of my "white" students wrote about her experiences trying to date and establish a relationship with a "black" classmate, especially when her father and many of her friends disapproved: "We tried to carry on a 'secret' relationship, but it just didn't work. We were always worried about who might see us out together. We both agreed it wasn't worth the trouble, so we called it off." Another student wrote that her mother was anxious to meet the college student she was dating until her mother learned he was "black," at which point she warned her daughter that "if you continue a relationship with the man you will lose everyone's respect. You will be considered the outcast of the family." This student decided to end her relationship with the man she was dating.

The response of avoidance is related to a second pattern that characterizes mixed contacts: Upon meeting, each party is unsure how the other views him or her or will act toward him or her. Thus the two parties become self-conscious about what they say and about their behavior:

- How was I to know which whites were good and which were bad? What litmus test could I devise? I distanced myself from everyone white, watching, listening, for hints of latent prejudice. But there were no formulas to follow. (McClain 1986:36)
- When you're black, you cannot help but second-guess everything.[14] Does this person like me because I'm black or because I'm me? Does this person hate me because I'm black or because I'm a jerk? You're always second-guessing, unless it's another black person. (King 1992:401)
- I just wish I could get over the fear I have of being rejected, the fear that nonwhites don't want people like me around, the fear they resent me for the color of my skin and what I have. (Northern Kentucky University student 1998)

For the stigmatized, the source of the uncertainty is not that everyone they meet views them in a negative

---

[13]The student decided to stay in class and learned that his preconceptions were wrong. "The class actually was an interesting class. The professor did not spend any more time than 'usual' on the topic of slavery. The professor knew the subject well and was a good instructor."

[14]With regard to second-guessing, one "black" student in my class wrote:
> When I was in high school, I noticed that I received a lot more attention from my art teachers than other students in my class. I didn't really know if it was because of my talents or because of my skin color. There was another student in my class who was white that I believed was better than me but the teachers always pushed me, not him, to enter the art competitions. There could have been several reasons why my teachers chose to enter me into art competitions rather than the white student, but whatever the reason, it turned into a situation of race. The white student and I never talked about this situation, but as the school year went on, we slowly stopped talking to each other.
> Finally, in college, I realized that my black heritage dominated the views others had to the point that they can only see me as "black." The fact that I was the only black in all my art classes in college probably did not help the fact that I was being labeled as an "exceptional black graphic designer." It was confusing at first because I didn't know if I was doing special art projects on campus (like the Black History Month posters) merely because I was black or because they liked my work. I think some people figured that I would do such projects related to African Americans because of my race classification.

light and treats them accordingly. Rather, the chance that they might encounter prejudice and discrimination gives them reason to be cautious about all encounters. According to the Kolts Report (a report written in response to the Rodney King beating, issued by Special Counsel James C. Kolts, a retired superior court judge, and his staff), there were 62 "problem" deputies out of a total of 8,000 members of the Los Angeles County Sheriff's Department. The investigators concluded that "nearly all deputies treat nearly all individuals, most of the time, with at least minimally acceptable levels of courtesy and dignity" (*Los Angeles Times* 1992:A18). Although only a small proportion of deputies (less than 1 percent) were identified as "problems," the cases of mistreatment were "outrageous enough and frequent enough to poison the well in some communities" (p. A18). Similarly, the majority of German citizens do not discriminate against Turks and other dark-skinned people in conscious or overt ways. Nevertheless, the German population includes some 6,000 skinheads who identify with Nazi ideology, a number large enough to make life dangerous for minorities in Germany.

A third pattern characteristic of mixed contacts is that normals often define accomplishments by the stigmatized—even minor accomplishments—"as signs of remarkable and noteworthy capacities" (Goffman 1963:14) or as evidence that they have met someone from the minority group who is an exception to the rule. In addition to defining the accomplishments of the stigmatized as something unusual, normals tend to interpret the stigmatized person's failings, both major and even minor (such as being late for a meeting, cashing a bad check, or leaving a small tip), as related to the stigma (pattern 4).

A fifth pattern noted with mixed contacts is that the stigmatized are likely to experience invasion of privacy, especially when people stare:

> I am tired of walking into restaurants, especially with a group of African-Americans, and having the patrons and proprietors act as if they were being visited by Martians.
>
> I'm tired of being out with my eight-year-old nephew and his best friend and never being completely at ease because I know they will act like little boys. Their curiosity will be perceived as criminal behavior. (Smokes 1992:14A)

If the stigmatized show their displeasure at such treatment, normals may treat such complaints as exag-

gerated, unreasonable, or much ado about nothing. They argue that everyone suffers discrimination in some way and that the stigmatized do not have the monopoly on oppression. The normals may announce that they are tired of the complaining and that perhaps the stigmatized are not doing enough to help themselves (Smokes 1992). Normals are apt to respond:

> I'm tired of them. (Smokes 1992:14A)
>
> Why can't they put past discrimination behind them? I think they've been whining too long, and I'm sick of it. (O'Connor 1992:12Y)
>
> My grandparents, when they came here from Italy, didn't ask anyone to speak Italian for them. They became American, and they never asked for anything. (Barrins 1992:12Y)

The discussion thus far may suggest that members of racial and ethnic minorities are passive victims who are at the mercy of the dominant group. This situation is not the case, however; minorities respond in a variety of ways to being treated as members of a category.

Responses to Stigmatization

In *Stigma: Notes on the Management of Spoiled Identity*, Goffman describes five ways in which the stigmatized respond to people who fail to accord them respect or who treat them as members of a category. First, they may attempt to correct the so-called failing. This response includes changing the visible cultural and physical characteristics believed to represent barriers to status and belonging. A person may undergo plastic surgery or do other things to alter the shape of nose, eyes, or lips, or may enroll in a school to change an accent. In her research for "Medicalization of Racial Features: Asian American Women and Cosmetic Surgery," Eugenia Kaw (1993) found that among Asian American women who choose to undergo cosmetic surgery, the most often requested kind of surgery is "double-eyelid" to make their eyes look bigger. Klaw found that these women were motivated to correct the perception that "'small, slanty' eyes and 'flat' nose" are associated with a passive, dull, and nonsocial personality.

The direct attempt to correct that characteristic defined as a failure is not unique to Asian American women. An Afro-German woman describes her efforts to change her physical appearance:

> When I was about thirteen I started to straighten my "horse hair" so that it would be like white people's

hair that I admired so much. I was convinced that with straight hair I would be less conspicuous. I would squeeze my lips together so that they appeared less "puffy." Everything, to make myself beautiful and less conspicuous. (Emde 1992: 103)

Turks who change their religion from Islam to Christianity, who Germanize their names, or who give up wearing head scarves to fit into German society have made direct attempts to correct their stigma. Often these kinds of responses are not easy to make, because such persons may be considered traitors to their racial or ethnic group (Safran 1986). People may experience conflicting reactions to their own visible changes, too:

> My wife is twenty-five years old, she is beginning to dress German now, she is becoming more and more beautiful in her German clothes. . . . [M]y friends grumble because my wife dresses so modern, they want her, if she's going to wear German clothes, to wear at least a headscarf and long skirts; why does everyone have to see from far away that my wife is a Turk[?] (Teraoka 1989:110)

The stigmatized also may respond in a second way. Instead of taking direct action and changing the "visible" attributes that normals define as failings, they may attempt an indirect response. That is, they may devote a great deal of time and effort to trying to overcome the stereotypes or appear as if they are in full control of everything around them. They may try to be perfect—to always be in a good mood, to outperform everyone else, or to master an activity ordinarily thought to be beyond the reach of or closed to people with such traits.

> [You have to be on guard all the time;] there is no way to get away from it. Because if you do something like close the door to your room, people will start saying, "Is she being angry? Is she being militant?" You can't even afford to be moody. A white girl can look spacey and people will say, "Oh, she's being creative." But if you walk around campus with anything but a big smile on your face, they'll wonder, "Why is she being hostile?" (Anson 1987:92)

The indirect response is common among black plumbers, electricians, building contractors, and other service workers (especially males) who must make house calls in predominantly white neighborhoods. One black contractor interviewed in *The New York*

Many black service workers who work in white neighborhoods consciously dress and act in ways to reassure whites that they are not "criminals" to be feared but are in the neighborhood for a purpose.

© Alan J. Duignan/LA Times

Times maintained that when doing business in white neighborhoods:

> "I pull out my briefcase, my notepad, and clipboard—I have no need for either one—and look and wave. You have to show you have a reason to be there" (Maher 1995:A1). Other precautions include: "Avoid working at night or showing up at a job early in the morning. Never linger inside houses or gaze at a resident's possessions. And always keep your tools at hand to allay suspicion." (p. A1)

As a third option, stigmatized people may respond by using their subordinate status for secon-

Protesting stigmatization: This Turkish woman is wearing a yellow star identifying herself as a Turk. The star recalls the Nazi period, when German Jews were forced to wear the Star of David.

© Coopet/Sipa Press

The hopeful symbolism of candles graces a demonstration by young Germans against xenophobia, or fear of foreigners.

© Rainer Unkel/Saba

dary gains, including for personal profit or as "an excuse for ill success that has come [their] way for other reasons" (Goffman 1963:10). If a black person, for example, levels a charge of racism and threatens to file a lawsuit in a situation in which he or she is justly sanctioned for poor work, academic, or other performance, then that person is using his or her status for secondary gains. Keep in mind, however, that the person is able to use his or her minority status in this way only because discrimination is commonplace in larger society.

A fourth response is to view discrimination as a blessing in disguise, especially for its ability to build character or for what it teaches a person about life and humanity.

Finally,[15] the stigmatized can condemn all of the normals and view them negatively:

> You build up these perceptions of whites, that whites are mean and vile, never trust a white person. (Anson 1987:127)

> This coldness here in Germany makes me sick; I am still homesick today, sometimes even more strongly than five years ago; homesickness is an illness, and the illness can only be cured in Turkey; but in Anatolia there is no work for me, no gain, no possibility to move up in life, to live like a human being, with a house and a steady income: I must stay in Germany for now, I must live with this illness; in this cold; the coldness here in Germany, that is its people. (Teraoka 1989:107)

Can the stigmatized and the normals ever trust one another? For an optimistic look at this question, see "Overcoming the Stigmatized/Normal Barrier."

[15]In looking over Goffman's five responses, it appears that he did not consider a sixth response: challenging the prejudice and discrimination.

Overcoming the Stigmatized/Normal Barrier

Goffman's ideas about "mixed contacts" are especially useful for analyzing first meetings between the stigmatized and the normals. Some stigmatized and normals, however, do manage to break through the barriers associated with stigma and get to know one another. Upon coming to know each other, one or both gain insight about the impact "race" and its historical legacy has had on them and their thinking. Some examples from Northern Kentucky University students illustrate:

- It all started when my twin sister Shelley let our family know that her new boyfriend she wanted us to meet was black. For the most part our family had no experience interacting with people of different races because we lived and grew up in an all-white neighborhood with maybe a few exceptions.

  It was "date night" for Shelley but it was also "meet night" for us. We all sat there on the couch together waiting with anticipation. As I sat there many things were running through my mind, most of which were strong generalizations I had about all blacks based on what I had learned about them from TV, my only source of knowledge. I was very nervous; my palms were sweaty, I was hot, and my stomach had butterflies swarming inside. Looking back, if I was this nervous just think how uncomfortable her new boyfriend had been. At this point I was drawing upon many false assumptions about "blacks," and basically the poor guy didn't stand a chance.

  It was somewhere around 6 or 7 o'clock and there was a gentle knock at the door. My sister strolled to the door quickly, realizing that it was her boyfriend. She escorted him into the living room and unfortunately, much to my surprise, he was a very clean-cut, nice-looking young man. She introduced him, "I would like you guys to meet Dave," and he replied, "How do you do." I thought to myself, "'Dave,' that's not a 'black' name," and, "How do you do, a black with manners." So already two of my assumptions about "blacks" were false and he had only been there

for about one minute. To my surprise, we actually started and had a conversation and got to know the real Dave, the "human being just like us." We all grew very fond of him. In fact, over time Dave and I have become great friends and I can honestly say that there is no other man I would want my sister, whom I love very much, to be with and spend the rest of her life. I know there is no one out there who would treat my sister with any more love and respect than my good friend and soon to be brother-in-law, Dave. Oddly enough, after getting to know Dave I found out that his biological mother was white and his biological father was an exchange student from Africa. That wasn't the real irony, however. That came when I found out that he was adopted and raised by white parents.

  The purpose of this paper is for me to express my concerns about our generalizations of others. This is a huge problem because everyone does it, whether or not they know it. The generalizations I made were based on "race," which made me a racist, something I never ever would have acknowledged myself as being. I thought that in order to be a racist you had to hate another race.

  It is upsetting to know that this wasn't the first time I had stereotyped and made inaccurate generalizations about someone. It had happened many times but I was mentally unaware that my thinking was based on inaccurate information. For example, when I see a news story about a robbery, rape, or murder, I would automatically assume that a "black" person was involved. Keep in mind while I'm sharing this with you that I am very embarrassed and ashamed of myself and the way I thought, but I think admitting this is the first step toward eliminating racist thinking. In the beginning it was hard and I had to tell myself in my head not to think or act a certain way, but now it is becoming automatic.

- Growing up in a predominantly white neighborhood, I was blinded by society's paradigms, and I couldn't see the problems facing all of us. I was raised in a Methodist, church-going family known by others as

loving, caring, considerate, and helpful. My parents raised me to believe that nobody is better than anyone else. No matter how much money your family has, where you live, and especially no matter what color your skin is, we are all children of God, and equal in His eyes. This is how my family, close friends, and I felt. I thought that this was how everyone thought. With the exception of those few ignorant people you'd see on TV, carrying their Nazi flags and hailing Hitler, I thought that the world was generally color-blind these days, and truly accepting of everyone. I now see that I was terribly naïve.

It was the fall of 1995. I had just graduated that spring from high school, and after a summer of work, parties, and innocent fun, I was off to Lexington. I was ready to take on the world. I had never really had much contact or involvement with black people. There were only 2 black students in my graduating class of 350, and while I knew both, I had never really "hung out" with either one. My parents didn't have any black friends. When we arrived at my dorm, my mom asked me if I was excited about meeting my roommate. See, I had friends that were also attending U.K., but we thought it would be better to live with new people and expand our horizons and, looking back, that was a great decision. When I got to my room, the door was locked. I figured that my new roommate hadn't moved in yet, but when I entered, I saw that he had moved in and just wasn't in the room yet. I started to put away my things when I noticed his stereo. It was a huge stereo with 2 gigantic speakers that came up to my chest. I began to thumb through some of his CDs. Snoop Doggy Dogg, 2 Pac, Dr. Dre, and other gangsta rap CDs filled his enormous box labeled "disks." As I put more of my stuff away, I noticed the posters that he had hung up. They were posters of rappers, basketball players, and Martin Luther King, Jr. I immediately knew that my new roommate was going to be "black." I kept telling myself that it was no big deal and that it would be great to meet someone new.

After I finished putting everything away, my family and I went out to get dinner. I hate to admit this, but the entire time we were eating I was thinking about my new black room-mate. "What if he doesn't like me? I've never had black friends, and I don't know how I'll handle my fears and ignorance!"

I was raised to believe that there weren't many things worse than being a racist, but I realized that at that moment I was being one. Even though I wasn't marching around with a shaved head and a Nazi flag, I was being racist. This only added to my mixed emotions by making me mad at myself and frustrated with my lack of an open mind.

After dinner and all of the good-byes with my family, I headed up to my dorm room. My stomach was sick, and it wasn't from dinner. I was absolutely terrified. Not only about meeting my new roommate, but living with him for the next year. When I reached my room, I turned the knob, and the door was un-locked. When I walked into the room I was absolutely shocked. Sitting there was a middle-aged couple, 2 little girls, and a young man who was now my new roommate. White, they were all white. I wasn't relieved or excited, just shocked. What was I thinking? How could I possibly have been so ignorant? I was so disap-pointed in my own pathetic general-izing and stereotyping that I felt sick. I had totally fallen victim to the race paradigm that has plagued our society for so long. It's embarrass-ing for me to be writing this, but I now realize how truly ignorant I was and how paradigms can shape our society today.

My roommate and I went on to become great friends, and we still are today. I can't tell you what would have happened had he been black, but I hope the outcome would have been the same.

• In sociology class we have talked about many types of discrimination. Discrimination happens daily throughout our lives. I am almost sure everyone has discriminated or been discriminated against. The kind of discrimination I'm going to write about is from the viewpoint of nonprejudiced discriminators. My friends and I always played basket-ball at the local park every Satur-day. The same people were always there and we always played against them. The area where we lived was almost all white, and only rarely did any of us come in contact with a person of a different race. One Saturday two black guys about the same age as us, which was sixteen or seventeen years old, pulled up and hopped out of the car with a basketball. One of the black guys was about six feet tall and probably weighed about 180 pounds. The other guy was slightly shorter but seemed to be the more muscular of the two. None of my friends or I stopped playing to ask if they would like to join us. No one said anything about it because we were all just hoping they wouldn't be there next Saturday. I don't really know why we didn't want to play with them. I don't think less of black people, but I feel that maybe society had shaped my thoughts of black peo-ple by myself believing in some of the stereotypes.

Saturday finally rolled around again. We all headed up to the park as usual to play some basketball. When we showed up the two black guys were already there shooting around. My five friends and I sat in the car for a minute to discuss if we should ask them to play with us. I spoke up and said, "I think we should go and play with them." Three of my friends really didn't want to play with them. I'm not really sure why. None of us had ever had a friend that was black or, for that matter, had every played a sport with a black person. I mean, there was only one black person in my whole high school. We finally de-cided to ask them to play with us. We wound up playing for two hours straight and no one said a word. After we were done playing we decided to go get something to drink. I asked the two black guys if they wanted to join us. The taller one spoke up and said, "Sure." We all introduced ourselves and started talking. The guys' names were Jason and Scott. The taller one was Jason and the other one was Scott. They were brothers. They told us they had just moved into town from Chicago. After hanging out with these guys for awhile we decided to go home. As we were leaving I asked Jason and Scott to play with us next week, and they said that they would. From then on we played basketball every Saturday with them for two years straight until they moved back to Chicago.

There was no real basis for the way my friends and I acted. It wasn't that we were prejudiced in any way, but we were discriminating against them because they were black. For some reason we didn't ask them to play with us right away. I know if they were white we would have asked them to play the first time they showed up at the basket-ball court. I feel ashamed of my actions and I know my friends felt the same way. I think the next time a situation of this sort comes up my friends and I will definitely act differently.

• I was sitting in my Ant 231 class here at Northern Kentucky Univer-sity trying to ignore the girl sitting next to me. She was rattling on about how her engagement was called off. She would not quit talk-ing and I could not hear the teacher, so I decided to leave. To my sur-prise she followed me out of the classroom, talking to my back the whole time. So not only did we both look bad in front of the professor, the whole class was disrupted as well. While I was standing in the hallway trying to organize my books and tune out Lori, another guy came out of the classroom and immedi-ately lashed out at Lori for talking. He said, "Maybe you don't under-stand this whole college thing, because talking during a lecture is a pretty high school thing to do!" I made a motion that I agreed with him by letting out a sigh, but he misinterpreted me and became offended. He turned to me and said, "Look, b—, just because I am black doesn't mean I don't want to learn anything!" By the way he was looking at me, I could tell that he was trying to see how I felt about him as a black man. I was in shock. I couldn't explain myself and I couldn't walk away. I just stood there looking at him. As I was standing there watching him return to class it became clear to me that many white people really do think that black people are less intelli-gent than they are. I thought to myself that it must be awful to live like that, to live knowing that be-cause you were born a different color you will live with certain stig-mas all your life and to know that no matter how much you learn or how nice you are, some people will

continued

Overcoming the Stigmatized/Normal Barrier continued

always be looking for your faults so that you fit their false perceptions, the perception that the white person created and the black person has to live with.

In the county I live in there is a population of 41,250 people. Of that number, 99 percent are white and one percent consists of black, American Indian/Alaskan Native, Asian/Pacific Islander, and Hispanic combined. I found these statistics amazing, and I asked my friends about how they felt about the black people that live in our area. After talking to them I came to the conclusion that they all characterized blacks in the same basic manner. They said that black people were loud, unintelligent, dirty, and troublesome, which

ultimately led me to believe that they did not come to this decision without some assistance. Instead, they followed what other people thought and did not try to find out for themselves what black people are really like. Otherwise, their stereotypical characteristics would have varied more noticeably from one person to the next. I also asked them how much contact they had with black people and all of them said that they had very limited contact with them. Most of them agreed that if they were encouraged as children to get to know black children then they wouldn't feel the way they do now.

I didn't really know such racism existed in my hometown until I reached high school. High school is

when teachers urge their students to voice their opinions, but they never offer their own opinion (especially on racism). The white residents and teachers in my town try and keep blacks in a certain section, and it reminds me of how the whites treat American Indians. There was not one black child in my school, nor was there a black person in my area. Without any black children to play with or adults to talk to, it is very hard to disprove any false perceptions that a person may have, whether it be a white person prejudiced against a black person or vice versa. Although racial discrimination doesn't occur only among black and white people, it is the most widely publicized discrimination.

Summary and Implications

In this chapter, we compared two systems of racial and ethnic classification. For the first time in its history, the U.S. system now allows people to identify with more than one racial category, but it still refuses to recognize

This photo of newly released slaves shows that people of different "races" (in this case, slaves and masters) have been producing offspring for some time. Note the wide range of physical characteristics. Based on physical characteristics alone, who would you label "black"? Which people in the photo are most likely to "blend into" the society upon release?

Courtesy of the Schomberg Center for Research in Black Culture

a multiracial category. Recently the German government failed to pass legislation offering automatic citizenship to people classified as "third-generation foreigners" and to "second-generation foreigners" with a parent who has lived in Germany since age 14.

Both changes are significant in their own ways. The change in the U.S. system represents a first step (albeit a very small one) toward acknowledging that racial purity is a myth and that people of different racial classifications have interconnected lives and histories. The change in the German system constitutes a first step toward recognizing that the "foreign population" is part of German society and has played an important role in rebuilding the German country since World War II.

We have also learned that the problem with classification schemes is that each category is assigned a status value, such that people in some categories are viewed and treated as superior to those in other categories. The concept of a "minority group" helps us to identify specific ways in which members of subordinate categories are systematically excluded, both consciously and unconsciously, from full participation in society and are denied equal access to positions of power, prestige, and wealth. Specifically, we have examined the roles played by racist ideology, prejudice, stereotypes, and discrimination in creating and reinforcing subordinate status.

On the surface, it might seem that if we end racial and ethnic classification, the categories would disappear and the inequalities would cease to exist. Such a solution, however, ignores a critical moral question: How do we deal with the legacy of classification? Keep in mind that the system of racial classification in the United States results from more than 200 years of laws and practices enforcing categorization. The German system of ethnic classification, which divides the population into ethnic Germans and foreigners, can be traced to 1945 and even to Hitler's quest to create a "master race." Ironically, in both countries we still need categories to monitor progress toward remedying inequalities resulting from classification.

The modest changes to the U.S. system and the attempt to change the German system offer some signs of hope for race and ethnic relations. Allowing people in the United States to identify themselves with more than one race is a first step toward acknowledging that racial categories are not clear-cut and that people from different "races" have produced offspring who logically belong to both "races." Once we acknowledge this intermingling in an official sense, the next step is to face the painful legacy of racial classification, which functioned to segregate biological relatives from one another based on their appearance.

Even if the legislation had passed, the proposed citizenship laws are not nearly as liberal as the U.S. citizenship policy, which gives automatic citizenship to anyone born on its soil. The failed legislation and accompanying debate did bring public attention to the fact that Germans who have lived in Germany all their lives are not fully part of German society.

* * *

At the time this book went to press, the German parliament passed new legislation that gives any child born on German soil automatic citizenship provided one parent has lived in the country for at least eight years. The law also allows foreigners who have lived in Germany for eight years to apply for citizenship. This new legislation sends the message that "foreigners" who have lived in Germany for a long time (in many cases for several generations) are part of German society. The legislation is also a necessary first step in overcoming the long-standing discrimination and prejudices that have worked against the smooth assimilation of immigrant communities into German society. This legislation takes effect on January 1, 2000.

Key Concepts

Use this outline to organize your review of the key chapter ideas.

Race/ethnic group
Minority groups
    Involuntary minorities
    Voluntary minorities
Assimilation
    Absorption assimilation
    Melting pot assimilation
Discrimination
    Individual discrimination
    Institutionalized discrimination
Ideology
    Racist ideology

Prejudice
    Stereotypes
    Selective perception
Discrimination — prejudice typology
    Nonprejudiced discrimination
    Prejudiced nondiscrimination
    Prejudiced discrimination
      Hate crimes
    Nonprejudiced nondiscrimination
Stigma
    Social identity
    Mixed contacts

internet assignment

Find the Web site "Interracial Voice." Write a review of this Web site. In your review, include an overview of the Web site (its purpose, content, and sponsors). How does it connect with the issue of racial classification in the United States?

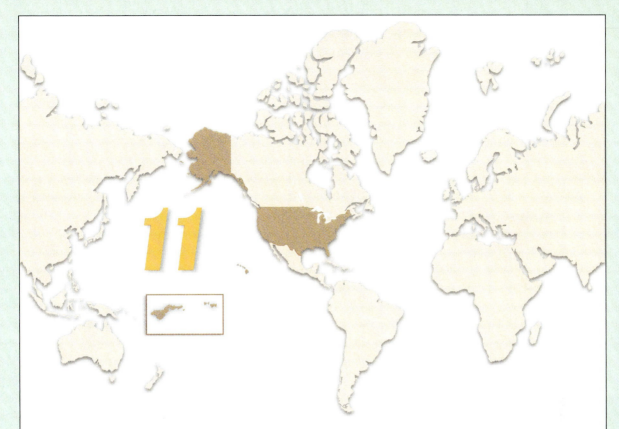

11

Gender

With Emphasis on American Samoa

Polynesian Cultural Center Laie, North Shore of O'ahu.
(Tami Dawson/Photo Resource Hawaii.)

AMERICAN SAMOA

Aunuu
Island

Tutuila Island

Pago Pago

Pacific Ocean

Olu Island

Tau Island

Manu'a Islands Group

In 1925, a 23-year-old named Margaret Mead who had never been west of Indiana traveled 8,000 miles by train across the United States to the West Coast and then by steamer to American Samoa to study adolescent girls (CPB/Polaroid Corporation 1981). She spent five weeks in Pago Pago learning the language and culture before going to Tau to study adolescent girls. Mead's research in Samoa and elsewhere helped us to see that if societies differ about how alike or unalike the two sexes should be, if they have different ideas about ideal behaviors and appearances for men and women, and if they differ in the tasks and responsibilities assigned to men and women, then we cannot explain male-female differences in biological terms alone. We must look to culture.

In 1925, Professor Franz Boaz wrote to his student Margaret Mead, who was doing research on adolescent girls in American Samoa, reminding her about the questions that most interested him: "One question that interests me very much is how the young girls react to the restraints of custom." In the United States, he noted, female adolescents often "display a strong rebellious spirit that may be expressed in sullenness or in sudden outbursts." Others rebel in more passive ways, by closing themselves off from others. Still others drown out their mental troubles by overdoing the social scene. Boaz questioned whether these reactions may occur in primitive societies or if the "desire for independence may be simply due to our modern conditions and to a more strongly developed individualism."

Boaz went on to write that

I am interested in the excessive bashfulness of girls in primitive society. I do not know whether you will find it [in Samoa]. It is characteristic of Indian girls of most tribes, and often not only in their relations to outsiders, but frequently within the narrow circle of the family. They are often afraid to talk and are very retiring before older people.

Another interesting problem is that of crushes among girls. For the older ones you might give special attention to the occurrence of romantic love, which is not by any means absent as far as I have been able to observe, and which, of course, appears most strongly where the parents or society impose marriages which the girls may not want. . . . (Reprinted in *Blackberry Winter: My Earlier Years* by Margaret Mead [1972])

Why Focus on American Samoa?

In this chapter, we consider the topic *gender*, which sociologists define as social distinctions based on culturally conceived and learned ideas about appropriate behavior and appearance for males and females. In exploring the concept of gender, we focus on American Samoa, a territory of the United States. We emphasize Samoa, because in 1925 a 23-year-old anthropologist named Margaret Mead traveled there in search of answers to Boaz's questions.

After spending five months observing and interviewing 68 adolescent girls between the ages of 9 and 20, Mead concluded that in Samoa adolescence "represented no period of crisis or stress. . . . The girls' minds were perplexed by no conflicts, troubled by no philosophical queries, beset by no remote ambitions. To live as a girl with many lovers as long as possible and then to marry in one's own village, near one's own relatives and to have many children, these were uniform and satisfying conditions" (Mead 1928:157).

In 1928, William Morrow and Company published *Coming of Age in Samoa,* in which Mead described in 294 pages why American and Samoan females experienced adolescence in such different ways. Her description of Samoan sexual customs became the most widely discussed and controversial part of her book (Marshall 1983). Mead maintained that, unlike their American counterparts, "Samoan children have complete knowledge of the human body and its functions" (p. 136) and, during adolescence, they come to possess "complete knowledge of sex, its possibilities and rewards" (p. 222). In fact, "we may say one striking difference between Samoan society and our own is the lack of specialization of feeling, and particularly of sex feeling, among the Samoans" (p. 215). Mead meant that casual, uncommitted sex and sexual experimentation, especially during adolescence, were the norms in

Samoa. Mead claimed that this practice was in sharp contrast to the United States, where a "tremendous fixation upon one individual leads the sheltered American girl to believe she has fallen 'in love' with the first man who kisses her" (p. 222).

We emphasize Samoa not only because of the controversy surrounding *Coming of Age in Samoa,* but also because Margaret Mead used Samoa as a vehicle to explore, reflect on, and critique gender relations in the United States.

Mead helped us to see that if societies differ in their views of how alike or unlike the two sexes should be, if they have different ideas about the ideal behaviors and appearances for men and women, and if they differ in the tasks and responsibilities assigned to men and women, then we *cannot* explain these differences in biological terms only. Instead, we must also look to culture.

Mead's vision of gender relations in Samoa and the United States remains important because the book,

which is still in print today, was translated into 16 languages. Since its publication, millions of college students and others around the world have read Mead's accounts of female adolescence and gender relations in Samoa. As we learn in this chapter, Mead's account of Samoa—whether right or wrong—and the controversy over "the real Samoa" probably reveal more about gender relations and politics in the United States than they tell us about American Samoa.

In this chapter, we explore the basic concepts that sociologists use to analyze the connection between gender and life chances. In outlining this connection, we examine how sociologists distinguish between sex (a biologically based classification scheme) and gender (a socially constructed phenomenon). In addition, we focus on the extent to which society is gender-polarized—that is, organized around the male-female distinction. Sociologists also seek to explain gender stratification and the mechanisms by which people learn and perpetuate their society's expectations about appropriate behavior and appearances for males and females.

Distinguishing Sex and Gender

Although many people use the words *sex* and *gender* interchangeably, the two terms do not have the exact same meaning. Sex is a biological concept, whereas gender is a social construct. In the following section, we pursue this distinction further, because it helps illustrate how the social differences between males and females develop.

Sex A biological concept determined on the basis of primary sex characteristics.

Primary sex characteristics The anatomical traits essential to reproduction.

Intersexed A broad term used by the medical profession to classify people with some mixture of male and female biological characteristics.

Sex as a Biological Concept

A person's **sex** is determined on the basis of **primary sex characteristics**, the anatomical traits essential to reproduction. Most cultures classify members of the population into two categories—male and female— largely based on what most people consider to be clear anatomical distinctions. Like race, however, even biological sex is not a clear-cut category, if only because some babies are born **intersexed.** The medical profession uses this broad term to classify people with some mixture of male and female biological characteristics.[1] Although we do not know how many intersexed babies are born each year, at Stanford Medical Center, as many as 20 births occur each year, for which "the doctors do not know whether to announce the arrival of a boy or girl" (Lehrman 1997:99).

If some babies are born intersexed, why does society not recognize an intersexed category? No such category exists because parents of intersexed children collaborate with physicians to assign their offspring to one of the two recognized sexes. Intersexed infants are treated with surgery and/or hormonal therapy. The rationale underlying medical intervention is the belief that the condition "is a tragic event which immediately conjures up visions of a hopeless psy-

---

[1]The intersexed group includes three very broad categories: (1) true hermaphrodites, persons who possess one ovary and one testis; (2) male pseudohermaphrodites, persons who possess testes and no ovaries, but some elements of female genitalia; and (3) female pseudohermaphrodites, persons who have ovaries and no testes but some elements of male genitalia.

chological misfit doomed to live always as a sexual freak in loneliness and frustration" (Dewhurst and Gordon 1993:A15).[2]

Adding a third category to include the intersexed would not do justice to the complexities of biological sex. French endocrinologists Paul Guinet and Jacques Descourt estimate that on the basis of variations in the appearance of external genitalia alone, the category "true hermaphrodite" may contain as many as 98 subcategories (Fausto-Sterling 1993). This variation tells us that " [n]o classification scheme could [do] more than suggest the variety of sexual anatomy encountered in clinical practice" (p. A15).

The picture becomes even more complicated when we consider that a person's primary sex characteristics may not match their sex chromosomes. Theoretically, one's sex is determined by two chromosomes: X (female) and Y (male). Each parent supposedly contributes one sex chromosome: the mother contributes an X chromosome and the father an X or a Y chromosome, depending on which one is carried by the sperm that fertilizes the egg. If the latter chromosome is a Y, then the baby will be a male. Although we cannot possibly know how many people's sex chromosomes do not match their anatomy, the results of mandatory "sex tests" of female athletes over the past 25 years have shown that such cases exist; indeed, a few women are disqualified from each Olympic competition and other major international competitions because they "fail" the tests (Grady 1992).[3]

Perhaps the most highly publicized, after-the-fact case is that involving Spanish hurdler Maria José Martinez Patino, who, although "clearly a female anatomically, is, at a genetic level, just as clearly a man" (Lemonick 1992:65). Upon giving her the test results, track officials advised her to warm up for the race but to fake an injury so as not to draw the media's attention to her situation (Grady 1992). Patino lost her right to com-

pete in amateur and Olympic events, but subsequently spent three years challenging the decision. The International Amateur Athletic Federation (IAAF) restored her status after deciding that her X and Y chromosomes gave her no advantage over female competitors with two X chromosomes (Kolata 1992; Lemonick 1992).[4]

As final evidence of the absence of a clear dividing line separating male from female, note the existence of transsexuals—people whose primary sex characteristics do not match the sex they perceive and know themselves to be. Those motivated to undergo surgery to become this sex are labeled "high-intensity" transsexuals (Bloom 1994). Although it is impossible to know how many such transsexuals exist, an estimated 25,000 people in the United States have had sex-change surgery (Brooke 1998). One surgeon with a practice in Trinidad, Colorado, who has performed 3,800 sex-change operations since 1969, noted that over the course of his career, only three patients requested that he reverse the change. The sex-change surgery phenomenon is not unique to the United States, as the charitable organization Interplast sends surgeons to "underdeveloped" countries to perform such operations. One surgeon, Don Laub, has performed 600 such sex-change (or sex-confirmation) operations since 1968 (Bloom 1994).

Why does no clear dividing line exist to separate everyone into one of the two biological categories, male and female? One answer lies with the biological mechanisms involved in creating males and females. In the first weeks of conception, the human embryo develops the potential to form a "female set of ovaries and a male set of testes." Approximately eight weeks into development, "a molecular chain of events orders one set to disintegrate." One week later, the embryo begins to develop an outside appearance that matches its external sex organs (Lehrman 1997:49). This complex chain of events may not be carried out "perfectly"; instead, it may be affected by any number of factors, including medications taken by the mother or environmental exposures.

In addition to primary sex characteristics and chromosomal sex, we use **secondary sex characteristics** to distinguish one sex from another. These physical traits are not essential to reproduction (breast development, quality of voice, distribution of facial and body

---

[2]Surgeon Linda Shortcliffe at Stanford Medical Center points out that these surgeries are very complex and "it's easier and less risky to do the work at about six months when the child is physically more mature, but if parents are extremely uncomfortable with their baby, [Shortcliffe] will operate within the first couple of weeks of life" (Lehrman 1997:49).

[3]In only one case in Olympic history, a man competed as a woman. In 1936, Nazi officials forced a German male athlete to enter the women's high jump competition. Three women jumped higher than he did (Grady 1992).

[4]Medical researchers hypothesize that a mutation in the Y chromosome in the case of anatomical females (who are genetically male) or a mutation in the X chromosome in the case of anatomical males (who are genetically female), respectively, suppresses or fails to suppress the production of excess testosterone.

Secondary sex characteristics Physical traits not essential to reproduction (breast development, quality of voice, distribution of facial and body hair, and skeletal form) that result from the action of so-called male (androgen) and female (estrogen) hormones.

hair, and skeletal form) but result from the action of so-called male (androgen) and female (estrogen) hormones. We use the term "so-called" because, although testes produce androgen and ovaries produce estrogen, the adrenal cortex produces androgen and estrogen in both sexes (Garb 1991). Like primary sex characteristics, none of these physical traits represents a clear dividing line by which to separate males from females. For example, biological females have the potential for the same hair distribution as biological males—follicles for a mustache, a beard, and body hair. Moreover, females produce both estrogen and androgen, a steroid hormone that triggers hair growth. In light of this information, we must ask, Why do women seem to have different patterns of facial and body hair growth than men? Before answering this question, we must consider the interrelationship between sex in the physical and reproductive sense and the concept of gender.

Gender as a Social Construct

Whereas sex is a biological distinction, **gender** is a social distinction based on culturally conceived and learned ideas about appropriate appearance, behavior, and mental or emotional characteristics for males and females (Tierney 1991). The terms **masculinity** and **femininity** signify the physical, behavioral, and mental or emotional traits believed to be characteristic of males and females, respectively (Morawski 1991).

To grasp the distinction between sex and gender, we must note that no fixed line separates maleness from femaleness. The painter Paul Gauguin pointed out this ambiguity in his observations about Maori men and women, which he recorded in a journal that he kept while painting in Tahiti in 1891. His observations were influenced by the norms regarding femininity around the turn of the century:

> Among peoples that go naked, as among animals, the difference between the sexes is less accentuated than in our climates. Thanks to our cinctures and corsets we have succeeded in making an artificial being out of

At the turn of the twentieth century, the artist Paul Gauguin observed that among Maori men and women the differences between the sexes was less accentuated such that there was something virile in the women and something feminine in the men.

Paul Gauguin, *Tahitian Women Inside a Hut*. The Hermitage Museum, St. Petersburg, Russia. Photo © Scala/Art Resource, NY.

woman. . . . We carefully keep her in a state of nervous weakness and muscular inferiority, and in guarding her from fatigue, we take away from her possibilities of development. Thus modeled on a bizarre ideal of slenderness . . . our women have nothing in common with us [men], and this, perhaps, may not be without grave moral and social disadvantages.

On Tahiti, the breezes from forest and sea strengthen the lungs, they broaden the shoulders and hips. Neither men nor women are sheltered from the rays of the sun nor the pebbles of the seashore. Together they engage in the same tasks with the same activity. . . . There is something virile in the women and something feminine in the men. ([1919] 1985:19–20)

Often we attribute differences between males and females to biology; in fact, they are more likely to be socially created. In the United States, for example, norms specify the amount and distribution of facial and body hair appropriate for females. It is deemed acceptable for women to have eyelashes, well-shaped eyebrows, and a well-defined triangle of pubic hair, but not to have hair above their lips, under their arms, on their inner thighs (outside the bikini line), or on their chin, shoulders, back, chest, breasts, abdomen, legs, or toes. Most men, and even women, do not realize that women work to achieve these cultural standards and that their compliance makes males and females appear more physically distinct in terms of this trait than they are in reality. We

Gender A social distinction based on culturally conceived and learned ideas about appropriate appearance, behavior, and mental or emotional characteristics for males and females.

Masculinity The physical, behavioral, and mental or emotional traits believed to be characteristic of males.

Femininity The physical, behavioral, and mental or emotional traits believed to be characteristic of females.

lose sight of the fact that significant but perfectly normal biological events—puberty, pregnancy, menopause, stress—contribute to the balance between two hormones, androgen and estrogen. Changes in the proportions of these hormones trigger hair growth that departs from societal norms about the appropriate amount and texture of hair for females. When women grow hair as a result of these events, they tend to think something is wrong with them, instead of seeing this development as a natural event. A "female balance" between androgen and estrogen is seen as one in which a woman's hair is consistent with these norms.[5]

The extreme measures taken by some women to eliminate facial and body hair are reflected in reports made by physicians who intervened after women were harmed by commercial treatments. For example, X-ray treatments—a popular choice between 1899 and 1940—caused adverse side effects, including wrinkling, scarring, discoloration, and cancerous growths.[6] A depilatory containing thallium acetate, a highly toxic substance that was popular in the 1960s, caused paralysis (Ferrante 1988).

Just as women strive to meet norms for facial and body hair, both men and women work to achieve the ideal standards of masculine and feminine beauty as portrayed in the media (for example, in magazines and on television) or as conveyed and reinforced in the ways people react to us. On a personal level, these ideal standards are not viewed objectively—as something created by people that varies across time and place. Despite all evidence to the contrary, we believe that facial and body hair *is* a masculine quality. In a similar vein, long hair on Samoan women does not simply signify feminine sexual attractiveness, it *is* feminine attractiveness. In this vein, Jeanette Mageo (1996) argues that ideal standards of beauty affect us personally because "what has personal significance is at least in part a product of how we are

At one time in Samoa, the transition from boyhood to manhood was marked by the long and painful process of tattooing the body from the waist to below the knee. This practice has not disappeared, as some present-day Samoan males choose to continue this tradition.
© David R. Austen/Stock Boston

regarded and treated by others. When a Samoan girl acts under constant threat of having her hair cut off or of being pulled home by her hair [because she is attracting male attentions], when her beauty is judged, at contests and elsewhere, by the length of her hair, the public symbol of hair cannot fail to touch her feelings" (p. 158).

Before the Christianization of Samoa, the transition from boyhood to manhood was accompanied by a "long and painful process of body tattooing, from the waist to below the knees" (Cote 1997:2). Tattooing did not merely signify manhood; it *was* manhood:

> . . . the man who was not tattooed . . . was not respected. . . . Until a young man was tattooed, he was considered in his minority. He could not think of marriage, and he was constantly exposed to taunts and ridicule, as being poor and of low birth, and having no right to speak in the society of men. But as soon as he was tattooed, he passed in his majority, and considered himself entitled to the respect and privileges of mature years. When a youth, therefore, reached the age of sixteen, he and his friends were all anxious that they should be tattooed. . . . On these occasions, six or a dozen young men would be tattooed at one time. . . . In two or three months the whole is completed. The friends of the young men are all the while in attendance with food. (Turner [1861] 1986:87–89)

This practice has not disappeared entirely, as many Samoan males still choose to tattoo their bodies in this fashion.

---

[5]Although "excess" hair sometimes warns a doctor to check for the presence of an underlying endocrine disorder, such as an ovarian tumor, in 99 percent of cases such hair is not associated with any pathology.

[6]The severity of damage caused by X-ray treatments is illustrated by the case of Mrs. A. E. C., one of many reported in the *Journal of the American Medical Association:*
Mrs. A. E. C., age thirty-two, with a history of treatment five years ago by the Tricho [X-ray] system in Boston. She had fifteen treatments on her chin for [superfluous hair]. She was treated every two weeks, and three areas on the underside of her chin were treated each time. Twice she was treated by the bookkeeper when the nurse was not in. Three years ago red blotches began to appear and they have continued. She shows definite signs of telangiectasis [a chronic dilation of blood vessels causing dark red blotches] and atrophy over an area of about three inches in diameter just below the sides of the chin. (American Medical Association 1929:286)

To this point, we have drawn a distinction between sex and gender. Although sociologists acknowledge that no clear biological markers exist with which to distinguish males from females, they would not argue that biological differences are nonexistent. Sociologists, however, are interested in the extent to which differences are socially induced. To put it another way, they study the actions that men and women take to accentuate the differences between them. As we see in the next section, these actions lead to gender polarization.

Gender Polarization

In *The Lenses of Gender,* Sandra Lipsitz Bem (1993) defines **gender polarization** as "the organizing of social life around the male-female distinction," so that a person's sex is connected to "virtually every other aspect of human experience, including modes of dress, social roles, and even ways of expressing emotion and experiencing sexual desire" (p. 192). To understand how life becomes organized around gender, we consider research by Alice Baumgartner-Papageorgiou.

In a paper published by the Institute for Equality in Education, Baumgartner-Papageorgiou (1982) summarizes the results of a study of elementary and high school students. In the study, she asked the students how their lives would be different if they were members of the opposite sex. Their responses reflect culturally conceived and learned ideas about sex-appropriate behaviors and appearances and about the imagined and real advantages and disadvantages to being male or female (Vann 1995). The boys generally believed that their lives would change in negative ways if they became girls. Among other things, they would become less active and more restricted in what they could do. In addition, they would become more conscious about tending to their appearance, finding a husband, and being alone and unprotected in the face of a violent attack:

- I would start to look for a husband as soon as I got into high school.

> **Gender polarization** "The organizing of social life around the male-female distinction," so that people's sex is connected to "virtually every other aspect of human experience, including modes of dress, social roles, and even ways of expressing emotion and experiencing sexual desire" (Bem 1993:192).

- I would play girl games and not have many things to do during the day.
- I'd use a lot of make-up and look good and beautiful. . . . I'd have to shave my whole body.
- I'd have to know how to handle drunk guys and rapists.
- I couldn't have a pocket knife.
- I would not be able to help my dad fix the car and truck and his two motorcycles. (pp. 2–9)

The girls, on the other hand, believed that if they became boys they would be less emotional, their lives would be more active and less restrictive, they would be closer to their fathers, and they would be treated as more than "sex objects":

- I would have to stay calm and cool whenever something happened.
- [I could sleep later in the mornings] since it would not take [me] very long to get ready for school.
- My father would be closer because I'd be the son he always wanted.
- I would not have to worry about being raped.
- People would take my decisions and beliefs more seriously. (pp. 5–13)

Although the Baumgartner-Papageorgiou study was published in 1982, these beliefs about how the character of one's life depends on one's sex seem to hold up across time, even among the college students enrolled in the author's introductory sociology classes. These students were asked to take a few minutes to write about how their lives would change in positive and negative ways as members of the other sex (see "Penalties and Privileges Associated with Being Male and Female"). The men generally believed that they would be more emotional and more conscious of their physical appearance and that their career options would narrow considerably. Some of their responses follow:

- I would be much more sensitive to others' needs and what I'm expected to do.
- I wouldn't always have to appear like I am in control of every situation. I would be comforted instead of always being the comforter.
- People would put me down for the way I look.
- I would be more emotional.
- I would worry more about losing weight instead of trying to gain weight.
- I probably wouldn't really feel any different, but people would see me as a female and respond

Penalties and Privileges Associated with Being Male and Female

In autumn 1998, students in the author's class were asked to list what they believe are the penalties and privileges associated with being male and female. Are there any penalties or privileges you would add or remove from the list? Why?

Penalties (Men)

Lower life expectancy

Breadwinner responsibility

Most dangerous occupations disproportionately filled with men

Higher insurance rates

More likely to pay for dates

Constraints on emotions

Expected and pressured to be more successful

Expected and pressured to be athletic

Expected to have a higher tolerance for discomfort

Less likely to get help when in trouble

Hard for men to behave and dress in ways considered feminine

Pressure to take initiative in asking women out and for their hand in marriage

Expected and pressured to take on role as protector

More difficult for men to get custody of children

Privileges (Men)

Better pay

Career choices not as likely to interfere with familial responsibilities

Greater career choices/job opportunities

Greater opportunities and respect in the world of sports

Fewer constraints on physical appearance

Less expensive to dress for success

More likely to be taken seriously

Less likely to have multiple sexual experiences evaluated harshly

More likely to be labeled as role models and heroes

More likely to experience independence at an earlier age

Penalties (Women)

Child-bearing experience (if unplanned and unwanted)

Lower pay

Fewer career choices

Career choices more likely to conflict with familial responsibilities

More time and attention paid to physical appearance

Strength and athletic ability less likely to be developed to full potential

Have to work harder to be taken seriously

Child more likely to carry father's name

More likely to be held responsible for housework and care of children

More likely to have to wait for men to ask them out and ask them to marry

Privileges (Women)

Child-bearing experience (when planned and desired)

Less likely to be considered a crime suspect because of reputation as the weaker sex

Fewer social costs for choosing not to work and for being supported by husband

Can use physical appearance to get "results" from men

More likely to receive help if in trouble

Easier for women to behave and dress in ways considered masculine (to a point)

Easier for women to freely express a wide range of emotions

---

accordingly. If I stayed in the construction program, I would have to fight the belief that men are the only real construction workers.

- My career options would narrow. Now I have many career paths to choose from, but as a woman I would have fewer.

- I would have to be conscious of the way I sit.

Notice that the first two responses suggest that some "feminine" traits would be a plus (being "more sensitive to others' needs" and being "comforted instead of

always being the comforter"). Both men and women can feel constrained by their gender roles.

The women in the class believed that as men they would have to worry about asking women out and about whether their major was appropriate. They also believed, however, that they would make more money, be less emotional, and be taken more seriously. Some of their responses follow:

- I would worry about whether a woman would say "yes" if I asked her out.

U.S. in Perspective

Gender Schemes in Educational Choices

This chart shows the college majors in which 70 or more percent of the degrees awarded go to males or to females. Note that 55 percent of degrees in all fields were awarded to women. If gender had no bearing on choice of major, 55 percent of degrees in each category would be awarded to women. Which fields are most disproportionately female? Disproportionately male? Why do you think 10 percent more women than men received a degree in 1994–1995?

	Bachelor's Degree		
	Men (%)	Women (%)	Total
Male-Dominated			
Computer and information sciences	71	29	24,404
Engineering	83	17	62,342
Engineering-related technologies	90	10	15,812
Precision production trades	70	30	353
ROTC and military technologies	89	11	27
Theological studies	75	25	5,578
Transportation and material moving	89	11	3,698
Female-Dominated			
Education	24	76	106,079
Health professions	18	82	79,885
Home economics	12	88	15,345
Law and legal studies	29	71	2,032
Library science	4	96	50
Psychology	27	73	72,083
Public administration and services	21	79	18,586
All Fields	45	55	1,160,134

Source: U.S. Department of Education (1998).

- I would earn more money than my female counterpart in my chosen profession.

- People would take me more seriously and not attribute my emotions to PMS.

- My dad would expect me to be an athlete.

- I'd have to remain cool when under stress and not show my emotions.

- I think that I would change my major from "undecided" to a major in construction technology.

Gender-schematic A term describing decisions that are influenced by a society's polarized definitions of masculinity and femininity rather than by criteria such as self-fulfillment, interest, ability, or personal comfort.

These comments by high school and college students show the extent to which life is organized around male-female distinctions. They also reveal that students' decisions about how early to get up in the morning, which subjects to study, whether to show emotion, how to sit, and whether to encourage a child's athletic development are gender-schematic decisions. Decisions and viewpoints about any aspect of life are considered **gender-schematic** if they are influenced by a society's polarized definitions of masculinity and femininity rather than by criteria such as self-fulfillment, interest, ability, or personal comfort. For example, college students make gender-schematic decisions about possible majors if they ask—even subconsciously—about the "sex" of the major and, if it matches their own sex, consider the major to be a viable option or, if the sex does not match, reject it outright (Bem 1993). (See "Gender Schemes in Educational Choices.")

Table 11.1 Sex Ratios for Various Age Groups in the United States and American Samoa

The data show the number of males per every 100 females for various age groups in Samoa and the United States. So, for example, in American Samoa there are 107 males age 4 and younger for every 100 females of that age. Note that the number of males declines from age 15 on in the United States. Also notice how the ratio of males to females varies from one age group to the next in American Samoa. Why do you think these ratios appear in Samoa?

| | Males per 100 Females | |
Age	Samoa	United States
4 and younger	107	105
5–9	108	105
10–14	116	105
15–19	103	102
20–24	97	103
25–29	91	100
30–34	106	99
35–39	98	99
40–44	110	98
45–49	121	96
50–54	122	94
55–59	121	92
60–64	102	89
65–69	94	84
70–74	111	77
75 and older	86	57

Source: U.S. Bureau of the Census (1992, 1998).

Even sexual desire between men and women is organized around male-female characteristics, such as height and age, that are unrelated to reproduction. Bem (1993) argues that

> neither women nor men in American society tend to like heterosexual relationships in which the woman is bigger, taller, stronger, older, smarter, higher in status, more experienced, more educated, more talented, more confident, or more highly paid than the man; they do tend to like heterosexual relationships in which the man is bigger, taller, stronger, and so forth, than the woman. (p. 163)

The negative consequences of channeling sexual desire according to age differences so that the woman in the relationship is usually younger than her partner becomes evident when we consider that the median age at first marriage for women is 24 and for men is 25.9 (Centers for Disease Control and Prevention 1998). In the United States, the average life expectancy for women is 6.2 years longer than that for men (the difference in life expectancy can be partly explained by the fact that men tend to hold the most hazardous jobs

in society). This practice of men marrying younger women, in combination with differences in life expectancy, means that women who marry can expect to live a significant portion of their lives as widows (see Table 11.1).

In addition to sexual desire between men and women being influenced strongly by gender-polarized ideas about such things as height and age, emotions toward persons of the same sex are affected as well. In Chapter 4, we learned that **social emotions** are internal bodily sensations that we experience in relationships with other people and that feeling rules are norms specifying appropriate ways to express those sensations. These other people may be boyfriends, spouses, parents, same-sex friends, teachers, and so on. When students in the author's class were asked to comment on social emotions or "internal bodily sensations" that they had felt and expressed toward someone of the

Social emotions Internal bodily sensations experienced in relationships with other people.

Feeling Rules and Same-Sex Friends

Of the 48 students attending a Race and Gender class at Northern Kentucky University in autumn 1998, 46 indicated that they knew a same-sex person for whom they felt strong affection and who made them feel "alive." However, 39 students said that it was important that their feelings of intense pleasure not be interpreted as sexual. Why do you think our society offers no vocabulary separate from the one we use to describe romantic relationships to describe strong same-sex friendships? Why are students in this class fearful that their feelings might be interpreted as sexual? What does that fear suggest about American culture?

I do have one male friend who I really look forward to seeing. It's just because we have been friends since kindergarten and we get along so well because we know each other the way you can only know someone who you grew up with for 15 years. We can make each other laugh easier than any of our other friends. I don't have any feelings towards him that I would feel guilty about, but even though we have been friends a long time, I still don't know that I would be comfortable saying something as simple as "I can't wait to see you." That's just how it is.

I have a good friend from high school whom I have a blast with. We are almost like brothers and I do feel alive when I am with him. I do not think that this person would construe these feelings as sexual. We were best friends for over four years and always partied together and had a great time together. It seemed like we are in our own little world at times, but we are always able to come back to reality. I don't feel guilty about feeling this way but I have not ever told him how I feel, yet something tells me that he felt the same way. I don't think, however, that I could ever tell him that "I

can't wait to see him again" because of the way our society has construed this to seem sexual.

I know someone of the same sex around whom I feel these feelings. I think that men and women do have feelings for the same sex but are often afraid to show those feelings because of how they have been raised. I have to say that I am afraid to say, "You mean a lot to me," or whatever. I think society has portrayed these feelings to be reserved for people of opposite sexes and that it is not 100 percent correct to share such feelings with the same sex. If I would consult with this person, I don't think that they would take it sexually. Honestly, I don't feel guilty about feeling this way because they are my friends and I love them. Sometimes I just need my female friends more than my male friends. Females understand me more.

My best friend and I have been friends for ten years. We were roommates for a couple of years and have always been close. I just love everything about her. Sometimes I look at her and I can't help but stare; it just makes me feel warm to be around her. For a while I thought maybe something was wrong with me because I thought about her so much and loved to be around her. Now, though, I realize that we have a bond like sisters, probably even closer. She can do no wrong in my eyes. I feel closer to her than I do to my own family and in some ways closer than my fiancé. I am not gay or bisexual, but some women just fascinate me and I enjoy their company so much that I cannot stop smiling and staring when I'm with them. I want to sit next to them and hold them, but I restrain myself most of the time and ration out my affection so that it is not misconstrued.

My best friend and I have never had a problem expressing our feelings. We have been friends since the third grade. I remember when we were

about 14 or 15 we were sitting together and I was scratching her arm. It's something she always liked and asks me to do. I've never had a problem with this—I'm an affectionate person—but some people do. Other girls that witnessed this began to tell people we were lesbians or at least acted like it. We both knew this was stupid and chalked it up to them being jealous of our friendship. To this day, I am so happy when she is around. I love her dearly and have told her so many times. I've also told her that I hope other people have the chance to have such a wonderful friend and feel this bond. It's never been weird for us.

I've had the same best friend since the fourth grade. The intense pleasure we discussed in class was present almost every time we hung out together. We would have slumber parties a lot and all that I remember is that whenever she spent the night we would laugh for hours (nonstop). Nothing could stop us from laughing at everything and anything. We were so happy we didn't want to go to sleep. We could stay up all night talking and laughing. It wasn't until high school that these sleep-overs stopped, and not until I had a boyfriend (and she had a boyfriend) that I felt comfortable telling her something we already knew, that I loved her. Now whenever I think she needs to be reminded, I tell her, but without a boyfriend as proof that it didn't "mean anything" I probably never would have told her.

One of my best friends is someone I love to hang out with. He's been my friend since the seventh grade and we are great friends even today. We've hung out a lot then and we still do. I have told people how I feel about him but not in a serious way. My friends and I tend to play as if we are gay at times and I reveal my feelings toward him in that joking

same sex, most indicated that other people made them feel uncomfortable and defensive about such feelings (see "Feeling Rules and Same-Sex Friends"). Specifically, students were asked to think of a same-sex friend

with whom they most liked to hang out with, either now or in the past. Although no physical/sexual component existed in the relationship, the students felt sensations of being "alive" and experienced intense

manner. Sometimes I feel weird saying it and I'm always afraid they'll burn me for it (make fun of me) but it's just that I have fun all the time around him. I've never told him this though. I think I feel too uncomfortable about that. I'm always afraid he'll think of me differently if he knew what I felt about him. I have fun and I love his company but I just can't say it.

I totally agree with you on this issue. I feel this way about my friend Steve. I think he is awesome. I just love to be around him. I have never thought about it but now that I do, I feel this way. I have said to him on several occasions that I love being around him and spending time with him. What is strange enough is he feels the same about me. We just agree on everything. It is great. I remember one time when we didn't go out together one weekend, he called and said he missed being around me. I think that is cool.

I have a same-sex friend that I truly enjoy being around. I think that if I express my intense feelings for my best friend to another person, she would probably consider it to be sexually oriented. I don't feel guilty for loving my friend, but I do feel guilty using words such as "love," "intense," etc., to describe our relationship. These words make me sound as if I am gay. I have never told my best friend that I love her and appreciate her, mainly because I am truly afraid of her reaction to such a statement. I am afraid that she would believe that I was in love with her sexually. I would never want to jeopardize my friendship by expressing simply what I feel!

For the last ten years, both my job and my social environment have left me with a very broad perspective on friendship/social interactions, so the notion of intense same-sex friendships is not a novelty to me. My best friend Jeff is a same-sex friend for whom I have intense

feelings. I often tell him how dear his friendship is to me, and now that he lives in Japan, I often include a statement of how dearly I miss him in my letters to him. Before he left for Japan, we had many late-nighters in which I would actually sleep over in the same bed because we'd been partying or because I was simply too sleepy to drive home. I had no fear that anyone would laugh at this friendship or find it strange, but I also never cared what anyone thought. I plan on visiting him in Japan and, because living situations are so small there, we'll probably end up sleeping together again. And yes, if it is an issue, I am straight.

I have a set of friends that I absolutely love hanging out with, but I have one friend in particular that, when I'm around him, I feel really great. If I told him that I liked his company that much, he would most likely call me a fag. When I read the question "Do you feel guilty [for your feelings]," at first I thought that I did not. However, as I am getting further into this I am beginning to feel weird. I have never said anything to him about how I feel, I guess because I'm a guy and guys just don't do that, generally speaking. What would I say, anyway? "Hey there buddy, I was just thinking about you and I realized that I really love your company. What do you say we have a couple beers tonight?" I doubt it.

I have intense (nonsexual) feelings about my close girlfriends. Two of my closest friends are so much fun to be around. When we go to dance clubs and stuff, we dance almost as if we were gay. It's so much fun. I tell them how much fun I have with them, specifically, and they tell me the same. I don't feel like I'm sexually out of the norm or anything. I know I'm totally heterosexual. It's just that sometimes I'll even go with them when my boyfriend insists that I don't go. This shows the intense "want" or "need" to be with them.

It's a unique way of bonding that I need to fulfill my self-image and social acceptance.

I have experienced intense feelings for a Guatemalan woman. When this woman moved into our strictly Anglo-Saxon sphere, I was elated. I sought out both her and her company, and her conversation excited me so that I felt a deep and intense excitement just being in her presence. I couldn't wait to see her, talk to her, and just to be in the same room with her. I felt aroused with the pleasure of living at those times. I never told her, as I felt she would back away from me if I did and more than anything, I wanted her in my life. Even though I knew there was nothing wrong with my feelings, I felt she would misconstrue them. That threat of losing her was so painful that it kept me from ever being honest. I found it hard to breathe, I stuttered and couldn't get my tongue around a single word that made sense. It was love of who she was, her personality, and experiences.

I do know a same-sex friend with whom I have such an intense relationship, and until today I thought I was just weird. I have had a very close friendship with this girl for about seven years and I've always felt this way about her. Whenever we're together, I feel so energized and so "healthy." I feel an almost uncomfortable closeness to her sometimes. For example, the other day we were watching TV and for some reason I just felt the urge to curl up next to her and snuggle. There was no feeling of sexual arousal in any way at all; it was just a type of closeness. It's almost like the same closeness I feel when I snuggle up to my mom, almost a motherly bond. I have never ever told her of these feelings because it would just come out so homosexual and it's nothing like that at all.

Source: Northern Kentucky University students, 1998.

pleasure when interacting with that person. Everyone except two persons in the class knew a same-sex friend for whom they felt this way. The students were also asked, "Have you ever told that person you feel this way?" (that is, "I love being around you" or "I feel so much energy when I am around you"). A sample of their answers to these questions are given in "Feeling Rules and Same-Sex Friends." Of the 48 students in class that day, 81.2 percent (39) indicated that it was important to understand that their intense feelings for the same-sex friend were not sexual. Sixteen students (32.7 percent) also indicated that they had not or could not tell this friend how they felt.

Social Emotions and the Mead–Freeman Controversy

As was pointed out in the introduction to the chapter, Mead maintained that "one striking difference between Samoan society and our own is the lack of specialization of feeling, and particularly of sex feeling, among the Samoans" (Mead 1928:215). According to Mead, Samoan feeling rules prevent adolescent sexual feelings and energies from being directed in gender-polarized ways—that is, only toward a particular opposite-sex person or even a few special opposite-sex persons. When Mead questioned 30 adolescent girls in detail about their sexual history and experimentation, she learned that 12 (40 percent) claimed they had engaged in heterosexual activities; 17 (56 percent) claimed they had engaged in homosexual activity; and 22 (73 percent) indicated experience with masturbation (Mead 1928:285). Mead concluded that in Samoa "masturbation is all but a universal experience, beginning at the age of six or seven. Theoretically it is discontinued with the beginning of heterosexual activity. . . . Among grown boys and girls casual homosexual practices also supplant it to a certain extent" (p. 136).[7] Homosexual relations between girls were described by Mead as being casual, never assuming any long-time importance, and "regarded as pleasant and natural diversion" (p. 147). Long-term and intense heterosexual passions were also rare, as Samoans "rate romantic fidelity in terms of days or weeks at most" (p. 155). Mead argued that, because Samoans possess "complete knowledge of sex, its possibilities and its rewards, they are able to count its true value" (p. 222). They do not reserve sex activity for just "important relationships" or "regard relationships as important [simply] because they are productive of sex satisfaction" (p. 222). Mead further concluded that "familiarity with sex and the recognition of sex as an art have produced a scheme of personal relations in which there are no neurotic pictures, no frigidity, no impotence, except as the temporary result of severe illness, and the capacity for intercourse only once in a night is counted as senility" (p. 151). "The Samoan girl never tastes the rewards of

romantic love as we know it, nor does she suffer as an old maid who has appealed to no lover or found no lover appealing to her, or as the frustrated wife in a marriage which had not fulfilled her high demands" (p. 211). Finally, Mead stated, "this acceptance of a wider range of sexual activities as normal provides a cultural atmosphere in which frigidity and psychic impotence do not occur and in which a satisfactory sex adjustment in marriage can always be established" (p. 223).

Mead's descriptions and conclusions regarding the nature of sexual activity and experimentation in Samoa included many comparative references to the adolescent sexual experience in America. In the latter case, she maintained, "secrecy, ignorance, guilty knowledge, [and] faulty speculations" result "in grotesque conceptions" of the physical facts of sex, and adolescents possess "a knowledge of the bare physical facts of sex without a knowledge of the accompanying excitement" (p. 216). In *Coming of Age in Samoa,* Mead wrote that our "records of maladjusted children are full of cases where children have misunderstood the sex act, have interpreted it as a struggle accompanied by anger, or as chastisement, have recoiled in terror from one highly charged experience" (p. 222).

According to Mead, everything about Samoan culture and the structure of its society fosters "the sunniest and easiest of attitudes toward sex." This society has eliminated "strong emotions," including jealousy, interest in competition, and "strong attachments to one person, be they parents or lovers."

Derek Freeman challenged Mead's motives, research methods, and conclusions about many aspects of Samoan life.[8] After decades of research, Freeman concluded that Samoan life possessed a Puritanical, aggressive, rank-conscious dark side. The most highly publicized part of his critique revolved around Mead's portrayal of Samoan sexual activity. In this regard, Freeman found the Samoans to be "a people who traditionally value virginity highly and so disapprove of premarital promiscuity as to exercise a strict surveillance over the comings and goings of adolescent girls" (p. 228). In fact, Freeman argued that "the cult of virginity is probably carried to a greater extreme than in any culture known to anthropology" (Freeman 1996:250).[9]

---

[7]Based on the experiences of the 30 adolescent girls interviewed by Mead, we can conclude that she exaggerated the universal nature of some sexual activities. Seventy-three percent hardly qualifies as "all but a universal experience."

[8]Freeman claims that he knew as early as November 1943 that he "would one day face the responsibility of writing a refutation of Mead's Samoan findings," a task he began in 1945 and that is still ongoing.

[9]Curiously, Freeman, the Samoans, and others who have critiqued Mead's research have not chosen to address Mead's account of near-universal homosexual experience and masturbation. Rather, the focus has been on proving her wrong about heterosexual activity.

In 1996, at the age of 86, Fa'apua'a Fa'amu, Mead's foremost Samoan friend of 1926, stepped forward after she learned that Margaret Mead had become famous by describing Samoa, "entirely incorrectly," as a "place where the 'community' does not attempt to 'curb' the sexual activity of adolescents" (Freeman 1996:xii). The 86-year-old woman "confessed" in tears to the Secretary for Samoan Affairs of the Government of American Samoa that she and her friend Fofoa had told "their American inquisitor the very antithesis of the truth" (p. xii). "We said we were out with boys. She failed to realize that we were just joking and must have been taken in by our pretenses . . . we just fibbed and fibbed to her" (p. viii).

If Mead was fooled—and doubt persists today as to whether she was—can we learn anything about feeling rules and gender relations from *Coming of Age in Samoa*? Regardless of whether her sources told the truth, Margaret Mead undoubtedly traveled to Samoa with a clear research agenda. She needed to find a "negative instance"—one exception to a supposed rule—to prove that adolescence need not be a time of inevitable emotional turmoil, stress, and rebellion, as it was for female adolescents in the United States. She found that instance in Samoa, where adolescence was portrayed largely as painless, smooth, untroubled, unstressed, and "perhaps the pleasantest time the Samoan girl will ever know." For Mead, this "negative instance" had to be a place that contrasted sharply with the United States. Perhaps she therefore needed to "imagine" a society where women were knowledgeable about the human body and could express sexual desire free of constraints and guilt.

As one professor of Samoan studies noted, "this may sound a bit harsh but I think Margaret Mead was talking about her own self. I think she brought her ideas. She was looking for a frame. But she brought her own ideas and her own problems from the U.S. of America and from her own self." The Secretary of Samoan Affairs agreed with the assessment: "I think she made up her mind that she was going to write that theory according to what she believed and not according to what the people are living here" (Australian Broadcasting Corporation and Discovery Channel 1988).

Compliance and Resistance to Gender Polarization

If we simply think about the men and women we encounter every day, we quickly realize that people of the same sex vary in the extent to which they meet their society's gender expectations. Some people con-

Jennifer Miller, a bearded lady for the Coney Island Sideshow, decided when she was 20 years old that shaving was about shame and hiding and that the only reason to shave would be to conform to an ideal standard of femininity.

© Deanne Fitzmaurice/San Francisco Chronicle

form to gender expectations; others do not. This variability, however, does not stop most people from using their society's gender expectations to evaluate their own and others' behavior and appearances in "virtually every other aspect of human experience, including modes of dress, social roles, and even ways of expressing emotion and experiencing sexual desire" (Bem 1993:192).

For 10 years, the author has collected response papers from students at Northern Kentucky University asking them to explain how gender ideals have shaped personal experiences in their lives. A reading of hundreds, even thousands, of these responses reveals that most students share personal experiences in which they (1) learn about gender ideals, (2) attempt to change behaviors and feelings that deviate from gender ideals, (3) give in to or comply with gender ideals, but maintain regrets about society's superficial standards, or (4) "come out of the closet" and openly challenge gender ideals. Here the word "closet" does not refer only to homosexuality, but rather to any situation in which someone hides personal information that contradicts gender ideals. "Compliance and Resistance to Gender Ideals" provides examples of the students' various types of personal experiences.

The accounts suggest that human beings do not passively accept gender ideals regarding masculinity

Compliance and Resistance to Gender Ideals

People do not always passively accept their society's gender ideals. Some people do not just comply with gender ideals; they also resist and even find ways to subvert these ideals. We might even argue that the power of gender ideals is reflected in people's compliance and resistance to them. Some examples of various ways that Northern Kentucky University students respond to gender ideals are listed below.

LEARN ABOUT GENDER IDEALS

- *When my son was about three years old, he wanted me to paint his nails. He thought it was really cool to have colored fingernails and toenails. He loves it! Later, at the age of five, he entered kindergarten. He really wanted me to paint his nails green so he could show off at school. I felt like I had to warn him how others might react. I told him that a lot of people think that nail polish is for girls only, and they might make fun of him. He looked up at me with a self-assured expression and said, "It's not just for girls. I'm a boy." In other words if boys were capable of painting their fingernails, then how could it just be for girls, and why would anyone think he was a girl if his fingernails were painted? He didn't understand why anyone would believe differently.*

- *In the past I often wore hand-me-downs from my brothers. Clothing was never an issue. This changed when I began to realize that unless I wanted to be ostracized, my mode of dress would have to conform to that of the other girls. In high school I began to realize how important my appearance and behavior were to being accepted. In the bathrooms the upperclassmen would show the freshmen how to throw up after lunch in order to stay thin. Anorexia and bulimia were widespread, and we were very ruthless and materialistic in our criticism of others. It suddenly became extremely important to me to be accepted. I began to practice to walk with a sway in my hips and to flirt with my eyes lowered and head tilted. I started to wear make-up to accentuate my feminine features. I learned that there was a difference in the way the girls from my all-girls' school behaved from all the boys across the street. I began to giggle, something I never really did growing up. I used to laugh out loud, but once in high school, I joined the other girls in laughing in high-pitched tones. Giggling was definitely a learned way of expressing emotions. Gender polarization has had its effects upon me. I now wear the color pink and, I am sorry to say, have been known to giggle upon occasion. I work as an amateur model and actress in Cincinnati and Columbus, banking on the male-female distinction.*

ATTEMPT TO CHANGE BEHAVIORS AND FEELINGS THAT "DEVIATE" FROM GENDER IDEALS

- *There is something that drove me crazy. The feeling I experienced and the warmth I felt toward this woman were extraordinary, but so wrong. I didn't have anywhere to turn and no one to talk to. I did not want to tell anyone who or what I had done. Society said it was wrong to feel and act this way, but why did I feel so good and think about her so much? I nearly drove myself crazy and ran as far as I could. I joined the Army to punish myself. Every day I hated what I had done and I wanted to hurt myself to make it go away. I hoped that by surrounding myself with men I would forget what I had done and find some man that would help make me into a "normal heterosexual woman." I would hurt my parents and I would lose my friends if they knew what I was. I felt so alone and so outcast. The only way I thought I could be forgiven was through pain and punishment.*

GIVE IN AND COMPLY WITH GENDER IDEALS BUT WITH REGRETS ABOUT SOCIETY'S SUPERFICIAL STANDARDS

- *In elementary school I was a "tomboy" and I wore my hair very short, never wore skirts or dresses, and always wore Nike high-top basketball shoes. Etched into my memory is the first time I remember feeling self-conscious about being a tomboy. I was in the third grade and was taking part in a physical fitness test in the gym. The test included the number of sit-ups we could do in a minute or the time it took us to run a certain distance. Needless to say, I scored higher than every boy. My male gym teachers looked at me with the strangest expression as if I had broken some unwritten rule. The boys said things like, "I can't believe we got beat by a girl." Others said, "Yeah, but she's not really a normal girl." From this point on I began to feel like an outcast. No girls really wanted to be my friends because I liked to do "boy" things. I was always excluded from their groups. I was never invited to any slumber parties. No boys wanted to befriend me either. Before too long, even my family tried to make me more feminine. Around the fifth grade, my aunt encouraged me to grow my hair long. My cousin sug-*

and femininity. In fact, most people find ways to subvert these ideals through deception, secret agreements with others, impression management, or outright challenges to the ideals (Mageo 1992). These student accounts also show that people can resist ideals even as they use them to evaluate themselves and others. We might argue that resistance, compliance, and the strain of simultaneously complying yet resisting speak to the importance of gender ideals in shaping our lives.

Our discussion has focused on how Northern Kentucky University students have complied with and rejected norms regarding appearances and behavior appropriate to the opposite sex. This examination shows that people cannot be divided into clear-cut categories. It is important to recognize that not every society divides people into "opposite" sexes—male or female. In American Samoa and some other areas of the Pacific Islands, there is a third gender—*fà-afafine*.

gested that I shave my legs. My grandmother bought me dresses, skirts, and hair bows. By then I was tired of the pressure. I was convinced I was a freak and that I must change to fit in and be accepted. I was determined to achieve a more feminine ideal. I said good-bye to my days as a tomboy and began to act like all other girls my age. I believe my experience in elementary school changed me forever. By being told that I was not feminine enough when I was young, I now constantly strive for this ideal even more than most women. I'm not saying that I don't like who I am today. I am quite comfortable with who I turned out to be. I often wonder, however, where I would be now if I had stayed with my old ways. Perhaps I would be the star of a college basketball team. I will never know because society transformed me into what it wanted me to be, a woman striving for feminine perfection.

- I distinctly remember when I first felt freakish about my facial hair. It was eleven years ago and I was a sophomore in high school. I had a huge crush on the star of our basketball team. Because I was painfully shy, I could never talk to him. After I had asked a few people about him, they sensed my crush and decided to tell him. One day a friend of mine told me she had asked him whether he might be interested in me and he said, "No way" and called me Blackbeard. She asked me what he could be talking about, obviously not noticing for herself, and I didn't have a clue at first. I honestly had never paid attention to the new dark hair growing on my face. After lunch I was still baffled, but it finally dawned on me that he was talking about the hair on my face. I was so embarrassed

and ashamed that I didn't think I was ever going to be able to face anyone ever again. Since that day I have bleached the hair on my face. I am sad that our culture resented something natural as unnatural. I'm a slave to creme bleach every month, and for what? Acceptance from men? No one has ever commented about my facial hair since then because I have conformed to the societal standards.

COME "OUT OF THE CLOSET" AND OPENLY CHALLENGE GENDER IDEALS, AND PAY THE PRICE

- I went on a blind date and met a girl who seemed too good to be true. Everything went well for the next three weeks. We hadn't been very intimate so when she asked me to come over to her house and stay with her, I jumped at the opportunity. I entered her house to find candles lit and soft music playing. We began to kiss and I noticed that she hadn't shaved her legs in a while. Disgusting, but under the circumstances I tried to block it out. As we got more involved I was surprised to find a lot of hair under her arms. I quickly pulled away. She asked what was wrong and after I told her she laughed at me. Needless to say I felt like I was going to throw up. I put my clothes on and quickly exited her house. On the way home I kept thinking how sick it made me that she didn't shave. I decided that I was never going to talk to her again. Sometime later she called me to explain. She told me that she didn't think women should have to shave and that she hadn't shaved for at least a year. She believed I should respect and even like her more for making a stand. Making a stand towards

what, I thought? Anyway, I told her that I thought it was disgusting and I told her I didn't want to see her again.

- In 1997 at a church retreat I met this guy, Jake, who did not care what other people thought of him. The first thing I noticed about him was his clothing. He wore both men and women's clothes. After talking with Jake for a while, I discovered that he was gay. He said the reason he wore women's clothes is that it made him feel more feminine. He did not like the fact that most people did not believe in mixing genders. You could be either male or female, but not both. Last Halloween Jake and I attended the same party. All of the guys were wearing scary costumes and the girls were wearing animal and dancer costumes. They wore the costume that society thinks fits the gender they were born into. Not Jake. He wore a sheet draped over himself, flowers in his hair, and sandals on his feet. People at the party were making fun of him and saying that he looked like a girl and that they didn't want to talk to him. Jake is a good example of a gender nonconformist because he did not conform to what other people wanted him to be. He would rather live his life as a little of both genders and be happy. A person is born into a gender and has a choice whether or not to conform to the expectations put in front of him/her. Jake went against the dominant paradigm and he paid the price. He lost out on many friends because both males and females could not accept that he did not act his gender.

A Third Gender?

Jeannette Mageo (1992) begins her article, "Male Transvestitism and Cultural Change in Samoa," by describing the guests attending a wedding shower in Samoa. Of approximately 40 "women," six were *fà-afafines*, guests who are not biologically female but who had taken on the "way of women" in dress, mannerism, appearance, and role. The closest word we have to this idea in American society is "transvestite." During that party, the *fà-afafines* staged a beauty contest in which each sang and danced a love song. Such beauty contests are well known in Samoa, and the winner "is sometimes the 'girl' who gives the most stunningly accurate imitation of real girls, such that even Samoans would be at a loss to tell the difference; sometimes the winner is the most brilliant comic" (Mageo 1998:213). Often the *fà-afafines* imitate popular foreign female vocalists, such as Whitney Houston or Madonna.

As Mageo notes, early Christian missionaries to Samoa did not mention transvestites in their written accounts of Samoan society. Given that the missionaries were preoccupied with documenting the sexual habits of the Samoans, it seems certain that they would have mentioned the *fà-afafines'* existence. This omission suggests that transvestitism was a marginal practice in pre-Christian Samoa. How then did it become commonplace among males, especially in urban areas? Mageo's answer is considerably more complex than the partial summary of her research presented here suggests. To fully appreciate her answer, one would have to spend weeks reading about Samoan culture, as *fà-afafines* cannot be truly understood apart from the larger societal context and broader history. Mageo (1992, 1998) argues that *fà-afafines* could not have become commonplace unless something about Samoan society supported gender blurring. In addressing this point, she notes that "on a personal level Samoans do not distinguish sharply between men and women, boys and girls." For example, "boys and girls take equal pride in their skills in fights; pre-Christian personal names are often not marked for gender, and outside school little boys and girls still wear much the same clothing" (1998:451).

Another practice that encourages the presence of *fà-afafines* relates to separation of boys and girls. Once Samoan boys reach the age of five or six, they begin spending the majority of their time in the company of other boys; at this point, they are prohibited from flirting with the girls. At the same time, "close and physically affectionate relations with same-sex people are established practices. In Samoa, as in much of the Pacific, boys may walk about hand-in-hand or with an arm draped around their comrade, and so may girls" (Mageo 1992:452).[10] (See "Hand-holding as an Expression of Affection (Not Homosexuality) Between Same-Sex Friends.")

Samoans also make a clear distinction between situations in which they must show respect and those in which they may engage in **ula,** highly sexualized entertainment including joking, jesting, and imitating. Although this kind of practice supports the existence of *fà-afafines,* other factors related to colonization and Christianization help to explain how *fà-afafines* became

transformed from a rare occurrence to a commonplace one.

At one time the *ula* was institutionalized, in the form of ceremonies involving young girls who were part of a village or organization, known as *aualuma.* The Christian missionaries sought (and succeeded on some levels) to change Samoan sexual customs, including the *ula.* These missionaries "saw Samoan sexual relations as practically without rules (ironically in much the same way Margaret Mead described them). They did not understand that "these relations reflected a set of rules invisible to missionaries. . . . Missionaries could not believe that a society in which young girls were encouraged to dance naked before torches and sing songs about sexual body parts could have any sexual morality at all." Because girls abandoned this role, Mageo makes the case that *fà-afafines* are both "stand-ins for bygone *aualuma* girls" and reminders of how girls are not to behave (1998:454).

Another factor that may account for the widespread emergence of the *fà-afafines* relates to changes in the position of and opportunities open to men in Samoa (see Table 11.2). Specifically, these changes are connected to the gradual and ongoing decline of the *aumaga,* an organization of younger and older men without titles. At one time, the *augama* was considered the "strength of the village" (Mead 1928:34), serving "as a village police force or an army reserve" (Mageo 1992:444). It took responsibility for the heavy work, whether that be "on the plantation, or fishing, [or] cooking for the chiefs" (Mead 1928:34). A system of mass education introduced by the missionaries, the shift away from an agricultural-based economy, along with the introduction of new technology and a wage-based economy eventually transformed the *augama,* in the process removing an important source of apprenticeship and status for Samoan males. This loss of status has been confounded by an unemployment rate of 12 percent. Moreover, when we consider that the total working-age population of Samoa is approximately 14,400, and that the two largest employers are the tuna canneries employing 4,282 and the Samoan government employing 4,000, we can see the problems of status that many males might experience (Infonautics Corporation 1998; U.S. Department of the Interior 1998). This economic situation has left the average man without a clear sense of status in Samoan society. For some men, becom-

---

Ula The Samoan practice in which individuals engage in highly sexualized entertainment including joking, jesting, and imitating.

---

[10]It is important to note that homosexuality is viewed as transitional and not as a "life-long" choice. Young people are expected to marry and have children.

U.S. in Perspective

Hand-holding as an Expression of Affection (Not Homosexuality) Between Same-Sex Friends

Countries highlighted in gold are those East Asian and Pacific countries where hand-holding between people of the same sex is not viewed as a sign of homosexuality. The hand-holding may be viewed as acceptable among men, women, and young people. Countries where hand-holding between same-sex friends is considered a sign of homosexuality are shaded in dark blue. (Data is not available for countries appearing white.)

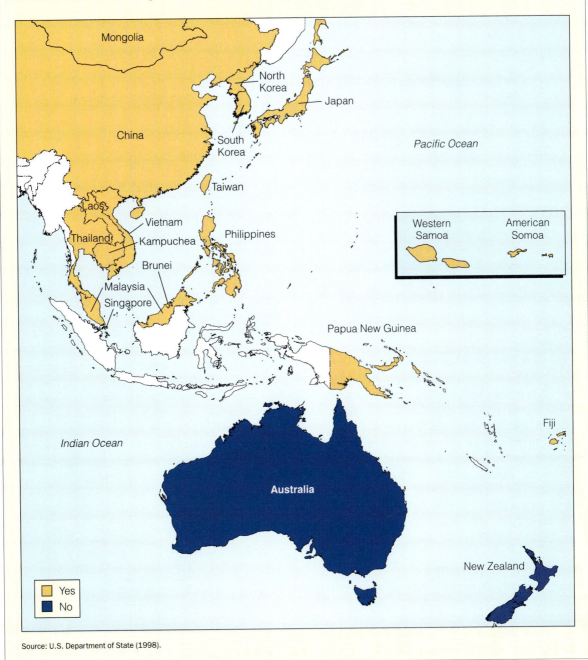

Source: U.S. Department of State (1998).

Table 11.2   Occupational Categories in Which Samoan Men and Women Are Disproportionately Overrepresented

In Samoa, women make up 41.2 percent of the labor force and men make up 58.7 percent. If occupations were filled without regard for gender, 41.2 percent of the people in each category would be female and 58.7 percent would be male.

Top Five Categories in Which Samoan Women Are Overrepresented

Occupation	Percentage of Female Workers
Secretaries, stenographers, and typists	94
Health service occupations (excluding home care)	90
Personal service occupations	72
Finance, insurance, and real estate	71
Machine operators, assemblers, and inspectors	65

Top Five Categories in Which Samoan Men Are Overrepresented

Occupation	Percentage of Male Workers
Construction trades	99
Mechanics and repairers	98
Forestry and fisheries	98
Transportation and material moving occupations	98
Precision production, craft, and repair occupations	90

Source: U.S. Department of the Interior (1992).

ing *fà-afafines* offers them an opportunity to step out of their roles as males and assume the status of well-known female impersonators.[11]

Gender Stratification

In Chapter 9, we learned that social stratification is the system used by societies to classify people into categories. When sociologists study stratification, they examine how the category into which people are classified affects their perceived social worth and their life chances. They are particularly interested in how individuals who possess one type of a characteristic (male reproductive organs versus female reproductive organs) are regarded and treated as more valuable than persons who possess the other type. Sociologist Randall Collins (1971) offers a theory of sexual stratification to analyze this phenomenon.

Economic Arrangements

Collins' theory of sexual stratification is based on three assumptions: (1) people use their economic, political, physical, and other resources to dominate others; (2) any change in the distribution of resources across a society alters the structure of domination; and (3) ideology is used to justify one group's domination over another. In

the case of males and females, males in general are physically stronger than females are. Collins argues that, because of differences in strength between men and women, the potential for coercion by males exists in every encounter of males with females.[12] He maintains that the ideology of sexual property—which he defines as the "relatively permanent claim to exclusive sexual rights over a particular person" (1971:7)—lies at the heart of sexual stratification and that women have historically been viewed and treated as men's sexual property.

According to Collins, the extent to which women are viewed as sexual property and subordinate to men historically has depended—and still depends—on two important and interdependent factors: women's access to agents of violence control, such as the police,[13] and

[11]While *fà-afafines* are included in activities reserved for females, such as wedding showers, and are accepted as entertainers, we should not assume that their lives are stress-free. For example, male relatives are often very unhappy with those who have become *fà-afafines,* in some cases beating them up.

[12]Note the word *potential*, which means "possible." Strength differences give men the potential to control women with physical force. Obviously, most men do not take advantage of this potential.

[13]Collins (1971) argues that women must have access to agents of violence control if they are to be men's equals. In other words, when women find themselves in a disadvantaged situation, they must be able to neutralize men's physical strength with support from the outside.

women's position relative to men in the labor market. On the basis of these factors, Collins identifies four historical economic arrangements: low-technology tribal societies, fortified households, private households, and advanced market economies.

These four economic arrangements are ideal types; the reality is usually a mixture of two or more types. Each arrangement is broadly characterized by distinct relationships between men and women. We begin our discussion by analyzing the first type, characteristic of low-technology tribal societies.

Low-technology tribal societies include hunting-and-gathering societies with technologies that do not permit the creation of surplus wealth—that is, wealth beyond what is needed to meet the basic needs (food and shelter). In such societies, sex-based division of labor is minimal because of an emphasis on collective welfare and the belief that all members must contribute if the group is to survive. In precontact Samoa,[14] "the basic economic tasks for men were agriculture, carpentry, hunting, and fishing; and for women, weaving, tapa making, and oil making" (Cote 1997:2). While all Samoan men and women learned the basic economic tasks deemed appropriate to their sex, they also specialized in certain areas, such as fishing or pigeon hunting, boat building, wood carving (for men), or producing the most valuable kinds of mats, cloth, medicines, or oils (for women).

Some evidence suggests that in hunting-and-gathering societies, women perform more menial tasks and work longer hours than men. Men hunt large animals, for example, while women gather most of the food and hunt smaller animals. In the Samoa described by Margaret Mead, the women performed routine agricultural work (weeding, transplanting, and gathering crops). Women also did "the routine fishing for octopuses, sea eggs, jellyfish, crabs, and other small prey" (p. 49).

Because almost no surplus wealth exists, marriage between men and women from different families does little to increase a family's wealth or political power. Consequently, daughters are not treated as "property" in the sense that they are not used as bargaining chips to achieve such aims. In comparison with fortified households and private household arrangements, women in low-technology tribal societies have more "freedom of self-determination, more social value, and higher status economic roles" (Cote 1997:8).

---

[14]*Precontact Samoa* is a term used to describe Samoa before colonization. Significant contact with Europeans occurred around 1830.

Fortified households include preindustrial arrangements characterized by no police force, militia, national guard, or other peacekeeping organization. Therefore, the household acts as an armed unit, and the head of the household represents its military commander. Fortified households "may vary considerably in size, wealth, and power, from the court of a king or great lord . . . down to households of minor artisans and peasants" (Collins 1971:11). All fortified households, however, share one common characteristic: the presence of a nonhouseholder class consisting of propertyless laborers and servants. In the fortified household, "the honored male is he who is dominant over others, who protects and controls his own property, and who can conquer others' property" (p. 12). Men treat women as sexual property in every sense: daughters are bargaining chips for establishing economic and political alliances with other households; male heads of household have sexual access to female servants; and women (especially in poorer households) bear many children, who eventually become an important source of labor. In this system, women's power depends on their relationship to the dominant men.

Private households emerge with a market economy, a centralized, bureaucratic state, and the establishment of agencies of social control that alleviate the need for citizens to take the law into their own hands. Thus private households exist when the workplace is separate from the home, men are still heads of households but assume the role of breadwinner (as opposed to military commander), and women remain responsible for housekeeping and child rearing. Men, as heads of households, control the property; it is a relatively

---

Low-technology tribal societies Hunting-and-gathering societies with technologies that do not permit the creation of surplus wealth.

Fortified households Preindustrial arrangements characterized by no police force, militia, national guard, or other peacekeeping organization; the household is an armed unit, with the head of the household as its military commander.

Private households The economic arrangements that exist when the workplace is separate from the home, men are still heads of households but assume the role of breadwinner (as opposed to military commander), and women remain responsible for housekeeping and child rearing.

Table 11.3	Women's Earnings as Percent of Men's Earnings, 1979–1996		
Year	Hourly	Weekly	Annual
1979	64.1	62.5	59.7
1980	64.8	64.4	60.2
1981	65.1	64.6	59.2
1982	67.3	65.4	61.7
1983	69.4	66.7	63.6
1984	69.8	67.8	63.7
1985	70.0	68.2	64.3
1986	70.2	69.2	65.2
1987	72.1	70.0	65.2
1988	73.8	70.2	66.0
1989	75.4	70.1	68.7
1990	77.9	71.9	71.6
1991	78.5	74.2	69.9
1992	80.3	75.8	70.8
1993	80.4	77.1	71.5
1994	80.6	76.4	72.0
1995	80.8	75.5	71.4
1996	81.2	75.0	73.8

Source: U.S. Department of Labor (1998).

new practice in the United States, for example, to put a house or credit in the names of both husband and wife. Moreover, men monopolize the most desirable and important economic and political positions under the private households arrangement.

According to Collins, a decline in the number of fortified households, the separation of work from home, a smaller family size, and the existence of a police force to which women can appeal in cases of domestic violence give rise to the notion of romantic love as an important ingredient in a marriage. In the marriage market, men offer women economic security because they dominate the important, high-paying positions. Women offer men companionship and emotional support and strive to be attractive—that is, to achieve the ideals of femininity, which may include possessing an 18-inch waist or removing most facial and body hair. At the same time, women try to act as

Advanced market economies Economic arrangements that offer widespread employment opportunities for women as well as for men.

sexually inaccessible as possible, because they offer sexual access to men only in exchange for economic security.

Advanced market economies offer widespread employment opportunities for women. Although women remain far from being men's economic equals, some now can enter relationships with men by offering more than an attractive appearance; they can provide an income and other personal achievements. Having more to offer, women can demand that men be physically attractive and meet the standards of masculinity. This situation may explain why more commercial attention has been given in the past decade to males' appearance—as evidenced by advertisements related to body building, hair styles, and male skin and cosmetic products.

Collins' classic theory of sexual stratification suggests that men and women cannot be truly equal in a relationship until women become men's economic equals (see Table 11.3). Many forces in society operate to perpetuate and maintain economic inequality. In 1990, for example, the United States terminated Dependency and Indemnity Compensation to spouses of deceased veterans upon the spouses' remarriage (*Samoa Daily News* 1998b). Of course, the veteran must have died from a disability, disease, or injury related to military service to warrant this compensation. The new policy removes a source of income from the surviving spouse, which in turn puts him or her at an economic disadvantage relative to the new spouse. Given that the benefits are a form of compensation for a life "sacrificed" in the line of duty, one might argue that the surviving spouse who is usually female and who supported the veteran's profession deserves compensation regardless of marital status.

In addition to addressing structural inequalities that put women at a disadvantage in the labor force, we must consider inequalities connected with housework and child-rearing responsibilities. In the United States, research shows consistently that among married couples in which both partners earn the same income and share egalitarian ideals, the husbands spend considerably less time than the wives in preparing meals, taking care of children, shopping, and performing other household tasks (Almquist 1992). One study showed that among "equal-earning" couples, the male partner's share of the housework is about 36 percent (Kamo and Cohen 1998). In the area of child-rearing responsibilities, simply consider that among single-parent households, single mother households outnumber single father households by 5 to 1.

Mechanisms of Perpetuating Gender Ideals

As mentioned at the onset of this chapter, people vary in the extent to which they conform to their society's gender expectations. This fact, however, does not prevent us from using gender expectations to evaluate our own and other people's behavior. For many people, failure to conform (whether deliberate or reluctant) serves as a source of intense confusion, pain, and pleasure. Sociologists are therefore interested in identifying the mechanisms by which individuals learn and perpetuate a society's gender expectations. To address this issue, we examine three important factors: socialization, situational constraints, and ideologies.

Socialization

In Chapter 5, we learned that socialization is a learning process that begins immediately after birth and continues throughout life. Through this process, "newcomers" develop their human capacities, acquire a unique personality and identity, and internalize (that is, take as their own and accept as binding) the norms, values, beliefs, and language they need to participate in the larger community. The socialization process may be direct or indirect. It is indirect when children learn gender expectations by observing others' behavior, such as the jokes or stories they hear about men and women, the reactions that significant others show to those who violate gender expectations, and the portrayals of men and women in magazines, books, and on television (Raag and Rackliff 1998). Socialization is direct when significant others intentionally convey the societal expectations to children, as in the following examples. Socialization theorists argue that an undetermined but significant portion of male-female differences are products of the ways in which males and females are treated during the socialization process.

Child development specialist Beverly Fagot and her colleagues observed how toddlers in a play group interacted and communicated with one another and how teachers responded to the children's attempts to communicate with them at ages 12 months and 24 months (Fagot et al. 1985). Fagot found no real sex differences in the interaction styles of 12-month-old boys and girls: All of the children communicated by gestures, gentle touches, whining, crying, and screaming. The teachers, however, interacted with the toddlers in gender-polarized ways. They were more likely to respond to girls when the girls communicated in gen-

Through the socialization process children learn gender-appropriate behavior. This four-year-old boy has rolled his hair up in curlers. Should someone tell him that curlers are for girls only or should he be allowed to play?

Courtesy of Yvonne Dupont Dressman

tle, "feminine" ways and to boys when the boys communicated in assertive, "masculine" ways. That is, the teachers tended to ignore assertive acts by girls and to respond to assertive acts by boys. Thus, by the time these toddlers reached two years of age, their communication styles showed quite dramatic differences.

Children's toys and celebrated images of males and females figure prominently in the socialization process, along with the ways in which adults treat children. Barbie dolls, for example, have been marketed for more than 30 years and currently are available in 67 countries. Executives at Mattel consider Barbie to be an **aspirational doll**—that is, a role model for the child. Barbie accounts for approximately half of all toy sales by Mattel (Boroughs 1990; Cordes 1992; Morgenson 1991; Pion 1993). An estimated 95 percent of girls between ages 3 and 11 in the United States have Barbie dolls, which come in several different skin colors. Market analysts attribute

Aspirational doll A doll considered to be a role model for a child.

Mattel's success to the fact that "they generally have correctly assessed what it means to a little girl to be grown-up" (Morgenson 1991:66). For boys, G.I. Joe became the first "action figure" toy on the market, being launched in 1964. It thrived for 12 years until 1976, when the line was canceled. G.I. Joe was reintroduced again in 1978 (Hasbro Toys 1998). Keep in mind that this is only one in a long line of action figures, including Transformers, Micronauts, Star Wars, Power Rangers, X-Men, Street Fighter, Bronze Bombers, and Mortal Kombat. The popularity of G.I. Joe generated several lines of comic books, 750 different action figures and vehicles, a motion picture, and cartoons; the G.I. Joe logo also appears on school supplies, video games, card games, lunch boxes, posters, and party supplies (Son 1998).

We might conclude that if we could change socialization experiences, then behavior should change accordingly. The Christian missionaries assigned to Samoa must have recognized this principle, as they sought to "destroy most of the social institutions that guided young Samoans through childhood to adulthood" (Cote 1997:7). Among other things, these missionaries sought to end the practice of tattooing (discussed earlier in this chapter), and they targeted the *aualuma* group of unmarried adolescent girls who "lived together" and "supported one another emotionally." The *aualuma* carried out village work projects and entertained visiting parties. Instead of living with the *aualuma,* unmarried girls were brought to live with the pastors and their wives. Here they learned how to sew and cook according to European standards (Cote 1997:8). By introducing mass education, the missionaries also changed the role of the *aumaga.* Instead of learning skills as members of the *aumaga,* boys studied inside the classroom, which prepared them to work for wages instead of for the village as a whole.

Structural or Situational Constraints

Situational theorists agree with socialization theorists that the social and economic differences between men and women cannot be explained by their biological makeup. In their view, these differences are caused by structural or situational constraints. **Structural constraints** are the established and customary rules, policies, and day-to-day practices that affect a person's life

chances. An example is occupations segregated by sex, so that women tend, more often than men, to be concentrated in low-paying, low-ranking, or dead-end jobs. (See Tables 11.2 and 11.3.) (Recall the discussion in Chapter 9 of occupations in which U.S. women are disproportionately underrepresented and overrepresented and of the inequality in men's and women's earnings.) The established and customary practices that put women at a disadvantage in the labor market are many and complex. Examples of such practices include the following:

- Limiting their job search to positions that are considered sex-appropriate

- Choosing specialties and fields that require working with children and young adults, that involve supervising other women, or that are otherwise considered feminine (for example, becoming a professor of social work rather than a professor of mathematics or computer sciences)

- Steering males and females into different gender-appropriate assignments and offering them different training opportunities and chances to move into better-paying jobs

- Choosing part-time jobs and disrupting careers to meet child-bearing and -rearing responsibilities

These structural constraints reinforce expectations about gender. For example, the different social and physical demands and skills required to perform the jobs held by men and women function toward "channeling their motivations and their abilities into either a stereotypically male or a stereotypically female direction" (Bem 1993:135). Sociologist Renee R. Anspach's research observing nurses and physicians working in neonatal care units illustrates vividly how one's position in a social structure can channel behavior in stereotypically male or female directions.

THE CASE OF PHYSICIANS AND NURSES Anspach spent 16 months conducting field research (observing and holding interviews) in two neonatal intensive care units (NICUs). Among other things, she found that nurses (almost all of whom were female) and physicians (usually male) used different criteria to answer the question, "How can you tell if an infant is doing well or poorly?" Physicians tended to draw on so-called objective (technical or measurable) information and immediate perceptual cues (skin color, activity level) obtained during routine examination:

Structural constraints The established and customary rules, policies, and day-to-day practices that affect a person's life chances.

Well, we have our numbers. If the electrolyte balance is OK and if the baby is able to move one respirator setting a day, then you can say he's probably doing well. If the baby looks gray and isn't gaining weight and isn't moving, then you can say he probably isn't doing well.

The most important thing is the gestalt. In the NICU, you have central venous pressure, left atrial saturations, temperature stability, TC (transcutaneous) oxymeters, perfusions (oxygenation of the tissues)—all of this adds in. You get an idea, when the baby looks bad, of the baby's perfusion. The amount of activity is also important—a baby who is limp is doing worse than one who's active. (Anspach 1987:219–220)

Although immediate perceptual and measurable signs were likewise important to the nurses, Anspach found that the nurses also considered interactional clues, such as the baby's level of alertness, ability to make eye contact, and responsiveness to touch:

I think if they're doing well they just respond to being human or being a baby. . . . Basically emotionally if you pick them up, the baby should cuddle to you rather than being stiff and withdrawing. Do they quiet when held or do they continue to cry when you hold them? Do they lay in bed or cry continuously or do they quiet after they've been picked up and held and fed. . . . Do they have a normal sleep pattern? Do they just lay awake all the time really interacting with nothing or do they interact with toys you put out, the mobile or things like that, do they interact with the voice when you speak? (p. 222)

Anspach concluded that the differences between nurses' and physicians' responses to the question "How can you tell if an infant is doing well or poorly?" could be traced to their daily work experiences. In the division of hospital labor, nurses interact more with patients than do physicians. Also, doctors and nurses have access to different types of knowledge about infants' condition, which correspond to our stereotypes of how females and males manage and view the world. Because physicians have only limited amounts of daily interaction and contact with infants, they tend to rely on perceptual and technological (measurable) cues. Nurses, on the other hand, remain in close contact with infants throughout the day; consequently, they are more likely to consider interactional cues as well as perceptual and technological ones.

Anspach (1987) suggests that one's position in the division of labor "serves as a sort of interpretive lens through which its members perceive their patients and predict their futures" (p. 217). Her findings suggest

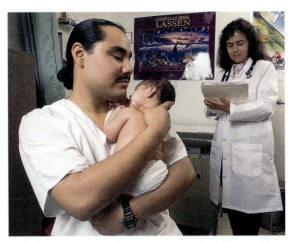

One's position in the social structure can channel behavior in stereotypically male or female directions. The job description of nurse, for example, requires the person in that position to interact more with patients than physicians do. Consequently, the nurse is more likely to consider interactional clues in evaluating a patient's medical condition.
© Anne Dowie

that when physicians make life-and-death decisions about whether to withdraw or continue medical care, they should collaborate with NICU nurses so that they can consider interactional as well as technological and immediate perceptual cues. Anspach's findings also suggest that if nurses' experiences and opinions counted more in medical diagnosis, we might see a corresponding increase in the prestige and salary associated with this largely female position.

The various structural constraints tend to push women into work roles that emphasize personal relationships and nurturing skills and that offer them control over family-oriented and feminine products/services. These work roles help to perpetuate and reinforce stereotypes that women and men are better suited to the jobs that we have come to define as masculine and feminine. Structural constraints may have also played some role in the drastically different views of Samoan life described by Mead and Freeman. To understand the potential role of structural/situational constraints in this case, we must ask whether being female affected Mead's choices and access to information. Likewise, we must ask whether being male affected Freeman's research.

Situational Constraints on Mead and Freeman

Recall that Margaret Mead was a 23-year-old female anthropologist when she went to American Samoa in

Mead drew conclusions about Samoan sexual customs from her discussions with adolescent females.

© Corbis-Bettmann

1925. She was greatly influenced by Franz Boaz,[15] who was worried that cultures—even in the most remote areas of the world—were disappearing "before the onslaught of modern civilization." Boaz believed that the "work of recording these unknown ways of life had to be done now—*now*—or they would be lost forever" (Mead 1972:137). He enlisted Mead to help in this effort. He decided that she should do research on adolescence, and he influenced her decision about where to conduct this research. Mead wanted to study people living on remote Polynesian Islands, but Boaz suggested Mead study a Native American culture because he believed it too dangerous for Mead to study outside the United States (Mead 1972). The two ultimately compromised, with Mead choosing American Samoa, a Pacific Island territory governed by the U.S. Department of Navy and a place in which a steamship went back and forth to and from the United States every three weeks (Mead 1973:132).

On August 21, 1925, Mead arrived in Pago Pago (the capital city). She stayed there in a rundown hotel for six weeks, learning the language and adjusting to the food and the Samoan way of life. From Pago Pago, she traveled to the island of Tau to begin her research on adolescents. Between November 1925 and June 1926, Mead lived in and worked out of the home of a U.S. Navy pharmacist and his family. She chose to live apart from the people under study, an unusual decision for an anthropologist.

This home base gave Mead easy access to 60 adolescent girls who became her main informants and who were free to visit her at all hours. The anthropologist developed close friendships with several of her key informants. Because Mead was a woman, she was denied access to some segments of Samoan society, including village chiefs and village council meetings.

Derek Freeman was also 23 years old when he first traveled to Western Samoa (not American Samoa) to do research between 1940 and 1943. During his stay, a village chief adopted the anthropologist as his son and Freeman became a honorary village chief, putting him in an "exceptionally favorable position to pursue research into the realities of Samoan life" (Freeman 1996). After studying a settlement of 400 people and its political structure for several years, Freeman realized that much of Margaret Mead's conclusions did not apply to the people he was studying. He returned to Western Samoa more than 20 years later in 1965 to collect facts to refute Mead's work. He resumed

[15]Franz Boaz is considered to be the "father" of modern anthropology.

his position as chief and took part in village council deliberations.[16]

Sexist Ideology

In Chapters 2 and 10, we learned that ideologies are ideas that support the interests of the dominant group but that cannot withstand scientific investigation. They are taken to be accurate accounts and explanations of why things are as they are. On closer analysis, however, we find that ideologies are at best half-truths, based on misleading arguments, incomplete analysis, unsupported assertions, and implausible premises.

Sexist ideologies are structured around three notions:

1. People can be classified into two categories, male and female.

2. A close correspondence exists between a person's primary sex characteristics and characteristics such as emotional activity, body language (see "Gender Polarization and Body Language"), personality, intelligence, the expression of sexual desire, and athletic capability.

3. Primary sex characteristics are so significant that they explain and determine behavior and social, economic, and political inequalities that exist between the sexes.

The **evolutionary view** of sex differences is one example of sexist ideology. It holds that human societies progress in stages from primitive (simple) to civilized (advanced), with each stage being characterized by a gradually more complex form of social organization. The evolutionary view "supported Anglo-Americans' definition of themselves as a superior race." In evolutionary terms, the less civilized the races, the more alike the two sexes are in terms of physical characteristics and the less its members are able to control sexual impulses (Newman 1996). According to this view, "the white woman's superiority to nonwhite peoples was due to her moral purity and sexual restraint, and was the basis of America's supposedly more advanced gender relations" (p. 244). Although the white woman was considered superior to all nonwhites, she was nevertheless

conceived as the weaker of the two sexes (Wishart 1995:2).

At the time that Margaret Mead was conducting her research in Samoa, the dominant ideology was an evolutionary view of gender relations. Mead's research undermined the evolutionary view because she argued and showed that male-female differences could not be simply biological in origin; they must be culturally determined. (Mead's research was also ideologically driven, but the ideology driving it was not sexist.)

Sexist ideologies are so powerful that "almost everyone has difficulty believing that behavior they have always associated with 'human nature' is not human nature at all but learned behavior of a particularly complex variety" (Hall 1959:67). Another example of a sexist ideology is the belief that men are prisoners of their hormones, making them powerless in the face of female nudity or sexually suggestive dress or behavior. Still another sexist ideology claims that men are not capable of forming relationships with other men that are as meaningful as those formed between women. Since the 1980s, dozens of books have been written by men to refute these stereotypes (Shweder 1994).

We might also add a fourth notion underlying sexist ideology: People who behave in ways that depart from ideals of masculinity or femininity are considered deviant, in need of fixing, and subject to negative sanctions ranging from ridicule to physical violence. This ideology is reflected in military policy toward homosexuals, for example. According to a U.S. Department of Defense (1990) directive:

> Homosexuality is incompatible with military service. The presence of such members adversely affects the ability of the Armed Forces to maintain discipline, good order, and morale; to foster mutual trust and confidence among the members; to ensure the integrity of the system of rank and command; to facilitate assignment and worldwide deployment of members who frequently must live and work under close conditions affording minimal privacy; to recruit and retain members of the military services; to maintain the public acceptability of military services; and, in certain circumstances, to prevent breaches of security. (p. 25)

[16]Freeman claimed that his study of a Western Samoan culture rather than the American Samoan culture studied by Mead made no difference, because the Samoans in Tau had assured him that life was essentially the same in both places. Freeman also argued that it did not matter that his research took place 40 years later, because he assumed that Samoan culture had not changed in any significant way since 1925.

Evolutionary view The idea that human societies progress in stages from primitive to civilized, with each stage being characterized by a gradually more complex form of social organization.

Gender Polarization and Body Language

Socialization theory may help explain the differing norms governing body language for males and females. According to women's studies professor Janet Lee Mills, norms governing male body language suggest power, dominance, and high status, whereas norms governing female body language suggest submissiveness, subordination, vulnerability, and low status. Mills argues that these norms are learned and that people give them little thought until someone breaks them, at which point everyone focuses on the rule breaker. Such norms can prevent women from conveying a sense of security and control when they are in positions that demand these qualities, such as lawyer, politician, or physician.

Mills suggests that women face a dilemma: "To be successful in terms of femininity, a woman needs to be passive, accommodating, affiliative, subordinate, submissive, and vulnerable. To be successful in terms of the managerial or professional role, she needs to be active, dominant, aggressive, confident, competent, and tough" (Mills 1985:9).

Janet Lee Mills is famous for this pose, which she captions, "Could you say no to this woman?" In it she violates many traditional female behaviors with relaxed posture: arms, legs positioned away from body; direct, confrontational eye contact; no affiliative smile.

Mills demonstrates the "power spread," another typical high-status male pose, with hands behind head, elbows thrust out, legs in a "broken-four" position, and an unaffiliative facial expression. Women dressed for success appear shocking in this pose.

Mills recruited Richard Friedman, Assistant to the President at the University of Cincinnati, to model for these photographs. Friedman does not appear shocking in the same pose, because many male executives conduct business from a similar position.

Ah, but doesn't Mills look feminine in this typically feminine pose with canted head, affiliative smile, ankles crossed, and hands folded demurely?

And doesn't Friedman look ridiculous in the same pose?

Power is often wielded in postural alignment, gesture, and use of objects. Mills holds power, along with papers on which attention is focused; Friedman signals deference with lowered gaze and constricted body.

Now Friedman holds power, along with papers, with lower limbs spread away from his body; Mills signals deference with an attentive pose, submissively bowed head, and a hand-to-mouth gesture of uncertainty.

In this typical office scene, Friedman holds power with an authoritative stance, one hand in pocket and the other at midchest, straight posture, and head high; Mills is submissive with canted head, smile, arms and hands close to her body. Note Friedman's wide, stable stance and Mills's unstable stance. Many women tend to slip into a similar posture when talking to a shorter male authority figure.

Now the tables are turned. Friedman defers to authority in a feminine, subordinate posture with scrunched-up spine, constricted placement of arms and legs, canted head, and smiling attentiveness.

All photos courtesy of Janet Lee Mills.

Navy SEALS, an elite unit, are taught to work in small units and to depend on each other. These trainees hug each other to retain body heat after a grueling night-time cold-water conditioning session. Would the presence of gays in the military disrupt this kind of training? Might the presence of two men who share an intense friendship also affect training?

© Rick Rickman/Matrix

No scientific evidence, however, supports this directive. In fact, whenever Pentagon researchers (with no links to the gay and lesbian communities and with no ax to grind) have found evidence that runs contrary to this directive, high-ranking military officials have generally refused to release the information or found the information unacceptable and directed researchers to rewrite the reports. For example, when researchers found that sexual orientation is unrelated to military performance and that men and women known to be gay or lesbian displayed military suitability that is as good as or better than that of men and women believed to be heterosexual,[17] U.S. Deputy Undersecretary of Defense Craig Alderman, Jr., wrote the researchers that the "basic work is fundamentally misdirected" (Alderman 1990:108). He explained that the researchers should determine whether being a homosexual was connected to being a security risk, not to determine whether homosexuals were suitable for military service. Although the researchers found no data to support a connection between sexual orientation and security risks, Alderman maintained that the findings were not relevant, useful, or timely. (See "Ideology to Support Military Policy.") The research that Alderman dismissed would have gone unnoticed, had Congressman Gerry Studds and House Arms Subcommittee Chairwoman Patricia Schroeder not insisted upon its release.

This example shows the role that ideologies play in setting policy. In this case, the ideologies are that homosexuality is incompatible with military service and that being homosexual represents a security risk to the United States. The case of the military also alerts us to the fact that other variables, such as a person's sexual orientation, interact with biological sex to affect the experience of being male or female in different ways. To examine this interaction, we turn to the work of sociologists Floya Anthias and Nira Yuval-Davis, who have written about the interconnection among gender, race, ethnicity, and country (the state).

Gender, Ethnicity, Race, and the State

Ethgender refers to people who share (or are believed by themselves or others to share) the same sex, race, and ethnicity. This concept acknowledges the combined (but not additive) effects of gender, race, and ethnicity on life chances. Ethgender merges two ascribed statuses into a single social category. In other words, a person is not just a resident of Samoa but an ethnic Samoan[18]—a white Samoan (Geschwender 1992). To complicate matters, people of a particular ethgender have a legal relationship with a country or state—that is, an individual may be a citizen, refugee (temporary worker), or national. In a legal sense, people born in American Samoa are classified as U.S. nationals, not citizens.[19]

We use the term *country* or *state* here to mean a governing body organized to manage and control specified activities of people living in a given territory. The governing body is almost always male-dominated. For example, in the 50-year history of the Samoan legislature, only five women have been elected to office. In the last election (November 1998), only 5 of 50 candidates competing to fill 20 seats in the Samoan House of Representatives were women (*Samoa Daily News* 1998d). The U.S. Congress is similarly male-dominated.

Ethgender People who share (or are believed by themselves or others to share) the same sex, race, and ethnicity.

[17]"Believed to be homosexual" is an important consideration, because the military currently follows a "don't ask, don't tell policy," and because some recruits remain in the "closet."

[18]Some 94 percent of people in Samoa are classified as Samoan.

[19]U.S. nationals, however, can apply for U.S. citizenship (*Samoa Daily News* 1998c).

Ideology to Support U.S. Military Policy

People who oppose the presence of gays and lesbians in the military stereotype homosexuals as sexual predators who are waiting to pounce on a heterosexual person while he or she is showering, undressing, or sleeping. Opponents seem to believe that any same-sex person is attractive to a gay or lesbian person. But as one gay ex-midshipmen notes, "Heterosexual men have an annoying habit of overestimating their own attractiveness" (Schmalz 1993:B1). The excerpt from Pentagon research included here shows that no evidence exists to support such a stereotype.

Those who resist changing the traditional policies support their position with statements of the negative effects on discipline, morale, and other abstract values of military life. Buried deep in the supporting conceptual structure is the fearful imagery of homosexuals polluting the social environment with unrestrained and wanton expressions of deviant sexuality. It is as if

persons with nonconforming sexual orientations were always indiscriminately and aggressively seeking sexual outlets. All the studies conducted on the psychological adjustment of homosexuals that we have seen lead to contrary inferences. The amount of time devoted to erotic fantasy or to overt sexual activity varies greatly from person to person and is unrelated to gender preference. In one carefully conducted study, homosexuals actually demonstrated a lower level of sexual interest than heterosexuals.

Homosexuals are like heterosexuals in being selective in their choice of partners, in observing rules of privacy, in considering appropriateness of time and place, in connecting sexuality with the tender sentiments, and so on. To be sure, some homosexuals are like some heterosexuals in not observing privacy and propriety rules. In fact, the manifold criteria that govern sexual interest are identical for homosexuals and heterosexuals, save for only one criterion: the gender of the sexual partner.

Age, gender, kinship, class membership, marital status, size and shape, social role, posture, manners, speech, clothing, interest/indifference signaling, and other physical and behavioral criteria are all differentiating cues. They serve as filters to screen out undesirable or unsuitable potential sex partners. With such an array of cues, many (in some cases, all) potential objects of interest are rejected. For most people, only a small number of potential partners meet the manifold criteria. Whether in an Army platoon or in a brokerage office, people are generally selective in their choice of intimate partners and in their expression of sexual behavior. Heterosexuals and homosexuals alike employ all these variables in selecting partners, the only difference being that the latter include same-gender as a defining criterion, the former include opposite-gender.

Source: Sarbin and Karols (1990:37).

Of the 435 members that make up the House of Representatives, only 56 (12.8 percent) are women. Of the 100 voting members of the Senate, 9 (9 percent) are women.

Sociologists Floya Anthias and Nira Yuval-Davis (1989) give special attention to women, their ethnicity, and the state. They argue that "women's link to the state is complex," and women "are a special focus of state control because of their role in human reproduction—women carry and give birth to children who become the state's citizens and future labor force." Anthias and Yuval-Davis maintain that the state's policies and political debate reflect its concerns about the kinds of babies (that is, their race and ethnicity) to which women give birth and about the ways in which these babies become socialized. They identify five areas of women's lives over which the state may choose to exercise control. One should not conclude, however, that women meekly accept the policies and programs imposed by the state. In fact, women often form organizations and work to change state policies. Examples of such organizations are the National Orga-

nization for Women (NOW) and the Pan Pacific and South East Asia Women's Association of American Samoa (PPSEAWA).

1. Women as Biological Reproducers of Babies of a Particular Ethnicity or Race As factors that can underlie a state's population control policies, Anthias and Yuval-Davis (1989) name "fear of being 'swamped' by different racial and ethnic groups" or fear of a "demographic holocaust"—that is, trepidation that a particular racial or ethnic group will die out or become too small to hold its own against other ethnic groups (p. 7). Such policies can range from physically limiting numbers of a particular racial or ethnic group deemed undesirable to actively encouraging the "right kind" of women to produce more children.[20] Policies that limit numbers include immigration control (limiting or excluding members of certain ethnic groups from entering a

---

[20]Governments can also institute policies that encourage women to reproduce by offering tax incentives, maternity leave, and other benefits.

country and subsequently producing children). The August 1998 call to reinstate a policy barring women who are classified as "foreigners" and who are six months pregnant from entering Samoa represents an example of such a policy (*Samoa Daily News* 1998a). It is intended to reduce the likelihood that such women will take advantage of Samoa's medical services or stay in Samoa after the baby's birth. Other population control policies include physical expulsion, extermination, forced sterilization, and massive birth control or family planning campaigns.

2. Women as Reproducers of the Boundaries of Ethnic or National Groups   In addition to implementing policies affecting the numbers and kinds of babies born, states may institute policies that define who the parents should be. Examples of such policies include laws prohibiting sexual relationships between men or women of different races or ethnicities, laws specifying marriage if the child is to be recognized as legitimate, and laws connecting the child's ethnic and legal status to the ethnicity of the mother or father. Over its more than 200-year history as an independent country, the United States has employed a bewildering variety of laws specifying the racial identity of babies born to parents who belonged to different racial categories. Keep in mind that the U.S. government does not recognize "multiracial people" (Knepper 1995:123); thus it has prohibited multiracial identity. As late as 1980, any person of mixed parentage was "classified according to the race of the nonwhite parent, and mixtures of nonwhite races [were] classified according to the race of the father" (U.S. Bureau of the Census 1993:21).[21]

3. Women as Transmitters of Social Values and Culture
The state can institute policies that either encourage women to be the main socializers of their offspring or leave socialization in the hands of the state. For example, it may tie welfare payments to nonemployment (or under welfare reform to employment) so that mothers are forced to stay home with the children. Some countries have generous maternity leave policies; others have no policies at all. Sometimes state leaders become concerned that children of particular ethnic or racial groups are not learning the cultural

values or language that they need to succeed in the dominant culture. This concern may motivate them to fund programs such as preschools that expose particular kinds of children to the necessary personal, social, and learning skills.

4. Women as Symbols of Urgent Issues   Political leaders often use images of women and men to symbolize the most urgent issues that they believe the country faces. In wartime, for example, the country may be represented as "a loved woman in danger or as a mother who lost her sons in battle" (Anthias and Yuval-Davis 1989:9–10). Men are called to battle so as to protect the women and children. Often the leaders present the image of a woman who meets the culture's ideal of femininity and who belongs to the dominant ethnic group. Sometimes political leaders evoke images of women of a certain ethnic or racial group as the source of a country's problems, such as Hispanic women who produce many children or African American welfare mothers with no economic incentives to practice birth control. A check of the facts often reveals that such images are unfounded. For instance, in "Fertility Among Women on Welfare: Incidence and Determinants," sociologist Mark R. Rank (1989) maintains that "it is impossible to calculate with any precision the fertility rate of women on public assistance" (p. 296) because the data available have serious flaws. For one thing, women move in and out of welfare, and we simply do not keep track of their fertility during these periods. "There is no way of judging whether the fertility rate of women on welfare is high or low" (p. 296) relative to the fertility rate of other women.

5. Women as Participants in National, Economic, and Military Struggles   States may implement policies governing the roles that women and men can assume in crises, notably in war. Historically, women have played supportive and nurturing roles, even in situations in which they have been exposed to great risks. In most countries, women are not drafted; they volunteer to serve. If they are drafted, the state defines acceptable military roles. If women fight, they often do so as part of special units or in an unofficial capacity.

Regardless of their official roles, women are affected by war. Since World War II, civilians have suffered an estimated 80 percent of all war-related deaths and casualties (Schaller and Nightingale 1992). Women are killed, taken prisoner, tortured, and raped. Even so, they often are not trained formally and systematically to fight. As a result, women occupy a different position than men during war.

---

[21]A special exception applied to persons of mixed Negro and Indian descent. In such cases, no matter what the father's race, the person was classified as Negro "unless the Indian ancestry very definitely predominates or unless the individual is regarded as Indian in the community" (p. 21). There were exceptions to this rule.

The fact that women's combat roles are limited does not mean that women are incapable of combat. In fact, during the Civil War women disguised themselves as men so they could serve in the war.

© Corbis-Bettmann

Through its military institutions, the state even establishes policies that govern male soldiers' sexual access to women outside military bases, both in general and in times of war. During the war in Bosnia, for example, the Serbs captured some women and sent them to places resembling concentration camps in which many were raped; they also kept other women in brothel-like houses and hotels. During World War II, Japanese military authorities forcibly recruited[22] between 60,000 and 200,000 women—mostly Korean but also Chinese, Taiwanese, Filipino, and Indonesian—to work as sex slaves in army brothels in the war zone (Doherty 1993; Hoon 1992). They referred to these captives as "comfort women."

In *Let the Good Times Roll: Prostitution and the U.S. Military in Asia,* Saundra Pollock Sturdevant and Brenda Stoltzfus (1992) examine "the sale of women's sexual labor outside U.S. military bases" (p. vii). They present evidence that the U.S. military helps regulate prostitution; that retired military officers own some of the clubs, massage parlors, brothels, discotheques, and hotels; and that the military provides the women with medical care so as to prevent the spread of sexually transmitted diseases between the soldiers and the women. As an example of the complex relationship between military men and local women, consider that thousands of Filipino women who live near the Subic Bay Naval Base filed a class action suit against the United States in 1993, arguing that the United States has moral and legal responsibilities to support the estimated 8,600 children fathered by U.S. servicemen stationed at Subic Bay.[23] In the 1940s, some 14,371 U.S. Marines were stationed in Samoa at the Tutuola naval station alone; another 25,000 to 30,000 troops were stationed in Western Samoa. In some locations, the servicemen outnumbered the Samoan population. A great deal of sexual interaction took place between Samoan women and U.S. troops. Although we will never know exactly how many illegitimate babies were fathered by American soldiers, the numbers are significant. "One mission society reported that in Upolu alone there were 1,200 known instances of illegitimate children by American soldiers" (Stanner 1953:327).

In evaluating the role of the military in the women's lives, we must consider that many poor women who live near the bases see a relationship with a U.S. recruit as their only way out of poverty and a desperate situation. The Subic Bay situation is therefore more complicated than the story presented by the media. To clarify this point, we will use an example from a student at Northern Kentucky University.

A young male student, perhaps 23 or 24 years old, stopped the author after a class in which we had discussed the Subic Bay base closing. He explained that when he was in the Navy and his ship docked at Subic Bay, it seemed as if the entire town turned out to welcome the ship. Local women were everywhere. Whereas many recruits visited prostitutes for one-night stands, others fell in love with local women. Usually

---

[22]*Recruit* is the word used by the Japanese to describe how they managed to bring Korean women to war zones. Documents show that the women were acquired through civilian agents, village raids, and applying pressure on Korean school administrations to provide girls (Hoon 1992).

[23]Under current U.S. immigration law, Filipino children of servicemen are not permitted to migrate to the United States with sponsorship from any American, as can the children of servicemen stationed in South Korea, Thailand, Cambodia, and Laos (Lambert 1993).

the commanding officers attempted to discourage permanent relationships. Although some might criticize the officers' actions, they might have been anticipating the difficulties ahead for a recruit who took the woman as his wife and then continued life "at sea," which would leave her alone. Unfortunately, the complexity of the relationship between the military and the local populations is often overlooked.

Summary and Implications

In this chapter, we distinguished between sex (a biological distinction) and gender (a social construction). We considered the problems associated with gender—culturally conceived and learned ideas about appropriate behavior and appearance for males and females.

Sociologists find gender a useful concept, not because all people of the same sex look and behave in uniform ways, but because a society's gender expectations are central to all people's lives, whether they conform rigidly or resist. For many people, failure to conform to gender expectations, even if they fail deliberately or conform only reluctantly, represents a source of intense confusion, pain, and/or pleasure (Segal 1990).

Unfortunately, many people equate the notion of gender polarization with the idea that women should abandon traditional vocations, such as housewife and mother, in favor of careers. Furthermore, many critics point to "gender awareness" and the corresponding push toward sexual equality as the cause of family breakdown.

From a sociological perspective, gender awareness and the goal of sexual equality should strengthen the family. How can this development arise? The information in this chapter helps us to see a number of important points related to this issue. If men and women did not feel constrained to select partners according to age (that is, if the typical woman were not generally several years younger than the typical male partner), the sex ratio would be more balanced, especially in old age. Society would include fewer widows or, at the very least, women who married would tend to spend fewer years alone.

If men and women did not feel constrained to meet artificial standards of beauty, the personal energy and financial resources channeled toward obtaining cosmetics, clothes, cosmetic surgery, depilatories, and diet products could be spent in other, more socially useful ways.

If women were paid a salary equal to that received by men, perhaps they would be less vulnerable to poverty in the event of separation, divorce, or the death of a spouse.

If men's and women's self-images, aspirations, and life chances were less constrained by gender scripts, men might choose to stay home and take care of children rather than pursue a full-time career. Similarly, women would not face the no-win situation associated with choosing between a family and career. As things stand today, if a married woman works and has no children, she is viewed as selfish; if she stays home and raises a family, she is considered "underemployed" at best; if she works and raises a family, people wonder how she can possibly do the job right; if she is divorced with children, she is responsible for the breakdown of the family unit; if she does not marry, she is considered a spinster.

Finally, consider the manner of conveying sexual intentions. If women did not feel constrained about communicating sexual concerns and interests (or lack of interest) and if men were more sensitive to these gender constraints on women, perhaps fewer unwanted pregnancies and abortions would occur. Whether a person is pro-choice or pro-life, he or she would agree that 1.5 million abortions[24] per year is an unacceptable—even alarming—number. The number of abortions suggests that honest communication between women and men about the consequences of sex does not take place for a significant number of couples. This failure to communicate is also reflected in the number of unplanned pregnancies, which are as high as 81.7 percent among women aged 15–19 (Forrest 1994).

Perhaps these constraints explain why Margaret Mead needed to "find" a society in which women "possess complete knowledge of sex, its possibilities and its rewards" and in which women did not fall in love with the first man who has kissed them. How many problems can be traced to men and women who form relationships based on sexual attraction? Whether Mead actually captured the reality of Samoan life is in some ways irrelevant when we consider that she was looking for a real or imagined alternative to the United States. In other words, American Samoa functioned for Mead as a mirror through which Americans could see themselves and another alternative. Her hope was that Americans might then envision new ways of reforming their social institutions to support gender ideals and inequality (Newman 1996).

---

[24]This number seems especially high when we consider that approximately 4.0 million births occur each year. For every 40 live births, 15 abortions are performed.

Key Concepts

Use this outline to organize your review of the key chapter ideas.

Sex
    Primary sex characteristics
    Secondary sex characteristics
Gender
    Gender polarization
    Social emotion
    Gender-schematic decisions
    Femininity
    Masculinity
Intersexed
Feminist

Sexual stratification
    Low-technology tribal societies
    Fortified households
    Nonhouseholder class
    Private households
    Advanced market economies
Perpetuating sexual stratification
    Socialization
    Situational constraints
    Sexist ideologies
    Institutionalized discrimination
Ethgender
State

internet assignment

In this chapter we learned that although men as a group, especially white males, have an economic advantage relative to women in the labor market, they are concentrated in the most hazardous occupations. Find the Web site of the U.S. Bureau of Labor Statistics and look for information that shows disadvantages to males, females, or both.

Economics and Politics

With Emphasis on the United States

ATM in Seattle. (Tim Crosby/Gamma Liaison.)

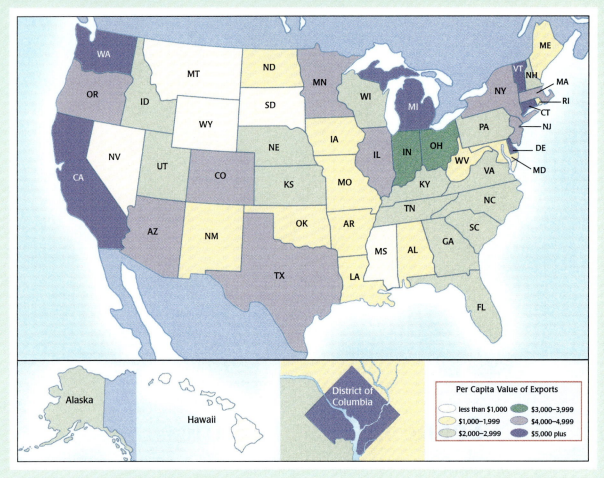

Sources: International Trade Administration (1998); U.S. Census Bureau (1997).

In 1997, the 50 states *exported* $687.6 billion in merchandise trade only to the world. This map shows the per capita value of exports for each state. In this case, per capita value is calculated by dividing the number of people in a state's work force by the dollar value of exports for that state. Which states have the largest per capita exports? The smallest per capita exports? Why do you think this is the case?

The following statements by President Clinton of the United States highlight the American belief that democracy (a political arrangement) and capitalism (an economic arrangement) go hand-in-hand.

- As I have told audiences from Santiago to Shanghai: unless nations deepen their democracies, unless the citizens of each nation feel they have a stake in their economy, they will resist reforms necessary for recovery. Unless citizens are empowered with the tools to master economic change, they will feel they are its victims, not its victors. So we will continue to offer them a better way. We will with our efforts to generate greater economic growth, to seek freer and fairer trade, and to promote liberty and democracy. (Clinton 1998c)

- The United States seeks a stable, economically dynamic and democratic Africa with which it can trade, in which it can invest profitably, and with which it can work to advance U.S. interests globally. This goal will be impossible to achieve unless and until Africa is fully integrated into the global economy. (White House 1998a)

- Today, we are drawing up the blueprints for a new economic age— blueprints not for starting big institutions but rather for freeing small entrepreneurs. We have the honor of designing the architecture for a global electronic marketplace, a marketplace that includes all nations, a marketplace with stable laws, strong protection for consumers, serious incentives for competition, and includes all people and nations. (Clinton 1998d)

- Today, 34 of the 35 nations in this hemisphere [the Americas] embrace democracy and free markets after decades of civil wars and unrest [have] given way to peace and stability. U.S. exports to this region are growing faster than to any other region; indeed 42% of all U.S. exports went to Free Trade Area of the Americas (FTAA) countries in 1997. The region is becoming a cornerstone for U.S. security and prosperity in [the] 21st century. (Clinton 1998b)

- We do not seek to impose our vision on others, but we are convinced that certain rights are universal—not American rights or European rights or rights for developed nations, but the birthrights of people everywhere, now enshrined in the United Nations Declaration of Human Rights—the right to be treated with dignity; the right to express one's opinions, to choose one's own leaders, to associate freely with others, and to worship, or not, freely, however one chooses. (Clinton 1998a)

- From the time our Administration took office in 1993, we have believed it vital to our future that the United States look not only to the West, but, as a Pacific power, to the East as well, and forge a strong Asia-Pacific community for the 21st century. Central to that effort are the APEC Leaders Forum, which just concluded, and the strengthening of our bonds with Japan and Korea, two of our strongest allies for promoting democracy, securing peace, and building prosperity. (Clinton 1998b)

Why Focus on the United States?

In this chapter, we consider the economy and politics together. We follow this course because a country's economy and government are fundamentally intertwined. That is, governments—whatever their form—design policies to manage economic activity and to distribute wealth. A government's economic policies are shaped by its visions of the ideal economy. As we will learn later, that vision usually falls somewhere on a continuum between two competing ideals: capitalism and socialism.

We focus on the United States for two reasons: because it possesses the largest economy in the world, and because after the Cold War ended in 1989 it became the world's only superpower. The **Cold War** is the name given to the massive military buildup that took place between 1945 and 1989 as the Soviet Union and the United States sought to contain the spread of one another's economic and political systems. In the case of the United States, its economic system was capitalism and its political system was democracy; in the case of the Soviet Union, the economic system was socialism and the political system was communism. The Cold War's fierceness becomes readily evident when we realize that during this period the number of nuclear weapons in the world grew from zero to 50,000,[1] and the United States and the Soviet Union became involved in as many as 120 proxy wars. The best known of these proxy wars were fought in Korea, Vietnam, and Afghanistan.

Today, many Americans believe that because their country "won" the Cold War, its political and economic systems "won" as well. In light of this history, President Clinton's statements about the connection between democracy and capitalism carry an even greater force and sense of conviction. His statements reflect a fundamental belief shared by many Americans: that democracy and capitalism are inseparable, and that each needs the other to exist. In theory, capitalism's emphasis on private ownership of property, the pursuit of personal profit, free competition, and consumer choice corresponds with democratic principles including freedom of choice, freedom of expression, and election of leaders who are accountable to the voters.

[1]The massive amount of nuclear waste generated to manufacture 50,000 weapons will remain radioactive for 240,000 years (see Chapter 2).

This chapter explores two major social institutions: the economy and politics. In discussing the economy, we pay special attention to three revolutions that have shaped economic activity. Those revolutions are the Agricultural Revolution, the Industrial Revolution, and the Information Revolution. We compare and contrast two major economic systems—capitalism and socialism. In addition, we explore the global economy and some characteristics of the U.S. economy. In the second half of the chapter, we delve into politics, paying special attention to three major political systems: democracy, communism, and totalitarianism. We also consider some of the essential characteristics of the U.S. political system.

Cold War The massive military buildup that took place between 1945 and 1989 as the Soviet Union and the United States sought to contain the spread of one another's economic and political systems.

Social institutions Widely recognized and relatively predictable and persistent patterns of behavior and interactions that have emerged over time to coordinate and channel human activity toward meeting basic human and societal needs.

Social Institutions

Sociologists classify both economic and political systems as major social institutions. **Social institutions** are widely recognized and relatively predictable and persistent patterns of behavior and interactions that have emerged over time to coordinate and channel human activity toward meeting basic human and societal needs. (This definition does not imply that the coordinating effort lacks flaws, if only because the process of meeting these needs causes certain groups to benefit at the expense of others.) In the case of economic institutions, those needs relate to the production, distribution, and consumption of goods and services. In the case of political institutions, they relate to the use of and access to the power that is essential to articulating and realizing individual, local, regional, national, international, or global interests and agendas.

In the chapters that follow we consider other important social institutions: the family (Chapter 13), education (Chapter 14), and religion (Chapter 15). Other institutions are also covered in this textbook, such as the criminal justice system (in Chapter 8), the medical system (in Chapter 6), and the media (in Chapters 3 and 6).

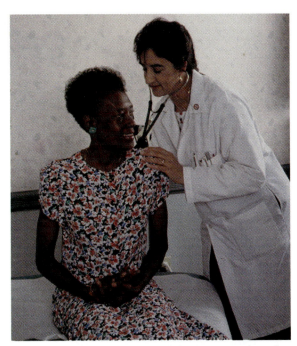

The service sector of the economy includes activities performed for others that result in no tangible product, such as medical care.

© Myrleen Ferguson/PhotoEdit

The Economy

A society's **economic system** is the institution that coordinates human activity in the effort to produce, distribute, and consume goods and services. **Goods** include any product that is manufactured, grown, or extracted from the earth, such as food, clothing, housing, automobiles, coal, computers, and so on. **Services** include activities performed for others that result in no tangible product, such as entertainment, transportation, financial advice, medical care, spiritual counseling, and education. Three major, ongoing revolutions have shaped the economic system: the Agricultural Revolution, the Industrial Revolution, and the Information Revolution.

The Agricultural Revolution

Many agricultural revolutions have taken place over the course of human history. The first important agricultural revolution took place approximately 10,000 years ago with the domestication of plants and animals. "Domestication" means that plants and animals were brought under human control. Instead of hunting for and gathering wild grains and available vegetation, people planned when they would plant and harvest crops. This planning allowed for a predictable food source and made it possible to produce more grain than people needed to just survive. The availability of excess grain facilitated the domestication of cattle, oxen, and

other animals, because enough food became available for animals as well as people. Domestication reduced the need to hunt for animals, and it offered people an additional source of power (other than human muscle) to transport heavy loads, to guard sheep, and so on.

Another major agricultural revolution occurred around 5000 B.C. with the invention of the scratch plow, "with a forward-curving wooed blade for cutting the soil, and a backward-curving pair of handles" by which farmers could direct oxen. Domestication and the invention of the plow are arguably the most fundamental inventions in the history of agriculture, because they created food surpluses (Burke 1978:9). The existence of a food surplus enables a society to support nonfood occupations, such as potters, weavers, bakers, musicians, craftsmen, and so on. Figure 12.1 shows how food surpluses triggered a chain of events that changed society in fundamental ways. Other innovations such as the windmill, the tractor, pesticides, and genetic engineering also triggered "agricultural revolutions" in that they dramatically increased crop yields, altered the number of people needed to produce food, and further ensured predictable food supplies.

The Industrial Revolution

In Chapter 6, we learned that an increase in population size (achieved through food surplus) intensifies the demands for resources. This transformation, in turn, stimulates the development of more efficient methods for producing goods and services. That is, as population size and density increase, society "advances steadily towards powerful machines, towards great concentrations of forces and capital, and consequently to the extreme **division of labor**" (Durkheim [1933] 1964:39),

---

Economic system An institution that coordinates human activity in the effort to produce, distribute, and consume goods and services.

Goods Any product that is manufactured, grown, or extracted from the earth, such as food, clothing, housing, automobiles, coal, computers, and so on.

Services Activities performed for others that result in no tangible product, such as entertainment, transportation, financial advice, medical care, spiritual counseling, and education.

Division of labor An arrangement where work is broken down into specialized tasks, with each task being performed by a different set of persons. Not only do the tasks themselves become specialized, but the parts and materials needed to manufacture products also come from many geographical regions.

Figure 12.1   Plant and Animal Domestication and the Plow: Effects on Civilization

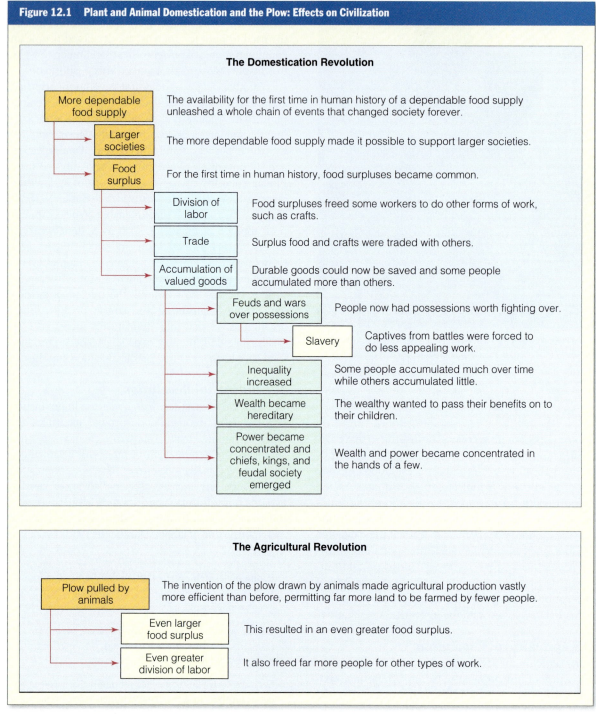

The Domestication Revolution

More dependable food supply
The availability for the first time in human history of a dependable food supply unleashed a whole chain of events that changed society forever.

Larger societies
The more dependable food supply made it possible to support larger societies.

Food surplus
For the first time in human history, food surpluses became common.

Division of labor
Food surpluses freed some workers to do other forms of work, such as crafts.

Trade
Surplus food and crafts were traded with others.

Accumulation of valued goods
Durable goods could now be saved and some people accumulated more than others.

Feuds and wars over possessions
People now had possessions worth fighting over.

Slavery
Captives from battles were forced to do less appealing work.

Inequality increased
Some people accumulated much over time while others accumulated little.

Wealth became hereditary
The wealthy wanted to pass their benefits on to their children.

Power became concentrated and chiefs, kings, and feudal society emerged
Wealth and power became concentrated in the hands of a few.

The Agricultural Revolution

Plow pulled by animals
The invention of the plow drawn by animals made agricultural production vastly more efficient than before, permitting far more land to be farmed by fewer people.

Even larger food surplus
This resulted in an even greater food surplus.

Even greater division of labor
It also freed far more people for other types of work.

Source: From *Sociology Timeline.* Copyright © 1995, IdeaWorks, Inc.™ Portions copyright © 1995, Board of Curators, University of Missouri.

where work is broken down into specialized tasks, with each task being performed by a different set of persons. Not only do the tasks themselves become specialized, but the parts and materials needed to manufacture products also come from many geographical regions.

To satisfy its growing demand for resources, the West vigorously colonized much of Asia, Africa, and the Pacific in the late nineteenth and early twentieth centuries. Western governments forced the local populations in these regions to cultivate and harvest crops and extract minerals and ores for export. Thus the Industrial Revolution was not unique to Western Europe and the United States, but rather forced people from even the most seemingly isolated and remote regions of the planet into a worldwide division of labor that continues today.

One fundamental feature of the Industrial Revolution was **mechanization,** the addition of external sources of power, such as oil or steam, to hand tools and modes of transportation. This innovation trend turned the spinning wheel into a spinning machine, the hand loom into a power loom, and the blacksmith's hammer into a power machine. It replaced wind-powered sailboats with steamships and horse-drawn carriages with trains. On a societal level, it changed the nature of work. The new forms of energy supported machines, factories, and mass production (see Chapter 1), thereby reducing the physical requirements needed to produce goods. As a result, industrialization transformed individual workshops into factories, craftspeople into machine operators, and manual production into machine production. Products previously designed as unique entities and assembled by a few people became standardized and could be assembled by many workers, each performing a specific task in the overall production process. Factory workers became wage laborers, meaning that they sold their labor for a price to an employer rather than working for themselves or a household.

The Post-industrial Society and the Information Revolution

In the 1950s, sociologist Daniel Bell argued that the modern industrial economy was being replaced by an economy centered around the creation of information and delivery of services rather than around the production of goods. In post-industrial society, the majority of workers are employed in service, managerial, technical, professional, research and development, or other information-based jobs—not manual jobs.

The invention of, and improvements in, computers and telecommunication technologies such as satellites and fiber optics has accelerated the shift away from a manufacturing-centered economy and toward an information- and service-based economy (see Figure 12.2). In particular, the computer—which was originally developed in the 1940s—benefited greatly from the development of the silicon chip, which can process millions of bits of information in a second or less. The silicon chip has reduced computer size and cost, making widespread use of these machines possible. Other inventions, such as fiber optics and satellites, have virtually eliminated the barriers of time and place, increasing the speed at which people can transmit images, voices, and data.

One fundamental feature of the Industrial Revolution was mechanization. It replaced muscle-generated power with power generated from external sources such as steam and oil.

© Corbis-Bettmann

Major Economic Systems

We can classify the economies of the world as falling somewhere on a continuum between capitalism and socialism. No economy in the world fully realizes capitalist or socialist principles. Nevertheless, the capitalist and socialist principles described here represent **economic ideals,** the standards against which real economic systems can be compared.

Capitalism

Capitalism is an economic system in which the raw materials and the means of producing and distributing goods and services remain privately owned. That is, capitalist systems are characterized by private ownership of the means of production. These systems

---

Mechanization The addition of external sources of power, such as oil or steam, to hand tools and modes of transportation.

Economic ideals The standards against which real economic systems can be compared.

Capitalism An economic system in which the raw materials and the means of producing and distributing goods and services are privately owned. Capitalist systems are characterized by private ownership of the means of production and are profit-driven, free of government interference, and consumer-driven.

Figure 12.2   Percentage of Work Force Employed in Manufacturing by State, 1996

Although the percentage of the work force employed in manufacturing is 15.3 percent for the United States as a whole, the percentages vary across the 50 states. Which states have the lowest percentage of workers employed in manufacturing? Which states have the highest percentage? Why do you think this is the case?

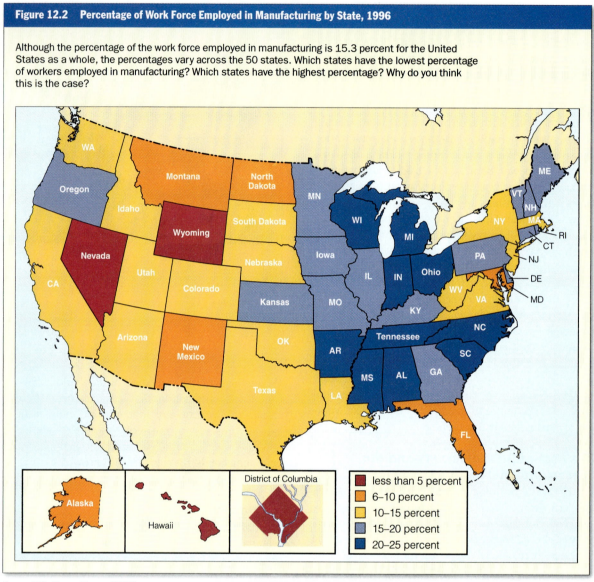

District of Columbia

	less than 5 percent
	6–10 percent
	10–15 percent
	15–20 percent
	20–25 percent

Source: U.S. Census Bureau (1997).

are profit-driven, free of government interference, and consumer-driven. **Private ownership** means that individuals (rather than workers, the government, or communal groups) own the raw materials, machines,

Private ownership A situation in which individuals (rather than workers, the government, or communal groups) own the raw materials, machines, tools, labor, trucks, buildings, and other inputs needed to produce and distribute goods and services.

tools, labor, trucks, buildings, and other inputs needed to produce and distribute goods and services. *Profit-driven* is the most important characteristic of capitalist systems. In such a system, the owners of the means of production and distribution are "driven" to continually increase and maximize profits. This profit motive ensures that owners, in their quest to maximize their return on investment, make the most effective use of labor and resources. The production and distribution process is consumer-driven because businesses depend on consumers "choosing" to purchase the goods and services. Thus consumers ultimately control the market and owners therefore have no choice except to produce goods and services that meet consumer demand.[2]

Ideally, capitalist systems remain free of government regulations and other types of interference. Instead, they are governed by the **laws of supply and demand**—that is, "as demand for an item increases, prices rise. When manufacturers respond to the price increase by producing a larger supply of that item, this increases competition and drives the price down" (Hirsch, Kett, and Trefil 1993:455). Although most economic systems in the world are classified as capitalist, including the U.S. system, no system fully realizes capitalist principles.[3]

For example, although most Americans consider the United States to be a capitalist country, capitalist principles do not apply to the operation of the defense industry. In this market, a single customer (the Pentagon) describes the weapons or other technologies it desires and contracts with a company to manufacture them. If this industry operated on capitalist principles, the supplier would invest its own money and resources to create and develop the product, and it then would compete with other companies to sell the product to the buyer (Lambert 1992). Because capitalist principles do not apply to defense contractors, they do not have to control costs, understand customer needs, or implement customer-oriented marketing or sales strategies.

Socialism

In contrast to capitalism, **socialism** is an economic system in which the raw materials and the means of producing and distributing goods and services are collectively owned. That is, public ownership—rather than private ownership—is an essential characteristic of this system. Socialists reject the idea that markets regulate

The New York Stock Exchange, located on Wall Street, is a well-known symbol of capitalism.
© Reuters/Mark Cardwell/Archive Photos

themselves through the laws of supply and demand, and government therefore plays a central role in regulating and controlling economic activity. In socialism's most extreme form, the pursuit of personal profit is forbidden. In less extreme forms, profit-making activities

---

[2]Although pursuit of profit is a largely self-centered activity, advocates of capitalism consider it socially legitimate because they assume that what is profitable for entrepreneurs and corporations and their stockholders also benefits society as a whole. Quality and cost-effectiveness are enhanced when competition exists among those who produce or distribute the same goods and services. Ideally, well-informed consumers "vote" for products with their purchases. Manufacturers and providers that cannot sustain consumers' interest and that cannot match or surpass their competitors in cost and quality will be forced to improve, or eventually they will go out of business.

[3]Many American politicians, economic analysts, and others complain that the United States is at an unfair advantage in the marketplace because it competes against economies whose governments interfere with the free exchange of goods and services.

Laws of supply and demand Natural laws regulating capitalist economies such that "as demand for an item increases, prices rise. When manufacturers respond to the price increase by producing a larger supply of that item, this increases competition and drives the price down" (Hirsch 1993:455).

Socialism An economic system in which the raw materials and the means of producing and distributing goods and services are collectively owned.

These workers are part of a cooperative association of farmers who work land owned by the state of Cuba. This photograph is a common image associated with socialist economic systems.
© Steve Cagan/Impact Visuals

World system theorists classify Haiti as a peripheral economy because most of the jobs that connect its people to the global economy are low-paying ones in industries such as textiles.
© Julio Etchart/Impact Visuals

are permitted as long as they do not interfere with larger collective goals.

As with capitalism, no economic system fully realizes socialist principles. The People's Republic of China, Cuba, and North Korea are classified as socialist, but they nevertheless permit varying degrees of personal wealth-generating activities that are closely watched and subject to state control.

World System Theory

Immanuel Wallerstein (1984) is the sociologist most frequently associated with world system theory, a modern theory concerned with capitalism. He has written extensively about the ceaseless expansion of a single market force—capitalism—that created the world-economy. The world-economy is not a new invention; it is at least 500 years old and continues to evolve. The world-economy, which currently encompasses more than 200 countries and thousands of cultures, is interconnected by a single division of labor. In this system, economic transactions transcend national boundaries (see Figure 12.3). Although each government seeks to shape the global market in ways that benefit its "own" corporations and national interests, no single political structure (world government) or national government holds authority over the system of production and distribution.

How has capitalism come to dominate the global network of economic relationships? One answer to this question can be found in the ways in which capitalists

respond to changes in the economy, especially to economic stagnation. Historically, five important responses have existed, all designed to generate profits through economic growth:

1. Lowering production costs by hiring employees who will work for wages (for example, by busting unions, buying out workers' contracts, or offering early retirement plans), by introducing labor-saving technologies (such as automating the production process), or by moving production facilities out of high-wage zones and into lower-wage zones inside or outside the country

2. Creating a new product that consumers "need" to buy, such as the videocassette recorder, computer, fax machine, or cellular phone

3. Improving an existing product and thus making previous versions obsolete

4. Expanding the outer boundaries of the world economy and creating new markets

5. Redistributing wealth to enable more people to purchase products and services

Concrete examples of each response can be found in Chapter 7 (see page 195).

As a result of these five responses to economic stagnation, capitalism has spread steadily throughout the globe and facilitated the interconnections between regional and national economies. In addition, every country of the world has come to play one of three different and unequal roles in the global economy: core, peripheral, or semiperipheral.

The Roles of Core, Peripheral, and Semiperipheral Economies

Core economies include the wealthiest, most highly diversified economies with strong, stable governments. The G-7 countries (Japan, Germany, France, the United States, Canada, United Kingdom, and Italy) are examples of core economies that collectively absorb nearly two-thirds of the other countries' exports (see Chapter 2 for a description of U.S. manufacturing operations in Mexico). Core economies import raw materials from the poorest countries and make use of free-trade zones around the world. The overwhelming majority of the *Fortune* 500 corporations, those "great global enterprises that make the key decisions—about what people eat and drink, what they read and hear, what sort of air they breathe and water they drink, and, ultimately, which societies will flourish and which city blocks will decay" (Barnet 1990:59)—have their headquarters in countries with core economies. The sales of these corporations exceed the gross national products (GNPs) of many countries (Currie and Skolnick 1988). When economic activity weakens in core economies, the other economies suffer because exports decline and price levels fall.

Peripheral economies are not very diversified, as they rely on a few or even a single commodity, such as coffee, peanuts, or tobacco, or a single mineral resource, such as tin, copper, or zinc (Van Evera 1990). These economies have a dependent relationship with core economies that traces its roots to colonialism. Peripheral economies operate on the fringes of the world economy. Most of the jobs that connect their workers to the global economy are low-paying and require few skills. In the midst of their widespread and chronic poverty, one finds islands of economic activity, including off-shore manufacturing zones, highly vulnerable extractive and single-commodity economies, and tourist zones.

Between the core and the periphery are the **semiperipheral economies,** characterized by moderate wealth (but extreme inequality) and moderately diverse economies. They exploit peripheral economies and are in turn exploited by core economies. By this definition, the Gulf countries of Bahrain, Kuwait, Oman, Qatar, Saudi Arabia, and the United Arab Emirates are semiperipheral economies. The core economies rely on them for cheap oil, and they depend on peripheral economies and other semiperipheral economies to supply cheap labor. The extent of that reliance becomes clear when we consider that 68 percent of the total labor force in the six Gulf countries comprises foreign workers. Egypt, Jordan, Yemen, Sudan, Bangladesh, India, Pakistan, and Sri Lanka are the major countries from which labor flows to these Gulf countries (Omran and Roudi 1993).

According to Wallerstein, semiperipheral economies play an important role in the world-economy because they are politically stable enough to provide useful places for capitalist investment if wage and benefit demands become too great in core economies. Consider that the U.S. Department of Commerce has identified 10 emerging markets, all of which can be classified as semiperipheral economies: Chinese Economic Area (which includes Hong Kong and Taiwan), India, Indonesia, South Korea, Argentina, Brazil, Mexico, Poland, Turkey, and South Africa.

The U.S. Economy

According to the *World Factbook* (U.S. Central Intelligence Agency 1997), the United States has "the most powerful, diverse, and technologically advanced economy in the world." In addition "U.S. firms are at or near the forefront in technological advances" in all economic sectors, especially in computers and software technologies and in medical, aerospace, and military equipment. The United States has the highest per capita GDP in the world—$28,600. Its economy can be classified as capitalist or market-oriented, in that "private individuals and business firms make most of the decisions" and the government purchases needed goods and services from private enterprises (as opposed to government-owned ones). The *World Factbook* also points out that, compared with firms in other core economies, American companies have greater flexibility in decisions to expand, lay off workers, and develop new products (U.S. Central Intelligence Agency 1997). This section examines some of the other major characteristics of the U.S. economy.

Core economies The wealthiest, most highly diversified economies with strong stable governments.

Peripheral economies Economies that rely on a few or even a single commodity, such as coffee, peanuts, or tobacco, or a single mineral resource, such as tin, copper, or zinc.

Semiperipheral economies Economies characterized by moderate wealth (but extreme inequality) and moderately diverse economies. Semiperipheral economies exploit peripheral economies and are in turn exploited by core economies.

Figure 12.3   Top Five Export Markets for the United States, 1997

This chart shows the five largest export markets for each state. It offers one example of how economic relationships transcend national borders in the world economy. Overall, which countries are the most important export markets? Find your home state. Are you surprised at which countries are the top export markets? Why?

| | Five Largest Export Markets | | | | |
	1	2	3	4	5
Alabama	Canada	Mexico	Japan	Korea	UK
Alaska	Canada	Japan	Korea	Belgium	UK
Arizona	Mexico	Japan	Netherlands	Malaysia	Canada
Arkansas	Canada	Russia	Hong Kong	Mexico	Japan
California	Japan	Canada	Mexico	Korea	Taiwan
Colorado	Japan	Mexico	Italy	UK	Canada
Connecticut	Canada	Japan	UK	Korea	Germany
Delaware	Canada	Belgium	Japan	Mexico	Brazil
Dist. of Col.	Taiwan	Saudi Arabia	UK	Egypt	Kuwait
Florida	Brazil	Canada	Venezuela	Columbia	Mexico
Georgia	Canada	Mexico	Japan	UK	Netherlands
Hawaii	Japan	Canada	New Zealand	Singapore	Korea
Idaho	Japan	Canada	UK	Germany	Singapore
Illinois	Canada	Japan	Mexico	UK	Germany
Indiana	Canada	Mexico	Japan	UK	France
Iowa	Canada	Mexico	Japan	UK	Korea
Kansas	Japan	Canada	Mexico	Venezuela	Korea
Kentucky	Canada	Japan	France	Mexico	UK
Louisiana	Japan	Canada	Belgium	Netherlands	Taiwan
Maine	Canada	Malaysia	Singapore	Japan	Korea
Maryland	Canada	UK	Mexico	Belgium	Germany
Massachusetts	Canada	Japan	UK	Germany	Netherlands
Michigan	Canada	Mexico	Japan	Belgium	Germany
Minnesota	Canada	Japan	Netherlands	Mexico	UK
Mississippi	Canada	Russia	Mexico	Honduras	UK
Missouri	Canada	Mexico	Belgium	Japan	Spain

Source: International Trade Administration (1998).

Primary sector (of the economy) Economic activities that generate or extract raw materials from the natural environment.

Secondary sector (of the economy) Economic activities that transform raw materials into manufactured goods.

Tertiary sector Economic activity related to delivering services, including the creation and distribution of information.

A Strong Tertiary Sector

We can think of an economy as being composed of three sectors: the primary sector, the secondary sector, and the tertiary sector. The **primary sector** includes economic activities that generate or extract raw materials from the natural environment. Mining, fishing, growing crops, raising livestock, drilling for oil, and planting, clearing, or gathering forest products are examples. The **secondary sector** consists of economic activities that transform raw materials from the primary sector into manufactured goods. The **tertiary sector**

| | \multicolumn{5}{c}{Five Largest Export Markets} | | | | |
	1	2	3	4	5
Montana	Canada	Belgium	Japan	Mexico	China
Nebraska	Japan	Canada	Korea	Mexico	Venezuela
Nevada	Canada	Japan	UK	Mexico	Philippines
New Hampshire	Canada	Germany	UK	Japan	Ireland
New Jersey	Canada	Japan	Korea	Netherlands	Germany
New Mexico	Philippines	Malaysia	Korea	Taiwan	Mexico
New York	Canada	Japan	Switzerland	UK	Mexico
North Carolina	Canada	Mexico	Japan	UK	Germany
North Dakota	Canada	Belgium	Germany	Mexico	Australia
Ohio	Canada	France	Mexico	Japan	UK
Oklahoma	Canada	Mexico	Japan	Venezuela	Netherlands
Oregon	Japan	Canada	Taiwan	Korea	Philippines
Pennsylvania	Canada	UK	Japan	Mexico	Brazil
Rhode Island	Canada	UK	Japan	Mexico	Germany
South Carolina	Canada	Mexico	Germany	Japan	UK
South Dakota	Canada	UK	Taiwan	Netherlands	Japan
Tennessee	Canada	Mexico	Japan	UK	China
Texas	Mexico	Canada	Japan	Singapore	UK
Utah	UK	Japan	Canada	Malaysia	Germany
Vermont	Canada	UK	Japan	Germany	France
Virginia	Canada	Japan	Belgium	Korea	Germany
Washington	Japan	UK	Canada	Korea	China
West Virginia	Canada	UK	Japan	Netherlands	Italy
Wisconsin	Canada	Japan	UK	Germany	France
Wyoming	Canada	Netherlands	Japan	Belgium	UK

encompasses economic activity related to delivering services, including the creation and distribution of information.

One way to determine the relative importance of each sector to an economy is to determine how much it contributes to the **gross domestic product (GDP)**, the monetary value of the goods and services that a nation's work force produces over the course of a year (or some other time period). In 1997, agriculture accounted for 1.6 percent of the U.S. GDP and manufacturing accounted for 17.0 percent. The tertiary or

service sector accounted for 76.5 percent of GDP (see Figure 12.4).

The importance of the tertiary sector to the economy is reflected in the list of occupations projected to experience the largest growth in terms of number of jobs created between 1996 and 2006 (Bureau of Labor

Gross domestic product (GDP) The monetary value of the goods and services that a nation's work force produces over the course of a year (or some other time period).

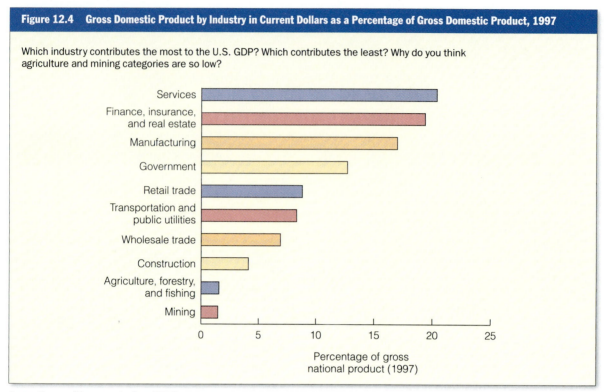

Figure 12.4   Gross Domestic Product by Industry in Current Dollars as a Percentage of Gross Domestic Product, 1997

Which industry contributes the most to the U.S. GDP? Which contributes the least? Why do you think agriculture and mining categories are so low?

Percentage of gross national product (1997)

Source: Bureau of Economic Analysis (1998).

Statistics 1998e). The 10 occupations estimated to have the largest job growth are cashiers (530,000 jobs), systems analysts (520,000), general managers and top executives (467,000), registered nurses (411,111), retail salespersons (408,000), truck drivers (404,000), home health aides (378,000), teacher's aides and educational assistants (370,000), nurse's aides and orderlies (333,000), and receptionists and information clerks (318,000).

The importance of the tertiary sector is also apparent in consumption patterns. The average U.S. household earns $39,928 (before taxes). Of that amount, $1,927 is spent on food (eaten away from home), $1,841 on health care, $1,813 on entertainment, and $3,223 on personal insurance and pensions (Bureau of Labor Statistics 1998f). The point is someone has to deliver the services that consumers demand.

Economist Robert Lawrence (1998) argues that because most Americans work in the service sector of the economy, they are largely protected from losing their jobs to global competition. According to Lawrence, this protection exists because "the essence of a service is that it has to be produced in the same place it is consumed" (p. 3). Others predict that the Internet has the potential

Technologies such as the Internet allow people to view and purchase products and services 365 days a year, 7 days a week, 24 hours a day.

© Neil Ricklen/PhotoEdit

to "to become the United States' most active trade vehicle within a decade, creating millions of high paying jobs. In addition, Internet shopping may revolutionize retailing by allowing consumers to sit in their homes and buy a wide variety of products and services from all over the world" (White House 1997).

Dominance of Large Corporations

A **monopoly** exists when a single producer dominates a market. Monopolies are illegal in the United States, because a company that does not face competition can set prices without any challenge. Another kind of marketplace domination is **oligopoly,** a situation in which a handful of producers dominate a market. For example, three foreign-headquartered corporations control 90 percent of technical, medical, and scientific publishing in the United States (Carvajal 1998).

The largest U.S.-based corporations[4] employ hundreds of thousands of people and operate on a multinational and even truly global scale. In 1997, General Motors (GM) was the largest corporation in the world as measured by revenues; its revenues amounted to $178.1 billion in that year (Fortune 1998). The company employs 608,000 people, making it the third largest employer among the Global 500,[5] and it has a "global presence in over 190 countries" (General Motors 1999).

Many of the world's largest companies are **conglomerates;** that is, they own "smaller" corporations acquired through merger or acquisition. GM is an example of a conglomerate. It is the parent company for a variety of smaller firms such as General Motors Acceptance Corporation (one of the world's largest financial service companies), Hughes Electronics Corporation (the world's largest producer of commercial communications satellites and a major player in the aerospace and defense systems industry), General Motors Locomotive Group, Allison Transmission (the world's largest designer and producer of heavy-duty transmissions), General Motors International Operations (vehicle design, production, and marketing), and Delphi Automotive Systems (General Motors 1999).

One way that corporations grow in size and influence is through mergers. In 1998, some of the largest mergers in corporate history were proposed. For example, Citicorp merged with Travelers Group to form Citigroup Incorporated, offering consumers "one-stop shopping for all their banking, investing, insurance, and other financial needs" (Online Newshour 1998). NationsBank merged with Bank America to create the first coast-to-coast bank in the United States; German-based Daimler-Benz merged with U.S.-based Chrysler Corporation; and Exxon Corporation merged with Mobil Corporation to create the world's largest oil company. Other mergers that occurred in 1998 spanned the entertainment and media, health care, pharmaceuticals, defense, insurance, and retail industries (Online Newshour 1998; Farber 1999).

Decline in Union Membership

Organized labor or unions seek to maintain and sometimes improve workers' wages, benefits, and working conditions. They also strive to ensure that the workplace remains secure for the future generation of workers. Despite these lofty goals, union membership in the United States has declined from a high of 35 percent of the work force in the 1950s to 14.5 percent of the work force in 1995 (Employment Policy Foundation 1998). The drop in membership is connected to many factors, including the following:

- The declining significance of the manufacturing sector in the overall economy (the traditional base of union membership)
- Increasing percentages of females in the work force (who work in jobs that are traditionally non-union positions)
- Increasing global competition, which adds pressure to keep wages down and union influence at a minimum (see Figure 12.5)

In spite of their dwindling rolls, unions remain a powerful interest group, as evidenced by their money contributions to pro-labor candidates running for elected office. Taken together, all union groups contributed an estimated $1 billion to candidates running for elected office in 1996, a presidential election year (Potter 1997). The AFL-CIO alone contributed $35 million (Online Newshour 1997). These funds come from mandatory dues paid by union members. Membership-generated contributions are controversial because unions tend to support Democratic candidates, even though one-third of their members support Republican candidates (Potter 1997).

Monopoly A situation in which a single producer dominates a market.

Oligopoly A situation in which a few producers dominate a market.

Conglomerates Large corporations that own "smaller" corporations acquired through merger or acquisition.

---

[4]Thirty-five percent (175) of the 500 hundred largest corporations in the world are based in the United States (Fortune 1998).

[5]The U.S. Postal Service is the largest employer in the world, with 898,384 employees, followed by Wal-Mart with 825,000 employees (Fortune 1998).

Figure 12.5 Average Hourly Compensation in U.S. Dollars for Production Workers in Manufacturing in Selected Countries, 1997

Which country has the highest average hourly wages? Which country the lowest? How might these differences help to explain declining union membership in the United States?

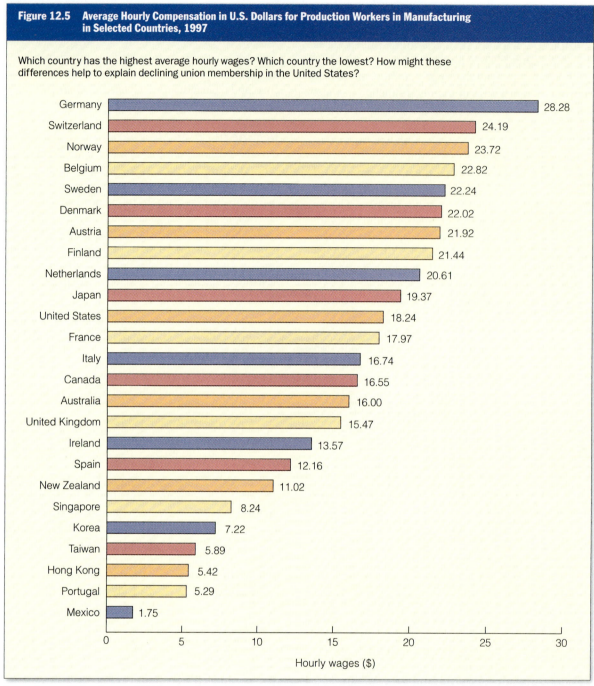

Source: Bureau of Labor Statistics (1998d).

Flexible Work Schedules

In 1997, more than 21.5 million individuals age 16 years and older who worked in U.S. nonagricultural industries did some job-related tasks at home as part of their primary job. That number represents 16 percent of the total U.S. work force. Among the 8 million people who hold second jobs, 37 percent (3 million) performed some work at home. Almost 60 percent of those who worked at home used a computer as part of their jobs and 70 percent used a telephone line (Bureau of

Table 12.1 Flexible Schedules: Full-Time Wage and Salary Workers by Occupation and Industry, May 1997

Which occupation/industry category has the highest percentage of workers with flexible work schedules? Which category has the lowest? What are the advantages and disadvantages of flexible versus nonflexible work schedules?

Occupation and Industry	Percentage
Managerial and professional specialty	38.9
Executive, administrative, and managerial	42.4
Professional specialty	35.5
Mathematical and computer scientists	59.0
Natural scientists	64.5
Teachers, college and university	64.7
Technical, sales, and administrative support	30.4
Technicians and related support	30.8
Sales occupations	41.0
Administrative support, including clerical	23.1
Service occupations	21.5
Private household	40.5
Protective service	16.6
Service, except private household and protective	21.8
Food service	22.7
Health services	17.6
Cleaning and building service	16.3
Personal service	29.1
Precision production, craft, and repair	17.6
Mechanics and repairers	18.3
Construction trades	17.7
Other precision production, craft, and repair	16.6
Operators, fabricators, and laborers	14.6
Machine operators, assemblers, and inspectors	10.3
Transportation and material moving	22.1
Handlers, equipment cleaners, helpers, and laborers	13.5

Source: Bureau of Labor Statistics (1998c).

Labor Statistics 1998a, 1998b). Persons in management and professional occupational categories are most likely to work at home.

In 1997, 25 million (27.6 percent) of the 90.5 million full-time workers in the United States took advantage of flexible work schedules. Surprisingly, that rate did not vary for workers based on sex, race, or age. Some occupational categories, however, were more likely than others to enjoy this flexibility (see Table 12.1).

Computer and Information Technology

Without doubt, computer technology is "altering the form, nature, and future course of the American economy, increasing the flow of products, creating entirely new products and services, altering the way firms respond to demand, and launching an information highway that is leading to the globalization of the product and financial markets" (McConnell 1996).

We will consider the banking industry to illustrate how computers have transformed the workplace and the economy. Automatic teller machines (ATMs) were first introduced in the early 1970s. They enable banks to deliver services 24 hours per day, 7 days per week, in an amazing array of locations, such as shopping centers, stadiums, airports, convenience stores, and any other locations where people gather. ATMs also allow people to make on-line loan applications, receive mini-financial and -bank statements, change currencies, purchase

Figure 12.6   Employment in Commercial Banks, Seasonally Adjusted, and Number of ATM Transactions, 1986–1996

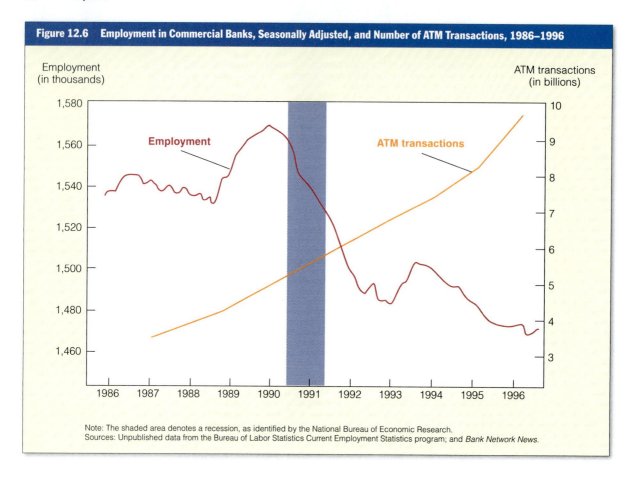

Note: The shaded area denotes a recession, as identified by the National Bureau of Economic Research.
Sources: Unpublished data from the Bureau of Labor Statistics Current Employment Statistics program; and *Bank Network News*.

travelers checks, and cash checks. These devices are not only convenient—allowing customers to complete transactions by machine (without a teller)—but also cost-effective. According to one estimate, teller-mediated transactions cost $1.07 per transaction; in contrast, an ATM-mediated transaction costs only $0.27 per transaction. As a result of this cost difference, as ATM transactions have increased, the number of employees in the banking industry has decreased (see Figure 12.6).

Through the advent of ATMs, computers now allow consumers to carry out financial transactions 365 days per year and 24 hours per day. In addition, they have expanded comsumers' options about when and where to conduct such transactions without bringing a corresponding increase in the number of workers needed to process the transactions.

Automate To use the computer as a means to increase workers' speed and consistency or as a source of surveillance.

Computers also affect the way in which workers perform their jobs. In Chapter 7, we looked at two broad ways in which computers are being used in the workplace: as automating tools and as informating tools. In her book, *In the Age of the Smart Machine,* social psychologist Shoshana Zuboff (1988) found that management can make a conscious choice about applying computers as automating tools or informating tools.

To **automate** means to use the computer as a means to increase workers' speed and consistency or as a source of surveillance (for example, by checking on workers or by keeping precise records of the number of keystrokes per minute or customer service exchanges). "Smart" equipment such as automatic timers, cash registers that calculate change, time clocks that monitor the speed at which orders are filled—anything that does the thinking for employees, "watches" them, or pushes them to produce—represents examples of computers that serve as automating tools.

To **informate** means to empower workers with decision-making tools such as employee-scheduling software intended to ensure that a sufficient number of employees are scheduled for the busiest times and shifts. Other examples of decision-making tools include software to keep track of payroll, sales, inventory, and purchasing. On the surface, the software may appear to be doing the work for managers. Keep in mind, however, that managers must interpret the results and use that information to make decisions. In any event, workers who use computers as informating tools experience work very differently than those who use computers as automating tools.

Political Systems and Power

A society's **political system** is the institution that regulates the use of and access to the power that is essential to articulating and realizing individual, local, regional, national, international, or global interests and agendas. **Power** is the probability that an individual can achieve his or her will even against another individual's opposition (Weber 1947). That probability increases if the individual can force people to obey his or her commands or if the individual has authority over others. **Authority** is legitimate power in which people believe that the differences in power are just and proper—that is, people view a leader as being entitled to give orders. Max Weber identified three types of authority—traditional, charismatic, and legal-rational.

Traditional authority relies on the sanctity of time-honored norms that govern the selection of someone to a powerful position (chief, king, queen) and that specify responsibilities and appropriate conduct for the individual selected. People comply because they believe they are accountable to the past and have an obligation to perpetuate it. (Their reasoning is apt to be, "It has always been like that.") To give up past ways of doing things is to renounce a heritage and an identity (Boudon and Bourricaud 1989).

Charismatic authority derives from the exceptional and exemplary qualities of the person who issues the commands. That is, charismatic leaders are obeyed because their followers believe in and are attracted irre-

sistibly to the leaders' vision. These individuals, by virtue of their special qualities, have the ability to unleash revolutionary changes; they can ask their followers to behave in ways that depart from rules and traditions.

Charismatic leaders often emerge during times of profound crisis (such as economic depressions or wars), because in these situations people are most likely to be drawn to someone with exceptional personal qualities who offers them a vision of a new order. A charismatic leader is more than popular, attractive, likable, or pleasant, however; a merely popular person, "even one who is continually in our thoughts" (Boudon and Bourricaud 1989:70), is not someone for whom we would break all previous ties and give up our possessions. Charismatic leaders may successfully demand that their followers make extraordinary personal sacrifices, cut themselves off from ordinary worldly connections, or devote their lives to achieving a vision that the leaders have outlined.

The source of the charismatic leader's authority, however, does not rest with the ethical quality of the command or vision. Adolf Hitler, Franklin D. Roosevelt, Mao Zedong, and Winston Churchill were all charismatic leaders. Each assumed leadership of a country during turbulent times. Likewise, each conveyed a powerful vision (right or wrong) of his country's destiny.[6]

Informate To use the computer to empower workers with decision-making tools, such as employee-scheduling software intended to ensure that a sufficient number of employees are scheduled for the busiest times and shifts.

Political system The institution that regulates the use of and access to power that is essential to articulating and realizing individual, local, regional, national, international, or global interests and agendas.

Power The probability that an individual can achieve his or her will even against another individual's opposition.

Authority Legitimate power in which people believe that the differences in power are just and proper—that is, people view a leader as being entitled to give orders.

Traditional authority A type of authority that relies on the sanctity of time-honored norms that govern the selection of someone to a powerful position (chief, king, queen) and that specify responsibilities and appropriate conduct for the individual selected.

Charismatic authority A type of authority that derives from the exceptional and exemplary qualities of the person who issues the commands.

[6]A description such as the following is typical of charismatic leaders: "He had a powerful sense of both his nation's destiny and his own. . . . [He] saw no difference between the two. Difficult, egocentric, vainglorious, he demanded complete loyalty from those beneath him but did not always bestow comparable loyalty on those above him. . . . His belief in his own vision and fate was so strong that few other men dared challenge it" (Halberstam 1986:111).

Queen Elizabeth II, Martin Luther King, Jr., and Al Gore possess power. But that power is grounded in three very different types of authority. In the case of Queen Elizabeth II authority is traditional; for Martin Luther King, Jr., it was charismatic; and for Al Gore it is legal-rational.

(left) AP/Wide World; *(center)* Archive Photos; *(right)* Ron Sachs/CNP/Archive Photos

Charismatic authority is a product of the intense relationships between leaders and followers. From a relational point of view, then, charisma is a highly unequal "power-relationship between an inspired guide and a cohort of followers" (Boudon and Bourricaud 1989:70) who believe wholeheartedly in the guide's promises and visions.

Over time, charismatic leaders and their followers come to constitute an "emotional community" devoted to achieving a goal and sustained by a belief in the leader's special qualities. Weber argues, however, that eventually the followers must be able to return to a normal life and to develop relationships with one another based on something other than their connections to the leader. Attraction and devotion cannot sustain a community indefinitely, if only because the object of these emotions—the charismatic leader—is mortal. Unless the charisma that bonds a community becomes routinized, the community may disintegrate from either exhaustion or a void in leadership. Routinized charisma develops as the community establishes procedures, rules, and traditions to regulate the members' conduct, recruit new members, and ensure the orderly transfer of power.

Ultimately, charismatic authority must come to rest on legal-rational grounds—that is, it must be grounded in the position, not in personal qualities. **Legal-rational authority** rests on a system of impersonal rules that formally specifies the qualifications for occupying a powerful position. These rules also regulate the scope of power and the conduct appropriate to someone holding a particular position. In cases of legal-rational authority, people comply with commands, decisions, and directives because they believe that those who have issued them have earned the right to rule (see Chapter 7 and the "Power Elite Model" section later in this chapter).

Forms of Government

Government is the organizational structure that directs and coordinates people's involvement in the political and economic activities of a country or some other territory, such as a city, county, or state (see "A Framework for Global Electronic Commerce"). In this chapter, we consider two major forms of government: democracy and totalitarianism.

Democracy

Democracy is a system of government in which power is vested in the citizen body and in which members of that citizen body participate directly or indirectly in the

Legal-rational authority A type of authority that rests on a system of impersonal rules that formally specifies the qualifications for occupying a powerful position.

Government The organizational structure that directs and coordinates people's involvement in a county's or some other territory's (city, county, state) political activities.

Democracy A system of government in which power is vested in the citizen body, and in which members of that citizen body participate directly or indirectly in the decision-making process.

U.S. in Perspective

A Framework for Global Electronic Commerce

The U.S. government has prepared a report listing strategies to help accelerate the growth of global commerce across the Internet. The report reflects the results of widespread consultation with industry, consumer groups, and the Internet community. The recommendations in this framework are guided by five major principles. Read over the recommendations. Do they reflect socialist or capitalist principles? Are there any other principles that should guide Internet policy?

PRINCIPLES

1. The private sector should lead. The Internet should develop as a market-driven arena, not a regulated industry. Even where collective action is necessary, governments should encourage industry self-regulation and private-sector leadership where possible.

2. Governments should avoid undue restrictions on electronic commerce. In general, parties should be able to enter into legitimate agreements to buy and sell products and services across the Internet with minimal government involvement or intervention. Governments should refrain from imposing new and unnecessary regulations, bureaucratic procedures, or new taxes and tariffs on commercial activities that take place via the Internet.

3. Where governmental involvement is needed, its aim should be to support and enforce a predictable, minimalist, consistent, and simple legal environment for commerce. Where government intervention is necessary, its role should be to ensure competition, protect intellectual property and privacy, prevent fraud, foster transparency, and facilitate dispute resolution—not to regulate.

4. Governments should recognize the unique qualities of the Internet. The genius and explosive success of the Internet can be attributed in part to its decentralized nature and to its tradition of bottom-up governance. Accordingly, the regulatory frameworks established during the past 60 years for telecommunication, radio, and television may not fit the Internet. Existing laws and regulations that may hinder electronic commerce should be reviewed and revised or eliminated to reflect the needs of the new electronic age.

5. Electronic commerce on the Internet should be facilitated on a global basis. The Internet is a global marketplace. The legal framework supporting commercial transactions should be consistent and predictable, regardless of the jurisdictions in which particular buyers and sellers reside.

Source: http://www.whitehouse.gov/WH/New/Commerce/summary.html.

---

decision-making process. The size of the citizen body usually makes direct participation impossible. Decision making usually takes place indirectly through elected representatives. This indirect form is known as representative democracy. Representative democracies hold free elections; every citizen has the right to vote. Candidates and parties can campaign in opposition to the party holding power, and the choice of candidates is not limited to a single party. In addition, when a majority votes to change the party in power, an orderly and peaceful transition in government occurs. In democracies elected representatives legislate; vote taxes; control the budget; and debate, support, question, discuss, criticize, and oppose government policies.

Democratic forms of government extend basic rights to all of their citizens (and legal residents). These rights include freedom of speech, movement, religion, press, and assembly (that is, the right to form and belong to parties and other associations) as well as freedom from "arbitrary arrest and imprisonment" (Bullock 1977:210–211).

In assessing whether a form of government is a democracy, it is important to consider who has the right to vote. At one time or another, many governments classified as democratic have excluded some portions of its populations from the decision-making process on the basis of race, sex, income, property, criminal status, mental health, religion, age, and other characteristics (Creighton 1992:430–31).

Totalitarianism

Totalitarianism is a system of government characterized by (1) a single ruling party led by a dictator, (2) an unchallenged official ideology that defines a vision

Totalitarianism A system of government characterized by (1) a single ruling party led by a dictator, (2) an unchallenged official ideology that defines a vision of the "perfect" society and the means to achieve that vision, (3) a system of social control that suppresses dissent and opposition, and (4) centralized control over the media and the economy.

of the "perfect" society and the means to achieve that vision, (3) a system of social control that suppresses dissent and opposition, and (4) centralized control over the media and the economy. Ideological goals vary, but may include overthrowing capitalist and foreign influences (China under Mao Zedong), creating the perfect race (Nazi Germany under Hitler), or meeting certain economic and development goals (China's Great Leap Forward).[7] Whatever the goals, the leaders, military, and secret police intimidate and mobilize the masses to help the state meet them (see Chapter 8 for an example—the Cultural Revolution).

Totalitarian governments are products of the twentieth century because a technology exists that allows a few people in power to control the behavior of the masses and the information the masses hear. Many of the governments labeled as totalitarian have followed Communist principles.[8] Traditionally, Communist governments instituted polices that outlawed private ownership of property, supported the equal distribution of wealth, and offered status and power to the working class (proletariat). Communist leaders mobilized the "masses" (that is, members of working class, peasants, minorities) to help them realize these principles and to bring about economic changes that benefited everyone—not just the elite and middle classes.[9]

The case of the Soviet Union under the leadership of Joseph Stalin represents one example of how Communist ideals were put into practice and helps to explain why most Communist governments eventually collapsed. Stalin believed that if the Soviet Union did not industrialize, it could not compete with European nations that were expanding their intelligence territories and fighting to gain influence. He maintained that the Soviet Union was 50 years behind these countries and that the gap must be closed within 10 years. To accomplish these goals, this leader initiated a reign of terror. Stalin forced millions of peasants to work in factories, seized millions of private land holdings, and created massive state-owned agricultural collectives. In addition, he established a brutal system of repression (characterized by secret police, forced labor camps, mass deportations, and purges) as means of controlling or eliminating anyone who opposed him, or even thought of opposing him. This system was so relentless and brutal that more than 20 million people died as a result of it. Because Stalin controlled and censored all information, the Soviet people learned only about his vision of socialism and heard only good things about his policies. In addition, Stalin strengthened control over the economy so that state bureaucrats decided what should be produced, how much should be produced, how much the products should cost, and where they should be distributed.[10]

> Individual factory managers not only were not free to produce what they wished but could not ship their output to whomsoever they chose, could not fire unwanted or inefficient workers, could not refuse to accept output shipped to them under the plan directive, and could not even spend their factory's money to buy unauthorized electric light bulbs. . . .
>
> From what we know now about how anarchic the planning process was, the wonder is not that the Soviet economy has given out but that it went on as long as it did. . . . The remarkable achievements in space exploration, and the high level of war material were obtained by giving them absolute priority and close supervision, unlike the lack of attention accorded to "unimportant" goods like buttons, toilet paper, needles, thread, and diapers, which therefore regularly disappeared from the market. (Heilbroner 1990:4)

[7]See Chapter 8 for an overview of the Great Leap Forward.

[8]Until 1989, after democracy, the first form of government about which American students learned was communism. For the most part, students learned that communism was a major alternative to democracy and the greatest threat to representative democracy (Reitman 1998).

[9]Historian Eric Hobsbawm's (1997:28) thoughts about communism are useful for assessing its historical impact:

> Communism was a great cause. It was a cause for which people were prepared to sacrifice themselves in the interests of a greater cause. This as we know has great dangers; they were also prepared to sacrifice others. Nevertheless, let us not knock the sheer moral effect of this.
>
> I was told that the thing that impressed Mandela most about the Communist Party of South Africa was that . . . the whites didn't expect to get anything out of it, they expected to do it for the great cause of the liberation of other people. . . .
>
> What I would hope is that in the future nobody would be able to get the kind of power to commit the excesses and crimes in the name of idealism which were committed in these [communist] dictatorships.

[10]It is difficult to comprehend how millions of people could have suffered and died under Stalin's system without rising up to stop him. Even the best investigators offer no satisfactory explanation. "The question Why? remains unanswered. Perhaps the only answer is 'Because. Period'" (Tolstaya 1991:4).

How does a single person often gain control of an entire country? The political culture must be amenable to personal leadership. Authoritarian leaders or a small group of people receive some outside support from another government that expects to benefit from the change (Buckley 1998). During the Cold War, for example, the United States supported anti-Communist dictators and the Soviet Union supported anti-capitalist dictators. Between 1945 and 1989, the foreign and domestic policies of the United States and Soviet Union were shaped by Cold War dynamics. Virtually every U.S. policy—from the 1949 Marshall Plan to the covert aid given to the Contras during the Reagan administration—was influenced in some way by America's professed desire to protect the world from Soviet influence and the spread of Communism, even to the point of supporting anti-democratic, brutally repressive totalitarian regimes (McNamara 1989).

How does the military gain control of a country? Again, many factors are involved. Military leaders (supported by the military rank and file) may take power when a country is experiencing severe instability due to war, economic crisis, or fraudulent elections. During these times, the military is best positioned to take power because it is often the "only organized force and disciplined force capable of exercising power" (Buckley 1998). The military transition to power may take place more readily if significant segments of the population have come to depend on the military to earn a livelihood or for protection. After taking power, military leaders win civilian support by promising to stay in power for only a short time or until things return to normal. Once in power, however, the military and its leaders do not easily give up their power (Buckley 1998).

The U.S. Political System

Although the United States is classified as a representative democracy, not everyone who is eligible to vote participates in the electoral process.

Participation in the Political Process

Only 65.9 percent of Americans eligible to vote were registered voters in 1996. This percentage represented a decline from 74.3 percent who were registered to vote in 1968. Figure 12.7 shows the percentage of Americans by age and race who reported being registered to vote in 1966–1996. It is striking that only 35.7 percent of people classified as Hispanic and 32.4 percent of those aged 18–24 reported being registered voters (U.S. Census Bureau 1997).

Of course, not everyone who is registered to vote actually visits the polling stations. Only 54.2 percent of Americans reported that they voted in the 1996 election. Again, Hispanics stand out as the category least likely to vote, with 26.7 percent reporting that they voted. Likewise, people between the ages of 18 and 24 were the least likely age group to vote (32.4 percent).

In light of these voting and registration statistics, we might question whether elected officials represent everyone or just those groups who are likely to vote. We might also question whether elected officials represent all voters or whether they serve the powerful constituents who donate generously to their campaigns. In this regard, we can take two major perspectives on power sharing and influence: the power elite model and the pluralist model.

Power Elite Model

The **power elite** are those few people who occupy such lofty positions in the social structure of leading institutions that their decisions have consequences affecting millions of people worldwide. For the most part, the source of this power is legal-rational and resides not in the personal qualities of those in power, but rather in the positions that the power elite have come to occupy. "Were the person occupying the position the most important factor, the stock market would pay close attention to retirements, deaths, and replacements in the executive ranks" (Galbraith 1958:146).

The amount of power wielded by the elite over the lives of others reflects the nature and the quality of the tools that they can use, by virtue of their position, to rule, control, and influence others. These tools might include weapons, surveillance equipment, and specialized modes of communication. According to C. Wright Mills, since World War II, rapid advances in technology have allowed power to become concentrated in the

> **Power elite** Those few people who occupy such lofty positions in the social structure of leading institutions that their decisions have consequences affecting millions of people worldwide.

Figure 12.7   Percentage of Americans by Selected Age and Race Categories Who Reported Being Registered to Vote, 1966–1996

Study the two line graphs. Do you notice any patterns related to age or race? Between 1966 and 1996, do you see any interesting changes in the percentages of people registered in each race and age category? How might you explain these fluctuations?

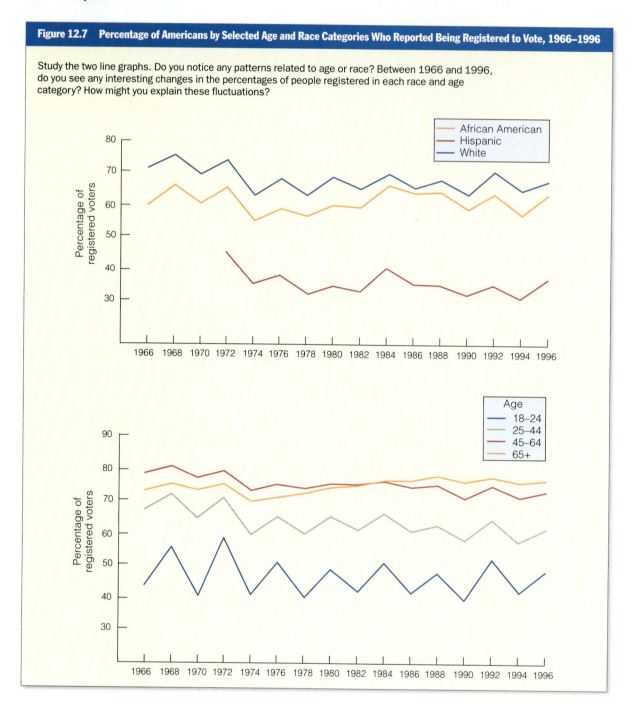

hands of a few; those with access to such power can exercise an extraordinary influence over not only their immediate environment, but also millions of people, tens of thousands of communities, entire countries, and the globe.

In writing about the power elite, Mills does not focus on any single individual, but rather discusses those who occupy the highest positions in the leading institutions in the United States. According to Mills, the leading institutions are the military, corporations

Millions of Military Telephones

In the last five years the Bell System has furnished millions of telephones for war, including 1,325,000 head sets for air and ground forces and more than 1,500,000 microphones. . . . Also more than 1,000,000 airplane radio transmitters and receivers . . . 1,000,000 miles of telephone wire in cables . . . a vast quantity of switchboards, gun directors and secret combat equipment. That helps to explain why we are short of all kinds of telephone facilities here at home.

BELL TELEPHONE SYSTEM

For one measure of the extent to which government, military, and corporate interests became intertwined during World War II, we can look at a Bell Telephone System advertisement published in July 30, 1945, issue of Life *magazine.*

Property of AT&T Archives. Reprinted with permission of AT&T.

(especially the 200 or so largest American corporations), and government. "The power to make decisions of national and international consequence is now so clearly seated in political, military, and economic institutions that other areas of society seem off to the side and, on occasion, readily subordinated to these" (Mills 1963:27).

The origins of these institutions' power can be traced to World War II, when the political elite mobilized corporations to produce the supplies, weapons, and equipment needed to fight the war. U.S. corporations, which were left unscathed by the war, were virtually the only companies in the world able to offer the services and products that the war-torn countries needed for rebuilding.[11] The interests of the government, the military, and corporations became further intertwined when the political elite decided that a permanent war industry was needed to contain the spread of Communism. Thus, over the past 45 or 50 years, these three institutions have become deeply and intricately interrelated in hundreds of ways.

Because the military, government, and corporations are so interdependent and because decisions made by the elite of one realm affect the elite of the other two sectors, Mills believes that everyone has a vested interest in cooperation. Shared interests cause those who occupy the highest positions in each realm to interact with one another. Out of necessity, then, a triangle of power has emerged. We should not assume, however, that the alliance among the three sectors is untroubled, that the powerful in each realm share the same mind-set, that they know the consequences of their decisions, or that they are joined in a conspiracy to shape the fate of a country or the globe:

> At the same time it is clear that they know what is on each other's minds. Whether they come together casually at their clubs or hunting lodges, or slightly more formally at the Business Advisory Council or the Committee for Economic Development or the Foreign Policy Association, they are definitely not isolated from each other. Informal conversation elicits plans, hopes, and expectations. There is a community of interest and sentiment among the elite. (Hacker 1971:141)

Mills gives no detailed examples of the actual decision-making process that takes place at the power elite level. Instead, he focuses on understanding the consequences of this alliance rather than elucidating the details of the decision-making process or assessing the players' motives.

Mills acknowledges that the power elite are not totally free agents, subject to no controls. A chief executive officer of a major corporation is answerable to unions, the Occupational Safety and Health Administration, the Food and Drug Administration, or perhaps other regulatory bodies. Pentagon officials

[11]As Stalin moved to consolidate his power in Eastern Europe, Japan and the countries of Western Europe—their populations demoralized, their economies in ruins, and their infrastructures devastated—had little choice but to accept help from the United States in the form of the Marshall Plan.

The protest organized by Greenpeace represents one example of the kinds of activities special interest groups engage in to make their positions known to elected officials.

© AP/Wide World

Pluralist model A model that views politics as an arena of compromise, alliances, and negotiation among many competing and different special-interest groups and power as something that is dispersed among those groups.

Special-interest groups Groups composed of people who share an interest in a particular economic, political, and social issue and who form an organization or join an existing organization with the goal of influencing public opinion and government policy.

Political action committees (PACs) Committees that raise money to be donated to the political candidates most likely to support their special interests.

are subject to congressional investigations and budget constraints. Defense contractors must be aware of the Federal False Claims Act, which gives a share in the settlement to any employee who can prove that the contractor has defrauded the government (Stevenson 1991). The president of the United States is constrained by bureaucratic "red tape" and by a sometimes slow-moving, politically oriented Congress.[12] A pluralist model of power acknowledges these constraints and offers an alternative vision of how power is distributed and decisions are made in the United States.

Pluralist Model

A **pluralist model** views politics as an arena of compromise, alliances, and negotiation among many competing special-interest groups and power as something that is dispersed among those groups. **Special-interest groups** consist of people who share an interest in a particular economic, political, or social issue and who form an organization or join an existing organization with the goal of influencing public opinion and government policy. Some special-interest groups form **political action committees (PACs)** that raise money to be donated to the political candidates most likely to support their special interests. Examples of PACs listed among the top 50 contributors to political campaigns in 1998 include the National Rifle Association, Association of Trial Lawyers, Campaign for Working Families, United AutoWorkers, National Education Association, American Medical Association, National Automobile Dealer Association, United Parcel Service, Black America's Political Action Committee, Dairy Farmers of America, and National Association of Social Workers (Federal Election Commission 1998).

According to the pluralist model, no single interest group dominants the U.S. political system. Competing groups thrive and are able to express views, through opinion polls, unions, protest, e-mails, and PACs. "If a variety of interest groups are able to exercise power and influence in the system through a variety of means, then the system, with all its flaws, can be considered democratic" (Creighton 1992). One problem with the pluralist model is that, even though several thousand PACs operate in the U.S. political arena, we cannot conclude every special-interest group has the resources to organize and donate money to defend its interests.

---

[12]Mills questions, however, whether these constraints on the power elite have "much significance when weighed against the areas of unrestricted action open to [them]" (Hacker 1971:36).

The Role of Government

To determine the extent to which Americans believe that the government should play an active role in solving personal and social problems, we can study the answers given by a random sample of 5,000 people interviewed by the Gallup Organization in 1998. When asked "Which group in society should have the greatest responsibility for helping the poor?" 32 percent answered "government," but 68 percent named some other entity such as the poor themselves (28 percent), churches (14 percent), families and relatives of poor people (12 percent), or other private charities (6 percent). When respondents were given two choices regarding who should help the poor—the government or the poor themselves—65 percent named the government, and 29 percent named the poor (Gallup Organization 1998).

When asked "Do you think the U.S. government should or should not redistribute wealth by heavy taxes on the rich?" 45 percent answered that the government should tax the rich, while 51 percent believed that the government should not impose such taxes (Gallup Organization 1998). The answers to these questions suggest that Americans are divided about the extent to which government should become involved in people's lives, at least when it comes to helping the poor or otherwise redistributing wealth (Gallup Organization 1998).

Putting aside citizens' beliefs about government involvement, the fact is that the U.S. federal government is involved in its people's lives. Consider that 7.5 million civilian employees[13] carry out a variety of federal and state government services related to corrections, transportation, health care, mail delivery, space exploration, Social Security, trade, housing, recreation, and national defense (U.S. Census Bureau 1997). Moreover in 1997, 9.4 million people received unemployment benefits and 44.2 million people[14] received Social Security benefits. Clearly, government plays an important role in people's lives (Rosenbaum 1999).

Summary and Implications

This chapter focused on two major social institutions: government and politics. We learned that a country's economic and political systems are fundamentally intertwined. That is, governments, whatever their form, design policies to manage economic activity and to distribute wealth. A government's economic policies are shaped by its vision of the ideal economy.

We began this chapter with a series of statements reflecting an American belief that capitalism and democracy go hand-in-hand: that capitalism's emphasis on private ownership of property, the pursuit of personal profit, free competition, and consumer choice corresponds with democratic principles, including freedom of choice, freedom of expression, and elected leaders accountable to the voters. We focused on the United States because it possesses the largest economy in the world and because it became the world's only superpower after the Cold War ended in 1989. In addition, the United States presents an interesting case because many Americans believe that their country "won" the Cold War, and therefore that its political and economic systems—capitalism and democracy—defeated socialism and communism.

In exploring economic systems, we learned that they have been shaped by three major, overlapping revolutions: the Agricultural Revolution, the Industrial Revolution, and the Information Revolution. Each revolution is associated with major technologies that have had a profound impact on human relationships and organizations: the domestication of plants and animals and the invention of the plow (Agricultural Revolution), mechanization (Industrial Revolution), and the introduction of the computer chip, satellites, and fiber optics (Information Revolution).

It is important to keep in mind that the influence of the Industrial Revolution was not confined to only Western Europe and the United States. Instead, as part of this event the countries known as "industrialized" forced people from even the most seemingly isolated and remote regions of the planet into a worldwide division of labor that benefited the colonizers. The effects of the Industrial Revolution are evident today when we consider three different and unequal roles that countries occupy in the current global economy: core, peripheral, and semiperipheral.

Capitalism thrived during the Industrial Revolution, with socialism emerging as an alternative economic arrangement. From a socialist perspective, capitalism survived and flourished by sucking the blood of human labor. The drive for profit was viewed

[13]The federal government employed 2.8 million and the state governments employed 4.7 million civilian employees in 1997.

[14]Approximately 12 million of those receiving Social Security payments were younger than 65 years.

as a "boundless thirst—[a] werewolflike hunger—that takes no account of the health and life of the worker unless society forces it to do so" (Marx 1987:142). The thirst for profit "chases the [capitalist] over the whole surface of the globe" (Marx [1881] 1965:531) in search of the cheapest labor and raw materials.

Socialist and capitalist systems differ in terms of their attitudes toward the ownership of property, profit-making activities, and distribution of wealth. Each economic system became associated with a form of government that differed in the way it directed and coordinated political and economic activities. Capitalist systems encourage private ownership of the means of production, are profit-driven, remain free of government interference, and are consumer-driven. Socialist systems support collective ownership of property, forbid or control individual profit-making activities, and take an active role in managing the economy.

The period between 1945 and 1989, known as the Cold War, can be defined as a massive military buildup in which the Soviet Union and the United States sought to contain the spread of one another's economic and political systems. During the Cold War, the United States supported anti-Communist dictators and the Soviet Union supported anti-capitalist dictators. The foreign and domestic policies of both countries were dramatically shaped by Cold War dynamics. In the United States, virtually every policy—from the 1949 Marshall Plan to the diversion of aid to the Contras during the Reagan administration—was influenced in some way by America's professed desire to protect the world from Soviet influence and the spread of Communism, even to the point of supporting anti-democratic, brutally repressive totalitarian governments.

In addition to its role as a superpower, the United States is the largest and most powerful economy in the world. U.S. domestic and international policies support democratic forms of government and economies that adhere to capitalist principles. It appears that capitalism has "won"—at least for now. Around the world, other countries continue to struggle, however. Global-level organizations such as the World Bank and the International Monetary Fund may consent to work with "troubled economies" provided their governments implement various "corrective measures," including (1) devaluing currency to reduce the money supply, (2) supporting market mechanisms rather than government price controls, subsidies, and credits, (3) eliminating public sectors of the economy, and (4) providing incentives for private-sector development (Kabir 1995).

Key Concepts

Use this outline to organize your review of the key chapter ideas.

Social Institutions
   Economic Systems
   Political Systems
Economic Ideals
   Capitalism
      Private Ownership
      Law of Supply and Demand
   Socialism
Economic Sectors
   Primary Sector
   Secondary Sector
   Tertiary Sector
   Goods
   Services
World System Theory
   Core Economies
   Peripheral Economies
   Semiperipheral Economies
Gross Domestic Product

Division of Labor
Mechanization
Monopoly
Oligarchy
Conglomerate
Automate
Informate
Governments
   Democracy
   Totalitarianism
Power
   Power Elite Model
   Pluralistic Model
      Special interest groups
      Political Action Committees
Authority
   Traditional
   Charismatic
   Legal-rational

internet assignment

Find the CIA's *World Factbook* on the Internet. Select a region of the world (Africa, Asia, Europe, Latin America). How many countries have economies in which more than half the population work in the area of agriculture? In manufacturing?

Population and Family Life

With Emphasis on Brazil

Street scene in São Paolo, Brazil. (© Paulo Frieman/Sygma.)

North
2.59

North East
27.2

Central West
58.6

South
38

South East
67

Rio de Janeiro
Federal Unit
291

São Paulo
Federal Unit
126

Inset figures represent number of persons per square kilometer.

Brasilia ★

Highway Branch 364

Rio de Janeiro
Population 9.8 million

São Paulo
Population 16.1 million

Road to Disaster?

Highway BR364 was built by the Brazilian government in 1980 to link its densely populated Southeast region to its northern Amazon territories. The result was a sudden movement of poor people and economic "developers" into the rain forest and a sudden, appalling wave of forest destruction. Did the rapidly growing population of Brazil make this event inevitable? Is the virtual disappearance of the rain forest also unavoidable?

This chapter explores whether rapid population growth threatens the viability of life for families and for whole societies. Is humanity simply overwhelming the carrying capacity of the Earth, or are other factors at work? For example, with more than 50 percent of the cultivable land in Brazil owned by less than 1 percent of the country's landowners—and much of that land lying idle—was the destruction of the rain forests "inevitable"?

The chapter will also lead us to consider carefully what the concept "family" means.

There were nine children before me and twelve after me. . . . Out of this total of twenty-two, seventeen lived, but four died in infancy, leaving thirteen still to hold the family fort.

- Mine was a difficult birth, I am told. Both mother and son almost died. . . . After my birth Mother was sent to recuperate for some weeks and I was kept in the hospital while she was away. I remained there for some time, without name, for I wasn't baptized until my mother was well enough to bring me to church. (Brown 1992:85)

- One of the common experiences of people in their early forties, which they seem to need to talk about, is having parents who are in their sixties and seventies.... Ten years ago, when we were in our early thirties, we talked about our children—about pregnancies and births, bottles versus breast-feeding, how to get "them" to sleep through the night, and then about toilet training and schools. We were primarily parents, and our own parents were secondary subjects. They were just grandparents and in-laws who were or were not helpful or demanding, visiting, vacationing, or whatnot. Only in the last four or five years,

I realize, have I been telling and hearing stories about parents, usually with friends, but sometimes with people I have just met, if they are my age. (Sayre 1983:124)

- There were a couple of years' worth of thinking that went into the decision to have another child. We already had a school-age son, and for a long time I thought we'd just have one child, because working and raising a family and trying to have a normal kind of life, too, was kind of hard work. But after Erik was in school, and I saw him developing into a really neat little person, I kept trying to decide in my own mind whether to do it again. . . . I felt if we were going to add to the family, this would be the time. (Sorel 1984: 60–61)

- Rachel's day is long. She rises "when the sky is beginning to lighten," cooks breakfast, gets the children off to school and cleans the house. Then she sets off for the holding, two miles away up and downhill, where she tethers the animals to graze and gets down to planting, digging or weeding her corn, cassava and cowpeas, eating a snack in the field. On her way home she gathers whatever firewood she finds. Then she fetches water,

half an hour's walk away, with the return journey uphill. As the sun begins to set, she cooks the evening meal of [thick] maize porridge in a pot balanced on three stones, until it is as stiff as bread dough: "You are stirring solidly for an hour," she complains, "and it gets harder and harder until at the end the sweat is pouring off you." Getting the maize ground is another chore: twice a week she must trek two miles to the nearest neighbor who possesses a handmill. (Harrison 1987:438–439)

Why Focus on Brazil?

In this chapter, we focus on Brazil, which has the sixth largest population in the world and is the most populous country in South America, with 164 million people. Approximately 30 percent of Brazil's population is younger than 15 years old; 5.2 percent is more than 65 years old. In contrast, 22 percent of the U.S. population is younger than 15 years old; 13 percent is older than 65. These data tell us that Brazil has a relatively young population, whereas the United States has a relatively old population. To put it another way, in Brazil 1 in 20 people is older than 65 compared with 1 in 8 people in the United States. Obviously, this difference in age structure affects family composition, the nature of family relationships, and the family-related issues considered most pressing in each country.

In addition to examining the age structure, we will identify and compare other demographic characteristics that affect family composition, relationships, and issues in the United States and Brazil. Those demographic characteristics include the following:

1. Births, including the number of children born to a woman and the spacing between those births. The number can range from none to a child born in every year of a woman's reproductive cycle.
2. Deaths, including how and when (infancy through old age) a family member dies.
3. Employment, including how much and what kind of work each family member must do to sustain the family's standard of living.

Population and family life are paired together in this chapter because the factors that affect a country's population size, distribution, and age-sex composition also affect the "typical" family's composition, relationships, and experiences. Obviously, in comparison with the United States, fewer people in Brazil—a country in which 1 in 20 people is older than age 65—experience the need to talk about having parents in their sixties and seventies. Similarly, more families in Brazil will have to face and come to terms with a child's death, as 58 of every 1,000 live infants born die before the age of one each year. In the United States, the comparable figure is 9 of every 1,000 live infants born. Finally, in a society such as Brazil, where 73 percent of the rural population live in absolute poverty, we would expect that many people would work a "long day" simply trying to secure the basic resources that their families need to survive.

Sociologists who study population size, distribution, and age-sex structure are known as **demographers.** We devote the first half of this chapter to an overview of the key demographic concepts and theories. In the second half, we consider key demographic events: family composition, relationships, and experiences. One event that is central to demographic analysis is the Industrial Revolution.

Demographers   Sociologists who study population size, distribution, and age-sex structure.

Doubling time   The estimated number of years required for a country's population to double in size.

The Industrial Revolution and Family Events

As we learned elsewhere in this textbook (see Chapters 1, 6, and 7), the Industrial Revolution was not unique to Western Europe and the United States. In fact, this event forced people from even the most seemingly isolated and remote regions of the planet into a worldwide division of labor that continues through today (see "Company Towns: The Brazilian Experience"). Its effect was not uniform, but varied according to country and region of the world.

We can classify the countries of the world into two broad categories with regard to industrialization: the mechanized rich and the labor-intensive poor. Comparable but misleading dichotomies include developed and developing, industrialized and industrializing, and First World and Third World. These names are misleading because they suggest that a country is either industrialized or not industrialized. The dichotomy implies that a failure to industrialize is what makes a country poor, and it camouflages the fact that as Europe and North America plunged into industrialization, they took possession of Asia, Africa, and South America, establishing nonindustrial economies there that were oriented to their industrial needs, not to the needs of the colonized countries. The point is that labor-intensive poor countries were part of the Industrial Revolution from the beginning.

The World Bank, the United Nations, and other international organizations use a number of indicators to distinguish between mechanized rich and labor-intensive poor countries, including the following:

- **Doubling time,** the estimated number of years required for a country's population to double in size.

Table 13.1 Demographic Differences Between Labor-Intensive Poor and Mechanized Rich Countries

Labor-intensive poor countries are very different from mechanized rich countries with regard to indicators used by international organizations to classify countries. Here, data are presented for the 10 most populous countries in the world. Compare the U.S. figures with those for any of the labor-intensive poor countries. What contrasts do you see? How might each contrast affect the people living in that country?

	Population Doubling Time (years)	Infant Mortality (per 1,000 births)	Total Fertility (children born per woman)	Per Capita GNP Income (in U.S. $)	Annual per Capita Consumption of Energy (in kilograms of oil equivalent)
Labor-Intensive Poor					
Nigeria	23	72.6	6.3	1,250	109
Bangladesh	29	104.6	4.4	1,040	70
Pakistan	23	99.5	6.4	1,930	389
Brazil	37	58.0	2.4	5,580	1,589
India	34	76.3	3.4	1,360	324
Indonesia	40	65.0	2.7	3,090	207
China (mainland)	53	52.1	1.8	2,500	593
Mechanized Rich					
United States	89	7.9	2.1	25,850	11,236
Germany	(—)	6.3	1.5	16,580	5,683
Japan	217	4.3	1.6	20,200	6,263

Source: U.S. Central Intelligence Agency (1995).

- **Infant mortality**, the number of deaths in the first year of life for every 1,000 live births.

- **Total fertility**, the average number of children that women bear over their lifetime.

- **Per capita income**, the average income that each person in a country would receive if the country's gross national product were divided evenly among all its citizens.

- The percentage of the population engaged in agriculture.

- **Annual per capita consumption of energy**, the average amount of energy each person consumes over a year. A low per capita energy consumption suggests that, for the vast majority of people, work is labor-intensive rather than machine-intensive. In labor-intensive work, considerable physical exertion is required to produce food and goods.

When we use the term **labor-intensive poor** (or an equivalent term) to characterize a country, we mean that the country differs markedly on these and other indicators from countries considered to be industrialized (Stockwell and Laidlaw 1981; see Table 13.1). According to these measures, approximately 107 countries are low- or lower-income economies (labor-intensive poor), and 40 are high-income economies (mechanized rich) (Lutz 1994; see Figure 13.2).

Infant mortality The number of deaths in the first year of life for every 1,000 live births.

Total fertility The average number of children that women bear over their lifetime.

Per capita income The average income that each person in a country would receive if the country's gross national product were divided evenly among all its citizens.

Annual per capita consumption of energy The average amount of energy each person consumes over a year.

Labor-intensive poor A term used to characterize a country that differs markedly from industrialized countries on indicators such as doubling time, infant mortality, total fertility, per capita income, and annual per capita consumption of energy.

Company Towns: The Brazilian Experience

The need to provide accommodation at new sites of economic activity has long been integral to the development process, from pithead villages, industrial and philanthropic settlements (Pullman, Port Sunlight) to socialist planning in eastern Europe (Nova Huta, Dunaujvaros). Single enterprise communities continue to be established in the Third World, most commonly where resource endowments requiring substantial capital investment occur in environments lacking established settlement. Such "company towns" are particularly associated with agribusiness, lumbering, mining, and industries dependent on bulky raw materials or cheap energy. Their precise form varies from temporary camps exploiting ephemeral resources such as timber, or at project construction sites; tracts of company-built workers' housing grafted onto existing settlements; to complete "new towns" set into existing urban networks or at the resource frontier.

In Brazil the earlier company towns were associated with nineteenth-century mines and water-powered textile mills (Figure 13.1a). In [the region of Brazil called] Minas Gerais, the little cotton town of Biribiri was laid out at a remote waterfall in the Serra do Espinhaço, consisting of the mill, church, store, schoolhouse and a neat square of 40 houses. The British-owned St. John del Rey gold-mining company, established in 1834, built a town for its workers at Nova Lima where, until the 1950s, there was social segregation between British managers and foremen, and Brazilian laborers. Railway building fostered the construction of company housing adjacent to railyards and workshops, as at Divinópolis.

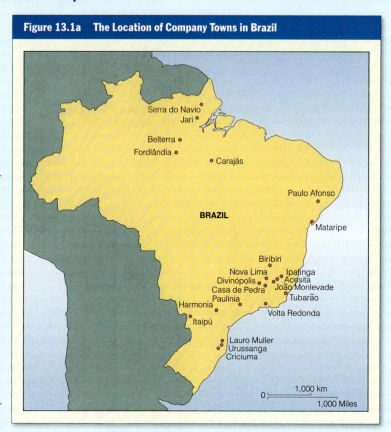

Figure 13.1a   The Location of Company Towns in Brazil

The (Brazilian) steel industry has created several company towns where plant size, coupled with raw material requirements, compelled companies to build steel mills and substantial settlements at green field sites. The iron ores of Minas Gerais generated steel towns in the Doce Valley, at Monlevade (1935), Acesita (1944) and Ipatinga (1956). Brazil's class company town is Volta Redonda, established at a green field site in the Paríba valley in 1941, to service the country's first coke-fired steelworks. The (company) Cia. Siderúrgica Nacional, initially built 2,000 houses, adding a further 2,000 by 1952. It also provided churches, hotels, cinemas, sports facilities, hospitals, primary and secondary schools and a technical school to train employees. The town is socially zoned, with workers' housing near the plant, and more substantial properties for technicians and managers on the hill-

The official statistics used to identify labor-intensive poor and mechanized rich countries are counts of how many people have experienced an event. For example, infant mortality is an annual count of how many infants die for every 1,000 children born. A high infant mortality rate tells us that the death of an infant is a common experience for many families.

Industrialization and Brazil

Brazil has a powerful economy (the tenth largest in the world). Moreover, after the United States, it is the world's largest exporter of agricultural and food products. Its major exports include soybeans, coffee, tobacco transport equipment, footwear, orange juice, iron ore, and steel products. The benefits that we

sides. Population grew from 200 in 1941 to over 30,000 in 1954 and 180,000 in 1980. In addition, the company built small settlements at its iron and coal mines (Casade Pedra, Lauro Muller, etc.).

Other company towns have been associated with manganese (Serra do Navio), paper (Harmonia), oil refining (Matairpe and Paulínia), and hydroelectric projects (Paulo Afonso and Itaipú). In Amazonia there have been abortive plantation settlements at Fordiândia, Belterra, and Jari.

The domination of these settlements by a single enterprise influences their economy, population and townscape. The company provides employment, controls rents, zones lands, and supplies essential utilities, social services, recreational facilities, and sometimes retail provision. Populations have grown rapidly, and tend to be immigrant, youthful and relatively affluent; employment opportunity is male-dominated. Town plans are simple, uniform in layout and date, with standardized housing which varies only in size and site with employee status.

Brazil's most recent company town is at the iron mine of Carajás (Figure 13.1b). Its sponsor, the Cia. Vale do Rio Doce had built earlier towns at its mines in Minas Gerais and port of Tubarão (Espírito Santo). Carajás was established in 1980, for a planned workforce of 2,000 to be housed in single men's flats and family houses, with the company providing essential services. The townsite is screened from the mine, and includes leisure areas and protected forest reserves. Though carefully planned, Carajás demonstrates some of the difficulties engendered by company towns, being located in an ecologically sensitive area, offering very little employment for women, and attracting spontaneous migrants to unplanned adjacent settlements.

When a private company founds an isolated company town, what implications does that have for the people who will need to come to work there? What sorts of advantages and disadvantages is a company town likely to present to them? What sorts of social ties may be broken? What new ones reinforced? What sorts of personal opportunities might be diminished?

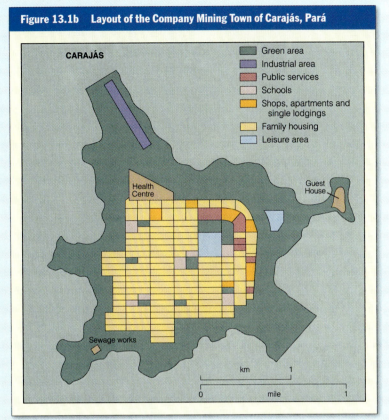

Figure 13.1b   Layout of the Company Mining Town of Carajás, Pará

CARAJÁS

- Green area
- Industrial area
- Public services
- Schools
- Shops, apartments and single lodgings
- Family housing
- Leisure area

Health Centre

Guest House

Sewage works

Source: From "Company Towns: The Brazilian Experience" by John Dickenson in *Atlas of World Development*, edited by Tim Unwin. New York: John Wiley & Sons Ltd. Copyright © Developing Areas Research Group. Reprinted by permission of the Developing Areas Research Group of the Royal Geographical Society (with the Institute of British Geographers) and John Wiley & Sons, Ltd.

associate with industrialization, however, have bypassed the majority of Brazilians, who continue to live at or below subsistence level. Brazil is a country in which poverty is widespread and chronic (not the result of some temporary misfortune). Sixty percent of its population can be classified as poor (Calsing 1985).

The World Bank classifies Brazil as a country with a middle-income to upper-middle economy. Nevertheless, as many as 60 percent of Brazil's people live in extreme poverty. The unequal distribution of wealth is reflected by the fact that the most affluent 20 percent of the population claims 63 percent of Brazil's total household income, whereas the least affluent 20 percent of

Figure 13.2   The Highest Per Capita Incomes Are Found in the North

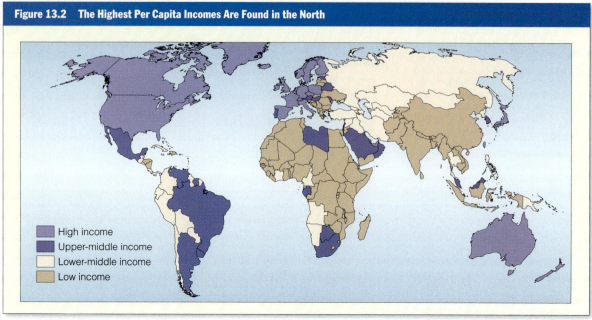

High income
Upper-middle income
Lower-middle income
Low income

Source: Adapted by John R. Weeks in 1996 from data in the World Bank (1994), Table 1.

the population holds only 3.1 percent of all household income (Romero 1996).

In Brazil, the effects of industrialization vary across and within its five major regions:

- The North (the Amazon Basin, covering half of the country and characterized by vast reaches of largely uninhabited tropical forest).

- The Northeast (semiarid scrubland prone to periodic drought and massive flooding; heavily settled and poor).

- The South (rich farmland and pasture lands and large modern cities with a large and relatively prosperous population).

- The Southeast (huge, densely populated urban centers, including the city and state of São Paulo).

- The Central West (home of one of the earth's major ecological frontiers and some of the largest cattle ranches in South America, sparsely populated). The major city in this region is Cuiaba, which is the site of a satellite tracking station maintained by NASA (Beresky 1991).

Approximately 25 percent of Brazil's people live in the Southeast, which is dominated by the metropolises of São Paulo[1] and Rio de Janeiro—two of the world's 20 largest cities and centers of multinational commercial, industrial, and agricultural activity (see the

chapter-opening map). As many as 30 percent of the people in these two cities live in urban squatter settlements called *favelas*,[2] in makeshift dwellings constructed from cardboard, metal, or wood, with inadequate sewers, running water, and electricity.

Approximately one-third of Brazil's people, most of whom are descendants of slaves who worked the Portuguese sugarcane plantations between the sixteenth and nineteenth centuries, live on the drought-stricken and exhausted land in the northeastern part of the country. Even after the abolition of slavery in 1888, the plantation agricultural system, which was oriented toward export, persisted. Today, the most productive land in the Northeast continues to be used to grow

[1]According to Latin American Studies scholar Warren Dean (1991), "São Paulo has to be explained. The rest of Brazil and indeed the rest of Latin America [have] not experienced that city's and state's breakthrough of sustained economic development based on diversified and technologically complex industrial production" (p. 649).

[2]Many of the people who live in these squatter settlements have been pushed off the surrounding plantation land because of mechanization and have come to the cities in search of a better life. Latin American Studies scholar Thomas G. Sanders (1988) maintains that outsiders stereotype *favelas* as centers of crime, prostitution, and extreme family disorganization, but that most studies show the opposite. The vast majority of *favelas* dwellers are "honest, employed, hardworking, and have high aspirations for their children. They live in *favelas* because of low cost, location near their workplaces, and often lack of viable alternative housing solutions" (p. 5).

sugarcane, soybeans, and other export crops, leaving the peasants with no land on which to grow subsistence crops.

Over the years, landless peasants from the Southeast and Northeast have migrated to overcrowded cities in search of work or have sought land in the sparsely populated interior of the country. The interior houses the Amazon forest, the world's largest tropical jungle and rain forest. The rate of migration to the Amazon region accelerated when the Brazilian government began constructing a network of highways and roads in the 1960s to connect the Amazon region with the rest of the country. The network of roads opened the land to foreign and Brazilian investors, individuals living on the fringes of the cities, and the landless and unemployed people from the Northeast. To convert forest to pasture and farmland, the settlers cut the trees and other vegetation, allowed them to dry, and then set them afire during the dry season, which runs from June to October. (Brazilian and U.S. scientists, monitoring this practice via satellite, counted 170,000 fires in 1987.) The ash from the burning trees and other foliage acts as a fertilizer for a few years; after that time, however, the land can no longer support crops. It is then abandoned to cattle ranchers, whose herds graze it for a few years until the land becomes totally exhausted (Simons 1988).

Although isolated geographically from the rest of the country, the Amazon region was not uninhabited; it was occupied by indigenous people, rubber tappers, nut gatherers, and others whose forest-centered livelihood was disrupted by the highway construction and subsequent human migration. The forest dwellers, especially the indigenous people, suffered (and still suffer) cultural and physical extinction at the hands of mining companies that extract mineral wealth buried beneath the rain forest, lumber companies in search of rare jungle trees, cattle ranchers that graze their herds on deforested land, government development projects, and land-hungry peasants. Although the government has set aside land for these peoples, the forest dwellers face great pressure to abandon their language and culture and enter the dominant society. Unfortunately, if they enter society, they do so as landless peasants, low-paid laborers, or beggars (Caufield 1985).

This background information on Brazil sets the stage for thinking about industrialization's wide-ranging and uneven consequences. To help us grasp these consequences, we turn to a model that outlines historical changes in birth and death rates in Western Europe and the United States, especially those affected by the Industrial Revolution. This model relies on the theory of demographic transition. After describing this model, we discuss how well it applies to non-Western countries such as Brazil.

The Theory of Demographic Transition

In the 1920s and early 1930s, demographers observed birth and death rates in various countries and noticed a pattern: Both birth and death rates were high in Africa, Asia, and South America; death rates were declining while birth rates remained high in Eastern and Southern Europe; and birth rates were declining and death rates were low in Western Europe and North America. Demographers observed that Western Europe and North America had the following sequence with regard to birth and death rates (see "How Demographers Measure Change"):

1. Birth and death rates remained high until the middle of the eighteenth century, at which time death rates began to decline.

2. As the death rates decreased, the population grew rapidly, because more births than deaths occurred. The birth rates began to decline around 1800.

3. By 1920, both birth and death rates had dropped below 20 per 1,000 (see Figure 13.3).

On the basis of these observations, demographers put forth the theory of the demographic transition. That is, they proclaimed that a country's birth and death rates are linked to its level of industrial or economic development.

Note that this model documents the general situation—it should not be construed as a detailed description of the experiences of any single country. Even so, we can say that all countries have followed or are following the essential pattern of the demographic transition, although they differ with regard to the timing of the declines and the rate at which their populations increase after death rates begin to fall. The theory of the demographic transition includes more than this pattern, however; it also seeks to explain the events that caused birth and death rates to drop in the mechanized rich countries (with the exception of Japan). The factors that underlie changes in birth and death rates in the labor-intensive poor countries, however, are fundamentally different from those that caused such changes in the so-called developed or industrialized countries.

U.S. in Perspective
How Demographers Measure Change

To discuss change, demographers must specify a time period (a year, a decade, a century) over which they keep track of how many times an event (a birth, a death, a move) occurs. The accompanying table shows the number of births and deaths that occurred in Brazil and the United States in 1997.

The simplest way to express change is in absolute terms—that is, to state the number of times that an event occurred. (In 1997, there were an estimated 3,552,353 births in Brazil and 3,892,478 births in the United States.) Expressing change in absolute terms is not very useful, however, for making comparisons between countries that have different population sizes. Consequently, for comparative purposes demographers calculate *rates* of births, deaths, and migrations, usually per 1,000 people in the population. Rates are calculated by dividing the number of times that an event occurs by the size of the population at the onset of the year and then multiplying that figure by 1,000. The 1997 death rate for Brazil is calculated as follows:

$$\frac{1,443,351}{169,806,557} \times 1,000 = 8.5$$

When the total number of people in the population is used as the denominator, the rates are called *crude rates*. Sometimes demographers wish to know how many births, deaths, or moves occur within a specific segment of the population (among males or among females aged 15 to 64). In these cases, the denominator is the number of people in that segment of the population. For example, there are 55,769,122 women between the ages of 15 and 64 in Brazil. The age-specific birth rate for these women thus is calculated as follows:

$$\frac{3,552,353}{55,769,122} \times 1,000 = 63.7$$

	Mid-1997 Population	Births	Deaths	Births per 1,000 Population	Deaths per 1,000 Population
Brazil	169,806,557	3,552,353	1,443,351	20.9	8.5
United States	270,311,756	3,892,478	2,378,736	14.4	8.8

Source: U.S. Central Intelligence Agency (1998).

Stage 1: High Birth and Death Rates

For most of human history—the first 2 to 5 million years—populations grew very slowly, if at all. World population remained less than 1 billion until A.D. 1800, at which point it began to grow explosively. By 1930, the world's population had increased to 2 billion; during the next 50 years, it increased by another 2.5 billion. During the 1980s, the world's population increased by approximately 750 million to reach its present size of 6.0 billion. Demographers speculate that growth until 1800 was slow because **mortality crises**—frequent and violent fluctuations in the death rate caused by war, famine, and epidemics—were a regular feature of life.

Stage 1 of the demographic transition is often referred to as the stage of high potential growth: If something happened to cause the death rate to decline—for example, improvements in agriculture, sanitation, or medical care—the population would increase dramatically. In this stage, life is short and brutal; the death rate almost always exceeds 50 per 1,000. When mortality crises occur, the death rate seems to have no limit. Sometimes half of the population is affected, as when the Black Plague struck Europe, the Middle East, and Asia in the middle of the fourteenth century and recurred periodically for approximately 300 years. Within 20 years of its onset, the plague killed an estimated three-fourths of the people in the affected populations.[3]

---

[3]The medieval Italian writer Giovanni Boccaccio ([1353] 1984) recorded his impressions of the event:

> It began in both men and women with certain swellings either in the groin or under the arm-pits [and] began to spread indiscriminately over every part of the body; and after this, the symptoms of the illness changed to black . . . almost all died after the third day. (p. 728)

Mortality crises Frequent and violent fluctuations in the death rate caused by war, famine, and epidemics.

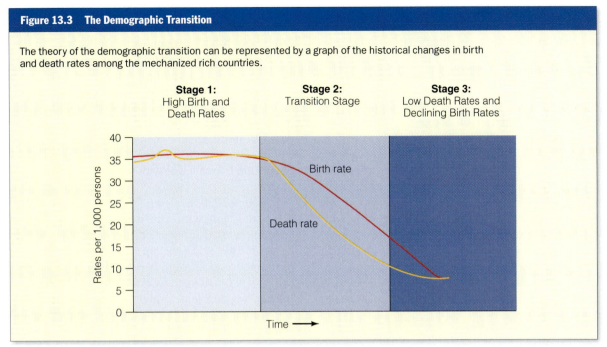

Figure 13.3   The Demographic Transition

The theory of the demographic transition can be represented by a graph of the historical changes in birth and death rates among the mechanized rich countries.

Stage 1:
High Birth and
Death Rates

Stage 2:
Transition Stage

Stage 3:
Low Death Rates and
Declining Birth Rates

Birth rate

Death rate

Rates per 1,000 persons

Time

Source: U.S. Central Intelligence Agency (1998).

Another mortality crisis—but one that has not received as much attention as the Black Plague—affected the indigenous populations of North America when the Europeans arrived in the fifteenth century. A large proportion of the native population died because they had no resistance to diseases such as smallpox, measles, tuberculosis, and influenza, which the colonists brought with them. Others were killed outright by colonists because they refused to work as slaves on plantations. Historians continue to debate what proportion of the native population died as a result of this contact, with their estimates ranging between 50 and 90 percent.

Thus, in stage 1, average life expectancy at birth remained short—perhaps between 20 and 35 years—with the most vulnerable groups being women of reproductive age, infants, and children younger than age 5.[4] It is believed that women gave birth to large numbers of children and that the crude birth rate was about 50 per 1,000, the highest rate possible for human beings. Families remained small, however, because one in three infants died before reaching age 1, and another died before reaching adulthood. If the birth rate had not remained high, the society would have become

extinct. Demographer Abdel R. Omran (1971) estimates that in societies in which life expectancy at birth is 30 years, each woman must have an average of seven live births to ensure that two children survive into adulthood. She must bear six sons to ensure that at least one survives until the father reaches age 65 (if the father lives that long).

In Western Europe before 1650, high mortality rates were associated closely with food shortages and famines. Even when people did not die directly from starvation, they died from diseases that preyed on their weakened physical state. Thus Thomas Malthus ([1798] 1965), a British economist and an ordained Anglican minister, concluded that "the power of population is so superior to the power in the earth to produce subsistence for man, that premature death must in some shape or other visit the human race" (p. 140). According to Malthus, **positive checks** served to keep population size in line with the food supply. He defined positive checks as events that increase mortality, including epidemics of infectious and parasitic disease, war, and famine. Malthus believed that the only moral

---

[4]The high infant and child mortality rate pulled down the average life expectancy at birth; many people managed to live well beyond age 30.

Positive checks Events that increase mortality, including epidemics of infectious and parasitic disease, war, and famine.

way to prevent populations from growing beyond a size that could not be supported by the food supply was delayed marriage and celibacy. He regarded any other method—such as infanticide, homosexuality, or sterility caused by sexually transmitted diseases—as immoral.

Stage 2: The Transition Stage

Around 1650, mortality crises became less frequent in Western Europe; by 1750, the death rate had begun to decline slowly in that region. This decline was triggered by a complex array of factors associated with the onset of the Industrial Revolution. The two most important factors were as follows: (1) increases in the food supply, which improved the nutritional status of the population and increased its ability to resist diseases, and (2) public health and sanitation measures, including the use of cotton to make clothing and new ways of preparing food. The following excerpt elaborates on these trends:

> The development of winter fodder for cattle was important; fodder allowed the farmer to keep his cattle alive during the winter, thereby reducing the necessity of living on salted meats during half of the year. . . . [C]anning was discovered in the early nineteenth century. This method of food preservation laid the basis for new and improved diets throughout the industrialized world. Finally, the manufacture of cheap cotton cloth became a reality after mid-century. Before then, much of the clothes were seldom if ever washed, especially among the poor. A journeyman's or tradesman's wife might wear leather stays and a quilted petticoat until they virtually rotted away. The new cheap cotton garments could easily be washed, which increased cleanliness and fostered better health. (Stub 1982:33)

Contrary to popular belief, advances in medical technology had little influence on death rates until the turn of the twentieth century—well after improvements in nutrition and sanitation had caused dramatic decreases in deaths due to infectious diseases.

Over a 100-year period, the death rate fell from 50 per 1,000 to less than 20 per 1,000, and life expectancy at birth increased to approximately 50 years of age. As death rates declined, fertility remained high. It may

even have increased temporarily, because improvements in sanitation and nutrition enabled women to carry more babies to term. With the decrease in the death rate, the **demographic gap**—the difference between birth rates and death rates—widened, and population size increased substantially. **Urbanization,** an increase in the number of cities and growth in the proportion of the population living in cities, accompanied the unprecedented increases in population size. (As recently as 1850, only 2 percent of the world's population lived in cities with populations of 100,000 or more.)

Around 1880, however, fertility began to decline. The factors that caused birth rates to drop are unclear and subject to debate among demographers. One thing is certain: The decline was not caused by innovations in contraceptive technology, because the methods available in 1880 had been available throughout history. Instead, the decline in fertility seems to be associated with several other factors. First, the economic value of children declined in industrial and urban settings, as children no longer represented a source of cheap labor but rather became an economic liability to their parents. Second, with the decline in infant and childhood mortality, women no longer had to bear a large number of children to ensure that a few survived. Third, a change in the status of women gave them greater control over their reproductive lives and made child bearing less central to women's lives. Scholars disagree, however, about the specific conditions under which women are able to control their reproductive lives.

Stage 3: Low Death Rates and Declining Birth Rates

Around 1930, both birth and death rates fell to less than 20 per 1,000, and the rate of population growth slowed considerably. Life expectancy at birth surpassed 70 years, an unprecedented age. The remarkable successes in reducing infant, childhood, and maternal mortality rates permitted accidents, homicides, and suicide to become the leading causes of death among young people. The reduction of the risk of dying from infectious diseases ensures that people who would have died of infectious diseases in an earlier era can survive into middle age and beyond, when they face an elevated risk of dying from degenerative and environmental diseases (heart disease, cancer, strokes, and so on). For the first time in history, persons 50 years of age and older account for more than 70 percent of the annual deaths. Before stage 3, infants, children, and young women accounted for the largest share of deaths (Olshansky and Ault 1986).

Demographic gap The difference between birth rates and death rates.

Urbanization An increase in the number of cities and the proportion of the population living in cities.

As death rates decline, disease prevention becomes an important issue. The goal is to live not only a long life but a "quality life" (Olshansky and Ault 1986; Omran 1971). As a result, people become conscious of the link between their health and their lifestyle (sleep, nutrition, exercise, and drinking and smoking habits). In addition to low birth and death rates, stage 3 is distinguished by an unprecedented emphasis on consumption (made possible by advances in manufacturing and food production technologies).

Theoretically, all of the mechanized rich countries now occupy stage 3 of the demographic transition. At one time some sociologists and demographers maintained that the so-called developing countries would follow this model of development as they industrialized. The nature of industrialization in labor-intensive poor countries is so fundamentally different from the version that occurred in the mechanized rich countries, however, that the former countries are unlikely to follow the same path.

The "Demographic Transition" in Brazil

One reason that the nature of industrialization in labor-intensive poor countries is so fundamentally different from that of mechanized rich countries relates to the fact that most labor-intensive poor countries were once colonies of mechanized rich countries. As noted earlier, the mechanized rich countries established economies oriented toward their own industrial needs, rather than the needs of the countries they colonized. In the case of Brazil, the Portuguese forced the indigenous peoples to grow crops and mine metals and materials for export to the mother country, but they did not allow the Brazilian people to develop and establish their own native industries. For more than a century following its independence from Portugal in 1822, Brazil possessed a one-crop, export-oriented economy that was dominated first by sugarcane, then by rubber, and finally by coffee. As is usually the case with one-product export economies, Brazil's economy experienced cycles of boom and bust.

Consider the rubber boom. Between 1900 and 1925, the Brazilian city of Manaus in the Amazon was the rubber capital of the world, supplying 90 percent of the world's demand for this product. The profits were reaped by a handful of Amazon rubber barons, who controlled huge plantations on which Indians and peasants from northeastern Brazil worked under conditions resembling slavery. The rubber boom came to an end when Henry Wickham, a British businessman, smuggled 70,000 rubber plant seedlings out of Brazil to plant in Singapore. Eventually, Asian rubber priced Brazilian rubber out of the world market (Revkin 1990).

Even today, multinational corporations employ workers in developing countries to perform low-skill, low-paying, labor-intensive work whose products are exported to people in the developed countries. The remnants of colonization help explain why the model of the demographic transition does not apply to labor-intensive poor countries. When compared with mechanized rich countries, labor-intensive poor countries differ on several characteristics: They have a faster decline in death rates, relatively high birth rates despite declines in the death rate, a more rapid increase in population size, and unprecedented levels of rural-to-urban and rural-to-rural migration.

Death Rates

Death rates in the labor-intensive poor countries declined much more rapidly than they did in the mechanized rich countries; only 20 to 25 years (rather than 100 years) elapsed before the death rate fell from 50 per 1,000 to less than 10 per 1,000. Demographers attribute the relatively rapid decline to cultural diffusion (see Chapter 4). That is, the labor-intensive poor countries imported some Western technology—pesticides, fertilizers, immunizations, antibiotics, sanitation practices, and higher-yield crops—which caused an almost immediate decline in the death rates. Brazil, for example, imported DDT to eradicate the mosquitoes that carried diseases such as yellow fever and malaria. In addition, its government has developed antidotes for snake, spider, and scorpion bites which it has distributed to clinics around the country. Among other things, the government is also working to stamp out Chagas' disease and schistosomiasis, serious disorders caused by parasites in the blood (Nolty 1990).

The swift decline in death rates has caused the populations in labor-intensive poor countries to grow very rapidly. Some demographers believe that such countries may be caught in a **demographic trap**—the point at which population growth overwhelms the environment's carrying capacity:

> Once populations expand to the point where their demands begin to exceed the sustainable yield of

Demographic trap The point at which population growth overwhelms the environment's carrying capacity.

As the example of Brazil demonstrates, the issue of population growth is not simply a question of sheer numbers but also involves the carrying capacity of the land.

© Alain Keeler/Sygma

local forests, grasslands, croplands, or aquifers, they begin directly or indirectly to consume the resource base itself. Forests and grasslands disappear, soils erode, land productivity declines, water tables fall, or wells go dry. This in turn reduces food production and incomes, triggering a downward spiral. (Brown 1987:28)

Nepal (located between China and India) and Costa Rica (in Central America) represent two cases of the demographic trap. Because of population pressures, the peasants of Nepal have been forced to cultivate steep and forested hillsides, and the women have no choice but to collect feed for livestock and wood for cooking and heating from these sites. As the forests recede, the daily journey for fodder and fuel grows longer and cultivation becomes more difficult. As a result, Nepalese family incomes have dropped and diets have deteriorated. In fact, the malnutrition rates in the Nepal villages are correlated strongly with deforestation rates (Durning 1990).

In Costa Rica, a country once covered by tropical forest, the demographic trap is fueled by population growth and land policies of the past two decades, which favor a small number of cattle ranch owners (perhaps 2,000) and disregard the needs of peasants who have become landless under these policies. Approximately half the nation's cultivable land is used to raise cattle, an industry that requires little labor and hence provides few employment opportunities for landless peasants. "The rising tide of landlessness has spilled over into expanding cities, onto the fragile slopes, and into the forests, where families left with little choice accelerate the treadmill of deforestation" (Durning 1990:146).

The point at which population growth overwhelms the environment's carrying capacity is rarely caused by population pressures alone. In 1980, the Brazilian government built highway BR364 to link São Paulo with the Amazon state of Rondônia. Significant numbers of landless peasants were offered 100 acres of free land in the Amazon forest, provided that they agree to clear the land and construct a house. Unfortunately, much of the land turned out to be sandy and unfit to sustain crops; millions of acres of forest land were therefore destroyed without resolving the land problem. Technically, one could argue that the population growth in the Amazon overwhelmed its capacity to feed people. On the other hand, critics of the plan argue that government officials sent peasants blindly into the jungle, unprepared for the difficulty in making a living there. In addition, the 100-acre plots were divided without reference to terrain (many plots consisted of rocky hills, and others had no source of water). Most importantly, the chronic landlessness flows from the system of land distribution, not a lack of land: 50 percent of the cultivable land in Brazil is owned by less than 1 percent of all landowners, and much of that land remains idle. Thus, in the final analysis, it is land distribution policies—not population pressures per se—that send peasants into the hillsides, forests, and grasslands (Sanders 1986).

Birth Rates

Birth rates, although still high, are beginning to show signs of decline in most developing countries. The demographic gap remains wide, however (see Figure 13.4). It is not clear exactly which factors have caused total fertility to decline in developing countries. Sociologist Bernard Berelson (1978) has identified some important "thresholds" associated with industrialization and declines in fertility:

1. Less than 50 percent of the labor force is employed in agriculture.

2. At least 50 percent of persons between the ages of 5 and 19 are enrolled in school.

3. Life expectancy is at least 60 years.

4. Infant mortality is less than 65 per 1,000 live births.

5. Eighty percent of the females between the ages of 15 and 19 are unmarried.

Most of these conditions have been met in Brazil. Total fertility has declined during the past two decades from 5.75 children to 2.3 children per woman (Brooke

Figure 13.4 Crude Birth and Death Rates: United States Versus Brazil

These graphs show that as the death rate in each country declined, the demographic gap was much wider and more persistent in Brazil than in the United States. The U.S. graph shows dramatically how the 1950s earned the name of "baby boom." Around what five-year mark did Brazil's birth rate start a dramatic decline? Note the differences in scales between the two graphs. What was the approximate numerical difference between birth and death rates for each country as of 1990?

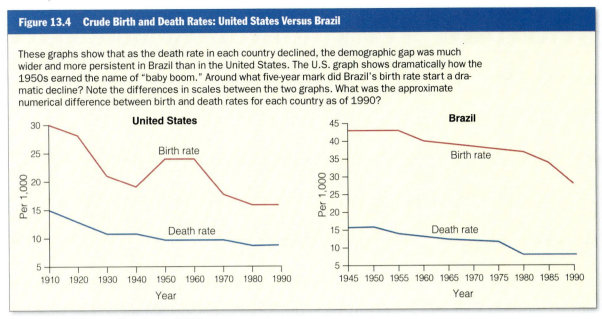

Source: *Statistical Abstracts of Latin America* (1989).

1989; U.S. Bureau of the Census 1996).[5] The factors identified by Berelson may indeed be responsible for this decline, along with other factors such as changes in government family policy and the near-universal access to television. During the 1960s, the Brazilian government had established pronatalist policies designed to increase the size of the population, with particular emphasis on bolstering the population in its Amazon states. Families who had children received tax breaks and maternity bonuses (U.S. Department of the Army 1983). Finally, however, the government abandoned these policies because they only served to increase the number of people living in already densely populated regions of the country. A review of the results of the 1996 Brazil Demographic and Health Survey shows that women do not yet have control over the number and timing of births. Nevertheless, 70.3 percent are using some method of birth control, the most popular being the pill (the most *ever* used), followed by female sterilization (the one most women *currently* use). (See Tables 13.2 and 13.3.)

Wider access to television has played some role in declining fertility in Brazil. Since 1960, the percentage of households with television sets has risen from 5 to 72 percent, and many Brazilians can now watch pro-

grams that encourage new norms. For example, the popular Brazilian soap operas feature small, consumer-oriented families. When large families are part of the drama, they usually are depicted as poor and miserable (Brooke 1989).

The speed at which death rates declined in relation to birth rates in developing countries constitutes only one of several important differences between the mechanized rich countries and the labor-intensive poor countries. Two other important differences are the rate at which population is increasing and migration, the movement of people from one area to another.

Population Growth

The rate at which a population increases is tied to a number of factors, including its sex-age composition. Generally, population size increases faster in countries with a disproportionate number of young adults (men and women of reproductive age) than in countries with a disproportionate number of middle-aged and older people.

A population's age and sex composition is commonly depicted as a **population pyramid**, a series of horizontal bar graphs, each of which represents a

[5]This decline has not been as dramatic in rural areas, where total fertility is 3.5 children per woman (Brazil Demographic and Health Survey 1998).

Population pyramid A series of horizontal bar graphs, each representing a different five-year age cohort.

Table 13.2   Fertility Preferences Among Brazilian Women with Children

What percentage of Brazilian women wanted their first child at the time it was conceived? What percentage wanted no more children at the time they had their third child?

Percentage Distribution of Births in the Five Years Preceding the Survey, by Planning Status, According to Birth Order

Planning Status	Birth Order				
	1	2	3	4+	All
Wanted then	61.0	55.2	44.4	31.5	50.5
Wanted later	31.8	29.0	22.5	15.4	26.1
Wanted no more	8.8	15.1	31.0	51.4	22.3
Missing	0.7	0.6	2.1	1.4	1.0
Total	100	100	100	100	100
(N)	(1,888)	(1,476)	(853)	(1,121)	(5,318)

Source: Brazil Demographic and Health Survey (1998).

Table 13.3   Knowledge, Ever-Use, and Current Use of Methods Among Brazilian Women in a Marital Union (percent)

Which contraceptive do women know the most about? Is that contraceptive the one most women currently use? What is the most ever-used contraceptive? What is the contraceptive most women currently use? Why do you think "male sterilization" is not used more frequently given the high percentage of women who use "female sterilization"?

Method	Know Method	Ever Used	Currently Using
Any modern method	99.9	91.4	70.3
Pill	99.4	78.9	20.7
IUD	78.5	3.8	1.1
Injection	88.5	10.4	1.2
Norplant	12.2	0.0	0.0
Vaginal methods	41.1	1.4	0.1
Condom	98.9	38.7	4.4
Female sterilization	96.8	40.1	40.1
Male sterilization	79.5	2.8	2.5
Any traditional method	91.4	39.2	6.1
Abstinence	87.5	23.0	3.0
Withdrawal	58.0	25.8	3.1
Other	3.0	1.7	0.3
Any method	99.9	93.6	75.7

Note: Vaginal methods include diaphragm, foam, tablets.
Source: Brazil Demographic and Health Survey (1998).

different five-year age cohort. (A **cohort** is a group of people who share a common characteristic or life event; in this case, it includes everyone born in a specific five-

Cohort A group of people who share a common characteristic or life event.

year period.) Two bar graphs are constructed for each cohort: one for males and another for females. The bars are placed end to end, separated by a line representing zero. Typically, the left-hand side of the pyramid depicts the number or percentage of males that make up each age cohort, and the right-hand side depicts the number or percentage of females. The graphs are stacked

Figure 13.5   Population Pyramids

The two graphs show (a) the expansive pyramid characteristic of Brazil and (b) the stationary pyramid characteristic of the United States.

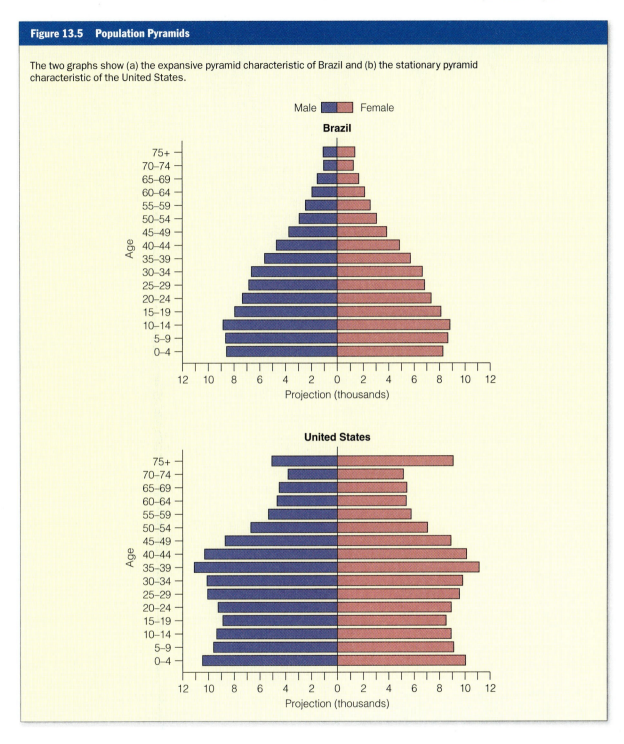

according to age; the age 0–4 cohort forms the base of the pyramid and the 75+ age cohort is at the apex (see Figure 13.5). The population pyramid allows us to view the relative sizes of the age cohorts and to compare the relative number of males and females in each cohort.

The population pyramid offers a snapshot of the number of males and females in the various age cohorts at a particular time. Generally, a country's population pyramid approximates one of three shapes—expansive, constrictive, or stationary.

Expansive pyramids, characteristic of labor-intensive poor countries, are triangular; they are broadest at the base, and each successive bar is smaller than the one below it (see Figure 13.5). The relative sizes of the age cohorts in expansive pyramids show that the population is increasing in size and composed disproportionately of young people. **Constrictive pyramids,** which characterize some European societies —most notably Switzerland and western Germany— are narrower at the base than in the middle. This shape shows that the population is composed disproportionately of middle-aged and older people. **Stationary pyramids,** which are characteristic of most developed countries, are similar to constrictive pyramids, except that all age cohorts are roughly the same size and the fertility rate matches the replacement level (see Figure 13.5).

Demographers study age-sex composition to place birth and death rates within a broader context. For example, between 1990 and 1996, the world's population increased by an estimated 491 million people. A significant proportion of this annual increase can be attributed to two countries—the People's Republic of China and India. Even though 90 countries had higher birth rates than India and 134 countries had higher birth rates than China (U.S. Bureau of the Census 1991), these two countries nevertheless accounted for 35.2 percent (172 million) of the increase; China's population grew by 76 million people, and India's grew by 96 million. This large share of the overall growth can be

attributed to the fact that both countries initially had large populations (1.21 billion people live in China and 952 million live in India) and to the fact that more than 25 percent of the Indian and Chinese populations are of child-bearing age—between age 15 and 49 (U.S. Bureau of the Census 1996).

Age-sex composition helps explain why death rates in labor-intensive poor countries are the same as those in mechanized rich countries (and sometimes even lower). Brazil has an official death rate of 8.5 per 1,000; the United States has an official death rate of 8.8 per 1,000. Thirteen percent of the U.S. population, however, is older than age 65, compared with 5 percent of Brazil's population. If the chances of survival were truly equal in the two countries, the death rate should be substantially higher in the mechanized rich country because a greater percentage of the population is older and thus at higher risk of death.

Another reason that death rates are higher in the United States than in Brazil is that in many labor-intensive poor countries, especially in rural areas, an unknown number of deaths are never registered. Consequently, a gap exists between statistics and reality. According to researchers Marilyn K. Nations and Mara Lucia Amaral (1991), "for all official purposes death occurs in Brazil only when the event is registered by surviving family members" (p. 207).

Migration

Migration is a product of two factors: **push factors,** the conditions that encourage people to move out of an area, and **pull factors,** the conditions that encourage people to move into a particular area. "On the simplest level [and in the absence of force] it can be said that people move because they believe that life will be better for them in a different area" (Stockwell and Groat 1984:291). Some of the most common push factors include religious or political persecution, discrimination, depletion of natural resources, lack of employment opportunities, and natural disasters (droughts, floods, earthquakes, and so on). Some of the most common factors that pull people into an area are employment opportunities, favorable climate, and tolerance.

Migration can be classified into two broad categories: international and internal.

INTERNATIONAL MIGRATION International migration involves the movement of people between countries. Demographers use the term *emigration* to denote the departure of individuals from one country and the term *immigration* to denote the entrance of individuals into a new country. Unless their countries are severely under-

---

Expansive pyramids Population pyramids characteristic of labor-intensive poor countries that are broadest at the base with each successive bar smaller than the one below it, showing that the population is increasing in size and composed disproportionately of young people.

Constrictive pyramids Population pyramids characteristic of some European societies, that are narrower at the base than in the middle, showing that the population is composed disproportionately of middle-aged and older people.

Stationary pyramids Population pyramids characteristic of most developed countries that are similar to constrictive pyramids, except that all age cohorts in the population are roughly the same size and the fertility rate matches the replacement level.

Push factors The conditions that encourage people to move out of an area.

Pull factors The conditions that encourage people to move into a particular area.

populated, most governments restrict the numbers of foreign people who are allowed to enter. The United States, for example, followed a policy of open migration until 1822, at which time the government imposed restrictions.

Three major flows of intercontinental (and, by definition, international) migration occurred between 1600 and the early part of the twentieth century:

- The massive exodus of European peoples to North America, South America, Asia, and Africa to establish colonies and commercial ventures, in some cases eventually displacing native peoples and establishing independent countries (this trend affected the United States, Brazil, Argentina, Canada, New Zealand, Australia, and South Africa)

- The smaller flow of Asian migrants to East Africa, the United States (including Hawaii, which did not become a state until 1959), and Brazil, where they provided cheap labor for major transportation and agricultural projects

- The forced migration of some 11 million Africans by Spanish, Portuguese, French, Dutch, and British slave traders to the United States, South America, the Caribbean, and the West Indies

In all the Americas, Brazil imported the greatest number of African slaves (Skidmore 1993).

These three migration flows are responsible for the highly diverse ethnic composition of Brazil: "There are few other [countries] on earth where such a wide spread of skin tones, from whitest white to yellow to tan to deepest black, are grouped under one nationality" (Nolty 1990:6). The mixing process began when the Portuguese sailors who colonized Brazil intermingled (often forcibly) with native women and African slave women. After it abolished slavery in 1888, the Brazilian government encouraged Europeans and Asians to immigrate into the country to replace slave labor. Between 1884 and 1914, a large number of Japanese and Italians entered Brazil as indentured servants; they worked on large plantations until they had earned enough to buy their way out of servitude.

---

[6]Ninety-four cities in the world have populations exceeding 2 million. Of these cities, 39 are in mechanized rich nations and 55 are in labor-intensive poor countries. By the year 2000, 34 additional cities will be added to this list; all but 4 of these new cities will be in developing countries. The proportion of the population that lives in the largest cities (2 million or more) varies considerably by country. Almost 25 percent of Mexico's population lives in Mexico City; about 20 percent of Brazil's population lives in São Paulo and Rio de Janeiro. In contrast, only 9 percent of the U.S. population lives in its largest city, New York City.

INTERNAL MIGRATION   In contrast to international migration, internal migration involves movement within the boundaries of a single country—from one state, region, or city to another. Demographers use the term *in-migration* to denote the movement of people into a designated area and the term *out-migration* to denote the movement of people out of a designated area.

One major type of internal migration is the rural-to-urban movement (urbanization) that accompanies industrialization.[6] Urbanization in labor-intensive poor countries, however, differs substantially from that occurring in mechanized rich societies:

> The world has never seen such extremely rapid urban growth. It presents the cities, especially in the developing countries, with problems new to human experience, as well as old problems—urban infrastructure, food, housing, employment, health, education—in new and accentuated forms. Many developing countries will have to plan for cities of sizes never conceived of in currently developed countries. High population growth in developing countries, whatever other factors enter the process, is inseparable from this phenomenon. (Rusinow 1986:9)

The unprecedented growth of urban areas stems from a number of causes. First, the growth of cities in the mechanized rich countries kept a closer pace to the number of jobs created during the industrialization process a century and a half ago. Many Europeans who were pushed off the land were able to emigrate to sparsely populated places like North America, South America, South Africa, New Zealand, and Australia. If people who fled to other countries in the eighteenth and nineteenth centuries had been forced to make their living in European cities, the conditions there would have been much worse than they actually were:

> Ireland provides the most extreme example. The potato famine of 1846–1849 deprived millions of peasants of their staple crop. Ireland's population was reduced by 30 percent in the period 1845–1851 as a joint result of starvation and emigration. The immigrants fled to industrial cities of Britain, but Britain did not absorb all the hungry Irish. North America and Australia also received Irish immigrants. Harsh as life was for these impoverished immigrants, the new continents nonetheless offered them a subsistence that Britain was unable to provide. . . . [T]here are no longer any new worlds to siphon off population growth from the less industrialized countries. (Light 1983:130–131)

In Brazil, the problem of urbanization is compounded by the fact that many migrants who come to the cities depart from some of the most economically precarious sections of Brazil. In fact, most rural-to-urban migrants are not pulled into the cities by employment opportunities, but rather are forced to move there because they have no alternatives. In the Northeast region of Brazil, peasants are pushed out because droughts and floods occur, because land is deforested and exhausted, and because the land is concentrated in the hands of a few people. When these migrants come to the cities, they face not only unemployment, but also a shortage of housing and a lack of services (electricity, running water, waste disposal). One distinguishing characteristic of cities in labor-intensive poor countries is therefore the prevalence of slums and squatter settlements, which are much poorer and larger than even the worst slums in the mechanized rich countries. A survey of living conditions in São Paulo, the largest city in Brazil and the third largest in the world, sponsored by the International Institute for Labour Studies, found that between 31.0 and 42.3 percent of families had housing needs. That is, these families' homes were constructed with makeshift materials or they lacked enough space for ordinary family activities such as sleeping, cooking, bathing, and laundry. An estimated 35.7 percent of families were classified as "bad," "very poor," or "poor" with regard to labor market status. An estimated 74 percent of the population in São Paulo were found to be lacking in at least one of four areas: housing, education, employment, or income (SEADE Foundation 1994).

Another type of massive internal migration—rural-to-rural—is taking place in many labor-intensive poor countries. We know very little about the extent of this migration (except that it may involve as many as 370 million people worldwide) or its impact on the communities that the migrants enter and leave. We do know, however, that in this kind of migration peasants move to increasingly more marginal or fragile lands. Generally, this change of scenery does not improve their social and economic status. In other words, peasants do not move to find *better* jobs than those they have; they move to find *any* kind of work or to search for land (Feder 1971).

In Brazil, an estimated 10.5 million workers without land "migrate all over the country . . . invading any empty patch which may seem unclaimed" (Cowell 1990:137). Some of the most desperate migrate into regions of the Amazon states. These peasants do not settle in the wilderness, clear the ground of trees and rocks, and become prosperous farmers, however. More often than not, they barely grow enough food to survive. This situation is not unique to Brazil. Sociologist Alfredo Molano (1993) estimates that in Colombia, settlers—"fleeing violence, political persecution, or economic deprivation—go to the forests looking for another way of finding subsistence" (p. 43). The desperate activities of migrants in search of wood for cooking and land on which to grow food and set up households results in a level of deforestation that surpasses the number of acres taken by the lumber industry (Semana 1993).

Industrialization, whatever its form, has been a major influence on the basic structure of family life. In this section, we have identified various trends, such as the relatively rapid decline in death rates coupled with relatively slow decline in birth rates, that obviously affect family size and composition. Likewise, an increase in life expectancy is associated with changes in child-bearing experiences (a decrease in infant and maternal mortality, for example) and the increased probability that parents will live to see their children into adulthood. In light of the fact that changes affect the structure of the family, we should not be surprised to learn that definitions of and ideas about the "typical" family reflect these changes.

The Structure of Family Life

The structure of family life is clearly affected by birth rates, death rates, life expectancy, and migration. As mentioned earlier, many Brazilian families will have to face and come to terms with an infant's death if 58 of every 1,000 infants die before the age of one year. In the United States, the comparable death rate for newborns is 9 of every 1,000 infants. At the same time, we cannot easily generalize about the Brazilian family. The sheer diversity of experiences makes it impossible to make blanket statements about how industrialization and its impact on births, deaths, work experiences, and migration affect family life. Consider the differences among three Brazilian families with whom sociologist Hélène Tremblay (1988) stayed as part of her 60-country, 118-family research circuit to learn about how families around the world live. Excerpts from her diaries about how each family begins the day give us some insights into the impossibility of generalizing about family life in a specific country. Consider the Yanomami family of Amazonia, which lives in a *shabono,* a communal hut that houses 15 fire sites, one for each family unit. This family consists of a mother (one of three wives, two of whom live elsewhere), a father, and three children (ages 18 months, 4 years, and 7 years).

- Mother plucks a few bananas hanging over her head and throws them on the coals along with some palm leaves stuffed with nutmeats. Breakfast will soon be ready. After feeding her children [she] joins the other women on their way to fish. [Father] is off hunting with the other men, taking advantage of the final days of the dry season. (p. 3)

Compare the Yanomami family's experience with those of two other Brazilian families:

- The Mariano de Souza Caldas family of Guriri lives in a mud hut on a government-issued land plot. The smell is appalling because they live just fifty meters from the city garbage dump. The family consists of a pregnant mother, a father, and ten children who range in age from eleven months to seventeen years old. The father gently pushes his sleeping son and gets out of bed. He moves the baby's crib, which is blocking the doorway . . . and heads for the well at the bottom of the yard. . . . He showers and rinses his mouth . . . will milk the cows . . . and will spend all morning combing the dumps in search of items that he can recycle for money. (p. 23)

- The Aratanha family of Rio de Janeiro lives in a fenced-in, highly guarded high rise with many amenities (soccer field, tennis courts, a gym). The family consists of a mother and father (both of whom are physicians) and two children (ages eighteen months and three years). They have a maid who helps care for the children. The father gets out of bed quickly without waking his wife . . . [and] leaves very early to avoid the traffic jams. . . . [The mother is] grateful for her new car. It saves her half an hour every morning. (p. 31)

In the sections that follow, we consider how the changes in life expectancy, which include changes in birth and death rates, lead to changes in family function and composition and in the status of children. We also consider how these factors lead to changes in the division of labor between men and women and to changes in the reasons why people marry. Finally, we discuss how these demographic factors affect the definition of family.

The Consequences of Long Life

Since the turn of the century, the average life expectancy at birth has increased by 28 years in the mechanized rich countries and by 20 years (or more) in the labor-intensive poor countries. In *The Social Consequences of Long Life*, sociologist Holger R. Stub (1982) describes at least four ways in which gains in life expectancy have altered the composition of the family since 1900.

First, the chance that children will lose one or both parents before they reach 16 years of age has decreased sharply. There was a 24 percent chance of such an occurrence in 1900; today, the probability is less than 1 percent. At the same time, parents can expect that their children will survive infancy and early childhood. In 1900, 250 of every 1,000 children born in the United States died before age 1; 33 percent did not live to age 18. Today, only 10 of every 1,000 children born die before they reach age 1; fewer than 5 percent die before reaching age 18.

Second, the potential length of marriages has increased. Given the mortality patterns in 1900, newly married couples could expect their marriage to last an average of 23 years before one partner died (if we assume they did not divorce). Today, assuming that they do not divorce, newly married couples can expect to be married for 53 years before one partner dies. This structural change may be one of several factors underlying the current high divorce rates. At the turn of the century, death nearly always intervened before a typical marriage had run its natural course. Now, many marriages run out of steam with decades of life remaining for each spouse. When people could expect to live only a few more months or years in an unsatisfying relationship, they would usually resign themselves to their fate. But the thought of 20, 30, or even 50 more years in an unsatisfying relationship can provoke decisive action at any age (Dychtwald and Flower 1989:213). According to Stub, divorce dissolves today's marriages at the same rate that death did 100 years ago.

Third, people now have more time to choose and get to know a partner, settle on an occupation, attend school, and decide whether they want children. Moreover, an initial decision made in any one of these areas is not final. The amount of additional living time enables individuals to change partners, careers, or educational and family plans—a luxury not shared by their turn-of-the-century counterparts. Stub (1982) argues that the midlife crisis is related to long life because many people "perceive that there yet may be time to make changes and accordingly plan second careers or other changes in life-style" (p. 12).

Finally, the number of people surviving to old age has increased. (In countries where fertility is low or declining, the proportion of old people in the population—not merely the number—is increasing as well.) In 1970, approximately 25 percent of people in their late fifties had at least one surviving parent; in 1980, 40 percent had a surviving parent. Even more astonishing, in 1990, 20 percent of people in their early sixties and 3 percent of people in their early seventies had at least one surviving parent (Lewin 1990). Although a small number of people have always lived to age 80 or 90, "[t]here is no historical precedent for the aging of our population. We are in the midst of a new phenomenon" (Soldo and Agree 1988:5).

Much has been written about the growing numbers of people older than age 65 in the world, especially in

Living Arrangements and Quality of Life of Older Americans

Many assume that health among the elderly has improved because these people, as a group, are living longer. Others hold a contradictory image of the elderly, seeing them as dependent and frail. The truth actually lies somewhere in between. Poor health is not as prevalent as many assume. In 1992, about 3 in every 4 noninstitutionalized persons aged 65 to 74 considered their health to be good. Two in three aged 75 or older felt similarly.

On the other hand, as more people live to the oldest ages, there may also be more who face chronic, limiting illnesses or conditions, such as arthritis, diabetes, osteoporosis, and senile dementia. These conditions result in people becoming dependent on others for help in performing the activities of daily living. With age comes increasing chances of being dependent. For instance, while 1 percent of those aged 65 to 74 years lived in a nursing home in 1990, nearly 1 in 4 aged 85 or older did. And among those who were not institutionalized in 1990–1991, 9 percent aged 65 to 69 years—but 50 percent aged 85 or older—needed assistance performing everyday activities such as bathing, getting around inside the home, and preparing meals. (See graph below.)

As more and more people live long enough to experience multiple, chronic illnesses, disability, and dependency, an increasing number of relatives in their fifties and sixties will face the concern and expense of caring for them. The parent-support ratio gives us an approximate idea of things to come. This ratio equals the number of persons aged 85 and older per 100 persons aged 50 to 64. Between 1950 and 1993, the ratio tripled from 3 to 10. Over the next six decades, it could triple yet again, to 29.

Of the 2.2 million Americans who died in 1991, 1.6 million (or 7 of 10) were elderly. Seven of 10 elderly deaths could be attributed to either heart disease, cancer, or stroke. Although death rates from heart disease have declined for the elderly since the 1960s, this malady remains the leading cause of death among them. Death rates from cancer, on the other hand, have increased since 1960.

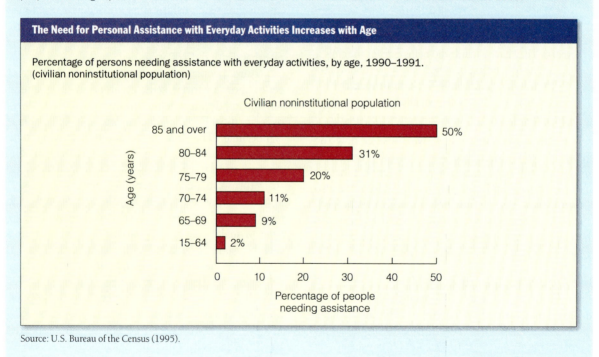

The Need for Personal Assistance with Everyday Activities Increases with Age

Percentage of persons needing assistance with everyday activities, by age, 1990–1991. (civilian noninstitutional population)

Civilian noninstitutional population

- 85 and over: 50%
- 80–84: 31%
- 75–79: 20%
- 70–74: 11%
- 65–69: 9%
- 15–64: 2%

Age (years)

Percentage of people needing assistance

Source: U.S. Bureau of the Census (1995).

terms of the challenges of caring for the disabled and the frail elderly. Yet, the emphasis on this segment of the older population should not obscure the fact that the elderly are a rapidly changing, heterogeneous group. They differ according to gender, age (a 30-year difference separates persons aged 65 from individuals in their nineties), social class, and health status. Most older people today are in relatively good health; in the United States, only 5 percent of persons aged 65 and older live in nursing homes (Eckholm 1990). (See "Living Arrangements and Quality of Life of Older Americans.")

In most countries, including the United States, female relatives care for most disabled and frail elderly persons (Stone et al. 1987; Targ 1989). In the United States, approximately 72 percent of the caregivers are women. Of this 72 percent, 22.7 percent are spouses,

28.9 percent are daughters, and 19.9 percent are daughters-in-law, sisters, grandmothers, or some other female relative or nonrelative (Stone et al. 1987). The major issue faced by even the closest of families is how to meet the needs of the disabled and frail elderly so as not to "constrain investments in children, impair the health and nutrition of younger generations, impede mobility," or impose too great a psychological, physical, or time-demand stress on the caregivers. This issue may be even more problematic in developing countries, where the presence of older dependents "may alter the level of subsistence of other members in the family" (United Nations 1983:570). In subsistence cultures—even those that display considerable respect toward the elderly—"the decrepit elderly may be seen as too much of a burden to be supported during periods of deprivation and may be killed, abandoned, or forsaken" (Glascock 1982:55). Although many programs already exist and more are being developed to support caregivers and address the issues associated with caring for the disabled and frail elderly, most efforts remain in the pilot stages. "It is going to take as many as fifteen or twenty years to sort out what works" (Lewin 1990:A11).

The Status of Children

The technological advances associated with the Industrial Revolution decreased the amount of physical exertion and time needed to produce food and other commodities. As human muscle and time became less important to the production process, children lost their economic value. In nonmechanized, extractive economies, children represent an important source of cheap and unskilled labor for the family. This fact may partly explain why the states in Amazonian Brazil have some of the highest total fertility rates found anywhere in the world (Butts and Bogue 1989). The total fertility rate in rural Acre is 8.03, a level that demographers have defined as approximating the human capacity to reproduce. (See "The Economic Role of Children in Labor-Intensive Environments.")

Middle- and upper-class children in mechanized rich economies are likely to live in settings that strip them of whatever potential economic contribution they might make to the family economy (Johansson 1987). In these economies, the family's energies shift away from production of food and other necessities and toward the consumption of goods and services.

The U.S. Department of Agriculture estimates that the average family earning an annual income of at least $56,700 will spend $346,980 to house, feed, clothe, transport, and supply medical care (not covered by insurance) to a child until he or she is 22 years old (*American Demographics* 1996), with annual expenses ranging from $5,100 to $10,510 (Wark 1995). These figures represent a basic sum; the costs grow even higher when we include extras (such as summer camps, private schools, sports, and music lessons). Even if children go to work when they reach their teens, they often use that income to purchase items for themselves and do not contribute to household expenses. Demographer S. Ryan Johansson (1987) argues that in the mechanized rich economies, couples who choose to have children bring them into the world to provide intangible, "emotional" services—love, companionship, an outlet for nurturing feelings, enhancement of dimensions of adult identity—rather than economic services.

Urbanization and Family Life

Urbanization encompasses two phenomena: the migration of people from rural areas to cities so that an increasing proportion of the population comes to reside in cities (see "The Growth of Major Urban Centers"), and a change in the ties that bind people to one another. (See the discussion of mechanical and organic solidarity in Chapter 6.) In mechanized rich countries, people who live in urban settings usually are not part of self-sufficient economic units composed of family members. Instead, they have opportunities to make contact with people outside the family network, and they come to depend on people and institutions other than the family (such as the workplace, hospitals and clinics, counseling services, schools, day-care centers, and the media) to meet their various needs. People's relationships with most of the individuals they encounter in the course of a day are transitory, limited, and impersonal. The family does not cease to be important to people's lives, however. Instead, it becomes important in a different way—less as an economic unit and more as a source of personal support. Most sociological research on the American family shows that the great majority of people have some family members living near them, that they interact with those relatives regularly, and that they define these interactions as meaningful and important (Goldenberg 1987).

Urbanization in itself, however, does not necessarily reduce the economic function of the family. When studying the urban poor in Brazil, Janice Perlman (1967) found that the majority of migrants to Rio de Janeiro (67 percent) came to the city in the company of relatives or went to the homes of relatives upon arrival; many made every effort to live as close to one another as possible. As a result, some neighborhoods became dominated by large kin networks. Perlman's research also showed that family networks are extremely useful

The Economic Role of Children in Labor-Intensive Environments

Children often are an important source of cheap labor, especially for nonmechanized industries. The childhood experiences of Chico Mendes, the son of an Amazon rubber tapper, illustrate the economic role of children in the kind of labor-intensive work that takes place in rubber plantations.

Chico was an important Brazilian grassroots environmentalist and union activist dedicated to preserving the Amazon and to improving rubber tappers' economic standard of living as compared with those of cattle ranchers and rubber barons. He was murdered in 1988 by Amazon cattle ranchers. Indians and rubber tappers are now working together against those who misuse or otherwise destroy the resources of the forests.

> Childhood for Chico Mendes was mostly heavy work. . . . If all his siblings had lived to adulthood, Chico would have had seventeen brothers and sisters. As it was, conditions were so difficult that by the time he was grown, he was the oldest of six children—four brothers and two sisters.
>
> When he was five, Chico began to collect firewood and haul water. A principal daytime occupation of young children on the seringal has always been lugging cooking pots full of water from the nearest river. . . . Another chore was pounding freshly harvested rice to remove the hulls. A double-ended wooden club was plunged into a hollowed section of tree trunk filled with rice grains, like an oversize mortar and pestle. Often two children would pound the rice simultaneously, synchronizing their strokes so that one club was rising as the other descended.
>
> By the time he was nine, Chico was following his father into the forest to learn how to tap. . . . Well before dawn, Chico and his family would rise. . . . [T]hey would grab the tools of their trade—a shotgun, . . . a machete, and a pouch to collect any useful fruits or herbs found along the trail. They left before dawn, because that was when the latex was said to run most freely. . . .
>
> As they reached each tree, the elder Mendes grasped the short wooden handle of his rubber knife in two hands, with a grip somewhat like that of a golfer about to putt. . . . It was important to get the depth [of the cut] just right. . . . A cut that is too deep strikes the cambium [generative tissue of the tree] and imperils the tree; a cut that is too shallow misses the latex-producing layer and is thus a waste. . . .
>
> Chico learned how to position a tin cup, or sometimes an empty Brazilian nut pod, just beneath the low point of the fresh cut on a crutch made out of a small brand. . . . The white latex immediately began to dribble down the slash and into the cup. . . .
>
> Chico quickly adopted the distinctive, fast stride of the rubber tapper. . . . The pace has evolved from the nature of the tapper's day. The 150 and 200 rubber trees along an estrada are exasperatingly spread out. A simple, minimal bit of work is required at each tree, but there is often a 100-yard gap between trees. Thus, a tapper's morning circuit can be an 8 to 11-mile hike. And that is just the morning. Many tappers retrace their steps in the afternoon to collect the latex that has accumulated from the morning cuts.
>
> By the time Chico and his father came full circle on an estrada and returned to the house, it would be close to midday, time for a lunch. . . . Then, they would retrace their steps on the same trail, to gather the latex that had flowed from the trees during the morning. Only rarely would they return home before five o'clock. By then they would be carrying several gallons of raw latex in a metal jar or sometimes in a homemade, rubber-coated sack.
>
> The very best rubber is produced when this pure latex is immediately cured over a smoky fire.
>
> The smoking of the latex was done in an open shed that was always filled with fumes from the fire. The smokier the fire, the better. Usually, palm nuts were added to the flames to make the smoke extra thick. Chico would ladle the latex onto a wooden rod or paddle suspended over a conical oven in which the nuts and wood were burned. As the layers built up on the rod, the rubber took on the shape of an oversize rugby football. The smoking process would continue into the evening. After a day's labor of fifteen hours or more, only six or eight pounds of rubber were produced. The tappers often developed chronic lung diseases from exposure to the dense, noxious smoke.
>
> Toward December, with the return of the rainy season, the tappers stopped harvesting latex and began collecting Brazil nuts.
>
> During the rainy season, the Mendeses often crouched on their haunches around a pile of Brazil nut pods, hacking off the top of each one with a sharp machete blow, then tossing the loose nuts onto a growing pile. For tappers in regions with Brazil nut trees, the nuts can provide up to half of a family's income. One tree can produce 250 to 500 pounds of nuts in a good season and some tappers' trails pass enough trees to allow them to collect more than three tons of nuts each season. While that might sound like a potential windfall, even as late as 1989, tappers received only three or four cents a pound for the harvest—which later sold for more than $1 a pound at the export docks.
>
> When the Mendeses were not harvesting latex or nuts, they tended small fields of corn, beans, and manioc.

Source: From *The Burning Season* by Andrew Revkin, pp. 69–76. Copyright © 1990 by Andrew Revkin. Reprinted by permission of Houghton Mifflin Company.

to people living in precarious financial conditions. She found that 63 percent of the people she studied had gotten their first job through friends or kin.

Anthropologist Claudia Fonseca (1986) also confirmed the significance of family relationships in Brazil. She studied 68 mothers from a shantytown of 750 squatters living on an empty lot in a middle-class neighborhood in southern Brazil. She found that 50 percent of the mothers had sent a child away to live under the care of another person while they recovered from illness or financial setbacks. Most often, this caretaker was a relative.

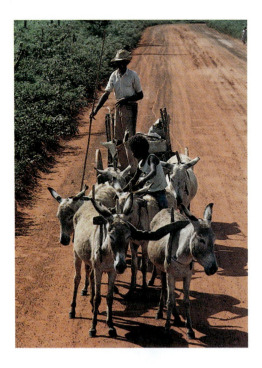

The effects of industrialization on family life are too complex to be captured by sweeping generalizations. Brazil, for example, includes regions where children remain sources of cheap labor in a nonmechanized economy. Yet it also includes cities where middle-class children, like their counterparts in the United States, play the role of consumers in a mechanized rich country.

© Stephanie Maze/Woodfin Camp & Associates (left); © Paulo Fridman/Sygma (right)

Judith Goode (1987) has written extensively about Colombian street children, and Thomas Sanders (1987a, 1987b) has studied Brazilian street children. Both groups of children are frequently misperceived by government officials and middle-class people as abandoned by their parents, as delinquents, and as a reflection of widespread family pathology among squatters. Both writers conclude, however, that street children (usually between the ages of 10 and 15) tend to come from poverty-stricken families "who are trying to maximize their household incomes by utilizing the money-earning capacities of their children in such activities as shining shoes, carrying packages, watching cars, and begging" (Sanders 1987b:1). Goode (1987) writes:

> Attempts to classify types of income for street children often distinguish between those activities which are legal, such as performing services as shoeshine boys, street vendors, and car watchers; those which are more marginal, such as begging and singing on buses; and those which are illegal and anti-social, such as purse-snatching, picking pockets, and other forms of theft. (p. 6)

The Division of Labor Between Men and Women

In 1984, sociologist Kingsley Davis published "Wives and Work: The Sex Role Revolution and Its Consequences." In this article, Davis identified "as clear and definite a social change as one can find" (p. 401):

Between 1890 (the first year for which reliable data exist) and 1980, the proportion of married women in the labor force rose from less than 5 percent to more than 50 percent. Davis found that this pattern holds true for virtually every industrialized country except Japan.[7] He attributes this change to the Industrial Revolution and its effect on the division of labor between men and women (see Figure 13.6). Keep in mind that Davis's theory is intended to explain how industrialization in the mechanized rich countries affected the division of labor between males and females; it does not systematically address how industrialization affects the division of labor between males and females in labor-intensive poor countries.

THE DIVISION OF LABOR AND THE BREADWINNER SYSTEM   Before industrialization—that is, for most of human history—the workplace comprised the home and the surrounding land. Under these circumstances, the division of labor was based on sex. In nonindustrial societies (which includes two major types, hunting-and-gathering and agrarian societies), men provided raw materials through hunting or agriculture, and women processed these materials. Women also worked in agriculture and provided some raw materials by gathering food and hunting small animals, and they took care of the young.

[7]Japan followed an accelerated model: It underwent a rapid decline in mortality followed by an equally rapid decline in fertility. The rapid decline in fertility was due to high rates of abortion.

U.S. in Perspective
The Growth of Major Urban Centers

Most people in the United States are probably still accustomed to thinking of Europe and the United States as leading cultural forces in the world. Yet we also associate culture and innovation with great cities.

Consider this table and map showing the world's 15 largest cities, ranked by size. In what regions and countries of the world are the largest cities concentrated today? Is this concentration a strong reason to expect that one or

more of these parts of the world have become, or are about soon to become, great centers of world culture? Why?

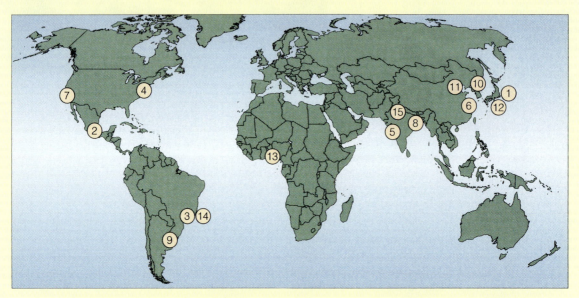

Rank	City, Country	Population 1996 (thousands)	Annual Growth Rate 1990–1995 (%)	Percentage Increase Between 1975 and 1995	Population of City as Percentage of Total Population*
1	Tokyo, Japan	26,959	1.45	36.36	21.56
2	Mexico City, Mexico	16,562	1.81	47.4	18.17
3	São Paulo, Brazil	16,533	1.84	64.56	10.40
4	New York City, United States	16,332	0.34	2.85	6.11
5	Mumbia (Bombay), India	15,138	4.24	120.79	1.63
6	Shanghai, China	13,584	0.36	18.71	1.11
7	Los Angeles, United States	12,410	1.60	39.03	4.65
8	Calcutta, India	11,923	1.81	51.15	1.28
9	Buenos Aires, Argentina	11,801	1.15	29.07	33.95
10	Seoul, South Korea	11,609	1.92	70.52	25.85
11	Beijing, China	11,299	0.87	32.23	0.93
12	Osaka, Japan	10,609	0.23	7.77	8.48
13	Lagos, Nigeria	10,287	5.68	211.73	9.21
14	Rio de Janeiro, Brazil	10,181	1.00	29.63	6.40
15	Delhi, India	9,948	3.85	124.76	1.07

*Denotes percentage of the total population of the country in which the corresponding city is located.

The Industrial Revolution separated the workplace from the home and altered the division of labor between men and women. It destroyed the household economy by removing economic production from the home and taking it out of the women's hands:

> The man's work, instead of being directly integrated with that of wife and children in the home or on the surrounding land, was integrated with that of non-kin in factories, shops, and firms. The man's economic role became in one sense more important to the family, for he was the link between the family and the wider market economy, but at the same time his personal participation in the household diminished. His wife, relegated to the home as her sphere, still performed the parental and domestic duties that women had always performed. She bore and reared children, cooked meals, washed clothes, and cared for her husband's personal needs, but to an unprecedented degree her economic role became restricted. She could not produce what the family consumed, because production had been removed from the home. (Davis 1984:403)

Davis called this new economic arrangement the "breadwinner system." From a historical point of view, this system is not typical. Rather, it is peculiar to the middle and upper classes and is associated with a particular phase of industrialization—the stage lasting from the point at which agriculture loses its dominance to the point where only 25 percent of the population still work in agriculture. In the United States, "the heyday of the breadwinner system was from about 1860 to 1920" (p. 404). This system has been in decline for some time in the United States and other industrialized countries, but it tends to recur in countries undergoing a particular phase of development.

Davis asks, "Why did the separation of home and workplace lead to this system [in most mechanized rich countries]?" The major reason, he believes, is that women had too many children to engage in work outside the home. This answer is supported by the fact that family size, in the sense of the number of living members, reached its peak from the mid-1800s to the early 1900s. During this period, infant and childhood mortality declined but the old norms favoring large families persisted.

THE DECLINE OF THE BREADWINNER SYSTEM The breadwinner system did not last long because it placed too much strain on husbands and wives and because a number of demographic changes associated with industrialization worked to undermine it. The strains stemmed from several sources. Never before had the

While media headlines about street children in Brazil focus on the violent gangs, most street children try to find work to help increase their families' household incomes.
© Sean Sprague/Panos Pictures

roles of husband and wife been so distinct. Never before had women played less than a direct, important role in producing what the family consumed. Never before had men been separated from the family for most of their waking hours. Never before had men had to bear the sole responsibility of supporting the entire family. Davis regards these events as structural weaknesses in the breadwinner system. Given the presence of these weaknesses, the system needed strong normative controls to survive. "The husband's obligation to support his family, even after his death, had to be enforced by law and public opinion, illegitimate sexual relations and reproduction had to be condemned, divorce had to be punished, and marriage had to be encouraged by making the lot of the 'spinster' a pitiful one" (p. 406).

Davis maintains that the normative controls collapsed because of the strains inherent in the breadwinner system and because of demographic and social changes that accompanied industrialization. These changes included decreases in total fertility, increases in life expectancy, increases in the divorce rate, and increases in opportunities for employment perceived as suitable for women.

DECLINES IN TOTAL FERTILITY The decline in total fertility began before married women entered the labor force in large numbers. This fact led Davis to conclude that the decline itself changed women's lives

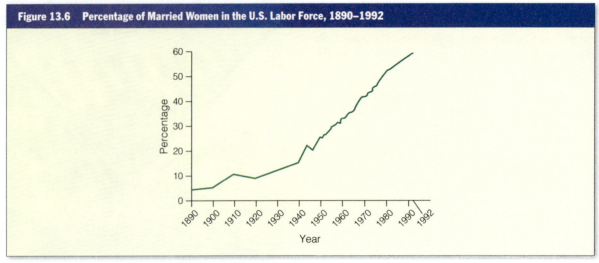

Figure 13.6   Percentage of Married Women in the U.S. Labor Force, 1890–1992

Sources: Adapted from U.S. Bureau of the Census as reported by Davis
(1984:398–399) and U.S. Bureau of the Census (1993:400).

so that they had the time to work outside the home, especially after their children entered school. During the 1880s, total fertility among white women was approximately 5.0; in the 1930s, it averaged 2.4; during the 1970s, it averaged 1.8. Not only did the number of children decrease, but reproduction ended earlier in women's lives and the births were spaced closer together than they had been in earlier years. (The median age of mothers at the last birth was 40 years in 1850; by 1940, it had fallen to 27.3 years.) Davis attributes the changes in child-bearing patterns to two sources: the forces of industrialization, which changed children from an economic asset to an economic liability, and the "desire to retain or advance one's own and one's children's status in a rapidly evolving industrial society" (p. 408).

INCREASED LIFE EXPECTANCY   With the relatively short life expectancy in 1860 and the relatively high age at which women had their last child (age 40), the average woman was dead by the time her last child left home. By 1980, given the changes in family size, spacing of children, and age of last pregnancy, the average woman could expect to live 33 years after her last child left home. As a result, child care came to occupy a smaller proportion of a woman's life. In addition, although life expectancy has increased for both men and women, on average women can expect to live longer than men. In 1900, women outlived men by about 1.6 years on average; in 1980, they outlived men by approximately 7 years. Yet, because brides tend to be three years younger on average than their husbands,

married women can expect to live past their husbands' death for approximately 10 years. In addition, the distorted sex ratio caused by males' earlier death decreases the probability of remarriage. Although few women think directly about their husbands' impending death as a reason for working, the difference in mortality remains a background consideration.

INCREASED DIVORCE   Davis traces the rise in the divorce rate to the breadwinner system, specifically to the shift of economic production outside the home:

> With this shift, husband and wife, parents and children, were no longer bound together in a close face-to-face division of labor in a common enterprise. They were bound, rather, by a weaker and less direct mutuality—the husband's ability to draw income from the wider economy and the wife's willingness to make a home and raise children. The husband's work not only took him out of the home but also frequently put him into contact with other people, including young unmarried women [who have always worked] who were strangers to his family. Extramarital relationships inevitably flourished. Freed from rural and small-town social controls, many husbands either sought divorce or, by their behavior, caused their wives to do so. (pp. 410–411)

Davis notes that an increase in the divorce rate preceded married women's entry into the labor market by several decades. He argues that once the divorce rate reaches a certain threshold (higher than a 20 percent chance of divorce), more married women seriously consider seeking employment to protect themselves in case of divorce. When both husband and wife participate in the labor force, the chances of divorce increase even more. In that case, both partners interact with people who are strangers to the family. Moreover, they live in three different worlds, only one of which they share.

INCREASED EMPLOYMENT OPPORTUNITIES FOR WOMEN Davis believes that married women are motivated to seek work because of changes in child-bearing experiences, increases in life expectancy, the rising divorce rate, and the inherent weakness of the bread-winner system. They tend to act on this motivation as their opportunities to work increase. With improvements in machine technology during industrialization, productivity increased, the physical labor required to produce goods and services decreased, and wages rose in industrialized societies. (Chapter 2 explains how industrialization proceeds in labor-intensive poor countries.) As industrialization matures, it brings a corresponding increase in the kinds of jobs perceived as suitable for women (traditionally, nursing, clerical and secretarial work, and teaching).

The two-income system is not free of problems, however. First, the present system "lacks normative guidelines. It is not clear what husbands and wives should expect of each other. It is not clear what ex-wives and ex-husbands should expect, or children, cohabitants, friends, and neighbors. Each couple has to work out its own arrangement, which means in practice a great deal of experimentation and failure" (Davis 1984:413). Second, although the two-income system gives wives a direct role in economic production, it also requires that they work away from the home, a situation that complicates child care. Third, even in the two-income system, the woman remains primarily responsible for domestic matters. Davis maintains that women bear this responsibility because men and women are unequal in the labor force. (On the average, women earn 66 cents for every dollar

As reflected in the Ms. Foundation for Women's "Take Our Daughters to Work" campaign, the growing importance of employment and careers to women in industrialized societies represents a profound social change as well as an economic one. Here, a nine-year-old girl is spending a day at work with her mother, a state supreme court justice.

© Marty Lederhandler/AP/Wide World Photos

earned by men.) As long as women make an unequal contribution to the household income, they will do more work around the house.

The problems of the new system are stressed further by the fact that a large percentage of married women (almost 40 percent) do not work outside the home. Davis believes that this lack of participation reflects the psychosocial costs of employment to married women. The dilemma that women face can be summarized as follows: if married women work and have no children, they are selfish; if they stay home and raise a family, they are considered underemployed at best; if they work and raise a family, people wonder how they can possibly do either job right.

Summary and Implications

We have waited until the end of the chapter to offer a definition of the family, because the challenge of defining the family is complicated by historic events that affect its composition, size, and function (see Table 13.4). Simply consider the ways in which the division of labor changed because of declines in total fertility and increased life expectancy (especially increases in female life expectancy relative to men) and the subsequent effect on family size and composition. The challenge of defining the family is

further complicated by the fact that an amazing variety of family arrangements exists worldwide—a variety reflected in the numerous norms that specify how two or more people can become a family. These norms include those governing the number of spouses that a person can have, the way a person should select a spouse, the ideal number and spacing of children, the circumstances under which offspring are considered legitimate (or illegitimate), the ways in which people trace their descent,

Table 13.4   Effects of Industrialization on Key Episodes in Family Life

Key Episode	Before Industrialization	Mature Phase of Industrialization
Child-bearing Experiences		
Birth rates (annual)	30–50 births/1,000 people	Fewer than 20 births/1,000 people
Total fertility	9	2
Spacing	Entire reproductive life (15–49)	Short segment of reproductive life
Maternal mortality (annual)	More than 600 deaths/100,000 births	30 deaths/100,000 births
Chance of dying from pregnancy-related causes	1 in 21	1 in 10,000
Chances of Survival		
Death rate (annual)	501 deaths/1,000 people	Fewer than 10 deaths/1,000 people
Life expectancy	20–30	70 years
Major causes of death	Epidemic sources, deficiency diseases, malnutrition, parasite diseases	Degenerative and human-made
Nature of Labor		
Livelihood	Labor-intensive agriculture	Information and services
Source of power	Human muscles and other sources of power	Machines and created power
Economy	Subsistence-oriented	Consumption-oriented
Residence	Rural	Urban

Table 13.5   Distribution of Family and Nonfamily Households in the United States, 1960 and 1992 (percent)*

Total Households	1960 (53,021,000)	1992 (95,669,000)
I. Family household		
A. Nuclear		
Married couple with children	52.18	26.46
Married couple with no children	22.00	21.64
Father-child(ren)	2.33	3.16
Mother-child(ren)	8.48	12.22
B. Extended†		
Grandparent(s), grandchildren, both parents	N.A.	0.52
Grandparent(s), grandchildren, mother only	N.A.	1.82
Grandparent(s), grandchildren, father only	N.A.	0.15
Grandparent(s), grandchildren, no parents	N.A.	0.91
II. Nonfamily household		
One person living alone	14.88	25.06
Institutions†	3.61	3.48
III. Other living arrangements§	6.71	3.46

* Most recent data.

Note: Percentages do not total 100 because of inconsistencies and overlaps in the categories defined by the U.S. Bureau of the Census; N.A. means "not available."

†Only those extended family categories used by the U.S. Bureau of the Census.

‡Institutions include those living in correctional institutions, nursing homes, and juvenile centers.

§Other living arrangements include college dormitories, military quarters, emergency shelters, and other unrelated persons living together.

Sources: U.S. Bureau of the Census (1961:40; 1993:64–65).

Table 13.6   Distribution of Family and Nonfamily Arrangements in Brazilian Households, 1960 and 1984 (percent)*

Family and Nonfamily Household	1960*	1984†
I. Family household	92.8	93.2
A. Nuclear	68.9	70.3
Married couple with children	54.1	46.5
Married couple with no children	8.5	12.8
Father or mother, child—family	6.3	11.0
B. Extended	22.6	14.1
Married couple with children, unmarried and relatives	13.1	6.6
Married couple, no children and other relatives	2.9	1.6
Other extended family (no married couple, only father or mother, child—family and relatives)	6.6	5.9
C. Complex	1.5	8.7
Married couple with children unmarried and nonrelatives	1.5	8.7
II. Nonfamily household	7.2	6.8
Men or women living alone	5.3	5.5
Other nonfamily	1.9	1.3
Total percentage	100	100
Absolute values in thousands	13,532	31,075

*For the 1960 census, the family-household classification did not include maids and lodgers. At other times only maids are not included.

†1984 is the last year for which data are available. The rural population of the Central West region was not considered in 1976, nor was the North region considered in 1984.

Source: Goldani (1990:527). Original sources: Brazilian Population Censuses, tapes of 1 percent in 1970 and 75 percent in 1980; and Household Survey, 1976 and 1984 tapes.

and the nature of the parent-child relationship over the child's and parent's lives. In light of this variability, we should not be surprised to learn that it is difficult to construct a definition of family and that "no general theory or universal model of the family can be formulated" (Behnam 1990:549).

Most official definitions of family emphasize blood ties, adoption, or marriage as criteria for membership, and the function of procreation and socialization of offspring. The U.S. Bureau of the Census (1993) uses the term *family* to mean "a group of two or more persons related by birth, marriage, or adoption and residing together in a household" (p. 5). This definition has been in effect since 1950. Between 1930 and 1950, however, the definition of family revolved around a head of the household and reflected living arrangements in a more rural America:

a private family comprises a family head and all other persons in the house who are related to the head by blood, marriage, or adoption, and who live together and share common housekeeping arrangements. The term "private household" is used to

include the related family members (who constitute the private family) and the lodgers, servants, or hired hands, if any who regularly live in the home. (U.S. Bureau of the Census 1947:2)

Until the mid-1970s, the official definition of family in Brazil included the following criteria:

(1) [that] family is synonymous with legal marriage, (2) marriage lasts until a spouse dies; (3) the husband is the breadwinner and the sole earner; (4) the wife is a full-time home-maker and her work has no [formal] economic value; (5) the husband is the legal head of the family. (Goldani 1990:525)

The Brazilian government's official definition of what constitutes a family has changed several times in the past 25 years, and the preceding definition of family has now been abandoned constitutionally. For example, the 1988 constitution recognized a stable union between a man and a woman as a family and children living with one parent as a family.

Changes in the official definition of family reflect dramatic changes in the size and composition of families that have occurred in both the United States and

Family Portraits

Can a single image represent all that we mean by "family"? What ideas of "family" are conveyed by these pictures? The examples shown here are just a few of the many types of arrangements possible. Many sociologists are greatly interested in the concept of family and problems of families, in all their complexity. At an undergraduate level, this inquiry takes the form of a course often called Marriage and Family.

Among the 118 photographs on board the Voyager 1 and Voyager 2 spacecraft was this one entitled "Family Portrait."

© Nina Leen/Life Picture Service

A Mexican family celebrating a birthday.

© Chip and Rosa Maria de la Cueva Peterson

A father with his adopted children.

© J. Patrick Forden/Sygma

An African mother and her children (absent the father, who lives and works in the city).

© Betty Press/Woodfin Camp & Associates

Lesbian parents with their child.

© Chris Maynard/Gamma Liaison

Brazil in the past 20 to 30 years (see Tables 13.5 and 13.6). Both the contemporary U.S. and Brazilian definitions reflect changes in the division of labor between men and women.

Nevertheless, the current definitions fail to capture certain demographic realities that are beyond any one individual's control (see "Family Portraits"). For example, according to the official definitions of a family, two elderly widows who live together and offer one another emotional support and financial commitment do not qualify as a family, nor would a couple who "adopt" an elderly man and commit to care for him. Likewise, older people who choose to assume caregiver roles to people in need (foster grandchildren, the sick and dying) do not qualify as family. On the surface, this distinction may not seem important—until we consider that official definitions of family can be used to deny family-related benefits to people who do not fit the definition of family. Such benefits include insurance coverage, housing, time off from work, inheritances (especially in the absence of a will), and opportunities to cash in individual retirement accounts without penalty to help a "family member" in need.

In this vein, a number of recent court decisions in the United States have ruled that criteria other than socially recognized bonds of blood, marriage, or adoption must be considered in defining whether a person belongs to a family and thus is entitled to family-related benefits. These factors, which are part of any healthy family environment, include exclusivity, longevity, emotional support, and financial commitment (Gutis 1989a, 1989b). Clearly, at least with regard to official definitions of family, it is not especially useful to think about the family in terms of specific membership (that is, son, daughter, mother, father) or a specific function, such as procreation or socialization. Instead, we would do better to take a contextual view of the family and consider how key demographic events transform its character in fundamental ways (the transformation may have positive or negative consequences). The official definition, however, drives policies that can alleviate or exacerbate those consequences.

Key Concepts

Use this outline to organize your review of key chapter ideas.

Demography
Mechanized rich
Labor-intensive poor
    Doubling time
    Infant mortality
    Total fertility
    Per capita income
    Annual per capita consumption of energy
Theory of demographic transition
    Demographic gap
Demographic trap
Population pyramid
    Cohort
    Expansive pyramid
    Stationary pyramid
    Constructive pyramid

Migration
    International migration
        Emigration
        Immigration
    Internal migration
        In-migration
        Out-migration
    Push factors
    Pull factors
Mortality crisis
Urbanization

internet assignment

Go to Pennsylvania State University's Population Research Institute Web site (http://www.pop.psu.edu/Demography /demography.html). Browse the links connected with this Population Research Institute and links to other Web sites. Select a document that adds to the information in this chapter. Write a brief description of the information, and explain how it further clarifies a demographic concept covered in this chapter.

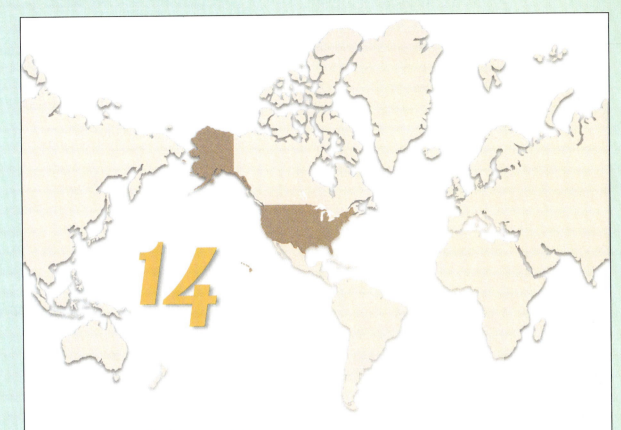

14

Education

With Emphasis on Public Education in the United States

High school graduation ceremony. (Mike Mazzaschi/Stock Boston.)

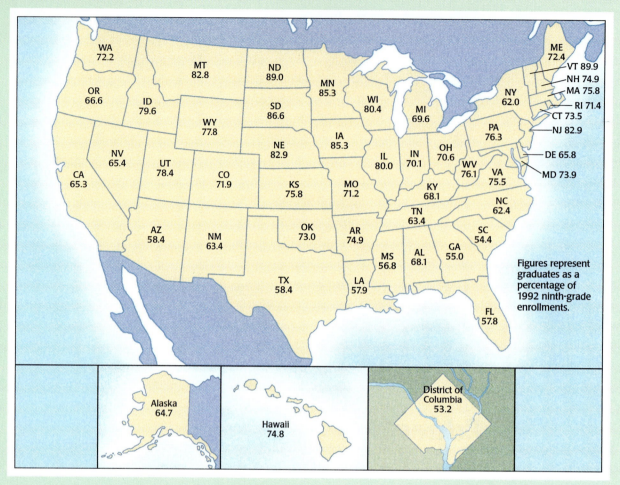

Figures represent graduates as a percentage of 1992 ninth-grade enrollments.

Source: National Center for Education Statistics

This map shows the 1995–1996 public high school graduation rate for the 50 states. The numbers represent graduates as a percentage of 1992 ninth-grade enrollments. Overall, 68.1 percent of ninth graders enrolled in fall 1992 graduated from high school four years later. Which state had the highest graduation rate? Which state had the lowest? Why do you think that 31.9 percent of ninth graders in the United States did not graduate four years later? Why do you think some states manage to graduate 89 percent but other states graduated only 55 percent? The chapter explores the U.S. system of public education and asks whether something about that system encourages almost one-third of ninth graders to "drop out" before graduation.

The Congress declares that the National Education Goals are the following:

- All children in America will start school ready to learn.
- The high school graduation rate will increase to at least 90 percent.
- All students will leave grades 4, 8, and 12 having demonstrated competency over challenging subject matter, including English, mathematics, science, foreign languages, civics and government, economics, arts, history, and geography.
- Every school in America will ensure that all students learn to use their minds well, so they may be prepared for responsible citizenship, further learning, and productive employment in our nation's modern economy.
- The nation's teaching force will have access to programs for the continued improvement of their professional skills and the opportunity to acquire the knowledge and skills needed to instruct and prepare all American students for the next century.
- United States students will be first in the world in mathematics and science achievement.
- Every adult American will be literate and will possess the knowledge and skills necessary to compete in a global economy and exercise the rights and responsibilities of citizenship.
- Every school in the United States will be free of drugs, violence, and the unauthorized presence of firearms and alcohol and will offer a disciplined environment conducive to learning.

- Every school will promote partnerships that will increase parental involvement and participation in the social, emotional, and academic growth of children.

These national goals, to be achieved by the year 2000, were first announced by President George Bush in September 1989 at the Governors Conference. In April 1994, President Bill Clinton signed the Goals 2000 Act, which sets aside $700 million in federal funds for states and school districts that implement programs aimed at meeting guidelines related to these goals.

Notice the preoccupation with preparing students to compete in a global economy and with creating persons prepared for responsible citizenship. These preoccupations suggest that the system of education functions to teach children (and adults) the skills they need to adapt to their environment and to make decisions in their everyday lives that benefit humankind. In the United States, the ongoing belief that the school system does not achieve these functions drives efforts to reform or restructure the system. Ironically, while schools have been treated as the primary source of many of the country's problems, they are also treated as the solution to those problems.

Why Focus on the United States?

In this chapter we focus on public education in the United States, the system in which 89 percent of students are enrolled. Throughout U.S. history, it seems as if public education has always been in a state of crisis and under reform. Between 1880 and 1920, for example, the public was concerned about whether schools were producing qualified workers for the growing number of factories and businesses and whether they were instilling a sense of national identity (that is, patriotism) in immigrant and ethnically diverse student populations. When the United States became involved in major wars—World War I, World War II, the Korean conflict, and the Vietnam War—the public voiced concerns about whether the schools were turning out recruits physically and mentally capable of defending the American way of life.

When the Soviet *Sputnik* satellites were launched in the mid-1950s, Americans were forced to consider the possibility that their public schools were not educating students in mathematics and the sciences as well as the Soviet schools were. In the 1960s, civil rights events forced Americans to question whether the public schools were offering children from less advantaged ethnic groups and social classes the knowledge and skills to compete economically with children in more advantaged groups. Beginning in the late 1970s and continuing throughout the 1980s and the 1990s, many critics, both at home and abroad, have maintained that the U.S. system of education is not adequate for meeting the challenges associated with being part of a global economy. Many employers claim that they are unable to find enough workers with the level of reading, writing, mathematical, and critical thinking skills needed to function adequately in the workplace. Reform efforts have focused on increasing "seat time," or the amount of time Americans devote to academic activities, either in school or at home. Many studies have shown that U.S. students devote considerably less time to academic activities in comparison to foreign counterparts. (See Table 14.1.) Many reformers believed that if we increased the time spent on academic activities, academic achievement would increase as well.

In the 1990s, the focus shifted away from "time" and quantity to the quality of the curriculum and instruction, with attempts being made to measure its effectiveness. Goals 2000 supported educational reform on the state and local levels. Reform efforts are to be guided by five fundamental principles:

- All students can learn.
- Lasting improvements depend on school-based leadership.
- Simultaneous top-down and bottom-up reform is necessary.
- Strategies must be locally developed, comprehensive, and coordinated.
- The entire community must be involved in developing strategies for system-wide improvement. (U.S. Department of Education 1998a)

What Is Education?

In the broadest sense, **education** includes those experiences that train, discipline, and develop the mental and physical potentials of the maturing person. An experience that educates may be as commonplace as reading a sweater label and noticing that it was made in Taiwan or as intentional as performing a scientific experiment to learn how genetic makeup can be altered deliberately through the use of viruses. In view of this definition and the wide range of experiences it encompasses, we can say that education begins when people are born and ends when they die.

Education Those experiences that train, discipline, and develop the mental and physical potentials of the maturing person.

Informal education Education that occurs in a spontaneous, unplanned way.

Formal education A systematic, purposeful, planned effort intended to impart specific skills and modes of thought.

Schooling A program of formal and systematic instruction that takes place primarily in classrooms but also includes extracurricular activities and out-of-classroom assignments.

Sociologists make a distinction, however, between formal and informal education. **Informal education** occurs in a spontaneous, unplanned way. Experiences that educate informally occur naturally; they are not designed by someone to stimulate specific thoughts or interpretations or to impart specific skills. Informal education takes place when a child puts her hand inside a puppet, then works to perfect the timing between the words she speaks for the puppet and the movement of the puppet's mouth.

Formal education encompasses a purposeful, planned effort intended to impart specific skills and modes of thought. Formal education, then, is a systematic process (for example, military boot camp, on-the-job training, programs to stop smoking, classes to overcome fear of flying) in which someone designs the educating experiences. We tend to think of formal education as consisting of enriching, liberating, or positive experiences, but it can also include impoverishing and narrowing situations (such as indoctrination or brainwashing). In any case, formal education is considered a success when the people instructed internalize (or take as their own) the skills and modes of thought that those who design the experiences seek to impart.

This chapter is concerned with a specific kind of formal education—schooling. **Schooling** is a program of formal and systematic instruction that takes place

Table 14.1   Comparison of Taiwanese, Japanese, and American First Graders and Fifth Graders

During the late 1970s and throughout the 1980s, education critics focused on the amount of time that Americans spent on academic activities relative to foreign counterparts. One of the most publicized studies compared the time American, Taiwanese, and Japanese first and fifth graders spent on academic activities inside and outside the classroom.

	Americans	Taiwanese	Japanese
Homework			
Minutes per day doing homework (Monday–Friday)			
First graders	14 min	77 min	37 min
Fifth graders	46 min	114 min	57 min
Minutes doing homework (Saturday)			
First graders	7 min	83 min	37 min
Fifth graders	11 min	73 min	29 min
Classroom			
Percentage of time devoted to academic activities			
First graders	69.8%	85.1%	79.2%
Fifth graders	64.5%	91.5%	87.4%
Hours per week devoted to academic interests			
First graders	19.6 hr	40.4 hr	32.6 hr
Fifth graders	19.6 hr	40.4 hr	32.6 hr

Source: Stevenson, Lee, and Stigler (1986).

primarily in classrooms but also includes extracurricular activities and out-of-classroom assignments. In its ideal sense, "[e]ducation must make the child cover in a few years the enormous distance traveled by mankind in many centuries" (Durkheim 1961:862). More realistically, schooling represents the means by which those who design and implement programs of instruction seek to pass on the values, knowledge, and skills that they define as important for success in the world. What constitutes an ideal education—the goals that should be achieved, the material that should be covered, the best techniques of instruction—is elusive and debatable. Conceptions vary according to time and place; they differ according to whether schools are viewed primarily as mechanisms by which the needs of a society are met or as the means by which students learn to think independently, thus becoming free from the constraints on thought imposed by their family, peers, culture, and nation.

Social Functions of Education

Sociologist Emile Durkheim believed that education functions to serve the needs of society. In particular, schools exist to teach children the skills they need to adapt to their environment. To ensure this end, the state (or other collectivity) reminds teachers "constantly of the ideas, the sentiments that must be impressed" on children if they are to adjust to the society in which they must live. Educators must agree upon and pass on a sufficient "community of ideas and sentiments without which there is no society." Without this sense of community, "the whole nation would be divided and would break down into an incoherent multitude of little fragments in conflict with one another" (Durkheim 1968:79, 81). Logic underscores efforts to use the schools as mechanisms for meeting the needs of society—whether the goal is to instill a love of country, train a skilled labor force, take care of children while their parents work, or teach young people to drive.

Another, quite different conception views education as a liberating experience that releases students from the blinders imposed by the accident of birth into a particular family, culture, religion, society, and time in history. To meet this goal, schools should be designed to broaden students' horizons so they will become aware of the conditioning influences around them and will learn to think independently of any authority. Under this conception, schools function as agents of change and progress.

These two aims are not necessarily contradictory if what benefits the group also liberates the individual. For example, democracies and the free market system require an informed public that is capable of

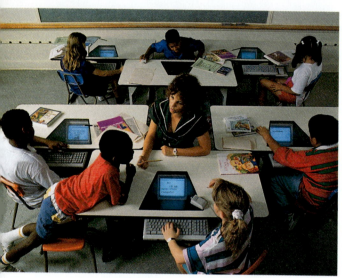

Schooling can serve different purposes. Training in the use of computers, for example, may be designed primarily to equip students with the skills they need to adapt to their environment. It can also be used to broaden students' intellectual horizons and encourage creativity and independent thinking. Have you been asked to use computers in any of your previous education? Apart from improving your "computer skills," were the assignments designed to broaden your intellectual perspective or social opportunities? How?

© James Wilson/Woodfin Camp & Associates

independent thought. Most sociological research, however, suggests that schools are more likely to be designed to meet the perceived needs of society rather than to liberate minds. Despite this intention, a significant percentage of the population in every country seems to be **functionally illiterate**—that is, some people do not possess the level of reading, writing, and calculating skills needed to adapt to the society in which they live. In fact, to many critics of the U.S. educational system, illiteracy in the United States has reached crisis proportions. To evaluate the "literacy crisis," we need to define illiteracy, distinguish among kinds of literacy, and see how the United States compares to other countries when it comes to preparing its students to be educated workers and citizens.

Functionally illiterate A term describing people who do not possess the level of reading, writing, and calculating skills needed to adapt to the society in which they live.

Illiteracy The inability to understand and use a symbol system, whether it is based on sounds, letters, numbers, pictographs, or some other type of character.

Illiteracy in the United States

In the most general and basic sense, **illiteracy** is the inability to understand and use a symbol system, whether it is based on sounds, letters, numbers, pictographs, or some other type of character. Although the term *illiteracy* is used traditionally in reference to the inability to understand letters and their use in reading and writing, as many kinds of illiteracy exist as there are symbol systems—computer illiteracy, mathematical illiteracy (or innumeracy), scientific illiteracy, cultural illiteracy, and so on.

If we confine our attention merely to languages, of which between 3,500 and 9,000 (including dialects) exist, we can see that the potential number of literacies is overwhelming (Ouane 1990) and that people cannot possibly be literate in every symbol system. If a person speaks, writes, and reads in only one language, by definition he or she is illiterate in perhaps as many as 8,999 languages. Such a profound level of illiteracy rarely presents a problem, however, because usually people need to know and understand only the language of the environment in which they live.

This point suggests that illiteracy is a product of one's environment—that is, people are considered illiterate when they cannot understand or use the symbol system of the surrounding environment in which they wish to function. Examples of this illiteracy include not being able to use a computer, access information, read a map to find a destination, make change for a customer, read traffic signs, follow the instructions to assemble an appliance, fill out a job application, or comprehend the language of people in that environment.

In the United States (and in all countries, for that matter), some degree of illiteracy has always existed, but conceptions of what people needed to know to be considered literate have varied over time. At one time, people were considered literate if they could sign their names and read the Bible. At another time, a person who had completed the fourth grade was considered literate. The National Literacy Act of 1991 defines literacy as "an individual's ability to read, write, and speak English and compute and solve problems at levels of proficiency necessary to function on the job and in society, to achieve one's goals, and to develop one's knowledge and potential" (U.S. Department of Education 1993a:3).

In 1988, the U.S. Congress requested that the Department of Education define literacy in the context of the new economic order and attempt to estimate the number of illiterate Americans. The project involved a representative sample of 26,091 adults (13,600 were interviewed, and 1,000 were surveyed in each of 11 states). In addition, researchers interviewed 1,100 federal and state prison inmates. The project, which was completed in 1993, revealed that 21 to 23 percent of those contacted "demonstrated skills in the lowest level of prose, document, and quantitative proficiencies." Another 25 to 28 percent performed at Level 2.[1] All of these respondents fell below the Level 3 criteria: They could not (1) write a brief letter explaining an error made on a credit card bill, (2) identify information from a bar graph, or (3) use a calculator to find the difference between the regular price and the sale price of an item in an advertisement.

The fact that almost half of the adult population functions at the two lowest levels leads social critics—most notably U.S. government officials and business leaders—to identify the school system as one major source of the problem. The high estimates of illiteracy, even among individuals who complete high school, has inspired people to ask how so many students could attend school for 10 to 12 years without acquiring enough reading, writing, and problem-solving skills to deal effectively with the work-related problems encountered in the new kinds of entry-level jobs. This situation has prompted many critics to examine the ways in which schools in countries considered as major economic competitors to the United States—most notably European and Pacific Rim countries—educate their students.

Insights from Foreign Education Systems

Sociologists John W. Meyer and David P. Baker (1996) argue that one legacy of the Reagan-Bush presidencies is internationalization of U.S. educational policy. Apparently, an increasingly competitive world economic climate and accompanying concerns about world position prompted critics to compare the U.S. system of education with those of foreign countries on a host of attributes, including amount of homework, parental support, teachers' salaries, classroom environment, and amount of time children spent watching television. The assumption underlying such comparisons is that the way a society structures the education experience is connected to education-related outcomes such as test scores, literacy levels, and work skills.

Meyer and Baker (1996) point out that the International Education Association (IEA) has collected cross-national data on educational achievement since the 1950s, but only in the past 10 to 15 years have the media focused on the findings. They maintain that, more often than not, the media emphasize that the test scores of U.S. students are "not especially high" or are "very poor" in comparison with students in other countries. Usually, U.S. scores are suggested to be lower than they actually are, because the media reports rank (for example, as "eleventh of fourteen countries," "last," or "next to last"). The reports fail to mention that average test scores for many countries ranking lower and higher are not measurably different from the U.S. average. For example, whereas U.S. ninth graders averaged 67 percent on the IEA science test, that score is not measurably different from those achieved by Spain (67.5) or Scotland (67.9), two countries that technically rank higher than the United States.

In addition, the media and education critics focus considerable attention on how U.S. students compare with students in selected countries—those currently considered to be major economic competitors (Germany, Japan) or countries expected to emerge as major economic competitors (China, South Korea). As Meyer and Baker point out, a disproportionate focus is also placed on tests in which U.S. students do less well than on tests in which they do better than students in other countries. The fact that students in 20 countries score measurably lower than U.S. nine-year-olds in reading attracts much less attention than the fact that U.S. nine-year-olds rank ninth out of ten countries in mathematics (see Table 14.2).

In the next sections of this chapter, we examine the historical background of the U.S. system of education and consider some of its general and distinguishing characteristics.

The Development of Mass Education in the United States

The United States was the first country in the world to embrace the concept of mass education. In doing so, it broke with the European view that education should

[1]According to the authors of this study:

The approximately 90 million adults who performed in Levels 1 and 2 did not necessarily perceive themselves as being "at risk." Across the literacy scales, 66 to 75 percent of the adults in the lowest level and 93 to 97 percent in the second lowest level described themselves as being able to read or write English "well" or "very well." Moreover, only 14 to 25 percent of the adults in Level 1 and 4 to 12 percent in Level 2 said they get a lot of help from family members or friends with everyday prose, document, and quantitative literacy tasks. It is therefore possible that their skills, while limited, allow them to meet some or most of their personal and occupational literacy needs. (U.S. Department of Education 1993a:XV)

Table 14.2   International Distribution of Academic Achievement Relative to the United States, 1991–1992

This table is better than many other public reports on the quality of American education. It does not exaggerate the differences between U.S. and foreign students. In some tables we see only one column of data, "rank," which does not tell us whether the differences between the United States and the countries ranking higher or lower are really significant. In addition, the media that filter the data before we see it often fail to give headline attention to areas in which U.S. students do very well. Rather than giving us the entire picture, they give us selected parts of it that exaggerate the failure of U.S. students.

What measurable differences do you see in this table between the United States and other countries? What specific strengths or weaknesses does the table suggest in U.S. education?

Subject and Age	Number of Countries Performing			Number of Countries in the Study	U.S. Rank
	Measurably Higher Than the United States	Not Measurably Different from the United States	Measurably Lower Than the United States		
Mathematics (1991)					
9-year-olds	7	2	0	10	9
13-year-olds	12	1	1	15	13
Science (1991)					
9-year-olds	0	7	2	10	3
13-year-olds	5	7	1	14	11
Reading (1991–1992)					
9-year-olds	1	1	20	23	2
14-year-olds	3	15	6	23	9

Source: U.S. Department of Education (1996).

be limited to an elite few (for example, the top 5 percent or only individuals who could afford it). In 1852, Massachusetts legislators passed a law making elementary school mandatory for all children. Within 60 years, all of the states had passed compulsory attendance laws.

A number of factors other than the legal mandate encouraged parents to comply with attendance laws, however. First, as the pace of industrialization increased, jobs moved away from the home and neighborhoods and into factories and office buildings. As mechanization increased, apprenticeship opportunities gradually dried up. With the disappearance of family farms and businesses, parents could no longer train their children because familiar skills were becoming obsolete. Thus, with the home and work environments operating independently of each other, parents were no longer available to oversee their children. Second, a tremendous influx of immigrants to the United States between 1880 and 1920 created a large labor pool—a surplus—from which factory owners could draw workers, thereby eliminating the need for child laborers. This combination of events created an environ-

ment in which children had no place to go except to school.

At least two prominent features of early American education have endured to the present: textbooks modeled after catechisms and single-language instruction.

Textbooks

The most vocal early educational reformers, such as Benjamin Rush, Thomas Jefferson, and Noah Webster, insisted that schools represented an important mechanism by which a diverse population could acquire a common culture. They believed that the new "perfectly homogeneous" American was one who studied at home (not abroad) and who used American textbooks. To use Old World textbooks "would be to stamp the wrinkles of decrepit age upon the bloom of youth" (Webster, quoted in Tyack 1966:32).

The first textbooks in the United States were modeled after catechisms, short books covering religious principles written in question-and-answer format. Each question had one answer only; in repeating this answer, the student strictly adhered to the question's wording (Potter and Sheard 1918). This format discouraged readers from behaving as active learners "who could frame questions, interpret materials, and reflect on the significance of what was presented. . . . No premium was placed on generating and inventing ideas or arguing about the truth or value of what others had written" (Resnick 1990:18). The reader's job was to memorize the "right" answers for the questions. Under this model, not surprisingly, textbooks tended to be written in such a way that the primary reason to read them was to find the "right" answers to the accompanying questions.

The influence of the catechisms on learning today becomes evident whenever students are assigned to read a chapter and answer the list of questions at the end. Many students discover that they do not need to read the material to answer the questions; instead, they can simply skim the text until they find key words that correspond to the question wording, and then copy the surrounding sentences.

In the past few years, some educators have questioned the value of textbooks as a learning tool (see "The Content of U.S. Math Lessons"). Although some schools have abandoned textbooks, especially at the elementary level, we do not know the scope of this trend. Educators in the field of elementary reading seem to be leading the way by increasingly abandoning traditional textbooks modeled after catechisms in favor of children's literature books and daily writing projects (Richardson 1994). Historical novels and edited volumes that include essays, various perspectives, and "real" or authentic literature are among the alternatives to textbooks.

Single-Language Instruction

The "peopling of America is one of the great dramas in all of human history" (Sowell 1981:3). It involved the conquest of the native peoples, the annexation of Mexican territory along with many of its inhabitants (who lived in what is now New Mexico, Utah, Nevada, Arizona, California, and parts of Colorado and Texas), and an influx of millions of people from practically every country in the world. School reformers—primarily people of Protestant and British background—saw public education as the vehicle for "Americanizing" a culturally and linguistically diverse population, instilling a sense of national unity and purpose, and training a competent work force. As Benjamin Rush argued, "Let our pupil . . . be taught to love his family but let him be taught, at the same time, that he must forsake and even forget them, when the welfare of his country requires it" (Rush, quoted in Tyack 1966:34). To meet

these nation-building objectives, people had to abandon their native languages and learn to speak a common language. Consequently, students were taught in English, the language of the established elite.

Although the United States is hardly unique in pressuring its people to abandon their native tongues and learn to speak a common language, it is probably the only country in the world that does not emphasize learning at least one other language. Education critic Daniel Resnick believes that the absence of serious foreign language instruction contributes to the parochial nature of American schooling. The almost exclusive attention to a single language has deprived students of the opportunity to appreciate the connection between language and culture and to see that language is a tool that enables them to think about the world.

According to Resnick, the focus on a single language "has cut students off from the pluralism of world culture and denied them a sense of powerfulness in approaching societies very different from their own" (1990:25). It also denies students from non-English-speaking heritages the means of reflecting on and fully appreciating their ancestors' lives.

In the United States, English became the language people needed to speak if they were to enter the mainstream of society and have a chance at upward mobility. In this respect, English became positioned against the language and even regional accents of parents and grandparents. It was the language "through which one's ethnic self was converted into one's acquired American personality . . . [and] as an instrument through which one could hide and mask ethnic (and regional) origins" (Botstein 1990:63). Noah Webster, an early influential education reformer, wrote in the preface of his spelling book that the United States must promote a "uniformity and purity of language" and "demolish those odious distinctions of provincial dialects which are subject to reciprocal ridicule in different states" (quoted in Tyack 1966:33).

Fundamental Characteristics of Contemporary American Education

A number of characteristics distinguish American education from other systems of education. They include the availability of college, the lack of a uniform curriculum, funding that varies by state and community, the belief that schools can be the vehicle for solving a variety of social problems, and ambiguity of purpose and value.

U.S. in Perspective

The Content of U.S. Math Lessons

This graph shows that 87 percent of eighth-grade U.S. math lessons observed were rated as low in quality. No lessons were rated as high in quality. In November 1996, *Newsweek* education reporter Pat Wingert explained in an interview with *MacNeil/Lehrer Newshour* (*Online Newshour* 1996a) correspondent Elizabeth Farnsworth why the ratings were so low relative to Japan and Germany, two countries regarded as major economic competitors with the United States:

FARNSWORTH: What did this study find American schools are doing wrong?

WINGERT: Basically, we seem to be doing a lot of things wrong. This study says that American teachers spend more time teaching math and science. They give out more homework, which means they're correcting more homework. They're covering more topics in a year than your typical teachers around the world. But they're teaching in a very dry way. They present a problem to the class and say, "I'm going to try to do this kind of fraction problem." Then they have kids do fraction problems at their desks. Then they give them fraction problems for homework.

That's not the most effective way to do it. What [the researchers] saw in the videotapes in Japan, for example, was that the Japanese teachers were teaching the way that

American reform documents have been saying for years that we should be doing it. They give kids problems in a more realistic way. You know: "You have this piece of property. You have to divide it in half. How would you do it?"

And then they have the kids sit there and really struggle. It's not an easy answer. It's something that they may have to work out for 20

minutes, not sure they have the right answer. Then they have a class discussion where kids will get up and say, "I think this is the way to do it." That kind of discussion gets kids emotionally involved in the answer. Brain research says the more emotional you are about something that you're learning, the better you're going to remember it. So those kids are getting excited. They're fighting

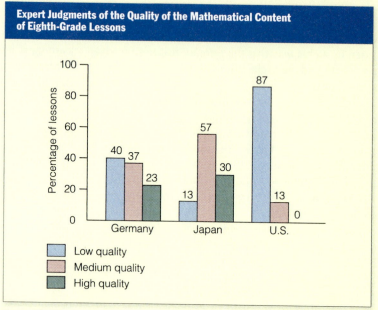

Expert Judgments of the Quality of the Mathematical Content of Eighth-Grade Lessons

Source: Third International Mathematics and Science Society Study; unpublished tabulations, Videotape Classroom Study, University of California–Los Angeles, 1996.

The Availability of College

One of the most distinctive features about the U.S. education system is that in theory anyone can attend a college if he or she has graduated from high school or has received a general equivalency degree (GED). As a result, the United States has the world's highest postsecondary enrollment ratio. Sixty-seven percent of 1997 high school graduates were enrolled in college in fall 1997 (U.S. Bureau of Labor Statistics 1998).[2] This figure varies depending on racial classification. For example, whereas 67.5 percent of high school graduates classified as white and 65.5 percent classified as Hispanic enrolled in college, only 59.6 percent of high

[2]This statement does not mean that the 67 percent who enroll out of high school go on to complete college.

over their answer. Then the teacher finally gets up and says, "Well, I think this is the best way. Now watch while I do this." And all the kids that have been sitting there struggling suddenly say, "He's right! That is the right way." And they remember it as a result.

It was very frustrating to the researchers to realize, when they talk to the teachers in American classrooms, that the teachers are very aware of what math reform documents had said and thought that they were doing it. Yet, when they videotaped, the researchers [saw] they're not doing it. [The teachers] don't realize they're not doing it.

So what do we do now to get them to understand that we still have big changes to make? How do we get American teachers now to say, after spending all this time doing all this homework, we're still not doing better? How do we keep them on track to fight this?

FARNSWORTH: You have an example to show us, don't you?

WINGERT: Right. [Wingert shows an example.] This example was handed out by the Department of Education. This shows how a typical math problem is taught in the United States. And as you can see, basically once you learn the method or the fast way of doing it, it doesn't

really draw you in. Once you know the method, you just do it.

[Wingert shows a second example.] [In a better] question, you have to read what the question is about, and then you have to sit there and look at the graph and struggle. Several of us were looking at this before the show and had a discussion among us. Was this the right answer? Is this the right answer? Is this what they're trying to say? You have to sit there and think about it, so you get drawn in more. You have to think more, apply more.

FARNSWORTH: Is part of the problem there's too much being covered and not enough in this different way?

WINGERT: Right. Most math teachers follow their textbooks, and most textbooks are written in a way so that teachers in California and teachers in Texas and teachers in New York will all buy them. The result is we're packing way too much into each textbook. We're covering way too many topics. What the Japanese would do, for example, is start teaching kids fractions and [continue teaching] fractions every day for multiple days. They would build on that knowledge. In a typical American classroom kids might get a little fractions this year in the fall and maybe a little more in the spring. By spring, of course, they've forgotten some of the stuff

that they learned in the fall. Then next year they might learn a little more, but the teacher has to take them back to teach them again. We keep going over the same ground again and again. We're wasting time.

FARNSWORTH: Does this imply mostly changes in teacher training?

WINGERT: Yes.

FARNSWORTH: There may be too much content, and there need to be changes in the way teachers approach it?

WINGERT: There needs to be reorganization of the content. It needs to be done in a more efficient way that builds on knowledge. Teachers need to be retrained. Teachers need to see videotapes or see real live teachers teaching the way that they should be teaching. It's very hard to change, just like it's hard to change the way that you're parenting. So teachers need to see real-life examples. They need to see a teacher in front of them, doing it the right way.

Does this interview give any clues about other changes, outside the classroom—for example, at administrative levels or among publishers—that might lead to improvements in U.S. schools?

Source: From "Education Report Card" (November 21), *Online Newshour.* Copyright © 1996. Reprinted by permission.

school graduates classified as black matriculated in a postsecondary program (U.S. Bureau of Labor Statistics 1998).

To examine another indicator of relatively easy access to college, consider that 20 percent of U.S. four-year colleges and universities accept students regardless of what courses they took, what grades they earned in high school, or what scores they received on ACT or SAT tests. Almost all colleges and universities offer remedial courses for those students who lack skills necessary to do college-level work. Thirty percent of all freshman students who entered college in fall 1995 took one or more remedial courses in reading, writing, or mathematics (U.S. Department of Education 1998c).

In most other countries, education beyond high school is available to a smaller percentage of the

U.S. in Perspective
How Many Get Their Bachelor's Degree?

This graph shows the percentages of males and females earning bachelor's or equivalent degrees in selected countries as of 1991. Spend some time considering what you can learn from it about specific countries that interest you and about comparisons between the United States and other countries. Here are some questions you might ask:

1. Which countries educate lower percentages of college graduates than does the United States? Which educate higher percentages? What do the more striking differences suggest?

2. Is it necessarily a good thing for a country to have the highest percentage of college graduates? Are there ways in which a high percentage might reflect or conceal important weaknesses in a country's overall educational system?

3. What differences do you see in the proportion of men versus women completing college in various countries? What might account for these differences in several countries? Would you consider an equal proportion of male and female graduates ideal? How so? How perhaps not?

4. The text says that "the United States has the world's highest postsecondary *enrollment* ratio." Is that claim inconsistent with the data in this graph? Why?

Sources: U.S. Department of Education (1996); Organization for Economic Co-operation and Development, unpublished data.

Table 14.3   Range of Tuition at Four-Year Institutions,* 1995–1996

About how much per year did most private and public college students pay for their 1995–1996 tuition? How much more would you add for other expenses, such as books and transportation? Consider your own college tuition. Where do you fit on this chart? How do you compare with your peers?

	Number of Colleges	Proportion of Total Enrollment (%)
Private Institutions		
$20,000 or more	36	3.2
15,000–19,999	142	12.5
10,000–14,999	427	37.7
5,000– 9,999	386	34.1
2,000– 4,999	129	11.4
Less than $2,000	13	1.1
Total	1,133	100.0
Public Institutions		
$5,000 or more	21	4
3,000–4,999	172	32.2
2,000–2,999	170	31.9
Less than $2,000	170	31.9
Total	533	100.0

*Includes only those institutions that provided final or estimated 1994–1995 tuition and fees by September 1, 1995.
Source: College Board (1996).

population (see "How Many Get Their Bachelor's Degree?"). In England and Wales, 16-year-olds who get low scores on the General Certificate of Secondary Education Test cannot take the advanced courses required for college. Similarly, scores on the French baccalaureate examination determine who can attend college (Chira 1991). From this perspective, the educational opportunities in the United States are admirable. A college education is open to everyone (who can find money to pay the tuition) regardless of previous educational failures; it is not reserved for a privileged few (see Table 14.3). This policy reflects the American belief in equal opportunity to compete at whatever point in life one wishes to enter the game: "We may not all hit home runs, the saying goes, but everyone should have a chance at bat" (Gardner 1984:28).

At the same time, the unrestricted right to a college education seems to be connected with a decline in the value of a high school education and the effort put into achieving that level of education. As high school enrollments increased over time to include the middle class and eventually the poor and minority groups, the perceived value of the high school diploma declined and the perceived need for a college degree increased. As

As the United States has made it easier for people to enter college, it has made secondary schools less rigorous. Is this choice a wise trade-off? Should high school diplomas be based on stricter and more specific standards of accomplishment with regard to verbal and mathematical literacy? For example, should all students be required to pass algebra? How might such requirements have affected you and other students you knew in high school?

© Valrie Massey/Photo 20-20

college enrollments increased and a college degree came to be defined as important for success, the high school no longer represented the last stop before young people entered the mainstream labor force. Consequently, high schools faced less pressure to ensure that their graduates had acquired the skills necessary for literacy in the workplace. As a result, the high school curriculum became less important and less rigorous (Cohen and Neufeld 1981).

Defenders of U.S. public schools argue that a less rigorous curriculum is a necessary consequence of trying to educate all citizens: accommodating everyone requires a reduction in the standards. The consequences of such a policy are nevertheless clear: The high school diploma loses its value, and the goal of achieving equality through compulsory and free education is undermined. Mass education accomplished through using social promotion (passing students from one grade to another on the basis of age rather than academic competency), awarding high school diplomas to people with fifth-grade reading abilities, and issuing certificates of achievement to those who fail minimum competency tests is not the same as educating everyone. True equality in education is achieved only if everyone has an opportunity to earn a valued degree.

Differences in Curriculum

No uniform curriculum exists in the United States. Each state sets its own broad curriculum requirements for kindergarten through high school; each school, in turn, interprets and implements these requirements. Consequently, even with state guidelines, the textbooks, assignments, instructional methods, staff qualifications, and material covered vary across the schools within each state. In addition to curriculum differences across states, students in the same school are usually grouped or "tracked" on the basis of tests or past performance. For example, students enrolled in standard diploma (versus college preparatory, honors, or advanced) tracks often take fewer mathematics courses and different kinds of mathematics (general math instead of algebra) to meet a state's mathematics requirement. They take English composition rather than creative writing to meet the state English requirement.

As one indicator of the extent to which a national curriculum is resisted, consider that the National Commission on Excellence in Education (NCEE) recommends that high school students complete four years of English, three years of mathematics, three years of science, and three years of social studies. Only 19.8 percent of the public school districts have graduation requirements that meet NCEE recommendations (U.S. Department of Education 1998b).[3] (See "Percentage of Public School Districts with Graduation Requirements That Meet or Exceed NCEE Guidelines.") Forty-five percent of private high schools have graduation requirements that meet or exceed NCEE guidelines (U.S. Department of Education 1998b).

Although most other countries also track students at some point in their careers, all of their students are exposed to a core curriculum, even if everyone cannot assimilate the material. The Japanese, for example, believe that a standardized curriculum is the only way to ensure that everyone has an equal chance at the rewards brought by education (Lynn 1988). "They put nearly their entire population through twelve tough years of basic training" (Rohlen 1986:38). Although the Japanese system includes three tracks at the secondary level—academic, specialized vocational, and comprehensive—all Japanese students take classes in foreign language, social studies, mathematics, science, health and physical education, and fine arts. Japanese schools do not have gifted or self-paced learning programs (Rohlen 1986). They do not teach some students general mathematics and others algebra; everyone is taught algebra. In elementary school, all Japanese children learn to read music and play a wind instrument and a keyboard instrument (Rohlen 1986). Likewise, all German students take a foreign language, German, physics, chemistry, biology, mathematics, and physical education.

Another source of differences in curricula around the United States arises from the fact that some minority group members are more likely to take or be assigned to less demanding classes. For example, in 1994, 61.6 percent of high school graduates classified as white took algebra II, compared with 43.7 percent of graduates classified as black. Only 39.2 percent of students classified as Native American took algebra II, but 67 percent of high school graduates classified as Asian took algebra II (U.S. Department of Education 1996).

---

[3]These figures apply to the 1993–1994 school year, the most recent year for which data are available.

U.S. in Perspective

Percentage of Public School Districts with Graduation Requirements That Meet or Exceed NCEE Guidelines

Which states have the fewest percentage of school districts meeting NCEE guidelines? Which states have the most? How might you explain the resistance to NCEE recommendations?

(The National Commission on Excellence in Education (NCEE) recommends that students complete four years of English, three years of math, three years of science, and three years of social studies.)

Source: U.S. Department of Education, National Center for Education Statistics, *Schools and Staffing Survey (SASS)*, 1993–94 (Teacher Questionnaire).

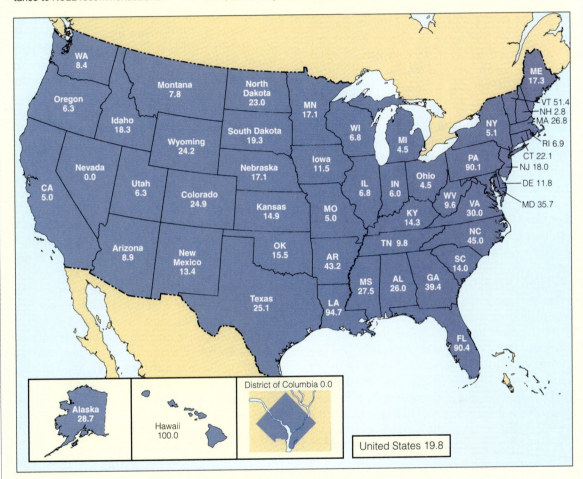

E. D. Hirsch, Jr. (1993), author of a best-selling and controversial[4] book, *Cultural Literacy: What Every*

[4]Hirsch's book is controversial because it attempts to identify and define concepts that every American should know. According to Hirsch, "this common knowledge or collective memory allows people to communicate, to work together, and to live together. It forms the basis for communities, and if shared by enough people, it is a distinguishing characteristic of a national culture" (Hirsch, Kett, and Trefil 1993:ix).

American Needs to Know, maintains that the nonuniform and diverse nature of the U.S. curriculum has created one of the most unjust and inegalitarian school systems in the developed world.

It happens that the most egalitarian elementary-school systems are also the best. . . . The countries that achieve these results tend to teach a standardized curriculum in early grades. Hungarian, Japanese, and Swedish children have a systematic grounding in shared knowledge. Until third graders learn what third graders are supposed to know, they do not pass on to fourth grade. No pupil is allowed to escape the knowledge net. (Hirsch 1989:32)

Surprisingly, little evidence indicates that tracking students into remedial or basic courses contributes to intellectual growth, corrects academic deficiencies, prepares students for success in higher tracks, or increases interest in learning. Instead, the special curricula exaggerate and widen differences among students and perpetuate beliefs that intellectual ability varies according to social class and ethnic group (Oakes 1986a, 1986b).

Differences in Funding

A 1993 report on education in 24 countries sponsored by the Paris-based Organization for Economic Cooperation and Development (OECD) concluded that no country has greater educational disparities between rich and poor than the United States (Sanchez 1993). In the United States, schools differ in regard to not only curriculum requirements, but also funding. Elementary and secondary schools receive approximately 7 percent of their funding from the federal government, 49.4 percent from the state government, and 43.8 percent from local and other intermediate sources, primarily property taxes (U.S. Department of Education 1993b). Such heavy reliance on state revenue is problematic, because the less wealthy states generate less tax revenue than do the wealthier states. Of the 24 countries studied by the OECD, the United States ranked third in spending per student at the secondary level and first at the primary level (Celis 1993a). On the other hand, the size of school funding disparities between the poorest and wealthiest states is considerably greater in the United States than that observed in other industrialized countries.

For example, the 12 Canadian provinces spend different amounts of money per student; expenditures range from a high of $6,500 per year per student to a low of $3,000. In Canada, a difference of about $3,500 separates the richest province from the poorest province. In the United States, however, state-level expenditures ranging from a high of $9,565 per student to a low of $3,867 yield a difference of $5,698 between the richest and poorest states (U.S. Department of Education 1995). (See "Percentage of School

Funding That Comes from Local and Intermediate Sources.")

Likewise, an even heavier reliance on local revenue is problematic, because it causes funding disparities among schools within states. In this regard, the courts in at least 28 states have evaluated (or are in the process of evaluating) claims that methods of financing have helped create unequal school systems within the state or have ruled that the methods of school financing are unconstitutional (Celis 1992, 1993b). A dramatic case in point is Kentucky. In response to a 1986 lawsuit filed by 66 mostly rural school districts that challenged the state's system of funding education, the Kentucky Supreme Court declared on June 8, 1989, that it found the entire state system of public education deficient and unconstitutional. The court mandated that every aspect of the public school system be reconsidered and that a new system be created no later than April 15, 1990. "The court concluded that a school system in which a significant number of children receive an inadequate education or ultimately fail is inherently inequitable and unconstitutional" (Foster 1991:34). To remedy this inequity, the court turned the academic success of all students into a constitutional obligation and required the state legislature to devise a system to ensure that every student was "learning at the highest level of which he or she is capable" (Foster 1991:36). In 1990, the legislature responded by passing the Kentucky Education and Reform Act (KERA), which set into motion the restructuring of education's rules, roles, and relationships. The sociological significance of KERA is that the state recognized that inequality could be corrected only by a complete rethinking and overhaul of the delivery of education to students.

Education-Based Programs to Solve Social Problems

The United States uses education-based programs to address a variety of social problems, including parents' absence from the home, racial inequality, drug and alcohol addictions, malnutrition, teenage pregnancy, sexually transmitted diseases, and illiteracy. Although all countries support education-based programs that address social problems, the United States is unique in that education is viewed as the primary solution to many of its problems. In the United States,

U.S. in Perspective

Percentage of School Funding That Comes from Local and Intermediate Sources

For which states is local and intermediate funding the most important source of school funding? For which states is it the least important source? Look at the opening map on page 426. Is there a connection between the percentage of funding and the graduation rate?

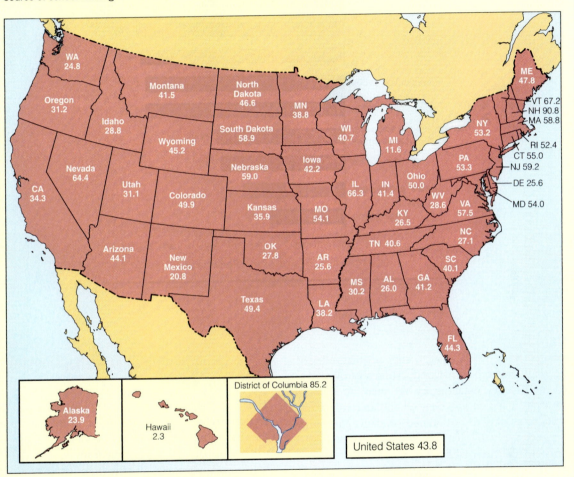

WA 24.8

Oregon 31.2

Montana 41.5

Idaho 28.8

North Dakota 46.6

MN 38.8

ME 47.8

VT 67.2

NH 90.8

MA 58.8

Wyoming 45.2

South Dakota 58.9

WI 40.7

MI 11.6

NY 53.2

RI 52.4

CT 55.0

NJ 59.2

Nevada 64.4

Utah 31.1

Nebraska 59.0

Iowa 42.2

PA 53.3

CA 34.3

Colorado 49.9

IL 66.3

IN 41.4

Ohio 50.0

WV 28.6

VA 57.5

DE 25.6

MD 54.0

Kansas 35.9

MO 54.1

KY 26.5

Arizona 44.1

New Mexico 20.8

OK 27.8

AR 25.6

TN 40.6

NC 27.1

SC 40.1

MS 30.2

AL 26.0

GA 41.2

Texas 49.4

LA 38.2

FL 44.3

Alaska 23.9

Hawaii 2.3

District of Columbia 85.2

United States 43.8

[t]he process became familiar: discover a social problem, give it a name, and teach a course designed to remedy it. Alcoholism? Teach about temperance in every school. Venereal disease? Develop courses in social hygiene. Youth unemployment? Improve vocational training and guidance. Carnage on the highways? Give driver education classes to youth. Too many rejects in the World War I draft? Set up programs in health and physical education. (Tyack and Hansot 1981:13)

The importance of these programs notwithstanding, the schools alone cannot solve such complex problems. Other programs must be implemented in concert with education-based programs.

Ambiguity of Purpose and Value

As noted earlier, in comparison with the people of other countries, Americans tend to be ambivalent about the purpose and value of an education. In

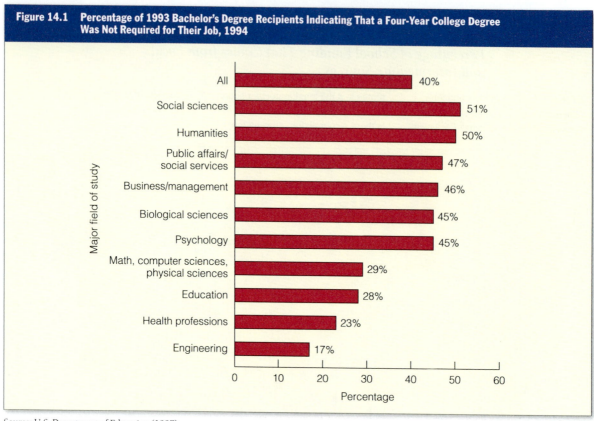

Figure 14.1   Percentage of 1993 Bachelor's Degree Recipients Indicating That a Four-Year College Degree Was Not Required for Their Job, 1994

Source: U.S. Department of Education (1997).

general, the U.S. public strongly supports mass education and the right to a college education regardless of a student's past academic history. Yet, many Americans view elementary school, high school, and college as merely something to be endured; students count the days until they are "out." The source of this ambivalence could be connected to the fact that most Americans tend to equate education with increased job opportunities and higher salaries. The connection is not always realized on a personal level, however. As many as 40 percent of college graduates are underemployed—that is, they do not believe a college degree is required for the job they obtained (U.S. Department of Education 1998c; see Figure 14.1).

Open college enrollments, diverse and special curricula, unequal funding, the problem-solving burden, and a national ambivalence toward education explain in part the graduation rate observed in the U.S. education system. In the next section, we look more closely at the role of the classroom environment.

A Close-Up View: The Classroom Environment

The classroom is where schooling takes place. In this section, we examine the operation of the classroom: the curriculum to which many students are exposed, the practice of tracking, testing of students, and problems faced by teachers. The section focuses on practices that lead to boredom and failure and that ultimately undermine the time and energy devoted to academic pursuits. Certainly, some schools in the United States achieve stimulating classroom environments. The problem is that there are not enough of them.

The Curriculum

Teachers everywhere in the United States teach two curricula simultaneously—a formal curriculum and a hidden curriculum. The various academic subjects—mathematics, science, English, reading, physical education, and so on—make up the **formal curriculum.** Students do not learn in a vacuum, however. As teachers instruct students and as students complete their assignments, other activities are going on around them. Social anthropologist Jules Henry (1965) maintains that these other important activities represent the hidden curriculum. The **hidden curriculum** consists of the other activities that go on as students learn the subject matter and the "lessons" that those other activities convey to students about the value and meaning of what they are learning. The teaching method, the types of assignments and tests, the tone of the teacher's voice, the attitudes of classmates, the number of students absent, the frequency of the teacher's absences, and the number of interruptions during a lesson are all factors in the classroom as students learn the formal curriculum. These so-called extraneous events function to convey messages to students not only about the value of the subject, but also about the values of society, the place of learning in their lives, and their role in society.

HIDDEN CURRICULUM: THE CASE OF SPELLING BASEBALL Henry (1965) uses typical classroom scenes (acquired from thousands of hours of participant observation), such as a session of "spelling baseball," to demonstrate the seemingly ordinary process by which a hidden curriculum is transmitted and to show how students become exposed simultaneously to the two curricula. Although Henry observed this scene in 1963, his observations hold even today:

> The children form a line along the back of the room. They are to play "spelling baseball," and they have lined up to be chosen for the two teams. There is much noise, but the teacher quiets it. She has selected a boy and a girl and sent them to the front of the room as team captains to choose their teams. As the boy and girl pick the children to form their teams, each child chosen takes a seat in orderly succession around the room. Apparently they know the game well. Now Tom, who has not yet been chosen, tries to call attention to himself in order to be chosen. Dick shifts his position to be more in the direct line of vision of the choosers, so that he may not be overlooked. He seems quite anxious. Jane, Tom, Dick, and one girl whose name the observer does not know, are the last to be chosen. The teacher even has to remind the choosers that Dick and Jane have not been chosen.

> The teacher now gives out words for the children to spell, and they write them on the board. Each word is a pitched ball, and each correctly spelled word is a base hit. The children move around the room from base to base as their teammates spell the words correctly.

> The outs seem to increase in frequency as each side gets near the children chosen last. The children have great difficulty spelling "August." As they make mistakes, those in the seats say, "No!" The teacher says, "Man on third." As a child at the board stops and thinks, the teacher says, "There's a time limit; you can't take too long, honey." At last, after many children fail on "August," one child gets it right and returns, grinning with pleasure, to her seat. . . . The motivation level in this game seems terrific. All the children seem to watch the board, to know what's right and wrong, and seem quite keyed up. There is no lagging in moving from base to base. The child who is now writing "Thursday" stops to think after the first letter, and the children snicker. He stops after another letter. More snickers. He gets the word wrong. There are frequent signs of joy from the children when their side is right. (Henry 1965:297–298)

According to Henry, learning to spell is not the most important lesson that students learn from this exercise. They are also learning important cultural values from the way in which spelling is being taught—that is, they are learning to fear failure and to envy success. In exercises like spelling baseball, "failure is paraded before the class minute upon minute" (p. 300), and success is achieved after others fail. And, "since all but the brightest children have the constant experience that others succeed at their expense they cannot but develop an inherent tendency to hate—to hate the success of others" (p. 296).

In an exercise such as spelling baseball, students also learn to be absurd. *To be absurd,* as Henry defines it, means to make connections between unrelated things or events and not to care whether the connections are appropriate or inappropriate. From Henry's point of view, spelling baseball teaches students to be absurd

Formal curriculum The various subjects such as mathematics, science, English, reading, physical education, and so on.

Hidden curriculum All the things that students learn along with the subject matter.

Jules Henry maintains that when American children are sent to the blackboard, especially to do math problems, they learn to "fear failure" and "envy success." If they get the problem wrong, a classmate will be asked to correct the mistake; hence, success is achieved at the expense of another's failure.

© John Ficara/Woodfin Camp & Associates

because no logical connection exists between learning to spell and baseball. "If we reflect that one could not settle a baseball game by converting it into a spelling lesson, we see that baseball is bizarrely irrelevant to spelling" (p. 300). Yet, most students participate in classroom exercises like spelling baseball without questioning their purpose. Although some children may ask, "Why are we doing this? What is the point?" and may be told, "So you can learn to spell" or "To make spelling fun," few children challenge the purpose of this activity even further. Students go along with the teacher's request and play the game as if spelling is related to baseball because, according to Henry, they are terrified of failure and want so badly to succeed.

Henry argues that classroom activities such as spelling baseball prepare students to fit into a competitive and consumption-oriented culture. Because the U.S. economy depends on consumption, the country benefits if its citizens purchase nonessential goods and services. The assignments that children undertake in school do not prepare them to question false or ambiguous statements made by advertisers; schools do not properly prepare demanding individuals to "insist that the world stand up and prove that it is real" (p. 49). Henry argues that this sort of training—this hidden curriculum—makes possible an enormous amount of selling that otherwise could not take place:

[I]n order for our economy to continue in its present form people must learn to be fuzzy-minded and impulsive, for if they were clear-headed and deliberate, they would rarely put their hands in their pockets. . . . If we were all logicians the economy [as we know it] could not survive, and herein lies a terrifying paradox, for in order to exist economically as we are we must . . . remain stupid. (p. 48)

According to Henry, students who have the intellectual strength to see through absurd assignments such as spelling baseball and who find it impossible to accept such assignments as important may rebel against the system, refuse to do the work, drop out, or eventually think of themselves as stupid. We cannot know how many students perform poorly in school because they cannot accept the manner in which subjects are taught. The United States has diverse public school systems; for instance, different schools teach reading in different ways. Yet, the vast majority of Americans, especially middle- and lower-class students, typically learn to carry out these kinds of assignments in the manner described here. In addition, students, especially young ones, cannot articulate what they do not like about school; many come to believe that school is not for them and think that they have failed, rather than believing that school has failed them.

Tracking

Most schools in the United States arrange students in instructional groups according to similarities in past academic performance or performance on standardized tests. In elementary school, this practice is often known as ability grouping; in middle school and high school, it is called streaming or tracking. Under such a sorting and allocation system, students may be assigned to separate instructional groups within a single classroom; they may be separated across the entire array of subjects (that is, college preparatory versus general studies courses); or they may be sorted with regard to selected subjects such as mathematics, science, and English (that is, advanced placement, honors, regular, basic).

The following rationales underlie ability grouping, streaming, or tracking (hereafter referred to as "tracking"):

1. Students learn better when they are grouped with those who learn at the same rate. The brighter students are not held back by the slower learners, and the slower learners receive the extra time and

special attention needed to correct academic deficiencies.

2. Slow learners develop more positive attitudes when they do not have to compete with the more academically capable.

3. Groups of students with similar abilities are easier to teach.

Actual research, however, suggests that tracking has a positive effect on high-track students, negative effects on low-tracked students, and no noticeable effects on middle-track or regular track students (Hallinan 1996).

THE EFFECTS OF TRACKING Sociologist Jeannie Oakes (1985) investigated how tracking affected the academic experiences of 13,719 middle school and high school students in 297 classrooms and 25 schools across the United States.

> The schools themselves were different: some were large, some very small; some in the middle of cities; some in nearly uninhabited farm country; some in the far West, the South, the urban North, and the Midwest. But the differences in what students experienced each day in these schools stemmed not so much from where they happened to live and which of the schools they happened to attend but, rather, from differences within each of the schools. (p. 2)

Oakes's findings were consistent with the findings of hundreds of other studies of tracking in terms of how students were assigned to groups, how they were treated, how they viewed themselves, and how well they did:

Placement: Poor and minority students were placed disproportionately in the lower tracks.

Treatment: The different tracks were not treated as equally valued instructional groups. Clear differences existed with regard to the quality, content, and quantity of instruction and to classroom climate as reflected in the teachers' attitude and in student–student and teacher–student relationships. Low-track students were consistently exposed to inferior instruction—watered-down curricula and endless repetition—and to a more rigid, more emotionally strained classroom climate.

Self-image: Low-track students did not develop positive images of themselves because they were identified publicly and treated as educational

To teachers, students' boredom may seem to be a problem of individual motivation or even laziness. But another possibility is that the social environment of the classroom, including some teaching styles, promotes a lack of responsiveness on the part of students.

© Bob Daemmrich/The Image Works

discards, damaged merchandise, or unteachable. Overall, among the average and the low-track groups, tracking seemed to foster lower self-esteem and promote misbehavior, higher dropout rates, and lower academic aspirations. In contrast, placement in a college preparatory track had positive effects on academic achievement, grades, standardized test scores, motivation, educational aspirations, and attainment. "And this positive relationship persists even after family background and ability differences are controlled" (Hallinan 1988:260).

Achievement: The brighter students tended to do well, regardless of the academic achievements of the students with whom they learn.

These findings were reflected in the written answers that teachers and students gave to various questions asked by Oakes and her colleagues. For example, when teachers were asked about the classroom climate, high-track teachers tended to reply in positive terms:

> There is a tremendous rapport between myself and the students. The class is designed to help the students in college freshman English composition. This

makes them receptive. It's a very warm atmosphere. I think they have confidence in my ability to teach them well, yet because of the class size—32—there are times they feel they are not getting enough individualized attention. (Oakes 1985:122)

Low-track teachers replied in less positive terms:

> This is my worst class. Kids [are] very slow—underachievers and they don't care. I have no discipline cases because I'm very strict with them and they are scared to cross me. They couldn't be called enthusiastic about math—or anything, for that matter. (p. 123)

Clear differences also appeared in the high-track and low-track students' responses to the question, "What is the most important thing you have learned or done so far in this class?" The replies of high-track students centered around themes of critical thinking, self-direction, and independent thought:

> The most important thing I have learned in this [English] class is to loosen up my mind when it comes to writing. I have learned to be more imaginative. (p. 87)

> The most important thing I have learned in this [math] class is the benefit of logical and organized thinking; learning is made much easier when the simple processes of organizing thoughts have been grasped. (p. 88)

Low-track students were more likely to give answers that centered around themes of boredom and conformity:

> I think the most important is coming into [math] class and getting out folders and going to work. (p. 89)

> To be honest, nothing. (p. 71)

> Nothing I'd use in my later life; it will take a better man than I to comprehend our world. (p. 71)

Although many educators have recognized the problems associated with tracking, efforts to detrack have

collided with demands for curriculum differentiation from politically powerful parents of high-achieving or identified "gifted" students who insist their children must get something more than the "other" students (Wells and Oakes 1996). As a result, tracking persists "even though many educators and policymakers acknowledge that students in the low and middle tracks are not held to high enough standards and thus are not adequately prepared for either college or the transition to work" (Wells and Oakes 1996:137).

TEACHERS' EXPECTATIONS AND SELF-FULFILLING PROPHECIES   Tracking can also create self-fulfilling prophecies by affecting teachers' expectations of the academic potential and abilities of students placed in each track. The idea of a self-fulfilling prophecy is a deceptively simple, yet powerful concept that originated from an insight by William I. and Dorothy Swain Thomas: "If [people] define situations as real, they are real in their consequences" ([1928] 1970:572). A **self-fulfilling prophecy** begins with a false definition of a situation. The false definition, however, is assumed to be accurate, and people behave as if it were true. In the end, the misguided behavior produces responses that confirm the false definition (Merton 1957).

A self-fulfilling prophecy in education can occur if teachers and administrators assume that some children are "fast," "average," or "slow" learners and consequently expose them to "fast," "average," and "slow" learning environments. Over time, real differences in quantity, quality, and content of instruction cause many students to actually become (and believe that they are) "slow," "average," or "fast." In other words, the prediction or prophecy of academic ability becomes an important factor in determining academic achievement:

> The tragic, often vicious, cycle of self-fulfilling prophecies can be broken. The initial definition of the situation which has set the circle in motion must be abandoned. Only when the original assumption is questioned and a new definition of the situation is introduced, does the consequent flow of events [show the original assumption to be false]. (Merton 1957:424)

In the tradition of symbolic interactionism, Robert Rosenthal and Lenore Jacobson designed an experiment to test the hypothesis that teachers' positive expectations about students' intellectual growth can

---

Self-fulfilling prophecy   A concept that begins with a false definition of a situation, which people assume to be accurate and thus behave as if it were true. The misguided behavior produces responses that confirm the false definition.

become a self-fulfilling prophecy and lead to increases in students' intellectual competence.[5, 6]

Rosenthal and Jacobson's experiment took place in an elementary school called Oak School, a name given the school to protect its identity. The student body came mostly from lower-income families and was predominantly white (84 percent); 16 percent of the students were Mexican Americans. Oak School sorted students into ability groups based on reading achievement and teachers' judgments.

At the end of a school year, Rosenthal and Jacobson gave a test, purported to be a predictor of academic "blooming," to those students who were expected to return the next year. Just before classes began in the fall, all full-time teachers were given the names of the white and Hispanic students from all three ability groups who had supposedly scored in the top 20 percent. The teachers were told that these students "will show a more significant inflection or spurt in their learning within the next year or less than will the remaining 80 percent of the children" (Rosenthal and Jacobson 1968:66). Teachers were also told not to discuss the scores with the students or the students' parents. Actually, the names given to teachers were chosen randomly: the differences between the children earmarked for intellectual growth and the other children existed only in the teachers' minds. The students were retested after one semester, at the end of the academic year, and after a second academic year.

Overall, intellectual gains, as measured by the difference between successive test scores, were greater for those students who had been identified as "bloomers" than they were for those students who were not identified as such. Although "bloomers" benefited in general, some benefited more than others: first- and second-graders, Hispanic children, and children in the middle

track showed the largest increases in test scores. The "bloomers" received no special instruction or extra attention from teachers—the only difference between them and the unidentified students was the teacher's belief that the "bloomers" bore watching. Rosenthal and Jacobson speculated that this belief was communicated to "bloomers" in very subtle and complex ways, which they could not readily identify:

> To summarize our speculations, we may say that by what she said, by how and when she said it, by her facial expressions, postures, and perhaps by her touch, the teacher may have communicated to the ["bloomers"] that she expected improved intellectual performance.
>
> It is self-evident that further research is needed to narrow down the range of possible mechanisms whereby a teacher's expectations become translated into a pupil's intellectual growth. (p. 180)

The research on tracking and teachers' expectations shows that the learning environment affects academic achievement and that tracking and expectations are two mechanisms that contribute to the unequal distribution of knowledge. Because teachers draw on test results to form their expectations about students' academic potential, and because academic personnel place students in different ability groups based on test results (along with teacher evaluations), tests represent another mechanism that contributes to the unequal distribution of knowledge and skills.

Various practices within schools clearly contribute to the unequal distribution of knowledge and skills. We cannot blame teachers and other school personnel entirely for this outcome, however. Teachers do not have exclusive control over the classroom environment and cannot single-handedly create students who are interested in learning. For many teachers, the environment in which they work can make teaching problematic.

Problems Faced by Teachers

Teachers' jobs are complex; teachers are expected to undo learning disadvantages generated by larger inequalities in the society and to handle an array of discipline problems. More than 50 percent of elementary and high school teachers surveyed in a recent Gallup poll responded that discipline is a very serious or fairly serious problem in their school, as are uncompleted homework assignments, cheating, stealing, drugs and alcohol, truancy, and absenteeism. When compared

---

[5]Rosenthal and Jacobson were influenced by animal experiments in which trainers' beliefs about the genetic quality of the animals affected the animals' performances. When trainers were told that an animal was genetically inferior, the animal performed poorly; when trainers were told that an animal was genetically superior, the animal's performance was superior. This difference arose despite the fact that no such genetic differences distinguished the animals defined as dull or bright.

[6]In one of the few research studies on track mobility (track reassignments) during high school, sociologist Maureen T. Hallinan (1996) found that there is "considerably more mobility than is typically assumed" in tracking (p. 983). Thirty percent of the students she studied changed English tracks (19 percent moved upward; 11 percent moved downward) and 11 percent changed math tracks (7 percent moved upward; 4 percent moved downward).

Figure 14.2   Student Problems as Seen by Teachers

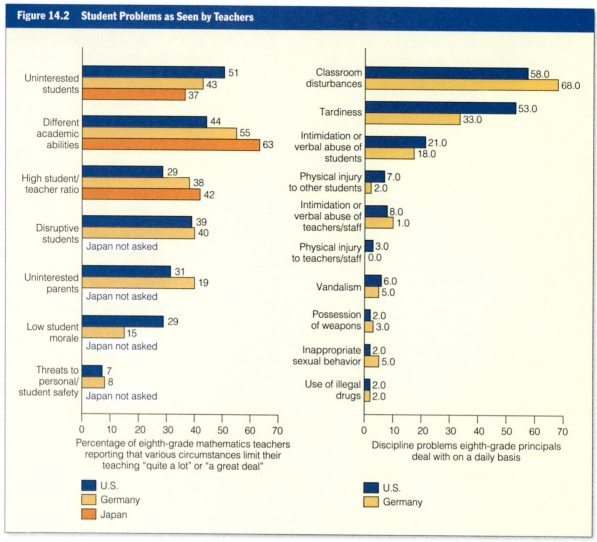

Percentage of eighth-grade mathematics teachers reporting that various circumstances limit their teaching "quite a lot" or "a great deal"

Discipline problems eighth-grade principals deal with on a daily basis

Primary source: Third International Mathematics and Sciences Study, unpublished tabulations, U.S., German, and Japanese teacher surveys, Westat, 1996; secondary source, National Center for Education Statistics (1996).

with their counterparts in Germany and Japan, U.S. teachers name "uninterested students," "uninterested parents," "low student morale," "tardiness," and "intimidation or verbal abuse of teachers/staff" as problems that limit teaching effectiveness and disrupt the learning environment (see Figures 14.2 and Figure 14.3).

In addition to these problems, U.S. teachers work in environments that discourage systematic learning outside the classroom and collaboration with other teachers. Psychologist Harold Stevenson (1992) finds that Asian schools plan extracurricular activities after school hours. During this time, students learn com-

puter skills and teachers do not have to devote classroom time to this subject. In addition, Stevenson finds that Asian teachers work together very closely in preparing lesson plans. The level of collaboration is equivalent to that needed to put on a theatrical production. In contrast, teachers in the United States prepare lesson plans on their own. Asian teachers have more time to collaborate because they teach for approximately 60 percent of the school day; in the remaining time, they discuss ideas with other teachers. American teachers, on the other hand, remain in the classroom for at least 85 percent of the school day.

Figure 14.3   Percentage of High School Seniors Who Reported Being Victimized at School During the Previous 12 Months: 1976–1996

Many critics present schools as more dangerous places than they were 10 or even 20 years ago. U.S. Department of Education data suggest that this assertation is not the case. What does the chart say about the 1990s? Which year (or years) seems to have been the most "dangerous"?

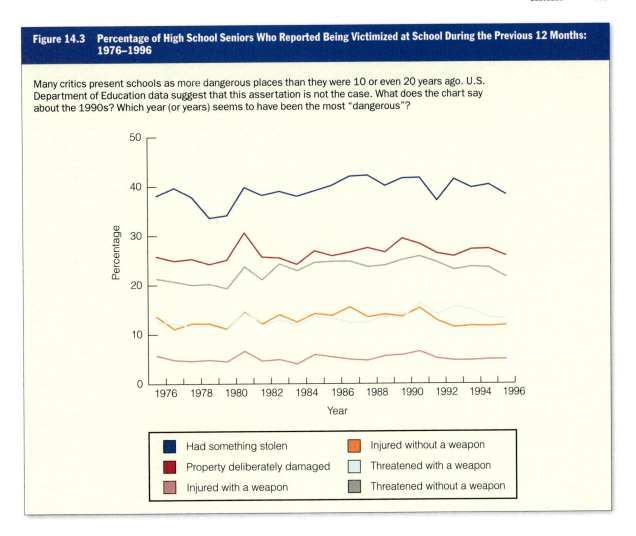

The job of teaching is further complicated by the social context of education. In the United States, teachers must deal with students from diverse family and ethnic backgrounds and with a student subculture that values and rewards athletic achievement, popularity, social activities, jobs, cars, and appearance at the expense of academic achievement. We examine these aspects of social context in the next section.

The Social Context of Education

Sociologist James S. Coleman (the 1991 president of the American Sociological Association) has studied both family background and the adolescent student subculture. Both factors affect the classroom atmosphere and the learning experience.

Family Background

Coleman (1966) was the principal investigator behind *Equality of Educational Opportunity,* popularly known as the Coleman Report. The U.S. government supported this project under the directive of the 1964 Civil Rights Act, which prohibited discrimination for reasons of color, race, religion, or national origin in public places (restaurants, hotels, motels, and theaters); mandated the desegregation of public schools (racial segregation in public schools was ruled unconstitutional by the Supreme Court in 1954); and forbade discrimination in employment. Coleman's intent was to examine the degree to which public education is segregated and to explore inequalities of educational opportunity in the United States. Coleman and his six colleagues surveyed 570,000 students and 60,000 teachers, principals, and school superintendents in 4,000 schools across the United States. Students filled out questionnaires about their home background and educational aspirations and took standardized achievement tests intended to measure verbal ability, nonverbal ability, reading comprehension, mathematical ability, and general knowledge. Teachers, principals, and superintendents answered questionnaires about their backgrounds, training, attitudes, school facilities, and curricula.

Coleman found that a decade after the Supreme Court's famous desegregation decision in 1954 (*Brown v. Board of Education*), U.S. schools were still largely segregated: 80 percent of white children attended schools that were 90 to 100 percent white, and 65 percent of black students attended schools that were more than 90 percent black. Almost all students in the South and the Southwest attended schools that were 100 percent segregated. Although Mexican Americans, Native Americans, Puerto Ricans, and Asian Americans also attended primarily segregated schools, they were not segregated from whites to the same degree that blacks were. The Coleman Report also found that white teachers taught black children, but that black teachers did not teach whites: approximately 60 percent of the teachers who taught black students were black, whereas 97 percent of the teachers who taught white students were white. When the characteristics of teachers of the average white student were compared with those of teachers of the average black student, the study found no significant differences in professional qualifications (as measured by degree, major, and teaching experience).

Coleman found sharp differences across racial and ethnic groups with regard to verbal ability, nonverbal ability, reading comprehension, mathematical achievement, and general information as measured by the standardized tests. The white students scored highest, followed by Asian Americans, Native Americans, Mexican Americans, Puerto Ricans, and African Americans.

Contrary to Coleman's expectations, on average no significant differences in quality existed between schools attended predominantly by the various ethnic groups and schools attended by whites. (Quality was measured in terms of age of buildings, library facilities, laboratory facilities, number of books, class size, expenditures per pupil, extracurricular programs, and the characteristics of teachers, principals, and superintendents.) Surprisingly, any variations in school quality that Coleman found did not have much effect on the students' test scores.

Test scores were affected, however, by family background and other students' attributes. The average

minority group member was likely to come from an economically and educationally disadvantaged household and was likely to attend school with students from similar backgrounds. Fewer of his or her classmates would complete high school, maintain high grade-point averages, enroll in college preparatory curricula, or be optimistic about their future. Coleman also found some support for the idea that "[t]he higher achievement of all racial and ethnic groups in schools with greater proportions of white students is largely, perhaps wholly, related to effects associated with the student body's educational background and aspirations" (1966:307, 310). This finding does not mean that a white environment provides something magical. The Coleman Report examined the progress of blacks who had participated in school integration programs and found that their scores were higher than those of their counterparts who attended schools with members of the same social class. The important variable is the social class of one's classmates, not ethnicity:

> Taking all these results together, one implication stands out above all: That schools bring little influence to bear on a child's achievement that is independent of his background and general social context; and that this very lack of an independent effect means that the inequalities imposed on children by their home, neighborhood, and peer environment are carried along to become the inequalities with which they confront adult life at the end of school. (1966:325)

Coleman's findings that school expenditures are not an accurate predictor of educational achievement (as measured by standardized tests) were cited in support against allocating additional funds to the public school system. Yet, the finding that schools *do not* make a difference does not mean that schools *cannot* make a difference. A more accurate interpretation of Coleman's results is that schools, as currently structured, have no significant effect on test scores; this conclusion implies that the educational system needs restructuring.

Coleman's findings about the composition of the student body and about the higher test scores earned by economically disadvantaged black students in predominantly middle-class schools were used to support busing as a means of achieving educational equality. Although Coleman initially supported this policy, he later retracted his endorsement after busing hastened "white flight," or the migration of middle-class white Americans from cities to suburbs. This migration only intensified the racial segregation in city and suburban

Once the fond symbol of rural schooling, in recent years the school bus has become instead a symbol of unsuccessful court-ordered integration. Behind the failure of busing to achieve integration is the larger fact that the middle class finds one means or another to segregate its children educationally from the children of the lower class.

© George Goodwin/The Picture Cube

A number of studies indicate that the home environment is an important variable — though not the only one — in explaining academic success.

© George Goodwin/Monkmeyer Press

schools. As the ratio of white to black students dropped sharply, the positive effects of desegregation proved to be short-lived. As a result, economically and educationally disadvantaged blacks were sent from their deficient schools into equally deficient lower-class and lower-middle-class white neighborhoods. Coleman adamantly maintained that court-ordered busing alone could not achieve integration:

> With families sorting themselves out residentially along economic and racial lines, and with schools tied to residence, the end result is the demise of the common school attended by children from all economic levels. In its place is the elite suburban school . . . the middle-income suburban school, the low-income suburban school, and the central-city schools of several types — low-income white schools, middle-income white schools, and low- or middle-income black schools. (1977:3–4)

The findings of this study do not imply that a person remains trapped by his or her family background. Indeed, Coleman never claimed that family background explains all of the variations in test scores. He did claim, however, that it was the single most important factor in his study.

School segregation appears to have changed very little over the past three decades. In 1968, the federal government reported that 76 percent of black students and 55 percent of Hispanic students attended predominantly minority schools (schools in which 50 percent or more of the students are black, Asian, Native American, and/or Hispanic). In 1991, the Harvard Project on School Desegregation found those figures to be 66 percent for blacks and 74.3 percent for Hispanics. In some states, such as Illinois, Michigan, New York, and New Jersey, more than 50 percent of the schools are composed of 90 to 100 percent minority students. As in the 1960s, black and other minority students are significantly more likely to find themselves in schools in which overall academic achievement is undervalued and low. The Harvard Project recommended that busing — the most widely used strategy for integrating schools — be supplemented by other strategies, such as finding ways to integrate neighborhoods and enforcing desegregation laws (Celis 1993b).

Many subsequent studies have confirmed the importance of family background in educational achievement (Hallinan 1988). For example, the International Association for the Evaluation of Educational Achievement tested students in 22 countries on six subjects. The association found that the "home environment is a most powerful factor in determining the level of school achievement of students, student interest in school learning, and the number of years of schooling the children will receive" (Bloom 1981:89; Ramirez and Meyer 1980). Nevertheless, in this international study, in the Coleman study, and in other studies, home background (defined in terms of parents' ethnicity, income, education, and occupation) explains only about 30 percent of the variation in students' achievement. Thus factors other than socioeconomic status must affect academic performance:

> In most if not all societies, children and youth learn more of the behavior important for constructive participation in the society outside of school than within. This fact does not diminish the importance of school but underlines the nation's dependence on the home, the working place, the community institutions, the peer group and other informal experiences to furnish a major part of the education required for a child to be successfully inducted into society. Only by clear recognition of the school's special responsibilities can it be highly effective in educating its students. (Tyler, quoted in Purves 1974:74c)

In view of these findings, schools have a special responsibility not to duplicate the inequalities outside the school. Over the past three decades, researchers have found that "schools exert some influence on an individual's chances of success, depending on the extent to which they provide equal access to learning" (Hallinan 1988:257–258). Unfortunately, as we have learned, several characteristics of U.S. education and practices within the schools—hidden curricula, self-fulfilling prophecies, and tracking—work to perpetuate social and economic inequalities. Now we turn to another problematic phenomenon that teachers confront daily—a student value system that de-emphasizes academic achievements.

Adolescent Subcultures

Around the turn of the century—the early decades of late industrialization—fewer than 10 percent of teenagers 14 to 18 years of age attended high school in the United States. Young people attended elementary school to learn the three Rs; they then learned the skills needed to make a living from their parents or from their neighbors. As the pace of industrialization increased, jobs increasingly moved away from the home and the neighborhood and into factories and office buildings. Parents no longer trained their children, because the skills they knew were becoming outdated and obsolete. Children therefore came to expect that they would not make a living in the same way as their parents. In short, as the economic focus in the United States shifted from predominantly farm and small-town work environments to the factory and office, the family became less involved in the training of its children and, by exten-

sion, less involved in children's lives. The transfer of work away from the home and neighborhood removed opportunities for parents and children to work together. Under this new arrangement, family occasions became events that were consciously arranged to fit everyone's work schedule.

According to Coleman, this shift in training from the family to the school cut adolescents off from the rest of society and forced them to spend most of the day with members of their own age group. Adolescents came "to constitute a small society, one that has most of its important interactions within itself, and maintains only a few threads of connection with the outside adult society" (Coleman, Johnstone, and Jonassohn 1961:3).

Coleman surveyed students from 10 high schools in the Midwest to learn about adolescent society. He selected schools representative of a wide range of environments: five schools were located in small towns, one in a working-class suburb, one in a well-to-do suburb, and three in cities of varying sizes. Also, one of the schools was an all-male Catholic school. Coleman was interested in the adolescent status system, a classification of achievements resulting in popularity, respect, acceptance into the crowd, praise, awe, and support, as opposed to isolation, ridicule, exclusion from the crowd, disdain, discouragement, and disrespect. To learn about this system, he asked students questions similar to the following:

- How would you like to be remembered—as an athlete, as a brilliant student, as a leader in extracurricular activities, or as the most popular student?

- Who is the best athlete? The best student? The most popular? The boy the girls go for most? The girl the boys most go for?

- What person in the school would you like most to date? To have as a friend? What does it take to get in with the leading crowd in this school?

Based on the answers to these and other questions, Coleman identified a clear pattern common to all 10 schools. "Athletics was extremely important for the boys, and social success with boys [accomplished through being a cheerleader or being good-looking] was extremely important for girls" (1960:314). Coleman found that girls in particular did not want to be considered as good students, "for the girl in each grade in each of the schools who was most often named as best student has fewer friends and is less often in the leading crowd than is the boy most often named as

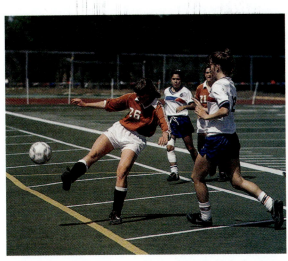

Coleman's research on adolescent subcultures, published in 1961, is considered a classic study. Do his conclusions about the importance for boys of sports and the importance for girls of success with boys still hold true in U.S. high schools? Has the emergence of female sports or other factors changed things for a significant number of either sex? How could you go about researching this issue? What hypotheses might you explore?

© Bob Daemmrich/Stock, Boston

best student" (Coleman 1960:338). A popular boy could often be a good student or dress well or have enough money to meet social expenses, but to truly be admired he must also be a good athlete. Coleman also found that the peer group had more influence over and exerted more pressure on adolescents than did their teachers, and a significant number of adolescents were influenced more by the peer group than by their parents.

Why does the adolescent society penalize academic achievement in favor of athletic and other achievements? Coleman maintained that the manner in which students are taught contributes to their lack of academic interest: "They are prescribed 'exercises,' 'assignments,' 'tests,' to be done and handed in at a teacher's command" (1960:315). This type of academic work does not require creativity but rather conformity. Students show their discontent by choosing to become involved in and acquiring things they can call their own—athletics, dating, clothes, cars, and extracurricular activities. Coleman, Johnstone, and Jonassohn (1961) noted that this reaction is inevitable given the passive roles that students play in the classroom:

> [One] consequence of the passive, reactive role into which adolescents are cast is its encouragement of

irresponsibility. If a group is given no authority to make decisions and take action on its own, the leaders need show no responsibility to the larger institution. Lack of authority carries with it lack of responsibility; demands for obedience generate disobedience as well. But when a person or group carries the authority for his own action, he carries responsibility for it. In politics, splinter parties which are never in power often show little responsibility to the political system; a party in power cannot show such irresponsibility. . . . An adolescent society is no different from these. (p. 316)

Athletics is one of the major avenues open to adolescents, especially males, in which they can act "as a representative of others who surround [them]" (1961:319). Others support this effort, identify with the athletes' successes, and console athletes when they fail. Athletic competition between schools generates an internal cohesion among students as no other event can. "It is as a consequence of this that the athlete gains so much status: he is doing something for the school and the community" (p. 260).

Coleman argues that, because athletic achievement is widely admired, everyone with some ability will try to develop this talent. With regard to the relatively unrewarded arena of academic life, "those who have most ability may not be motivated to compete" (p. 260). This reward structure suggests that the United States does not draw into the competition everyone who has academic potential.

Coleman's findings should deliver the message that the peer group represents a powerful influence on learning, but they should not leave the impression that the peer group's world does not overlap with the family or the classroom. In fact, it seems more appropriate to consider the interrelationships of the multiple contexts of students' lives. From data collected during interviews and observations with 54 ethnically and academically diverse youth in four urban desegregated high schools in California, educators Patricia Phelan, Ann Locke Davidson, and Hanh Cao Yu (1991, 1993; Phelan and Davidson 1994) generated a model of the interrelationships between students' family, peer, and school worlds. The Students' Multiple Worlds Model describes the ways in which sociocultural aspects of students' worlds (for example, norms, values, beliefs, expectations, actions) combine to affect their thoughts and actions with respect to school and learning. These researchers were particularly concerned with understanding students' perceptions of boundaries and borders between worlds and identifying adaptation

Students' Multiple Worlds Model and Typology

Many factors affect students' academic performances, including peer groups, the classroom environment, and family background. Education researchers Patricia Phelan, Ann Locke Davidson, and Hanh Cao Yu made an important attempt to understand the interrelatedness of students' worlds. Rather than compartmentalizing aspects of students' lives, their typology gives us a way of looking more holistically at the processes young people use to manage, more or less successfully, the transitions between their various contexts.

CONGRUENT WORLDS/SMOOTH TRANSITIONS

These students describe values, beliefs, expectations, and normative ways of behaving as similar across their worlds. Moving from one setting to another is harmonious and uncomplicated, and boundaries are easily managed. This statement does not mean that students act exactly the same way or discuss the same things with teachers, friends, and family members, but rather that commonalties among worlds override differences. Students who exhibit this pattern say that their worlds are merged by their common sociocultural components rather than bounded by conspicuous differences. Although many of these youths are white, upper-middle-class, and high-achieving, this situation is not always the case. Some minority students describe little difference across their worlds and experience transitions as smooth. Likewise, academically average students can also exhibit patterns that fit this type.

DIFFERENT WORLDS/BORDER CROSSINGS MANAGED

For some adolescents, differences in family, peer, or school worlds (with respect to culture, ethnicity, socioeconomic status, or religion) require them to adjust and reorient as movement among contexts occurs. For example, a

student's family world may be dominated by an all-encompassing religious doctrine in which values and beliefs are contrary to those found in school and peer worlds. For other students, home and neighborhood are viewed as starkly different than school, particularly for students of color who are bused. And for still other students, differences between peers and family are dominant themes. Regardless of any differences, students in this category are able to use strategies that enable them to manage crossings successfully (in terms of what is valued in each setting). This statement does not mean that crossings are always easy, that they are made in the same way, or that they cannot result in personal and psychic consequences. It is not uncommon for high-achieving minority youth to exhibit patterns common to this type.

DIFFERENT WORLDS/BORDER CROSSINGS DIFFICULT

Students in this category define their family, peer, or school worlds as distinct. They say they must adjust and reorient as they move across worlds and among contexts. Unlike students who manage to make adjustments successfully, however, these students have not learned, mastered, or been willing to adopt all of the strategies necessary for successful transitions. For example, a student may perform poorly in a class in which the teacher's interaction style, the student's role, or the learning activity are oppositional to what takes place within the student's peer or family worlds. Other students in this category describe their comfort and ease at school and with peers but are essentially estranged from their parents. In these cases, parents' values and beliefs are frequently more traditional, more religious, or more constrained than those of their children, making the students' adaptation to their home world difficult and

conflictual. Other youth say that the socioeconomic circumstances of their families work against their full engagement in school. For youth who exhibit patterns common to this type, border crossing involves friction and discomfort and in some cases is possible only under particular conditions. This pattern often includes adolescents on the brink between success and failure, involvement and disengagement, commitment and apathy. These are some of the students for whom classroom and school climate conditions can mean the difference between staying in school and dropping out.

DIFFERENT WORLDS/BORDER CROSSINGS RESISTED

In this category, the values, beliefs, and expectations across students' worlds are so discordant that students perceive borders as insurmountable and actively or passively resist transitions. Border crossing is frequently so painful for these students that, over time, they develop reasons and rationales to protect themselves against further distress. In such cases, borders are viewed as insurmountable, and students actively or passively resist attempts to embrace other worlds. For example, some students say that school is irrelevant to their lives; others immerse themselves fully in the world of their peers. Rather than moving from one setting to another, blending elements of all, these students remain constrained by borders they perceive as rigid and impenetrable. Although low-achieving students (seemingly unable to profit from school and classroom settings) are typical of this type, high-achieving students who do not connect with peers or family also exhibit patterns in this type.

Source: Phelan, Davidson, and Yu (1991, 1993, 1994); synthesized by Patricia Phelan, University of Washington, Bothell.

strategies that students employ as they move from one context to another. Although they found a good deal of variety in students' descriptions of their worlds and in their perceptions of boundaries, the investigators also uncovered four distinctive patterns that students follow as they move between and adapt to different contexts and settings (see "Students' Multiple Worlds Model and Typology"). This typology does not divide students along ethnic, achievement, or gender lines, but rather focuses on the congruency of students' worlds and the borders that they face. In other words, youth of the same ethnicity or students who achieve at the same level can be found within any of the four types.

Summary and Implications

We began this chapter by examining public high school graduation rates. In 1995–1996, the United States had a 68.1 percent graduation rate. Thus 30 percent of the 1992 ninth-grade class did not graduate during the 1995–1996 school year. For some states, the graduation rate was as high as 89.9 percent; for others, it was as low as 54.4 percent.

The low graduation rate does not mean that 30 percent of all students leave high school and never receive a high school diploma. We know, for example, that 720,000 people took the GED exam in 1997,[7] and that more than 20 million have taken the GED test since 1949—the year the tests were first offered. Seventy percent of those who took the exam in 1997 were younger than 30, and 66 percent of those who took the test did so with the intention of attending college.

Dropping out of school clearly does not limit individuals' opportunities to eventually get a high school diploma through an alternative means or to pursue education beyond high school, as at least 90 percent of colleges recognize the GED as valid (American Council on Education 1998). The most recent data show that 5 percent of young adults between the ages of 18 and 24 have earned a high school diploma via the GED (Thomas 1997). This fact leads us to ask, "What happens to the other 25 percent?" Clearly, many acquire their GED after age 24.

When "dropout rates" are so high—one in every four students nationwide—sociologists view it as their obligation to look beyond individual motivation for answers. In this chapter, we have tried to explore the system of American education—its strengths and its shortcomings—to ascertain what supports such a high "dropout" rate.

First, in the United States, a college education is open to anyone (who can find money to pay the tuition) regardless of previous educational failure. College is not just an option for a privileged few. As almost all colleges and universities accept the GED as a legitimate alternative to the high school diploma, high school students realize that, even if they drop out, they can always obtain a GED.

Second, in theory the United States makes a connection between jobs and education, but only one-third of college graduates see a connection between their degree and their job description. One in three college graduates can therefore convey a message to their children and thereby fuel a belief that education is really unrelated to one's job. In the end, employers value the "pieces of paper" more than the content of one's educational experience.

As a country, the United States resists a national curriculum that defines what every graduate from high school or college should know. On the one hand, this decision allows each school to work out the curriculum that best fits the needs of its community or region. On the other hand, it leaves people with the impression that no subject or information is so valuable that everyone should be exposed to it. This invitation to chaos can leave many people (students and teachers alike) frustrated about the purpose and value of education.

Other factors that contribute to the high dropout rate are related to inequalities in learning opportunities that are affected by tracking, funding, family background, and self-fulfilling prophecies. In addition, some groups (such as women and minorities) do not gain the same economic and job opportunities from their educational achievements as do other groups.

Relative to other systems of education, the U.S. system seems to promote a curriculum that makes learning less rewarding and interesting for many people. In the United States, high-track students are usually exposed to the most innovative and creative teaching methods, while their average- and low-track

---

[7]The number of people who take the GED exam each year is at record levels. The record year is 1996, in which 758,000 people completed the test (Ponessa 1997).

Title III—State and Local Education Systemic Improvement

SEC. 301. FINDINGS

The Congress finds that—

(1) all students can learn and achieve to high standards and must realize their potential if the United States is to prosper;

(2) the reforms in education from 1977 through 1992 have achieved some good results, but such reform efforts often have been limited to a few schools or to a single part of the educational system;

(3) leadership must come from teachers, related services personnel, principals, and parents in individual schools, and from policymakers at the local, State, tribal, and national levels, in order for lasting improvements in student performance to occur;

(4) simultaneous top-down and bottom-up education reform is necessary to spur creative and innovative approaches by individual schools to help all students achieve internationally competitive standards;

(5) strategies must be developed by communities and States to support the revitalization of all local public schools by fundamentally changing the entire system of public education through comprehensive, coherent, and coordinated improvement in order to increase student learning;

(6) parents, teachers, and other local educators, and business, community, and tribal leaders must be involved in developing systemwide improvement strategies that reflect the needs of their individual communities;

(7) State and local education improvement efforts must incorporate strategies for providing all students
and families with coordinated access to appropriate social services, health care, nutrition, and early childhood education, and child care to remove preventable barriers to learning and enhance school readiness for all students;

(8) States and local educational agencies, working together, must immediately set about developing and implementing such systemwide improvement strategies if our Nation is to educate all children to meet their full potential and achieve the National Education Goals described in title I;

(9) State and local systemic improvement strategies must provide all students with effective mechanisms and appropriate paths to the work force as well as to higher education;

(10) businesses should be encouraged—

(A) to enter into partnerships with schools;

(B) to provide information and guidance to schools based on the needs of area businesses for properly educated graduates in general and on the need for particular workplace skills that the schools may provide;

(C) to provide necessary education and training materials and support; and

(D) to continue the lifelong learning process throughout the employment years of an individual;

(11) schools should provide information to businesses regarding how the business community can assist schools in meeting the purposes of this Act;
(12) institutions of higher education should be encouraged to enter into partnerships with schools to provide information and guidance to schools on the skills and knowledge graduates need in order to enter and successfully complete postsecondary education, and schools should provide information and guidance to institutions of higher education on the skills, knowledge, and preservice training teachers need, and the types of professional development educators need in order to meet the purposes of this Act;

(13) the appropriate and innovative use of technology, including distance learning, can be very effective in helping to provide all students with the opportunity to learn and meet high standards;

(14) Federal funds should be targeted to support State and local initiatives, and to leverage State and local resources for designing and implementing systemwide education improvement plans;

(15) all students are entitled to participate in a broad and challenging curriculum and to have access to resources sufficient to address other education needs; and

(16) quality education management services are being utilized by local educational agencies and schools through contractual agreements among local educational agencies or schools and businesses providing quality education management services.

Source: U.S. Department of Education (1998a).

counterparts are not. Jules Henry argues that the U.S. curriculum is designed to teach three major lessons: to fear failure, to envy success, and to be absurd. In addition, U.S. teachers seem to work in environments that discourage systematic learning outside the classroom and collaboration with other teachers. In effect, these teachers can neither assume that students learn anything outside the classroom nor collaborate with colleagues to coordinate learning or to determine what students are learning in other courses.

The education reforms associated with Goals 2000 recognized that "simultaneous top-down and bottom-up education reform is necessary to spark creative and innovative approaches by individual schools to help all students achieve internationally competitive standards" (U.S. Department of Education 1998a). It is interesting that the U.S. Congress has instituted Goals 2000 policies that emphasize change on the local level (see "Title III—State and Local Education Systemic Improvement" for an overview of how Goals 2000 should be realized).

Key Concepts

Use this outline to organize your review of the key chapter ideas.

Education
   Informal education
   Formal education
     Schooling
Illiteracy
Formal curriculum

Hidden curriculum
   To be absurd
Tracking/ability grouping
Self-fulfilling prophecy
Status system

internet assignment

Find the report *The Condition of Education, 1998*. Review this report and find data that reflect the strengths of American education. In addition, find data that reflect the weaknesses of American education.

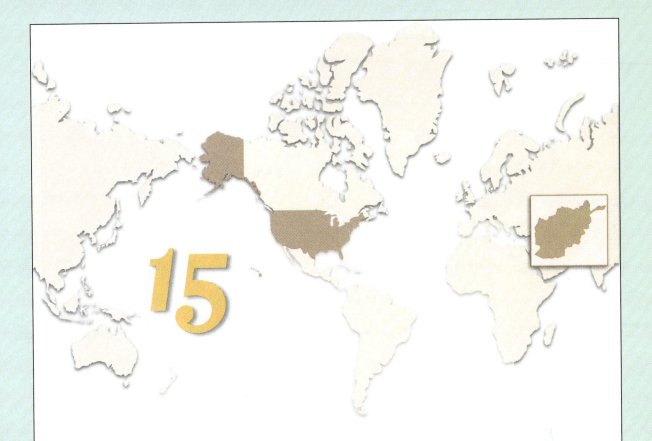

15

Religion

With Emphasis on the Islamic State of Afghanistan

Mountainous terrain in Afghanistan. (© Roger Lemoyne/Gamma-Liaison.)

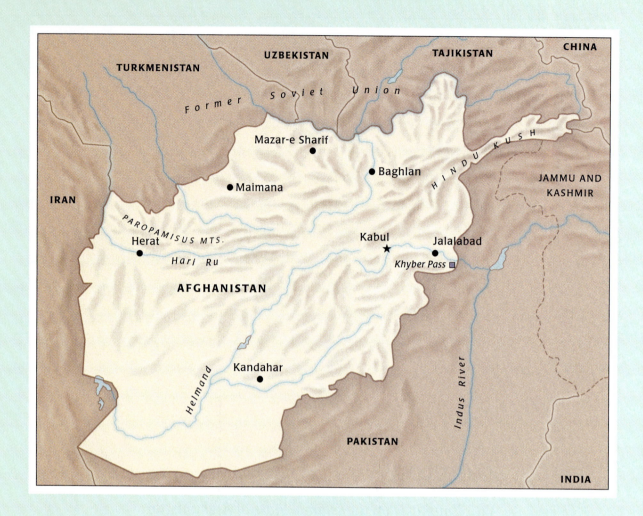

The following labels appear on the map:

TURKMENISTAN   UZBEKISTAN   TAJIKISTAN   CHINA

Former Soviet Union

Mazar-e Sharif

Baghlan

HINDU KUSH

JAMMU AND KASHMIR

Maimana

IRAN

PAROPAMISUS MTS.

Herat

Kabul   Jalalabad

Hari Ru   Khyber Pass

AFGHANISTAN

Helmand

Kandahar

Indus River

PAKISTAN

INDIA

The Historical Context

Before the nineteenth century, mountainous Afghanistan lay in the path of invaders from China, Persia (ancient Iran), and the Indian subcontinent. In the nineteenth and twentieth centuries, the country became a battleground for the British and Russian empires and, after World War II, for the United States and the Soviet Union. In 1979, the Soviet Union invaded Afghanistan to support a secular government. U.S. observers called Afghanistan "the Soviet Union's Vietnam," a title that history has borne out.

Supported materially by the United States, Afghanistan's military resistance to the Soviets was mobilized in large part through religious institutions that proclaimed a "holy war." The Soviet exit in 1989 left a ravaged country full of religiously charged armed rival factions. When the Taliban government took power in 1996, it justified many of its new policies on religious Islamic grounds. Westerners tend to see such policies as simply fanatical and irrational and to overlook the histories that led up to them. They also tend to disregard the parallels that

can be drawn between the role of religion in Afghan society and its significant role in our own.

What is religion? What is its connection to society? How have outspoken religious leaders recently become so influential in government in many parts of the world, including the United States?

In explaining how sociologists view and study religion as an aspect of social life, this chapter helps us make sense of events in Afghanistan.

In the early morning hours of September 27, 1996, the Taliban captured Kabul, the capital city of Afghanistan. Here is how the event was reported by Mark Austin of Independent Television News (ITN):

MARK AUSTIN, ITN: On a hill overlooking Kabul, these are Afghanistan's new soldiers of God, praying they say for peace and stability in a country that's known only conflict for nearly two decades. But below them is a battle-torn city where the fear of war is fast being replaced by a fear of repression. It's symbolized by the white flag of the Taliban militia, heavily armed religious students who patrol the streets, enforcing their vision of Islamic law. The penalties for disobedience: flogging or even death. Their first edict, women must not work, must not be seen uncovered on the streets. Men must grow beards and pray five times a day. The only sounds from the radio, Islamic prayer and poetry. All music and entertainment is banned here. Television shops are being closed down, TV's and video recorders destroyed, tapes hung from trees.

[Taliban] SPOKESMAN: We will confiscate it and destroy it stage by stage.

MARK AUSTIN: At the gates of the presidential palace, we took tea with one group of militia men who told us their goal was a pure Islamic society, free of crime and corruption. But when we toured the palace itself, they proudly showed us works of art they destroyed.

[Taliban] SPOKESMAN: The painting is against Islam.

MARK AUSTIN: After seventeen years of war and suffering, what this city is now experiencing is the most extreme brand of Islam anywhere in the world. The Taliban takeover may have brought temporary peace of a kind, but for the people of Kabul, it's peace at a price. These are the child victims of the Taliban assault on Kabul, appalling injuries caused by shelling and rocket fire. But their tragedy is compounded by the imposition of strict Islamic laws. Eighty percent of the nurses and 40 percent of the doctors here are women, and now most are too frightened even to leave their homes. These are the hands of one of the city's top surgeons. She won't risk being identified but says it's almost as if women no longer exist.

SURGEON: I can't go to my job. I can't help my people because they said the women must sit in the houses, and they can't go outside. It's really bad for us. I'm very sorry, and I want to leave this country.

MARK AUSTIN: Many are already leaving Kabul. Aid workers say more than 100,000 have fled in the last few days. Reports of arrests and beatings abound in this city, and for the women here, the veil conceals the fear that children do not hide. The Taliban are urging people to stay, but there's a sense of panic, and the exodus continues, leaving those who remain to come to terms with life under new rulers with new rules and an existence that many women here say is taking them back to the dark ages. (*Online NewsHour* 1996)

Why Focus on Afghanistan?

In this chapter, we focus on Afghanistan. At the time of this writing, two-thirds of the country was under the control of an Islamic group known as the Taliban. The Taliban rebels received headline news coverage beginning in 1994 (but especially in 1996, after gaining control of the Afghan capital, Kabul). Much of the coverage focused on the Taliban's brutal execution of the ex-Afghan President Najibullal and his brother and its imposition of a strict and harsh version of Islamic law on the local populations. The news focus on this Islamic group, to the neglect of larger historical context, is consistent with how the media cover Islam in general.

The images of repression and violence are consistent with the "snapshot" images that many Americans associate with the Muslim religion, or Islam. For many, Islam evokes images of the November 1995 explosion of a U.S. military training and communication center in Riyadh, the capital of Saudi Arabia; the 1993 World Trade Center bombings in New York City; the 1988 bombing of Pan Am Flight 103 over Lockerbie, Scotland; and the stepped-up airport security in the wake of the Gulf War with Iraq in 1991.

Whatever the event, one idea seems to dominate the coverage: some Islamic group—the Movement of Islamic Amal, the Islamic Jehad organization, the Hezbollah (Party of God), or now the Taliban—is responsible. In other words, each event becomes reduced to the actions of religious fanatics acting solely from "primitive and irrational" religious conviction. This focus masks more important political, geographic, and economic factors and causes viewers to lose sight of the larger questions: For example, why has Afghanistan experienced civil war for the past 20 years? Is the Taliban's version of Islamic law consistent with "Islamic principles"? How did the Taliban rise to power? In an October 9, 1996, U.S. Department of State press statement, spokesman Nicholas Burns suggested that the situation in Afghanistan somehow goes beyond religion:

There's a great interest in what happens in Afghanistan, because it is a pivotal country in terms of where it is situated in that part of the world, and I've just listed some of the issues that are of concern to us [such as terrorism, narcotics, human rights, and the treatment of women]. So, we're going to maintain a close view of the events. We'll have contact with the Taliban and others, and we'll just have to proceed on a step-by-step basis to see if we can establish better and closer contacts. (Burns 1996)

In this chapter, we examine religion from a sociological perspective. Such a perspective is useful because it allows us to step back and view, in a detached way, an often emotion-charged subject. Detachment and objectivity are necessary if we wish to avoid making sweeping generalizations about the nature of religions, such as Islam, that are unfamiliar to many of us.

When sociologists study religion, they do not investigate whether God or some other supernatural force exists, whether certain religious beliefs are valid, or whether one religion is better than another. Sociologists cannot study such questions because they adhere to the scientific method, which requires them to study only observable and verifiable phenomena. Instead, they investigate the social aspects of religion, focusing on the characteristics common to all religions, the types of religious organizations, the functions and dysfunctions of religion, the conflicts within and between religious groups, the way in which religion shapes people's behavior and their understanding of the world, and the way in which religion is intertwined with social, economic, and political issues.

Here we begin with a definition of religion. Defining religion is a surprisingly difficult task and one with which sociologists have been greatly preoccupied.

What Is Religion? Weber's and Durkheim's Views

Figure 15.1 shows the "major religions" of the world. But what makes something a religion? In the opening sentences of *The Sociology of Religion*, Max Weber (1922) states, "To define 'religion,' to say what it is, is not possible at the start of a presentation such as this. Definition can be attempted, if at all, only at the conclusion of the study" (p. 1). Despite Weber's keen interest and his extensive writings about religious activity, he could offer only the broadest of definitions: religion encompasses those human responses that give meaning to the ultimate and inescapable problems of existence—birth, death, illness, aging, injustice, tragedy, and suffering (Abercrombie and Turner 1978). To Weber, the hundreds of thousands of religions, past

Figure 15.1 Major Religions of the World

The map shows where certain large religious groups predominate around the world. There is much overlap, however—for example, millions of Catholics in the United States, millions of Muslims in India, many Protestants in Latin America. Other distinct religions do not appear on the map because they do not dominate a single large geographic area—for example, Judaism, Sikhism, Zoroastrianism (the Parsis). The term *syncretism* refers to compatible combinations of belief systems, such as Confucianism, Buddhism, Taoism, and Shinto in China and Japan (see "The World's Major Non-Christian Religions" later).

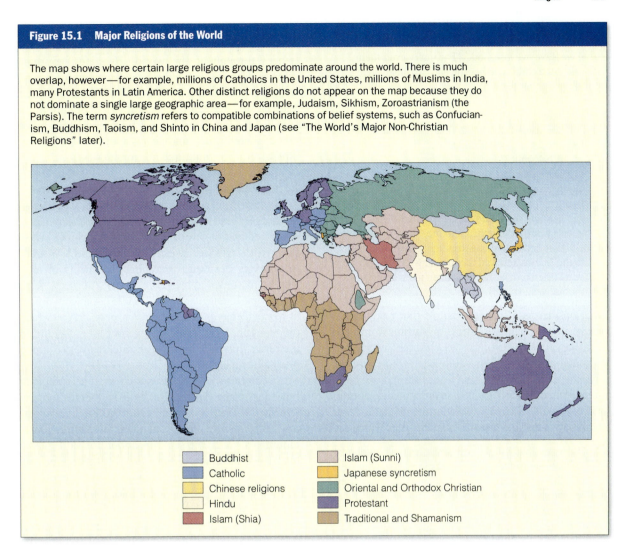

- Buddhist
- Catholic
- Chinese religions
- Hindu
- Islam (Shia)
- Islam (Sunni)
- Japanese syncretism
- Oriental and Orthodox Christian
- Protestant
- Traditional and Shamanism

and present, represented a rich and seemingly endless variety of responses to these problems. In view of this variety, he believed that no single definition could hope to capture the essence of religion.

Like Max Weber, Emile Durkheim believed that nothing is as vague and diffused as religion. In the first chapter of his book *The Elementary Forms of the Religious Life*, Durkheim ([1915] 1964) cautions that when studying religions, sociologists must assume that "there are no religions which are false" (p. 3). Like Weber, Durkheim believed that all religions are true in their own fashion—all address the problems of human existence, albeit in different ways. Consequently, he said, those who study religion must first rid themselves of all preconceived notions of what religion should be. We cannot attribute to religion the charac-

teristics that reflect only our own personal experiences and preferences.

In *The Spiritual Life of Children*, psychiatrist Robert Coles (1990) recounts his conversation with a 10-year-old Hopi girl, which illustrates Durkheim's point. The conversation reminds us that if we approach the study of religion with preconceived notions, we will lose many insights about the nature of religion in general:

> "The sky watches us and listens to us. It talks to us, and it hopes we are ready to talk back. The sky is where the God of the Anglos lives, a teacher told us. She [the teacher] asked where our God lives. I said, I don't know. I was telling the Truth! Our God is the sky, and lives wherever the sky is. Our God is the sun and the moon, too; and our God is our [the Hopi]

people, if we remember to stay here [on the consecrated land]. This is where we're supposed to be, and if we leave, we lose God." [The interviewer then asked the child whether she had explained all of this to the teacher.]

"No."

"Why?"

"Because — she thinks God is a person. If I'd told her, she'd give us that smile."

"What smile?"

"The smile that says to us, You kids are cute, but you're dumb; you're different — and you're all wrong!"

"Perhaps you could have explained to her what you've just tried to explain to me."

"We tried that a long time ago; our people spoke to the Anglos and told them what we think, but they don't listen to hear us; they listen to hear themselves." (p. 25)

Consider as a second example that many critics view the *hijab,* or modest dress of Muslim women, as evidence that the women are severely oppressed. Although women in the Middle East certainly do not have the same rights as men, critics should not be so quick to assume that the *hijab* is *the* source of oppression, especially when we consider the view that some Muslim women hold toward American dress customs:

> If women living in western societies took an honest look at themselves, such a question [as to why Muslim women are covered] would not arise. They are the slaves of appearance and the puppets of male chauvinistic society. Every magazine and news medium (such as television and radio) tells them how they should look and behave. They should wear glamorous clothes and make themselves beautiful for strange men to gaze and gloat over them.
>
> So the question is not why Muslim women wear *hijab,* but why the women in the West, who think they are so liberated, do not wear *hijab*? (Mahjubah 1984)

The conversation between Cole and the Hopi child and the discussion of *hijab* show that preconceived notions of what constitutes religion and uninformed opinions about the meaning of religious symbols and practices can close people off to a wide range of religious beliefs and experiences.

Sacred A term describing everything that is regarded as extraordinary and that inspires in believers deep and absorbing sentiments of awe, respect, mystery, and reverence.

In formulating his ideas about religion, Durkheim remained open to the many varieties of religious experiences throughout the world. He identified three essential features that he believed were common to all religions, past and present: (1) beliefs about the sacred and the profane, (2) rituals, and (3) a community of worshipers. Thus Durkheim defined religion as a system of shared rituals and beliefs about the sacred that bind together a community of worshipers.

Beliefs About the Sacred and the Profane

At the heart of all religious belief and activity stands a distinction between two separate and opposing domains: the sacred and the profane. The **sacred** includes everything that is regarded as extraordinary and that inspires in believers deep and absorbing sentiments of awe, respect, mystery, and reverence. These sentiments motivate people to safeguard what is sacred from contamination or defilement. To find, preserve, or guard that which they consider sacred, people have gone to war, sacrificed their lives, traveled thousands of miles, and performed other life-endangering acts (Turner 1978).

Definitions of what is sacred vary according to time and place. Sacred things may include objects (chalices, sacred documents, and books), living creatures (cows, ants, birds), elements of nature (rocks, mountains, trees, the sea, sun, moon, or sky), places (churches, mosques, synagogues, birthplaces of religious founders), days that commemorate holy events, abstract forces (spirits, good, evil), persons (Christ, Buddha, Moses, Mohammed, Zarathustra, Nanak), states of consciousness (wisdom, oneness with nature), past events (the crucifixion, the resurrection, the escape of the Jews from Egypt, the birth of Buddha), ceremonies (baptism, marriage, burial), and other activities (holy wars, just wars, confession, fasting, pilgrimages).

Durkheim ([1915] 1964) maintains that the sacredness springs not from the item, ritual, or event itself, but rather from its symbolic power and from the emotions that people experience when they think about the sacred thing or when they are in its presence. These emotions are so strong that believers feel part of something larger than themselves and become outraged when others behave inappropriately in the presence of the sacred thing.

Ideas about what is sacred are such an important element of religious activity that many researchers classify religions according to the type of phenomenon that their followers consider sacred. One such typology consists of three categories: sacramental, prophetic, and mystical religions (Alston 1972).

In **sacramental religions,** followers seek the sacred in places, objects, and actions believed to house a god or a spirit. These locations may include inanimate objects (relics, statues, crosses), animals, trees, plants, foods, drink (wine, water), places, and certain processes (such as the way in which people prepare for a hunt or perform a dance). Examples of a sacramental religion include the various forms of Native Spirituality. An excerpt from the "Statement of Walter Echo-Hawk Before the United States Commission on Civil Rights" illustrates the sacramental qualities:

> First, I should tell you something about traditional or tribal religion, as native religion is vastly different from the Judeo-Christian religions most of us are familiar with. Because native religions are so different, the religion of the "redman" has never been understood by the non-Indian soldiers, missionaries, or government officials. (1) It is important to note that there are probably as many native religions as there are Indian tribes in this country. (2) None of these religions or religious tenets [has] been reduced to writing in a holy document such as the Bible or Koran.[1] (3) None of these religions have man-made churches in the Judeo-Christian sense; rather the native religions are practiced in nature, at sacred sites, or in temporary religious structures—such as a tepee or sweat lodge. (4) The religious beliefs are tied to nature, the spiritual forces of nature, the natural elements, and the plants and creatures which make up the environment. . . . Natives are dependent upon all these things in order to practice their many religious ceremonies, rituals and religious observances. (Echo-Hawk 1979:280)

In **prophetic religions,** the sacred revolves around items that symbolize significant historical events or around the lives, teachings, and writings of great people. Sacred books, such as the Christian Bible, the Muslim Koran, and the Jewish Torah hold the records of these events and revelations. In the case of historical events, God or some other higher being is believed to be directly involved in the course and the outcome of the event (a flood, the parting of the Red Sea, the rise and fall of an empire). In the case of great people, the lives and inspired words of prophets or messengers reveal a higher state of being, "the way," a set of ethical principles, or a code of conduct. Followers seek to imitate this life. Some of the best-known prophetic religions include Judaism as revealed to Abraham in Canaan and to Moses at Mount Sinai, Confucianism (founded by Confucius), Christianity (founded by Jesus Christ), and Islam (founded by Mohammed).

In **mystical religions,** followers seek the sacred in states of being that, at their peak, can exclude all awareness of one's existence, sensations, thoughts, and surroundings. In such states, the mystic becomes caught up so fully in the transcendental experience that earthly concerns seem to vanish. Direct union with the divine forces of the universe assumes the utmost importance. Not surprisingly, mystics tend to become involved in practices such as fasting or celibacy to separate themselves from worldly attachments. In addition, they meditate to clear their minds of worldly concern, "leaving the soul empty and receptive to influences from the divine" (Alston 1972:144). Buddhism and philosophical Hinduism are two examples of religions that emphasize physical and spiritual discipline as a means of transcending the self and earthly concerns.

Keep in mind that the distinctions among sacramental, prophetic, and mystical religions are not clear-cut. In fact, most religions incorporate or combine elements of other religions. Consequently, the majority of religions cannot be classified into a single category, although one category often predominates.

According to Durkheim ([1915] 1964), the sacred encompasses more than the forces of good: "There are gods [that is, satans] of theft and trickery, of lust and war, of sickness and of death" (p. 420). Evil and its various representations, however, are generally portrayed as inferior and subordinate to the forces of good: "in the majority of cases we see the good victorious over evil, life over death, the powers of light over the powers of darkness" (p. 421). Even so, Durkheim considers superordinary evil phenomena to fall under the category "sacred" (as he defines it) because they are endowed with special powers and serve as the object of rituals (confessions, baptisms, penance, fasting, exorcism) designed to overcome or resist their negative influences.

Sacramental religions Religions in which the sacred is sought in places, objects, and actions believed to house a god or spirit.

Prophetic religions Religions in which the sacred revolves around items that symbolize significant historical events or around the lives, teachings, and writings of great people.

Mystical religions Religions in which the sacred is sought in states of being that, at their peak, can exclude all awareness of one's existence, sensations, thoughts, and surroundings.

---

[1]To understand the nature of Native American religion, one must necessarily look to the native practitioners themselves for information.

Religious beliefs, doctrines, legends, and myths detail the origins, virtues, and powers of sacred things and describe the consequences of mixing the sacred with the profane. The **profane** encompasses everything that is not sacred, including things opposed to the sacred (the unholy, the irreverent, the contemptuous, the blasphemous) and things that stand apart from the sacred, although not in opposition to it (the ordinary, the commonplace, the unconsecrated, the temporal, the bodily) (Ebersole 1967). Believers often view contact between the sacred and the profane as being dangerous and sacrilegious, as threatening the very existence of the sacred, and as endangering the fate of the person who made or allowed the contact. Consequently, people take action to safeguard sacred things by separating them in some way from the profane. For example, some people refrain from speaking the name of God in frustration; others believe that a woman must cover her hair or her face during worship and that a man must remove his hat during worship.

The distinctions between the sacred and the profane do not mean that a person, object, or idea cannot pass from one domain to another or that something profane cannot ever come into contact with the sacred. Such transformations and contacts are authorized through rituals—the active and most observable side of religion.

Rituals

In the religious sense, **rituals** are rules that govern how people must behave in the presence of the sacred to achieve an acceptable state of being. These rules may take the form of instructions detailing the appropriate context for worship, the roles of various participants, acceptable dress, and the precise wording of chants, songs, and prayers. Participants must follow instructions closely so as to achieve a specific goal, whether it

be to purify the participant's body or soul (confession, immersion, fasting, seclusion), to commemorate an important person or event (pilgrimage to Mecca, Passover, the Last Supper), or to transform profane items into sacred items (water to holy water, bones to sacred relics). During rituals, behavior is "coordinated to an inner intention to make contact with, or to participate in, the invisible world or to achieve a desired state" (Smart 1976:6).

Rituals can be as simple as closing one's eyes to pray or having one's forehead be marked with ashes; they can also be as elaborate as fasting for three days before entering a sacred place to chant, with head bowed, a particular prayer for forgiveness. Although rituals are often enacted in sacred places, some are codes of conduct aimed at governing the performance of everyday activities—sleeping, walking, eating, defecating, washing, dealing with the opposite sex.

The Taliban has received worldwide attention for the code of conduct that it has imposed on those under its rule and for the severe punishments issued to those who violate these codes. For example, women who appear in public must be covered from head to toe; they are not permitted to leave their homes without an acceptable reason or to work outside the home even as health care workers or as distributors of international aid. Members of the Taliban have sought out and destroyed things they consider un-Islamic, including tape recorders, cassettes, toys, TVs, and car radio antennas (Amnesty International 1996a, 1996b).

According to Durkheim, the nature of the ritual is relatively insignificant. Rather, the important element is that the ritual be shared by a community or worshipers and evoke certain ideas and sentiments that help individuals feel part of something bigger than themselves.

Community of Worshipers

Durkheim uses the word **church** to designate a group whose members hold the same beliefs with regard to the sacred and the profane, who behave in the same way in the presence of the sacred, and who gather in body or spirit at agreed-on times to reaffirm their commitment to those beliefs and practices. Obviously, religious beliefs and practices cannot be unique to an individual; they must be shared by a group of people. If not, then the beliefs and practices would cease to exist when the individual who held them died or if he or she chose to abandon them. In the social sense, religion is inseparable from the idea of church. The gathering and the sharing create a moral community and allow worshipers to share a common identity. The gathering, however, need

---

Profane A term describing everything that is not sacred, including things opposed to the sacred and things that stand apart from the sacred, although not in opposition to it.

Rituals Rules that govern how people must behave in the presence of the sacred to achieve an acceptable state of being.

Church A group whose members hold the same beliefs with regard to the sacred and the profane, who behave in the same way in the presence of the sacred, and who gather in body or spirit at agreed-on times to reaffirm their commitment to those beliefs and practices.

It is a challenge to see through one's own preconceptions about what is "right" in everyday life. A Western woman may look on the cover-up Muslim hijab as a sign of sexual oppression. The Muslim woman may look on the Western woman's garb as the result of media pressures on her to display herself as a sex object for men.

© Evan Agostini/Gamma-Liaison (left); © Anne Dowie (right)

not take place in a common setting. When people perform a ritual on a given day or at given times of day, the gathering is spiritual rather than physical.

Durkheim ([1915] 1964) uses the term *church* loosely, acknowledging that it can assume many forms: "Sometimes it embraces an entire people . . . sometimes it embraces only a part of them . . . sometimes it is directed by a corps of priests, sometimes it is almost completely devoid of any official directing body" (p. 44). Sociologists have identified at least five broad types of religious organizations (communities of worshipers): ecclesiae, denominations, sects, established sects, and cults. As with most classification schemes, the categories overlap on some characteristics because the classification criteria for religions are not always clear.

ECCLESIAE An **ecclesiae** is a professionally trained religious organization governed by a hierarchy of leaders, which claims everyone in a society as its members. Membership is not voluntary; it is the law. Consequently, considerable political alignment exists between church and state officials, so that the ecclesiae represents the official church of the state. Ecclesiae formerly existed in England (the Anglican church), France (the Roman Catholic church), and Sweden (the Lutheran church). The 1987 Afghanistan constitution stated that the religion of Afghanistan is Islam. Today

Islam is the official religion of Bangladesh and Malaysia, Iran has been an Islamic republic since the Ayatollah Khomeini took power in 1979, and Saudi Arabia is a monarchy based on Islamic law (*World Almanac and Book of Facts 1991* 1990).

Individuals are born into ecclesiae; newcomers to the society are converted; dissenters often are persecuted. Those who do not accept the official view tend to emigrate or occupy the most marginal status in the society. The ecclesiae claims to be the one true faith and often does not recognize other religions as valid. In its most extreme form, it directly controls all facets of life. In Afghanistan, for example, non-Muslims must conform to Islamic legal and unwritten restrictions and are prohibited from proselytizing (Kurian 1992).

DENOMINATIONS A **denomination** is a hierarchical organization in a society in which church and state usually remain separate; it is led by a professionally trained

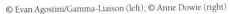

Ecclesiae A professionally trained religious organization governed by a hierarchy of leaders, which claims as its members everyone in a society.

Denomination A hierarchical organization in a society in which church and state are usually separate, led by a professionally trained clergy.

"Here's the church, here's the steeple, . . . " the child's rhyme says. Is building a church simply a practical necessity? Or is there some underlying idea in Christianity that tends to make Christian communities feel the need to build large structures?

© Michael Dwyer/Stock, Boston

clergy. In contrast to an ecclesiae, a denomination is one of many religious organizations in the society. For the most part, denominations tolerate other religious organizations; they may even collaborate to address and solve some problems in the society. Although membership is considered to be voluntary, most people who belong to denominations did not choose to join them. Rather, they were born to parents who are members. Denominational leaders generally make few demands on the laity, and most members participate in limited and specialized ways. For example, they may choose to send their children to church-operated schools, attend church on Sundays and religious holidays, donate money, or attend church-sponsored functions. The leaders of a denomination do not oversee all aspects of the members' lives. Yet, even though laypeople vary widely in lifestyle, denominations frequently

Sect A small community of believers led by a lay ministry, with no formal hierarchy or official governing body to oversee the various religious gatherings and activities.

Established sects Religious organizations, resembling both denominations and sects, composed of renegades from denominations or ecclesiae that have existed long enough to acquire a large following and widespread respectability.

attract people of particular races and social classes—that is, their members are drawn disproportionately from specific social and ethnic groups.

Eight major religious denominations exist in the world—Buddhism, Christianity, Confucianism, Hinduism, Islam, Judaism, Taoism, and Shinto—each dominant in different areas of the globe (see "The World's Major Non-Christian Religions"). For example, Christianity predominates in Europe, North and South America, New Zealand, Australia, and the Pacific Islands; Islam predominates in the Middle East and North Africa; Hinduism predominates in India (see "The Visibility of Major Christian Denominations by State").

SECTS AND ESTABLISHED SECTS   A **sect** is a small community of believers led by a lay ministry, with no formal hierarchy or official governing body to oversee the various religious gatherings and activities. Sects typically are composed of people who broke away from a denomination because they came to view it as corrupt. They then created the offshoot in an effort to reform the religion from which they separated.

All of the major religions encompass splinter groups that have sought, at one time or another, to preserve the integrity of their religion. In Islam, for example, the most pronounced split occurred 1,300 years ago, approximately 30 years after the death of the Prophet Mohammed. The dividing issue related to Mohammed's successor. The Shia maintained that the successor should be a blood relative of Mohammed; the Sunni believed that the successor should be selected by the community of believers and need not be related by blood. After Mohammed's death, the Sunni (encompassing the great majority of Muslims) accepted Abu-Bakr as the caliph (successor). The Shia supported Ali, Mohammad's first cousin and son-in-law, and they called for the overthrow of the existing order and a return to the pure form of Islam. Today Shiaism represents the dominant religion in the Islamic Republic of Iran (95 percent), and Sunni Muslim dominates in the Islamic Republics of Pakistan (77 percent) and Afghanistan (84 percent). These divisions within Islam have existed for so long that Sunni and Shia have become recognized as **established sects,** renegades from denominations or ecclesiae that have existed long enough to acquire a large following and widespread respectability. In some ways, established sects resemble both denominations and sects. As you might expect, several divisions exist within each established sect, as splinter groups attempted to

The Visibility of Major Christian Denominations by State

This map shows the number of people in the state in relation to the number of Catholic, Baptist, Methodist, Lutheran, and Christian churches. This number is a rough indicator of the influence of these denominations. That is, the greater the number of people per church, the less pervasive their influence in that state. In which states are these denominations least influential? In which states are they most influential?

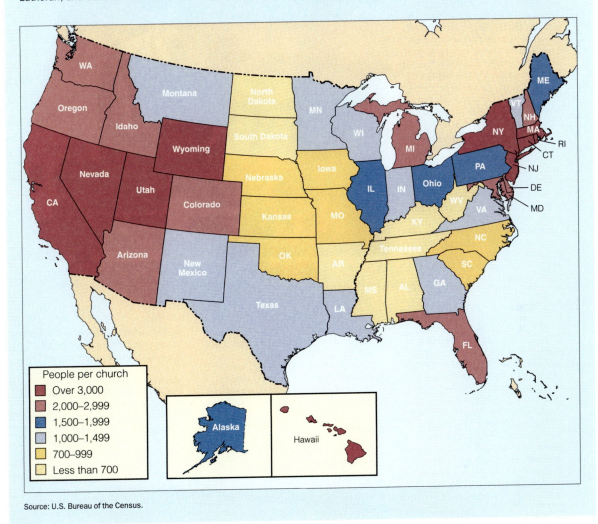

People per church
- Over 3,000
- 2,000–2,999
- 1,500–1,999
- 1,000–1,499
- 700–999
- Less than 700

Source: U.S. Bureau of the Census.

reform some policy, practice, or position held by the religious organization from which they separated.

Similarly, several splits have occurred within the Christian churches. In 1054, for example, the Eastern Orthodox churches rejected the pope as the earthly deputy of Christ and questioned the papal claim of authority over all Catholic churches in the world. The Protestant religions owe their origins to Martin Luther (1483–1546), who also challenged the papal authority and protested many of the practices of the Roman Catholic church.[2] Divisions also exist within various Protestant sects and between Catholics. Offshoots of the Roman Catholic church, for example, include Maronites, Greek Catholics, Greek Orthodox, Jacobites, and Gregorians.

[2]This protest was dubbed the Reformation because it involved efforts to reform the Catholic church and to cleanse it of corruption, especially with regard to paying for indulgences (the forgiveness of sins upon saying specific prayers or performing specific good deeds at the order of a priest). Luther believed that a person is saved not by the intercession of priests or bishops, but by private and individual faith (Van Doren 1991).

The World's Major Non-Christian Religions

BUDDHISM

Buddhism has 307 million followers. It was founded by Siddhartha Gautama, known as the Buddha (Enlightened One), in southern Nepal in the sixth and fifth centuries B.C. The Buddha achieved enlightenment through meditation and gathered a community of monks to carry on his teachings. Buddhism teaches that meditation and the practice of good religious and moral behavior can lead to Nirvana, the state of enlightenment, although before achieving Nirvana one is subject to repeated lifetimes that are good or bad depending on one's actions (karma). The doctrines of the Buddha describe temporal life as featuring "four noble truths": existence is a realm of suffering; desire, along with the belief in the importance of one's self, causes suffering; achievement of Nirvana ends suffering; and Nirvana is attained only by meditation and by following the path of righteousness in action, thought, and attitude.

The famous Daibutsu in Kamakura, Japan.

© Frilet/Sipa Press

CONFUCIANISM

A faith with 5.6 million followers, Confucianism was founded by Confucius, a Chinese philosopher, in the sixth and fifth centuries B.C. Confucius's sayings and dialogues, known collectively as the Analects, were written down by his followers. Confucianism, which grew out of a strife-ridden time in Chinese history, stresses the relationship between individuals, their families, and society, based on li *(proper behavior) and* jen *(sympathetic attitude). Its practical, socially oriented philosophy was challenged by the more mystical precepts of Taoism and Buddhism, which were partially incorporated to create neo-Confucianism during the Sung* dynasty *(A.D. 960–1279). The overthrow of the Chinese monarchy and the Communist revolution during the twentieth century have severely lessened the influence of Confucianism on modern Chinese culture.*

HINDUISM

A religion with 648 million followers, Hinduism developed from indigenous religions of India in combination with Aryan religions brought to India in c. 1500 B.C. and codified in the Veda and the Upanishads, the sacred scriptures of Hinduism. Hinduism is a term used to broadly describe a vast array of sects to which most Indians belong. Although many Hindu reject the caste system—in which people are born into a particular subgroup that determines their religious, social, and work-related duties—it is widely accepted and classifies society at large into four groups: the Brahmins or priests, the rulers and warriors, the farmers and merchants, and the peasants and laborers. The goals of Hinduism are release from repeated reincarnation through the practice of yoga, adherence to Vedic scriptures, and devotion to a personal guru. Various deities are worshiped at shrines; the divine trinity, representing the cyclical nature of the universe, are Brahma*

People are not born into sects, as they are with denominations; theoretically, they convert. Consequently, newborns are not baptized; they choose membership later in life, when they are considered able to decide for themselves. Sects vary on many levels, including the degree to which they view society as religiously bankrupt or corrupt and the extent to which they take action to change people in society.

Cults Very small, loosely organized groups, usually founded by a charismatic leader who attracts people by virtue of his or her personal qualities.

CULTS   Generally, **cults** are very small, loosely organized groups, usually founded by a charismatic leader

the creator, Vishnu the preserver, and Shiva the destroyer.

ISLAM

Islam has 840 million followers. It was founded by the prophet Muhammad, who received the holy scriptures of Islam, the Koran, from Allah (God) c. A.D. 610. Islam (Arabic for "submission to God") maintains that Muhammad is the last in a long line of holy prophets, preceded by Adam, Abraham, Moses, and Jesus. In addition to being devoted to the Koran, followers of Islam (Muslims) are devoted to the worship of Allah through the Five Pillars: the statement "There is no god but God, and Muhammad is his prophet"; prayer, conducted five times a day while facing Mecca; the giving of alms; the keeping of the fast of Ramadan during the ninth month of the Muslim year; and the making of a pilgrimage at least once to Mecca, if possible. The two main divisions of Islam are the Sunni and the Shiite; the Wahabis are the most important Sunni sect, while the Shiite sects include the Assassins, the Druses, and the Fatimids, among countless others.

JUDAISM

Stemming from the descendants of Judah in Judea, Judaism was founded c. 2000 B.C. by Abraham, Isaac, and Jacob and has 18 million followers. Judaism espouses belief in a monotheistic God, who is creator of the universe and who leads His people, the Jews, by speaking through prophets. His word is revealed in the Hebrew Bible (or Old Testament), especially in that part known as the Torah. The Torah also contains, according to Rabbinic tradition, a total of 613 biblical commandments, including the Ten Commandments, which are explicated in the Talmud. Jews believe that the human condition can be improved, that the letter and the spirit of the Torah must be followed, and that a Messiah will eventually bring the world to a state of paradise. Judaism promotes community among all people of Jewish faith, dedication to a synagogue or temple (the basic social unit of a group of Jews, led by a rabbi), and the importance of family life. Religious observance takes place both at home and in temple. Judaism is divided into three main groups who vary in their interpretation of those parts of the Torah that deal with personal, communal, international, and religious activities: the Orthodox community, which views the Torah as derived from God, and therefore absolutely binding; the Reform movement, which follows primarily its ethical content; and the Conservative Jews, who follow most of the observances set out in the Torah but allow for change in the face of modern life. A fourth group, reconstructionist Jews, rejects the concept of the Jews as God's chosen people, yet maintains rituals as part of the Judaic cultural heritage.

SHINTO

Shinto, with 3.5 million followers, is the ancient native religion of Japan, established long before the introduction of writing to Japan in the fifth century A.D. The origins of its beliefs and rituals are unknown. Shinto stresses belief in a great many spiritual beings and gods, known as kami, who are paid tribute at shrines and honored by festivals, and reverence for ancestors. While there is no overall dogma, adherents of Shinto are expected to remember and celebrate the kami, support the societies of which the kami are patrons, remain pure and sincere, and enjoy life.

TAOISM

Both a philosophy and a religion, Taoism was founded in China by Lao-tzu, who is traditionally said to have been born in 604 B.C. Its number of followers is uncertain. It derives primarily from the Tao-te-ching, which claims that an ever-changing universe follows the Tao, or path. The Tao can be known only by emulating its quietude and effortless simplicity; Taoism prescribes that people live simply, spontaneously, and in close touch with nature and that they meditate to achieve contact with the Tao. Temples and monasteries, maintained by Taoist priests, are important in some Taoist sects. Since the Communist revolution, Taoism has been actively discouraged in the People's Republic of China, although it continues to flourish in Taiwan.

Look back at several of these descriptions. What are some sacramental, prophetic, and mystical characteristics they mention? Suppose you were asked to write a brief piece describing Christianity in the same ways that the other major religions are described here. What would you say?

Source: Reprinted with permission of Macmillan General Reference USA, a Simon & Schuster Macmillan Company, from The New York Public Library Desk Reference, second edition. A Stonesong Press book. Copyright © 1989, 1993 by The New York Public Library and The Stonesong Press, Inc. The name, "The New York Public Library," is a trademark and the property of The New York Public Library, Astor, Lenox, and Tilden Foundations.

who attracts people by virtue of his or her personal qualities. Because the charismatic leader plays such a central role in attracting members, cults often dissolve after the leader dies. For this reason, few cults last long enough to become established religions. Even so, a few manage to survive, as evidenced by the fact that the major world religions began as cults. Because cults form around new and unconventional religious practices, outsiders tend to view them with considerable suspicion.

Cults vary according to their purpose and to the level of commitment that the cult leaders demand of converts. They may draw members by focusing on highly specific, but eccentric interests, such as astrology, UFOs, or transcendental meditation. Members may be attracted by the promise of companionship, a

The tokens of devotion left by "pilgrims" to the grave of James Dean illustrate the difficulty of making absolute distinctions between the sacred and the profane.

© James F. Hopgood

cure from illness, relief from suffering, or enlightenment. A cult may meet infrequently and strictly voluntarily (such as at conventions or monthly meetings). In some cases, however, the cult leaders may require members to break all ties with family, friends, and jobs and thus to come to rely exclusively on the cult to meet all their needs.

A Critique of Durkheim's Definition of Religion

Durkheim's definition of religion relies on the most outward, most visible characteristics of religion. Critics argue, however, that the three essential characteristics—beliefs about the sacred and the profane, rituals, and a community of worshipers—are not unique to religious activity. This combination of characteristics, they say, applies to many gatherings—sporting events, graduation ceremonies, reunions, and political rallies—and many political systems (for example, Marxism, Maoism, and fascism). On the basis of these characteristics alone, it is difficult to distinguish between an assembly of Christians celebrating Christmas, a patriotic group supporting the initiation of a war

> **Civil religion** "Any set of beliefs and rituals, related to the past, present and/or future of a people (nation), which are understood in some transcendental fashion" (Hammond 1976:171).

against another country, and a group of fans eulogizing James Dean or Elvis Presley. In other words, religion is not the only unifying force in society that incorporates these three elements defined by Durkheim as characteristic of religion. Civil religion represents another such force that resembles religion as Durkheim defined it.

Civil Religion

Civil religion is "any set of beliefs and rituals, related to the past, present and/or future of a people (nation), which are understood in some transcendental fashion" (Hammond 1976:171). A nation's beliefs (such as individual freedom or equal opportunity) and rituals (parades, fireworks, singing of the national anthem, 21-gun salutes, and so on) often assume a sacred quality. Even in the face of internal divisions based on race, ethnicity, region, religion, or gender, national beliefs and rituals can still inspire awe, respect, and reverence for the country. These sentiments are most notably evident on national holidays that celebrate important events or people (such as Presidents' Day, Martin Luther King, Jr., Day, or Independence Day), in the presence of national monuments or symbols (the flag, the Capitol, the Lincoln Memorial, the Vietnam Memorial), and at times of war or other national crises.

Often political leaders appeal to these sentiments to win an election, to legitimize their policies, to rally a nation around a cause that requires sacrifice, or to motivate a people to defend their country, as President George Bush attempted to do in his State of the Union Address on January 7, 1991:

> I come to this house of the people to speak to you and all Americans, certain that we stand at a defining hour.
>
> Halfway around the world, we are engaged in a great struggle in the skies and on the seas and sands. We know why we're there. We are Americans—part of something larger than ourselves.
>
> For two centuries, we've done the hard work of freedom. And tonight we lead the world in facing down a threat to decency and humanity.
>
> For two centuries, America has served the world as an inspiring example of freedom and democracy. For generations, America has led the struggle to preserve and extend the blessings of liberty. And today, in a rapidly changing world, American leadership is indispensable. Americans know that leadership brings burdens, and requires sacrifice.
>
> But we also know why the hopes of humanity turn to us. We are Americans; we have a unique responsibility to do the hard work of freedom. And when we do, freedom works.

One can argue that the Cold War[3] between the United States and the former Soviet Union elevated each country's economic and political systems to the level of a "religion." During the Cold War, relations between the United States and the Soviet Union fell short of direct, full-scale military engagement; even so, as many as 120 wars were fought in so-called Third World countries. In many of these wars, the United States and the Soviet Union supported opposing factions with weapons and other military equipment, combat training, medical supplies, economic aid, and food. Three of the best-known proxy wars were fought in Korea, Vietnam, and Afghanistan.

Soviet and American leaders justified their direct or indirect intervention in these proxy wars on the grounds that it was necessary to contain the spread of the other side's economic and political system, to protect national and global security, and to prevent the other side from shifting the balance of power in favor of its system.

In the mid-1960s, President Lyndon Johnson justified U.S. involvement in Vietnam in this way:

> If we allow the Communists to win in Vietnam, it will become easier and more appetizing for them to take over other countries in other parts of the world. We will have to fight again some place else—at what cost, no one knows. That is why it is vitally important to every American family that we stop the Communists in South Vietnam. (Johnson 1987:907)

For their part, the Soviets justified their involvement on the grounds that they were blocking "the export of 'counterrevolution,' or actions by Western powers that would hinder the historic progress of socialism" (Zickel 1991:681). They defined it as their "internationalist duty" to defend their socialist allies and to give military and economic support to wars of national liberation in Third World countries. Such actions were rationalized according to "the Marxist belief that workers around the world are linked by a bond that transcends nationalism" (pp. 999–1000).

Since the end of World War II, the foreign and domestic policies of both the United States and the former Soviet Union have been shaped by Cold War dynamics. In the United States, for example, virtually every policy—from the 1949 Marshall Plan to covert aid to the Contras during the Reagan administration—was influenced in some way by America's professed desire to protect the world from the Soviet influence and the spread of Communism. Robert S. McNamara (1989), U.S. Secretary of Defense under Presidents Kennedy and Johnson, remarked that "on occasion after occasion, when confronted with a choice between support of democratic governments and support of anti-Soviet dictatorships, we have turned our backs on our traditional values and have supported the antidemocratic" and brutally repressive and totalitarian regimes (p. 96).

As this brief overview of the Cold War's history reveals, each of the two governments' economic and political beliefs assumed a sacred quality that unified and motivated each side to sacrifice millions of human lives at home and abroad in the name of those principles. The larger point is that the traits cited by Durkheim as characteristics of religion apply to other events, relationships, and forces within society that many people would not define as religious.

Narrower, less inclusive definitions of religion than that proposed by Durkheim are problematic as well. Suppose that we defined religion as the belief in an ever-living god. This definition would exclude polytheistic religions such as Hinduism, which has more than 640 million adherents. It would also exclude religions in which a deity plays little or no role, such as Buddhism, which has more than 300 million followers. Thus narrow definitions of religion offer no improvement relative to broad ones.

Despite its shortcomings, Durkheim's definition of religion remains one of the best and most widely used. No sociologist with any standing in the discipline can study religion without encountering and addressing Durkheim's definition. The question "What is religion?" is not just a sociological question; it is also a question asked by those governments that guarantee their residents freedom of religion (see "Freedom of Religious Expression in U.S. Prisons").

Besides proposing a definition of religion, Durkheim wrote extensively about the functions of religion. His work laid the foundation for the functionalist perspective of religion.

The Functionalist Perspective

Some form of religion appears to have existed for as long as humans have lived (at least 2 million years). In view of this fact, functionalists maintain that religion must serve some vital social functions for the individual and

---

[3]The Cold War is the name given to the political tension and military rivalry that existed between the United States and the former Soviet Union from approximately 1945 until its symbolic end on November 9, 1989—the date on which the Berlin Wall "fell." The Cold War included an arms race, in which each side competed to match and then surpass any advances made by the other in the number and technological quality of weapons.

U.S. in Perspective
Freedom of Religious Expression in U.S. Prisons

The following selection is an excerpt of a statement by Larry F. Taylor, warden at the Federal Correctional Institution in Lompoc, California, speaking before the U.S. Commission on Civil Rights.

Thank you, Commissioner Freeman. Mr. Chairman, first of all, let me state that Lompoc is a maximum security institution designed to accommodate about 1,200 offenders [and] . . . a minimum security camp for 420 inmates serving relatively short sentences.

Persons confined in the major institutions are serving relatively long sentences for drug offenses, trafficking in drugs, bank robbery, homicides, hijacking, kidnapping, and all the kinds of offenses that you would expect to find at a major institution.

We have a mixed racial group: about one-third of our population black, one-third white, one-third Chicano, and about 100 Indians at any one time, Native American Indians.

I think it's important to say that the exercise of religion in an institution of this type cannot be absolute

and I think it's subject to reasonable regulations designed to protect the welfare of the staff and the inmates, the welfare of the community, control and discipline of the inmates, proper exercise of institutional authority in scheduling activities, etc., and reasonable economic considerations.

I don't think, for example, we can provide clergy for every kind of religion represented in our facility; we can't provide facilities for every type of religion represented in our facility. There are some difficult religious issues which a prison administrator has to face each day: first of all, what is a religion? Thank goodness I don't have to decide what a religion is. You've heard the list of religions that are represented in the Bureau of Prisons. That's decided for me by general counsel and by several court cases. But another difficult issue is what are ceremonies, artifacts, and symbols that are mandated by certain recognized religions? And can we make those available to the inmate population?

Areas of concern for us have to do primarily with security. Can we operate a safe humane institution

and yet allow a number of these kinds of things into the institution?

Of course, our goal is to try to be as flexible as possible and allow the greatest amount of flexibility in exercising religion wherever we can. We are concerned about diets and, you know, it is easy to prepare 1,600 meals three times a day if we're preparing the same thing for everybody; but if you have seven or eight different groups who need, who have different kinds of diet requirements, then that task becomes much more difficult. When certain requirements are accepted or recognized, the question of having the necessary experts there is still to advise us on proper preparation. This has been particularly troublesome with the Jewish inmates and also troublesome with Muslim inmates who have special diet requirements.

We have a similar problem with what is considered proper religious wearing apparel. Special clothing also becomes a security problem when going in and out of our visiting room. Head gear and items of this sort need to be carefully searched when a man enters or leaves the visiting room.

---

for the group. On the individual level, people embrace religion in the face of uncertainty: they draw on religious doctrine and ritual to comprehend the meaning of life and death and to cope with misfortunes and injustices (war, drought, illness). Life would be intolerable without reasons for existing or without a higher purpose to justify the trials of existence (Durkheim 1951). Try to imagine, for example, how people cope with the immense devastation and destruction resulting from decades of war. In Afghanistan, after Soviet troops invaded the country in 1979, troops attacked civilian populations, burned village crops, killed livestock, used lethal and nonlethal chemical weapons, planted an estimated 10 million mines, and engaged in large-scale, high-altitude carpet bombing.

In the countryside it was standard Soviet practice to bombard or even level whole villages suspected of harboring resistance fighters. Sometimes women,

children, and old men were rounded up and shot. This devastation of towns and villages forced many civilians to seek refuge in Kabul, whose prewar population of less than one million swelled to nearly two million. (Kurian 1992:5)

Even after the Soviets withdrew from Afghanistan in 1989, the civil war continued as various political parties competed to fill the power vacuum. Table 15.1 summarizes the tragic results of 20 years of war in this country. In light of this situation, is it any wonder that the Afghan people turn to religion?

A crisis like [war] makes a Muslim search within himself for values to counter the bestiality of man. We look around at what they have done, and the only thing you can say is: Thank God, who is, in a situation like this, the only refuge for those who cannot understand why man would do this to his fellow man. (Ibrahim 1991:A7)

Mr. Cripe indicated one example of Rastafarians, where we've had some difficulty searching their hair for drugs and things of that nature. Symbols often cause racial tension within our institutions for various reasons. The recognition of holy days and when they should be celebrated is another sensitive issue. Are the inmates allowed off work during those holy days? If one group has the day off from work, then all groups would like their days recognized, and be off from work on that day.

Scheduling of prayer hours so they do not interfere with counts and other institutional functions is also important. Chanting, for example, isn't very popular early in the morning, yet, some of our inmates believe that they must chant at sunrise. In one case at Lompoc, one man assaulted another man because of the noise that was being made and the invasion of privacy.

Special instruments used in the practice of religion is another area that we have to be concerned about. For Native American Indians we allow drums into the institution and although we haven't in the past, we are now looking at allowing a peace pipe into the institution.

Incidentally, in the past when the peace pipe was allowed, there was a conflict in which the pipe disappeared. About forty or fifty inmates decided that that pipe was not going to be returned to its proper place. As a result, there was a great potential for violence and disruption of prison routine because of that incident. Many times, it's a question of what is required of a man's religion as opposed to what he prefers. A question of individual preference over mandated requirements. Clergy don't always agree, courts don't always agree, and certainly prison administrators don't agree on what's required.

One thing that's important to remember: whatever we do for one religious group, we must be willing to do for all religious groups, and so wherever we give a little in one area, we also have to be willing to give in another area. I think the primary barriers to the exercise of religion as far as the Native Americans are concerned is the lack of knowledge, lack of knowledge on my part, lack of knowledge of many administrators who are charged with the responsibility of running the prisons. What is the Indian religion? The same lack of understanding was true of Muslims

in the sixties when we first started running into those kind of issues. The Indian religious issue is further complicated by the lack of written definition. I think most of the Indian religion is passed down by word of mouth and only after making some effort and doing some research have I become a little more comfortable with some of the requirements of the Indian religion within our institutions. I think there is a lack of documented history. The Indian religion is a combination of cultural, medical, recreation, and social needs. It serves all those purposes as far as I can understand and so, sometimes, what may seem to be a social event may have, indeed, some very serious religious significance for Indians.

Warden Taylor is happy not to be the one who must "decide what religion is." What is it that he must decide? How is that difficult? What parallels do you see between what he must do and what governments must do to maintain civility or law when a society admits the possibility of more than one religion?

Source: U.S. Commission on Civil Rights (1979).

In addition to turning to religion in the face of intolerable circumstances, people rely on religious beliefs and rituals to help them achieve a successful outcome (the birth of a healthy child, a job promotion) and to gain answers to questions of meaning: How did we get here? Why are we here? What happens to us when we die? According to Durkheim, people who have communicated with their god or with other supernatural forces (however conceived) report that they gain the inner strength and the physical strength to endure and to conquer the trials of existence.

It is as though [they] were raised above the miseries of the world. . . . Whoever has really practiced a religion knows very well . . . these impressions of joy, of interior peace, of serenity, of enthusiasm which are, for the believer, an experimental proof of his beliefs. (Durkheim [1915] 1964:416–417)

Religion functions in several ways to promote group unity and solidarity. First, the shared doctrine and rituals create emotional bonds among believers. Second, all religions strive to raise individuals above themselves—to help them achieve a life better than they would lead if they were left to their own impulses. In this sense, religion offers ideas of proper conduct that carry over into everyday life. When believers violate this code of conduct, they feel guilt and remorse. Such feelings, in turn, motivate them to make amends. Third, although observances of many religious rituals function to alleviate individual anxieties, uncertainties, and fears, they also establish, reinforce, or renew social relationships, binding individuals to a group. Finally, religion functions as a stabilizing force in times of severe social disturbances and abrupt change. During such times, many regulative forces in society may break down. In the absence of such regulative forces, people are more likely to

Table 15.1 Profile of the Islamic State of Afghanistan	
Population	16 million
World Food Program	Feeds, aids 1.8 million people
Refugees	6.0 million fled the country;
	2.0 million displaced internally;
	3.0 million repatriated
Deaths as a result of war	500,000 military; 1.5 million civilian (since 1979)
Life expectancy	42 years
Malnutrition	15 to 20 percent of children under the age of 7
Damage from war	19,000/22,000 of the estimated villages bombarded by Soviet air force
Mines	10 million implanted mines
Access to drinking water	5 percent rural; 40 percent urban
Maternal mortality	11,700 women die for every 100,000 child births
Infant mortality	182/1,000; 257/1,000 die before age 5

Sources: U.S. Central Intelligence Agency (1995), Carter Center (1996), and United Nations (1996).

turn to religion in search of a force that will bind them to a group. This tie helps people think less about themselves and more about some common goal (Durkheim 1951), whether that goal is to work for peace or to participate more fervently in armed conflicts. One could argue that the Taliban hoped to restore order after the chaos wrought by war.

The fact that religion functions to meet individual and societal needs, combined with the fact that people create sacred objects and rituals, led Durkheim to reach a controversial but thought-provoking conclusion: The something out there that people worship is actually society.

Society as the Object of Worship

If we operate under the assumption that all religions are true in their own fashion and that the variety of religious responses is virtually endless, we find support for Durkheim's conclusion that people create everything encompassed by religion—gods, rites, sacred objects. That is, people play a fundamental role in determining what is sacred and how people should act in the presence of the sacred. Consequently, at some level, people worship what they (or their ancestors) have created. This point led Durkheim to conclude that the real object of worship is society itself—a conclusion that many critics cannot accept (Nottingham 1971). (See "Salman's Portrait of Jesus.")

Let us give Durkheim the benefit of the doubt, however, and ask, Is there anything about the nature of society that makes it deserving of such worship? In

reply to this question, Durkheim gave what sociologist W. S. F. Pickering (1984) calls a "virtual hymn to society, a social Gloria in Excelsis" (p. 252). Durkheim maintained that society transcends the individual life because it frees people from the bondage of nature (as in nature and nurture). How does it accomplish this task? Chapter 5 presented cases showing the consequences of extreme isolation, neglect, and limited social contact. Such cases make it clear that "it is impossible for a person to develop without social interaction" (Mead 1940:135). In addition, studies of mature and even otherwise psychologically and socially sound persons who experience profound isolation—astronauts orbiting alone in space, prisoners of war placed in solitary confinement, individuals who volunteer to participate in scientific experiments in which they are placed in deprivation tanks—show that when people are deprived of contact with others, they lose a sense of reality and personal identity (Zangwill 1987). The fact that we depend so strongly on society supports Durkheim's view that for the individual "it is a reality from which everything that matters to us flows" (Durkheim, cited in Pickering 1984:252).

Durkheim, however, does not claim that society provides us with perfect social experiences:

Society has its pettiness and it has its grandeur. In order for us to love and respect it, it is not necessary to present it other than it is. If we were only able to love and respect that which is ideally perfect, . . . God Himself could not be the object of such a feeling, since the world derives from Him and the world

Salman's Portrait of Jesus

The religious painter and illustrator Warner Salman (pictured on the right) created this image of Jesus Christ in 1940. The number of times this image has been reproduced on "church bulletins, calendars, posters, book marks, prayer cards, tracts, buttons, stickers and stationary" is more than 500 million (Grimes 1994). This number translates into an image for every 10 people alive on the planet. Given that Jesus was born in Bethlehem, a town in the Middle Eastern part of the world, is the Sallman image the most accurate representation of how Jesus might have looked? Now re-evaluate Durkheim's belief that "at some level people worship what they or others have created."

AP/Wide World

is full of imperfection and ugliness. (quoted in Pickering 1984:253)

Durkheim observed that whenever a group of people (no matter what kind) has strong conviction, that conviction almost always takes on a religious character. Religious gatherings and affiliations become ways of affirming convictions and mobilizing the group to uphold them, especially when it is threatened. As an example of this mobilizing function, consider that after the Soviet Union invaded Afghanistan in December 1979, Afghan freedom fighters known as the *mujahidin* resisted, making it impossible for the Soviet Union to establish control outside the major urban centers. For example, Afghanistan's mountainous terrain ensured that the Khyber Pass was a key position in Britain's nineteenth-century colonial strategies. It also made it extremely difficult for massive, highly mechanized Soviet forces to find small bands of *mujahidin* moving on foot (see the chapter-opening photo). Analogous problems worked against the United States in Vietnam.

A Critique of the Functionalist View of Religion

To claim that religion functions as a strictly integrative force is to ignore the long history of wars between different religious groups and the many internal struggles among factions within the same religious group. For example, although the Afghan *mujahidin* opposed the Soviet occupation and its secular government, many competing factions existed within the *mujahidin*. After the Soviets withdrew, the former *mujahidin* resistance commanders became the major power brokers, and each took control of different cities outside Kabul. At this point, as has happened in the past, the same rugged terrain that made it impossible for the Soviets to gain control over the entire country made it difficult for any one internal group to consolidate power. Tribal elders and religious students, in turn, have tried to wrest control from rebel commanders (U.S. Central Intelligence Agency 1995). The following is a partial list of the various political parties now operating in Afghanistan: Islamic Society, Islamic Party (which has three factions), Islamic Union for the Liberation of Afghanistan, Islamic Revolutionary Movement, Afghanistan National Liberation Front, National Islamic Front (which has two major factions), Islamic Unity Party, Islamic Movement, and the Taliban (Religious Student Movement).

If religion were entirely an integrative force, then the religious symbols that unite a community of worshipers would not bind them so strongly that they would be willing to destroy persons who did not belong to or support their religion or version of a religion. The Amnesty International (1996c) document, *Afghanistan: Grave Abuses in the Name of Religion*, outlines human rights violations by the Taliban committed in the name of religion. Those abuses include "indiscriminate killings, arbitrary and unacknowledged detention of civilians, physical restrictions on women for reasons of their gender, the beating and ill-treatment of women, children and detainees, deliberate and arbitrary killings, amputations, stoning and executions."

The functionalist perspective, then, tends to overemphasize the constructive consequences associated with religions' unifying, bonding, and comforting functions. Strict functionalists who focus only on the consequences that lead to order and stability tend to overlook the fact that religion can also unify, bond, and comfort believers in such a way that it supports war and other forms of conflict between ingroups and outgroups. The conflict perspective, on the other hand, acknowledges the unifying, comforting functions of religion, but views such functions as ultimately problematic.

The Conflict Perspective

Scholars who view religion from the conflict perspective focus on how religion turns people's attention away from social and economic inequality. This perspective stems from the work of Karl Marx, who believed that religion was the most humane feature of an inhumane world and that it arose from the tragedies and injustices of human experience. He described religion as the "sigh of the oppressed creature, the sentiment of a heartless world, and the soul of soulless conditions. It is the opium of the people" (Pelikan and Fadiman 1990:80). According to Marx, people need the comfort of religion to make the world bearable and to justify their existence. In this sense, he said, religion is analogous to a sedative.

Even though Marx acknowledged the comforting role of religion, he focused on its repressive, constraining, and exploitative qualities. In particular, he conceptualized religion as an ideology that justifies the status quo, rationalizing existing inequities or downplaying their importance. This aspect of religion is especially relevant with regard to the politically and economically disadvantaged. For them, said Marx, religion serves as a source of false consciousness. That is, religious teachings encourage the oppressed to accept the economic, political, and social arrangements that constrain their chances in this life because they are promised compensation for their suffering in the next world.

Consider how material published in 62 languages by the Watchtower Bible and Tract Society (1987) and distributed worldwide describes life in God's Kingdom:

> God's Kingdom will bring earthly benefits beyond compare, accomplishing everything good that God originally purposed for his people to enjoy on earth. Hatreds and prejudices will cease to exist. . . . The whole earth will eventually be brought to a garden-like [paradise]. . . . No longer will people be crammed into huge apartment buildings or rundown slums. . . . People will have productive, satisfying work. Life will not be boring. (pp. 3–4)

This kind of ideology led Marx to conclude that religion justifies social and economic inequities and that religious teachings inhibit protest and revolutionary change. He went so far as to claim that religion would be unnecessary in a truly classless society—that is, a propertyless society providing equal access to the means of production. In the absence of material inequality, no exploitation and no injustice would occur; these experiences cause people to turn to religion. In sum, Marx believed that religious doctrines turn people's attention away from unjust political and economic arrangements and that they rationalize and defend the political and economic interests of the dominant social classes. For some contemporary scholars, this legitimating function explains the fact that most religions allow only a specific category of people— men—to be leaders and to handle sacred items. A letter written to the editor of *Christianity Today* in reaction to the article "Women in Seminary: Preparing for What?" shows how the "facts" of the Bible can be used to explain and justify such inequalities:

> If the Lord meant for a woman to lead the church in such roles as preacher, elder, pastor, minister, prophet, priest, et cetera, why didn't he provide early Christians with a scriptural prototype? Where in Scripture can a woman priest be found? A woman (literary) prophet? A woman apostle? A woman elder or pastor? Could it be the Lord didn't intend for a woman to serve in any of these positions? It makes me wonder. (*Christianity Today* 1986:6)

As a more extreme example, in the name of Islam, Taliban leaders placed severe restrictions on female dress, employment, and access to education. The restrictions were so severe that the Save the Children Fund, a humanitarian organization that has been working in Afghanistan and with Afghan refugees in Pakistan since the mid-1970s, suspended its operations in western Afghanistan (Save the Children Fund 1996).

Sometimes believers may twist religion in ways that serve the interests of dominant groups. During the days of slavery, for example, some Christians prepared special catechisms for slaves to study. The following questions and answers were included in such catechisms:

Q: What did God make you for?

A: To make a crop.

Q: What is the meaning of Thou shalt not commit adultery?

A: To serve our heavenly Father, and our earthly Master, obey our overseer, and not steal anything. (Wilmore 1972:34)

Often political leaders use religion to unite their country in war against another. Both Iraqi President Saddham Hussein and U.S. President George Bush invoked the name of God to rally people behind their causes in the Persian Gulf War of 1991:

> In the name of God, the merciful, the compassionate: Our armed forces have performed their holy war duty of refusing to comply with the logic of evil, imposition and aggression. They have been engaged in an epic, valiant battle that will be recorded by history in letters of light. (spokesperson for the Iraqi Government, quoted in "Iraqi Message" 1991:Y1)

> This we do know: Our cause is just. Our cause is moral. Our cause is right. May God bless the United States of America. (Bush 1991:A8)

A Critique of the Conflict Perspective of Religion

The major criticism leveled at Marx and at the conflict perspective is that religion is not always the sign of the oppressed creature. On the contrary, the oppressed have often used religion as a vehicle for protesting or working to change social and economic inequities. **Liberation theology** represents one such approach. Liberation theologians maintain that they have a responsibility to demand social justice for the marginalized peoples of the world, especially landless peasants and the urban poor, and to take an active role at the grassroots level to bring about political and economic justice. Ironically, this doctrine is inspired by Marxist thought, in that it advocates raising the consciousness of the poor and teaching them to work together to obtain land and employment and to preserve their cultural identity.

Sociologist J. Milton Yinger (1971) identifies at least two interrelated conditions under which religion can become a vehicle of protest or change. In the first

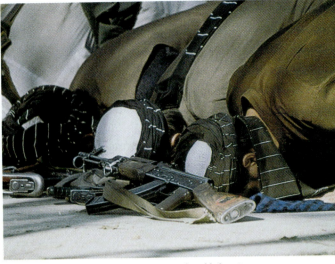

Although many Westerners become uncomfortable hearing Muslims or others describe armed struggle in terms of "holy war," U.S. government officials and politicians have a long tradition of claiming that fighting wars on behalf of the United States involves some "sacred duty." The similar language reflects a similar underlying frame of mind.

© Corbis-Bettmann

condition, a government or other organization fails to achieve clearly articulated ideals (such as equal opportunity, justice for all, or the right to bear arms). In the second condition, a society becomes polarized along class, ethnic, or sectarian lines. In such cases, disenfranchised or disadvantaged groups may form sects or cults and use seemingly eccentric features of the new religion to symbolize their sense of separation (p. 111) and to rally their followers to fight against the establishment or the dominant group. In the United States, one religion that emerged in reaction to society's failure to ensure equal opportunity was the Nation of Islam.

In the 1930s, black nationalist W. D. Farad, who went by a variety of names including Farad Mohammed, founded the Nation of Islam and began preaching in the Temple of Islam in Detroit. (When Farad disappeared in 1934, he was replaced by his chosen successor, Elijah Mohammed.) Farad taught that the white

Liberation theology A doctrine that maintains that organized religions have a responsibility to demand social justice for the marginalized peoples of the world, especially landless peasants and the urban poor, and to take an active role at the grassroots level to bring about political and economic justice.

The Nation of Islam was organized in the United States in response to widespread exclusion of African Americans from mainstream society. Despite its representation in the media as a violent organization, the Nation of Islam focuses on empowering members through religious training that defines moral behavior and through small-business enterprise.

© Tannenbaum/Sygma

man was the personification of evil and that black people were Muslim but that their religion had been stripped away after they came to America as slaves. In addition, he taught that the way out was not through gaining the "devil's" (that is, the white man's) approval but through self-help, discipline, and education. Members received an X to replace their slave names (hence Malcolm X). In the social context of the 1930s, this message was very attractive:

> You're talking about Negroes. You're talking about niggers, who are the rejected and the despised, meeting in some little, filthy, dingy little [room] upstairs over some beer hall or something, some joint that nobody cares about. Nobody cares about these people. . . . You can pass them on the street and in 1930, if they don't get off the sidewalk, you could have them arrested. That's the level of what was going on. (National Public Radio 1984a)

The Nation of Islam is only one example of a religious organization working to improve life for African Americans. Historically, African American churches have reached out to millions who have felt excluded from the system (Lincoln and Mamiya 1990).

They have demonstrated their capability for empowerment, emancipating black people from the ravages of generations of slavery and equipping them to recover a sense of identity, to forge ties of communal loyalty, to help themselves and others and to create cultural expressions that embody their hopes and aspirations. (Forbes 1990:2)

For example, African American churches were instrumental forces in achieving the overall successes of the civil rights movement. In fact, some observers argue that the movement would have been impossible if the churches had not become involved (Lincoln and Mamiya 1990).

Religion in a Polarized Society

As Durkheim observed, whenever a group of people holds strong convictions, those convictions almost always take on a religious character. Religion serves as a vehicle around which convictions can be affirmed and around which a group can rally to uphold those convictions.

Although the full story of Afghanistan is very complicated and lies beyond the scope of this chapter, we can identify at least one factor that has contributed to major polarizations among the Afghan peoples: geographic location, which has rendered Afghanistan of strategic importance to several foreign powers.

As the chapter-opening map shows, Afghanistan is landlocked, and two-thirds of its terrain is mountainous. China, Iran, Pakistan, and three former Soviet republics (Tajikistan, Turkmenistan, and Uzbekistan) border the country. Given its location, it is not surprising that many foreign governments evinced interest in Afghanistan. Consequently, the factional fighting in Afghanistan has been backed by foreign powers.

The Cold War between the United States and the Soviet Union made Afghanistan a focus of those two countries' conflict. When the Soviet Union invaded Afghanistan in 1979 and put its Afghan supporters in charge, the United States supported the Afghan freedom fighters known as the *mujahidin,* funneling money through Pakistan. Barnett Rubin (1996) argues that no one paid more for the U.S. Cold War victory than Afghanistan and its people. The political maneuverings of Afghan and foreign leaders "may inspire cynicism or repulsion, but millions of unknown people scarified their homes, their land, their cattle, their health, their families, with barely hope of success or reward, at least in this world" (p. 21).

The cases of the Nation of Islam and Afghanistan show that the larger social, economic, and political context must be incorporated into any discussion of the role of religion in shaping human affairs. This line of investigation was particularly interesting to Max Weber, who examined how religious beliefs direct and are used to justify economic activity.

Max Weber: The Interplay Between Economics and Religion

Max Weber was interested in understanding the role of religious beliefs in the origins and development of **modern capitalism,**

> a form of economic life which involved the careful calculation of costs and profits, the borrowing and lending of money, the accumulation of capital in the form of money and material assets, investments, private property, and the employment of laborers and employees in a more or less unrestricted labor market. (Robertson 1987:6)

In his book *The Protestant Ethic and the Spirit of Capitalism,* Weber (1958) asked why modern capitalism emerged and flourished in Europe rather than in China or India (the two dominant world civilizations at the end of the sixteenth century). He also asked why business leaders and capitalists in Europe and the United States were overwhelmingly Protestant. To answer these questions, Weber studied the major world religions and some of the societies in which these religions were practiced. He focused on understanding how norms generated by different religious traditions influenced the adherents' economic orientations and motivations.

On the basis of his comparisons, Weber concluded that a branch of Protestant tradition—Calvinism—supplied a "spirit" or ethic that supported the motivations and orientations required by capitalism. Unlike other religions that Weber studied, Calvinism emphasized **this-worldly asceticism**—a belief that people are instruments of divine will and that God determines and directs their activities. Consequently, people glorified God when they accepted a task assigned to them and carried it out in exemplary and disciplined fashion and when they did not indulge in the fruits of their labor (that is, when they did not use money to eat or drink or otherwise relax to excess). In contrast, Buddhism, a religion that Weber defined as the Eastern parallel and opposite of Calvinism, "emphasized the basically illusory character of worldly life and regarded release from the contingencies of the everyday world as the highest religious aspiration" (Robertson 1987:7).

The Calvinists conceptualized God as all-powerful and all-knowing; they also emphasized **predestination,** the belief that God has foreordained all things, including the salvation or damnation of individual souls. According to this doctrine, people could do nothing to change their fate. To compound matters, only relatively few people were destined to attain salvation.

Weber maintained that such beliefs created a crisis of meaning among adherents as they tried to determine how they should behave in the face of their predetermined fate. Such pressures led them to look for concrete signs that they were among God's chosen people, destined for salvation. Consequently, accumulated wealth became an important indicator of whether one was among the chosen. At the same time, this-worldly asceticism "acted powerfully against the spontaneous enjoyment of possessions; it restricted consumption, especially of luxuries" (Weber 1958:171). Frugal behavior encouraged people to accumulate wealth and make investments, important actions for the success of capitalism.

This calculating orientation was not an official part of Calvinist doctrine per se. Rather, this orientation grew out of and was supported by asceticism and predestination. Given this distinction, we must not misread the role that Weber attributed to the Protestant ethic in supporting the rise of a capitalistic economy. According to Weber, the ethic was a significant ideological force; it was not the sole cause of capitalism but "*one* of the causes of *certain aspects* of capitalism" (Aron 1969:204). Unfortunately, many people who encounter Weber's ideas overestimate the importance that he assigned to the Protestant ethic for achieving economic success, drawing a conclusion that Weber himself never reached: the reason that some groups and societies are disadvantaged is simply that they lack this ethic.

Modern capitalism "A form of economic life which involved the careful calculation of costs and profits, the borrowing and lending of money, the accumulation of capital in the form of money and material assets, investments, private property, and the employment of laborers and employees in a more or less unrestricted labor market" (Robertson 1987:6).

This-worldly asceticism A belief that people are instruments of divine will and that their activities are determined and directed by God.

Predestination The belief that God has foreordained all things, including the salvation or damnation of individual souls.

Finally, note that Weber was writing about the origins of industrial capitalism, not about the form of capitalism that exists today, which heavily emphasizes consumption and self-indulgence. He maintained that once established, capitalism would generate its own norms and become a self-sustaining force. In fact, Weber argued that "[c]apitalism produces a society run along machine-like, rational procedures without inner meaning or value and in which men operate almost as mindless cogs" (Turner 1974:155). In such circumstances, religion becomes an increasingly insignificant factor in maintaining the capitalist system. Some sociologists argue that industrialization and scientific advances cause society to undergo unrelenting secularization, a process in which religious influences become increasingly irrelevant not only to economic life, but also to most aspects of social life. Others argue that as religion becomes less relevant to economic and social life in general, a significant number of people take on a fundamentalist view—that is, they seek to reexamine their religious principles in an effort to identify and return to the most fundamental and basic principles (from which believers have departed) and to hold those principles up as the definitive and guiding blueprint for life.

Two Opposing Trends: Secularization and Fundamentalism

Secularization and fundamentalism are processes that have become increasingly popular in the recent past. Each has expanded in spite of the other's growth, or possibly in opposition to it. This section examines these two opposing trends.

Secularization

In the most general sense, **secularization** is a process by which religious influences on thought and behavior are reduced. It is difficult to generalize about the causes and consequences of secularization because they vary across contexts. Americans and Europeans associate secularization with an increase in scientific understanding and technological solutions to everyday problems of living. In effect, science and technology assume

Secularization A process by which religious influences on thought and behavior are reduced.

Subjective secularization A decrease in the number of people who view the world and their place in it from a religious perspective.

roles that were once filled by religious belief and practice. Most Muslims, on the other hand, do not attribute secularization to science or to modernization; indeed, many devout Muslims are physical scientists. In a National Public Radio interview, a Muslim geologist explained that Islam was compatible with science:

> I realized that there is something—there's a glue, there's a creator—behind all those marvelous facts of life in the field of geology, of chemistry. [I had] to think and rethink about my attitude towards life, and that there should be a creator and a very wise creator who has already given all of us laws, and produced all these facts of life. (National Public Radio 1984c)

From a Muslim perspective, secularization is a Western-imposed phenomenon—specifically, a result of exposure to what many people in the Middle East consider the most negative of Western values. This point is illustrated by the observations of a Muslim who attended college in Britain and by a Muslim artist:

> If I did not watch out [while I was in college], I knew that I would be washed away in that culture. In one particular area, of course, was exposure to a society where free sexual relations prevailed. There you are not subject to any control, and you are faced with a very serious challenge, and you have to rely upon your own strength, spiritual strength to stabilize your character and hold fast to your beliefs. (National Public Radio 1984b)

> You can't be against foreign influence because art and culture and science are things that you learn from people and [that] we exchange. It depends on what kind of foreign influence is being exerted. For example, if we speak of the United States, the U.S. has a tradition of science and of art and culture . . . which can be extremely useful and can enrich our own culture if there is an exchange. But the type of culture that we are getting from the U.S. at the moment is the *Dallas* series, the cowboy serials, crimes, commodity values, big cars, luxury, things like that. This kind of culture can't help. (National Public Radio 1984b)

Subjective secularization is a decrease in the number of people who view the world and their place in it from a religious perspective. In other words, paradigms shift from a religious understanding of the world, grounded in faith, to an understanding grounded in observable evidence and the scientific method. In the face of uncertainty, secular thinkers do not turn to religion or to a supernatural[4] power that

---

[4]The supernatural encompasses forces that can defy the laws of nature and affect the outcome of an event.

When Ben Franklin flew his kite in a lightning storm, was he intervening in divine events? Some people think so.

© North Wind Pictures

they have come to view as a distant, impersonal, and even inactive phenomenon. Consequently, they come to believe less in direct intervention by the supernatural and to rely more strongly on human intervention or scientific explanation:

> Consider the case of the lightning rod. For centuries, the Christian church held that lightning was the palpable manifestation of divine wrath and that safety against lightning could only be conforming to divine will. Because the bell towers of churches and cathedrals tended to be the only tall structures, they were the most common targets of lightning. Following damage or destruction of a bell tower by lightning, campaigns were launched to stamp out local wickedness and to raise funds to repair the tower. Ben Franklin's invention of the lightning rod caused a crisis for the church. The rod demonstrably worked. The laity began to demand its installation on church towers—backing their demands with a threat to withhold funds to restore the tower should lightning strike it. The church had to admit either that Ben Franklin had the power to thwart divine retribution or that lightning was merely a natural

phenomenon. (Stark and Bainbridge 1985: 432–433)

Considerable debate exists over the extent to which secularization is taking place. Data collected by the Gallup Organization over the past 20 years show little change in the degree of importance that Americans report assigning to religion. More than 90 percent have a religious preference, and almost 70 percent are members of one of the more than 250,000 places of worship within the United States (White House press release 1995); approximately 40 percent attend church weekly; and almost 60 percent state that religion is very important in their lives (Gallup and Castelli 1989). Polls also show that almost 80 percent of Americans are "sometimes very conscious of God's presence" and that more than 80 percent agree that "even today, miracles are performed by the power of God" (Gallup and Castelli 1989:58).

Fundamentalism

In "Popular Conceptions of Fundamentalism," anthropologist Lionel Caplan (1987) offers his readers an extremely clear overview of a complex religious phenomenon—**fundamentalism**, a belief in the timeless nature of sacred writings and a belief that such writings apply to all kinds of environments. In its popular usage, this term is applied to a wide array of religious groups in the United States and around the world, including the Moral Majority in the United States, Orthodox Jews in Israel, and various Islamic groups in the Middle East.

Religious groups labeled as fundamentalist are usually portrayed as "fossilized relics . . . living perpetually in a bygone age" (Caplan 1987:5). Americans frequently employ this simplistic analysis to explain events in the Middle East, especially the causes of the political turmoil that threatens the interests of the United States (including its need for oil). Such oversimplification misrepresents fundamentalism, however, and it cannot explain the widespread appeal of contemporary fundamentalist movements within several of the world's religions.

THE COMPLEXITY OF FUNDAMENTALISM   Fundamentalism is a more complex phenomenon than popular conceptions lead us to believe. It is impossible to define

Fundamentalism A belief in the timeless nature of sacred writings and a belief that such writings apply to all kinds of environments.

It was only in the 1950s that the phrase "under God" was inserted into the U.S. Pledge of Allegiance that most schoolchildren were required to recite each day. At the same time, the United States started printing "In God We Trust" on its coins. American demonization of Communism in the Cold War played a role in these decisions—over the objection of people who saw them as violations of the principle of separation between church and state.

© R. W. Carter/SuperStock

a fundamentalist in terms of age, ethnicity, social class, or political ideology, because fundamentalism appeals to a wide range of people. Moreover, fundamentalist groups do not always position themselves against those in power; in fact, they are equally likely to be neutral or to support existing regimes fervently. Perhaps the most important characteristic of fundamentalists is their belief that a relationship with God, Allah, or some other supernatural force provides answers to personal and social problems. In addition, fundamentalists often wish to "bring the wider culture back to its religious roots" (Lechner 1989:51).

Caplan (1987) identifies a number of other traits that seem to characterize fundamentalists. First, fundamentalists emphasize the authority, infallibility, and timeless truth of sacred writings as a "definitive blueprint" for life (p. 19). This characteristic does not mean that any definitive interpretation of sacred writings actually exists. Any sacred text has as many interpreta-

tions as there are groups that claim it as their blueprint. For example, even members of the same fundamentalist organization may disagree about the true meaning of the texts they follow.

Second, fundamentalists usually conceive of history as a "process of decline from an original ideal state, [and] hardly more than a catalog of the betrayal of fundamental principles" (p. 18). They conceptualize human history as a "cosmic struggle between good and evil": the good results from one's dedication to principles outlined in sacred scriptures, and the evil is an outcome of countless digressions from sacred principles. To fundamentalists, truth is not a relative phenomenon; it does not vary across time and place. Instead, truth is unchanging and knowable through the sacred texts.

Third, fundamentalists do not distinguish between the sacred and the profane in their day-to-day lives. Religious principles govern all areas of life, including family, business, and leisure. Religious behavior, in their view, does not take place only in a church, a mosque, or a temple.

Fourth, fundamentalist religious groups emerge for a reason, usually in reaction to a perceived threat or crisis, whether real or imagined. Consequently, any discussion of a particular fundamentalist group must include some reference to an adversary.

Fifth, one obvious concern for fundamentalists is the need to reverse the trend toward gender equality, which they believe is symptomatic of a declining moral order. In fundamentalist religions, women's rights often become subordinated to ideals that the group considers more important to the well-being of the society, such as the traditional family or the right to life. Such a priority of ideals is regarded as the correct order of things.

ISLAMIC FUNDAMENTALISM   In *The Islamic Threat: Myth or Reality?* professor of religious studies John L. Esposito maintains that most Americans' understanding of fundamentalism does not apply very well to contemporary Islam. The term *fundamentalism* has its roots in American Protestantism and the twentieth-century movement that emphasizes the literal interpretation of the Bible. Fundamentalists are portrayed as static, literalist, retrogressive, and extremist. Just as we cannot apply the term *fundamentalism* to all Protestants in the United States, so too we cannot apply it to the entire Muslim world, especially when we consider that Muslims make up the majority of the population in at least 45 countries.

Islamic fundamentalism includes the idea that it is wrong to lend money and require interest to be paid. Recently, Islamic banks have found ways to attract money that they can then make available for investment without breaking Islamic law. For example, depositors to the bank can be regarded as owning shares in the bank and receive a legitimate dividend on their ownership. To accommodate Muslims, some globally conscious non-Islamic banks now offer similar arrangements.

© V. Riviere/Sygma

Esposito believes that a more apt term is **Islamic revitalism** or *Islamic activism*. The form of Islamic revitalism may vary from one country to another, but seems to be characterized by the following themes:

A sense that existing political, economic, and social systems had failed; a disenchantment with and at times a rejection of the West; a quest for identity and greater authenticity; and the conviction that Islam provides a self-sufficient ideology for state and society, a valid alternative to secular nationalism, socialism, and capitalism. (Esposito 1992:14)

In "Islam in the Politics of the Middle East," Esposito (1986) asks, "Why has religion [specifically Islam] become such a visible force in Middle East politics?" He believes that Islamic revitalism represents a "response to the failures and crises of authority and legitimacy that have plagued most modern Muslim states" (p. 53). Recall that after World War I, France and Britain carved up the Middle East into nation-states, drawing the boundaries so as to meet the economic and political needs of Western powers. Lebanon, for example, was created in part to establish a Christian tie to the West; Israel was envisioned as a refuge for persecuted Jews when no country seemed to want them; the Kurds received no state; Iraq became virtually landlocked; and resource-rich territories were incorporated into states with very sparse

populations (Kuwait, Saudi Arabia, the Emirates). Their citizens viewed many of the leaders who took control of these foreign creations "as autocratic heads of corrupt, authoritarian regimes that [were] propped up by Western governments and multinational corporations" (p. 54).

When Arab armies from six states lost "so quickly, completely, and publicly" to Israel in 1967, Arabs were forced to question the political and moral structure of their societies (Hourani 1991:442). Had the leaders and the people abandoned Islamic principles or deviated too far? Could a return to a stricter Islamic way of life restore confidence to the Middle East and give it an identity independent of the West? Questions of social justice also arose. Oil wealth and modernization policies had led to rapid increases in population and urbanization and opened up a vast chasm between the oil-rich countries, such as Kuwait and Saudi Arabia, and the poor, densely populated countries, such as Egypt, Pakistan, and Bangladesh. Western capitalism, which was seen as one of the primary forces behind these trends, seemed blind to social justice, instead promoting unbridled consumption and widespread poverty. Likewise, Marxist socialism (a godless alternative) had failed to produce social justice. It is no wonder that the Taliban and other Muslim groups in Afghanistan reject Western capitalism and Marxist socialism because the disintegration of Afghanistan is a product of the Cold War between the United States and the former Soviet Union.

For many people, Islam offers an alternative vision for society. According to Esposito (1986), five beliefs guide Islamic activists (who follow many political persuasions, ranging from conservative to militant):

1. Islam is a comprehensive way of life relevant to politics, state, law, and society.

2. Muslim societies fail when they depart from Islamic ways and follow the secular and materialistic ways of the West.

3. An Islamic social and political revolution is necessary for renewal.

Islamic revitalism "A sense that existing political, economic, and social systems had failed; a disenchantment with and at times a rejection of the West; a quest for identity and greater authenticity; and the conviction that Islam provides a self sufficient ideology for state and society, a valid alternative to secular nationalism, socialism, and capitalism" (Esposito 1992:14).

Depicted here are American Muslims at prayer. How does this picture compare with the images of Islam that you are likely to see in the mass media?

© John Van Hasselt/Sygma

4. Islamic law must replace Western-inspired or Western-imposed laws.

5. Science and technology must be used in ways that reflect Islamic values to guard against the infiltration of Western values.

Muslim groups differ dramatically in their beliefs about how quickly and by what methods these principles should be implemented. Most Muslims, however, are willing to work within existing political arrangements; they condemn violence as a method of bringing about political and social change.

The information presented in this section points out the complex interplay of religion with political, economic, historical, and other social forces. Fundamentalism cannot be viewed in simple terms; rather, any analysis must consider the broader context. A focus on context allows us to see that fundamentalism can represent a reaction to many events and processes, including secularization, foreign influence, failure or crisis in authority, the loss of a homeland, and rapid change.

Summary and Implications

In this chapter, our examination of sociological concepts and perspectives showed that the civil wars in Afghanistan and the recent initiatives by the Taliban cannot be blithely reduced to the actions of fanatics acting from primitive and irrational religious convictions. Religion is important in understanding the nature of these civil wars, but its importance in elucidating their causes depends on understanding how religion is used rather than on knowing the religious groups to which the majority in each faction belongs.

Durkheim's ideas about the nature of religion, despite their shortcomings, allow us to consider religion's functions for the individual and the group. Most importantly, he viewed religion as a rich and seemingly endless variety of responses to problems of human existence. People embrace religion in the face of uncertainty to cope with misfortunes and injustices, when they have a great emotional investment in securing a successful outcome, and in an effort to find answers to agonizing questions about the meaning of life and death.

In addition, religion contributes to group unity and solidarity. Whenever members of a group have a strong conviction (no matter what kind of group and no matter what the conviction), the conviction almost always takes on a religious character. Religious gatherings and

affiliations become ways of affirming beliefs and mobilizing the group members to uphold their beliefs, especially if they are threatened. Religion can then unite a community of worshipers so strongly that they would be willing to destroy those who do not share their views. The convictions that religion is called on to legitimate may be either honorable or unprincipled.

Karl Marx was concerned about religion's repressive, constraining, and exploitative qualities. He believed that religious doctrines can be used to turn people's attention away from unjust political and economic arrangements and that they rationalize and defend the political and economic interests of the dominant classes. For example, those in political power may use religion to unite the society in war against another society or to dominate in some other way. Marx, however, ignored the possibility that the oppressed might use religion as a vehicle by which to protest or to work to change the existing social and economic inequities.

By outlining the role that religious beliefs played in the origins and development of modern capitalism, Max Weber alerts us to yet another way in which religion affects economic life. At the same time, Weber argued that once established, capitalism would generate its own value-rational logic, which in turn would make religion less relevant to economic activity. (See Chapter 7 for a discussion of value-rational action.) In fact, the advances in science and technology that accompany modern capitalism cause society to become secularized. Ironically, the same forces that support secularization processes also support the rise of fundamentalism.

The many uses to which religion are turned show that religion is a multifaceted and complex phenomenon that cannot be discussed in isolation. To under-

stand the wars in Afghanistan, we must understand the larger social, economic, and political context. With this information in hand, we can examine how religion shapes people's behavior and how it legitimates their actions. We can also see how "civil religion" motivated the United States and the Soviet Union to fight for their respective convictions in the same way that Islamic principles formed the foundation of the *mujahidin's* actions.

Key Concepts

Use this outline to organize your review of the key chapter ideas.

Religion
  Church
  Ecclesiae
  Denominations
  Sects
  Established sects
  Cults
Civil religion
Rituals
Sacred
  Sacramental religions
  Prophetic religions
  Mystical religions

Profane
Fundamentalism
Secularization
  Objective secularization
  Subjective secularization
Modern capitalism
  Predestination
  This-worldly asceticism
Class society

internet assignment

Find the Web site of the International Religious Foundation, an organization dedicated to promoting world peace through religious dialogue and cooperation. Select "Introduction" for more information on the project's purpose. The foundation has posted the scriptures of several of the major world religions indicating how each religion views (1) ultimate reality, (2) divine law, truth, and cosmic principles, (3) the purpose of human life, (4) life beyond death, and (5) the human condition—in addition to 16 other topics.

Also, find the "Description of Fifty-Seven Religions, Faith Groups, and Ethical Systems" maintained by the Ontario Con-

sultants on Religious Tolerance, an organization whose aims are to promote tolerance of minority religions, offer useful information on controversial religious topics, and expose hatred and misinformation about any religion.

Compare and contrast the views of Islam with those of one or two other religions. Are there any differences in views or principles that would represent grounds for violent conflict? Explain. What does your finding say about the role of religion in conflicts?

16

Social Change

With Emphasis on the Internet

Woman with laptop computer. (© David Lissy/The Picture Cube, Inc.)

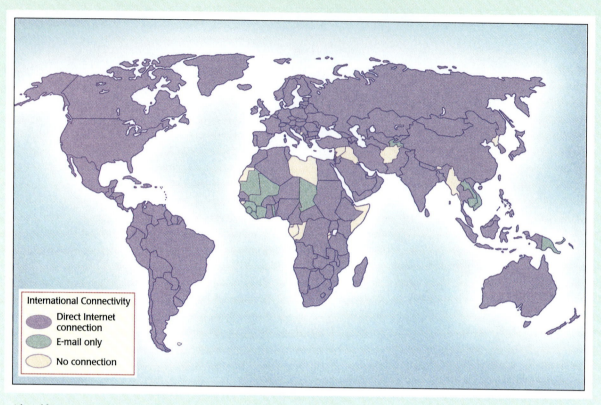

Adapted from a map Copyright ©1995 by Larry Landweber and the Internet Society. Unlimited permission to copy or use is granted subject to inclusion of the copyright notice.

How Worldwide Is the Web?

This map gives an idea of how "worldwide" the World Wide Web actually was as of June 1997, according to a copyrighted report on the Internet from Larry Landweber and the Internet Society. Countries shown in purple had Internet access via telephone and other media, such as the "backbone" data transmission lines and relay systems that now connect major computing centers around the world. Green countries had e-mail only.

Yellow countries had neither telephone e-mail nor access to the Internet. Of course, within each country that had connections, some regions and people had more and better connections than others.

This map focuses on power and opportunity in a time of rapid worldwide social change. This chapter considers the sociological study of social change in the context of technological innovation. It examines the vast changes at work today whose ultimate effects on people and the organization of social life we must work hard to foresee and understand.

One advantage of the Internet over the print world is that it allows authors to update information whenever they see fit. For example, you may be able to get an update to the data shown on the map at http://www.isoc.org.

Consider the following newspaper and magazine headlines:

"Third-World Pioneers Use Net to Promote Business"

"Internet Is Becoming an Essential Tool"

"Internet Snares First Criminal as FBI Traps Bank Robber in the Web"

"Teachers Explore Online Options"

"More Michigan Colleges Bring the Classroom to Cyberspace"

"Parishioners Flock Back to Fold as Some Churches Evangelize On-Line"

"Judge Turns Back Laws to Regulate Internet Decency"

"Are We Creating Internet Introverts?"

Most of us have seen or heard these kinds of headlines, and even if we have not accessed the Internet, we have heard many things about its potential and peril. Taken together, such headlines suggest that the Internet "is coming straight at us with the speed and momentum of a locomotive or at full steam. We can either prepare to jump on board or be mowed down by it" (Crawford 1995). Sociology offers a framework that can help us conceptualize these changes and place them in a larger social and historical context.

Why Focus on the Internet?

This chapter considers social change by focusing on the Internet—a technology that is expected to set into motion social, economic, and political changes equivalent to those revolutionary changes triggered by the printing press. The **Internet** is a vast network of computer networks connected to one another via special software and phone, fiber-optic, or other type of lines that has the potential to connect millions, even billions, of computers and their users on a global scale. It also has the potential to provide users with access to every word, image, or sound that has ever been written, produced, filmed, photographed, painted, or otherwise recorded (Berners-Lee 1996).

Although it is virtually impossible to catalog the myriad changes associated with (and expected to be associated with) the Internet, we can say that, for those who can afford it, the Internet will (1) speed up old ways of doing things, (2) give individuals access to the equivalent of a printing press, (3) allow people to bypass the formalized hierarchy devoted to controlling the flow of information, (4) change how students learn, and (5) provide a vehicle by which the lives of people around the world will become more intertwined as they seek to exchange information and mobilize to define and solve a wide range of social problems (see "Five Challenges to Build a Global Information Infrastructure"). In this regard, the Internet facilitates both "globalization-from-above" and "globalization-from-below."

Globalization-from-above connects those people around the world with educational, economic, and political advantages, excluding those who are not so advantaged. **Globalization-from-below,** however, involves interdependence at the grassroots level that aims to protect, restore, and nurture the environment; to enhance ordinary people's access to the basic resources they need to live a dignified existence; to democratize local, national, and transnational political institutions; and to ease tensions and prevent violent conflict between power centers and authority structures (Brecher, Childs, and Cutler 1993).

Five Challenges to Build a Global Information Infrastructure

Minneapolis, MN—In a speech before the United Nations' chief telecommunications organization, Vice President Gore challenged delegates representing more than 180 nations to use our newest technologies to preserve our oldest values.

"Four years ago, I asked you to help create a global information superhighway," Vice President Gore said. "Today, I thank you for what you have done to bring about the most stunning revolution the world has known, and I challenge you to build on this unprecedented opportunity by putting these new global networks to work helping people."

"Today, we can build on our progress and use these powerful new forces of technology to advance our oldest and most cherished values: to extend knowledge and prosperity to the most isolated inner cities at home, and the most remote rural villages around the world; to bring twenty-first-century learning and communication to places that don't even have phone service today; to share specialized medical technology that can save and improve lives; to deepen the meaning of democracy and freedom in this Internet age," he said.

The Vice President proposed five new challenges, which he characterized as a "Declaration of Interdependence."

First, he challenged the world community to improve access to technology so everyone on the planet is within walking distance of basic telecommunication services by the year 2005. For all our progress, 65% of the world's households still have no phone service.

Second, he challenged the world community to bridge language barriers by developing technologies with real-time digital translation so anyone on the planet can talk to anyone else. Such technologies could reduce the cost of doing business and increase international cooperation.

Third, he challenged the world community to create a global knowledge network of people working to improve the delivery of education, health care, agricultural resources, and sustainable development, and to ensure public safety. The Vice President challenged the education community to link together practitioners, academic experts, and not-for-profit organizations working on our most pressing social and economic needs.

Fourth, he challenged the world community to ensure that communica-

tions technology protects the free flow of ideas and supports democracy and free speech. We must continue to work to ensure that the global information infrastructure (GII) promotes this goal.

Fifth, he challenged the world community to create networks that allow every micro-entrepreneur in the world to advertise, market, and sell products directly to the world market. Such networks will enable entrepreneurs to keep more profits, provide information about world prices, develop technology as a business tool, increase the diversity of the global marketplace, and create jobs.

Additionally, the Vice President called on the world community to address the year 2000 computer problem, which, if not addressed, could pose serious problems for commerce and communications all over the world.

"We must ensure that the international system is ready for the year 2000—because one weak link in the system will hurt us all," Vice President Gore said. "Together, we must solve this problem."

Source: White House (1998).

Internet A vast network of computer networks connected to one another via special software and phone, fiber-optic, or other type of lines with the potential to connect millions, even billions, of computers and their users on a global scale.

Globalization-from-above A term describing the Internet's ability to connect those people around the world with educational, economic, and political advantages, excluding those who are not so advantaged.

Globalization-from-below Interdependence at the grassroots level that aims to protect the environment, enhance ordinary people's access to basic resources, democratize political institutions, and ease tensions and prevent violent conflict between power centers and authority structures.

Social change Any significant alteration, modification, or transformation in the organization and operation of social life.

Sociologists define **social change** as any significant alteration, modification, or transformation in the organization and operation of social life. Social change is an important topic in the discipline of sociology. In fact, sociology emerged as a discipline attempting to understand social change. Recall that the early sociologists were obsessed with understanding the nature and consequences of the Industrial Revolution—an event that triggered dramatic and seemingly endless changes in every area of social life. Over time, the discipline of sociology has evolved to encompass a wide range of rich concepts and theories that can be applied to the study of social change.

When sociologists study change, they first must identify the aspect of social life they wish to study that has changed or is undergoing change. The list of possible topics is virtually endless. Examples include changes in the division of labor; changes in the meaning of what constitutes deviance; changes in how people communicate; changes in the amount of goods and services that people produce, sell, or buy from others; and changes in the size of world population and in the average life span (Martel 1986). Upon identifying a topic, sociologists ask at least two key questions: What factor(s) cause that change? What are the consequences of that change for social life?

In this chapter, we examine the concepts and theories that sociologists use to answer these complex questions. We use the Internet as an example—specifically, some of the historical events that led to the invention of the Internet and some of the consequences of that invention for social life.

Causes and Consequences of Social Change

When we think about the factors that cause a specific social change, we usually cannot pinpoint a single factor as the source of an identified change. More often than not, change results from a sequence of events. An analogy may help clarify this point. Suppose that a wide receiver, after catching the football and running 50 yards, is tackled at the five-yard line by a cornerback. One could argue that the cornerback caused the receiver to fall to the ground. Such an account, however, would not fully explain what happened. For one thing, a tackle is not the act of one person; "there is present a simultaneous conflict of forces" (Mandelbaum 1977:54) between the tackler (who is attempting to seize and throw his or her weight onto the person with the ball) and the wide receiver (who is doing everything possible to elude the tackler's grasp). To complicate matters, the wide receiver and the tackler are each members of a team, and their teammates' actions help determine how the play develops and ends. Similarly, as we will learn in this chapter, the Internet is a product of a seemingly endless sequence of interrelated events, beginning most directly with the Cold War.[1]

Even though the story of the Internet is embedded within a virtually endless stream of interrelated events and consequences, we can nevertheless identify some key types of factors that resulted in the Internet's invention and the subsequent changes that this information technology triggered and has the potential to trigger: conflict, pursuit of profit, innovations, and revolutionary ideas. Sociologists define these factors as some of the most important agents and outcomes of change in general. Thus their importance for understanding social change goes beyond the Internet.

Conflict

As sociologist Lewis Coser (1973) points out in his essay "Social Conflict and the Theory of Social Change," conflict will always exist, if only because we never achieve a perfect "concordance between what individuals and groups within a system consider their just due and the system of allocation" (p. 120). Conflict occurs whenever a group takes action to increase its share of or control over wealth, power, prestige, or some other valued resource and when these demands are resisted by those who benefit from the current distribution system. In other words, those who gain control of valued resources strive to protect their own interests against efforts by those who lack control but hope to gain a share.

This fact of life has motivated Michael and Rhonda Hauben (1996) to draft a Declaration of the Rights of Netizens,[2] which argues that access to the Internet should be considered a human right and not something reserved for a privileged elite. Among the rights (and obligations) they identify are the right of equal times and quality of connection and the obligation to consider ideas posted according to their merit (see "Who Controls the Internet?").

Conflict, whether it involves violent clashes or public debate, is both a consequence and a cause of change. In general, any kind of change has the potential to trigger conflict between those who benefit from the change and those who stand to lose because of it. When the bicycle was invented in the 1840s, for example, horse dealers organized against it because it threatened their livelihoods. Some physicians declared that people who rode bicycles risked getting "cyclist sore throat" and "bicycle stoop." Church groups protested that bicycles would swell the ranks of "reckless" women (because bicycles could not be ridden sidesaddle).[3]

Conflict can lead to change as well. It can be a constructive and invigorating force that prevents a social system from becoming stagnant, unresponsive, or inefficient. Conflict can create new norms, relationships, and ways of thinking—consider the results of the antinuclear, civil rights, and women's movements. In addition, it can generate new and efficient technologies.

Ironically, war has generated advances in lifesaving medical technologies. During World War I,

---

[1]Of course, a thorough analysis of all causes that led to the invention of the Internet would also include the events that triggered the Cold War.

[2]A *netizen* is a citizen of the Internet.

[3]I read about this resistance to the bicycle many years ago while teaching a Sociology of Sport course. Although the example is very memorable, I cannot remember the source.

Who Controls the Internet?

On one level, the Internet does put the individual (who can afford it) at the center of information in that he or she can access information quickly and directly, bypassing hierarchies devoted to managing and granting access to information. On another level, however, individual users may not be able to entirely control their access to information. It would be naive to believe that the Internet provides access to information that governments, corporations, and other organizations and agencies wish to keep confidential or secret. Cyberspace law professor Stuart Biegel (1996) discounts the widely held belief

that no one is in charge of the Internet. In fact, many persons and groups are "in charge" in some capacity. They include:

- The government, which provided funds to "invent" the Internet and sets policies that affect funding, growth, and access
- Internet access providers, which act as the gatekeepers in that users pay them a fee to access the Internet
- Communication companies investing in the Internet infrastructure, such as the providers of the high-speed lines that route information

- Companies that create the hardware and software used to access the Internet and that will shape its future direction and applications
- Colleges and universities, whose faculty develop programs and post large quantities of noncommercial information
- The Internet Society, which was chartered by the U.S. government "for the purpose of commenting on policies and overseeing other boards and task forces dealing with network policy" (Biegel 1996)

---

many soldiers fighting on manure-covered farmlands contracted tetanus.[4] In addition, large numbers of soldiers were injured by machine gun shrapnel and bombshells. Physicians experimenting with antitoxins eventually found a cure for tetanus and made considerable advances in reconstructive surgery. Similarly, the kinds of injuries incurred during World War II motivated doctors to create a system of collecting and preserving blood plasma and to mass-produce an effective drug, penicillin, to treat wound infections (Colihan and Joy 1984). The Internet is another example of a technology whose origins are rooted in a most destructive form of conflict—war.

The Cold War Origin of the Internet

As discussed in Chapter 12, the *Cold War* is the name given to the political tension and military rivalry that existed between the United States and the former Soviet Union from approximately 1945 until its symbolic end on November 9, 1989, when the Berlin Wall fell. The Cold War included an arms race—that is, an intense and ongoing buildup by the Soviet Union and the United States in which each side competed to match and surpass any advances made by the other in the quantity and technological quality of weapons on the grounds that it was necessary to contain the spread of the other side's economic and political system. Both

countries created vast, entrenched armaments industries; both made the research and development of weapons (especially nuclear weapons)[5] their highest priorities.

U.S. government leaders had pulled together scientists from three sectors—military, industrial, and academic—to coordinate their research and expertise and thereby serve the Cold War effort. Because these scientists worked in offices and laboratories located all around the United States, Defense Department officials worried about the consequences of an attack on a military laboratory, defense contractor, or university site. Those in power realized they needed a computer network that would allow information stored at one site to be transferred to another site in the event of an attack, especially a nuclear attack. At the same time, the computer network had to be designed such that if one or more parts of the network failed or were knocked out by a bomb, the other parts could continue to operate. Such a design meant that no central control could exist for the network. After all, if central control is destroyed, the entire network crashes. The Internet began in the late 1960s as ARPANET (Advanced Research Projects Agency), linking four universities: the University of California–Los Angeles (UCLA), the University of California–Santa Barbara, the University of Utah, and Stanford University. Thus the Internet was originally designed to (1) transfer information from one site to

---

[4]Tetanus is an infectious disease characterized by muscular spasms and difficulty in opening the mouth (lockjaw). The tetanus bacterium is dangerous because it produces a toxin that affects the heart and the breathing muscles.

[5]During the Cold War, the number of nuclear weapons grew to 50,000.

Historical factors can spur or retard the pace of innovation. The Cold War stimulated enormous U.S. and Soviet governmental support for science and technological research, including the "conquest of space." When the statues of Lenin came down in Europe (this one was in East Germany), one driving motivation for innovation also disappeared. Other factors, such as growing global competition in trade, keep the pressure on.

© Regis Bossu/Sygma

another quickly and efficiently in the event of war and (2) create an information-sharing system absent central control.

Whether conflict will lead to reform (improvements or alterations in current practices) or to revolution (complete and drastic change) depends on a broad range of factors and contingencies. Sociologist Ralf Dahrendorf has identified and described some of these factors in his classic essay, "Toward a Theory of Social Conflict" (1973).

Structural Origins of Conflict

In his essay, Dahrendorf asks two questions: What is the structural source of conflict, and what forms can conflict take? Dahrendorf's answers rest on the following assumptions. First, in every organization that has a formal authority structure (a state, a corporation, the military, the judicial system, a school system), clear dichotomies exist between those who control the formal system of rewards and punishment (and thus have the authority to issue commands) and those who must obey those commands or face the consequences (loss of job, jail, low grades, and so on). Second, a distinction between "us" and "them" arises naturally from the unequal distribution of power. Using these assumptions, we can trace the structural origins of conflict to the nature of authority relations. That conflict can assume many forms. It can be mild or severe; "it can even disappear for limited periods from the field of vision of a superficial observer" (p. 111). As long as an authority structure exists, however, conflict cannot be abolished.

Dahrendorf outlines a three-stage model of conflict in which progression from one stage to another depends on many things. He does not claim to elucidate all possible courses followed by conflict. In fact, he reminds readers that the conflicts he names are some of the most obvious. The point is that conflict—its course and its resolution—is a complicated phenomenon encompassing many elements.

In the first stage, every authority structure contains at least two groups with opposite interests. Those with power have an interest in preserving the system; those without power have an interest in changing it. These opposing interests, however, remain below the surface until the groups (especially those without power) organize. "It is immeasurably difficult to trace the path on which a person . . . encounters other people just like himself, and at a certain point . . . [says] 'Let us join hands, friends, so that they will not pick us off one by one'" (Dahrendorf 1973:240).

Often a significant event makes seemingly powerless people aware that they share an interest in seeing the system changed. Vaclav Havel, the president of the Czech Republic, believes that the Chernobyl incident may have played an important role in bringing about the revolutions in central Europe. Chernobyl, located in Ukraine, was the site of a nuclear power plant meltdown in 1986, the most serious kind of accident that can occur at a nuclear power plant. Although Ukraine, Russia, and Belarus suffered the most radioactive contamination, areas as far away as Sweden were affected as well. Havel maintains that after the Chernobyl meltdown, people in what was then Czechoslovakia dared to complain openly and loudly to one another (Ash 1989).

Sometimes, too, people organize because they have nothing left to lose. As one East German scientist explained:

You don't need courage to speak out against a regime. You just need not to care anymore—not to care about being punished or beaten. I don't know why it all happened this year [in 1989]. We finally reached the point where enough people didn't care anymore what would happen if they spoke out. (Reich 1989:20)

In the second stage, if those without authority have opportunities to communicate with one another, the freedom to organize, the necessary resources, and a leader, then they will organize. At the same time, those in positions of authority often use the power of their positions to censor information, restrict resources, and undermine leaders.

One technology that will certainly enhance potential opportunities for people to communicate with one another at the grassroots level is the Internet. The word *potential* is appropriate because, although Internet technology has helped activists organize and pressure authority structures, activists have not yet tapped its full potential with regard to "delivering the message" and causing structural impact (Afonso 1997). Activist organizations with Web pages on the Internet range from the American Civil Liberties Union to Voters Telecommunications Watch.

In the third stage, organizations composed of individuals without power enter into a state of conflict with those in power. The speed and the depth of change depend on the capacity of the ruling group to stay in power and on the kind and degree of pressure exerted from below. The intensity of the conflict can range from heated debate to violent civil war, but it is always contingent on many factors, including opportunities for mobility within the organization and the ability of those in power to control the conflict. If those who lack authority are confident that eventually they will achieve such a position, the conflict is unlikely to become violent or revolutionary (see "Public Citizen's Greatest Hits"). If those in power decide that they cannot afford to compromise and mobilize all their resources to thwart protests, two results are possible. First, the protesters may believe that the sacrifices are too great and may withdraw. Alternatively, the protesters may decide to meet the "enemy" directly, in which case the conflict becomes bloody.

To this point, we have discussed conflict in very general terms. Another important agent of change is a specific kind of conflict, motivated by the pursuit of profit that is characteristic of capitalism, an economic system whose modern origins can be traced back 500 years.

The Pursuit of Profit

Karl Marx believed that an economic system—capitalism—ultimately caused the explosion of technological innovation and the enormous and unprecedented increase in the amount of goods and services produced during the Industrial Revolution. In a capitalist system (discussed in Chapter 12), profit is the most important measure of success. To maximize profit, the successful entrepreneur reinvests profits so as to expand consumer markets and to obtain technologies that allow products and services to be produced at the highest quality and the greatest cost-effectiveness.

The capitalist system acts as a vehicle of change because the instruments of production must be revolutionized constantly. Marx believed that capitalism was the first economic system capable of maximizing the immense productive potential of human labor and ingenuity. He also believed, however, that capitalism ignored too many human needs, and that too many people could not afford to buy the products of their labor. That is, capitalism survived and flourished by "sucking the blood" of living labor. The drive for profit (which Marx maintained is derived from the labor of those directly involved in the production process) is a "boundless thirst . . . [a] werewolflike hunger . . . [that] takes no account of the health and the length of life of the worker unless society forces it to do so" (Marx, cited in Carver 1987:142). The thirst for profit "chases the bourgeoisie over the whole surface of the globe" ([1881] 1965:531). Marx's theories influenced a group of contemporary sociologists—world system theorists—to write about capitalism as the agent of change underlying global interdependence.

World System Theory

Immanuel Wallerstein (1984) is the sociologist most frequently associated with world system theory, a modern theory concerned with capitalism. Since the early 1970s, he has been writing a four-volume work (three volumes have been published to date) about the ceaseless expansion, over the past 500 years, of a single market force—capitalism. According to Wallerstein, although stagnant periods have occurred and some countries (the Communist countries, for example) have tried to withdraw from the capitalist economy, no real contraction has occurred.

Hence, by the late nineteenth century, the capitalist world-economy included virtually the whole inhabited earth and it is presently striving to overcome the

Public Citizen's Greatest Hits

Public Citizen, founded by Ralph Nader in 1971, is a consumer advocacy organization with 150,000 members. It stands up against "thousands of special interest lobbyists in Washington," including auto makers, drug companies, energy companies, and so on. It accepts no government or corporate support. The following is a list of some of its most important victories—all achieved without bloodshed.

Source: Public Citizen (1996).

1998   Public Citizen sues the FDA to release the research protocol for the banned diet drug, Redux.

1994   Public Citizen successfully challenges the FDA's failure to restrict use of the drug Parlodel.

1993   Public Citizen lobbies successfully to end funding for the advanced breeder reactor.

1992   Public Citizen helps defeat an industry-backed bill that would restrict the use of the courts by consumers injured by defective products.

1991   Public Citizen publishes *They Love to Fly and It Shows,* a study of lobbyist-funded House travel during 1989–1990.

1990   Public Citizen publishes an account of the savings and loan scandal, entitled *Who Robbed America?*

1989   A regulation requiring air bags or passive seat belts in all cars takes effect after a 20-year battle.

1988   Public Citizen publishes *Worst Pills, Best Pills,* a guide to dangerous drugs for the elderly. By 1995, almost 2 million copies are sold.

1987   Public Citizen forces the chemical companies that produced Agent Orange to make all relevant documents public.

1986   Congress passes a law requiring health warnings on the labels of snuff and chewing tobacco, capping Public Citizen's two-year campaign.

1985   After a four-year battle by Public Citizen, the FDA requires aspirin labels to warn of the risk of Reye's syndrome.

1984   Public Citizen helps block legislation that would make it easier for manufacturers to use cancer-causing food additives.

1983   *Over the Counter Pills That Don't Work* is published.

1982   Oraflex, an arthritis drug that caused dozens of deaths and hundreds of injuries, is withdrawn from the market after Public Citizen's campaign.

1981   Public Citizen helps thwart President Reagan's attempt to bury the Consumer Product Safety Commission in the Commerce Department.

1980   Public Citizen plays a critical role in passage of the Superfund law to require cleanup of toxic waste dumps.

1979   In response to a 1977 petition by Public Citizen, the EPA bans DBCP, a pesticide that causes sterility in human males and cancer in laboratory animals.

1978   Congress passes the National Consumer Cooperative Bank bill, drafted by Public Citizen, providing $300 million in "seed money" for consumer cooperatives.

1977   Public Citizen mobilizes citizens, persuading President Carter to stop construction of the Clinch River breeder reactor.

1976   The FDA bans red dye no. 2, after Public Citizen's four-year campaign against the carcinogenic food dye.

1975   Public Citizen obtains a court order requiring the Labor Department to reconsider its inadequate rules to protect workers from 14 carcinogenic chemicals.

1974   Critical Mass Energy Project is founded to mobilize opposition to nuclear power.

1973   Congress Watch, the legislative lobbying arm of Public Citizen, is formed.

1972   The Litigation Group and the Tax Reform Research Group are founded.

1971   Public Citizen is founded by Ralph Nader.

technological limits to cultivating the remaining corners; the deserts, the jungles, the seas, and indeed the other planets of the solar system. (p. 165)

Wallerstein distinguishes between the terms *world economy* and *world-economy.* People who use the term **world economy** (without the hyphen) envision the world as consisting of 160 or so national economies that have established trade relationships with one another. In this vision, globalization is portrayed as a relatively new phenomenon and as a process in which the countries of the world have moved from relatively isolated, self-sufficient economies to economies that trade with one another to varying degrees. Although popular, this

> **World economy** An economy consisting of the 160 or so national economies that have established trade relationships with one another.

Computers started out as room-sized machines invented to perform mathematical calculations for the Department of Defense and the Census Bureau. Today the processing power needed to perform a variety of tasks can be packed into spaces that can fit on a person's lap or the dashboard of a car.

© Bettmann Archive (left); © Mark Richards/PhotoEdit (right)

conception of global interdependence is not very accurate. The more accurate term (and conception) is *world-economy*. The **world-economy** (with a hyphen) is not recent; it has been evolving for at least 500 years and continues to evolve today. People who use the hyphenated term envision a world (encompassing hundreds of countries and thousands of cultures) interconnected by a single division of labor. In the world-economy, economic transactions transcend national boundaries. Although the world-economy is not new, the giant multinational corporations of today "could not exist at their present scale if there were no computers. Computers are their global nervous systems; their way of keeping track of their billions of moving parts, keeping them synchronized and moving in the same direction for control purposes" (Mander 1997). Although each government seeks to shape the global market in ways that benefit its "own" corporations and national interests, no single political structure (world government) or national government has authority over the entire system of production and distribution.

The Role of Capitalism in the Global Economy

How has capitalism come to dominate the global network of economic relationships? One answer lies in the ways in which capitalists respond to changes in the

World-economy An economy encompassing hundreds of countries and thousands of cultures interconnected by a single division of labor.

economy, especially to economic stagnation. Historically, five important responses have been attempted, all designed to create economic growth. Here we will consider how the computer and the Internet affect those responses.

1. *Lowering production costs* by hiring employees who will work for lower wages (for example, by busting unions, buying out workers' contracts, or offering early retirement plans), by introducing labor-saving technologies (such as computerizing the production process), by moving production facilities out of high-wage zones and into lower-wage zones inside or outside the country, or by contracting services to be delivered via the Internet. For example, some large U.S. corporations hire computer programmers in India to write software and send it to them via the Internet (Associated Press 1996).

2. *Creating a new product that consumers "need" to buy* (such as the videocassette recorder, the computer, or the fax machine—products introduced in the late 1970s and early 1980s). In 1996, Sony and Fujitsu introduced "Web TV" with the aim of allowing Internet users to surf or browse the Web from an easy chair.

3. *Improving on an existing product and thus making previous versions obsolete.* Every improvement in a personal computer's central processing unit (CPU) makes the previous CPU obsolete. As one indicator of the speed at which a "new" CPU becomes obsolete, consider that the author used a 1992 reference book to research CPUs before writing the previous sentences. At the time of the book's print-

ing (Bear and Pozerycki 1992), the highest-grade CPU was a 386 model. Seven years later, four versions of a 486 and the Pentium I and II have appeared on the market, with an upgrade of the Pentium II due out in 1999 (Schwartz 1996).

4. *Expanding the outer boundaries of the world-economy and creating new markets.* The commercialization of the Internet, beginning in the mid-1990s, represents the attempt to expand the boundaries of the world-economy and/or establish new markets. As of mid-1996, an estimated 12.8 million hosts were active on the Internet; 25.8 percent were commercial hosts (Kantor and Neubarth 1996). The commercial Web sites describe a range of products, services, and entertainment activities, ranging from antihistamines to zoos.

5. *Redistributing wealth to enable more people to purchase products and services.* Henry Ford was the first to attempt this tactic on a large scale; in 1908, he came up with the revolutionary concept of paying workers a wage ($5 per day) large enough to allow them to purchase the products of their labor (Halberstam 1986). Because of the shortage of hard currency in Russia, the new states of Eurasia, and Eastern Europe, American, Western European, and Japanese executives have set up barter systems there (Holusha 1989). "Pepsico, for example, exports wooden chairs from Poland to its Pizza Hut franchises in the United States, and sells its soft drink to the Soviet Union in exchange for old submarines" (O'Sullivan 1990:22). Another example of this strategy is IBM's Global Campus, which is designed to help colleges and universities put their "campuses" online. As of early 1997, IBM had enrolled the 23-campus California State University system, 21 other U.S. colleges and universities, and a number of schools outside the United States. The purpose of the program is to help schools with start-up costs and to enhance existing distance-learning programs (and, not incidentally, to compete with its rival, Apple). The universities that have signed up with this pilot project have purchased or are considering future purchases of elements of the IBM Global Campus package (see Figure 16.1).

As a result of these five responses to economic stagnation, capitalism has spread steadily to encompass the globe and facilitate globalization-from-above. In addition, every country of the world has come to play one of three different and unequal roles in the global economy: core, peripheral, and semiperipheral (see Chapter 12). Their unequal roles in the global economy are also reflected in Internet access.

The chapter-opening map identified the "have" and "have not" countries of the world in terms of Internet access. If we use the number of telephones per 100 people as a rough indicator of the potential of a country's population to connect to the Internet, we can see that access reflects that country's role in the global economy (see "How Connected Are We?"). Obviously, the number of telephones speaks only to potential access as only some 30 percent of all people in the United States have access to a computer through home, work, or school.[6] In peripheral and semiperipheral economies, 85 percent or more of the people do not have a telephone. Thus, only a privileged elite have access to computers, and even for this group poor and intermittent phone service constrains Internet use. The commercial uses of the Internet by peripheral and semiperipheral economies reinforce and perpetuate unequal roles in the global economy. For example, the Internet is used to promote the tourist industry, attract jobs, sell the few products the country exports, and solicit contributions and other forms of aid.

Innovations

Innovation is the invention or discovery[7] of something new—an idea, a process, a practice, a device, or a tool. Innovations can be classified broadly in one of two categories—basic or improving. The distinction between the two is not always clear-cut.

Basic innovations are revolutionary, unprecedented, or ground-breaking inventions that represent the cornerstones for a wide range of applications.

---

[6]Not everyone, however, with access to a computer has access to a computer with a modem. A U.S. Census Bureau survey found that in 1993 (the last year for which data are available), only 35.3 percent of households with a computer had a telephone modem.

[7]A **discovery** is the uncovering of something that had existed before but had remained hidden, unnoticed, or undescribed.

Discovery The uncovering of something that had existed before but had remained hidden, unnoticed, or undescribed.

Innovation The invention or discovery of something new—an idea, a process, a practice, a device, or a tool.

Basic innovations Revolutionary, unprecedented, or ground-breaking inventions that represent the cornerstones for a wide range of applications.

Figure 16.1   IBM's Global Campus

This map shows campuses enrolled in the IBM Global Campus program as of early 1997. (For an update, try the Internet source of the data or search on "IBM Global Campus.") How would you explain the initially high concentration of campuses in California? Where in the United States and abroad would you expect more campuses to join soon? Is the program truly "global" now? If not, why does IBM call it "global"? How global is it likely to become? Why?

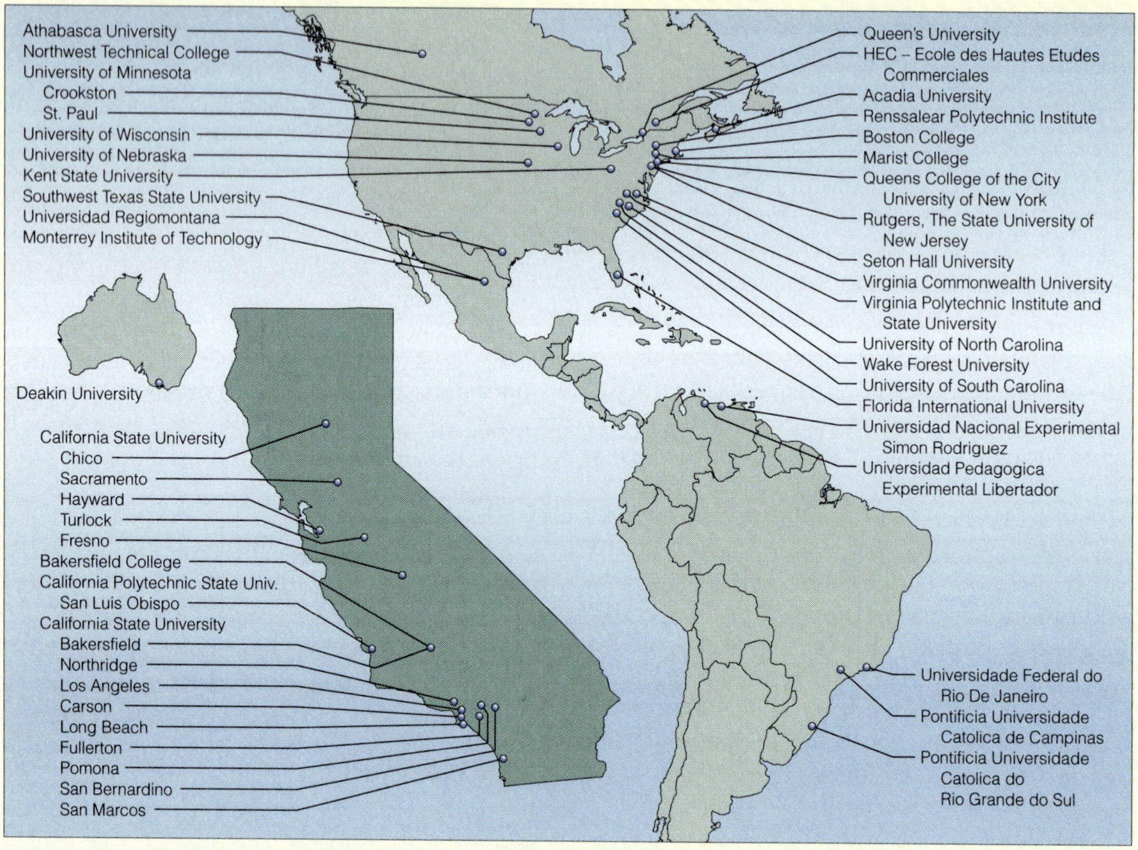

Athabasca University
Northwest Technical College
University of Minnesota
  Crookston
  St. Paul
University of Wisconsin
University of Nebraska
Kent State University
Southwest Texas State University
Universidad Regiomontana
Monterrey Institute of Technology

Deakin University

California State University
  Chico
  Sacramento
  Hayward
  Turlock
  Fresno
Bakersfield College
California Polytechnic State Univ.
  San Luis Obispo
California State University
  Bakersfield
  Northridge
  Los Angeles
  Carson
  Long Beach
  Fullerton
  Pomona
  San Bernardino
  San Marcos

Queen's University
HEC – Ecole des Hautes Etudes
  Commerciales
Acadia University
Rensselaer Polytechnic Institute
Boston College
Marist College
Queens College of the City
  University of New York
Rutgers, The State University of
  New Jersey
Seton Hall University
Virginia Commonwealth University
Virginia Polytechnic Institute and
  State University
University of North Carolina
Wake Forest University
University of South Carolina
Florida International University
Universidad Nacional Experimental
  Simon Rodriguez
Universidad Pedagogica
  Experimental Libertador

Universidade Federal do
  Rio De Janeiro
Pontificia Universidade
  Catolica de Campinas
Pontificia Universidade
  Catolica do
  Rio Grande do Sul

Source: http://ike.engr.washington.edu/igc/affiliate/html (1997).

Examples of basic innovations include the first mainframe computer, the first PC (personal computer),[8] computer languages, modems, disks, printers, and scanners.

Improving innovations, on the other hand, represent modifications of basic inventions that improve upon the originals—that is, to make them smaller,

Improving innovations Modifications of basic inventions that improve upon the originals—that is, to make them smaller, faster, less complicated, or more efficient, attractive, or profitable.

faster, less complicated (that is, user-friendly), or more efficient, attractive, or profitable. Each upgrade of a personal computer's CPU is an improving innovation that increases the power and speed of the machine in handling tasks such as making calculations, supporting software applications, and processing information.

Innovations are sociologically significant because they change the ways in which people think, solve problems, and relate to one another. For example, the

[8]The name of the first mainframe was ENIAC (Electronic Numerical Integrator and Computer). The name of the first PC was Altair (Noack 1997).

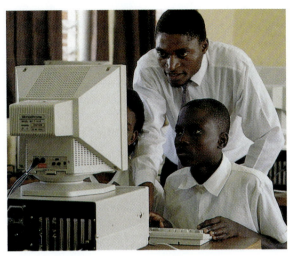

Only a small portion of the world's population has access to a computer. An even smaller share has access to the Internet.

© Chris Sattlberger/Panos Pictures

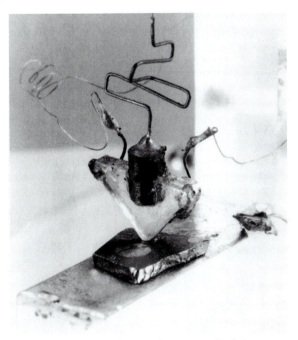

This first semiconductor (transistor) is an example of a basic innovation in that it became the cornerstone for a wide range of applications.

Property of AT&T Archives. Reprinted with permission of AT&T.

Internet has the potential to affect almost every area of life, including friendship, job hunting, finding an address, reading the news, adopting, and finding an apartment. The following descriptions of Web sites illustrate these broad capabilities.

- America's Job Bank

  http://www.ajb.dni.us/index/html

  America's Job Bank makes it easy to post your resume on the Internet, search for jobs according to category and geographic location, and look for job market information (that is, geographic profiles and occupational trends).

- Interactive Atlas

  http://www.mapquest.com

  Mapquest posts the Interactive Atlas, an on-line service that allows users access to county- and city-level maps from six continents. You can enter a street address along with its city, state, and ZIP code and access a map of that address and surrounding streets and landmarks. A one-time free registration is required, which allows you to save and store maps for later use. A graphical browser is required to access this site.

- CReAte Your Own Newspaper (CRAYON)

  http://crayon.net/

  CRAYON is a service maintained by Pressence Incorporated for managing news. After completing a free registration, users select the news sources from which they want to draw information. News sources are available at the international (such as *This Week in Germany*), national (such as *USA Today Nationwide*), and local (such as *Detroit News*) levels. Users may select information to be taken from specific sections, such as Sports or Weather, of the chosen papers. Users assign a name and motto to their paper and are given a uniform resource locator (URL) address where they can read their paper daily. A graphical browser is required to access this site.

- CyberFriends

  http://www.cyberfriends.com

  The purpose of CyberFriends is to provide an easy way for people sharing similar interests and professions to meet on the Internet. You simply fill out an application, which asks approximately 18 optional and required questions (age, sex, nationality, hobbies, personal philosophy, and so on). In two or three weeks, your application will be posted free of charge. In the meantime, you can search CyberFriends' listings for profiles and e-mail addresses for the names of people with whom you might like to correspond.

How Connected Are We?

This map shows about how many telephones there are per 1,000 people in each country, which provides a rough indicator of the potential for a country's general population to connect with the Internet. If and when the Internet can be accessed via cellular phones, poor countries may be able to reduce the infrastructure costs of connection that comes with traditional wire connections—one less barrier to access.

In the United States, economic class is reflected in unequal access: 24 percent of households with an annual income of less than $5,000 do not have access to telephone service, whereas almost all households with annual incomes of more than $20,000 have access. The data also show large differences in access to phones when households are classified as black, Hispanic, or white.

Suppose you were the Secretary of Communications Development for one of the poor, largely agricultural countries on the map on page 398. What sort of communications technology would be your first priority for the use of larger numbers of people? Would it be the telephone? Something else? If you wanted to dramatically increase the number of people with ready access to a telephone line, how might you try to make that happen? Are there certain groups of people that you would especially want to assist with telephone access? Who and why?

Sources: United Nations Development Programme (1993); U.S. telephone access data from *Statistical Abstract of the United States* (1995).

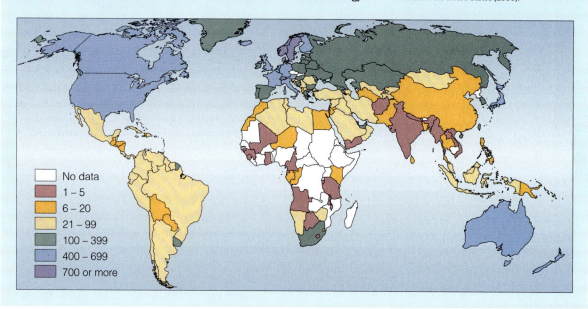

Legend:
- No data
- 1 – 5
- 6 – 20
- 21 – 99
- 100 – 399
- 400 – 699
- 700 or more

- Faces of Adoption

  http://www.ggw.org/cap

  Children Awaiting Parents (CAP) posts photographs and descriptions of children awaiting adoption. This independent, not-for-profit organization is dedicated to "uniting America's Children who have special challenges with adoptive families."

- Rent•net (Online Apartment Guide)

  http://www.rent•net

Cultural base The number of existing inventions.

This Web site allows apartment hunters to shop for apartments (unfurnished and furnished) in the 50 states, Canada, and 5 other foreign countries. Its database includes 1,157,000 apartment units. This site also offers a cost-of-living comparison between the current city and the destination city. In addition, it provides information on self-storage facilities and other relocation services.

The Cultural Base and the Rate of Change

Anthropologist Leslie White (1949) maintained that once a basic or an improving innovation has been invented, it becomes part of the **cultural base**, which he defines as the number of existing inventions. The size of the cultural base determines the rate of change.

White defined an **invention** as the synthesis of existing inventions. For example, the first airplane was a synthesis of many preexisting inventions, including the gasoline engine, the rudder, the glider, and the wheel. PocketNet, a product of AT&T's Wireless Data Division, combines cell-phone, e-mail, and fax technology to create a device that allows owners to send and receive e-mail and redirect messages to a fax machine (Pulver 1997).

White suggested that the number of inventions in the cultural base increases geometrically—1, 2, 4, 8, 16, 32, 64, and so on. (Geometric growth is equivalent to a state of runaway expansion.) He argued that if a new invention is to emerge, the cultural base must be large enough to support it. If the Wright brothers had lived in the fourteenth century, for example, they never could have invented the airplane because the cultural base did not contain the ideas, materials, and inventions to support its creation.

The seemingly runaway expansion, or increases in the volume, of new inventions prompted White to ask, Are people in control of their inventions, or do inventions control people? For all practical purposes, he believed that inventions control us. He cited two arguments to support this conclusion. First, he suggested that the old adage, "Necessity is the mother of invention," is naive; in too many cases, the opposite idea—that invention is the mother of necessity—is true. That is, an invention becomes a necessity because we find uses for that invention after it comes into being:

> We invent the automobile to get us between two points faster, and suddenly we find we have to build new roads. And that means we have to invent traffic regulations and put in stop lights [and build garages]. And then we have to create a whole new organization called the Highway Patrol—and all we thought we were doing was inventing cars. (Norman 1988:483)

Second, White (1949) argued that when the cultural base is capable of supporting an invention, then the invention will come into being whether people want it or not.[9] White supported this conclusion by pointing to **simultaneous-independent inventions**, situations in which the same invention is created by two or more persons working independently of one another at about the same time (sometimes within a few days or months). He cited some 148 such inventions—including the telegraph, the electric motor, the microphone, the telephone, the microscope, the steamboat, and the airplane—as proof that someone will make the necessary synthesis if the cultural base is ready to support a particular invention. In other words, the light bulb and the airplane would have been developed regardless of whether Thomas Edison and the Wright brothers (the people we traditionally associate with these inventions) had ever been born. According to White's conception, inventors may be geniuses, but they must also be born at the right place and the right time—that is, in a society with a cultural base sufficiently developed to support their inventions.

According to White's theory, if the parts are present, someone eventually will put them together. The implications are that people have little control over whether an invention should come into being and that they adapt to inventions after the fact. Sociologist William F. Ogburn (1968) calls the failure to adapt to a new invention *cultural lag*.

Cultural Lag

In his theory of cultural lag, Ogburn (1968) distinguishes between material and nonmaterial culture. Recall from Chapter 4 that material culture includes tangible creations or objects—including resources (oil, trees, land), inventions (paper, guns), and systems (factories, sanitation facilities)—that people have created or, in the case of resources such as oil, have identified as having the properties to serve a particular purpose. Nonmaterial culture, on the other hand, includes intangible creations such as beliefs, norms, values, roles, and language.

Although Ogburn maintains that both material and nonmaterial culture are important agents of social change, his theory of cultural lag emphasizes the material component, which he suggests is the more important of the two. Ogburn believes that one of the most urgent challenges facing people today is the need to

---

[9]Physicist Robert Oppenheimer, who led the research and development of the atomic bomb in the 1940s but opposed the building of the hydrogen bomb in the 1950s, would have agreed with White on this idea. Oppenheimer said, "It is a profound and necessary truth that the deep things in science are not found because they are useful; they are found because it was possible to find them" (1986:11).

Invention A synthesis of existing inventions.

Simultaneous-independent inventions Situations in which the same invention is created by two or more persons working independently of one another at about the same time.

adapt to material innovations in thoughtful and constructive ways. He uses the term **adaptive culture** to describe the portion of the nonmaterial culture (norms, values, and beliefs) that adjusts to material innovations. Such adjustments are not always immediate. Sometimes they take decades; sometimes they never occur. Ogburn uses the term **cultural lag** to refer to a situation in which adaptive culture fails to adjust in necessary ways to a material innovation.

Ogburn, however, is not a **technological determinist**—someone who believes that humans have no free will and are controlled entirely by their material innovations. Instead, he notes that people do not adjust to new material innovations in predictable and unthinking ways; rather, they choose to create them and only then choose how to use them (see "Pooling Resources and Talent to Create the Internet"). For example, according to Nobel Prize–winning physicist Isidor I. Rabi (1969), considerable choice, leadership, ingenuity, and planning went into the creation of the first and subsequent atomic bombs:

> With very few exceptions, we went around to the various laboratories and took every productive scientist from his laboratory; we took scientists from their teaching laboratories, and we put them into large laboratories for the war effort. The Radiation Laboratory at Cambridge was one of them; another was at Los Alamos, and so on. There they were, in very large numbers, devoted to making weapons, to the problems of war. This was to my knowledge the first great attempt at marrying the military and the scientific. (Rabi 1969:37–38)

Ogburn argues that if people have the power to create material innovations, they also have the power to destroy, ban, regulate, or modify their use. The challenge lies, however, in convincing people that they need to address an invention's potential disruptive consequences before those consequences have a chance to materialize. The U.S. government, for example, created the Internet as a way of managing the potentially devastating consequences of nuclear attacks. On the other hand, it has yet to fully address the problem of nuclear waste. As one critic of the nuclear weapons program commented as early as 1950, "Nuclear waste is like getting on a plane, and in mid-air you ask the pilot, how are we going to land? He says, we don't know—but we'll figure it out by the time we get there" (Bauerlein 1992:34).

In our discussion about innovations as a trigger of social change, we have emphasized material inventions (devices, tools, and equipment). Innovations can also consist of nonmaterial inventions, such as a revolutionary idea.

Revolutionary Ideas

In *The Structure of Scientific Revolutions,* Thomas Kuhn (1975) maintains that most people perceive science as an evolutionary enterprise. That is, over time scientists move closer to finding the solution to problems by building on their predecessors' achievements.

Kuhn takes issue with this evolutionary view, arguing that some of the most significant scientific advances have been made when someone breaks away from or challenges the prevailing paradigms. He defines **paradigms** as the dominant and widely accepted theories and concepts in a particular field of study. Paradigms gain their status not because they explain everything, but because they offer the "best way" of looking at the world at that time. The dominant paradigms of an academic discipline generally are recorded in its textbooks, which summarize the body of accepted theory and illustrate it with the most successful applications, the most exemplary observations, and the most supportive experiments. On the one hand, paradigms are important thinking tools; they bind a group of people with common interests into a scientific or national community. Such a community cannot exist without agreed-on paradigms. On the other hand, paradigms can act as blinders, limiting the kinds of questions that people ask and the observations that they make.

The explanatory value, and hence the status, of a paradigm is threatened by any **anomaly,** an observation or observations that it cannot explain. The existence of an anomaly alone, however, usually does not persuade people to abandon a particular paradigm.

Adaptive culture The portion of the nonmaterial culture (norms, values, and beliefs) that adjusts to material innovations.

Cultural lag A situation in which adaptive culture fails to adjust in necessary ways to a material innovation.

Technological determinist Someone who believes that human beings have no free will and are controlled entirely by their material innovations.

Paradigms The dominant and widely accepted theories and concepts in a particular field of study.

Anomaly An observation or observations that a paradigm cannot explain.

U.S. in Perspective

Pooling Resources and Talent to Create the Internet

Today, the United States leads the world in many of the technologies that make the information superhighway possible—semiconductors, fiber optics, high-speed switches, supercomputing, advanced software, and so on. One reason for this is the billions of dollars that the federal government, especially the Defense Department and NASA, has invested in information technology over the last 40 years. Many of the most important technological breakthroughs in computing and communications are due in part to a federal research grant or a federal contract.

The very first electronic computer, the ENIAC, was the product of federal funding. There are dozens of similar examples. A recent study by the National Research Council on the High-Performance Computing and Communications Program documented how federal research and development funding played a key role in the development of timesharing, computer networking, workstations, computer graphics, the mouse, the windows interface, VLSI circuit design, RISC computing, parallel computing, and digital libraries.

There has been a very effective partnership between government, academia, and industry. By providing funding, the federal government enables academic researchers to do the kind of long-term research that businesses can't afford to do. Even more important, that funding makes possible the training of the next generation of scientists and engineers. The research results and the talent from our universities are then applied by the private sector to develop the new products and new processes that they need to stay competitive in world markets.

The Internet provides an excellent example of how the federal government can work with universities and industry to develop new technologies that can create whole new industries—and thousands of new, high-paying jobs. The ARPANET, the precursor of the Internet, began as a research project funded by the Defense Department's Advance Research Project Agency in 1969. . . . In 1986, the National Science Foundation started the NSFNET program in order to expand the ARPANET and connect more university researchers to the network. The network grew in both size and speed. By the early 1990s, the NSFNET backbone network was operating at 45 megabits per second, almost 1,000 times faster than the original ARPANET.

In the past five years, the NSFNET has evolved into the Internet. . . . There is a similar story for the World Wide Web. . . . According to the latest figures, there is a new Web page being created every four seconds. The Web was developed at CERN, the European Laboratory for Particle Physics, in Switzerland. In 1993, researchers at the National Center for Supercomputing Applications at the University of Illinois, one of the supercomputer centers funded by the National Science Foundation, developed an easy-to-use software package called Mosaic, which made it possible to use the World Wide Web to view not only text, but also images, video, and audio. In less than a year, there were more than a million copies of Mosaic in use and Internet usage was skyrocketing. This new technology has made possible companies like Netscape and Spyglass and fundamentally changed the way the Internet is used.

Funding for the development of Mosaic came from the High-Performance Computing and Communications Initiative, a federal R&D program authorized by the High-Performance Computing Act of 1991, which then-Senator Al Gore introduced in 1988. That program provides more than $1 billion a year to ten participating agencies that fund research in universities, federal laboratories, and industry. The HPCC program has pooled the talent and resources available throughout the federal government to ensure effective coordination between agencies and ensure the efficient use of taxpayers' dollars. In many ways, the HPCC program is a model of how government, industry, and academia can work together to ensure that the U.S. stays at the leading edge of technology.

As the source indicates, this excerpt was derived from a White House press release. Naturally, the U.S. federal government argues that its programs benefit the United States. How does it describe its role here in relation to technological innovation and related social change? Is the argument convincing? Who, does it claim, are some specific beneficiaries of its actions? Do its actions favor some groups over others? Which?

Source: White House (1996).

According to Kuhn, before people discard old paradigms, someone must articulate an alternative paradigm that accounts convincingly for the anomaly. He hypothesizes that the people most likely to put forth new paradigms are those who are least committed to the old paradigms—the young and those new to a field of study.

A scientific revolution occurs when enough people in the community break with the old paradigm and change the nature of their research or thinking so as to favor the incompatible new paradigm. Kuhn considers a new paradigm incompatible with the one it replaces because it "changes some of the field's most elementary theoretical generalizations" (1975:85). The new

Because nuclear waste remains radioactive for 240,000 years, the EPA has mandated that a warning marker be designed to inform future generations about where waste has been stored or buried. What challenges are associated with creating such a marker? Do the markers shown in the photograph represent an acceptable solution?

© Rich Frishman/Tony Stone Images

paradigm causes converts to see the world in an entirely new light and to wonder how they could possibly have taken the old paradigm seriously. "[W]hen paradigms change, the world itself changes with them. Led by a new paradigm, scientists adopt new instruments and look in new places" (p. 111). To illustrate Kuhn's theory, one can argue that Internet technology is changing the paradigms governing education—specifically, the paradigms governing our conceptions of literacy.

The Internet and Changing Paradigms of Literacy

The Internet puts the individual at the center of information[10] in that he or she can access information quickly and directly, bypassing hierarchies devoted to managing and granting access to information (Crawford 1995). As one example, consider that we rely on the media for most of our information about world events. Under the current arrangements, reporters go to the source of the news, decide what is important, and then relay it to their viewers and readers for a price. Within the Internet, however, users can go to the source of the news and read the organizations' press releases, thereby avoiding the hierarchy that brings us an added layer of "filtered" news.

These characteristics of the Internet raise questions about the kinds of skills one needs to manage such a powerful information retrieval and distribution tool. In "Redefining Success: Public Education in the 21st Century," Yvonne Katz and Gay Chedester (1992) point out that as an increasing number of students come to school with notebook-size computers tapped into all the world's knowledge, teachers need to adopt a new literacy paradigm in which they cease being the brokers of information and begin teaching students the literacies of choice, connectedness, and creativity (see Table 16.1).

What is *literacy of choice*? The most obvious characteristic of the Internet is the sheer glut of useful and trivial information that it can deliver with the touch of a few keys. Because of its size, the documents posted on the Internet defy traditional approaches to indexing or categorization. If we do not give people the emotional and conceptual skills to sift through and choose among the resources available, they will simply be carried along in an avalanche of documents, forever surfing from site to site.[11] Two Northern Kentucky University students write about the emotional and conceptual challenges they face while doing research on the Internet:

> The amount of information on the Internet is massive. You can find something on almost every topic. However, you must be mentally and physically tough when searching for information. Sometimes you can feel totally overwhelmed to the point of panic. But don't worry—most people feel this way at one time or another. My advice is to try to relax and not feel that you have to look over or reach everything on a topic.

> Be patient with the Internet; don't spend too much time at any one site unless it is directly relevant to your task at hand. Look through the sites quickly and if the information isn't useful, keep going and don't look back. Above all, take an "I am in charge" attitude. Don't let the Internet overwhelm you. Maneuvering around the Internet is an acquired skill. The more time you spend at it, the more efficient you will become.

A person who possesses choice literacy, then, is someone who is not overwhelmed by information and has the emotional and conceptual ability to choose among the resources available.

---

[10]One reason that people have access to so much information via the Internet is that the Internet's strongest and most valuable quality is that most of the millions of documents available for access are formatted according to standard specification known as HTML (HyperText Markup Language). (Don't tune out because the concept of hypertext seems too technical.) Hypertext simply refers to a programming feature that connects documents to one another and allows readers to move quickly via linked keywords within and across documents located anywhere on the Internet.

[11]As of January 3, 1997, Lycos (a search engine) claimed to have indexed 60,434,860 Web pages.

Table 16.1   Year-to-Year Increases in Internet Access by School Characteristic, 1994–2000

This table shows that by the year 2000 most elementary and secondary schools expect to have Internet access. What types of schools were most likely to have Internet access in 1994? Which types of schools were the least likely? Do these numbers mean that at least 91 percent of all elementary and secondary students will have access to the Internet by 2000? Why or why not?

Schools Having Access to the Internet

School Characteristics	1994	1995	1996	1997	2000*
All public schools	**35**	**50**	**65**	**78**	**95**
Instructional level†					
Elementary	30	46	61	75	94
Secondary	49	65	77	89	98
Size of enrollment					
Less than 300	30	39	57	75	93
300 to 999	35	52	66	78	96
1,000 or more	58	69	80	89	97
Type of community					
City	40	47	64	74	94
Urban fringe	38	59	75	78	97
Town	29	47	61	84	95
Rural	35	48	60	79	95
Geographic region					
Northeast	34	59	70	78	98
Southeast	29	44	62	84	96
Central	34	52	66	79	92
West	42	48	62	73	96
Minority enrollment					
Less than 6 percent	—	52	65	84	95
6 to 20 percent	—	58	72	87	97
21 to 49 percent	—	54	65	73	98
50 percent or more	—	40	56	63	91
Students eligible for free or reduced-price lunch					
Less than 11 percent	—	62	78	88	97
11 to 30 percent	—	59	72	83	98
31 to 70 percent	—	47	58	78	93
71 percent or more	—	31	53	63	93

*Estimate of number of schools planning or expecting to have Internet access.
†Data for combined schools (those that span elementary and secondary grades) are included in the totals and in analyses by other school characteristics but are not shown separately.
Source: U.S. Department of Education, National Center for Education Statistics (1998).

The second type of literacy is the *literacy of connectedness*. The Internet owes its existence to the fact that millions of computers across the world are connected to one another. It is not run by some person or a group; it is not a prearranged library of information. Thus the literacy of connectedness means understanding that we are part of a community of information.

A third type of literacy is *creative literacy*. Creative literacy involves the ability to use others' information in creative ways and to contribute information that others can use. The success of the Internet depends on people being more than passive consumers of information; it has thrived because millions of people were willing to post helpful information for others to use. In other words, good citizens of the Internet do not just take information from it; they give something back. Creative literacy also involves the belief that ideas are works in progress and that authors are free to re-create them at any time. One Northern Kentucky University student expressed this idea as follows:

> The Internet presents a whole new way of thinking about information. My advice is to expect change and learn to be comfortable with it. People who post

information on the Internet can choose to take it off or revise it whenever they see fit. In the print world, authors might want to change their ideas but they have to wait until the next edition (if there is a next edition) to get it out there for the world to see. With the Internet, authors can revise their thinking and make changes on the spot. So don't be surprised if you find a document is gone or has changed in dramatic or subtle ways from one visit to the next.

Summary and Implications

Sociologists have identified at least four main agents of change: conflict over scarce and valued resources, capitalism, innovations, and revolutionary ideas. In this chapter, we have considered a major innovation—the Internet—billed by some as "a second printing press." We have studied the role that each agent of change has played in creating this technology and the changes (real and potential) associated with the Internet. The Internet represents the synthesis of a long trail of "existing inventions"—that is, it emerged as a product of conflict (the Cold War) and deliberate planning by government leaders, defense contractors, and academic researchers. Contrary to popular opinion, the Internet does not operate independently of controlling organizations. From a sociological point of view, the access to this technology is a scarce and valued resource that various organizations—governments, Internet access providers, communication companies, hardware and software companies, colleges and universities, Internet societies, and grassroots organizations—are attempting to control.

From a global perspective, a person's access to the Internet is shaped to a large extent by his or her position in the capitalist-driven economic order. People who live in core economies (especially the United States) are the most likely to have Internet access. Even within core economies, access to the Internet depends on one's ability to purchase access. From the perspective of a student whose school offers Internet access, it might seem that the Internet is "free." In fact, access to this resource is tied to the cost of tuition or a school district's ability to afford it.

By studying the four major agents of change, we can see the ways in which the character of the Internet has been shaped by the past and reflects current economic and political arrangements. On the one hand, the phenomenon of globalization-from-above shows the power of the Internet to perpetuate and intensify current arrangements. On the other hand, the Internet has the potential to change the way that people think and relate

to one another by breaking away from past arrangements. As an example, consider the Internet's effect on conceptions of literacy. Likewise, this technology could potentially invigorate globalization-from-below.

In studying social change, sociologists ask two key questions: What factor(s) cause that change? What are the consequences of that change for social life? With regard to the first question, we have considered how the four major triggers of change contributed to the Internet's emergence and ongoing development and how the Internet has inspired change in its own right. Keep in mind that we have covered concepts and theories in Chapters 1 through 14 that can help us conceptualize the consequences of the Internet. A sample of applications follow.

We can use the concept of **global interdependence** to think about the Internet as an invention capable of connecting the lives of people around the world and facilitating action directed at solving social problems on a global scale (see Chapter 1). We can use the questions and vocabulary of the **three theoretical perspectives** to frame an analysis of the Internet and its consequences (see Chapter 2). The **methods of social research** can help us evaluate research and conduct studies related to the Internet and its consequences (Chapter 3). We can use the concept of **cultural diffusion** to think about the Internet as a tool for promoting cultural exchange (Chapter 4). The concepts associated with **socialization** and **resocialization** help us consider the process by which "newcomers" learn about the Internet and come to accept (or reject it) as a tool for solving the problems of living (Chapter 5). We can draw on the concept of **solidarity** to frame an analysis of the Internet as a tool that binds (or fails to bind) people to one another (Chapter 6). The concept of **informal and formal aspects of organization** can help us think about the official and unofficial uses of the Internet in organizational settings (Chapter 7). With regard to **deviance, conformity,** and **social control,** we can envision the Internet

Access to computers in schools varies greatly even in the United States. One school offers a state-of-the art computer lab, another only the most minimal computer resources.

© A. Ramey/PhotoEdit (left); © David Young-Wolff/PhotoEdit (right)

as a mechanism of social control (for example, via crimestopper Web sites) or as a mechanism that needs to be controlled (Chapter 8). The sociological framework for analyzing **social stratification** as it relates to **social class, race,** and **gender** alerts us to the structural dynamics supporting inequality that affect people's access to the Internet (Chapters 9, 10, and 11). We can consider the relationship between a country's role in the worldwide **division of labor** and its citizens' access to the Internet (Chapter 12). The concepts of **literacy** can help us understand the Internet's role in changing definitions of literacy (Chapter 14). Finally, with regard to religion, we can consider the Internet's role in promoting **secularization** and **fundamentalism** (Chapter 15).

The larger point behind listing these possible approaches to studying the Internet is to show the power, vitality, relevance, and versatility of sociological concepts and theories for framing and explaining a variety of social phenomena and issues.

Key Concepts

Use this outline to organize your review of key chapter concepts.

Social change
    Globalization-from-above
    Globalization-from-below
Innovations
    Discoveries
    Basic innovations
    Improving innovations
    Inventions
    Simultaneous-independent innovations

Cultural lag
    Adaptive culture
    Technological determinism
Paradigms
World-economy versus world economy

internet assignment

Test your Internet searching and browsing skills. Find examples of how the Internet is changing the way that people solve everyday problems other than those described in this chapter (for example, finding an apartment, determining a location, obtaining news).

Key Concepts

Ability grouping The arranging of elementary school students into instructional groups according to similarities in past academic performance and/or on standardized test scores.

Achieved characteristics Attributes acquired through some combination of choice, effort, and ability.

Active adaptation A biologically based tendency to adjust to and resolve environmental challenges.

Adaptive culture The portion of nonmaterial culture (norms, values, and beliefs) that adjusts to material innovations.

Advanced market economics Economic arrangements that offer widespread employment opportunities for women as well as for men.

Alienation A state in which human life is dominated by the forces of human inventions.

Annual per capita consumption of energy The average amount of energy each person in a nation consumes over a year.

Anomaly An observation or observations that a paradigm cannot explain.

Anomie See *Structural strain.*

Apartheid A system of laws in which everyone in South Africa was put into a racial category and issued an identity card denoting his or her race.

Apartheid policies Policies centered on maintaining separate black and white areas.

Archival data Data that have been collected by other researchers for some other purpose.

Ascribed characteristic Any physical trait that is biological in origin and/or cannot be changed but to which people assign overwhelming significance.

Aspirational doll A doll considered to be a role model for a child.

Assimilation A process by which ethnic and racial distinctions between groups disappear.

Authoritarian government A political system in which no separation of powers exists; a single person (king, dictator), group (family, military, single party), or social class holds all the power.

Authority Legitimate power in which people believe that the differences in power are just and proper.

Automate To use the computer as a means to increase workers' speed and consistency or as a source of surveillance.

Back stage The region out of sight where individuals can do things that would be inappropriate or unexpected on the front stage.

Basic innovations Revolutionary, unprecedented, or ground-breaking inventions that represent the cornerstones for a wide range of applications.

Be absurd, to To make connections between unrelated things or events and not to care whether the connections are appropriate or inappropriate.

Beliefs Conceptions that people accept as true, concerning how the world operates and where the individual fits in relationship to others.

Boundaries The qualities marking some people off from others as a unified and distinctive group.

Bourgeoisie The owners of the means of production.

Bureaucracy An organization that uses the most efficient means to achieve a valued goal.

Capitalism An economic system in which the raw materials and the means of producing and distributing goods and services are privately owned. Capitalist systems are characterized by private ownership of the means of production and are profit-driven, free of government interference, and consumer-driven.

Caste system A system in which people are ranked on the basis of traits over which they have no control.

Catechisms Short books covering religious principles written in question-and-answer format.

Church A group whose members hold the same beliefs with regard to the sacred and the profane, who behave in the same way in the presence of the sacred, and who gather in body or spirit at agreed-on times to reaffirm their commitment to those beliefs and practices.

Civil religion "Any set of beliefs and rituals, related to the past, present and/or future of a people (nation), which are understood in some transcendental fashion" (Hammond 1976:171).

Claims makers People who articulate and promote claims and who tend to gain if the targeted audience accepts their claims as true.

Class A category that designates a person's overall status in society.

Classless society A propertyless society providing equal access to the means of production.

Class system A system in which people are ranked on the basis of merit, talent, ability, or past performance.

Cohort A group of people sharing a common characteristic or life event.

Cold War The massive military buildup that took place between 1945 and 1989 as the Soviet Union and the United States sought to contain the spread of one another's economic and political systems.

Collective memory The experiences shared and recalled by significant numbers of people.

Concepts Thinking and communication tools that are used to give and receive complex information efficiently and to frame and explain observations.

Confederate Someone who works in cooperation with a person conducting a research study.

Conflict According to Karl Marx, the major force that drives social change.

Conformists People who have not violated the rules of a group and are treated accordingly.

Conformity Behaviors and appearances that follow and maintain standards of a group, the acceptance of the cultural goals and the pursuit of these goals through legitimate means.

Conglomerates Large corporations that own "smaller" corporations acquired through merger or acquisition.

Connotation The set of associations that a word evokes.

Constrictive pyramids Population pyramids that are characteristic of some European societies that are narrower at the base than in the middle, showing that the population is composed disproportionately of middle-aged and older people.

Constructionist approach A sociological approach that focuses on the process by which specific groups, activities, conditions, or artifacts become defined as problems.

Content The cultural frameworks (norms, values, beliefs, material culture) that guide behavior, dialogue, and interpretations during interaction.

Context The larger historical circumstances that bring people together.

Control When the process of producing and acquiring the service or product is planned out in detail.

Control variables Variables suspected of causing spurious correlation.

Corporate crime Crime committed by a corporation as it competes with other companies for market share and profits.

Correlation A relationship between two variables such that a change in one variable is associated with a change in another.

Correlation coefficient A mathematical representative of the extent to which a change in one variable is associated with a change in another.

Counterculture A subculture that conspicuously challenges, rejects, or clashes with the central norms and values of the dominant culture.

Country A political entity, recognized by foreign governments, with a civilian and military bureaucracy to enforce its rules.

Crime Deviance that breaks the laws of society and is punished by formal sanctions.

Cults Very small, loosely organized groups, usually founded by a charismatic leader who attracts people by virtue of his or her personal qualities.

Cultural base The number of existing inventions.

Cultural genocide A form of ethnocentrism in which people of one society define the culture of another society not as merely offensive but as so intolerable that they attempt to destroy it.

Cultural lag A situation in which adaptive culture fails to adjust in necessary ways to a material innovation.

Cultural markers Distinctive characteristics that are used to clearly classify people into distinct cultural units.

Cultural relativism The perspective that a foreign culture should not be judged by the standards of a home culture and that a behavior or way of thinking must be examined in its cultural context.

Culture A way of life, especially general customs and beliefs of a particular group of people at a particular time.

Culture gap Ways of thinking and behaving that vary according to culture.

Culture shock The strain that people from one culture experience when they must reorient themselves to the ways of a new culture.

Dearth of feedback A situation in which not enough critical readers and listeners evaluate material before it is used by the popular media.

Democracy A system of government in which power is vested in the citizen body and in which members of that citizen body participate directly or indirectly in the decision-making process.

Demographers Sociologists who study population size, distribution, and age-sex structure.

Demographic gap The difference between birth rates and death rates.

Demographic trap The point at which population growth overwhelms the environment's carrying capacity.

Demography A subdiscipline within sociology that studies population trends.

Denomination A hierarchical organization in a society in which church and state are usually separate, led by professionally trained clergy.

Denotation A literal definition.

Dependent variable The behavior to be explained or predicted.

Deviance Any behavior or physical appearance that is socially challenged and/or condemned because it departs from the norms and expectations of a group.

Deviant subcultures Groups that are part of the larger society but whose members adhere to norms and values that favor violation of the larger society's laws.

Differential association A theory of socialization that explains how delinquent behavior is learned. It refers to the idea that "when persons become criminal, they do so because of contacts with criminal patterns and also because of isolation from anti-criminal patterns" (Sutherland and Cressey 1978:78).

Diffusion The process by which an idea, an invention, or some other cultural item is borrowed from a foreign source.

Discovery The uncovering of something that had existed before but had remained hidden, unnoticed, or undescribed.

Discrimination Intentional or unintentional unequal treatment of individuals or groups on the basis of attributes unrelated to merit, ability, or past performance.

Dispositional traits Personal or group traits such as motivation level, mood, and inherent ability.

Division of labor Work that is broken down into specialized tasks, with each task being performed by a different set of persons.

Documents Written or printed material used in research.

Dominant group The ethnic and racial group at the top of the hierarchy.

Double consciousness According to DuBois, "this sense of always looking at one's self through the eyes of others, of measuring one's soul by the tape of a world that looks on in amused contempt and pity." The double consciousness includes a sense of two-ness: "an American, a Negro; two souls, two thoughts, two unreconciled strivings; two warring ideals in one dark body, whose dogged strength alone keeps it from being torn asunder."

Doubling time The estimated number of years required for a country's population to double in size.

Downward mobility A loss of rank or status.

Dramaturgical model A model in which interaction is viewed as though it were theater, people as though they were actors, and roles as though they were performances presented before an audience in a particular setting.

Dysfunctions Disruptive consequences to society or to some segments of society.

Ecclesiae A professionally trained religious organization governed by a hierarchy of leaders which claims as its members everyone in a society.

Economic ideals The standards against which real economic systems can be compared.

Economic system An institution that coordinates human activity in the effort to produce, distribute, and consume goods and services.

Education Those experiences that train, discipline, and develop the mental and physical potentials of the maturing person.

Efficiency A corporation or organization offering the "best" products and services that allow consumers to get quickly from one state to another, that is, from hungry to full, from fat to thin, from uneducated to educated, and so on.

Engram Chemically formed entities in the brain that store in physical form a person's recollections of experiences.

Established sects Religious organizations, resembling both denominations and sects, composed of renegades from denominations or ecclesiae that have existed long enough to acquire a large following and widespread respectability.

Ethgender People who share (or are believed by themselves or others to share) the same sex and race and ethnicity.

Ethnic blending "Inter-ethnic unions (interbreeding) and shifts in ethnic affiliation" (Hirschman 1993:549).

Ethnic group A term that is used in reference to people who share or believe they share a national origin, cultural traits, or distinctive physical features.

Ethnicity A term that is used to classify people according to any number of attributes, including national origin, ancestry, distinctive and visible cultural traits (religious practice, dietary habits, style of dress, body ornaments, or languages), and/or socially important physical characteristics.

Ethnocentrism A viewpoint that uses one culture as the standard for judging the worth of foreign ways.

Evolutionary view The idea that human societies progress in stages from primitive to civilized, with each stage being characterized by a gradually more complex form of social organization.

Expansive pyramids Population pyramids characteristic of labor-intensive poor countries that are broadest at the base, with each successive bar smaller than the one below it, showing that the popula-tion is increasing in size and is composed disproportionately of young people. (See also *Population pyramid.*)

Externality costs Costs that are not figured into the price of a product but that are nevertheless paid by the consumer when using or creating a product.

Facade of legitimacy An explanation that members in dominant groups give to justify their actions.

Falsely accused People who have not broken the rules but who are treated as if they have done so.

Family Two or more people related to one another by blood, marriage, adoption, or some other socially recognized criteria.

Feeling rules Norms that specify appropriate ways to express internal sensations.

Femininity The physical, behavioral, and mental or emotional traits believed to be characteristics of females.

Folkways Norms that apply to the mundane aspects or details of daily life.

Formal curriculum The various subjects such as mathematics, science, English, reading, physical education, and so on. (See *Hidden curriculum.*)

Formal dimension (of an organization) The official, written guidelines, rules, regulations, and policies that define the goals and roles of the organization and its relationship to other organizations and integral parties.

Formal education A purposeful, planned effort intended to impart specific skills and modes of thought. (See *Informal education* and *Schooling.*)

Formal sanctions Definite and systematic laws, rules, regulations, and policies that specify (usually in writing) the conditions under which people should be rewarded or punished, and that define the procedures for allocating rewards and imposing punishments.

Fortified households Preindustrial arrangements characterized by no police force, militia, national guard, or other peacekeeping organization; the household is an armed unit with the head of the household as its military commander.

Front stage The region where people take care to create and maintain expected images and behavior.

Function The contribution of a part to the order and stability within the larger system.

Functionally illiterate A significant percentage of the population in every country that do not possess the level of reading, writing, and calculating skills needed to adapt to the society in which they live.

Fundamentalism A belief in the timeless nature of sacred writings and a belief that such writings apply to all kinds of environments.

Games Structured, organized activities that usually involve more than one person and a number of constraints concerning roles, rules, time, place, and outcome.

Gender A social distinction based on culturally conceived and learned ideas about appropriate appearance, behavior, and mental or emotional characteristics for males and females.

Gender polarization "The organizing of social life around the male-female distinction, so that people's sex is connected to virtually every other aspect of human experience, including modes of dress, social roles, and even ways of expressing emotion and experiencing sexual desire" (Bem 1993:192).

Gender schematic A term describing decisions that are influenced by society's polarized definitions of masculinity and femininity rather than by criteria such as self-fulfillment, interest, ability, or personal comfort.

Generalizability The extent to which findings can be applied to the larger population from which the sample is drawn.

Generalized other A system of expected behaviors, meanings, and viewpoints that transcend those of the people participating.

Global interdependence A state in which the lives of people around the world are intertwined closely and in which each country's problems—unemployment, employment, drug abuse, pollution, inequality, disease—are part of a larger global situation.

Globalization-from-above Connects people from around the world with educational, economic, and political advantages, excluding or pushing to the sidelines those who are not so advantaged.

Globalization-from-below Involves interdependence at the grassroots level that aims to protect, restore, and nurture the environment; enhance ordinary people's access to the basic resources they need to live a dignified existence.

Goods Any product that is manufactured, grown, or extracted from the earth, such as food, clothing, housing, automobiles, coal, computers, and so on.

Government The organizational structure that directs and coordinates peoples' involvement in a country's or some other territory's

(city, county, state) political activities.

Great changes Events whose causes lie outside ordinary people's characters or their immediate environments but profoundly affect their life chances.

Gross domestic product (GDP) The monetary value of the goods and services that a nation's work force produces over the course of a year (or some other time period).

Group Two or more people who share a distinct identity, feel a sense of belonging, and interact directly or indirectly with one another.

Hate crimes Actions aimed at humiliating members of a minority group and destroying their property or lives.

Hawthorne effect A phenomenon whereby observed persons alter their behavior when they learn they are being observed.

Hidden curriculum All the things that students learn along with the subject matter.

Households All related and unrelated persons who share the same dwelling.

Hypertext Software that allows readers to pick and choose among highlighted keywords (that is, important ideas) and follow links to related documents.

Hypothesis A trial explanation put forward as the focus of research that predicts how the independent and dependent variables are related and what outcome will occur.

Ideal type A deliberate simplification or caricature in that the characteristics emphasized exaggerate certain aspects of a bureaucracy, which makes them the subject of comparison.

Ideologies Fundamental ideas that support the interests of dominant groups.

Ideology A set of beliefs that are not challenged or subjected to scrutiny by the people who hold them.

Idiom A group of words that when taken together have a meaning different from the internal meaning of each word understood on its own.

Illiteracy The inability to understand and use a symbol system, whether it is based on sounds, letters, numbers, pictographs, or other type of character.

Impression management The process by which people in social situations manage the setting and their dress, words, and gestures to correspond to the impressions they are trying to make or the image they are trying to project.

Improving innovations Modifications of basic inventions that improve upon the originals—that is, to make them smaller, faster, less complicated, or more efficient, attractive, or profitable.

Independent variable The variable that explains or predicts the dependent variable.

Individual discrimination Any overt action on the part of an individual that depreciates someone from the outgroup, denies the outgroup opportunities to participate, or does violence to lives and property.

Infant mortality The number of deaths in the first year of life for every 1,000 live births.

Informal dimension Owner- or employee-generated norms that evade, bypass, do not correspond with, or are not systematically stated in official policies, rules, and regulations.

Informal education Education that occurs in a spontaneous, unplanned way.

Informal sanctions Spontaneous, unofficial expressions of approval or disapproval that are not backed by the force of law.

Informate To use the computer to empower workers with decision-making tools, such as employee-scheduling software intended to ensure that a sufficient number of employees are scheduled for the busiest times and shifts.

Information explosion A term describing the unprecedented increase in the volume of data made possible by the development of the computer and telecommunications.

Ingroups A group with which people identify and to which they feel closely attached, particularly when that attachment is founded on hatred from or opposition toward another group.

Innovation (as a response to structural strain) The acceptance of the cultural goals but the rejection of legitimate means to obtain these goals.

Innovation The development of something new—an idea, a practice, or a tool.

Issue A matter that can be explained only by factors outside an individual's control and immediate environment.

Institutionalized discrimination The established and customary ways of doing things in society; the unchallenged rules, policies, and day-to-day practices, that impede or limit minority members' achievements and keep them in a subordinate and disadvantaged position.

Institutionally complete Subcultures whose members do not interact with anyone outside their subculture to shop for food, attend school, receive medical care, or find companionship because the subculture satisfies these needs.

Intergenerational mobility A change (upward or downward) in rank over two or more generations.

Internalization The process in which people take as their own and accept as binding the norms, values, beliefs, and language needed to participate in the larger community.

Internal migration Movement within the boundaries of a single nation from one state, region, or city to another.

International migration Movement of people between countries.

Internet A vast network of computer networks connected to one another via special software and phone, fiber-optic, or other type of line with the potential to connect millions, even billions, of computers and their uses on a global scale.

Intersexed A broad term used by the medical profession to classify people with some mixture of male and female biological characteristics.

Interviews Face-to-face or telephone conversations between an interviewer and a respondent in which the interviewer asks questions and records the respondent's answers.

Intragenerational mobility Movement (upward or downward) during an individual's lifetime.

Invention A synthesis of existing inventions.

Involuntary minorities Ethnic and racial groups that were forced to become part of a country by slavery, conquest, or colonization.

Iron cage of rationality The irrationalities that rational systems generate.

Islamic activism See *Islamic revitalism.*

Islamic revitalism "A sense that existing political, economic, and social systems had failed; a disenchantment with, and at times a rejection of the West; a quest for identity and greater authenticity; and the conviction that Islam provides a self-sufficient ideology for state and society, a valid alternative to secular nationalism, socialism, and capitalism" (Esposito 1992:14).

Issues Matters that can be explained by factors outside an individual's control and immediate environment.

Labor-intensive poor A term used to characterize a country that differs markedly from industrialized countries on indicators such as doubling time, infant mortality, total fertility, per capita income, and annual per capita consumption of energy.

Latent dysfunctions Unintended, unanticipated disruptions to order and stability.

Latent functions The unintended, unrecognized, and unanticipated or unpredicted effects on order and stability.

Law of supply and demand Natural laws regulating capitalist economies such that "as demand for an item increases, prices rise. When manufacturers respond to the price increase by producing a larger supply of that item, this increases competition and drives the price down" (Hirsch 1993:455).

Liberation theology A doctrine that maintains that organized religions have a responsibility to demand social justice for the marginalized peoples of the world, especially landless peasants and the urban poor, and to take an active role at the grassroots level to bring about political and economic justice.

Life chances A critical set of potential social advantages.

Looking-glass self A way in which a sense of self develops in which people see themselves reflected in others' reactions to their appearance and behaviors.

Low-technology tribal societies Hunting-and-gathering societies with technologies that do not permit the creation of surplus wealth.

Manifest dysfunctions A part's expected or anticipated disruptions to order and stability.

Manifest functions A part's intended, recognized, expected, or predictable effects on order and stability.

Masculinity The physical, behavioral, and mental or emotional traits believed to be characteristics of males.

Master status of deviant An identification marking a rule breaker first as a deviant and then as having any other identification.

Material culture All physical objects that people have borrowed, discovered, or invented and to which they have attached meaning.

McDonaldization of society "The process by which the principles of the fast-food restaurant are coming to dominate more and more sectors of American society as well as the rest of the world" (Ritzer 1993:1).

Means of production The resources (land, tools, equipment, factories, transportation, and labor) essential to the production and distribution of goods and services.

Mechanical solidarity Social order and cohesion based on a common conscience or uniform thinking and behavior.

Mechanisms of social control All of the methods that people employ to teach, persuade, or force others to conform.

Mechanization The addition of external sources of power, such as oil or steam, to hand tools and modes of transportation.

Melting pot assimilation A process of cultural blending in which the groups involved accept many new behaviors and values from one another.

Methods of data collection The procedures used to gather relevant data.

Migration The movement of people from one area to another.

Minority groups Subgroups within a society that can be distinguished from members of the dominant groups by visible and identifying characteristics, including physical and cultural attributes.

Mixed contacts "The moments when stigmatized and normals are in the same 'social situation,' that is, in one another's immediate physical presence, whether in a conversation-like encounter or in the mere co-presence of an unfocused gathering" (Goffman 1963:12).

Modern capitalism "A form of economic life which involved the careful calculation of costs and profits, the borrowing and lending of money, the accumulation of capital in the form of money and material assets, investment, private property, and the employment of laborers and employees in a more or less unrestricted labor market" (Robertson 1987:6).

Monopoly A situation in which a single producer dominates a market.

Mores Norms that people define as essential to the well-being of the group.

Mortality crises Frequent and violent fluctuations in the death rate caused by war, famine, and epidemics.

Multinational corporation Enterprises that own or control production and service facilities in countries other than the one in which their headquarters are located.

Mystical religions Religions in which the sacred is sought in states of being that, at their peak, can exclude all awareness of one's existence, sensations, thoughts, and surroundings.

Nation A geographic area occupied by people who share a culture and history.

Nature Human genetic makeup or biological inheritance.

Negative correlation When one variable increases and the other variable decreases.

Negative sanction An expression of disapproval for noncompliance.

"Negatively privileged" property class Persons completely unskilled, lacking property, and dependent on seasonal or sporadic employment who constitute the very bottom of the class system.

Nonhouseholder class Property-less laborers and servants usually residing within fortified households. (See *Fortified households.*)

Nonmaterial culture Intangible creations or things that we cannot identify directly through the senses.

Nonparticipant observation A research technique involving detached watching and listening in which the researcher does not interact with study participants.

Nonprejudiced nondiscriminators (all-weather liberals) Persons who accept the creed of equal opportunity, and their conduct conforms to that creed.

Normals Those people who are in the majority or those who possess no discrediting attributes.

Norms Written and unwritten rules that specify behaviors appropriate and inappropriate to a particular social situation.

Nurture The environment or the interaction experiences that make up every individual's life.

Objectivity A state in which personal, subjective views do not influence one's opinons or behavior.

Obligations The relationship and behavior that a person enacting a role must assume toward others in a particular status.

Observation A research technique involving watching, listening to, and recording behavior and conversations as they happen.

Oligarchy Rule by the few, or the concentration of decision-making power in the hands of a few persons who hold the top positions in a hierarchy.

Oligopoly A situation in which a few producers dominate a market.

Operational definitions Clear, precise definitions and instructions about how to observe and measure variables.

Organic solidarity Social order based on interdependence and cooperation among people performing a wide range of diverse and specialized tasks.

Organization A coordinating mechanism created by people to achieve stated objectives.

Outgroup A group toward which members of an ingroup feel a sense of separateness, opposition, or even hatred. (See also *Ingroup*.)

Paradigms The dominant and widely accepted theories and concepts in a particular field of study.

Participant observation A research technique in which researchers interact directly with study participants.

Per capita income The average share of income that each person in a country would receive if the country's gross national product were divided evenly among all its citizens.

Play Voluntary and often spontaneous activity, with few or no formal rules, that is not subject to constraints of time.

Pluralist model A model that views politics as an arena of compromise, alliances, and negotiation among many competing and different special-interest groups and power as something that is dispersed among those groups.

Political action committees (PACs) Committees that raise money to be donated to the political candidates most likely to support their special interests.

Political parties Organizations oriented toward the planned acquisition of social power and toward influencing social action no matter what its content may be.

Political system The institution that regulates the use of and access to power that is essential to articulating and realizing individual, local, regional, national, international, or global interests and agendas.

Populations The total number of individuals, traces, documents, territories, households, or groups that could be studied.

Population (study of) A specialty within sociology that focuses on the number of people in and composition of various social groupings that live within specified boundaries and the factors that lead to changes in that social grouping's size and composition.

Population pyramid A series of horizontal bar graphs, each representing a different five-year age cohort.

Positive checks Events that increase mortality, including epidemics of infectious and parasitic disease, war, and famine.

Positive correlation When the value of one variable increases and there is a corresponding increase in the other variable.

Positive sanction An expression of approval and a reward for compliance.

"Positively privileged" social class Those individuals at the very top of the class system.

Predestination The belief that God has foreordained all things, including the salvation or damnation of individual souls.

Predictability For all intents and purposes, a state in which the service or product will be the same no matter where or when it is purchased.

Prejudice A rigid and usually unfavorable judgment about an outgroup that does not change in the face of contradictory evidence and that applies to anyone who shares the distinguishing characteristics of that group.

Prejudiced discriminators (active bigots) Persons who reject the notion of equal opportunity and profess a right, even a duty, to discriminate.

Prejudiced nondiscriminators (timid bigots) Persons who do not accept the creed of equal opportunity but refrain from discriminatory actions primarily because they fear the sanctions they may encounter if they are caught.

Primary groups Social groups characterized by face-to-face contact and strong ties among members.

Primary sector (of the economy) Economic activities that generate or extract raw materials from the natural environment.

Primary sex characteristics The anatomical traits essential to reproduction.

Private households The economic arrangements that exist when the workplace is separate from the home, men are still heads of households but assume the role of breadwinner (as opposed to military commander), and women remain responsible for housekeeping and child rearing.

Private ownership A situation in which individuals (rather than workers, the government, or communal groups) own the raw materials, machines, tools, labor, trucks, buildings, and other inputs needed to produce and distribute goods and services.

Probabilistic model A model in which the hypothesized effect does not always result from a hypothesized cause.

Profane A term describing everything that is not sacred, including those things opposed to the sacred and those things that stand apart from the sacred, although not in opposition to it.

Professionalization (within organizations) A trend in which organizations hire experts with formal training in a particular subject.

Proletariat Those individuals who must sell their labor to the bourgeoisie.

Prophetic religions Religions in which the sacred revolves around items that symbolize significant historical events or around the lives, teachings, and writings of great people.

Pull factors The conditions that encourage people to move into a particular area.

Pure deviants People who have broken the rules and are caught, punished, and labeled as outsiders.

Push factors The conditions that encourage people to move out of an area.

Quantification and calculation The ability of customers to easily evaluate a product or service with numerical indicators.

Race A term that refers to a group of people who possess certain distinctive physical characteristics.

Racial or **ethnic group** A group whose members believe (or outsiders believe) that they share a common national origin, cultural traits, or distinctive characteristics.

Racism An ideology that maintains that something in the biological makeup of a specific racial or ethnic group explains its subordinate or superior status.

Random sample A sample drawn in which every case in the population has an equal chance of being selected.

Rationalization A process whereby thought and action motivated by emotion, superstition, respect for mysterious forces, and tradition are replaced by thought and action grounded in the logical assessment of the most efficient ways to achieve a valued goal or end (known as value-rational action).

Rebellion The full or partial rejection of both the cultural goals and means of attaining those goals and the introduction of a new set of goals and means.

Reentry shock Culture shock in reverse; experienced upon returning home after living in another culture.

Reflexive thinking Stepping outside the self and observing and evaluating it from another's viewpoint.

Reliability The extent to which the operational definition gives consistent results.

Religion According to Emile Durkheim, a system of shared beliefs and rituals about the sacred that bind together a community of worshipers. (See also *Sacred.*)

Representative sample A research sample with the same distribution of characteristics as the population from which it is selected.

Research Fact gathering and explaining enterprise governed by strict rules.

Research design A plan for gathering data that specifies the population and method of data collection.

Research methods The various techniques used to formulate meaningful research questions and to collect, analyze, and interpret facts in ways that allow other researchers to check the results.

Research methods literate Knowing how to collect data worth entering into a computer and then how to interpret the resulting data.

Resocialization The process of discarding values and behaviors unsuited to new circumstances and replacing them with new, more appropriate values and norms.

Retreatism The rejection of both cultural goals and the means of achieving these goals.

Reverse ethnocentrism A type of ethnocentrism in which the home culture is regarded as inferior to a foreign culture.

Rights The behaviors that a person assuming a role can demand or expect from others.

Ritualism The rejection of cultural goals but a rigid adherence to the legitimate means of those goals.

Rituals Rules that govern how people must behave when in the presence of the sacred to achieve an acceptable state of being.

Role The behavior expected of a status in relationship to another status.

Role conflict A predicament in which the expectations associated with two or more roles in a role set contradict one another.

Role set An array of roles.

Role strain A predicament in which contradictory or conflicting expectations are associated with a person's role.

Role taking Stepping outside the self and imagining its appearance and behavior imaginatively from an outsider's perspective.

Sacramental religions Religions in which the sacred is sought in places, objects, and actions believed to house a god or a spirit.

Sacred A term describing everything that is regarded as extraordinary and that inspires in believers deep and absorbing sentiments of awe, respect, mystery, and reverence.

Sample A portion of the cases from a larger population.

Sampling frame A complete list of every case in the population.

Sanctions Reactions of approval and disapproval to others' behavior and appearances.

Scapegoat A person or a group that is assigned blame for conditions that cannot be controlled, threaten a community's sense of well-being, or shake the foundations of a trusted institution.

Schooling A program of formal and systematic instruction that takes place primarily in a classroom but also includes extracurricular activities and out-of-classroom assignments.

Scientific method An approach to data collection in which knowledge is gained through observation and its truth confirmed through verification.

Scientific revolution A condition that occurs when enough people in the community break with an old paradigm and change the nature of their research in favor of the incompatible new paradigm.

Secondary sector (of the economy) Economic activities that transform raw materials into manufactured goods.

Secondary sex characteristics Physical traits not essential to reproduction (breast development, quality of voice, distribution of facial and body hair, and skeletal form) that result from the action of so-called male hormones (androgen) and female hormones (estrogen).

Secondary sources or **archival data** Data that have been collected by other researchers for some other purpose.

Secret deviants People who have broken the rules but whose violation goes unnoticed, or, if it is noticed, prompts no one to enforce the law.

Sect A small community of believers led by a lay ministry, with no formal hierarchy or official governing body to oversee the various religious gatherings and activities.

Secularization A process by which religious influences on thought and behavior are reduced.

Selective perception The process in which prejudiced persons notice only those behaviors or events that support their stereotypes about an outgroup.

Self-administered questionnaire A set of questions given to respondents who read the instructions and fill in the answers themselves.

Self-fulfilling prophecy A concept that begins with a false definition of a situation. The false definition is assumed to be accurate, and people behave as if the definition were true. In the end, the misguided behavior produces responses that confirm the false definition.

Services Activities performed for others that result in no tangible product, such as entertainment, transportation, financial advice, medical care, spiritual counseling, and education.

Sex A biological concept determined on the basis of primary sex characteristics.

Sexist ideologies The ideologies that justify one sex's social, economic, and political dominance over the other.

Sexual property "The relatively permanent claim to exclusive sexual rights over a particular person" (Collins 1971:7).

Sick role A term coined by sociologist Talcott Parsons to represent the rights and obligations accorded people when they are sick.

Significant others People or characters who are important in a child's life in that they have considerable influence on that person's self-evaluation and encourage him or her to behave in a certain manner.

Significant symbols A word, gesture, or other learned sign used to convey a meaning from one person to another.

Simultaneous-independent inventions Situations in which the same invention is created by two or more persons working independently of one another at about the same time.

Situational factors Forces outside an individual's control, such as environmental conditions or bad luck.

Small groups Two to about twenty people who interact with one another in meaningful ways.

Social A quality of interaction that involves the ways of seeing, thinking, acting, and responding to others that are shaped by two factors: (1) forces outside the individuals or (2) the presence of other people who notice what is going on.

Social change Any significant alteration, modification, or transformation in the organization and operation of social life.

Social control Methods used to teach, persuade, or force a group's members, and even nonmembers, to comply with and not deviate from norms and expectations.

Social emotions Internal bodily sensations that we experience in relationships with other people.

Social facts Ideas, feelings, and ways of behaving "that possess the remarkable property of existing outside the consciousness of the individual" (Durkheim 1982:51).

Social identity The category to which a person belongs and the qualities that others believe, rightly or wrongly, to be "ordinary and natural" (Goffman 1963:2) for a member of that category.

Social institutions Widely recognized and relatively predictable and persistent patterns of behavior and interactions that have emerged over time to coordinate and channel human activity toward meeting basic human and societal needs.

Social interactions Everyday events in which at least two people communicate and respond through language and symbolic gestures to affect one another's behavior and thinking. In the process, the parties involved define, interpret, and attach meaning to the encounter.

Social mobility Movement from one class to another.

Social movements Organized, deliberate efforts by believers to transform, reform, or replace some element of culture, and in the process to convert nonbelievers to their position.

Social promotion Passing students from one grade to another on the basis of age rather than academic competency.

Social relativity The view that ideas, beliefs, and behavior vary according to time and place.

Social status A position in a social structure.

Social stratification The systematic process by which people are divided into categories that are ranked on a scale of social worth.

Social structure Two or more people interacting and interrelating in specific expected ways, regardless of the unique personalities involved.

Social trap A situation where people and the organizations they have created lead them into some direction or some set of relationships that later proves to be not only unpleasant or lethal, but also such that they see no easy way to avoid it.

Socialization A process by which people develop their human capacities and acquire a unique personality and identity and by which culture is passed from generation to generation.

Socialism An economic system in which the raw materials and the means of producing and distributing goods and services are collectively owned.

Society A large complex of human relationships; a system of interaction.

Sociological imagination The ability to connect seemingly impersonal and remote historical forces to the most basic incidents of an individual's life. The sociological imagination enables people to distinguish between personal troubles and public issues.

Sociological theory A set of principles and definitions that tell how societies operate and how people relate to one another and respond to the environment.

Sociology The systematic study of social interaction.

Solidarity The ties that bind people to one another in a society.

Special interest groups Groups composed of people who share an interest in a particular economic, political, and/or social issue and who form an organization or join an existing organization with the goal of influencing public opinion and government policy.

Spurious correlation A correlation that is coincidental or accidental because some third variable is related to both the independent and dependent variables.

State (1) A political entity recognized by foreign governments, with a civilian and military bureaucracy to carry out its policies, enforce its rules, and regulate other activities within its borders; (2) a governing body organized to manage and control specified activities of people living in a given territory.

Stationary pyramids Population pyramids characteristic of most developed countries that are similar to constrictive pyramids, except that all age cohorts in the population are roughly the same size and the fertility rate matches replacement level. (See also *Population pyramid* and *Constrictive pyramids.*)

Status group An amorphous group of persons held together by virtue of a lifestyle that has come to be expected of all those who wish to belong to the circle.

Status system A classification of achievements resulting in popularity, respect, and acceptance into the crowd versus disdain, discouragement, and disrespect.

Status value The value that some characteristics impart to those who possess them, causing that individual to be regarded and treated as more valuable or more worthy than persons who possess features from other categories.

Stereotypes Exaggerated and inaccurate generalizations about people who are members of an outgroup.

Stigmas Deeply discrediting statuses in that they overshadow all other statuses a person occupies.

Streaming The arranging of middle school and high school students into instructional groups according to similarities in past academic performance and/or on standardized test scores.

Structural constraints The established and customary rules, policies, and day-to-day practices that affect a person's life chances.

Structural strain Any situation in which (1) the valued goals have unclear limits, (2) people are unsure whether the legitimate means that society provides will allow them to achieve their valued goals, and (3) legitimate opportunities for meeting the goals remain closed to a significant portion of the population.

Structure of opportunities The chances available in a society to achieve a valued goal.

Structured interview An interview in which the wording and the sequence of questions are set in advance and cannot be changed during the interview.

Subcultures Groups that share in some parts of the dominant culture but have their own distinctive values, norms, language, or material culture.

Subjective secularization A decrease in the number of people who view the world and their place in it from a religious perspective.

Symbol Any kind of physical phenomenon to which people assign a name, meaning, or value.

Symbolic gestures Nonverbal cues such as tone of voice and other body movements that convey meaning from one person to another.

Technological determinist Someone who believes that human beings have no free will and are controlled entirely by their material innovations.

Technology The knowledge, skills, and tools used to transform resources into forms with specific purposes, and the skills and knowledge required to use them.

Territories Settings that have borders or that are set aside for particular activities.

Tertiary sector Economic activity related to delivering services, including the creation and distribution of information.

Theory A framework that can be used to comprehend and explain events.

Theory of the demographic transition A model that outlines historical changes in birth- and death rates among the mechanized rich countries and the factors underlying those changes. Some demographers have theorized that a country's birth- and death rates are linked to its level of industrial or economic development.

This-worldly asceticism A belief that people are instruments of divine will and that their activities are determined and directed by God.

Total fertility The average number of children that women bear over their lifetime.

Totalitarianism A system of government characterized by (1) a single ruling party led by a dictator, (2) an unchallenged official ideology that defines a vision of the "perfect" society and the means to achieve that vision, (3) a system of social control that suppresses dissent and opposition, and (4) centralized control over the media and the economy.

Total institutions Institutions in which people surrender control of their lives, voluntarily or involuntarily, to an administrative staff and carry out daily activities with others required to do the same thing.

Traces Materials or other evidence that yield information about human activity.

Tracking The arranging of middle school and high school students into instructional groups according to similarities in past academic performance and/or on standardized test scores.

Traditional A goal pursued because it was pursued in the past.

Traditional authority A type of authority that relies on the sanctity of time-honored norms that govern the selection of someone to a powerful position (chief, king, queen) and that specify responsibilities and appropriate conduct for the individual selected.

Trained incapacity The inability to respond to new and unusual circumstances or to recognize when official rules and procedures are outmoded or no longer applicable.

Transformative powers of history The concept that most significant historical events have dramatic consequences on people's opportunities and that events should be viewed in the context of time and space.

Troubles Personal problems and difficulties that can be explained in terms of individual characteristics such as motivation level, mood, personality, or ability.

Ula The Samoan practice in which individuals engage in highly sexualized entertainment including joking, jesting, and imitating.

Unit of analysis Who or what is to be studied in a research project.

Unprejudiced discriminators (fair-weather liberals) Persons who believe in equal opportunity but engage in discriminatory behaviors because it is to their advantage to do so or because they fail to consider the discriminatory consequences of some of their actions.

Unstructured interview An interview in which the question-answer sequence is spontaneous, open-ended, and flexible.

Upward mobility A gain in rank or status.

Urbanization An increase in the number of cities and the proportion of the population living in cities.

Urban underclass The group of families and individuals in the inner city who are outside the mainstream of the American occupational system and who consequently represent the very bottom of the economic hierarchy.

Validity The degree to which an operational definition measures what it claims to measure.

Value-rational action Action and thought grounded in the logical assessment of the most efficient ways to achieve a valued goal or end.

Value rationale A goal pursued because it is valued, and it is pursued with no thought of foreseeable consequences and often without consideration of the appropriateness of the means chosen to achieve it.

Values General, shared conceptions about what is good, right, appropriate, worthwhile, and important with regard to conduct, appearance, and states of being.

Variable Any trait or characteristic that can change under different conditions or that consists of more than one category.

Vertical mobility Movement that occurs when a change in class status corresponds to a gain or loss in rank.

Voluntary minorities Members of racial or ethnic groups that came to a country expecting to improve their way of life.

White-collar crime "Crimes committed by persons of respectability and high social status in the course of their occupations" (Sutherland and Cressey 1978:44).

Witch-hunt A campaign to identify, investigate, and correct behavior that undermines a group or a country.

World economy An economy consisting of the 160 or so national economies that have established trade relationships with one another.

World-economy An economy encompassing hundreds of countries and thousands of cultures interconnected by a single division of labor.

References

CHAPTER 1

Aasen, Larry. 1997. "Film Box Office Top 100 of 1996." http://www.leonardo.net/aasen/top1996.html

Albrow, Martin. 1990. "Globalisation, Knowledge and Society: Introduction." In Martin, William, and Mark Beittel. 1998. "Toward a Global Sociology? Evaluating Current Conceptions, Methods, and Practices." *The Sociological Quarterly* 39(1):139.

American Sociological Association. 1995. *Careers in Sociology*. Washington, DC.

Barnet, Richard J., and John Cavanagh. 1994. *Global Dreams*. New York: Simon & Schuster.

Behrangi, Samad. 1994. Quoted on page 27 in *International Rural Education Teacher and Literary Critic: Samad Behrangi's Life and Thoughts*, by M. Hussein Fereshtch. *Journal of Global Awareness* (Fall 1994), vol. 2, no. 1.

Berger, Peter L. 1963. *Invitation to Sociology: A Humanistic Perspective*. New York: Anchor.

Boden, Deirdre, Anthony Giddens, and Harvey L. Molotch. 1990. "Sociology's Role in Addressing Society's Problems Is Undervalued and Misunderstood in Academe." *The Chronicle of Higher Education* (February 21):B1, B3.

Brecher, Jeremy, John Brown Childs, and Jill Cutler. 1993. *Global Visions: Beyond the New World Order*. Boston: South End Press.

Canada News Wire. 1998. "Canada Faces Worst-Ever Blood Supply Crisis, Says Canadian Society for Transfusion Medicine." http://www.newswire.ca/releases/May1998/12/c2674.html

Carey, Peter. 1995. Quoted in *Interview*, by Kevin Bacon and Bill Davis. Glimmer Train Stories, Issue 14, Spring 1995. Glimmer Train Press.

Cassidy, John. 1997. "The Return of Karl Marx." *The New Yorker* (October 20, 27).

Center for International Financial Analysis and Research, Inc. 1992. *CIFAR's Company Handbook*. Princeton, NJ: The Center.

Chaliand, Gerard, and Jean-Pierre Rageau. 1995. *The Penguin Atlas of Diasporas*. New York: Penguin Group.

Charyn, Jerome. 1978. "Black Diamond." *The New York Review of Books* (August 17):41.

Conrad, Robert Edgar. 1996. "Slave Trade." Pages 127–128 in *Encyclopedia of Latin American History and Culture,* ed. B. A. Tenenbaum. New York: Scribner's.

Donnelly, Peter. 1996. "The Local and the Global: Globalization in the Sociology of Sport." *Journal of Sport and Social Issues* (August):239

DuBois, W. E. B. [1899] 1996. *The Philadelphia Negro: A Social Study*. Philadelphia: University of Pennsylvania Press.

———. [1903] 1996. *The Souls of Black Folk*. http://www.cc.columbia.edu/acis/bartleby/dubois/

Durkheim, Emile. [1888] 1978. *On Institutional Analysis,* ed. and trans. M. Traugott. Chicago: University of Chicago Press.

———. 1982. *The Rules of Sociological Method and Selected Texts on Sociology and Its Method,* ed. Steven Lukes, trans. W. D. Halls. New York: Macmillan Press Ltd.

El Diario. 1998. Blood Advertisements. 17 August 1998. Juarez, Mexico.

Engels, Frederick. [1883] 1993. "Frederick Engels' Speech at the Grave of Karl Marx." http://csf.Colorado.EDU/psn/marx/Archive/1883-Death/burial.htm. Transcribed by Mike Lepore.

Fletcher, Max E. 1974. "Harriet Martineau and Ayn Rand: Economics in the Guise of Fiction." *American Journal of Economics and Sociology* 33(4):367–379.

Freund, Julien. 1968. *The Sociology of Max Weber*. New York: Random House.

Gates, Henry Louis. 1995. "The Political Scene: Powell and the Black Elite." *The New Yorker* (September 25):64–80.

Giddens, Anthony. 1990. *The Consequences of Modernity*. In Martin, William, and Mark Beittel. 1998. "Toward a Global Sociology? Evaluating Current Conceptions, Methods, and Practices." *The Sociological Quarterly* 39(1):139.

Gordon, John Steele. 1989. "When Our Ancestors Became Us." *American Heritage* (December):106–221.

Gould, Stephen Jay. 1981. *The Mismeasure of Man*. New York: Norton.

Harvey, Jean, Genevieve Rail, and Lucie Thibault. 1996. "Globalization and Sport: Sketching a Theoretical Model for Empirical Analyses." *Journal of Sport and Social Issues* (August):258.

Holloway, Thomas H. 1996. "Immigration." Pages 239–242 in *Encyclopedia of Latin American History and Culture,* ed. B. A. Tenenbaum. New York: Scribner's.

International Trade Administration. 1997a. "U.S. Imports for Consumption: December 1997 and 1997 Year-to-Date." http://www.itadoc.gov/industry/otea/Trad...il/Latest-December/Imports/30/300210.html

———. 1997b. "U.S. Domestic Exports: December 1997 and 1997 Year-to-Date." http://www.ita.doc.gov/industry/otea/Trad...il/Latest-December/Exports/30/300210.html

Japan Surveillance Report. 1998. "AIDS Scandal." http://www.t3.rim.or.jp/~aids/survey_e.html

Kahn, Chuck. 1998. http://www.vex.net/~odin/Gross/

Konner, Melvin. 1997. "Take Only as Directed." *The New York Times Book Review* (October 17):28.

Krever Commission Report. 1997. http://www.hcsc.gc.ca/datapcb/communc.home/krever/k_main_e.html#Table

Lemert, Charles. 1995. *Sociology After the Crisis*. Boulder, CO: Westview Press.

Lengermann, Patricia M. 1974. *Definitions of Sociology: A Historical Approach*. Columbus, OH: Merrill.

Lewis, David Levering. 1993. *Biography of a Race, 1868–1919.* New York: A John Macrae Book/Holt.

Lewis, Paul. 1998. "Marx's Stock Resureges on 150-Year Tip." *The New York Times* (June 27):A17.

Lohr, Steve. 1993. "New Appeals to Pocketbook Patriots." *The New York Times* (January 22):23+.

———. 1997. "The Future Came Faster in the Old Days." *The New York Times* (October 5):A1.

Marcus, Steven. 1998. "Marx's Masterpiece at 150." *The New York Times Book Review* (April 26):39.

Martin, William, and Mark Beittel. 1998. "Toward a Global Society? Evaluating Current Conceptions, Methods, and Practices." *The Sociological Quarterly* 39(1):139.

Martineau, Harriet. [1837] 1968. *Society in America,* ed. and abridged by S. M. Lipset. Gloucester, MA: Peter Smith.

Marx, Karl. [1881] 1965. "The Class Struggle." Pages 529–535 in *Theories of Society,* ed. T. Parsons, E. Shils, K. D. Naegele, and T. R. Pitts. New York: Free Press.

Mills, C. Wright. 1959. *The Sociological Imagination.* New York: Oxford University Press.

Moore, William. 1966. "Global Sociology: The World as a Singular System." In Martin, William, and Mark Beittel. 1998. "Toward a Global Sociology? Evaluating Current Conceptions, Methods, and Practices." *The Sociological Quarterly* 39(1):139.

The New Columbia Encyclopedia. 1975. "Steamship." New York: Columbia University Press.

Ornstein, Robert, and Paul Ehrlich. 1989. *New World, New Mind.* New York: Touchstone.

Page, Clarence. 1996. *Showing My Color.* New York: HarperCollins.

Random House Encyclopedia. 1990. "European Imperialism in the 19th Century." New York: Random House.

Reich, Robert B. 1988. "Corporation and Nation." *The Atlantic Monthly* (May):76–81.

Reuters News Service. 1998. "Britain to import blood plasma due to 'mad cow' risk." http://www.nando.net/newsroom/ntn/health/022798/health22_19947_noframes.html

Schmitt, Eric. 1997. "Made in U.S.A: A Standard Undaunted in a Global War." *The New York Times* (December 6):A1.

Shelton, Rachel. 1998. "Concept Paper." Paper prepared for SOC 210-Race and Gender.

Terry, James L. 1983. "Bringing Women . . . In: A Modest Proposal." *Teaching Sociology* 10(2):251–261.

Texas A&M. 1998. "What Is Sociology?" http://www.tamu.edu/socdept/whtsoci.shtml

Thorne, Barrie. 1993. Quoted on page xxxi in *Down to Earth Sociology: Introductory Readings,* by James M. Henslin. New York: Free Press.

U.S. Census Bureau. 1996. "International Demographic Data." http://www.census.gov/ftp/pub/ipc/www/idbsum.html

U.S. Central Intelligence Agency. 1995. *World Factbook 1995.* http://www.odci.gov/cia/publications/95fact/index.html

U.S. Department of Commerce. 1993. *Globalization of the Mass Media.* Washington, DC: U.S. Government Printing Office.

U.S. Department of Defense. 1996. "Active Duty Military Personnel Strengths by Regional Area and by Country." http://web1.whs.osd.mil/mmid/military/309ab966.htm (June 30).

U.S. Department of the Interior, Bureau of Mines. *Minerals Yearbook. Vol. 7: Area Reports: International 1998.* Washington, DC: U.S. Government Printing Office.

U.S. Federal Communication Commission. 1997. "International Telecommunications." http://www.fcc.gov/Bureaus/CommonCarrier/Reports/FCCState_Link/socc.html

Variety. 1995. "Worldwide Rentals Beat Domestic Take." (February 13–19):280.

Webb, R. K. 1960. *Harriet Martineau, a Radical Victorian.* New York: Columbia University Press.

White House press release. 1995. "Miami International to Be Designated 'Reinvention Lab.'" Office of the Vice President, August 30.

Zuboff, Shoshana. 1988. *In the Age of the Smart Machine.* New York: Basic Books.

CHAPTER 2

Amirahmadi, Hooshang, and Weiping Wu. 1995. "Export Processing Zones in Asia." *Asian Survey* 35(9):828.

Bacon, David. 1997. "Evening the Odds: Cross-border Organizing Gives Labor a Chance." *The Progressive* 61(7):29.

———. 1995. "After NAFTA." *Environmental Action Magazine* 27(3):33.

Bearden, Tom. 1993. "Focus—Help Wanted (Interview with Anonymous Denver Woman on Her Use of Undocumented Worker for Child Care)." *MacNeil/Lehrer Newshour* (transcript #4548). New York: WNET.

Blakeslee, Sandra. 1994. "Tanning: A Failed Skin Restoration Project." *The New York Times* (December 7).

Blumer, Herbert. 1962. "Society as Symbolic Interaction." In *Human Behavior and Social Processes,* ed. A. Rose. Boston: Houghton Mifflin.

Bradsher, Keith. 1998a. "Making Threats and Losing Cash in G.M. Strike." *The New York Times* (July 8):A1.

———. 1998b. "Little Progress Seen in Strike Talks Between G.M. and U.A.W." (July 9):C2.

Brayshaw, Charles. 1998. "Freer Trade." *Twin Plant News* (May):35.

Brofenbrenner, Kate. 1997. "We'll Close! Plant Closings, Plant-Closing Threats, Union Organizing and NAFTA." *Multinational Monitor* 18(3). http://www.essential.org/monitor/hyper/mm0397.04.html

Carver, Terrell. 1987. *A Marx Dictionary.* Totowa, NJ: Barnes & Noble.

Case, Brendan M. 1996. "Cashing In on Immigration." *Los Angeles Times* (September 14):17.

Castillo, Arture. 1996. Personal correspondence (December 6).

Coalition for Justice in the *Maquiladoras.* 1996. http://www.pctvi.com/laamn/cjm.html

Cornejo, José A. Perez. 1988. "The Implications for the U.S. Economy of Tariff Schedule Item 807 and Mexico's Maquila Program." *Maquiladora Newsletter* 15(5):2–5.

Dillon, Sam. 1998. "A 20-Year G.M. Parts Migration to Mexico." *The New York Times* (June 24) C1.

Dodge, David. 1988. "Insights into the Mexicans." Pages 46–55 in *Fodor's Mexico,* ed. A. Beresky. New York: Fodor's Travel Publications.

Dowla, Asif. 1997. "Export Processing Zones in Bangladesh." *Asian Survey* 37(6):561.

Ehrenreich, Barbara, and Annette Fuentes. 1985. "Life on the Global Assembly Line." Pages 373–388 in *Crisis in American Social Institutions,* 6th ed., ed. J. H. Skolnick and E. Currie. Boston: Little, Brown.

Engels, Friedrich. [1881] 1996. "A Fair Day's Wage for a Fair Day's Work." http://www.idbsu.edu/surveyrc/Staff/jaynes/marxism/fairwage.htm

Essential Organization. 1996. "Executive Summary." *Monitoring Border and Environmental and Health Conditions Two Years After NAFTA.* http://www.essential.org/orgs/public_citizen/pctrade/borderexec

Fallows, James. 1993. "Commentator Says NAFTA Impact Will Be Minimal to Nil."

"Morning Edition." National Public Radio. September 17, 1993.

Gans, Herbert. 1972. "The Positive Functions of Poverty." *American Journal of Sociology* 78:275–289.

Garcia, Juan Ramon. 1980. *Operation Wetback: The Mass Deportation of Mexican Undocumented Workers in 1954.* Westwood, CT: Greenwood.

Glionna, John M. 1992. "The Paper Chase." *Los Angeles Times* (November 8):E1.

Goldberg, Raymond. 1995. "We're Creating Jobs in the Third World." *The New York Times* (August 1):A10.

Golden, Tim. 1995. "Mexicans Find Dreams Devalued." *The New York Times* (January 8):E5.

Greenhouse, Steven. 1997. "Voluntary Rules on Apparel Labor Proving Elusive." *The New York Times* (February 1):A1.

Haederle, Michael. 1992. "Composing a Message for the Ages: 'Keep Out!'" *Los Angeles Times* (October 5):A3.

Halberstam, David. 1986. *The Reckoning.* New York: Morrow.

Hamashige, Hope. 1995. "Wiring for Dollars." *Los Angeles Times* (February 9):D1

Herzog, Lawrence A. 1985. "The Cross-Cultural Dimensions of Urban Land Use Policy on the U.S.–Mexico Border: A San Diego–Tijuana Case Study." *Social Science Journal* (July):29–46.

Jacobson, Gary. 1988. "The Boom on Mexico's Border." *Management Review* (July):21–24.

Kraul, Chris. 1996. "State and Baja Team Up to Promote Business Deal." *Los Angeles Times* (July 25):A1.

Krause, Charles. 1996. "A Tale of Two Cities." *Online NewsHour: Across the Border.* http://www.pbs.org/newshour/ bb/ election/august96/sandiego_ 8-9.html

Lal, Barbara Ballis. 1995. "Symbolic Interaction Theories." *American Behavioral Scientist* 38(3):421.

Laredo/Webb County. 1997. "Ten Years of Economic Growth for Laredo/Webb County: 1987–1997." http://www. laredo-ldf.com/tenyears.html

Lee, Patrick, and Chris Kraul. 1993. "Uniqueness of *Maquiladora* Could Fade." *Los Angeles Times* (November 19):D1.

Lekachman, Robert. 1985. "The Specter of Full Employment." Pages 74–80 in *Crisis in American Institutions,* ed. J. H. Skolnick and E. Currie. Boston: Little, Brown.

Levi, Issaac A. 1995. "Peso Crisis Haunts Mexico." *USA Today* (December 20):4B.

Los Angeles Times. 1996. Advertisements (October 31).

Magaziner, Ira C., and Mark Patinkin. 1989. *The Silent War: Inside the Global Business Battles Shaping America's Future.* New York: Random House.

Marx, Karl. [1888] 1961. "The Class Struggle." Pages 529–535 in *Theories of Society,* ed. T. Parsons, E., and K. D. Shils.

Mata, Ruben. 1998. "Canadian *Maquilas.*" *Twin Plant News* (April):18.

McAllen Economic Development Corporation. 1995–1997. http://www.medc.org/

Mead, George H. 1934. *Mind, Self and Society.* Chicago: University of Chicago Press.

Meredith, Emily, and Garrett Brown. 1995. "The *Maquiladora* Health and Safety Support Network: A Case Study of Public Health Without Borders." *Social Justice* 22(4):85–87.

Meredith, Robyn. 1997. "The Brave New World of General Motors." *The New York Times* (October 26).

Merton, Robert K. 1967. "Manifest and Latent Functions." Pages 73–137 in *On Theoretical Sociology: Five Essays, Old and New.* New York: Free Press.

Morrow, David J. 1996. "Trials of Human Guinea Pigs." *The New York Times* (September 29):10.

Myerson, Allen R. 1995. "Greyhound: The Airline of the Road." *The New York Times* (January 18):C1.

———. 1997a. "Borderline Working Class." *The New York Times* (May 8):C1.

———. 1997b. "On the Road to Monterrey." *The New York Times* (December 18):C1.

The NAFTA. 1993. Washington, DC: U.S. Government Printing Office.

The New York Times. 1996. "Study Shows Danger to Skin and Gives Hope of a Savior." (January 25).

Noll, Cheryl L. 1992. "Mexican Maquiladora Workers: An Attitude Toward Working." *Southwest Journal of Business and Economics* (Spring):1–7.

O'Hare, William P. 1988. "The Rise of Poverty in Rural America." *Population Trends and Public Policy* (July). Washington, DC: Population Reference Bureau.

Parra, Angelo, Edmundo Elias Fernandez, and Carol S. Osmond. 1996. "*Maquila* Program: Key Features of the *Maquiladora* Program." *Twin Plant News* (July):21–24.

Partida, Gilbert A., and Cesar Ochoa. 1990. "Border Waste Program." *Twin Plant News* (April):29–31.

Patten, Mike. 1996. "Border University Offers Engineering Help to Maquilas." *Twin Plant News* (June):26–32.

Pearce, Jean. 1987. "Mexico Holds Special Place in the Heart of Japan: An Interview with Sergio Gonzalez Galvez, Ambassador of Mexico." *Business Japan* (November/December):47–49.

Pina, Rudy R. 1996. "Verifying Origin." *Twin Plant News* (April):39–40.

Preston, Julia. 1998. "Latin America Is Buffeted, But Seems Stronger Than in '94." *The New York Times* (September 1):A3.

Puente, Maria. 1997. "At the Border, Preventive Policing." *USA Today* (February 18):3A.

Risen, James. 1992. "U.S., Mexico, Canada Agree to Form Huge Common Market." *Los Angeles Times* (August 13):A1+.

Roderick, Larry M., and J. Rene Villalobos. 1992. "Pollution at the Border." *Twin Plant News* (November):64–67.

Rodriguez, Richard. 1992. *Days of Obligation: an Argument with My Mexican Father.* New York: Viking.

Rose, Kenneth J. 1988. *The Body in Time.* New York: Wiley.

Sanders, Thomas G. 1987. "Tijuana, Mexico's Pacific Coast Metropolis." *UFSI Reports* 38:1–8.

Sexton, Joe. 1995. "A Factory Reinvents the Sweatshop." *The New York Times* (May 29):Y18.

Smith, Hedrick. 1998. "A Fast-Track Reality Check: A Veteran Reporter Looks at the Not-So-Nifty Side of NAFTA." *Washington Monthly* 30(1):8.

Suro, Roberto. 1991. "Border Boom's Dirty Residue Imperils U.S.–Mexico Trade." *The New York Times* (March 31):Y1.

Sweatshop Watch. 1997. http://www. sweatshopwatch.org/swatch/industry/

Thelen, David. 1992. "Of Audiences, Borderlands, and Comparisons: Toward the Internationalization of American History." *Journal of American History* (September):432–451.

Tuleja, Tad. 1987. *Curious Customs: The Stories Behind 296 Popular American Rituals.* New York: Harmony.

Tumin, Melvin. 1964. "The Functionalist Approach to Social Problems." *Social Problems* 12:379–388.

Twin Plant News. 1990a. Advertisement: "Maquila's Multi-Billion Dollar Market: At Your Fingertips!" (July):53.

———. 1990b. Advertisement: "Shopping the Interior? Picture This. . . . " (February):7.

———. 1996a. Advertisement: "TPN at 11: A Look Ahead." (August):36–37.

———. 1996b. Advertisement: "Yucatan." (August):34.

———. 1998a. "Professional Daily Minimum Wages in Pesos." (April):35.

———. 1998b. "QA: The North American Free Trade Agreement. . . . " (May):9.

———. 1998c. "QA: There Has Been a Significant Increase in the Number of Maquilas. . . . " (June):9.

————. 1998d. *"Maquila* Scoreboard." (June):45.

Uchitelle, Louis. 1993. "Those High-Tech Jobs Can Cross the Border, Too." *The New York Times* (March 28):E4.

U.S. Bureau of Labor Statistics. 1998. "Real Earnings in May 1998." http://stats.bls.gov/newsrels.htm

U.S. Central Intelligence Agency. 1998. *World Factbook 1998.* http://www.odci.gov/cia.

U.S. Central Intelligence Agency. 1995. *World Factbook 1995.* http://www.odci. gov/cia/publications/95fact/index.html

U.S. Department of Justice. 1997. *INS Fact Book.*

U.S. Department of Labor, Bureau of Labor Statistics. 1998.

U.S. Department of State. 1998. *Background Notes: Mexico, March 1998.* http://www.state.gov/www/background_ notes/mexico_0398rsity_bgn.html

University of California, Davis Migration Dialogue. 1998. "Remittances: Mexico." http://migration.ucdavis.edu

Vargas, Lucinda. 1998. *"Maquila* Update." *Twin Plant News* (April):41.

Verhovek, Sam Howe. 1998a. "Trade Pact Brings Drug Searches and Traffic Jams." *The New York Times* (March 20):A10.

————. 1998b. "Torn Between Nations, Mexican-Americans Can Have Both." *The New York Times* (April 14):A12.

————. 1998c. "Benefits of Free-Trade Pact Bypass Texas Border Towns." *The New York Times* (June 23):A1.

————. 1998d. "Mexican Border Fence Friendlier Than a Wall." *The New York Times* (December 8):A1.

Weisman, Alan. 1986. *La Frontera: The United States Border with Mexico.* New York: Harcourt Brace Jovanovich.

White, George, and Andrea Maier. 1992. "A Closer Look at the Trade Agreement." *Los Angeles Times* (August 13):A7.

White, Leslie A. 1949. *The Science of Culture: A Study of Man and Civilization.* New York: Farrar, Straus.

Wilson, Patricia A. 1992. *Exports and Local Development: Mexico's New Maquiladoras.* Austin: University of Texas Press.

Working Mother. 1997. "The Best Companies for Working Mothers." (October):25–96.

Xenon Laboratories, Inc. 1998. "Universal Currency Converter." http://www.xc.net/cgi-bin/ucc/convert

Yim, Yong Soon. 1989. "American Perceptions of Korean-Americans: An Analytical Study of a 1988 Survey." *Korean and World Affairs* 13 (Fall):519–542.

CHAPTER 3

Allison, Anne. 1991. "Japanese Mothers and *Obentōs:* The Lunch-Box as Ideological State Apparatus." *Anthropological Quarterly* 64:195–208.

American Society of Microbiology. 1996. "Operation Clean Hands: Press Release." http://www.asmusa.org/pcsrc/och.htm

Bailey, William T., and Wade C. Mackey. 1989. "Observations of Japanese Men and Children in Public Places: A Comparative Study." *Psychological Reports* 65:731–734.

Cameron, William B. 1963. *Informal Sociology.* New York: Random House.

Collins, Randall. 1988. "Theoretical Continuities in Goffman's Work." Pages 41–63 in *Erving Goffman: Exploring the Interaction Order,* edited by P. Drew and A. Wooton. Boston: Northeastern University Press.

Dornbusch, Rudi. 1998. "What's the Weakest Link In the World Economy? Japan." *Business Week* (January 12):28.

Drew, Paul, and Anthony Wooton. 1988. "Introduction." Pages 1–13 in *Erving Goffman: Exploring the Interaction Order,* edited by P. Drew and A. Wooton. Boston: Northeastern University Press.

Dye, Lee. 1995. "Duplication of Research Isn't as Bad as It Sounds." *Los Angeles Times* (April 26):D5+.

Fallows, James. 1989. "The Real Japan." *New York Review of Books* (July 20):23–28.

Giddens, Anthony. 1988. "Goffman as a Systematic Social Theorist." Pages 250–279 in *Erving Goffman: Exploring the Interaction Order,* edited by P. Drew and A. Wooton. Boston: Northeastern University Press.

Goffman, Erving. 1959. *The Presentation of Self in Everyday Life.* New York: Anchor.

————. 1961. *Asylums: Essays on the Social Situation of Mental Patients and Other Inmates.* New York: Anchor.

————. 1981. *Forms of Talk.* Oxford: Blackwell.

————. 1983. "The Interaction Order." *American Sociological Review* (February):1–17.

Goodman, Roger. 1993. *Japan's International Youth: The Emergence of a New Class of School Children.* Oxford: Clarendon.

Gregg, Alan. 1989. Quoted in *Science and the Human Spirit: Contexts for Writing and Learning,* by Fred D. White. Wadsworth Publishing Company. Belmont, CA.

Hacker, Andrew. 1997. "Review of 'The New American Reality: Who We Are, How We Got There' by Reynolds Farley." *Contemporary Sociology* 26(4):478.

Hagan, Frank E. 1989. *Research Methods in Criminal Justice and Criminology.* New York: Macmillan.

Hamada, Tomoko. 1996. "Unwrapping Euro-American Masculinity in a Japanese Multinational Corporation." Pages 160–176 in *Masculinities in Organizations,* edited by Cliff Cheng. Thousand Oaks, CA: Sage Publications.

Hilts, Philip J. 1991. "U.S. Abandons Idea of Carrying Out Household Survey on Cases of AIDS." *The New York Times* (January 11):A10.

Horan, Patrick M. 1995. "Review of 'Working with Archival Data: Studying Lives.'" *Contemporary Sociology* (May):423–424.

International Trade Association. 1998. "U.S–Japan Merchandise Trade Data (1960–1997)." http://www.ita.doc.gov/ region/japan/usjmt.html

Ishii-Kuntz, Masako. 1992. "Are Japanese Families 'Fatherless'?" *Sociology and Social Research* 76(3):105–110.

Japan Ministry of Finance. 1998. "Japan's Trade Balance in Dollar Terms." http://www.jetro.go.jp/FACTS/ UA-HANDBOOK/13.html

Joseph, Michael. 1982. *The Timetable of Technology.* London: Marshal Editions.

Jussaume, Ramond A., Jr., and Yoshiharu Yamada. 1990. "A Comparison of the Viability of Mail Surveys in Japan and the United States." *Public Opinion Quarterly* 54:219–228.

Katzer, Jeffrey, Kenneth H. Cook, and Wayne W. Crouch. 1991. *Evaluating Information: A Guide for Users of Social Science Research,* 3d ed. New York: McGraw-Hill.

Klapp, Orrin E. 1986. *Overload and Boredom: Essays on the Quality of Life in the Information Society.* New York: Greenwood.

Kōji, Kata. 1983. "Pachinko." Page 143 in *Kodansha Encyclopedia of Japan.* New York: Kodansha International.

Kristof, Nicholas. 1997a. "Introducing a Child to the Culture of Shame." *The New York Times* (August 31):E3.

————. 1997b. "I Lived to Tell About Japan's Medical Miracle." *The New York Times* (November 30):7.

————. 1997c. "In Japan, Nice Guys (and Girls) Finish Together." *The New York Times* (April 12):7.

Lewis, Catherine C. 1988. "Japanese First-Grade Classrooms: Implications for U.S. Theory and Research." *Comparative Education Review* 32(2):159–172.

Lucky, Robert W. 1985. "Message by Light Wave." *Science* (November):112–113.

Lynd, Robert S., and Helen M. Lynd. [1929] 1956. *Middletown: A Study in Modern*

American Culture. New York: Harcourt, Brace & World.

Macer, Darryl. 1995. "Editorial: Why a New Journal?" *Eubois Journal of Asian and International Bioethics.* http://20bell.biol.tsukuba.ac.jp/~macer/ EJ51A.html

Marsa, Linda. 1992. "Scientific Fraud." *Omni* (June):39+.

Michael, Donald. 1984. "Too Much of a Good Thing? Dilemmas of an Information Society." *Technological Forecasting and Social Change* 25(4):347–354.

Miller, Matthew. 1998. "Don't Worry, Be Happy: Why the Huge Trade Deficits Ahead May Be Good News." *U.S. News & World Report* 124(7):53.

Mouer, Ross, and Yoshio Sugimoto. 1990. *Images of Japanese Society.* London: Kegan Paul International.

National Academy of Sciences. 1995. "Science Ethics Guide Updated, Expanded for Graduate Students." http://www2.nas.edu.whatsnew/20fe.html

Ogawa, Noahiro, and Robert D. Retherford. 1993. "Care of the Elderly in Japan: Changing Norms and Expectations." *Journal of Marriage and the Family* (August):585–597.

Paulos, John A. 1988. *Innumeracy: Mathematical Illiteracy and Its Consequences.* New York: Hill & Wang.

Rathje, William L., and Cullen Murphy. 1992. *Rubbish: The Archaeology of Garbage.* New York: HarperCollins.

Robinson, Richard D. 1985. "Another Look at Japanese–United States Trade Relations." *UFSI Reports* 19:1–7.

Roethlisberger, F. J., and William J. Dickson. 1939. *Management and the Worker.* Cambridge, MA: Harvard University Press.

Rossi, Peter H. 1988. "On Sociological Data." Pages 131–154 in *Handbook of Sociology,* edited by N. Smelser. Newberry Park, CA: Sage.

Ryūzo, Satō. 1991. "Maturing from We-ism to Global You-ism." *Japan Quarterly* 38:273–282.

Sanger, David E. 1992. "A Defiant Detroit Still Depends on Japan." *The New York Times* (February 27):A1+.

Schonberg, Harold C. 1981. "Sumō Embodies Ancient Rituals." *The New York Times:*B9.

Shotola, Robert W. 1992. "Small Groups." Pages 1796–1806 in *Encyclopedia of Sociology,* vol. 4, edited by E. F. Borgatta and M. L. Borgatta. New York: Macmillan.

Singleton, Royce A., Jr., Bruce C. Straits, and Margaret Miller Straits. 1993. *Approaches to Social Research,* 2d ed. New York: Oxford University Press.

Smith, Joel. 1991. "A Methodology for the Twenty-First Century Sociology." *Social Forces* 70(1):1–17.

Swazey, Judith P., Melissa S. Anderson, and Karen Seashore Lewis. 1993. "Ethical Problems in Academic Research." *American Scientist* (November–December):542–553.

Thayer, John E. III. 1983. "Sumo." Pages 270–274 in *Kodansha Encyclopedia of Japan,* vol. 7. Tokyo: Kodansha.

Thomas, Bill. 1992. "King Stacks." *Los Angeles Times Magazine* (November 15):31+.

Ting-Toomey, Stella. 1991. "Intimacy Expressions in Three Cultures: France, Japan, and the United States." *International Journal of Intercultural Relations* 15(1):29.

Totten, Bill. 1990. "Japan's Mythical Trade Surplus." *The New York Times* (December 9):F13.

Tuss, Paul, Jules Zimmer, and Hsiu–Zu Ho. 1995. "Causal Attributions of Underachieving Fourth-Grade Students in China, Japan, and the United States." *Journal of Cross-Cultural Psychology* (July):408–425.

Tyson, Laura. 1997. "Trade Deficits Won't Ruin Us." *The New York Times* (November 24):A17.

United Nations. 1998. "Population (in Thousands) for the Countries of the World: 1996." gopher://gopher.undp.org:70/00/ungophers/popin/wdtrends/pop.1996.txt

U.S. Central Intelligence Agency. 1997. *World Factbook 1997.* http://www.odci.gov/cia/publications/factbook/

U.S. Department of Commerce. 1998. *U.S. Foreign Trade Highlights 1997.* http://www.ita.doc.gov/industry/oten/usfth/+13.prn

U.S. Department of State. 1996. *Background Notes: Japan, August 1996.* gopher://dosfan.lib.uic.edu/ 1D-%3A22525%ABackground%20Ser

U.S. Federal Communications Commission. 1997. "International Telecommunications." http://www.fcc.gov/Bureaus/Common_Carrier/Reports/FCC-State_Link/socc.htm

Vogt, Paul W. 1996. "Dictionary of Statistics and Methodology: A Nontechnical Guide for the Social Sciences." http://www.burns.com/wcbspurcorl.htm

White, R. D., ed. 1989. *Science and the Human Spirit.* Belmont, CA: Wadsworth.

Whyte, William H. 1988. *City: Rediscovering the Center.* New York: Doubleday.

Winkin, Yves. 1989. "Erving Goffman (1922–1982)." Pages 223–225 in *International Encyclopedia of Communications,* edited by E. Barnouw, G. Berbner, W. Schramm, T. L. Worth, and L. Gross. New York: Oxford University Press.

The World Almanac and Book of Facts 1998. 1997. Mahwak, NJ: World Almanac Books.

CHAPTER 4

An, Heejung. 1997. "Sports Categories in Korea." http://www.itp.tsoa.nyu.edu/~student/heejung/sports5.htn

Behrangi, Samad. 1994. Quoted in "International Rural Education Teacher and Literacy Critic: Samad Behrangi's Life and Thought." *Journal of Global Awareness* 21(1):27–35.

Benedict, Ruth. 1976. Quoted in *The Person: His and Her Development Throughout the Life Cycle,* Lidz, Theodore. New York: Basic Books.

Berkhofer, Robert F., Jr. 1978. *The White Man's Indian: Images of the American Indian from Columbus to the Present.* New York: Knopf.

Berreby, David. 1995. "Unabsolute Truths: Clifford Geertz." *The New York Times Magazine* (April 9):44–47.

Bok, Lee Suk. 1987. *The Impact of U.S. Forces in Korea.* Washington, DC: National Defense University Press.

Breton, Raymond. 1967. "Institutional Completeness of Ethnic Communities and the Personal Relations of Immigrants." *American Journal of Sociology* 70:193–205.

Brown, Rita Mae. 1988. *Rubyfruit Jungle.* New York: Bantam.

Cambridge International Dictionary of English. 1995. "Culture." Cambridge: Cambridge University Press.

Cherni, Leigh. 1998. "Telephone Conversation with Andrea Simone Bowers." July 1 and July 14.

CNN/Sports Illustrated. 1998. "1998 Phoenix Mercury Roster." http://www.cnnsi.com/basketball/wnba/stats/1998/news.phoros.html

The Comenius Group. 1996. *The Weekly Idiom.* http://www.comenius.com/idiom/index.html

Cremation Association of North America. 1997. "Cremation Stats Nationwide." http://www.cremation.org/stats.htm

Doi, Takeo. 1986. *The Anatomy of Dependence,* trans. by J. Bester. Tokyo: Kodansha International.

Evinger, William, ed. 1995. *Directory of U.S. Military Bases Worldwide.* Phoenix, AZ: Oryx Press

Fallows, James. 1988. "Trade: Korea Is Not Japan." *Atlantic Monthly* (October):22–33.

Frontline. 1988. "American Game, Japanese Rules" (transcript #611). Boston: WGBH Educational Foundation.

Geertz, Clifford. 1995. *After the Fact: Two Countries, Four Decades, One Anthropologist.* Cambridge, MA: Harvard University Press.

Goodavage, Maria. 1996. "Active Adults Bring Their Pampered Pooches Along." *USA Today* (August 20):D1.

Gordon, Emily Fox. 1995. "Faculty Brat: A Memoir." *Boulevard* 10(1–2):1–17.

Gordon, Steven L. 1981. "The Sociology of Sentiments and Emotion." Pages 562–592 in *Social Psychology Sociological Perspectives,* edited by M. Rosenberg and R. H. Turner. New York: Basic Books.

Halberstam, David. 1986. *The Reckoning.* New York: Morrow.

Hannerz, Ulf. 1992. *Cultural Complexity: Studies in the Social Organization of Meaning.* New York: Columbia University Press.

———. 1993. "When Culture Is Everywhere: Reflections on a Favorite Concept." *Ethnos* 58 (1/2):95–111.

Henry, William A. 1988. "No Time for the Poetry: NBC's Cool Coverage Stints on the Drama." *Time* (October 3):80.

Herskovits, Melville J. 1948. *Man and His Works: The Science of Cultural Anthropology.* New York: Knopf.

Hochschild, Arlie R. 1976. "The Sociology of Feeling and Emotion: Selected Possibilities." Pages 280–307 in *Another Voice,* edited by M. Millman and R. Kanter. New York: Octagon.

———. 1979. "Emotion Work, Feeling Rules, and Social Structure." *American Journal of Sociology* 85:551–575.

Holt International Children's Services. 1998. http://www.holtintl.org/intro.html

Hughes, Everett C. 1984. *The Sociological Eye: Selected Papers.* New Brunswick, NJ: Transaction.

Hurst, G. Cameron. 1984. "Getting a Piece of the R.O.K.: American Problems of Doing Business in Korea." *UFSI Reports* 19.

Ingram, Erik. 1992. "Water Use Continues to Decline: Bay Area Districts Report Record Savings." *San Francisco Chronicle* (July):A15.

Institute of International Education. 1997. *Open Doors 1996/97: Report on International Educational Exchange.* Todd Davis, ed.

International Baseball Association. 1993. *International Baseball Association Membership List* (July). Indianapolis: International Baseball Association.

Jun, Suk-ho, and Daniel Dayan. 1986. "An Interactive Media Event: South Korea's Televised 'Family Reunion.'" *Journal of Communication* (Spring):73–82.

Kang, K. Connie. 1995. *Home Was the Land of Morning Calm: A Saga of a Korean-American Family.* Reading, MA: Addison-Wesley.

Kim, Bo-Kyung, and Kevin Kirby. 1996. Personal correspondence (April 25).

Kim, Choong Soon. 1989. "Attribute of 'Asexuality' in Korean Kinship and Sundered Koreans during the Korean War." *Journal of Comparative Family Studies* 20(3):309–325.

Koehler, Nancy. 1986. "Re-entry Shock." Pages 89–94 in *Cross-Cultural Reentry: A Book of Readings.* Abilene, TX: Abilene Christian University Press.

Korean Overseas Information Service. 1995–1996. *Tourism.* http://korea.emb. washington.dc.us/Kois/explore/Facts/tourism.html

Kristof, Nicholas D. 1995. "Where a Culture Clash Lurks Even in the Noodles." *The New York Times* (September 4):Y4.

———. 1998. "Big Macs to Go." *The New York Times Book Review* (March 22):18.

Lamb, David. 1987. *The Arabs: Journeys Beyond the Mirage.* New York: Random House.

Larson, James F., and Nancy K. Rivenburgh. 1991. "A Comparative Analysis of Australian, U.S., and British Telecasts of the Seoul Olympic Opening Ceremony." *Journal of Broadcasting and Electronic Media* 35(1):75–94.

Lee, Jennifer. 1994. "The Invisible Nation of Korean Emigrants." *Korean Culture* (Winter):39–40.

Lidz, Theodore. 1976. *The Person: His and Her Development Throughout the Life Cycle.* New York: Basic Books.

Linton, Ralph. 1936. *The Study of Man: An Introduction.* New York: Appleton-Century-Crofts.

Liu, Hsein-Tung. 1994. "Intercultural Relations in an Emerging World Civilization." *Journal of Global Awareness* 2(1):48–53.

Los Angeles Times. 1995. "Spotlight on South Korea." (October 5):D4.

Magaziner, Ira C., and Mark Patinkin. 1989. *The Silent War: Inside the Global Business Battles Shaping America's Future.* New York: Random House.

Mahmood, Cynthia K., and Sharon Armstrong. 1992. "Do Ethnic Groups Exist? A Cognitive Perspective on the Concept of Cultures." *Ethnology* 31(1):1–14.

Mendelsohn, Harold. 1964. "Listening to Radio." Pages 239–249 in *People, Society, and Mass Communications,* edited by Lewis Anthony Dexter and David Manning White. London: Collier-Macmillan.

Moran, Robert T. 1987. "Cross-Cultural Contact: What's Funny to You May Not Be Funny to Other Cultures." *International Management* 42 (July/August):74.

Murphy, Dean. 1994. "New East Europe Retailers Told to Put on a Happy Face." *Los Angeles Times* (November 26):A1.

Oberdorfer, Don. 1997. *The Two Koreas: A Contemporary History.* Reading, MA: Addison-Wesley.

Park, Myung-Seok. 1979. *Communication Styles in Two Different Cultures: Korean and American.* Seoul: Han Shin.

Perry, Charles. 1993. "The American Grain." *Los Angeles Times* (July 1):H12.

Peterson, Mark. 1977. "Some Korean Attitudes Toward Adoption." *Korea Journal* 17(12):28–31.

Protzman, Ferdinand. 1991. "As Marriage Nears, Germans in the Wealthy West Fear Cost in Billions." *The New York Times* (September 24):A6.

Reader, John. 1988. *Man on Earth.* Austin: University of Texas Press.

Redfield, Robert. 1962. "The Universally Human and the Culturally Variable." Pages 439–453 in *Human Nature and the Study of Society: The Papers of Robert Redfield,* vol. 1, edited by M. P. Redfield. Chicago: University of Chicago Press.

Reid, Daniel P. 1988. *Korea: The Land of the Morning Calm.* Lincolnwood, IL: Passport.

Rohner, Ronald P. 1984. "Toward a Conception of Culture for Cross-Cultural Psychology." *Journal of Cross-Cultural Psychology* 15(2):111–138.

Rokeach, Milton. 1973. *The Nature of Human Values.* New York: Free Press.

Rosenfeld, Jeffrey P. 1987. "Barking Up the Right Tree." *American Demographics* (May):40–43.

Sapir, Edward. 1949. "Selected Writings of Edward Sapir." In *Language, Culture and Personality,* edited by D. G. Mandelbaum. Berkeley: University of California Press.

Savada, Andrea, and William Shaw, ed. 1990. *South Korea: A Country Study.* Federal Research Division, Library of Congress.

Schudson, Michael. 1989. "How Culture Works: Perspectives from Media Studies on the Efficacy of Symbols." *Theory and Society* 18:153.

Shapiro, Laura. 1992. "In the American Grain." *The New York Times Book Review* (August 2):9.

Sims, Calvin. 1995. "Don't Cry, the Land Is Rich in Kims and Lees." *The New York Times* (November 15):A4.

Smith, Lynn. 1996. "Adoptees Search the World for Their Roots." *The New York Times* (June 17):A1.

Sobie, Jane Hipkins. 1986. "The Cultural Shock of Coming Home Again." Pages 95–102 in *The Cultural Transition: Human Experience and Social Transformation in the Third World and Japan*, edited by M. I. White and S. Pollack. Boston: Routledge & Kegan.

Stephens, Michael. 1995. "Incompatible Lives: An Astute Observer Ponders the Gulf Between Her American Mind, Korean Soul." *Los Angeles Times Book Review* (August 20):2.

Sterngold, James. 1991. "New Doubts on Uniting Two Koreas." *The New York Times* (May 30):C1.

Sumner, William Graham. 1907. *Folkways*. Boston: Ginn.

Tuleja, Tad. 1987. *Curious Customs: The Stories Behind 296 Popular American Rituals*. New York: Harmony.

UNESCO. 1995. *UNESCO Statistical Yearbook 1995*. Paris: UNESCO Publishing and Bernan.

U.S. Army. 1998. "Standard Installation Topic Exchange Service." http://ww.dmdc.osd.mil/sites/index.html

U.S. Bureau of the Census. 1998. *Statistical Abstract of the United States 1997*. U.S. Department of Commerce.

———. 1997. "Country of Origin and Year of Entry Into the U.S. of the Foreign Born, by Citizenship Status: March 1997." http://www.bls.census.gov/cps/pub/1997/for_born.htm

U.S. Department of Commerce. 1997. "Selected Data for Nonbank Foreign Affiliates, by Country and by Major Industry, 1994 and 1995, Table 18," in *Survey of Current Business*, vol. 77(10). Washington, DC: U.S. Government Printing Office.

U.S. Department of Defense. 1998. *Worldwide Military Strength as of March 31, 1997*. Washington, DC: U.S. Government Printing Office.

U.S. Department of Justice. 1995. "Immigrants Admitted in Fiscal Years 1993 and 1994 for Regions and Top 15 Countries of Birth, Table 4," in *INS Fact Book: Summary of Recent Immigration Data*. Washington, DC: U.S. Government Printing Office.

U.S. Department of State. 1998. "Korea International Adoption." http://travel.state.gov/adoption_korea.html

U.S. Department of Transportation. 1993. "Passenger Travel Between U.S. and Foreign Countries Distribution by U.S. and Foreign Flag Carriers Commercial Traffic Only, Table IIa," in *U.S. International Air Travel Statistics*. Washington, DC: U.S. Government Printing Office.

U.S. Federal Communications Commission. 1997. "International Telecommunications." http://www.fcc.gov/Bureaus/Common_Carrier/Reports/FCC-State_Link/socc.htm

U.S. International Trade Administration. 1997. "Select Destinations Visited by U.S. Resident Travelers 1995–1996." http://tinet.ita.doc.gov/view/f-1996-10-5...html?ti_cart_cookie=19980630.094016.31886

Visser, Margaret. 1988. *Much Depends on Dinner*. New York: Grove.

———. 1989. "A Meditation on the Microwave." *Psychology Today* (December):38–42.

Wallace, Charles P. 1994. "Singapore Affirms Flogging of American." *Los Angeles Times* (April 1):A5.

Wallace, F. C. 1952. "Notes on Research and Teaching." *American Sociological Review* (December):747–751.

Wallerstein, Immanuel. 1990. "Culture as the Ideological Battleground of the Modern World-System." *Theory, Culture, and Society* 7:31–55.

Werkman, Sidney L. 1986. "Coming Home: Adjustment of Americans to the United States After Living Abroad." Pages 5–18 in *Cross-Cultural Reentry: A Book of Readings*. Abilene, TX: Abilene Christian University Press.

Winchester, Simon. 1988. *Korea: A Walk Through the Land of Miracles*. New York: Prentice Hall.

WNBA. 1998. "Players." http://www.wnba.com/playerindex.html

World Monitor. 1992. "The Map: Batters Up!" (April):11.

———. 1993. "The Map: Hoop-la" (February):10–11.

Yeh, May. 1991. "A Letter." *Amerasia* 17(2):1–7.

Yoo, Yushin. 1987. *Korea the Beautiful: Treasures of the Hermit Kingdom*. Los Angeles: Golden Pond.

CHAPTER 5

Abu-Rabia, Salim. 1998. "The Learning of Arabic by Israeli Jewish Children." *Journal of Social Psychology* 138(2):165–171.

Al-Batrawi, Khaled, and Mouin Rabbani. 1991. "Break Up of Families: A Case Study in Creeping Transfer." *Race and Class* 32(4):35–44.

Aviezer, Ora et al. 1994. "Children of the Dream Revisited: 70 Years of Collective Early Child Care in Israeli Kibbutzim." *Psychological Bulletin* 116(1):99–116.

BBC News. 1998a. "The People with Nothing to Celebrate." http://news.bbc.co.uk.hi.english/events/i...at_50/israel_today/newsid_86000/86063.stm

———. 1998b. "Israel Celebrates Half-Century." http://news.bbc.co.uk/hi/english/events/i...at_50/israel_today/newsid_85000/85903.stm

———. 1998c. "Israel—Key Facts." http://news.bbc.co.uk/hi/english/events/i...at_50/israel_today/newsid_79000/79650.stm

Ben-David, Amith, and Yoav Lavee. 1992. "Families in the Sealed Room: Interaction Patterns of Israeli Families During SCUD Missile Attacks." *Family Process* 31(1):35–44.

Bourne, Jenny. 1990. "The Rending Pain of Reenactment." *Race and Class* 32(2):67–72.

Bunuel, Luis. 1985. Quoted on page 22 in *The Man Who Mistook His Wife for a Hat and Other Clinical Tales,* by Oliver Sacks. New York: Summit Books.

Chaliand, Gerard, and Jean-Pierre Rageau. 1995. *The Penguin Atlas of Diasporas*. New York: Viking.

Clinton, William Jefferson. 1996. "Remarks by the President on Announcing Middle East Peace Summit Meeting." Office of the Press Secretary (September 29).

Cooley, Charles Horton. 1909. *Social Organization*. New York: Scribner's.

———. 1961. "The Social Self." Pages 822–828 in *Theories of Society: Foundations of Modern Sociological Theory*, edited by T. Parsons, E. Shils, K. D. Naegele, and J. R. Pitts. New York: Free Press.

———. 1964. *Human Nature and the Social Order*. New York: Schocken.

Corsaro, William A. 1985. *Friendship and Peer Culture in the Early Years*. Norwood, NJ: Ablex.

Coser, Lewis A. 1992. "The Revival of the Sociology of Culture: The Case of Collective Memory." *Sociological Forum* 7(2):365–373.

Davis, Kingsley. 1940. "Extreme Isolation of a Child." *American Journal of Sociology* 45:554–565.

———. 1947. "Final Note on a Case of Extreme Isolation." *American Journal of Sociology* 3(5):432–437.

Durkheim, Emile. [1915] 1964. *The Elementary Forms of the Religious Life*, 5th ed., trans. J. W. Swain. New York: McMillan.

Dyer, Gwynne. 1985. *War.* New York: Crown.

Elbedour, Bastien, Center, 1997.

Ellin, Abby. 1998. "I Want to Be a Chairborne Ranger: Boot Camp for the Office." *The New York Times* (May 24):B9.

Elon, Amos. 1993. "The Jews' Jews." *New York Review of Books* (June 10):14–18.

Figler, Stephen K., and Gail Whitaker. 1991. *Sport and Play in American Life.* Dubuque, IA: Brown.

Freeman, Norman H. 1987. "Children's Drawings of Human Figures." Pages 135–139 in *The Oxford Companion to the Mind,* edited by R. L. Gregory. Oxford: Oxford University Press.

Freud, Anna, and Sophie Dann. 1958. "An Experiment in Group Upbringing." Pages 127–168 in *The Psychoanalytic Study of the Child,* vol. 6, edited by R. S. Eissler, A. Freud, H. Hartmann, and E. Kris. New York: Quadrangle.

Friedman, Thomas L. 1989. *From Beirut to Jerusalem.* New York: Farrar, Straus & Giroux.

Goffman, Erving. 1961. *Asylums: Essays on the Social Situation of Mental Patients and Other Inmates.* New York: Anchor.

Goldman, Ari L. 1989. "Mementos to Preserve the Record of Anguish." *The New York Times* (February 29):Y14.

Gorkin, Michael. 1986. "Countertransference in Cross-Cultural Psychotherapy: The Example of Jewish Therapist and Arab Patient." *Psychiatry* 49:69-79.

Greenburg, Joel. 1997. "Palestinian Census Ignites Controversy Over Jerusalem." *The New York Times* (December 11):A3.

Griffith, Marlin S. 1977. "The Influences of Race on the Psychotherapeutic Relationship." *Psychiatry* 40:27–40.

Grossman, David. 1988. *The Yellow Wind,* trans. by H. Watzman. New York: Farrar, Straus & Giroux.

———. 1998. "Fifty Is a Dangerous Age." *The New Yorker* (April 20):55.

Halbwachs, Maurice. 1980. *The Collective Memory,* trans. by F. J. Ditter, Jr., and V. Y. Ditter. New York: Harper & Row.

Hannerz, Ulf. 1992. *Cultural Complexity: Studies in the Social Organization of Meaning.* New York: Columbia University Press.

Hassassian, Manuel. 1998. "Historical Justice and Compensation for Palestinian Refugees." http://www.pna.net/peace/historical_justice_htm

Hellerstein, David. 1988. "Plotting a Theory of the Brain." *The New York Times Magazine* (May 22):17+.

Israeli Ministry of Foreign Affairs. 1998a. "Israel at 50: A Statistical Glimpse." http://www.israel-mfa.gov.il/facts/israel50.html

———. 1998b. "Centenary of Zionism, 1897–1997." http://www.israel-mfa.gov.il/mfa/zionism/aliya.html#ethiop

Jensen, Holger, 1998. "Holger Jensen: The U.N. Vote on 'Palestine'." http://www2.nando.net:80/newsroom/ntn/voices/071098/voices5_8207_noframes.html

Jewish Student Online Research Center. 1998. "The Jewish Popluation of the World." http://www.us-israel.org/jsource/Judaism/jewpop.html

Kagan, Jerome. 1988. Interview on "The Mind," Public Broadcast Service (transcript). Boston: WGBH Educational Foundation.

———. 1989. *Unstable Ideas: Temperament, Cognition, and Self.* Cambridge: Harvard University Press.

Klieman, Aharon. 1998. "Safe Passage: The Rocky Road to a Middle East Peace." *World Affairs* 160(3):126.

Liebes, Tamar and Shoshana Blum-Kulka. 1994. "Managing a Moral Dilemma: Israeli Soldiers in the *Intifada.*" *Armed Forces & Society* 21(1):45.

Mannheim, Karl. 1952. "The Problem of Generations." Pages 276–322 in *Essays on the Sociology of Knowledge,* edited by P. Kecskemeti. New York: Oxford University Press.

Mead, George Herbert. 1934. *Mind, Self and Society.* Chicago: University of Chicago Press.

Merton, Robert K. 1976. *Sociological Ambivalence and Other Essays.* New York: Free Press.

Montgomery, Geoffrey. 1989. "Molecules of Memory." *Discover* (December):46–55.

Nova. 1986. "Life's First Feelings." (February 11).

Ornstein, Robert, and Richard F. Thompson. 1984. *The Amazing Brain.* Boston: Houghton Mifflin.

Palestinian Central Bureau of Statistics. 1998. "Projected Mid-Year Population By Age Groups and Sex (1996)." http://www.pcbs.org/english/pop1.htm

Palestinian National Authority. 1998. "Palestinian Labor and Employment: An Introduction." http://www.pna.net/facta/pal_labor_employ.htm

Passia. 1998. "Distribution of the Palestinian People Worldwide 1996/1997." HTTP://WWW.PASSIA.ORG/publications/passia_diary/Agenda98/Palestine/5.htm

Patai, Raphael. 1971. "Zionism." Page 1262 in *Encyclopedia of Zionism and Israel.* New York: McGraw-Hill.

Pawel, Ernst. 1989. *The Labyrinth of Exile: A Life of Theodor Herzl.* New York: Farrar, Straus & Giroux.

Penfield, Wilder, and P. Perot. 1963. "The Brain's Record of Auditory and Visual Experience: A Final Summary and Discussion." *Brain* 86:595–696.

Peres, Judy. 1998. "A Human Mosaic." *Chicago Tribune* (May 9)

Piaget, Jean. 1923. *The Language and Thought of the Child,* trans. by M. Worden. New York: Harcourt, Brace & World.

———. 1929. *The Child's Conception of the World,* trans. by J. Tomlinson and A. Tomlinson. Savage, MD: Rowan & Littlefield.

———. 1932. *The Moral Judgment of the Child,* trans. by M. Worden. New York: Harcourt, Brace & World.

———. 1946. *The Child's Conception of Time,* trans. by A. J. Pomerans. London: Routledge & Kegan Paul.

———. 1967. *On the Development of Memory and Identity.* Worchester, MA: Clark University Press.

Rabin, Yitzhak. 1993. "Making a New Middle East: 'Shalom, Salaam, Peace': Views of Three Leaders." *Los Angeles Times* (September 14):A7.

Restak, Richard M. 1988. *The Mind.* New York: Bantam.

Rodgers, Walter. 1998. "Army Holds Israeli Society Together." http://www.cnn.com/WORLD/meast/9804/29/israel.glue/index.html

Rose, Peter I., Myron Glazer, and Penina M. Glazer. 1979. "In Controlled Environments: Four Cases of Intensive Resocialization." Pages 320–338 in *Socialization and the Life Cycle,* edited by P. I. Rose. New York: St. Martin's.

Rowley, Storer. 1998a. "In the Eyes of Children." *Chicago Tribune* (May 9).

———. 1998b. "After the Army, You Feel More Israeli." *Chicago Tribune* (May 9).

Rubinstein, Danny. 1988. "The Uprising: Reporter's Notebook." *Present Tense* 15:22–25.

Sacks, Oliver. 1989. *Seeing Voices: A Journey into the World of the Deaf.* Los Angeles: University of California Press.

Satterly, D. J. 1987. "Jean Piaget (1896–1980)." Pages 621–622 in *The Oxford Companion to the Mind,* edited by R. I. Gregory. Oxford: Oxford University Press.

Schmemann, Serge. 1996. "Rival Claims on Jerusalem Lie at Heart of Arab-Israeli Enmity." *The New York Times* (October 27):4.

Shipler, David. 1986. *Arab and Jew: Wounded Spirits in a Promised Land.* New York: Times Books.

Sichrovsky, Peter. 1991. *Abraham's Children: Israel's Young Generation.* New York: Pantheon.

Smooha, Sammy. 1980. "Control of Minorities in Israel and Northern Ireland." *Society for Comparative Study of Society and History* 10:256–280.

Spitz, Rene A. 1951. "The Psychogenic Diseases in Infancy: An Attempt at Their Etiological Classification." Pages 255–278 in *The Psychoanalytic Study of the Child,* vol. 27, edited by R. S. Eissler and A. Freud. New York: Quadrangle.

Steiner, George. 1967. *Language and Silence: Essays on Language, Literature, and the Inhuman.* New York: Atheneum.

Theodorson, George A., and Achilles G. Theodorson. 1979. *A Modern Dictionary of Sociology.* New York: Barnes & Noble.

Townsend, Peter. 1962. Quoted on pages 146–147 in *The Last Frontier: The Social Meaning of Growing Old,* by Andrea Fontana. Beverly Hills, CA: Sage.

U.S. Central Intelligence Agency. 1998. *World Factbook 1997.* http://www.odci. gov/cia/publications/95fact/index.html

Usher, Graham. 1991. "Children of Palestine." *Race and Class* 32(4):1–18.

CHAPTER 6

Altman, Lawrence K. 1986. "Anxiety on Transfusions." *The New York Times* (July 18):A1, B4.

———. 1995. "Long-Term Survivors May Hold Key Clues to Puzzle of AIDS." *The New York Times* (January 24):B1.

American Association of Blood Banks. 1998. "Facts About Blood." http://www.aabb. org/docs/facts.html

American National Red Cross. 1996. "Biomedical Services 1995–96." http://www.crossnet.org/biomed/ bio-fact.html

Appleby, Drew. 1990. "Faculty and Student Perceptions of Irritating Behaviors in the College Classroom." *J. Staff, Prog., & Org. Dev.* 8(1):41.

Balter, Michael. 1998. "Virus from 1959 Sample Marks Early Years of HIV." *Science* 279(5852):801.

Barr, David. 1990. "What Is AIDS? Think Again." *The New York Times* (December 1): Y15.

Bass, Thomas. 1992. "Interview: Thomas Adeoyte Lambo." *Omni* (February):71.

Bloor, Michael, David Goldberg, and John Emslie. 1991. "Research Note: Ethnostatics [sic] and the AIDS Epidemic." *British Journal of Sociology* 42(1):131–138.

Brooke, James. 1987. "In Cradle of AIDS Theory, a Defensive Africa Sees a Disguise for Racism." *The New York Times* (November 19):B13.

———. 1988. "In Africa, Tribal Hatreds Defy the Borders of State." *The New York Times* (August 28):E1.

Centers for Disease Control and Prevention. 1998. *HIV/AIDS Surveillance Report,* year-end edition. 9(2).

Clark, Matt, with Stryker McGuire. 1980. "Blood Across the Border." *Newsweek* (December 29):61.

Clarke, Thurston. 1988. *Equator: A Journey.* New York: Morrow.

Cohen, Jon. 1997. "The Rise and Fall of Project SIDA." *Science* 278(5343):1565.

———. 1998. "Uninfectable." *The New Yorker* (July 6):33.

Colby, Ron. 1986. Quoted in "Did Media Sensationalize Student AIDS Case?" by John McGauley. *Editor and Publisher* 119:19.

Conrad, Joseph. 1971. *Heart of Darkness,* revised and edited by R. Kimhough. New York: Norton.

Crossette, Barbara. 1998. "Surprises in the Global Tourism Boon." *The New York Times* (April 12):5.

De Cock, Kevin M., and Joseph B. McCormick. 1988. "Correspondence: Reply to HIV Infection in Zaire." *New England Journal of Medicine* 319(5):309.

Doyal, Lesley, with Imogen Pennell. 1981. *The Political Economy of Health.* Boston: South End Press.

Durkheim, Emile. [1933] 1964. *The Division of Labor in Society,* trans. by G. Simpson. New York: Free Press.

The Economist. 1981. "America the Blood Bank." (October 17):87.

———. 1983. "Vein Hopes, Mainline Profits." (January 22):63–64.

Eisenberg, David M., et al. 1993. "Unconventional Medicine in the United States." *The New England Journal of Medicine* 328(4):246–252.

Fox, Renée. 1988. *Essays in Medical Sociology: Journeys into the Field.* New Brunswick, NJ: Transaction.

French, Howard. 1997a. "The Anatomy of Autocracy: Mobutuism's Three Decades." *The New York Times* (May 17):A1.

———. 1997b. "Mobutu Sese Seko, Zairian Ruler, Is Dead in Exile in Morocco at 66." *The New York Times* (September 8): C22.

———. 1997c. "New Rules in Africa: Borders Aren't Sacred." *The New York Times* (October 18):A1.

———. 1998. "Congo Not Alone in Blocking Search for Killers of the Hutu." *The New York Times* (May 7):A1.

Frontline. 1993. "AIDS, Blood, and Politics." Boston: WGBH Educational Foundation and Health Quarterly.

Giese, Jo. 1987. "Sexual Landscape: On the Difficulty of Asking a Man to Wear a Condom." *Vogue* 177(June):227+.

Global Childnet. 1995. "Seattle and Minneapolis Named Alternative Medicine Centers." http://edie.cprost. sfu.ca/ gcnet/ISS4-05f.html

Goffman, Erving. 1959. *The Presentation of Self in Everyday Life.* New York: Anchor.

———. 1963. *Stigma: Notes on the Management of Spoiled Identity.* Upper Saddle River, NJ: Prentice Hall.

Gourevitch, Philip. 1998. "The Vanishing." *The New Yorker* (June 2):49.

Grmek, Mirkod. 1990. *History of AIDS: Emergence and Origin of a Modern Pandemic,* trans. by R. C. Maulitz and J. Duffin. Princeton, NJ: Princeton University Press.

Grover, Jan Zita. 1987. "AIDS: Keywords." *October* 43:17–30.

Halberstam, David. 1986. *The Reckoning.* New York: Morrow.

Harris, Robert, and Jenny Paxman. 1982. *A Higher Form of Killing.* New York: Hill & Wang.

Henderson, Charles. 1997. "Survey: Americans Are Well-Informed About AIDS." *AIDS Weekly Plus* (December 22):10.

———. 1998. "Epidemiology: U.S. Sees AIDS Rise Among Older Americans." *AIDS Weekly Plus* (February 9):14.

Hiatt, Fred. 1988. "Tainted U.S. Blood Blamed for AIDS' Spread in Japan." *Washington Post* (June 23):A29.

Hilts, Philip J. 1988. "Dispelling Myths About AIDS in Africa." *Africa Report* 33:27–31.

Hunt, Charles W. 1989. "Migrant Labor and Sexually Transmitted Disease: AIDS in Africa." *Journal of Health and Social Behavior* 30:353–373.

Hurley, Peter, and Glenn Pinder. 1992. "Ethics, Social Forces, and Politics in AIDS-Related Research: Experience in Planning and Implementing a Household HIV Seroprevalence Survey." *Milbank Quarterly* 70(4):605–628.

International Federation of Pharmaceutical Manufacturers Associations. 1981. Personal correspondence.

Irwin, Kathleen. 1991. "Knowledge, Attitudes and Beliefs About HIV Infection and AIDS Among Healthy Factory Workers and Their Wives, Kinshasa, Zaire." *Social Science and Medicine* 32(8):917–930.

Johnson, Diane, and John F. Murray. 1988. "AIDS Without End." *New York Review of Books* (August 18):57–63.

Kaptchuk, Ted, and Michael Croucher, with the BBC. 1986. *The Healing Arts: Exploring the Medical Ways of the World.* New York: Summit.

Kennedy, Meaghan, et al. 1998. "Sexual Behavior of HIV-Infected Women Reporting Recent Sexual Contact with Women." *Journal of the American Medical Association* (280):29.

Kerr, Dianne L. 1990. "AIDS Update: Ryan White's Death." *Journal of School Health* 60(5):237–238.

Kolata, Gina. 1989. "AIDS Test May Fail to Detect Virus for Years, Study Finds." *The New York Times* (June 1):Y1.

Kornfield, Ruth. 1986. "Dr., Teacher, or Comforter? Medical Consultation in a Zairian Pediatrics Clinic." *Culture, Medicine and Psychiatry* 10:367–387.

Kramer, Staci D. 1988. "The Media and AIDS." *Editor and Publisher* 121:10–11, 43.

Krause, Richard. 1993. Quoted in *A Dancing Matrix: Voyage Along the Viral Frontier,* by Robin Marantz Henig. New York: Knopf.

Kraut, Alan M. 1994. *Silent Travelers: Germs, Genes, and the "Immigrant Menace."* New York: Basic Books.

Kurian, George. 1992. *Atlas of the Third World,* 2d ed. New York: Facts on File.

Lamb, David. 1987. *The Africans.* New York: Random House.

Lasker, Judith N. 1977. "The Role of Health Services in Colonial Rule: The Case of the Ivory Coast." *Culture, Medicine and Psychiatry* 1:277–297.

Lippmann, Walter. 1976. "The World Outside and the Pictures in Our Heads." Pages 174–181 in *Drama in Life: The Uses of Communication in Society,* edited by J. E. Combs and M. W. Mansfield. New York: Hastings House.

Liversidge, Anthony. 1993. "Heresy! 3 Modern Galileos." *Omni* (June):43–51.

Mark, Joan. 1995. *The King of the World in the Land of the Pygmies.* Lincoln, NE: University of Nebraska Press.

McKinley, James Jr. (with Howard French) 1997. "Uncovering the Guilty Footprints Along Zaire's Long Trail of Death." *New York Times* (November 14):A1.

McNeill, William H. 1976. *Plagues and People.* New York: Anchor.

Meltzer, Milton. 1960. *Mark Twain: A Pictorial Biography.* New York: Bonanza.

Merton, Robert K. 1957. *Social Theory and Social Structure.* Glencoe, IL: Free Press.

Michael, Robert T., John H. Gagnon, Edward O. Lauman, and Gina Kolata. 1994. *Sex in America: A Definitive Survey.* New York: Little, Brown.

Naisbitt, John. 1984. *Megatrends: Ten New Directions Transforming Our Lives.* New York: Warner.

Newman, Richard, and Doug Podolsky. 1994. "Bad Blood." *U.S. News & World Report* 116(25):68.

The New York Times. 1996. "Pesticides Can Harm Immune Systems, Researchers Warn." (March 2):A16.

Noble, Kenneth B. 1989. "More Zaire AIDS Cases Show Less Underreporting." *The New York Times* (December 26):J4.

———. 1992. "As the Nation's Economy Collapses, Zairians Squirm Under Mobutu's Heel." *The New York Times* (August 30):Y4.

Olmos, David. 1993. "From Acupuncture to Yoga, Alternative Healing Gains Ground." *Los Angeles Times* (August 22):D3.

Panos Institute. 1989. *AIDS and the Third World.* Philadelphia: New Society.

Parsons, Talcott. 1975. "The Sick Role and the Role of the Physician Reconsidered." *Milbank Memorial Fund Quarterly: Health and Society* 53(1):257–278.

Peretz, S. Michael. 1984. "Providing Drugs to the Third World: An Industry View." *Multinational Business* 84 (Spring):20–30.

Postman, Neil. 1985. *Amusing Ourselves to Death.* New York: Penguin.

Radetsky, Peter. 1997. "Immune to a Plague." *Discover* 18(6):60.

Reuters Health Information. 1998. "Small Minority of HIV-Infected Subjects May Survive 25 Years or More." http://www.ama-assn.org:80/special/hiv/newsline/reuters/07143355.htm

Reuters Health Information. 1999. "Older HIV-Positive Patients Are Diagnosed at More Advanced Stages." HIV/AIDS Information Center. *Journal of the American Medical Association,* http://www.ama-assn.org

Richardson, Lynda. 1998. "AIDS Group Urges New York to Start Reporting of HIV." *The New York Times* (January 13):A1.

Rock, Andrea. 1986. "Inside the Billion-Dollar Business of Blood." *Money* (March):153.

Shilts, Randy. 1987. *And the Band Played On: Politics, People, and the AIDS Epidemic.* New York: St. Martin's.

Sontag, Susan. 1989. *AIDS and Its Metaphors.* New York: Farrar, Straus & Giroux.

Stein, Michael. 1998. "Sexual Ethics: Disclosure of HIV-Positive Status to Partners." *Archives of Internal Medicine* 158:253.

Stolberg, Sheryl. 1996. "Officials Find Rare HIV Strain in L.A. Woman." *Los Angeles Times* (July 5):A1+.

Swenson, Robert M. 1988. "Plagues, History, and AIDS." *The American Scholar* 57:183–200.

Thomas, William I., and Dorothy Swain Thomas. [1928] 1970. *The Child in America.* New York: Johnson Reprint.

Times Atlas of World History. 1984. Maplewood, NJ: Hammond.

Trumbull, Todd. 1993. "The Rise of Alternative Medicine." *Los Angeles Times* (August 22):D3.

Turnbull, Colin M. 1961. *The Forest People.* New York: Simon & Schuster.

———. 1962. *The Lonely African.* New York: Simon & Schuster.

———. 1965. *Wayward Servants.* New York: Doubleday.

———. 1983. *The Human Cycle.* New York: Simon & Schuster.

Twain, Mark. 1973. "King Leopold's Soliloquy on the Belgian Congo." Pages 41–60 in *Mark Twain and the Three R's,* edited by Maxwell Geismar. New York: International Publishers.

UNAIDS and World Health Organization. 1998a. "HIV in Site: Gateway to AIDS Knowledge." http://hivinsite.ucsf.edu/social/un/2098.3ce0.html

———. 1998b. "Global HIV/AIDS and STD Surveillance." http://www.who.ch/emc-hiv/global_report/slides/slide16.html

Urban Institute. 1996. "Why Teenagers Do Not Use Condoms." http://www.urban.org/periodc/prr25_2c.htm

U.S. Bureau of the Census. 1992. *U.S. Exports and General Imports by Hormonized Commodity by Country,* Report FT947/91-A. Washington, DC: U.S. Government Printing Office.

———. 1996a. "Merchandise Trade—Exports by Commodity." *National Trade Data Bank* (January 2). Washington, DC: U.S. Government Printing Office.

———. 1996b. "Merchandise Trade—Imports by Commodity." *National Trade Data Bank* (January 2). Washington, DC: U.S. Government Printing Office.

U.S. Bureau of Labor Statistics. 1997. "Data for Occupations Not Studied in Detail." http://stats.bls.gov/oco/oco2005.htm

U.S. Bureau for Refugee Programs. 1988. *World Refugee Report.* Washington, DC: U.S. Government Printing Office.

U.S. Central Intelligence Agency. 1995. *World Factbook 1995.* http://www.odci.gov/cia/publications/95fact/cg.html

———. 1998. *World Factbook 1997.* http://www.odci.gov/cia/publications/factbook/cg.html

U.S. Department of Health and Human Services. 1990. *HIV/AIDS Surveillance*

Report. Washington, DC: U.S. Government Printing Office.

———. 1992. *AIDS Knowledge and the Attitudes for January–March 1991: Provisional Data from the National Health Interview Survey.* No. 216. Washington, DC: U.S. Government Printing Office.

U.S. General Accounting Office. 1987. *AIDS: Information of Global Dimensions and Possible Impacts.* Washington, DC: U.S. Government Printing Office.

———. 1997a. *Abstracts of GAO Reports and Testimony, FY97.* Washington, DC: U.S. Government Printing Office.

———. 1997b. *Blood Supply: Transfusion-Associated Risks.* Washington, DC: U.S. Government Printing Office.

Von Bargen, Jennifer; Anne Moorman, and Scott Holmberg. 1998. "How Many Pills Do Patients with HIV Infection Take?" *Journal of the American Medical Association* 280:29.

Watson, William. 1970. "Migrant Labor and Detribalization." Pages 38–48 in *Black Africa: Its Peoples and Their Cultures Today,* edited by J. Middleton. London: Collier-Macmillan.

Whitaker, Jennifer Seymour. 1988. *How Can Africa Survive?* New York: Harper & Row.

Wilson, Mary E. 1996. "Travel and the Emergence of Infectious Diseases." *Emerging Infectious Disease* (April–June 1995). http://www.cdc.gov/ncidod/EID/vol1no2/wilson.htm

Witte, John. 1992. "Deforestation in Zaire: Logging and Landlessness." *The Ecologist* 22(2):58.

Wood, Chris; Sally Whittet, and Caroline Bradbeer. 1997. "HIV Infection and AIDS." *British Medical Journal* 315(7120):1433.

World Health Organization. 1988. "A Global Response to AIDS." *Africa Report* (November/December):13–16.

Yasuda, Yukuo. 1994. "Japanese Hemophiliacs Suffering from HIV Infection." http://www.nmia.com/~mdibble/japan2.html

Zuck, Thomas F. 1988. "Transfusion-Transmitted AIDS Reassessed." *New England Journal of Medicine* 318:511–512.

CHAPTER 7

Aldrich, Howard E., and Peter V. Marsden. 1988. "Environments and Organizations." Pages 361–392 in *Handbook of Sociology,* edited by N. J. Smelser. Newbury Park, CA: Sage.

Alimi, Siamak. 1995. "Witness Statement, Defence." http://www.mcspotlight.org/people/witnesses/employment/alimi_siamak.html

Barnet, Richard J., and John Cavanagh. 1994. *Global Dreams: Imperial Corporations and the New World Order.* New York: Simon & Schuster.

Barnet, Richard J., and Ronald E. Müller. 1974. *Global Reach: The Power of the Multinational Corporations.* New York: Simon & Schuster.

Beech, Alan. 1994. "Witness Statement, Defence." http://www.mcspotlight.org/people/witnesses/employment/beech_alan.html

Beech, Elizabeth. 1998. *Fast Food Figures.* http://www.beech.org/treehouse/collect/fastfood/book.shtml

Bernstein, Charles. 1997. "Links to the Community." Restaurants and Institutions. http://www.rimag.com/10/cb_com.htm

Beveridge, Dirk. 1997. "British Judge Rules for McDonald's in Libel Beef." *The Seattle Times* (June 19):http://www.seattletimes.com/extra/browse/html97/altmacd_061997.html

Blau, Peter M., and Richard A. Schoenherr. 1973. *The Structure of Organizations.* White Plains, NY: Longman.

Bloyd-Pashkin, Sharon. 1997. "McVerdict (Vegetarian Activists Lose Case Against McDonald's Corp.)." *Vegetarian Times* (September):24.

Brett, Adrian. 1993. "Witness Statement, Defence." http://www.mcspotlight.org/people/witnesses/employment/brett_adrian.html

Burger King. 1998. "Corporate Information." http://www.burgerking.com/company/fastfacts.htm

Camp, Dave. 1998. "What Employers Need to Know to Comply with the New Minimum Wage Law." http://www.house.gov/camp/minwage.htm

Chenevière, Alain. 1987. *Vanishing Tribes.* Garden City, NY: Doubleday.

Clark, Andrew. 1993. "Learning the Rules of Global Citizenship: Transnationals Need to Meet Their Challenges, or Be Overwhelmed by Them." *The World Paper* (February):5.

CNN Newstand Fortune. 1998. "Pools of Case; McMake Over; Exit Strategy." http://www.cnn.com/TRANSCRIPTS/9808/12/nsf.00.html

Coton, Ray. 1995. "Witness Statement, Defence." http://www.mcspotlight.org/people/witnesses/employment/coton_ray.html

Crecca, Donna. 1997. "Five Models for Success from Foodservice 2005." *Restaurants and Institutions.* http://www.rimag.com/09/five.htm

Crispell, Diane. 1995. "Why Working Teens Get into Trouble." *American Demographics.*

http://www.demographics.com/publications/ad/95_ad/9502_ad/9502ab06.htm

Dairy Queen. 1998. http://www.dairyqueen.com

de la Torre, Jose. 1995. "Multinational Expansion Underscores Management Challenges." *Los Angeles Times* (January 29):D2.

Dominos. 1998. http://www.dominos.com/info/1997.html

Fortune. 1998a. "5 Hundred Company Snapshot." http://cgi.pathfinder.com/cgi-bin/fortune/fortune500/csnap.cgi?r96=135

———. 1998b. "America's Most Admired Companies." http://cgi.pathfinder.com/cgi-bin/...dmired/hts/industry.hts?indcode=15

———. 1998c. "The World's Largest Coporations." (August 3):130.

Frank, Diane. 1997. "The New ROI In Point of Sale." *Datamation* 43(11):73.

Freund, Julien. 1968. *The Sociology of Max Weber.* New York: Random House.

Friedman, Thomas. 1996. "Big Mac I." *The New York Times* (December 8):E15.

Gandhi, Maneka. 1995. "Starving on Junk Food." *World Press Review* 42(9):47.

Gibney, Simon. 1993. "Witness Statement, Defence." http://www.mcspotlight.org/people/witnesses/employment/gibney_simon.html#burns

Gilo, Kim and Tom Welsh. 1997. "McDonald's Marketing of 'Local Burger' Sizzles." *Korea Herald* http://www.koreaherald.co.kr/kh1126/m1126b.html

Gregory, Neville George. 1994. "Witness Statement, Prosecution." http://www.mcspotlight.org/people/witnesses/animals/gregory.html

Gross, Edward, and Amitai Etzioni. 1985. *Organizations in Society.* Upper Saddle River, NJ: Prentice Hall.

Hardees Food Systems. 1998. http://www.hardeesrestaurants.com/hardees/main.html

Hoffman, Ken. 1998. "Mulan McNuggets Just Processed Parts." *Sun Herald Online* http://vh1459.infi.net:80/living/docs/drive072298.htm

Hollander, Stanley. 1991. "Ray Kroc." *The Reader's Companion to American History,* edition 1991:624.

Internal Revenue Service. 1998. *Statistics of Income Bulletin* (Fall). Washington, DC: U.S. Government Printing Office.

International Potato Center. 1998. "Globalization of French Fries." http://www.cipotato.org/ph&mkt/fries.htm

Jacobs, Paul. 1996. "UC Relishes Power of the Patent." *Los Angeles Times* (February 14):A1+.

Johnson, Brad. 1997. "The World According to French Fries." *Restaurants and Institutions* http://www.rimag.com/14/bj_fry.htm

Karp. Jonathan. 1996. "Food for Politics: McDonald's Opens in India's Prickly Market." *Far Eastern Economic Review* (October 24):72.

Keller, George M. 1986. "International Business and the National Interest." *Vital Speeches of the Day* (December 1):124–128.

Kennedy, Paul. 1993. *Preparing for the Twenty-First Century.* New York: Random House.

Khan, Rahat Nabi. 1986. "Multinational Companies and the World Economy: Economic and Technological Impact." *Impact of Science on Society* 36(141):15–25.

Koenig, Peter. 1997. "McRevelations." *New Statesman* 126(4341):20.

Leidner, Robin. 1993. *Fast Food, Fast Talk: Service Work and the Routinization of Everyday Life.* Berkeley, CA: University of California Press.

Leohnhardt, David. 1998. "McDonald's Tarnished Arches." *Business Week* (March 9):70.

Lepkowski, Wil. 1985. "Chemical Safety in Developing Countries: The Lessons of Bhopal." *Chemical and Engineering News* 63:9–14.

Lindeman, Teresa. 1998. "Checking It Twice: Drive-thru's Battle Mixed-up Orders." *Post-Gazette Online.* http://www.post-gazette.com:80/businessnews/19980820drive1.asp

London Greenpeace Group. 1986. "What's Wrong with McDonald's? Everything They Don't Want You to Know." http://www.mcspotlight.org/case/pretrial/factsheet.html#everything.

Lowe, Kimberly. 1997. "The Cost of Employee Theft." *Restaurants and Institutions.* http://www.rimag.com/07/rep_thft.htm

Lowe, Kimberly, and Erin Nicholas. 1997. "Chaos in a Crowded Market." *Restaurants and Institutions.* http://www.rimag.com/14/400main.htm

Mannix, Margaret. 1996. "A Big Whopper Stopper?" *U.S. News & World Report* 120(28):60.

Marquesee, Mike. 1994. "The Difference the McLibel Two Enjoy." *New Statesman & Society* (June 24), vol. 7, no. 308, page 12.

McDonald's Corporation. 1997. http://www.mcdonalds.com

———. 1998. *McDonald's 1997 Annual Report.*

McGurn, William. 1997. "Think Locally." *Far Eastern Economic Review* (November 20):69.

McNeely, C. L. 1996. "Review of *Global Dreams: Imperial Corporations and the New World Order.*" *Contemporary Sociology* 25(5):610–611.

Michels, Robert. 1962. *Political Parties,* trans. by E. Paul and C. Paul. New York: Dover.

Moskowitz, Milton. 1987. *The Global Marketplace.* New York: Macmillan.

Multinational Monitor. 1994. "The Corporate Hall of Shame" (December). http://www.essential.org/monitor/hyper/

———. 1995. "Harnessing the Law to Clean Up India" (July/August). http://www.essential.org/monitor/hyper/mm0795.09.html

National Cattlemen's Beef Association. 1998. "Economic Importance of the United States Cattle Industry." http://www.beef.org/beef/librpub/perirsch/ustotal.html

National Library of Medicine. 1998. "Joint Collection Development Policy: Human Nutrition and Food." http://www.nlm.nih.gov/pubs/cd_hum.nut.html

National Restaurant Association. 1998. "Limited-Service Outlook." http://www.restaurant.org/research/forecast/fc98-05.htm

Perrett, Kevin. 1995. "Witness Statement, Prosecution." http://www.mcspotlight.org/people/witnesses/employment/perrett.html

Personick, Martin. 1991. "Profiles in Safety and Health: Eating and Drinking Places." *Monthly Labor Review* (June):19.

Pizza Hut. 1997. "About Pizza Hut." http://www.pizzahut.com/lowtech/LowFunFacts.htm

Platt, John. 1973. "Social Traps." *American Psychologist* 28:641.

Preston, June. 1997. "Courts Just One Field of Battle Between Coke and Pepsi." *Reuters Business Report* (May 8).

Quinlan, Michael. 1998. "Letter to Shareholders." *McDonald's 1997 Annual Report.*

Restaurants and Institutions. 1998. "Dining on Wall Street." (July 15):58.

Ritzer, George. 1993. *The McDonaldization of Society.* Thousand Oaks: Pine Forge Press.

Roberts, David. 1996. "Witness Statement, Prosecution." http://www.mcspotlight.org/people/winesses/employment/roberts.html

Rousseau, Rita. 1997. "The Labor-Law Blues." *Restaurants and Institutions.* http://www.rimag.com/15/bus_lab.htm

Sadri. Mahmoud. 1996. "Book Review of Occidentalism: Images of the West." *Contemporary Sociology* (September):612.

Sekulic, Dusko. 1978. "Approaches to the Study of Informal Organization." *Sociologija* 20(1):27–43.

Sheridan, Margaret. 1997. "Shrink Rap." http://www.rimag.com/22/bus_ms.htm

———. 1998. "Head Count." http://www.rimag.com/801/bus_ms.htm

Snow, Charles P. 1961. *Science and Government.* Cambridge, MA: Harvard University Press.

Standke, Klaus-Heinrich. 1986. "Technology Assessment: An Essentially Political Process." *Impact of Science on Society* 36(141):65–76.

Steenhuysen, Julie. 1997. "McDonald's Hopes to Repeat Beanie Babie Success." http://mcspotlight.va.com.au/media/press/reuter_8may97.html

Stone, Ann. 1997a. "Lean? No Thanks." *Restaurants and Institutions.* http://www.rimag.com/10/lean.htm

———. 1997b. "Retention Span." *Restaurants and Institutions.* http://www.rimag.com/15/bus_as.htm

———. 1998. "Taking Credit." *Restaurants and Institutions.* http://www.rimag.com/808/bus-as.htm

Strother, Susan. 1998. "McDonald's Image Is Not So Golden on Convention Menu: Franchisees' Complaints." *The Orlando Sentinal* (March 15):H1.

Subway. 1998. http://www.subway.com/

Triconglobal. 1998. http://www.triconglobal.com/defaul2.htm

Union Carbide Annual Report. 1984. "After Bhopal." Danbury, CT: Union Carbide.

UN Food and Agriculture Organization. 1998. http://www.fao.org/

U.S. Department of Commerce. 1997. *Country Commercial Guide: India.* http://www.state.gov/www/about_state/business/com_guides/1997/

U.S. General Accounting Office. 1978. *U.S. Foreign Relations and Multinational Corporations: What's the Connection?* Washington, DC: U.S. Government Printing Office.

Veblen, Thorstein. 1933. *The Engineers and the Price System.* New York: Viking.

Waters. Jennifer. 1998. "Fractured Franchise." *Restaurants and Institutions.* http://www.rimag.com/807/bus_jw.htm

Watson, James. 1997. *Golden Arches East: McDonald's in East Asia.* Stanford, CA: Sanford University Press.

Weber, Max. 1947. *The Theory of Social and Economic Organization,* edited and trans. by A. M. Henderson and T. Parsons. New York: Macmillan.

Wendy's International, Inc. 1997. *Shareholder Report.*

Wexler, Mark N. 1989. "Learning from Bhopal." *The Midwest Quarterly* 31(1):106–129.

Williams, Malcolm T. 1995. Personal correspondence. December 21.

Young, T. R. 1975. "Karl Marx and Alienation: The Contributions of Karl Marx to Social Psychology." *Humboldt Journal of Social Relations* 2(2):26–33.

Zuboff, Shoshana. 1988. *In the Age of the Smart Machine: The Future of Work and Power.* New York: Basic Books.

CHAPTER 8

Author X. 1992. "Mao Fever—Why Now?" Trans. and adapted from the Chinese by R. Terrill. *World Monitor* (December):22–25.

Becker, Howard S. 1963. *Outsiders: Studies in the Sociology of Deviance.* New York: Free Press.

———. 1973. "Labelling Theory Reconsidered." In *Outsiders: Studies in the Sociology of Deviance.* New York: Free Press.

Becker, Jasper. 1996. *Hungry Ghosts: Mao's Secret Famine.* New York: Henry Holt and Company.

Bernstein, Richard. 1982. *From the Center of the Earth: The Search for the Truth About China.* Boston: Little, Brown.

———. 1997. "Horror of a Hidden Chinese Famine." *New York Times* (Feb. 5):B9.

Bernstein, Thomas P. 1983. "Starving to Death in China." *New York Review of Books* (June 16):36–38.

Best, Joel. 1989. *Images of Issues: Typifying Contemporary Social Problems.* New York: Aldine de Gruyter.

Bracey, Dorothy H. 1985. "The System of Justice and the Concept of Human Nature in the People's Republic of China." *Justice Quarterly* 2(1):139–144.

Broaded, C. Montgomery. 1991. "China's Lost Generation." *Journal of Contemporary Ethnography* (October):352–379.

Broadfoot, Robert. 1993. Quoted In D. Holley. "Ancient Power Steps into Asian Spotlight" (a special Pacific Rim edition of *World Report*). *Los Angeles Times* (June 15):H15.

Butterfield, Fox. 1976. "Mao Tse-Tung: Father of Chinese Revolution." *New York Times* (September 10):A13+.

———. 1980. "The Pragmatists Take China's Helm." *New York Times Magazine* (December 28):22–35.

———. 1982. *China: Alive in the Bitter Sea.* New York: Times Books.

Calhoun, Craig. 1989. "Revolution and Repression in Tiananmen Square." *Society* (September/October):21–38.

Carrel, Todd, and Richard Hornik. 1994. "A Chinese Gold Rush? Don't Hold Your Breath." *New York Times* (September 14): A15.

Chambliss, William. 1974. "The State, the Law, and the Definition of Behavior as Criminal or Delinquent." Pages 7–44 in *Handbook of Criminology,* edited by D. Glaser. Chicago: Rand McNally.

Chang Jung. 1991. *Wild Swans: Three Daughters of China.* New York: Simon & Schuster.

———. 1992. Quoted in "Literature of the Wounded," by Jonathan Mirsky. *New York Review of Books* (March 5):6.

Chinese Embassy. 1998. "Response of Chinese Embassy in U.K. to the Film *Return to the Dying Rooms.*" http://oneworld.org/news/partner_news/china_top.html

Chiu Hungdah. 1988. "China's Changing Criminal Justice System." *Current History* (September):265–272.

Clark, John P., and Shirley M. Clark. 1985. "Crime in China—As We Saw It." *Justice Quarterly* 2(1):103–110.

Collins, Randall. 1982. *Sociological Insight: An Introduction to Nonobvious Sociology.* New York: Oxford University Press.

Deng Xiaoping. 1995. Quoted in "Tide Turning Against China's 'Special Zones,'" by Rone Tempest. *Los Angeles Times* (November 3):A5.

Durkheim, Emile. [1901] 1982. *The Rules of Sociological Method and Selected Texts on Sociology and Its Method,* edited by S. Lukes and trans. by W. D. Halls. New York: Free Press.

Erikson, Kai T. 1966. *Wayward Puritans.* New York: Wiley.

Fairbank, John King. 1987. *The Great Chinese Revolution 1800–1985.* New York: Harper & Row.

———. 1989. "Why China's Rulers Fear Democracy." *New York Review of Books* (September 28):32–33.

Faison, Seth. 1997. "Chinese Revise Criminal Code, Not Its Essence." *New York Times* (Mar. 7):A1, A7.

Farley, Maggie. 1995. "China's Neighborly Capitalists." *Los Angeles Times* (December 7):A17.

Federal Bureau of Prisons. 1998. *Quick Facts.* http://www.bop.gov/fact0598.html#Inst

Feng Jicai. 1991. *Voices from the Whirlwind: An Oral History of the Chinese Cultural Revolution.* New York: Pantheon.

Goldman, Merle. 1989. "Vengeance in China." *New York Review of Books* (November 9):5–9.

Gould, Stephen Jay. 1990. "Taxonomy as Politics: The Harm of False Classification." *Dissent* (Winter):73–78.

Hareven, Tamara K. 1987. "Divorce, Chinese Style." *Atlantic Monthly* (April):70–76.

Henriques, Diana B. 1993. "Great Men and Tiny Bubbles: For God, Country and Coca-Cola." *New York Times Book Review* (May 23):13.

Holley, David. 1993. "Ancient Power Steps into Asian Spotlight" (a special Pacific Rim edition of *World Report*). *Los Angeles Times* (June 15):H1+.

Hong Kong Trade Development Council. 1996. "Will the U.S. Focus on Trade Deficit with China in 1996?" *Business Alert* (January 4). http://www.tdc.org.hk/alert/us9611.htm

Ignatius, Adi. 1988. "China's Birthrate Is Out of Control Again as One-Child Policy Fails in Rural Areas." *Asian Wall Street Journal Weekly* (July 18):18.

Jerome, Richard. 1995. "Suspect Confessions." *New York Times Magazine* (August 13):28–31.

Kitsuse, John I. 1962. "Societal Reaction to Deviant Behavior: Problems of Theory and Method." *Social Problems* 9 (Winter):247–256.

Kometani, Foumiko. 1987. "Pictures from Their Nightmare." *New York Times Book Review* (July 19):9–10.

Kristof, Nicholas D. 1989. "China Is Planning 2 Years of Labor for Its Graduates." *New York Times* (August 13):Y1.

Kwong, Julia. 1988. "The 1986 Student Demonstrations in China." *Asian Survey* 28(9):970–985.

Lamberth, John. 1996. "Report of John Lamberth, Ph.D." http://www.aclu.org/court/lamberth.html

Lemert, Edwin M. 1951. *Social Pathology.* New York: McGraw-Hill.

Leys, Simon. 1989. "After the Massacres." *New York Review of Books* (October 12):1719.

———. 1990. "The Art of Interpreting Nonexistent Inscriptions Written in Invisible Ink on a Blank Page." *New York Review of Books* (October 11):8–13.

Link, Perry. 1989. "The Chinese Intellectuals and the Revolt." *New York Review of Books* (June 29):38–41.

Liu Binyan. 1993. "An Unnatural Disaster," trans. by P. Link. *New York Review of Books* (April 8):3–6.

Liu Binyan, and Perry Link. 1998. "A Great Leap Backward?" *New York Times Review of Books* (Oct 8):19–23.

Liu Zaifu. 1989. Quoted in "The Chinese Intellectuals and the Revolt," by Perry Link. *New York Review of Books* (June 29):40.

Lubman, Stanley. 1983. "Comparative Criminal Law and Enforcement: China." Pages 182–193 in *Encyclopedia of Crime and Justice,* edited by S. H. Kadish. New York: Free Press.

Mao Zedong (Mao Tse-tung). 1965. "Report on an Investigation of the Peasant Movement in Hunan (March 1927)." In *Selected Works of Mao Tse-tung*. Peking: Foreign Language Press.

Mathews, Jay, and Linda Mathews. 1983. *One Billion: A China Chronicle*. New York: Random House.

Merton, Robert K. 1957. *Social Theory and Social Structure*. Glencoe, Ill.: Free Press.

Milgram, Stanley. 1974. *Obedience to Authority: An Experimental View*. New York: Harper & Row.

———. 1987. "Obedience." Pages 566–568 in *The Oxford Companion to the Mind*, edited by R. L. Gregory. Oxford: Oxford University Press.

Montalbano, William D. 1993. "Lifetime of Change in a Dozen Years" (a special Pacific Rim edition of *World Report*). *Los Angeles Times* (June 15):H2.

Mosher, Steven W. 1991. "Chinese Prison Labor." *Society* (November/December): 49–59.

National Council for Crime Prevention in Sweden. 1985. *Crime and Criminal Policy in Sweden*. Report no. 19. Stockholm: Liber Distribution.

Oxman, Robert. 1993a. "China in Transition." Interview on *MacNeil/Lehrer Newshour* (December 27), transcript #4828. New York: WNET.

———. 1993b. "China in Transition: Mao to Markets." Interview on *MacNeil/Lehrer Newshour* (December 28), transcript #4829. New York: WNET.

———. 1993c. "China in Transition: Status Report (Chinese Women)." Interview on *MacNeil/Lehrer Newshour* (December 29), transcript #4830. New York: WNET.

———. 1993d. "Focus, Olympic Hurdle." Interview on *MacNeil/Lehrer Newshour* (September 21), transcript #4759. New York: WNET.

———. 1994a. "China in Transition: Any Progress? (Human Rights)." Interview on *MacNeil/Lehrer Newshour* (January 31), transcript #4853. New York: WNET.

———. 1994b. "China in Transition: Taking the Plunge (Education in China; Going into Business in China)." Interview on *MacNeil/Lehrer Newshour* (January 4), transcript #4834. New York: WNET.

Personal correspondence. 1993. Comments by an anonymous reviewer.

Piazza, Alan. 1996. Quoted in "In China's Outlands, Poorest Grow Poorer," by P. E. Tyler. *New York Times* (October 26):A1, A4.

Rojek, Dean G. 1985. "The Criminal Process in the People's Republic of China." *Justice Quarterly* 2(1):117–125.

Rorty, Amelie Oksenberg. 1982. "Western Philosophy in China." *Yale Review* 72(1):141–160.

Schell, Orville. 1996. "China's 'Model' State Orphanages Serve as Warehouses for Death." *Los Angeles Times* (January 7):M2+.

Seymour, James D., and Richard Anderson. 1998. *New Ghosts, Old Ghosts: Prison and Labor Reform Camps in China*. New York: M.E. Sharpe.

Shaw, Victor N. 1996. *Social Control in China: A Study of Chinese Work Units*. Westport: Praeger Publishers.

Shipp, E. R., Dean Baquet, and Martin Gottlieb. 1991. "Slaying Casts a New Glare on Law's Uncertain Path." *New York Times* (June 23):A1+.

Simmons, J. L., with Hazel Chambers. 1965. "Public Stereotypes of Deviants." *Social Problems* 3(2):223–232.

Spector, Malcolm, and J. I. Kitsuse. 1977. *Constructing Social Problems*. Menlo Park, CA: Cummings.

Strebeigh, Fredited by 1989. "Training China's New Elite." *Atlantic Monthly* (April):72–80.

Sumner, William Graham. 1907. *Folkways*. Boston: Ginn.

Sutherland, Edwin H., and Donald R. Cressey. 1978. *Principles of Criminology*, 10th ed. Philadelphia: Lippincott.

Tannenbaum, Frank. 1938. *Crime and the Community*. New York: Ginn.

Tien H. Yuan. 1990. "Demographer's Page: China's Population Planning After Tiananmen." *Population Today* 18(9):6–8.

Tien H. Yuan, Zhang Tianlu, Ping Yu, Li Jingneng, and Liang Zhongtang. 1992. "China's Demographic Dilemmas." *Population Bulletin* 47(1):1–44.

Tobin, Joseph J., David Y. H. Wu, and Dana H. Davidson. 1989. *Preschool in Three Cultures: Japan, China and the United States*. New Haven, CT: Yale University Press.

Tyler, Patrick E. 1996a. "In China's Outlands, Poorest Grow Poorer." *New York Times* (October 26):1+.

———. 1996b. "Chinese Maltreatment at Orphanage." *New York Times* (January 9): A4.

———. 1996c. "U.S. Rights Group Asserts China Lets Thousands of Orphans Die." *New York Times* (January 6):1+.

United Nations. 1998. *1996 Demographic Yearbook*. New York: United Nations Publications.

U.S. Bureau of Justice Statistics. 1996a. *Criminal Victimization 1994: National Crime Victimization Survey*. http:// www.ndjrs.org/txtfiles/cv94.txt

———. 1996b. *Statistics About Crime and Victims*. http://www.ojp.usdoj.gov/bjs/ cvict.htm

U.S. Central Intelligence Agency. 1998. *World Factbook 1998: China*. http:// www.odci.gov/cia/publications/factbook/ ch.html

U.S. Department of Justice. 1998a. *Company President Pleads Guilty to Mail Fraud*. http://www.doj.gov/opa/pr/1998/July/ 340enr.html

———. 1998b. *USX Settles Federal, State Environmental Claims*. http://www.doj. gov/opa/pr/1998/August/358enr.html

U.S. Department of State. 1996. *China Country Commercial Guide 1995–1996*. http://www.usia.gov/abtusia/posts/ HK1/wwheh10html

Wang Ruowang. 1989. Quoted in "The Chinese Intellectuals and the Revolt," by Perry Link. *New York Review of Books* (June 29):40.

Wang Shuo. 1997. Quoted in "Bad Boy," by Jamie James. *The New Yorker* (April 21):50.

Williams, Terry. 1989. *The Cocaine Kids: The Inside Story of a Teenage Drug Ring*. Reading, Mass. Addison-Wesley.

Wu Han. 1981. Quoted in *Coming Alive: China After Mao*, by Roger Garside. New York: McGraw-Hill.

WuDunn, Sheryl. 1993. "Booming China Is a Dream Market for West." *New York Times* (February 15):A1+.

Xu, Xinyi. 1995. "The Impact of Western Forms of Social Control on China." *Crime, Law, and Social Change* 23:67–87.

Zhou, Joseph. 1995. Quoted in "The Impact of Western Forms of Social Control on China: A Preliminary Evaluation," by Xinyi Xu. *Crime, Law, and Social Change* 23:67–87.

CHAPTER 9

African National Congress. 1996. "The African National Congress on the Working Class Struggle for National Liberation." gopher://gopher.anc.orgnc/ history/bababenz.pak

African National Congress. 1998. http://www.anc.org.za/

Amnesty Committee. 1998. South African Truth and Reconciliation Commission. Http://www.truth.org.za/amnesty.html

Berreman, Gerald D. 1972. "Race, Caste, and Other Invidious Distinctions in Social Stratification." *Race* 13(4): 385–414.

Boraine, Alex. 1998. *A Message from the Deputy Chairperson of the TRC*. http:// www.truth.org.za/reading/talk2/no3.htm

Boudon, Raymond, and François Bourricaud. 1989. *A Critical Dictionary of Sociology,* selected and trans. by P. Hamilton. Chicago: University of Chicago Press.

Bryant, Adam. 1999. "American Pay Rattles Foreign Partners." *New York Times* (January 17):4.1.

Center on Budget and Policy Priorities. 1997. "Pulling Apart: A State-by-State Analysis of Income Trends." http://www.cbpp.org/pa-statelist.htm

Constitution of South Africa. 1996. "Preamble." gopher://gopher.anc.org.za/00/anc/misc/sacon96l.txt

Coser, Lewis A. 1977. *Masters of Sociological Thought,* 2nd ed., edited by R. K. Merton. New York: Harcourt Brace Jovanovich.

Crapanzano, Vincent. 1985. *Waiting: The Whites of South Africa.* New York: Random House.

Crystal, Graef. 1995. "Growing the Pay Gap." *Los Angeles Times* (July 23):D2.

Daley, Suzanne. 1998. "In Support of Apartheid: Poison Whiskey and Sterilization." *New York Times* (June 11):A3

Davis, Kingsley, and Wilbert E. Moore. 1945. "Some Principles of Stratification." Pages 413–445 in *Ideological Theory: A Book of Readings,* edited by L. A. Coser and B. Rosenberg. New York: Macmillan.

Drogin, Bob. 1995. "Apartheid Brutality on Trial." *Los Angeles Times* (February 20):A6+.

———. 1996. "South Africa Bringing Power to the People." *Los Angeles Times* (January 31):A1+.

DuPree, David. 1998. "Hitting the Salary Heights Can Come with a Painful Price." *USA Today* (February 26):10L.

Eiseley, Loren. 1990. "Man: Prejudice and Personal Choice." Pages 640–943 in *The Random House Encyclopedia,* 3rd ed. New York: Random House.

Federal Reserve Bank of New York. 1998. *Exchange Rates for the South African Rand, October 27, 1998.* http://www.dna.lth.se/cgi-bin/kurt/rates?ZAR+ALL

Frontline. 1985. "A Class Divided," transcript #309. Boston: WGBH Educational Foundation.

Gerth, Hans, and C. Wright Mills. 1954. *Character and Social Structure: The Psychology of Social Institutions.* London: Routledge & Kegan Paul.

Hattas, Riefaat. 1997. Quoted in *Youth Hearings.* Truth and Reconciliation Commission. http://www.truth.org/hrvtrans/children/hattas.htm

Howard, Judith A., and Jocelyn A. Hollander. 1997. *Gendered Situation, Gendered Selves: A Gender Lens on Social Psychology.* Thousand Oaks: Sage Publications.

Ignatieff, Michael. 1997. "Digging Up the Dead." *New Yorker* (November 10):84–93.

Jencks, Christopher. 1990. Quoted in "The Rise of the 'HyperPoor,'" by David Whitman. *U.S. News & World Report* (October 15):40–42.

Jones, L. Gregory. 1998. "How Much Truth Can We Take?" *Christianity Today* (Feb 9):18–24.

Keller, Bill. 1993. "South Africa's Wealth Is Luring Black Talent." *New York Times* (February 12):A1+.

———. 1994. "Mandela's Party Publishes Plan to Redistribute Wealth." *New York Times* (January 15):Y3.

Lamb, David. 1987. *The Africans.* New York: Vintage.

Lambert, Father Rollins. 1988. "A Day in the Life of Apartheid: The Editor's Interview with Father Rollins Lambert." *U.S. Catholic* 53:26–32.

Lewis, Anne C. 1996. "Average Teachers' Salaries Fall." *America Tomorrow.* http://www.heartofamerica.org/ati/ac170224.htm

Liebenow, J. Gus. 1986. "South Africa: Home, 'Not-So-Sweet,' Homelands." *UFSI Reports* no. 23.

Loy, John W., and Joseph F. Elvogue. 1971. "Racial Segregation in American Sport." *International Review of Sport Sociology* 5:5–24.

Mabuza, Lindiwe. 1990. "Apartheid: Far from Over." *New York Times* (June 20):A15.

Mandela, Nelson. 1990. "I Am the First Accused" (Rivonia Trial Statement 1964). *One Nation, One Country: The Phelps-Stokes Fund* 4 (May):17–45.

———. 1997. *Report by the President of the ANC, Nelson Mandela, to the 50th National Conference of the African National Congress.* http://www.anc.org.za/ancdocs/history/mandela/1997/sp971216.html

Marx, Karl. 1909. *Capital: A Critique of Political Economy,* vol. III, edited by F. Engles, trans. by E. Untermann. Chicago: Kerr.

———. [1895] 1976. *The Class Struggles in France 1848–1850.* New York: International.

Medoff, Marshall H. 1977. "Positional Segregation and Professional Baseball." *International Review of Sport Sociology* 12:49–56.

O'Hare, William P. 1996. "A New Look at Poverty in America." *Populations Bulletin* (September):2–48.

O'Hare, William P., and Brenda Curry-White. 1992. "Demographer's Page: Is There a Rural Underclass?" *Population Today* 20(3):6–8.

Passell, Peter. 1994. "South Africa's Huge Challenge: Bringing the Good Life to Blacks." *New York Times* (June 9):C2.

Ridgeway, Cecilia. 1991. "The Social Construction of Status Value: Gender and Other Nominal Characteristics." *Social Forces* 70(2):367–386.

Roberts, Margaret. 1994. "The Ending of Apartheid: Shifting Inequalities in South Africa." *Journal of the Geographical Association* (January 1):53–64.

Rockefeller Foundation. 1996a. "Where the American Public Would Set the Poverty Line." *Poverty Research Brief* no. 23. http://www.cdinet.com/Rockefeller/Briefs/brief23.html

———. 1996b. "White Poverty in America." *Poverty Research Brief* no. 28. http://www.cdinet.com/Rockefeller/Briefs/brief28.html

———. 1996c. "Working Their Way Out of Poverty?" *Poverty Research Brief* no. 11. http://www.cdinet.com/ Rockefeller/Briefs/brief11.html

Ross, Edward Alsworth. [1908] 1929. *Social Psychology: An Outline and Source Book.* New York: Macmillan.

Russell, Diana E. H. 1989. *Lives of Courage: Women for a New South Africa.* New York: Basic Books.

Seidman, Judy. 1978. *Ba Ye Zwa: The People Live.* Boston: South End Press.

Simpson, Richard L. 1956. "A Modification of the Functional Theory of Social Stratification." *Social Forces* 35:132–137.

South Africa Labour Development Research Unit. 1994. *South Africans Rich and Poor: Baseline Household Statistics.* Cape Town: University of Cape Town.

Statistics South Africa. 1998a. *Population Census, 1996 Announcement of Results.* http://www.css.gov.za/

———. 1998b. "Women and Men in South Africa." *Population Census, 1996 Announcement of Results.* http://www.css.gov.za/

Tumin, Melvin M. 1953. "Some Principles of Stratification: A Critical Analysis." *American Sociological Review* 18:387–394.

U.S. Bureau of the Census. 1992. *Money Income of Households, Families, and Persons in the United States: 1991.* Washington DC: U.S. Government Printing Office.

———. 1997. *Statistical Abstract of the United States 1997.* Washington DC: U.S. Government Printing Office.

———. 1995. *Statistical Abstract of the United States 1995.* Washington, DC: U.S. Government Printing Office.

———. 1998a. *1990 Census Lookup.* http://venus.census.gov/cdrom/lookup

———. 1998b. *The Official Statistics.* http://www.census.gov/

U.S. Central Intelligence Agency. 1997. *World Factbook 1997.* http://www.odci. gov/cia/publications/factbook/index.html

U.S. Department of Labor. 1995. *Employment, Hours, and Earnings, United States, 1990–1995.* Washington, DC: U.S. Government Printing Office.

———. 1998. *Employment and Earnings, United States, 1998.* Washington, DC: U.S. Government Printing Office.

U.S. Department of State. 1996. *Background Notes: South Africa, November 1994.* gopher://dosfan.lib.uic.edu/

Violations Commission. 1998. South African Truth and Reconciliation Commission. http://www.truth.org.za/hrv.htm

Wacquant, Loic J. D. 1989. "The Ghetto, the State, and the New Capitalist Economy." *Dissent* (Fall):508–520.

Wacquant, Loic J. D., and William Julius Wilson. 1989. "The Cost of Racial and Class Exclusion in the Inner City." *Annals of the American Academy* (January):8–25.

Weber, Max. 1982. "Status Groups and Classes." Pages 69–73 in *Classes, Power, and Conflict: Classical and Contemporary Debates,* edited by A. Giddens and D. Held. Los Angeles: University of California.

———. [1947] 1985. "Social Stratification and Class Structure." Pages 573–576 in *Theories of Society: Foundations of Modern Sociological Theory,* edited by T. Parsons, E. Shils, K. D. Naegele, and J. R. Pitts. New York: Free Press.

Weekend World: Johannesburg. 1977. "Lesson 8: Basic Economics at People's College" (April 24).

Wilson, Francis, and Mamphela Ramphele. 1989. *Uprooting Poverty: The South African Challenge.* New York: Norton.

Wilson, William Julius. 1983. "The Urban Underclass: Inner-City Dislocations." *Society* 21:80–86.

———. 1987. *The Truly Disadvantaged: The Inner City, the Underclass, and Public Policy.* Chicago: University of Chicago Press.

———. 1991. "Studying Inner-City Social Dislocations: The Challenge of Public Agenda Research" (1990 presidential address). *American Sociological Review* (February):1–14.

———. 1994. "Another Look at the Truly Disadvantaged." *Political Science Quarterly* 106(4):639–656.

Wirth, Louis. [1945] 1985. "The Problem of Minority Groups." Pages 309–315 in *Theories of Society: Foundations of Modern Sociological Theory,* edited by T. Parsons,

E. Shils, K. D. Naegele, and J. R. Pitts. New York: Free Press.

Wren, Christopher S. 1991. "South Africans Desegregate Some White Public Schools." *New York Times* (January 10):A1.

Yeutter, Clayton. 1992. "When 'Fairness' Isn't Fair." *New York Times* (March 24):A13.

CHAPTER 10

Alba, Richard D. 1992. "Ethnicity." Pages 575–584 in *Encyclopedia of Sociology,* vol. 2, edited by E. F. Borgatta and M. L. Borgatta. New York: Macmillan.

Anson, Robert Sam. 1987. *Best Intentions: The Education and Killing of Edmund Perry.* New York: Random House.

Atkins, Elizabeth. 1991. "For Many Mixed-Race Americans, Life Isn't Simply Black or White." *New York Times* (June 5):B8.

Barrins, Adeline. 1992. Quoted in "The Tallest Fence: Feelings on Race in a White Neighborhood." *New York Times* (June 21):Y12.

Breton, Raymond, Wsevolod W. Isajiw, Warren E. Kalbach, and Jeffrey G. Reitz. 1990. *Ethnic Identity and Equality: Varieties of Experience in a Canadian City.* Toronto: University of Toronto.

Carver, Terrell. 1987. *A Marx Dictionary.* Totowa, NJ: Barnes & Noble.

Castles, Stephen. 1986. "The Guest-Worker in Western Europe—An Obituary." *International Migration Review* 20(4): 761–778.

Cheliand, Gerard, and Jean-Pierre Rageau. 1995. *The Penguin Atlas of Diasporas.* New York: Penguin.

Cherni, Leigh. 1998. *Double Consciousness; A Crisis of Citizenship,* submitted to Johns Hopkins Citizenship Essay Contest.

Cohen, Roger. 1998. "The German 'Volk' Seem Set to Let Outsiders In." *New York Times* (Oct. 16):A4

Cornell, Stephen. 1990. "Land, Labour and Group Formation: Blacks and Indians in the United States." *Ethnic and Racial Studies* 13(3):368–388.

Crapanzano, Vincent. 1985. *Waiting: The Whites of South Africa.* New York: Random House.

Davis, F. James. 1978. *Minority–Dominant Relations: A Sociological Analysis.* Arlington Heights, IL: AHM.

Dunne, John Gregory. 1991. "Law and in Los Angeles." *New York Review of Books* (October 10):26.

Emde, Helga. 1992. "An 'Occupation Baby' in Postwar Germany." Pages 101–111 in *Showing Our Colors: Afro-German Women Speak Out,* edited by M. Opitz, K.

Oguntoye, and D. Schultz. Amherst: University of Massachusetts Press.

Encyclopedia of Latin American History and Culture. 1996. New York: Scribner's.

Faist, Thomas, and Hartmut Häubermann. 1996. "Immigration, Social Citizenship and Housing in Germany." *International Journal of Urban and Regional Geography* (March):83–98.

Fein, Helen. 1978. "A Formula for Genocide: Comparison of the Turkish Genocide (1915) and the German Holocaust (1939–1945)." *Comparative Studies in Sociology* 1:271–294.

Gallup Organization. 1997. "Special Reports: Black/White Realtions in the United States." (June 10) http://www.gallup.com/ special_reports/black-white.htm

———. 1998. "Special Reports: Black/White Relations in the United States." http://www.gallup.com/special_ reports/black-white.htm

German Government. 1998. http://www. bundesregierung.de/english/

Goffman, Erving. 1963. *Stigma: Notes on the Management of Spoiled Identity.* Upper Saddle River, NJ: Prentice Hall.

Gordon, Milton M. 1978. *Human Nature, Class, and Ethnicity.* New York: Oxford University Press.

Heilig, Gerhard, Thomas Büttner, Wolfgang Lutz. 1990. *Germany's Population: Turbulent Past, Uncertain Future. Population Bulletin,* vol. 45., no. 4 (December).

Herbert, Ulrich. 1995. "Immigration, Integration, Foreignness: Foreign Workers in Germany since the Turn of the Century." *International Labor and Working-Class History* (Fall):91–93.

Holmes, Steven A. 1997. "People Can Claim More Than One Race on Federal Forms." *New York Times* (Oct. 30):A1, 19.

Holzner, Lutz. 1982. "The Myth of Turkish Ghettos: A Geographic Case Study of West German Responses Towards a Foreign Minority." *Journal of Ethnic Studies* 9(4):65–85.

Houston, Velin Hasu. 1991. "The Past Meets the Future: A Cultural Essay." *Amerasia Journal* 17(1):53–56.

Ireland, Patrick R. 1997. "Socialism, Unification Policy, Disorder and the Rise of Racism in Eastern Germany." *International Migration Review* 31(4):541.

Jones, Tamara, and Hugh Pope. 1993. "Kurds Raid Turk Offices in Europe." *New York Times* (June 25):A1+.

Jopke, Christian. 1996. "Multiculturalism and Immigration: A Comparison of the United States, Germany, and Great Britain." *Theory and Society* 25(4):449–500.

Kaw, Eugenia. 1993. "Medicalization of Racial Features: Asian American Women and Cosmetic Surgery." *Medical Anthropology Quarterly* 74–89.

Kienbaum, Barbara, and Manfred Grote. 1997. "German Unification as a Cultural Dilemma: a Retrospective." *East European Quarterly*. 31(2):223.

King, Lloyd. 1992. "Lloyd King." Pages 397–401 in *Race: How Blacks and Whites Think and Feel About the American Obsession*, edited by Studs Terkel. New York: New Press.

Kramer, Jane. 1993. "Letter from Europe: Neo-Nazis: A Chaos in the Head." *New Yorker* (June 14):52–70.

Krell, Gert, Hans Nicklas, and Anne Ostermann. 1996. "Immigration, Asylum, and Anti-Foreigner Violence in Germany." *Journal of Peace Research,* 153–170.

Kurthen, Hermann. 1995. "Germany at the Crossroads: National Identity and the Challenges of Immigration." *International Migration Review* (Winter):914–937.

Lee, Sharon M. 1993. "Racial Classification in the U.S. Census: 1890–1990." *Ethnic and Racial Studies* 16(1):75–94.

Lieberman, Leonard. 1968. "The Debate over Race: A Study in the Sociology of Knowledge." *Phylon* 39 (Summer): 127–141.

Lock, Margaret. 1993. "The Concept of Race: An Ideological Construct." *Transcultural Psychiatric Research Review* 30:203–227.

Los Angeles Times. 1992. "Probe Finds Pattern of Excessive Force, Brutality by Deputies" (July 21):A18.

Maher, Adrian. 1995. "Black Tradesmen Face a Daily Wall of Suspicion." *New York Times* (March 20):A1, 20.

Mandel, Ruth. 1989. "Turkish Headscarves and the 'Foreigner Problem': Constructing Difference Through Emblems of Identity." *New German Critique* 46(Winter):27–46.

Marshall, Tyler. 1992. "Saying 'No' to Nazis in Germany." *Los Angeles Times* (December 5):A1, 13.

———. 1993. "Arson Attacks on Foreigners No Longer Big News in Germany." *Los Angeles Times* (July 17):A8.

Martin, Philip L., and Mark J. Miller. 1990. "Guests or Immigrants? Contradiction and Change in the German Immigration Policy Debate since the Recruitment Stop." *Migration World* 15(1):8–13.

McClain, Leanita. 1986. *A Foot in Each World: Essays and Articles,* edited by C. Page. Evanston, IL: Northwestern University Press.

McDowell, Jeanne. 1989. "He's Got to Have It His Way." *Time* (July 17):92–94.

McIntosh, Peggy. 1992. "White Privilege and Male Privilege: A Personal Account of Coming to See Correspondences Through Work in Women's Studies." Pages 70–81 in *Race, Class, and Gender: An Anthology,* edited by M. L. Andersen and P. H. Collins. Belmont, CA: Wadsworth.

Merton, Robert K. 1957. *Social Theory and Social Structure.* New York: Free Press.

———. 1976. "Discrimination and the American Creed." Pages 189–216 in *Sociological Ambivalence and Other Essays.* New York: Free Press.

New York Times. 1990. "Advertising: New Group Makes the Case for Black Agencies and Media" (October 26): C18.

———. 1998. "2 German Parties Reach Deal to Relax Law on Citizenship." (October 15):A11.

O'Brien, Peter. 1996. "Germany's Newest Aliens: The East Germans." *East European Quarterly*. 30(4):449.

O'Connor, Peggy. 1992. Quoted in "The Tallest Fence: Feelings on Race in a White Neighborhood." *New York Times* (June 21):Y12.

Ogbu, John U. 1990. "Minority Status and Literacy in Comparative Perspective." *Daedalus* 119(2):141–168.

Opitz, May. 1992a. "Recapitulation and Outlook." Pages 228–233 in *Showing Our Colors: Afro-German Women Speak Out,* edited by M. Opitz, K. Oguntoye, and D. Schultz. Amherst: University of Massachusetts Press.

———. 1992b. "Three Afro-German Women in Conversation with Dagmar Schultz: The First Exchange for This Book." Pages 145–164 in *Showing Our Colors: Afro-German Women Speak Out,* edited by M. Opitz, K. Oguntoye, and D. Schultz. Amherst: University of Massachusetts Press.

Page, Clarence. 1990. "Black Youth Need More Help, Not More Scorn." *Cincinnati Post* (March 9):A10.

———. 1996. *Showing My Color: Impolite Essays on Race and Identity.* New York: Harper Collins.

Poston, Dudley L., and Mei-Yu Yu. 1990. "The Distribution of the Overseas Chinese in the Contemporary World." *International Migration Review* (Fall):480–508.

Rawley, James A. 1981. *The Transatlantic Slave Trade: A History.* New York: Norton.

Reynolds, Larry T. 1992. "A Retrospective on 'Race': The Career of a Concept." *Sociological Focus* 25(1):1–14.

Safran, William. 1986. "Islamization in Western Europe: Political Consequences and Historical Parallels." *Annals* 485 (May):98–112.

Sallen, Herbert. 1995. "Jews in Germany Today." *Society* 32(4):53.

Sayari, Sabri. 1986. "Migration Policies of Sending Countries: Perspectives on the Turkish Experience." *Annals* 485 (May):87–97.

Schneider, Peter. 1989. "If the Wall Came Tumbling Down." *New York Times Magazine* (June 25):22+.

Segal, Aaron. 1993. *An Atlas of International Migration.* London: Zell.

Smokes, Saundra. 1992. "A Lifetime of Racial Rage Control Snaps with a Telephone Call." *Cincinnati Post* (May 13):14A.

Steele, Shelby. 1990. "A Negative Vote on Affirmative Action." *New York Times Magazine* (May 13):46–49+.

Teraoka, Arlene Akiko. 1989. "Talking 'Turk': On Narrative Strategies and Cultural Stereotypes." *New German Critique* 46:104–128.

Terkel, Studs. 1992. *Race: How Blacks and Whites Think and Feel About the American Obsession.* New York: New Press.

Thränhardt, Dietrich. 1989. "Patterns of Organization Among Different Ethnic Minorities." *New German Critique* 46:10–26.

The Times Atlas of World History. 1984. Maplewood, NJ: Hammond.

Toro, Luis Angel. 1995. "'A People Distinct from Others: Race and Identity in Federal Indian Law and the Hispanic Classification in OMB Directive No. 15." *Texas Tech Law Review* 26:1219–1274.

U.S. Bureau of the Census. 1994. *Current Population Survey Interviewing Manual.* Washington, DC: U.S. Government Printing Office.

U.S. Commission on Civil Rights. 1981. *Affirmative Action in the 1980s: Dismantling the Process of Discrimination (A Proposed Statement).* Clearinghouse Publication 65. Washington, DC: U.S. Government Printing Office.

U.S. Department of Justice. 1996a. *Hate Crime Report 1995.* http://www.fbi. gov/ucr

———. 1996b. Press release (October 31). http://www.usdoj.gov/gopherdata/ press_releases/previous/1996/Oct96/ 532cr.htm

———. 1996c. *Hate Crime Statistics 1996.* http://www.fbi.gov/ucr

———. 1997. "Justice Department Reaches Settlement with New Mexico Lender That Allegedly Discriminated Against Hispanics." (January 28) http://www. usdoj.gov/opa/pr/1997/January97/036cr. htm

———. 1998. "Justice Department Accuses City of Garland, Texas, of Discriminating Against Minority Applicants." (February 6) http://www.usdoj.gov/opa/pr/1998/February/048.htm.html

White, Jenny B. 1997. "Turks in the New Germany." *American Anthropologist* 99(4): 754–769.

Wilpert, Czarina. 1991. "Migration and Ethnicity in a Non-immigration Country: Foreigners in a United Germany." *New Community* 18(1):49–62.

Wirth, Louis. 1945. "The Problem of Minority Groups." Pages 347–372 in *The Science of Man,* edited by R. Linton. New York: Columbia University Press.

CHAPTER 11

Alderman, Craig, Jr. 1990. "10 February 1989 Memo for Mr. Peter Nelson." Page 108 in *Gays in Uniform: The Pentagon's Secret Reports,* edited by K. Dyer. Boston: Alyson.

Almquist, Elizabeth M. 1992. Review of "Gender, Family, and Economy: The Triple Overlap." *Contemporary Sociology* 21(3):331–332.

American Medical Association Bureau of Investigation. 1929. "The Tricho System: Albert C. Gryser X-Ray Method of Depilation." *Journal of the American Medical Association* 92:252.

Anspach, Renee R. 1987. "Prognostic Conflict in Life-and-Death Decisions: The Organization as an Ecology of Knowledge." *Journal of Health and Social Behavior* 28(3):215–231.

Anthias, Floya, and Nira Yuval-Davis. 1989. "Introduction." Pages 1–15 in *Woman-Nation-State,* edited by N. Yuval-Davis and F. Anthias. New York: St. Martin's.

Australian Broadcasting Corporation and Discovery Channel. 1988. *Margaret Mead and Samoa.* Produced by Cinetel Productions.

Baumgartner-Papageorgiou, Alice. 1982. *My Daddy Might Have Loved Me: Student Perceptions of Differences Between Being Male and Being Female.* Denver: Institute for Equality in Education.

Bem, Sandra Lipsitz. 1993. *The Lenses of Gender: Transforming the Debate on Sexual Inequality.* Binghamton, NY: Vail-Ballou.

Bloom, Amy. 1994. "The Body Lies." *New Yorker* (July 18):38–49.

Boroughs, Don L. 1990. "Valley of the Doll?" *U.S. News & World Report* (December 3):56–59.

Brooke, James. 1998. "Sex-Change Industry a Boon to Small City." *New York Times* (November 8):14.

Centers for Disease Control and Prevention. 1998. http://www.cdc.gov

Collins, Randall. 1971. "A Conflict Theory of Sexual Stratification." *Social Problems* 19(1):3–21.

Cordes, Helen. 1992. "What a Doll! Barbie: Materialistic Bimbo or Feminist Trailblazer?" *Utne Reader* (March/April):46, 50.

Cote, James. 1997. "A Social History of Youth in Samoa: Religion, Capitalism, and Cultural Disenfranchisement." *International Journal of Comparative Sociology* 38(3–4):217.

CPB/Polaroid Corporation. 1981. *Margaret Mead: Taking Note.* Produced, written, and directed by Ann Peck.

Dewhurst, Christopher J., and Ronald R. Gordon. 1993. Quoted in "How Many Sexes Are There?" *New York Times* (March 12):A15.

Doherty, Jake. 1993. "Conference to Focus on Plight of Wartime 'Comfort Women.'" *Los Angeles Times* (February 20):B3.

Fagot, Beverly, Richard Hagan, Mary Driver Leinbach, and Sandra Kronsberg. 1985. "Differential Reactions to Assertive and Communicative Acts of Toddler Boys and Girls." *Child Development* 56(6):1499–1505.

Fausto-Sterling, Anne. 1993. "How Many Sexes Are There?" *New York Times* (March 12):A15.

Ferrante, Joan. 1988. "Biomedical Versus Cultural Constructions of Abnormality: The Case of Idiopathic Hirsutism in the United States." *Culture, Medicine and Psychiatry* 12:219–238.

Forrest, J. D. 1994. "Epidemiology of Unintended Pregnancy and Contraceptive Use." *American Journal of Obstetrics and Gynecology* (May) part 2, vol. 170, no. 5, pages 1485–1489.

Freeman, Derek. 1983a. Quoted in McDowell, Edwin. "New Samoa Book Challenges Margaret Mead's Conclusions." *New York Times* (January 31):A1.

———. 1983b. Quoted in Elshtain, Jean. "Coming of Age in America: Why the Attack on Margaret Mead." *Progressive* (October):33.

———. 1996. *Margaret Mead and the Heretic: The Making and Unmaking of an Anthropological Myth.* Australia: Penguin Books.

Garb, Frances. 1991. "Secondary Sex Characteristics." Pages 326–327 in *Women's Studies Encyclopedia. Vol. 1: Views from the Sciences,* edited by H. Tierney. New York: Bedrick.

Gauguin, Paul. [1919] 1985. *Noa Noa: The Tahitian Journal,* trans. by O. F. Theis. New York: Dover.

Geschwender, James A. 1992. "Ethgender, Women's Waged Labor, and Economic Mobility." *Social Problems* 39(1):1–16.

Grady, Denise. 1992. "Sex Test of Champions." *Discover* (June):78–82.

Hall, Edward T. 1959. *The Silent Language.* New York: Doubleday.

Hasbro Toys. 1998. "G.I. Joe: Chronology." http://www.hasbrotoys.com/gijoe/crono.html

Hoon, Shim Jae. 1992. "Haunted by the Past." *Far Eastern Economic Review* (February 6):20.

Howard, Jane. 1984. *Margaret Mead: A Life.* New York: Simon and Schuster.

Infonautics Corporation. 1998. *American Samoa: Chapter 1. General Information.* http://www.elibrary.com

Kamo, Yoshinori, and Ellen L. Cohen. 1998. "Division of Household Work Between Partners: A Comparison of Black and White Couples." *Special Issue: Comparative Perspectives on Black Family Life,* vol. 1 *Journal of Comparative Family Studies* (Spring), vol. 29, no. 1, page 131(15).

Knepper, Paul. 1995. "Historical Origins of the Prohibition of Multiracial Legal Identity in the States and the Nation." *State Constitutional Commentaries and Notes: A Quarterly Review* 5(2):14–20.

Kolata, Gina. 1992. "Track Federation Urges End to Gene Test for Femaleness." *New York Times* (February 12): A1, B11.

Lambert, Bruce. 1993. "Abandoned Filipinas Sue U.S. Over Child Support." *New York Times* (June 21):A3.

Lehrman, Sally. 1997. "WO: Forget *Men Are from Mars, Women Are from Venus.* gender." *Stanford Today* (May/June):47.

Lemonick, Michael D. 1992. "Genetic Tests Under Fire." *Time* (February 24):65.

Lewin, Tamar. 1993. "At Bases, Debate Rages Over Impact of New Gay Policy." *New York Times* (December 24):A1+.

Mageo, Jeannette. 1992. "Male Transvestism and Cultural Change in Samoa." *American Ethnologist* 19(3):443.

———. 1996. "Hairdos and Don'ts: Hair Symbolism and Sexual History in Samoa." *Frontiers* 17(2):138.

———. 1998. *Theorizing Self in Samoa: Emotions, Genders, and Sexualities.* Ann Arbor: University of Michigan Press.

Marshall, Eliot. 1983. "A Controversy on Samoa Comes of Age." *Science* (March 4):1042.

Mead, Margaret. 1928. *Coming of Age in Samoa: A Psychological Study of Primitive*

Youth for Western Civilisation. New York: William Morrow.

———. 1972. *Blackberry Winter: My Earlier Years.* New York: William Morrow.

Mills, Janet Lee. 1985. "Body Language Speaks Louder Than Words." *Horizons* (February):8–12.

Morawski, Jill G. 1991. "Femininity." Pages 136–139 in *Women's Studies Encyclopedia. Vol. 1: Views from the Sciences,* edited by H. Tierney. New York: Bedrick.

Morgenson, Gretchen. 1991. "Barbie Does Budapest." *Forbes* (January 7):66–69.

Newman, Louise M. 1996. "Coming of Age, but Not in Samoa: Reflections on Margaret Mead's Legacy for Western Liberal Feminism." *American Quarterly* 48(2):223–272.

Pion, Alison. 1993. "Accessorizing Ken." *Origins* (November):8.

Raag, Tarja, and Christine Rackliff. 1998. "Preschoolers' Awareness of Social Expectations of Gender: Relationships to Toy Choices." *Sex Roles: A Journal of Research* 38(9–10):685.

Rank, Mark R. 1989. "Fertility Among Women on Welfare: Incidence and Determinants." *American Sociological Review* 54(4):296–304.

Samoa Daily News. 1998a. "Hospital Worried Over Costs of Foreigners' Childbirth." (August 11): http://www.ipacific.com/archive/1998/0811.txt.

———. 1998b. "Restoration of DIC Benefits for Unmarried Spouses." (September 14): http://www.ipacific.com/archive/1998/0914.txt.

———. 1998c. "Faleomavaega Dismisses Allegation He is Not Qualified." (October 9): http://www.ipacific.com/archive/1998/1009.txt.

———. 1998d. "Five Women Seek Elected Office in a Male Dominated Arena." (November 3): http://www.ipacific.com/archive/1998/1103.txt.

Sarbin, Theodore R., and Kenneth E. Karols. 1990. "Nonconforming Sexual Orientations and Military Suitability." Pages 6–49 in *Gays in Uniform: The Pentagon's Secret Reports,* edited by K. Dyer. Boston: Alyson.

Schaller, Jane Green, and Elena O. Nightingale. 1992. "Children and Childhoods: Hidden Casualties of War and Civil Unrest." *Journal of the American Medical Association* 268(5):642–644.

Schmalz, Jeffrey. 1993. "From Midshipman to Gay-Rights Advocate." *New York Times* (February 4):B1+.

Segal, Lynne. 1990. *Slow Motion: Changing Masculinities, Changing Men.* London: Virago.

Shweder, Richard A. 1994. "What Do Men Want? A Reading List for the Male Identity Crisis." *New York Times Book Review* (January 9):3, 24.

Son, Eugene. 1998. "G.I. Joe—A Real American FAQ." http:www.yojoe.com/faq/gifaq.txt.

Sturdevant, Saundra Pollock, and Brenda Stoltzfus. 1992. *Let the Good Times Roll: Prostitution and the U.S. Military in Asia.* New York: New Press.

Tattersall, Ian. 1993. "Focus—All in the Family" (*Homo sapiens* Exhibit at New York's American Museum of Natural History). Interview on *MacNeil/Lehrer Newshour,* July 12, transcript #4708. New York: WNET.

Tierney, Helen. 1991. "Gender/Sex." Page 153 in *Women's Studies Encyclopedia. Vol. 1: Views from the Sciences,* edited by H. Tierney. New York: Bedrick.

Turner, George. 1986 [1861]. *Samoa: Nineteen Years in Polynesia.* Apia; Western Samoa and Cultural Trust.

U.S. Bureau of the Census. 1992. "Social, Economic, and Housing Characteristics: American Samoa." *1990 Census of Population and Housing.*

———. 1993. *1990 Census of Population and Housing Content Reinterview Survey: Accuracy of Data for Selected Population and Housing Characteristics as Measured by Reinterview.* Washington, DC: U.S. Government Printing Office.

———. 1998. http://www.census.gov/

U.S. Department of Defense. 1990. "DOD Directive 1332.14." Page 19 in *Gays in Uniform: The Pentagon's Secret Reports,* edited by K. Dyer. Boston: Alyson.

U.S. Department of Education. 1998. *The Chronicle of Higher Education 1998–1999: Almanac Issue.* Washington, DC: U.S. Government Printing Office.

U.S. Department of the Interior. 1998. *United States Insular Areas and Freely Associated States.* http://www.doe.gov/oia/oiafacts.html#page2

———. 1992. http://www.doe.gov/

U.S. Department of Labor. 1998. "Women's Earnings as Percent of Men's, 1979–1996." http://www.dol.gov.dol/wb/public/wb_pubs/7996.html

U.S. Department of State. 1998. "Background Notes: East Asia and the Pacific." http://www.state.gov/www/background_notes/eapbgnhp.html

Vann, Elizabeth. 1995. "Implications of Sex and Gender Differences for Self: Perceived Advantages and Disadvantages of Being the Other Gender." *Sex Roles: A Journal of Research* 33(7–8):531.

Wishart, David. 1995. *The Roles and Status of Men and Women in Nineteenth Century Omaha and Pawnee Societies: Postmodernist Uncertainties and Empirical Evidence.* University of Nebraska Press. Lincoln.

CHAPTER 12

Barnet, Richard J. 1990. "Reflections: Defining the Moment." *New Yorker* (July 16):45–60.

Boudon, Raymond, and François Bourricaud. 1989. *A Critical Dictionary of Sociology,* selected and trans. by P. Hamilton. Chicago: University of Chicago Press.

Bregger, John. 1996. "Measuring Self-Employment in the United States." *Monthly Labor Review* (January/February):3–9.

Buckley, Alan D. 1998. "Comparing Political Systems." http://www.smc.edu/homepage/abuckley/ps2/ps2types.htm

Bullock, Alan. 1977. "Democracy." Pages 211–212 in *The Harper Dictionary of Modern Thought,* edited by A. Bullock, S. Trombley, and B. Eadie. New York: Harper & Row.

Bureau of Economic Analysis. 1998. "Gross Product by Industry." *National Accounts Data.* http://www.bea.doc.gov/bea/dn2/gposhr.htm

Bureau of Labor Statistics. 1998a. "Work at Home in 1997." *Labor Force Statistics from the Current Population Survey* (March 11). http://stats.bls.gov/news.release/homey.nws.htm

———. 1998b. "Employment Situation Summary." *The Employment Situation News Release* (December 4). http://stats.bls.gov/news.release/empsit.nws.htm

———. 1998c. "Workers on Flexible and Shift Schedules in 1997 Summary." *Labor Force Statistics from the Current Population Survey* (March 26). http://stats.bls.gov/news.release/flex.nws.htm

———. 1998d. "Hourly Compensation Costs in U.S. Dollars." *Foreign Labor Statistics.* http://stats.bls.gov/news.release/ichcc.t02.htm

———. 1998e. "The 10 Occupations with the Largest Job Growth, 1996–2006." *Employment Projections.* http://stats.bls.gov/news.release/ecopro.table7.htm

———. 1998f. "Consumer Expenditures in 1997." *Consumer Expenditure Surveys* (December 8). http://stats.bls.gov/news.release/cesan.nws.htm

———. 1998g. "Entrants Concentrated in Sales, Clerical and Service Occupations." *MLR: The Editor's Desk.* http://stats.bls.gov/opub/ted/1998/Dec/wk4/art01.htm

Burke, James. 1978. *Connections*. Boston: Little, Brown.

Carvajal, Doreen. 1998. "Book Publishers Seek Global Reach and Grand Scale." *New York Times* (October 19):C7.

Chehabi, H. E., and Juan J. Linz. 1998. *Sultanistic Regimes*. Johns Hopkins University Press web page. http://128.220.50.88/press/books/titles/s98/s98chsu.htm

Clinton, Bill. 1998a. "President Clinton: Promoting Human Rights in China." *The White House at Work* (June 29). http://www.whitehouse.gov/WH/Work/

———. 1998b. "President Clinton: Working to Strengthen the Global Economy, Promoting Security in Asia." *The White House at Work* (November 18). http://www.whitehouse.gov/WH/Work/111898.html

———. 1998c. "President Clinton: Promoting Exports, Creating New Jobs." *The White House at Work* (November 10). http://www.whitehouse.gov/WH/Work111098.html

———. 1998d. "President Clinton: Leading the World into the Information Age." *The White House at Work* (November 30). http://www.whitehouse.gov/WH/Work/113098.html

Creighton, Andrew L. 1992. "Democracy." Pages 430–434 in *Encyclopedia of Sociology, Volume 1*, edited by E. F. Borgatta and M. L. Borgatta. New York: MacMillan.

Currie, Elliot, and Jerome H. Skolnick. 1998. *America's Problems: Social Issues and Public Policy*, 2nd ed., edited by E. Etzioni-Halevy and A. Etzioni. New York: Basic Books.

Durkheim, Emile. [1933] 1964. *The Division of Labor in Society*, trans. by G. Simpson. New York: Free Press.

Employment Policy Foundation. 1998. "Union Membership as a Percent of Employment." http://epf.org/union1.htm

Farber, David. 1999. "Monster Mergers, Record Year." *CNBC & The Wall Street Journal* (December 29). http://www.msnbc.com

Federal Election Commission. 1998. "Top 50 PAC's-Receipts." http://www.fec.gov

Fortune. 1998. "The Global 500: The World's Largest Corporations." http://cgi.pathfinder.com/fortune/global500/intro.html

Galbraith, John K. 1958. *The Affluent Society*. Boston: Houghton Mifflin.

Gallup Organization. 1998. *A Gallup Poll Social Audit: Have/Have-Nots*. http://www.gallup.gov

General Motors. 1999. "Company Profile." http://www.gm.com/about/info/overview/facts.html

Hacker, Andrew. 1971. "Power to Do What?" Pages 134–146 in *The New Sociology: Essays in Social Science and Social Theory in Honor of C. Wright Mills*, edited by I. L. Horowitz. New York: Oxford University Press.

Halberstam, David. 1986. *The Reckoning*. New York: Morrow.

Heilbroner, Robert. 1990. "Reflections: After Communism." *The New Yorker* (September 10):91–100.

Hirsch, E. D., Jr., Joseph F. Kett, and James Trefil. 1993. *The Dictionary of Cultural Literacy*. New York: Houghton Mifflin.

Hobsbawm, Eric. 1997. Quoted in "At the Age of 80 This Most Pre-eminent of Historians Remains Convinced of the Moral Force of Communism" by J. Lloyd. *New Statesman* 126(June 6):28–31.

International Trade Administration. 1998. "State Exports to Countries and Regions." http://www.ita.doc.gov

Kabir, Bhuian Monoar. 1995. "Politico-economic Limitations and the Fall of the Military Authoritarian Government in Bangladesh." *Armed Forces & Society: An Interdisciplinary Journal* 21(4):533–573.

Lambert, Michael. 1992. "Defense Contractors Use Down-Sizing Not Diversification to Maintain Profits." *Aviation Week and Space Technology* (March 9):61–63.

Lawrence, Robert. 1998. Quoted in "One World, One Market." Online Newshour (August 11). http://www.pbs.org/newshour/bb/economy/july-dec98/globalization_8-11.htm

Marx, Karl. [1881] 1965. "The Class Struggle." Pages 529–535 in *Theories of Society*, edited by T. Parsons, E. Shils, K. D. Naegele, and J. R. Pitts. New York: Free Press.

McAneny, Leslie. 1996. "Public Confidence in Major Institutions Little Changed from 1995." *Gallup Poll Archives*. http://198.175.140.8/poll%5Farchives/1996/960606.htm

McConnell, Sheila. 1996. "The Role of Computers in Reshaping the Work Force." *Monthly Labor Review* (August):3–5.

McNamara, Robert S. 1989. *Out of the Cold: New Thinking for American Foreign and Defense Policy in the 21st Century*. New York: Simon & Schuster.

Mills, C. Wright. 1963. "The Structure of Power in American Society." Pages 23–38 in *Power, Politics and People: The Collected Essays of C. Wright Mills*, edited by I. L. Horowitz. New York: Oxford University Press.

Morisi, Teresa. 1996. "Commercial Banking Transformed by Computer Technology." *Monthly Labor Review* (August):30–36.

Newport, Frank. 1997. "Small Business and Military Generate Most Confidence in Americans." *Gallup Poll Archives*. http://198.175.140.8/poll%5Farchives/1997/970815.htm

Omran, Abdel R., and Farzaneh Roudi. 1993. "The Middle East Population Puzzle." *Population Bulletin* 48(1):1–40.

Online Newshour. 1997. "Working Issues." *Online Newshour Forum* (February 20). http://www.pbs.org/newshour

———. 1998. "Pooling Interest." *Online Newshour Forum* (May 26). http://www.pbs.org/newshour

Potter, Edward. 1997. "The State of the American Workplace." *Online Newshour Forum* (September 3). http://www.pbs.org/newshour

Reitman, Janet. 1998. "All the World's Governments: From Democrats to Dictators and Princes to Parliaments, Here's a Guide to Governing the World." *Scholastic Update* (February 23), vol. 130, no. 10, page 12(2).

Rosenbaum, David E. 1999. "Social Security: The Basics, With a Tally Sheet." *New York Times* (January 28):A19.

Stevenson, Richard W. 1991. "Nothrop Settles Workers' Suit on False Missile Tests for $8 Million." *New York Times* (June 25):A7.

Tolstaya, Tatyana. 1991. "In Cannibalistic Times." *New York Times Review* (April 11), pages 3–6.

U.S. Census Bureau. 1997. "State Rankings." http://www.census.gov/statab/ranks/pg16.txt

U.S. Central Intelligence Agency. 1997. *World Factbook*. http://www.odci.gov/cia/publications/factbook

Van Evera, Stephen. 1990. "The Case Against Intervention." *Atlantic Monthly* (July):72–80.

Wallerstein, Immanuel. 1984. *The Politics of the World Economy: The States, the Movements and the Civilizations*. New York: Cambridge University Press.

Weber, Max. 1947. *The Theory of Social and Economic Organization*, trans. by A. M. Henderson and T. Parsons. Glencoe, IL: Free Press.

White House. 1997. "Text of the President's Message to Internet Users." http://www.whitehouse.gov/WH/New/Commerce/message.html

———. 1998a. "U.S. Trade and Investment in Sub-Saharan Africa." *Issues* http://www.whitehouse.gov/Africa/trade2-plain.html

———. 1998b. "President Clinton: Shared Values and Interests with the Americas." *The White House at Work* (April 16). http://www.whitehouse.gov/WH/Work/041698.html.

Wiatrowski, William. 1994. "Small Businesses and Their Employees." *Monthly Labor Review* (October):29–35.

Zuboff, Shoshana. 1988. *In the Age of the Smart Machine.* New York: Basic Books.

CHAPTER 13

American Demographics. 1996. "Estimating Child-Raising Costs" (October 1996). http://www.marketingpower.com/Publications/FC/sample/sampFQ8.HTM

Atlas of World Development. 1994. New York: Wiley.

Behnam, Djamshid. 1990. "An International Inquiry into the Future of the Family: A UNESCO Project." *International Social Science Journal* 42:547–552.

Berelson, Bernard. 1978. "Prospects and Programs for Fertility Reduction: What? Where?" *Population and Development Review* 4:579–616.

Beresky, Andrew E., ed. 1991. *Fodor's Brazil: Including Bahia and Adventures in the Amazon.* New York: Fodor's Travel Publications.

Boccaccio, Giovanni. [1353] 1984. "The Black Death." Pages 728–740 in *The Norton Reader: An Anthology of Expository Prose,* 6th ed. edited by A. M. Eastman. New York: Norton.

Brazil Demographic and Health Survey. 1996. "Results from the Demographic and Health Survey." *Studies in Family Planning* 29(1):88–90.

Brooke, James. 1989. "Decline in Births in Brazil Lessens Population Fears." *New York Times* (August 8):Y1+.

Brown, Christy. 1992. "The Letter 'A.'" Pages 85–90 in *One World, Many Cultures,* edited by S. Hirschberg. New York: Macmillan.

Brown, Lester R. 1987. "Analyzing the Demographic Trap." *State of the World 1987: A Worldwatch Institute Report on Progress Toward a Sustainable Society.* New York: Norton.

Burke, B. Meredith. 1989. "Ceausescu's Main Victims: Women and Children." *New York Times* (January 16):Y15.

Butts, Yolanda, and Donald J. Bogue. 1989. *International Amazonia: Its Human Side.* Chicago: Social Development Center.

Calsing, Elizeu Francisco. 1985. "Extent and Characteristics of Poverty in Brazil. Estimation of Social Inequalities." *Revista Paraguaya de Sociologia* 22:29–53.

Caufield, Catherine. 1985. "A Reporter at Large: The Rain Forests." *The New Yorker* (January 14):41+.

Cowell, Adrian. 1990. *The Decade of Destruction: The Crusade to Save the Amazon Rain Forest.* New York: Holt.

Davis, Kingsley. 1984. "Wives and Work: The Sex Role Revolution and Its Consequences." *Population and Development Review* 10(3):397–417.

Dean, Warren. 1991. *Review of Coffee, Contention, and Change: In the Making of Modern Brazil,* edited by Mauricio A. Font and Charles Tilly. *Journal of Latin American Studies* 23(3):649–650.

Dickenson, John. 1994. "Company Towns: The Brazilian Experience." Pages 186–187 in *Atlas of World Development,* edited by T. J. Unwin. New York: Wiley.

Durning, Alan B. 1990. "Ending Poverty." Pages 135–153 in *State of the World 1990: A Worldwatch Institute Report on Progress Toward a Sustainable Society,* edited by L. Starke. New York: Norton.

Dychtwald, Ken, and Joe Flower. 1989. *Age Wave: The Challenges and Opportunities of an Aging America.* Los Angeles: Tarcher.

Eckholm, Erik. 1990. "An Aging Nation Grapples with Caring for the Frail." *New York Times* (March 27):A1+.

Feder, Ernest. 1971. *The Rape of the Peasantry: Latin America's Landholding System.* Garden City, NY: Anchor.

Fonseca, Claudia. 1986. "Orphanages, Foundlings, and Foster Mothers: The System of Child Circulation in a Brazilian Squatter Settlement." *Anthropological Quarterly* 59:15–27.

Glascock, Anthony P. 1982. "Decrepitude and Death Hastening: The Nature of Old Age in Third World Societies (Part I)." *Studies in Third World Societies* 22:43–66.

Goldani, Ana Maria. 1990. "Changing Brazilian Families and the Consequent Need for Public Policy." *International Social Science Journal* 42(4):523–538.

Goldenberg, Sheldon. 1987. *Thinking Sociologically.* Belmont, CA: Wadsworth.

Goode, Judith. 1987. "Gaminismo: The Changing Nature of the Street Child Phenomenon in Colombia." *UFSI Reports,* no. 28.

Gutis, Philip S. 1989a. "Family Redefines Itself, and Now the Law Follows." *New York Times* (May 28):B1.

———. 1989b. "What Makes a Family? Traditional Limits Are Challenged." *New York Times* (August 31):Y15+.

Harrison, Paul. 1987. *Inside the Third World: The Anatomy of Poverty,* 2d ed. New York: Viking Penguin.

Johansson, S. Ryan. 1987. "Status Anxiety and Demographic Contraction of Privileged Populations." *Population and Development Review* 13(3):439–470.

Lewin, Tamar. 1990. "Strategies to Let Elderly Keep Some Control." *New York Times* (March 28):A1, A11.

Light, Ivan. 1983. *Cities in World Perspective.* New York: Macmillan.

Malthus, Thomas R. [1798] 1965. *First Essay on Population.* New York: Kelley.

Molano, Alfredo. 1993. Quoted in "Colombia's Vanishing Forests." *World Press Review* (June):43.

Nations, Marilyn K., and Mara Lucia Amaral. 1991. "Flesh, Blood, Souls, and Households: Cultural Validity in Morality Inquiry." *Medical Anthropology Quarterly* 5(3):204–220.

Nolty, Denise, edited by 1990. *Fodor's '90 Brazil: Including the Amazon and Bahia.* New York: Fodor's Travel Publications.

Olshansky, S. Jay, and A. Brian Ault. 1986. "The Fourth Stage of the Epidemiologic Transition: The Age of Delayed Degenerative Diseases." *Milbank Quarterly* 64(3):355–391.

Degenerative Diseases." *Milbank Quarterly* 64(3):355–391.

Omran, Abdel R. 1971. "The Epidemiologic Transition: A Theory of the Epidemiology of Population Change." *Milbank Quarterly* 49(4):509–538.

Perlman, Janice. 1967. *The Myth of Marginality.* Berkeley: University of California Press.

Revkin, Andrew. 1990. *The Burning Season: The Murder of Chico Mendes and the Fight for the Amazon Rain Forest.* Boston: Houghton Mifflin.

Rock, Andrea. 1990. "Can You Afford Your Kids?" *Money* (July):88–99.

Romero, Simon. 1996. "Brazilian Income Distribution has Improved Under 'Real Plan.'" http://www.latinolink.com/biz/0906bbra.htm

Rusinow, Dennison. 1986. "Mega-Cities Today and Tomorrow: Is the Cup Half Full or Half Empty?" *UFSI Reports,* no. 12.

Sanders, Thomas G. 1986. "The Politics of Agrarian Reform in Brazil." *UFSI Reports,* no. 32.

———. 1987a. "Brazilian Street Children. Pt. I: Who They Are." *UFSI Reports,* no. 17.

———. 1987b. "Brazilian Street Children. Pt. II: The Public and Political Response." *UFSI Reports,* no. 18.

———. 1988. "Happiness Also Rises Up There: The *Favelas* of Rio." *UFSI Reports,* no. 2.

Sayre, Robert F. 1983. "The Parents' Last Lessons." Pages 124–142 in *Life Studies: A Thematic Reader,* edited by D. Cavitch. New York: St. Martin's.

SEADE Foundation. 1994. *Survey of Living Conditions in the Metropolitan Area of São Paulo,* Research Series 101. Geneva: International Institute for Labour Studies.

Semana. 1993. "Colombia's Vanishing Forests." *World Press Review* (June):43.

Simons, Marlise. 1988. "Man-Made Amazon Fires Tied to Global Warming." *New York Times* (August 12):Y1+.

Skidmore, Thomas E. 1993. "Bi-racial U.S.A. vs. Multi-racial Brazil: Is the Contrast Still Valid?" *Journal of Latin American Studies* 25:373–386.

Soldo, Beth J., and Emily M. Agree. 1988. "America's Elderly." *Population Bulletin* 43(3):5+.

Sorel, Nancy Caldwell. 1984. *Ever Since Eve: Personal Reflections on Childbirth.* New York: Oxford.

Stockwell, Edward G., and H. Theodore Groat. 1984. *World Population: An Introduction to Demography.* New York: Watts.

Stockwell, Edward G., and Karen A. Laidlaw. 1981. *Third World Development: Problems and Prospects.* Chicago: Nelson-Hall.

Stone, Robyn, Gail Lee Cafferata, and Judith Sangl. 1987. "Caregivers of the Frail Elderly: A National Profile." *Gerontologist* 27(5):616–626.

Stub, Holger R. 1982. *The Social Consequences of Long Life.* Springfield, IL: Thomas.

Targ, Dena B. 1989. "Feminist Family Sociology: Some Reflections." *Sociological Focus* 22(3):151–160.

Tremblay, Hélène. 1988. *Families of the World: Family Life at the Close of the Twentieth Century. Vol. 1: The Americas and the Caribbean.* New York: Farrar, Straus & Giroux.

United Nations. 1983. *World Population Trends and Policies: 1983 Monitoring Report,* vol. 1. New York: United Nations.

U.S. Bureau of the Census. 1947. *Statistical Abstract of the United States, 1947.* Washington, DC: U.S. Government Printing Office.

———. 1961. *Statistical Abstract of the United States, 1961.* Washington, DC: U.S. Government Printing Office.

———. 1991. *World Population Profile: 1991.* Washington, DC: U.S. Government Printing Office.

———. 1993. *Statistical Abstract of the United States, 1993.* Washington, DC: U.S. Government Printing Office.

———. 1995. "Sixty-five Plus in the United States." *Statistical Brief.* http://www.census.gov/

———. 1996. *World Population Profile: 1996.* Washington, DC: U.S. Government Printing Office.

———. 1998. "Population Pyramids for Brazil." http://www.census.gov/cgi-bin/ipc/idbpyry.pl

U.S. Central Intelligence Agency. 1992. *World Factbook 1992.* Washington, DC: U.S. Government Printing Office.

———. 1995. *World Factbook 1995: Brazil.* http://www.odci.gov/cia/publications/95fact/br.html

———. 1998. *World Factbook 1997: Brazil.* http://www.odci.gov/cia/ publications/97fact/br.html

U.S. Department of the Army. 1983. *Brazil, A Country Study,* 4th ed., edited by Richard F. Nyrop. Washington, DC: U.S. Government Printing Office.

U.S. Department of State. 1998. *Background Notes: Brazil, March 1998.* http://www.state.gov/www/background_notes/brazil_0398_bgn.html

van de Kaa, Dirk J. 1987. "Europe's Second Demographic Transition." *Population Bulletin* 42(1):1–59.

Wark, John T. 1995. "Raising Child to 22 Can Cost $265,000." *Detroit News* (October 19). http://detnews.com/menu/stories/20850.htm

Watkins, Susan C., and Jane Menken. 1985. "Famines in Historical Perspective." *Population and Development Review* 11(4):647–675.

Wilkie, James W., and Enrique Ochoa, eds. 1989. *Statistical Abstract of Latin America,* vol. 27. Los Angeles: University of California Press.

World Bank. 1990. *World Development Report, 1990.* New York: Oxford University Press.

———. 1994. *World Development Report, 1994: Infrastructure for Development.* New York: Oxford University Press.

CHAPTER 14

American Council on Education. 1998. *Center for Adult Learning and Education Credentials.* http://www.acenet.edu/programs/CALEC/GED

Bloom, Benjamin S. 1981. *All Our Children Learning: A Primer for Parents, Teachers and Other Educators.* New York: McGraw-Hill.

Botstein, Leon. 1990. "Damaged Literacy: Illiteracies and American Democracy." *Daedalus* 119(2):55–84.

Celis, William III. 1992. "A Texas-Size Battle to Teach Rich and Poor Alike." *New York Times* (February 12):B6.

———. 1993a. "International Report Card Shows U.S. Schools Work." *New York Times* (December 9):A1+.

———. 1993b. "Study Finds Rising Concentration of Black and Hispanic Students." *New York Times* (December 14):A1+.

Chira, Susan. 1991. "Student Tests in Other Nations Offer U.S. Hints, Study Says." *New York Times* (May 20):A1+.

Cohen, David K., and Barbara Neufeld. 1981. "The Failure of High Schools and the Progress of Education." *Daedalus* (Summer):69–89.

Coleman, James S. 1960. "The Adolescent Subculture and Academic Achievement." *American Journal of Sociology* 65:337–347.

———. 1966. *Equality of Educational Opportunity.* Washington, DC: U.S. Government Printing Office.

———. 1977. "Choice in American Education." Pages 1–12 in *Parents, Teachers, and Children: Prospects for Choice in American Education.* San Francisco: Institute for Contemporary Studies.

Coleman, James S., John W. C. Johnstone, and Kurt Jonassohn. 1961. *The Adolescent Society.* New York: Free Press.

Durkheim, Emile. 1961. "On the Learning of Discipline." Pages 860–865 in *Theories of Society: Foundations of Modern Sociological Theory,* vol. 2, edited by T. Parsons, E. Shils, K. D. Naegele, and J. R. Pitts. New York: Free Press.

———. 1968. *Education and Sociology,* trans. by S. D. Fox. New York: Free Press.

Foster, Jack D. 1991. "The Role of Accountability in Kentucky's Education Reform Act of 1990." *Education Leadership,* 34–36.

Gardner, John W. 1984. *Excellence: Can We Be Equal and Excellent Too?* New York: Norton.

Hallinan, M. T. 1988. "Equality of Educational Opportunity." Pages 249–268 in *Annual Review of Sociology,* vol. 14, edited by W. R. Scott and J. Blake. Palo Alto, CA: Annual Reviews.

———. 1996. "Track Mobility in Secondary School." *Social Forces* 74(3):983.

Henry, Jules. 1965. *Culture Against Man.* New York: Random House.

Hirsch, E. D., Jr. 1989. "The Primal Scene of Education." *New York Review of Books* (March 2):29–35.

———. 1985. *Cultural Literacy: What Every American Needs to Know.* New York: Houghton Mifflin.

Hirsch, E. D., Jr., Joseph F. Kett, and James Trefil. 1993. *The Dictionary of Cultural Literacy.* New York: Houghton Mifflin.

Lynn, Richard. 1988. *Educational Achievement in Japan: Lessons for the West.* London: Macmillan.

Merton, Robert K. 1957. *Social Theory and Social Structure.* Glencoe, IL: Free Press.

Meyer, John W., and David P. Baker. 1996. "Forming American Educational Policy with International Data: Lessons from the Sociology of Education." *Sociology of Education* (Extra Issue):123–130.

Oakes, Jeannie. 1985. *Keeping Track: How Schools Structure Inequality.* Binghamton, NY: Vail-Ballou.

———. 1986a. "Keeping Track. Part 1: The Policy and Practice of Curriculum Inequality." *Phi Delta Kappan* 67(September):12–17.

———. 1986b. "Keeping Track. Part 2: Curriculum Inequality and School Reform." *Phi Delta Kappan* 67(October):148–154.

Online NewsHour. 1996a. "Ready for Work" (March 27). http://www1.pbs.org/newshour

———. 1996b. "Education Report Card" (November 21). http://www1.pbs.org/newshour

Organization for Economic Co-operation and Development and Statistics Canada. 1995. *Literacy, Economy, and Society: Results of the First International Adult Literacy Survey.* Paris: OECD.

Ouane, Adama. 1990. "National Languages and Mother Tongues." *UNESCO Courier* (July):27–29.

Phelan, Patricia, and Ann Locke Davidson. 1994. "Looking Across Borders: Students' Investigations of Family, Peer, and School Worlds as Cultural Therapy." Pages 35–59 in *Pathways to Cultural Awareness: Cultural Therapy with Teachers and Students,* edited by George and Louise Spindler. Thousand Oaks, CA: Corwin.

Phelan, Patricia, Ann Locke Davidson, and Hanh Cao Yu. 1993. "Students' Multiple Worlds: Navigating the Borders of Family, Peer, and School Cultures." Pages 89–107 in *Renegotiating Cultural Diversity in American Schools,* edited by Patricia Phelan and Ann Locke Davidson. New York: Teachers College Press.

Phelan, Patricia, Ann Locke Davidson, and Hanh Thanh Cao. 1991. "Students' Multiple Worlds: Negotiating the Boundaries of Family, Peer, and School Cultures." *Anthropology and Education Quarterly* 22(3):224–250.

Ponessa, Jeanne. 1997. "GED Diploma Candidates Reach Record Level." *Education Week* (June 11): http://www.edweek.org/ew/vol-16/37ged.h16

Potter, J. Hasloch, and A. E. W. Sheard. 1918. *Catechizings for the Church and Sunday Schools,* 2d series. London: Skeffington.

Purves, Alan C. 1974. "Divergent Views on the Schools: Some Optimism Justified." *New York Times* (January 16):C74.

Ramirez, Francisco, and John W. Meyer. 1980. "Comparative Education: The Social Construction of the Modern World System." Pages 369–399 in *Annual Review of Sociology,* vol. 6, edited by A. Inkeles, N. J. Smelser, and R. H. Turner. Palo Alto, CA: Annual Reviews.

Resnick, Daniel P. 1990. "Historical Perspectives on Literacy and Schooling." *Daedalus* 119(2):15–32.

Richardson, Lynda. 1994. "More Schools Are Trying to Write Textbooks Out of the Curriculum." *New York Times* (January 31):A1+.

Rohlen, Thomas P. 1986. "Japanese Education: If They Can Do It, Should We?" *The American Scholar* 55:29–43.

Rosenthal, Robert, and Lenore Jacobson. 1968. *Pygmalion in the Classroom.* New York: Holt, Rinehart & Winston.

Sanchez, Claudio. 1993. Interview on National Public Radio, *Morning Edition* (December 8):11–13 (transcript). New York: WNET.

Sowell, Thomas. 1981. *Ethnic America: A History.* New York: Basic Books.

Stevenson, Harold. 1992. Interview on National Public Radio, *Morning Edition* (December 10):8–10 (transcript). New York: WNET.

Stevenson, Harold W., Shin-ying Lee, and James W. Stigler. 1986. "Mathematics Achievement of Chinese, Japanese, and American Children." *Science* 231: 693–699.

Third International Mathematics and Science Study. 1996. http://www.ed.gov/NCES/times/index.html

Thomas, David. 1997. "Dropout Rates Remain Stable Over the Last Decade." *U.S. Department of Education Press Release.* (December 17): http://www.ed.gov/Press-Releases

Thomas, William I., and Dorothy Swain Thomas. [1928] 1970. *The Child in America.* New York: Johnson Reprint.

Tyack, David. 1996. "Forming the National Character." *Harvard Educational Review* 36:29–41.

Tyack, David, and Elisabeth Hansot. 1981. "Conflict and Consensus in American Public Education." *Daedalus* (Summer):1–43.

U.S. Bureau of Labor Statistics. 1998. "College Enrollment and Work Activity of 1997 High School Graduates." *Labor Force Statistics from the Current Population Survey.* http://stats.bls.gov/news.release/hsgec.nws.htm

U.S. Department of Education. 1987. *Japanese Education Today.* Washington, DC: U.S. Government Printing Office.

———. 1993a. *Adult Literacy in America: A First Look at the Results of the National Literacy Survey.* Washington, DC: U.S. Government Printing Office.

———. 1993b. *Occupational and Educational Outcomes of Recent College Graduates 1 Year After Graduation: 1991.* Washington, DC: U.S. Government Printing Office.

———. 1995. *Digest of Education Statistics 1995,* by Thomas D. Snyder and Charlene M. Hoffman. NCES 95-029. Washington, DC: U.S. Government Printing Office.

———. 1996. *The Condition of Education, 1996,* by Thomas Smith. NCES 96-304. Washington, DC: U.S. Government Printing Office.

———. 1997. "Transition from College to Work." *The Condition of Education 1997, Indicator 31:* http://nces01.ed.gov/pubs/ce/c9731a01.html

———. 1998a. "Goals 2000: History." *Goals 2000: Reforming Education to Improve Student Achievement—April 30, 1998.* http://www.ed.gov/pubs/G2KReforming/g2ch1.html

———. 1998b. "Title III—State and Local Education Systemic Improvement." http://www.ed.gov/legislation/GOALS2000/TheAct/sec301.html

———. 1998c. "Transition from College to Work." *The Condition of Education 1998.* http://nces01.ed.gov/pubs/ce

Wells, Amy Stuart, and Jeannie Oakes. 1996. "Potential Pitfalls of Systematic Reform: Early Lessons from Research on Detracking." *Sociology of Education* (extra issue).

CHAPTER 15

Abercrombie, Nicholas, and Bryan S. Turner. 1978. "The Dominant Ideology Thesis." *British Journal of Sociology* 29(2):149–170.

Alston, William P. 1972. "Religion." Pages 140–145 in *The Encyclopedia of Philosophy,* vol. 7, edited by P. Edwards. New York: Macmillan.

Amnesty International. 1996a. "Afghanistan: AI Appeals to International Community to Take the Initiative to Promote and Protect Human Rights" (January 16). http://www.oneworld.org/amnesty/ai_afghan_jan18.html

———. 1996b. "Afghanistan: International Community Should Act Now to Prevent Possible Bloodbath" (September 26). http://www.oneworld.org/amnesty/press/afghanistan_sept26.html

———. 1996c. "Afghanistan: Taleban Take Hundreds of Civilians Prisoner" (October 2). http://www.oneworld.org/amnesty/press/afghanistan_oct12.html

Aron, R. 1969. Quoted in *The Sociology of Max Weber* by Julien Freund. New York: Random House.

Berger, Peter L. 1967. *The Sacred Canopy: Elements of a Sociological Theory of Religion.* New York: Doubleday.

Burns, Nicholas. 1996. "U.S. Department of State Daily Press Briefing" (September 30) http://www.state.gov/www/briefs/index/%20 Briefings%3A9609 %20Press %20Briefings%3A9960930%20Daily%20Briefing

Bush, George H. 1991. *State of the Union Address by the President of the United States* (January 29). Washington, DC: U.S. Government Printing Office.

Caplan, Lionel. 1987. "Introduction: Popular Conceptions of Fundamentalism." Pages 1–24 in *Studies in Religious Fundamentalism,* edited by L. Caplan. Albany: State University of New York Press.

Carter Center. 1996. "State of World Conflict Report: Afghanistan, 1994–1995." http://www.emory.edu/CARTER_CENTER/PUBS/SWCR9495/afghan.htm#summary

Christianity Today. 1986. "Letters" (October 17):6.

Coles, Robert. 1990. *The Spiritual Life of Children.* Boston: Houghton Mifflin.

Durkheim, Emile. [1915] 1964. *The Elementary Forms of the Religious Life,* 5th ed., trans. by J. W. Swain. New York: Macmillan.

———. 1951. *Suicide: A Study in Sociology,* trans. by J. A. Spaulding and G. Simpson. New York: Free Press.

Ebersole, Luke. 1967. "Sacred." Page 613 in *A Dictionary of the Social Sciences,* edited by J. Gould and W. L. Kolb. New York: UNESCO.

Echo-Hawk, Walter. 1979. "A Prepared Statement from Walter Echo-Hawk." Pages 280–287 in *Religious Discrimination: A Neglected Issue by U.S. Commission on Civil Rights.* Washington, DC: U.S. Government Printing Office.

Esposito, John L. 1986. "Islam in the Politics of the Middle East." *Current History* (February):53–57, 81.

———. 1992. *The Islamic Threat: Myth or Reality?* New York: Oxford University Press.

Forbes, James. 1990. "Up from Invisibility: Review of *The Black Church in the African American Experience* by C. Eric Lincoln

and Lawrence H. Mamiya." *New York Times Book Review* (December 23):1–2.

Gallup, George, Jr., and Jim Castelli. 1989. *The People's Religion: American Faith in the '90s.* New York: Macmillan.

Grimes, William. 1994. "The Man Who Rendered Jesus for the Age of Duplication." *New York Times* (October 12):B1.

Haddad, Yvonne. 1991. Interview with Bill Moyers on "Images of God in the Arab World," Public Broadcast Service (transcript). Boston: WGBH Educational Foundation.

Hammond, Phillip E. 1976. "The Sociology of American Civil Religion: A Bibliographic Essay." *Sociological Analysis* 37(2):169–182.

Hourani, Albert. 1991. *A History of the Arab Peoples.* Cambridge, MA: Belknap.

Ibrahim, Youssef M. 1991. "In Kuwait, Ramadan Has a Bitter Taste." *New York Times* (March 19):A1, A7.

Johnson, Lyndon B. 1987. Quoted in *America's History Since 1865* by James A. Henretta, W. Elliott Brownlee, David Brody, and Susan Ware. Chicago: Dorsey.

Kurian, George Thomas. 1992. *Encyclopedia of the Third World.* New York: Facts on File.

Lechner, Frank J. 1989. "Fundamentalism Revisited." *Society* (January/February): 51–59.

Lincoln, C. Eric, and Lawrence H. Mamiya. 1990. *The Black Church in the African American Experience.* Durham, NC: Duke University Press.

Mahjubah: The Magazine for Moslem Women. 1984 (July).

McNamara, Robert S. 1989. *Out of the Cold: New Thinking for American Foreign and Defense Policy in the 21st Century.* New York: Simon & Schuster.

Mead, George Herbert. 1940. *Mind, Self and Society,* 3d ed. Chicago: University of Chicago Press.

National Public Radio. 1984a. "Black Islam." *The World of Islam* (tape). Washington, DC: NPR.

———. 1984b. "Decay or Rebirth: The Plight of Islamic Art." *The World of Islam* (tape). Washington, DC: NPR.

———. 1984c. "Voices of Resurgence." *The World of Islam* (tape). Washington, DC: NPR.

The New York Public Library Desk Reference. 1989. New York: Simon & Schuster.

Nottingham, Elizabeth K. 1971. *Religion: A Sociological View.* New York: Random House.

Online NewsHour. 1996. "Veiled in Fear" (October 9). pbs.org/newshour/bb/

asia/july-dec96/afghan_background_10-9.html

Pelikan, Jaroslav, and Clifton Fadiman, eds. 1990. *The World Treasury of Modern Religious Thought.* Boston: Little, Brown.

Pickering, W. S. F. 1984. *Durkheim's Sociology of Religion.* London: Routledge & Kegan Paul.

Robertson, Roland. 1987. "Economics and Religion." Pages 1–11 in *The Encyclopedia of Religion.* New York: Macmillan.

Rubin, Barnett R. 1996. Quoted in *A Nation in Arms* by Karl E. Meyer. *New York Times Book Review* (August 11):21.

Save the Children Fund. 1996. "Restrictions on Women Lead to Withdrawal from Western Afghanistan." http://www.oneworld.org/scf/press_mar7.html

Smart, Ninian. 1976. *The Religious Experience of Mankind.* New York: Scribner's.

Spokesperson for the Iraqi government. 1991. Quoted in "Iraqi Message: 'Duty' Fulfilled." *New York Times* (February 26):Y1.

Stark, Rodney, and William S. Bainbridge. 1985. *The Future of Religion: Secularization, Revival and Cult Formation.* Berkeley: University of California Press.

Stavenhagen, Rodolfo. 1991. "Ethnic Conflicts and Their Impact on International Society." *International Social Science Journal* (February): 117–132.

Turner, Bryan S. 1974. *Weber and Islam: A Critical Study.* Boston: Routledge & Kegan Paul.

Turner, Jonathan H. 1978. *Sociology: Studying the Human System.* Santa Monica, CA: Goodyear.

United Nations. 1996. *General Assembly Fifty-First Session Agenda Item 21* (December 3). gopher://gopher.un.org/00/ga/docs/51/plenary/A51-704.EN

U.S. Central Intelligence Agency. 1995. *World Factbook 1995.* http://www.odci.gov/cia/publications/95fact/af.html

U.S. Commission on Civil Rights. 1979. *Religious Discrimination: A Neglected Issue.* Washington, DC: U.S. Government Printing Office.

Van Doren, Charles L. 1991. *A History of Knowledge: Past, Present, and Future.* New York: Carol.

Watchtower Bible and Tract Society. 1987. *Life in a Peaceful New World.* Brooklyn: Watchtower.

Weber, Max. 1922. *The Sociology of Religion,* trans. by E. Fischoff. Boston: Beacon.

———. 1958. *The Protestant Ethic and the Spirit of Capitalism,* 5th ed., trans. by T. Parsons. New York: Scribner's.

White House press release. 1995. "Remarks by the President on Religious Liberty in America" (July 12). http://www.whitehouse.gov/WH/EOP/OP/html/book3-plain.html

Wilmore, Gayraud S. 1972. *Black Religion and Black Radicalism*. Garden City, NY: Doubleday.

The World Almanac and Book of Facts 1991. 1990. New York: Pharos.

Yinger, J. Milton. 1971. *The Scientific Study of Religion*. New York: Macmillan.

Zangwill, O. L. 1987. "Isolation Experiments." Pages 393–394 in *The Oxford Companion to the Mind,* edited by R. L. Gregory. New York: Oxford University Press.

Zickel, Raymond E. 1991. *Soviet Union: A Country Study,* 2d ed. Washington, DC: U.S. Government Printing Office.

CHAPTER 16

Afonso, Carlos A. 1997. "The Internet and Social Strategies." *Corporate Watch.* http://www.corpwatch.org/trac/feature/feature1/Afonso.html

Ash, Timothy Garton. 1989. *The Uses of Adversity: Essays on the Fate of Central Europe.* New York: Random House.

Associated Press. 1996. "Third World Pioneers Use Net to Promote Business." *USA Today* (December 30). http://usatoday.com/80+USATODAY_ONLINE+USATODAY_ONLINE+NEWS+NEWS++internet

Barnet, Richard J. 1990. "Reflections: Defining the Moment." *New Yorker* (July 16):45–60.

Bauerlein, Monika. 1992. "Plutonium Is Forever: Is There Any Sane Place to Put Nuclear Waste?" *Utne Reader* (July/August):34–37.

Bear, John, and David M. Pozerycki. 1992. *Computer Wimp No More: The Intelligent Beginner's Guide to Computers.* Berkeley: Ten Speed Press.

Berners-Lee, Wright Tim. 1996. Quoted in "Seek and You Shall Find (Maybe)" by Steve G. Steinberg. *Wired* (May):111.

Biegel, Stuart. 1996. "Does Anyone Control the Internet?" *UCLA Online Institute for Cyberspace Law and Policy.* http://www.gse.ucla.edu/iclp/control.html

Brecher, Jeremy, John Brown Childs, and Jill Cutler. 1993. *Global Visions: Beyond the New World Order.* Boston: South End Press.

Carver, Terrell. 1987. *A Marx Dictionary.* Totowa, NJ: Barnes & Noble.

Colihan, Jane, and Robert J. T. Joy. 1984. "Military Medicine." *American Heritage* (October/November):65.

Coser, Lewis A. 1973. "Social Conflict and the Theory of Social Change." Pages 114–122 in *Social Change: Sources, Patterns, and Consequences,* edited by E. Etzioni-Halevy and A. Etzioni. New York: Basic Books.

Crawford, Jack. 1995. "Renaissance Two: Second Coming of the Printing Press?" http://www.lincoln.ac.nz/reg/futures/renaiss2.htm

Currie, Elliott, and Jerome H. Skolnick. 1988. *America's Problems: Social Issues and Public Policy,* 2d ed. Boston: Little, Brown.

Dahrendorf, Ralf. 1973. "Toward a Theory of Social Conflict." Pages 100–113 in *Social Change: Sources, Patterns, and Consequences,* 2d ed., edited by E. Etzioni-Halevy and A. Etzioni. New York: Basic Books.

Halberstam, David. 1986. *The Reckoning.* New York: Morrow.

Hauben, Michael, and Rhonda Hauben. 1996. *Proposed Declaration of the Rights of Netizens.* http://www.columbia.edu/~rh120/netizen-rights.txt

Holusha, John. 1989. "Eastern Europe: Its Lure and Hurdles." *New York Times* (December 18):Y25+.

Kantor, Andrew, and Michael Newbarth. 1996. "Off the Charts: The Internet 1996." *Internet World* (December):46–57.

Katz, Yvonne, and Gay Chedester. 1992. "Redefining Success: Public Education in the 21st Century." *Catalyst.* gopher://borg.lib.vt.edu/00/catalyst/v22n3/katz.v22n3

Kuhn, Thomas S. 1975. *The Structure of Scientific Revolutions.* Chicago: University of Chicago Press.

Mandelbaum, Maurice. 1977. *The Anatomy of Historical Knowledge.* Baltimore: Johns Hopkins University Press.

Mander, Jerry. 1997. "The Net Loss of the Computer Revolution." *Corporate Watch.* http://www.corpwatch.org/trac/feature/feature1/mander.htm

Martel, Leon. 1986. *Mastering Change: The Key to Business Success.* New York: Simon & Schuster.

Marx, Karl. [1881] 1965. "The Class Struggle." Pages 529–535 in *Theories of Society,* edited by T. Parsons, E. Shils, K.

D. Naegele, and J. R. Pitts. New York: Free Press.

Nader, Ralph, James Love, and Andrew Saindon. 1995. "Project Censored's Top Ten Censored Stories of 1995." *Consumer Project on Technology* (July 14). http://censored.sonoma.edu/ProjectCensored/Stories1995.html

Noack, David. 1997. "The Origin of PCs." *Internet World* (February):49–54.

Norman, Donald A. 1988. Quoted in "Management's High-Tech Challenge." *Editorial Research Report* (September 30):482–491.

Ogburn, William F. 1968. "Cultural Lag as Theory." Pages 86–95 in *Culture and Social Change,* 2d ed., edited by O. D. Duncan. Chicago: University of Chicago Press.

Oppenheimer, Robert. 1986. Quoted in *The Making of the Bomb* by Richard Rhodes. New York: Touchstone.

O'Sullivan, Anthony. 1990. "Eastern Europe." *Europe* (September):21–22.

Public Citizen. 1996. http://www.citizen.org/

Pulver, Jeff. 1997. "Unwired." *Internet World* (February):100–102.

Rabi, Isidor I. 1969. "The Revolution in Science." Pages 28–66 in *The Environment of Change,* edited by A. W. Warner, D. Morse, and T. E. Cooney. New York: Columbia University Press.

Reich, Jens. 1989. Quoted in "People of the Year." *Newsweek* (December 25):18–25.

Schwartz, Bruce. 1996. "Weigh Pros and Cons Before Beefing Up the PC." *USA Today* (November 18):15E.

United Nations Development Programme. 1993. *Human Development Report.* New York: Oxford University Press.

U.S. Bureau of the Census. 1995. *Statistical Abstract of the United States 1995.* RIANE Publishing Company.

U.S. Department of Education. 1998. "Issue Brief: Internet Access in Public Schools." http://nces.ed.gov/pubs98/98031.html

Wallerstein, Immanuel. 1984. *The Politics of the World-Economy: The States, the Movements and the Civilizations.* New York: Cambridge University Press.

White, Leslie A. 1949. *The Science of Culture: A Study of Man and Civilization.* New York: Grove.

White House. 1996. "The Internet." http://www.whitehouse.gov/WH/EOP/OVP/24hours/internet.html

Index